AF370784

Applications of GNSS Reflectometry for Earth Observation

Applications of GNSS Reflectometry for Earth Observation

Editors

Nereida Rodriguez-Alvarez
Mary Morris

MDPI • Basel • Beijing • Wuhan • Barcelona • Belgrade • Manchester • Tokyo • Cluj • Tianjin

Editors
Nereida Rodriguez-Alvarez
Jet Propulsion Laboratory
California Institute
of Technology
Pasadena
USA

Mary Morris
Jet Propulsion Laboratory
California Institute
of Technology
Pasadena
USA

Editorial Office
MDPI
St. Alban-Anlage 66
4052 Basel, Switzerland

This is a reprint of articles from the Special Issue published online in the open access journal *Remote Sensing* (ISSN 2072-4292) (available at: www.mdpi.com/journal/remotesensing/special_issues/GNSS_Reflectometry).

For citation purposes, cite each article independently as indicated on the article page online and as indicated below:

LastName, A.A.; LastName, B.B.; LastName, C.C. Article Title. *Journal Name* **Year**, *Volume Number*, Page Range.

ISBN 978-3-0365-2066-7 (Hbk)
ISBN 978-3-0365-2065-0 (PDF)

Contents

About the Editors

Nereida Rodriguez-Alvarez

Nereida Rodriguez-Alvarez received her Ph.D. degree in remote sensing from the Universitat Politecnica de Catalunya (UPC), Barcelona, Spain, in December 2011. Since 2018, she has been with the Planetary Radar and Radio Science Group, Jet Propulsion Laboratory, Pasadena, CA, USA. She specializes in the use of bistatic radar signals for sensing the Earth, i.e., the use of signals of opportunity through Global Navigation Satellite System–Reflectometry (GNSS–R), and has widely contributed to the retrieval of land geophysical parameters and ocean surface winds. While at JPL, Nereida has made relevant accomplishments on wetlands, polar sea ice, and floods, for which she received the NASA Early Career Public Achievement Award in 2019. Currently, Dr. Rodriguez-Alvarez continues to explore applications of GNSS–R in Earth science, while she expands her knowledge to planetary science to sense the properties of Solar System bodies.

Mary Morris

Mary Morris joined JPL's Microwave Systems Technology group in March 2017. Her research interests lie at the intersection of remote sensing and atmospheric science, with particular expertise in microwave radiometry, GNSS–R, and tropical cyclones. She is currently a member of NASA's Cyclone Global Navigation Satellite System (CYGNSS) science team and the USAF Compact Ocean Wind Vector Radiometer (COWVR) instrument science team.

Preface to "Applications of GNSS Reflectometry for Earth Observation"

Global Navigation Satellite System–Reflectometry (GNSS–R) has proven an effective and valuable technique that has helped the scientific community learn more about the Earth system. By observing reflected Global Positioning System (GPS) signals, a unique set of data has been built up that complements other Earth observations; the forward-scattered signals can be related to properties of the Earth's surface and therefore indirectly related to geophysical phenomena. GNSS–R observations have enabled many studies of the atmosphere and hydrosphere, which is important for developing a better understanding of Earth system dynamics. Airborne and ground-based measurements enable smaller-scale studies of observation sensitivity before scaling up to spaceborne experiments. The availability of data from satellite missions—such as CYclone Global Navigation Satellite System (CYGNSS), TechDemoSat-1 (TDS-1), and the Soil Moisture Active Passive (SMAP) mission working in reflectometry mode (SMAP-R)—have made a significant impact on the scientific return of the GNSS–R measurements. Data from these spaceborne missions demonstrate the capabilities of GNSS–R and allow us to explore new applications.

This Special Issue provides a number of high-quality peer-reviewed papers that add valuable knowledge to the GNSS–R field. Authors have contributed with their works using aircraft, ground-based, and satellite data. Multiple applications have been addressed: soil moisture from CYGNSS, soil moisture at multiple GPS bands (L1 and L5), biomass, wetlands, flood inundation, ocean winds, and snow and ice thickness. Additionally, the Special Issue presents new analysis: the first evidence of swell and mesoscale ocean eddy signatures, the first evidence of ionospheric plasma depletions, target detection above the Earth's surface, winds near tropical oceanic precipitation systems in the presence of lightning, first evaluations of topography, and the polarimetric sensitivity of SMAP–R to crop growth. Many of these works include novel algorithms employing the use of neural networks. This special issue also incorporates publications dedicated to calibration and validation of GNSS-R, such as investigations on the topographical accuracy in the prediction of the GPS reflection points, the spatial resolution of GNSS–R signals, and the geolocation, calibration, and surface resolution of CYGNSS GNSS-R land observations. Additionally, a method is presented to untangle the incoherent and coherent scattering components in GNSS–R observations, along with the novel applications enabled by this capability.

We would like to thank the authors that have contributed to this Special Issue and acknowledge the quality and the relevance of their work. The works published amplify the applications of the GNSS–R field and increase our knowledge of the Earth system.

Nereida Rodriguez-Alvarez, Mary Morris
Editors

Article

Snow and Ice Thickness Retrievals Using GNSS-R: Preliminary Results of the MOSAiC Experiment

Joan Francesc Munoz-Martin [1,*], Adrian Perez [1], Adriano Camps [1], Serni Ribó [2,3], Estel Cardellach [2,3], Julienne Stroeve [4,5], Vishnu Nandan [4], Polona Itkin [6,7], Rasmus Tonboe [8], Stefan Hendricks [9], Marcus Huntemann [10], Gunnar Spreen [10] and Massimiliano Pastena [11]

[1] CommSensLab—UPC, Universitat Politècnica de Catalunya—BarcelonaTech, and IEEC/CTE-UPC, 08034 Barcelona, Spain; adrian.perez.portero@upc.edu (A.P.); camps@tsc.upc.edu (A.C.)

[2] Institute of Space Sciences (ICE, CSIC), Campus UAB, 08193 Barcelona, Spain; ribo@ice.csic.es (S.R.); estel@ice.csic.es (E.C.)

[3] Institut d'Estudis Espacials de Catalunya IEEC, 08034 Barcelona, Spain

[4] Centre for Earth Observation Science, 535 Wallace Building, University of Manitoba, Winnipeg, MB R3T 2N2, Canada; j.stroeve@ucl.ac.uk (J.S.); Vishnu.Nanda@umanitoba.ca (V.N.)

[5] CPOM, University of College London, Gower Street, London WC1E 6BT, UK

[6] Department of Physics and Technology, UiT The Arctic University of Norway, 8514 Tromsø, Norway; polona.itkin@uit.no

[7] Cooperative Institute for Research in the Atmosphere, Colorado State University, Fort Collins, CO 80523, USA

[8] Danish Meteorological Institute, Lyngbyvej 100, DK-2100 Copenhagen Ø, Denmark; rtt@dmi.dk

[9] Alfred Wegener Institute, Helmholtz Centre for Polar and Marine Research, Klussmanstr 3D, 27570 Bremerhaven, Germany; stefan.hendricks@awi.de

[10] Institute of Environmental Physics, University of Bremen, Otto-Hahn-Allee 1, 28359 Bremen, Germany; marcus.huntemann@uni-bremen.de (M.H.); gspreen@awi-polarstern.de (G.S.)

[11] European Space Agency Directorate of Earth Observation Keplerlaan 1, 2201 Noordwijk, The Netherlands; Massimiliano.pastena@esa.int

* Correspondence: joan.francesc@tsc.upc.edu; Tel.: +34-626-253-955

Received: 13 October 2020; Accepted: 3 December 2020; Published: 10 December 2020

Abstract: The FSSCat mission was the 2017 ESA Sentinel Small Satellite (S^3) Challenge winner and the Copernicus Masters competition overall winner. It was successfully launched on 3 September 2020 onboard the VEGA SSMS PoC (VV16). FSSCat aims to provide coarse and downscaled soil moisture data and over polar regions, sea ice cover, and coarse resolution ice thickness using a combined L-band microwave radiometer and GNSS-Reflectometry payload. As part of the calibration and validation activities of FSSCat, a GNSS-R instrument was deployed as part of the MOSAiC polar expedition. The Multidisciplinary drifting Observatory for the Study of Arctic Climate expedition was an international one-year-long field experiment led by the Alfred Wegener Institute to study the climate system and the impact of climate change in the Arctic Ocean. This paper presents the first results of the PYCARO-2 instrument, focused on the GNSS-R techniques used to measure snow and ice thickness of an ice floe. The Interference Pattern produced by the combination of the GNSS direct and reflected signals over the sea-ice has been modeled using a four-layer model. The different thicknesses of the substrate layers (i.e., snow and ice) are linked to the position of the fringes of the interference pattern. Data collected by MOSAiC GNSS-R instrument between December 2019 and January 2020 for different GNSS constellations and frequencies are presented and analyzed, showing that under general conditions, sea ice and snow thickness can be retrieved using multiangular and multifrequency data.

Keywords: GNSS-R; sea-ice; arctic; snow

1. Introduction

FSSCat was the winner of the "2017 ESA Sentinel Small Satellite (S^3) Challenge" and the overall winner of the 2017 "Copernicus Masters" competition [1]. The mission consists of two "federated" 6U ($10 \times 20 \times 30$ cm^3) CubeSats (^{3}Cat5/A and ^{3}Cat-5/B) in support of the Copernicus Land and Marine Environment services [2], which were successfully launched on 3 September 2020 onboard the Vega SSMS (Small Spacecraft Mission Service) maiden flight in a Sun Synchronous Orbit of 550 km height. The goal of the ESA S^3 Challenge was to develop, in a record time of one year, a novel small satellite mission to complement the Sentinel family missions. One of the instruments onboard FSSCat is the Flexible Microwave Payload-2 (FMPL-2) [3], a dual Global Navigation Satellite System reflectometer (GNSS-R) and microwave radiometer instrument implemented using a Software Defined Radio.

The GNSS signals produce a coherent reflection over the ice [4] that allows for the precise determination of the ice edge [5]. Moreover, other studies show the potential of GNSS-R measurements to infer sea-ice thickness, when the scattering is coherent and the complex reflected signals are downloaded to ground [6,7], based on the reflection between the sea-ice and the sea water underneath. However, this hypothesis has not been confirmed by in-situ measurements, as sea-ice thickness has been only retrieved at L-band by radiometers [8,9].

As part of the validation of the FMPL-2 GNSS-R instrument, the PYCARO-2 (P(Y) & C/A ReflectOmeter-2), an evolved version of PYCARO, [10] was designed to take part in the Multidisciplinary drifting Observatory for the Study of Arctic Climate (MOSAiC) expedition [11,12], the largest campaign ever aiming to study the Arctic Ocean.

The objective of MOSAiC was to understand the evolving Arctic climate system and the role it plays in changing the global climate. MOSAiC was a multidisciplinary campaign to study the Arctic Ocean from different points of view, and involving different teams: Ice, Ocean, Atmosphere, Ecology, Bio-geochemical systems, Satellite Remote Sensing, Modeling, and Aircraft teams. Most of them deployed a large suite of sensors to measure different parameters, such as ice thickness evolution, snow properties, the topography beneath the ice floe, etc. Most of the instruments run for over a year on top of a drifting ice floe. The diversity of simultaneous measurements and the extension of the campaign in time makes it an exceptional opportunity to cross-check and use the data retrieved as model validation for other sensors such as scatterometers, GNSS-R, or radiometer sensors on satellites. Here, we will focus on the MOSAiC remote sensing site and its ancillary measurements, which include:

- Three Remote Sensing Sites (alternating measurements every 2–3 weeks), with the following instruments: L-, C-, X-, Ka-, and Ku-band microwave scatterometers, P- to L-band, C-, X-, Ku-, K-, and W-band microwave radiometers, IR and hyperspectral cameras, and two multiconstellation and multiband GNSS-R instruments. The GNSS-R instruments (Figure 1) are part of a joint effort from the Institut d'Estudis Espacials de Catalunya (IEEC): Institute of Space Sciences (ICE, CSIC), and Universitat Politècnica de Catalunya (UPC) IEEC sections.
- Regular transects (1 km length over different sea-ice types) to measure total sea ice thickness, snow depth and density, Ku- and Ka-band radar backscatter, L-band radiometry, and additionally surface albedo with the returning insolation from spring onwards.
- Other ice, snow, ocean, and atmospheric measurements.

This work describes the data collected by one of the GNSS-R instruments developed by IEEC/UPC. Section 2 describes the instrument, the experiment set up, and the ground truth data used, Section 3 describes the Interference Pattern captured by the instrument, produced by the sea-ice floe, modeled as a four-layer dielectric, and Section 4 analyzes the results of the data retrieved by the PYCARO-2 IEEC/UPC instrument and its comparison to the four-layer model is presented in Section 3.

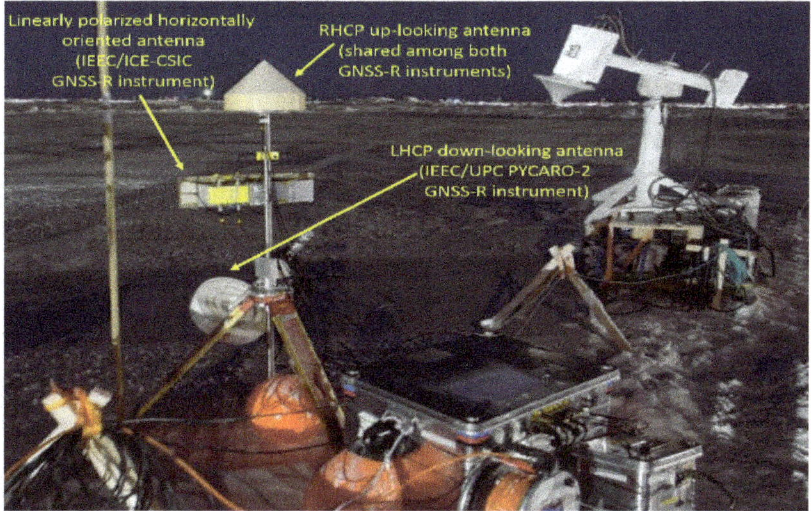

Figure 1. IEEC's GNSS-R instruments: IEEC-ICE/CSIC GNSS-R instrument is formed by the zenith antenna and by the 5 × 1 patch array pointing to the horizon. IEEC/CTE-UPC GNSS-R instrument is formed by the zenith antenna and the 45° incidence angle LHCP antenna. [Credits: Gunnar Spreen].

2. Instrument Description and Experiment Setup

The GNSS-R instruments belonging to the Remote Sensing site are divided in two. One of them is the ICE/CSIC instrument, which is a GNSS-R instrument working at linear polarization to collect the reflected signals off the sea-ice. Its aim is to provide sea-ice properties from scattered GNSS signals. The other one is the IEEC/UPC instrument (named PYCARO-2), which operates at Left hand Circular Polarization (LHCP) and aims to support the reflectivity modeling activities in support to the FMPL-2 data processing.

Both instruments share a common power and data interface, physical enclosure, and the Right Hand Circular Polarization (RHCP) antenna used for time and geo-reference the instrument. The primary objective of both GNSS-R instruments is to provide ground-truth data for reflection modeling and originally also to support the calibration/validation of the FMPL-2 [3] onboard the FSSCat mission [1,2]. This manuscript focuses only on the data retrieved by the IEEC/UPC PYCARO-2 instrument [13], which is equipped with an antenna working at the same polarization (LHCP) as the FMPL-2 instrument onboard FSSCat.

2.1. Circular Polarization GNSS-R Instrument

PYCARO-2 is an evolved version of the PYCARO instrument, which was the first GNSS-R instrument collecting polarimetric reflected Galileo signals from a stratospheric balloon (Bexus 19 in 2014 [14]). PYCARO was also the payload of the UPC ^{3}Cat-2 CubeSat [15]. PYCARO-2 is able to track Beidou B1/B2 in addition to GPS L1/L2 and Galileo E1/E5, as well as GLONASS L1/L2.

The down-looking antenna is a LHCP choke-ring antenna, tilted 45° downwards to maximize the power of the reflected signals, as most GNSS satellites reach ~50° elevation angle in polar regions. The antennas have a very smooth radiation pattern, as shown in Figure 2a and their configuration (height and pointing) can be seen in Figure 2b.

The instruments were shipped to Tromsø, Norway, on July 2019, from where the Polastern icebreaker carrying the MOSAiC expedition departed on 20 September 2019. The Remote Sensing (RS) site was set up during October 2019, and the instruments were successfully installed on 24 October 2019, as shown in Figure 2c. Despite being installed in October, several ice cracks (Figure 2d) during

November and December forced a relocation of the RS site, including the PYCARO-2 instrument. The final location of the instrument for the winter season was set in late December 2019.

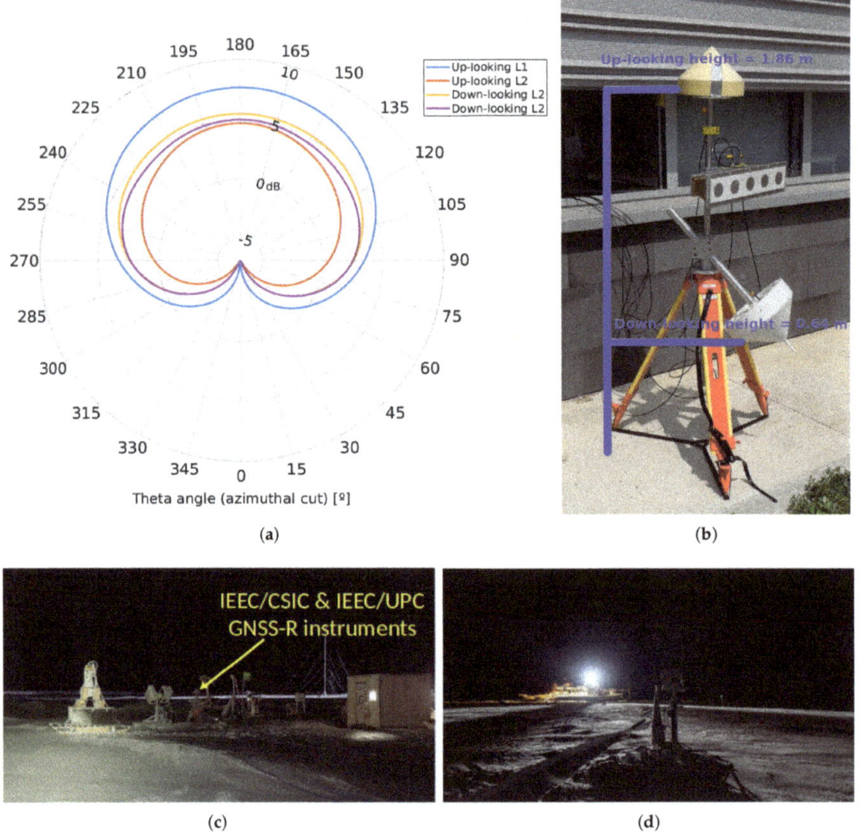

(a)

(b)

(c)

(d)

Figure 2. (**a**) Up-looking and down-looking antenna pattern; (**b**) antenna distribution and heights of each antenna; (**c**) MOSAIC Remote Sensing site with GNSS-R instruments (orange arrow), and (**d**) MOSAiC ice cracks during November–December 2019 [Credits: Gunnar Spreen and Stefan Hendricks].

2.2. Ground-Truth Data

In the vicinity of the GNSS instrument, different types of measurements were collected of both snow depth, and sea ice thickness. Measurements were repeated every few weeks by the combination of a Magnaprobe snow probe and broadband electromagnetic induction sensors (Geophex GEM-2). The Magnaprobe was used to retrieve snow thickness, with a precision better than 1 cm [16]. The GEM-2 was used to measure the combined thickness of the snow and the sea ice, as described by [17]. Finally, both measurements are combined to retrieve snow and sea-ice thickness separately. In this study, we use the quick-view GEM-2 thicknesses based on the 18 kHz in-phase channel and processed directly onboard and quality controlled against manual drill-hole observations. We chose this channel out of the available frequency set of the GEM-2 for optimal retrieval of sea ice thickness with an accuracy better than 10 cm.

The snow depth was also measured using an avalanche probe in the area surrounding the GNSS-R instrument mast. Figure 3a shows the evolution of the mean and standard deviation of the snow thickness in the area surrounding the GNSS instrument. Figure 3b shows a broad distribution of snow depths as a Probability Distribution Function (PDF) and points to a highly inhomogeneous snow surface, with developed snow dunes in two different time instants (blue, 12 January 2020, and green, 7 February 2020). Finally, Figure 3c shows the PDF of the sea-ice thickness measurements.

It can be observed that, for the period between December 2019 and January 2020, the PDF and the avalanche probe measurements are consistent, which is not the case for February 2020. The instrument performance is compared to these measurements, from where consistent ground-truth data is available. Furthermore, ice core drillings were performed during the first days of December, showing that the ice core during the week of the 21st of December 2019, was around 0.8 m and 1 m thick. The same ice core drillings were conducted in the time period shown in Figure 3c to validate the measurements taken by the Magnaprobe.

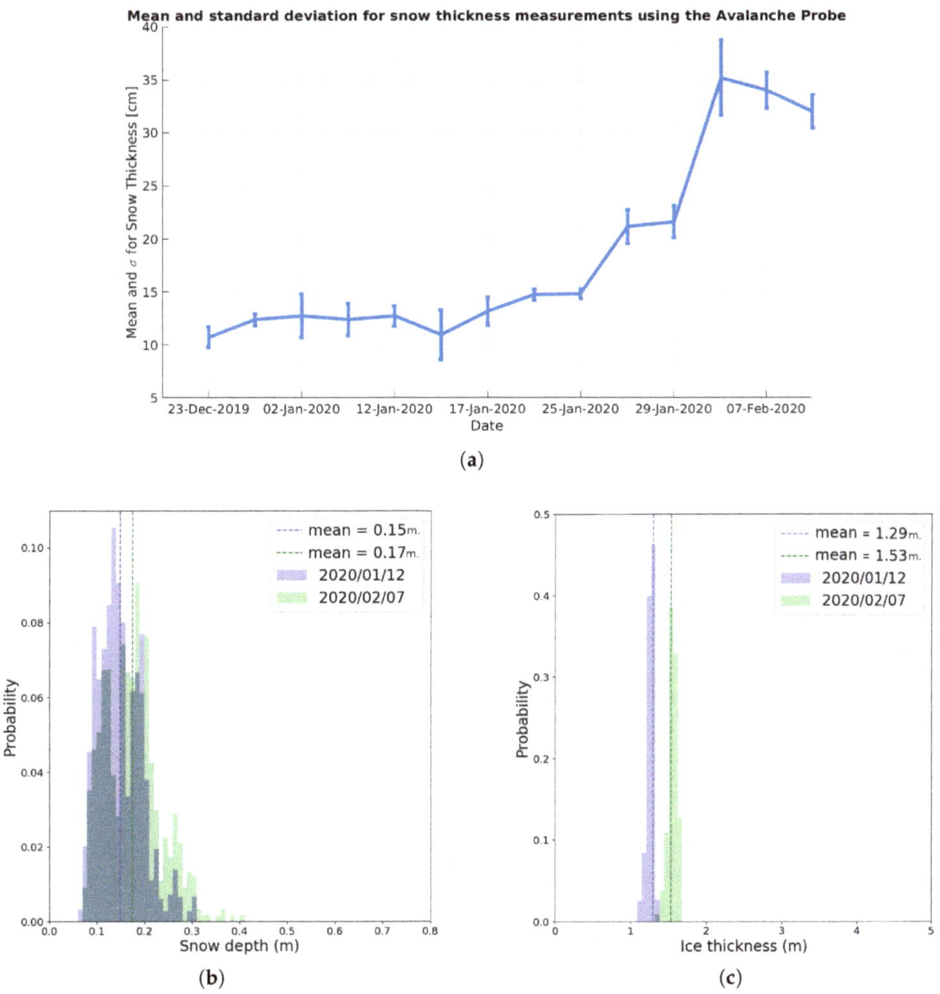

Figure 3. (**a**) Avalanche probe measurements in the vicinity of the GNSS-R instrument; (**b**) Snow depth and (**c**) Sea-ice thickness PDF for different dates in the level ice area adjacent to Remote Sensing site measured by Magnaprobe and GEM-2.

3. Theoretical Background: IPT Applied to the Ice Floe

In navigation receivers, multipath is avoided as much as possible as it degrades positioning accuracy. For ground-based GNSS-R instruments, the signal collected by a GNSS antenna is a combination of the incident wave, i.e., the one coming directly from the GNSS satellite at RHCP and the signal reflected over a certain surface. The combination of the direct and reflected signals creates strong reflectivity fluctuations or fringes, called an "Interference Pattern". In the PYCARO-2 case, the reflection of the GNSS signal on the sea ice floe produced the same type of fluctuations.

The Interference Pattern Technique (IPT) was conceived by [18] to extract a number of geophysical parameters from the Earth by means of ground-based instruments. Most of its applications are soil moisture [19], vegetation height [20], water level [21] monitoring, sea state [22], or snow depth [23] estimation. However, most of those works are based on a three-layer model, with the first layer being air, a second thin layer of substrate (i.e., snow, crop field), and finally, a third layer which is a semi-infinite reflective surface (i.e., land). Interfering reflections from multiple layers of snow and ice in Antarctica were analyzed in [24], however, according to the authors' knowledge, the IPT has never been applied over an ice floe, which contains four dielectric layers: air, a snow layer on top of the sea ice, the sea ice itself, and the sea water underneath.

To study the Interference Pattern (IP) fringes created by the multiple reflections in the floe, the instrument retrieves the signal-to-noise ratio (or peak of the Delay-Doppler Map minus noise background measured a few lags before the correlation peak [25]) once every 10 s, collecting thousands of reflections from different bands and constellations. The total amount of reflections collected was 300.000 for Beidou B1D1, 125.000 for Beidou B2D1, 425.000 for Galileo E1C, 290.000 for Galileo E5b, 275.000 for GPS L1 C/A, and 235.000 for GPS L2 P(Y). The data used to perform this study was collected from 20 December 2019 to 27 January 2020. However, due to several operational procedures of the instrument, most of the data was obtained between 15 and 27 January 2020. This study is focused on the analysis of this period of data, from which results and methodology will then be applied to the rest of the campaign.

3.1. Four-Layer IPT Model: Theoretical Definition

The scattering geometry is modeled as shown in Figure 4, where the dielectric constant of each media as in Section 6.2 of [26], and in particular, the sea-ice dielectric constant model is based in [27], but for the direct model. This study is performed assuming the dielectric constant of the substrate is not varying. Furthermore, simulations have been performed to prove that the effect of varying the dielectric constant of the substrate produced a negligible effected as compared to the variation of either the snow or the ice thickness. In addition, the properties selected (i.e., snow density, ice type, and ice temperature) used to compute the dielectric constants shown in Figure 4 are the average values of what MOSAiC researchers' measured. Therefore, following the procedures described in [28], the power received by either the zenith-looking or the 45°-looking antenna is proportional to (see Figure 2a).

$$P(\theta) \propto \left| E_i(\theta_i) + E_r(\theta_i) \right|^2 = |E_0|^2 \cdot \left| F_n(\theta_{dir}) + F_n\left(\theta_{ref}\right) \cdot R(\epsilon_r, \theta_i) \cdot e^{j\Delta\phi} \right|^2, \tag{1}$$

where E_i is the incident electric field, E_r is the reflected electric field, $F_n(\theta)$ is the antenna voltage pattern (amplitude and phase) for the up-looking and down-looking signals arriving to either the up-looking or down-looking antennas, with different patterns for each one (see Figure 2a), θ_{dir}, and θ_{ref} are the off-boresight arriving angles of the direct and reflected signals, θ_i is the incidence angle of the signal arriving from a particular GNSS spacecraft, E_{0_i} is the direct signal, the one that would be received if there were no interference, $R(\epsilon_r, \theta)$ is the reflection coefficient as defined in (3), and $\Delta\phi$ is the phase associated to the geometry. $\Delta\phi$ is given by

$$\Delta\phi = \frac{4\pi}{\lambda} \cdot h \cdot sin(\theta), \tag{2}$$

where λ is the electromagnetic wavelength, depending on the frequency band (L1 or L2), h is the antenna height, and θ is the satellite elevation angle. The reflectivity on top of the snow layer is computed iteratively as

$$R_i = e^{-(\frac{4\pi\sigma}{\lambda})^2} \frac{r_{i,i+1} + r_{i+1,i+2} \cdot e^S \cdot e^{j2\Psi}}{1 + r_{i+1,i+2} \cdot r_{i+1,i+2} \cdot e^S \cdot e^{j2\Psi}}, \tag{3}$$

where $r_{i,i+1}$ is the Fresnel coefficients between layers i and $i+1$, for the four-layer model being 1 = air, 2 = snow, 3 = ice, and 4 = saline water; S (see (4)) accounts for the reflectivity decrease due to the roughness at the interface [18], and Ψ is the phase produced by the reflection between the different layers, as detailed in (5)

$$S = -8 \cdot \left(\frac{\pi\sigma}{\lambda} \sqrt{\varepsilon_{r_{i+1}} - \varepsilon_{r_i} \cdot \cos^2(\theta)} \right)^2 \tag{4}$$

$$\Psi = \frac{2\pi}{\lambda} t_{i+1} \cdot \sqrt{\varepsilon_{r_{i+1}} - \varepsilon_{r_i} \cdot \cos^2(\theta)} \tag{5}$$

where t_{i+1} is the thickness of the $i+1$ layer, ε_{r_i} is the permittivity of the i layer, and σ the surface roughness of the interface.

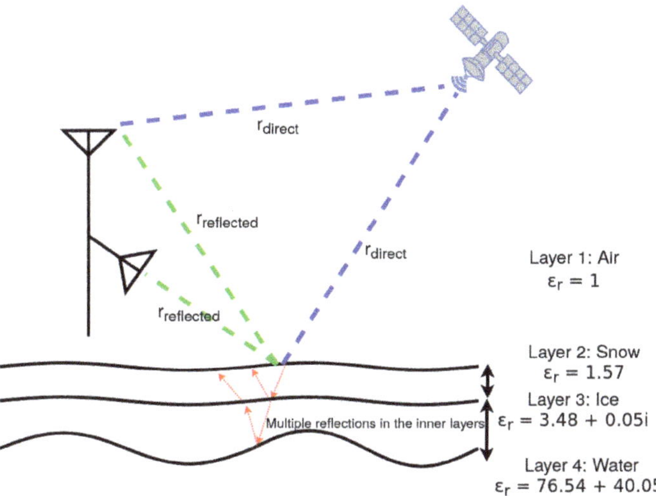

Figure 4. Four-model layer applied to PYCARO-2 instrument in MOSAiC campaign.

3.2. Four-Layer IP Model: PYCARO-2 Case

Each of the different surface permittivity values (ε_{r_i}) for this study have been modeled. The snow has been modeled as dry snow, as per in-situ measurements during the expedition, with a density of 296 kg/m^3, and $-25\,°$C [29]. The ice has been modeled as multiyear ice, with a mean temperature of $-25\,°$C. Note that simulations have been carried out at different temperatures, and the impact of varying the mean temperature is negligible in front of the interference pattern produced by variations on the layer thickness. Finally, the water layer has been modeled as saline water (32 psu) and at a temperature of $-1.7\,°$C [29].

3.2.1. Interference Pattern in the RHCP Zenith-Looking Antenna

This study is focused on two frequencies: 1575.42 MHz, used by GPS L1 C/A code and Galileo E1C; and 1207.14 MHz, used by Galileo E5b and Beidou B2D1 signals. As a first example, the four-layer IP model simulation results are shown in Figure 5a for the up-looking RHCP antenna at 1575.42 MHz for different values of snow, ice thickness, and surface roughness. Note that the antenna heights used to simulate the model are the ones indicated in Figure 2b. Moreover, the contribution to the interference pattern from the reflected path to the up-looking antenna starts to vanish at elevation angles larger than 25° due to the antenna pattern (see Figure 2a). Therefore, the model is only presented up to 30°. Note that the presented IPT is the normalized version of $P(\theta)$ from Equation (1) (i.e., normalized by $|E_0|^2$).

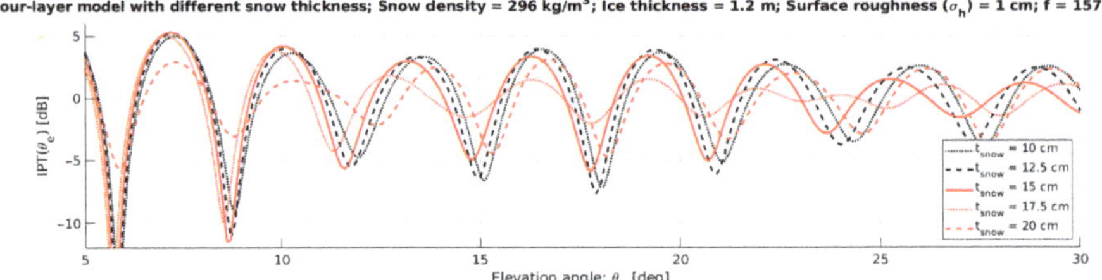
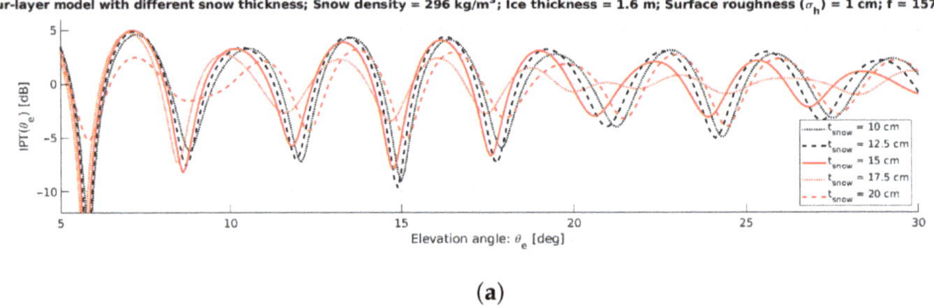

(a)

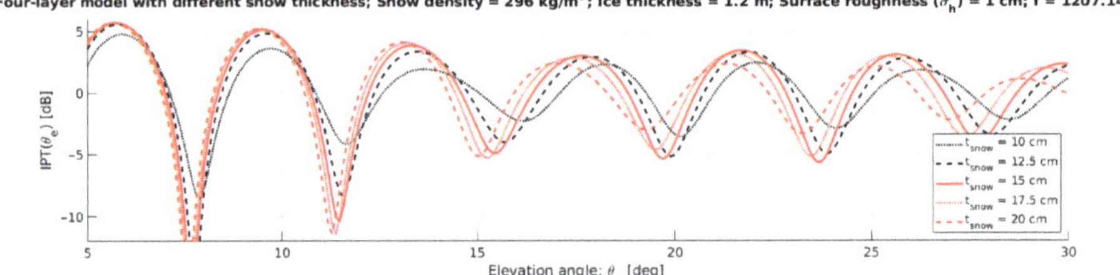
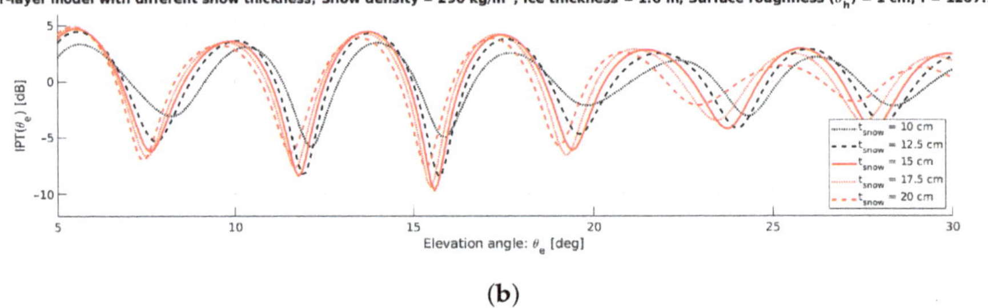

(b)

Figure 5. Four-layer model at RHCP for five different snow layer thickness at two predefined ice thickness, (**top**) 1.2 m, and (**bottom**) 1.6 m. Simulation at 1575.42 MHz (**a**) and 1207.14 MHz (**b**). Assuming a surface roughness between layers of 1 cm.

As it can be seen, the position of the different peaks as well as their shape depends on the thickness of the snow layer. The position of both peaks and notches depends mainly on the snow thickness: from 10 to 15 cm thickness, the position of the notches shifts towards the left (lower elevation angles), but for a thicker snow layer (20 cm) the notch position moves towards higher elevation angles (e.g., the notch at 15° of elevation in Figure 5a). In addition, there is also a small dependence on the ice thickness. Comparing the top (ice thickness = 1.2 m) and bottom (ice thickness = 1.6 m) plots in Figure 5a, the notch position is shifted slightly towards lower elevation angles, where thickness variation has less impact in the notch position.

A similar phenomenon happens at 1207.14 MHz (i.e., Galileo E5b and Beidou B2D1) where, depending on both the snow and the ice thickness in the four-layer model, the position where the minima occur fluctuates. As shown in Figure 5b, the four-layer model is also affected by the snow thickness in both the peaks' shape and the notch position. However, the ice thickness has a larger impact on the notch position as compared to the 1575.42 MHz case as, due to the longer wavelength, the penetration depth into the ice is larger.

Finally, Figure 6 presents a set of curves representing different values of both ice and snow thickness at 1575.42 MHz and 1207.14 MHz. The 1207.14 MHz signal is more sensitive to variations in the ice thickness than to snow thickness (10 to 15 cm), in terms of the positions of peaks and notches. In the 1575.42 MHz case, the nulls' position in the IP move ~0.5° in elevation, while in the 1207.14 MHz case they move ~1.5°.

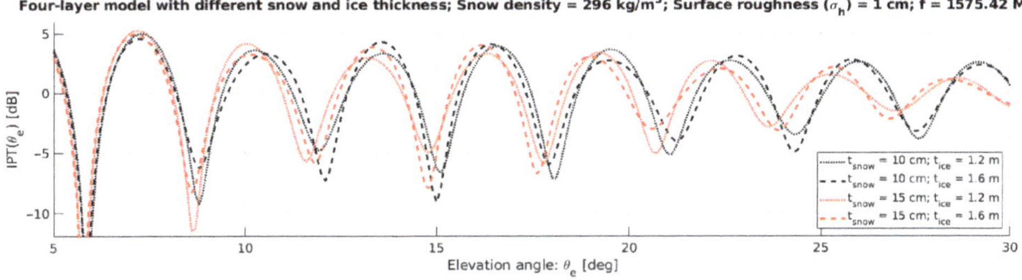

Figure 6. Comparison of the RHCP four-layer model at 1575.42 MHz (**top**) and 1207.14 MHz (**bottom**), using four different combinations of snow and ice thickness.

3.2.2. Interference Pattern in the LHCP Down-Looking Antenna

As the LHCP signal is received by the 45°-tilted down-looking antenna, its reception is not limited by the lower back lobes of the radiation pattern of the RHCP antenna. For this reason, model predictions are presented for elevations from 5° to 60°. Figure 7a presents the IP at LHCP and 1575.42 MHz. Using the same parameters as for the RHCP signal case, the sensitivity to snow thicknesses is even larger than for the RHCP case (Figure 5a). Note that around 30° and 45° of elevation, the ripples produced by the IP have the same shape for different snow thickness, i.e., the curve for 10 cm has the same shape as that for 20 cm, both for an ice thickness of 1.2 m and 1.6 m.

Figure 7b presents the IP at LHCP and 1207.14 MHz. As compared to Figure 5b, the simulated IPs are now sensitive to both ice and snow thickness variations. Comparing both IPs (Figure 7b top and Figure 7b bottom) around 20° of elevation, it can be appreciated that the valleys are modulated by the snow thickness, and their depth (amplitude) and shape depend on the ice thickness.

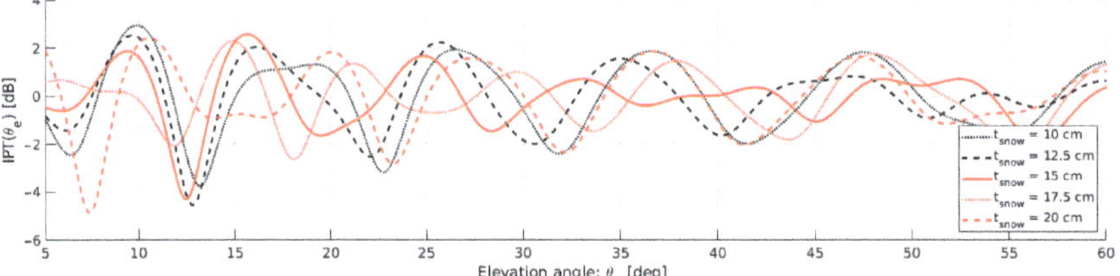

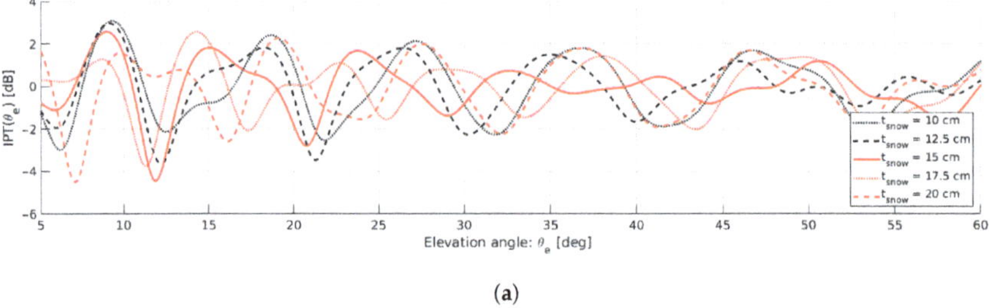

(a)

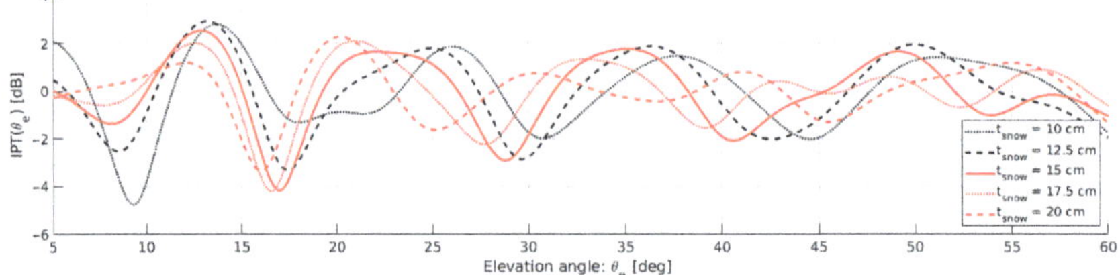

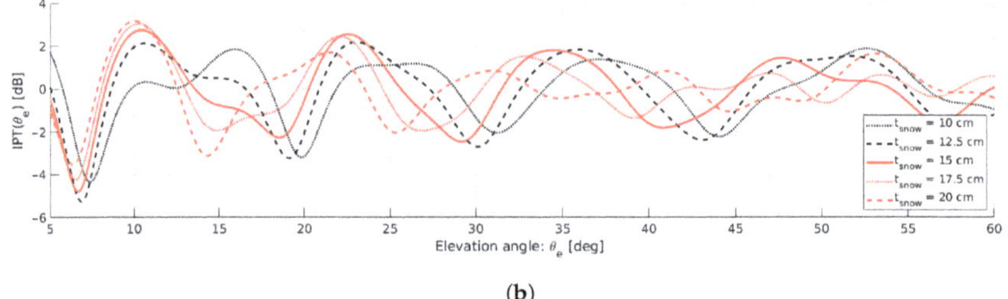

(b)

Figure 7. Four-layer model at LHCP for five different snow thickness, at two predefined ice thickness, 1.2 m (top), and 1.6 m (bottom). Simulation at 1575.42 MHz (**a**) and 1207.14 MHz (**b**). Assuming a surface roughness of 1 cm.

Comparing the overlaid curves from Figure 7a,b and for different ice thicknesses, in Figure 8 it can be clearly seen that the ice thickness variation does not affect the interference pattern produced between 30° and 42.5° (R2 in Figure 8) of elevation, although at low elevation angles, between 10° and 30°, the IP is sensitive to both snow and ice thickness. Note that, for the sake of simplicity, two ice thickness are shown, but simulations have been carried out for the complete range between 0.6 m. and 1.8 m. with steps of 0.1 m., showing that ice thickness variations are not affecting the interference pattern produced between 30° and 42.5°.

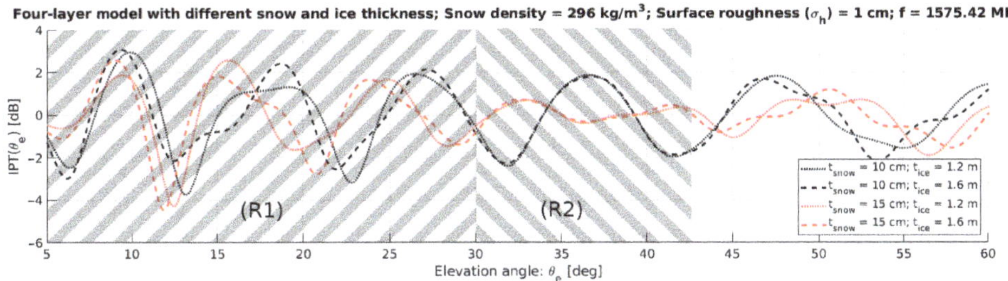

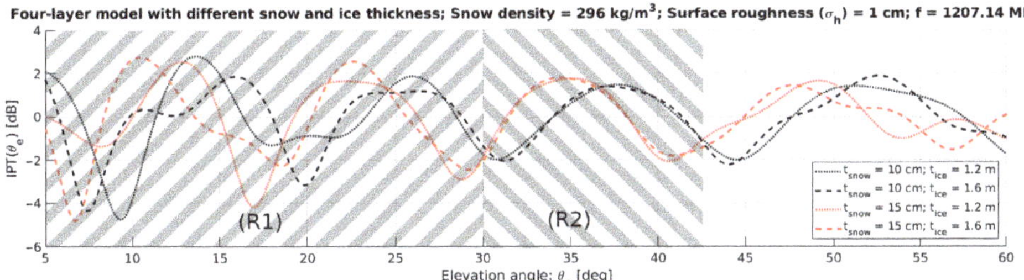

Figure 8. Comparison of the LHCP four-layer model at 1575.42 MHz and 1207.14 MHz, using four different combinations of snow and ice thickness. IPT in region 1 (R1) is affected by ice and snow thickness, while IPT in region 2 (R2) is unaffected by ice-thickness variations.

In order to summarize, the IPT with the four-layer model is sensitive to changes in the snow and ice thickness in both RHCP and LHCP signals. However, at 1575.42 MHz, the RHCP has a lower sensitivity to ice thickness. In the 1207.14 MHz case, as the wavelength is longer and thus able to penetrate more into the ice, it shows more sensitivity to ice thickness variations. The LHCP case shows a mixture of both phenomena depending on the elevation angle: for low elevation angles, the IPT is affected by both snow and ice thickness, but for angles between 30° and 42.5°, the IPT mainly depends on the snow thickness. Note that this model does not take into account any antenna orientation, as antenna pattern of the received signal is compensated for.

4. Data Analysis

The PYCARO-2 instrument receives the direct and reflected GNSS signals. In this section, the waveforms' signal-to-noise ratio (SNR) are presented as radar plots for the different frequencies and constellations. The fringes produced by the reflection geometry can be easily detected by the angle of arrival of the signal (i.e., azimuth and elevation angles).

Figure 9a shows the IPT corresponding to Galileo E1C code, and Figure 9b to the GPS L1 C/A code. It can be clearly identified, in both RHCP signals, where the reflections are coming from, located in the azimuth range from ~180° to ~300°. However, in the LHCP signal collected by the down-looking antenna, other interference patterns can be identified in both GPS and Galileo, most of them grouped around 220° and 250°. Moreover, at 1207.14 MHz (Figure 9c,d) Galileo E5B and Beidou B2D1 signals are also exhibiting an IP. It can be seen that the pattern can be appreciated around 210° and 260° of azimuth, whereas for other angles the fringes are slightly changed or even lost.

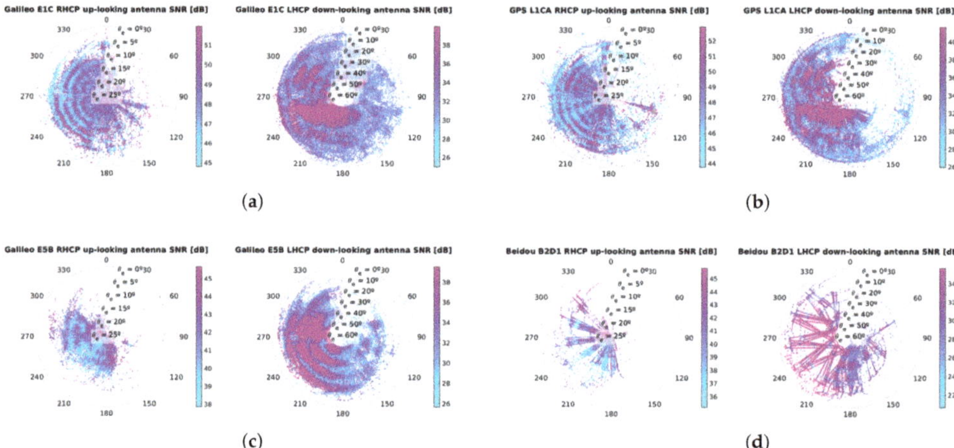

Figure 9. Radar plot of the SNR for (**a**) Galileo E1C, (**b**) GPS L1 C/A, (**c**) Galileo E5b signal, and (**d**) Beidou B2D1 during January 2020.

Because of the antenna pattern of the 45° down-looking antenna, most of the IPs are detected around 220° and 250°, where the fringes are stable at both frequencies. In order to compare the measurements from PYCARO-2 and the four-layer model described in Section 3, the signals have been filtered, and only those received in the 220°–250° of azimuth are used.

In order to ease the visualization of the signal in a 2D form, each SNR measurement retrieved by the instrument has been binned depending on the elevation angle of the transmitting satellite. Each measurement has been divided depending on the receiving antenna (i.e., RHCP or LHCP) and band (1575.42 MHz or 1207.14 MHz). Figure 10 shows, after calibration and azimuth filtering, a 2D histogram of the PYCARO-2 measurements overlaid with the RHCP (a) and the LHCP (b) four-layer model for different snow and ice thicknesses. In order to ease the representation and comparison of the interference pattern, the model curve has been rescaled to have its range in the same span as the PYCARO-2 measurement.

Note that, as seen in the previous section, both snow and ice thickness have an impact on the 1575.42 MHz band. However, it is not large enough to clearly distinguish which model parametrization is closer to the observed pattern. For the 1207.14 MHz case, the number of reflections received for low elevation angles is very small, and therefore, the ripples caused by the interference pattern cannot be clearly identified.

When moving to the LHCP case, as seen in Figure 10b, the retrieved signal is noisier at 1575.42 MHz, where there is only a small range of elevation angles (20° to 35°) with ripples present. However, at 1207.14 MHz, the signal presents notches that can be easily identified as the ones described by the red curve in Figure 10b. In this case, the signal at 1207.14 MHz is able to penetrate deeper into the substrate (i.e., the snow), and therefore it is slightly less sensitive to snow variations. As shown in Figure 3b,c, the snow depth has a larger dispersion than the sea-ice thickness.

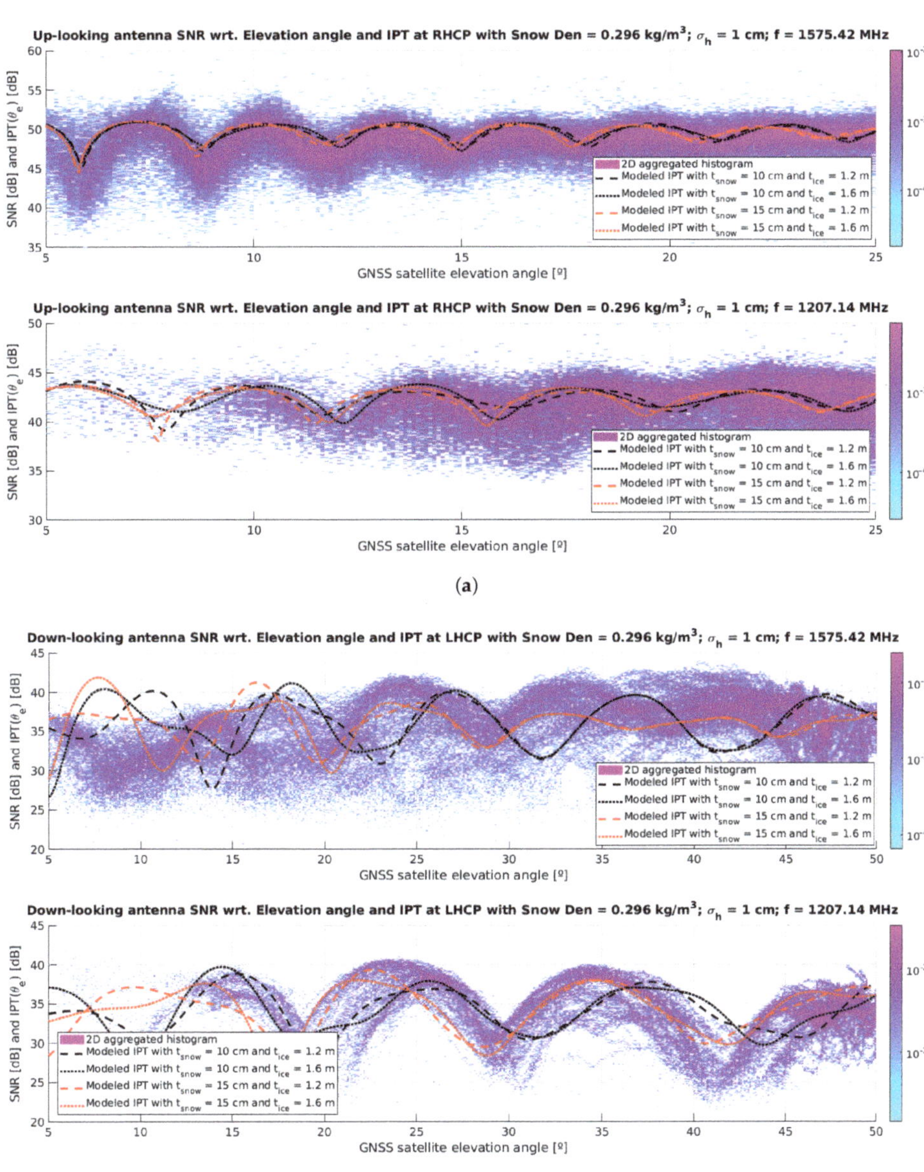

Figure 10. Aggregated 2D histogram (density plot) for the up-looking RHCP signal SNR for both 1575.42 MHZ and 1207.14 MHz; and IPT for RHCP (**a**) and LHCP (**b**) signal for different snow and ice thickness.

4.1. Snow and Ice Thickness Retrievals: Results and Discussion

In order to compare the model and the 2D histogram, the median curve shown in Figure 11 has been computed over a running window with a width of 0.5° in elevation angle. In order to compare those curves and the model, a nonlinear least square (NLS) minimization is performed in different steps. Note that the error function used for the process is the mean-squared error (MSE) of the SNR (dB).

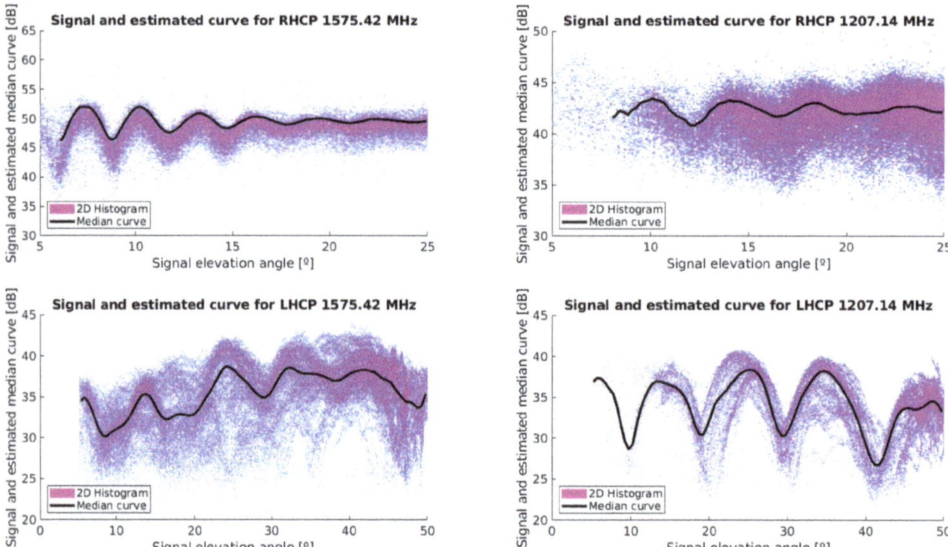

Figure 11. 2D histograms (density plot) and median curve (in black) for the received signals in GPS L1 C/A, Galileo E1C, Galileo E5b, and Beidou B2D1.

In order to properly retrieve snow and ice thickness, a first NLS minimization is performed for the LHCP signal at elevation angles between 30° and 42.5°, where the minimization parameter is the snow thickness. As seen in the previous section, the LHCP signal around 30° and 42.5° elevation angle is insensitive to ice thickness variations but not to snow thickness variations. In this case, the notch position of the different ripples can be used to estimate the snow thickness.

When the snow thickness is retrieved, a second NLS minimization is performed for the RHCP up-looking signal, between 5° and 25° of elevation. In this second case, the variable used for the second minimization is the ice thickness, using the snow thickness retrieved by the first NLS minimization. Note that in the RHCP case, both snow and ice thickness have an impact on the notch position, and therefore it is very advisable to first retrieve the snow thickness, as in the LHCP case the signal is mainly sensitive to the snow. Moreover, as the RHCP signal is coming from the up-looking antenna, the IPT can only be applied to very low elevation angles, where the antenna pattern has similar values for both the direct and the reflected signal. In this case the reflected signal is also RHCP, as the incidence angle is smaller than the Brewster angle.

4.1.1. Error Function Analysis and Ambiguity Removal

Both the snow and ice thickness IP produce periodic notches. Therefore, almost equal interference patterns are generated by different combinations of snow and ice thickness. To resolve this uncertainty, data from the two frequency bands (1575.42 MHz and 1207.14 MHz) can be used. The error function with respect to snow and ice thickness is shown in Figure 12. It can be clearly seen that the periodicity of the error function at 1575.42 MHz and 1207.14 MHz are different. In our case, there is a point where both error functions exhibit a minimum at the same time (blue dashed line). This point is the actual snow-ice thickness.

In the case of snow thickness (Figure 12, top), the signal presents a periodicity every 10 cm, and therefore both signals will not present a minimum at the same time in the observation range (i.e., 5 to 35 cm). Therefore, the actual snow thickness can be estimated by overlapping the two normalized error functions. By looking to the sum of both normalized curves, the minimum value is the actual thickness (14.4 cm), which is very close to the measurements by the MOSAiC Remote Sensing team, which were 14.5 ± 0.2 cm.

As for the ice thickness case, as shown in Figure 12 (bottom), a minimum in the MSE function is present every ~20 cm. Similar to the snow thickness case, the periods are not the same for each band. However, in this case, as the observation range (i.e., 0.5 to 2.5 m) is larger, there are different combinations of thickness values producing a minimum at the same time: 0.58 m. and 1.24 m. Since some a priori knowledge of the ice is required to resolve the ambiguity, we rely on the measured value of 1.21 m on 15 January 2020. Therefore, from both possible solutions, the correct value is likely 1.24 m.

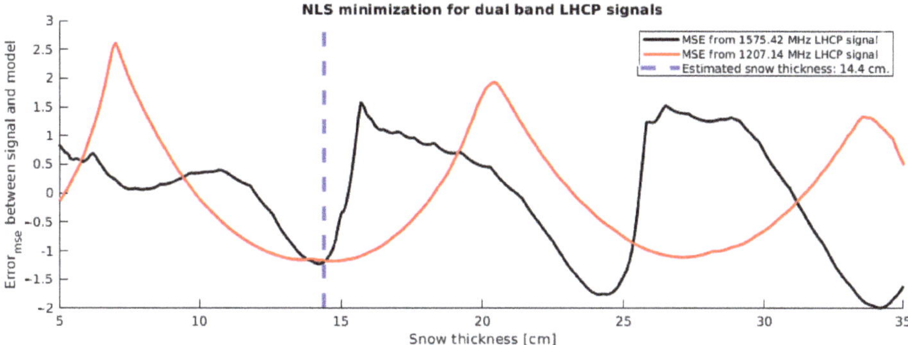

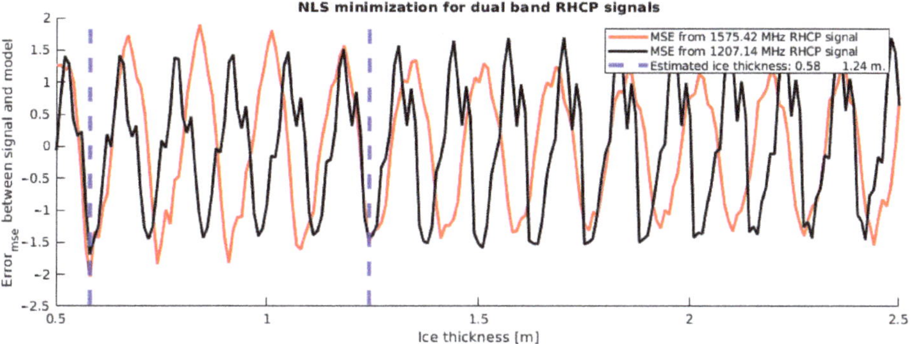

Figure 12. NLS minimization output for the data set between the 15th and the 27th of January 2020. MSE is represented for both snow and ice thickness, for different signal frequencies and polarization, and it has been normalized (z-score normalization) to ease its visualization. Note that the exact place where both signals present a minimum at the same thickness is represented by the blue dashed line thickness and the ice thickness have an impact on the notch positions; therefore, it is very advisable to first retrieve the snow thickness. As in the LHCP case, the signal is mainly sensitive to the snow.

4.1.2. Model and Measured Signal Overlay

Once the thickness parameters have been retrieved, they are used as input for the four-layer IPT model. The modelled IPT is then overlaid on top of the PYCARO-2 data, as shown in Figure 13. Note that the IPT model has been normalized in range (i.e., added a bias and multiplied by a scale factor) in order to better compare it with the SNR values of PYCARO-2.

It can be noticed that the IPT curves are not 100% coincident with PYCARO-2 data in all elevation angles. Despite that, the estimated thickness values are extracted from an MSE minimization process, therefore giving a minimum error for all elevation angles with respect to the median curve used.

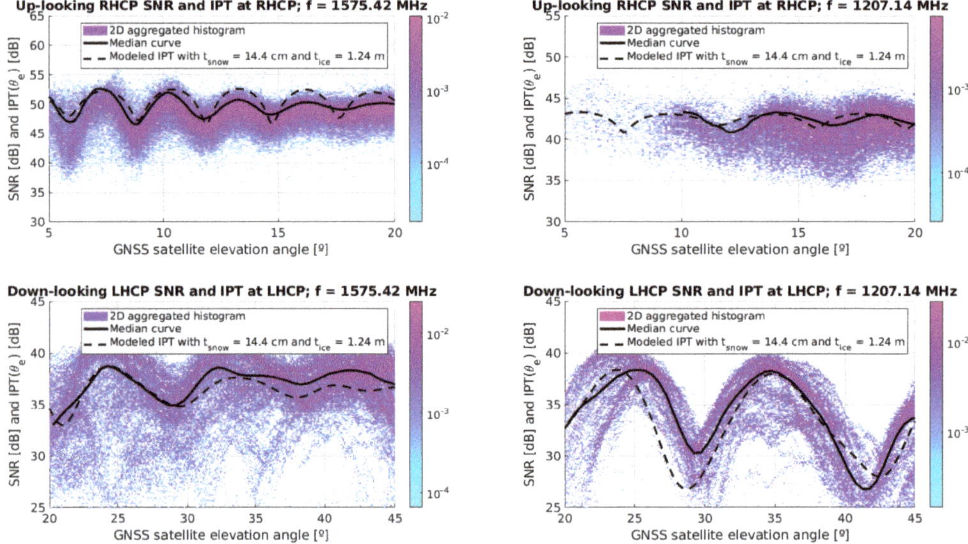

Figure 13. Modeled IPT using the retrieved snow and ice thickness overlaid on top of the signal retrieved by the PYCARO-2 instrument in the MOSAiC campaign during January 2020. Both the median curves and the modeled interference patterns are represented in black.

Note that in the 1575.42 MHz RHCP signal the notches are also very clear, and therefore we can check that the IPT model and the median curve present the same notches at 6°, 8°, 12°, 14.5°, and 17.5° of elevation angle. In the same way, the median curve and the IPT model for the 1207.14 MHz RHCP signal is presenting two notches at the same position, around 12° and 15.5° of elevation angle.

In order to validate the model for other data sets, another period of data has been used, between the 23rd and 27th of December 2019. In this case, the snow thickness measured from MOSAiC researchers was 11–13 cm, and the estimated ice thickness by means of ice core drilling performed some days before, on 20 December 2019, was around 90 cm. Figure 14 shows the error functions for both ice and snow thickness. In this case, for the minimization process at LHCP, both curves exhibit a minimum at 12 cm of snow thickness, being consistent with the measurements carried out during the campaign. Furthermore, the error curves for RHCP present once more a combination of two values with a minimum of 0.79 m and 1.68 m. In this case, thanks to the knowledge of the actual ice floe thickness, the selected solution is 0.79 m.

The overlaid version is presented in Figure 15. In this case, the notches at 1207.14 MHz, both RHCP and LHCP, are well-matched, as well as the minima and maxima at 1575.42 MHz, LHCP. Despite that, the modeled IPT for the RHCP signal at 1575.42 MHz, and the actual measured data by PYCARO-2 has some mismatch. Even then, the snow and ice thicknesses selected are the ones presenting a minimum MSE between the median curve and the IPT curve.

Even though the estimated ice thickness in this second case differs by 10 cm from the actual measured value, the accuracy of the GEM-2 instrument is 10 cm as well. In addition, as pointed out by the MOSAiC researchers, the ice floe is not evenly growing. To prove that, underwater pictures were taken by a remotely operated vehicle (ROV) during MOSAiC expedition. Furthermore, it is known that small (e.g., 10 cm) and local (meter-scale) undulations are present during the sea-ice freeze up [30], and therefore a single GEM-2 measurement cross-validated using an ice drilling may not be representative of the overall sea-ice thickness but an approximation. It is important to note that this second example of measurements was conducted during December, which is the time period when the ice floe is patched together to form new ice. In this case, some regions may present thinner or thicker spots of ice, which can affect the GNSS-R data measured by PYCARO-2. This phenomena may not be present in the data presented for January, as the different ice sheets have been already merged together.

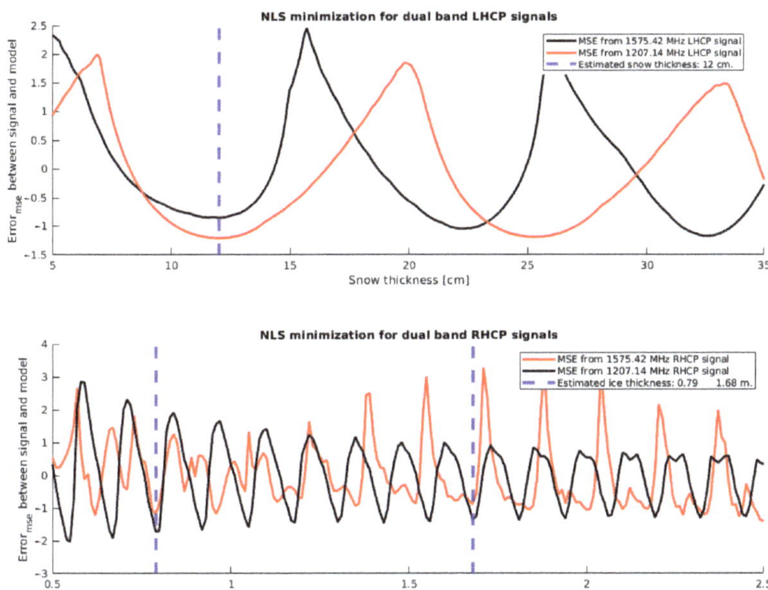

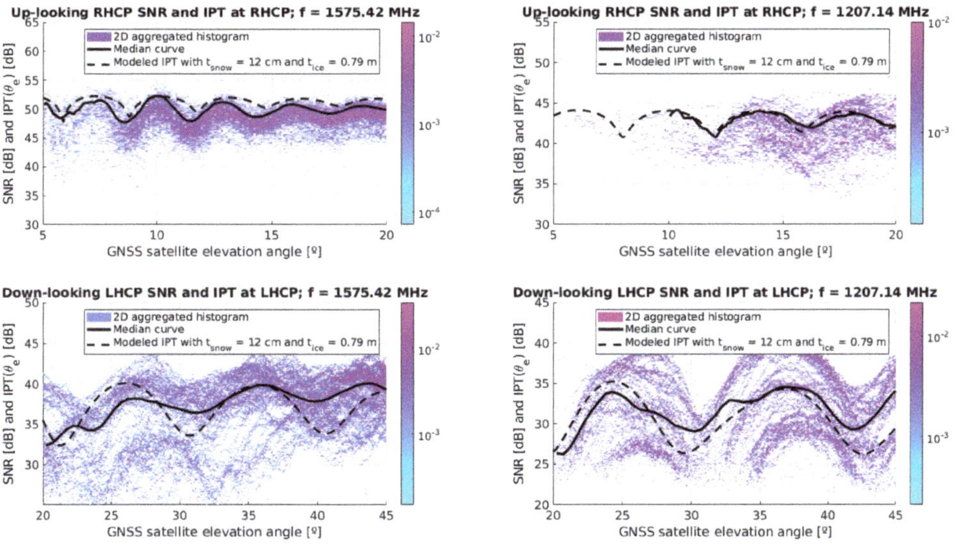

Figure 14. NLS minimization output for the data set between the 23rd and the 27th of December 2019. MSE is represented for both snow and ice thickness, for different frequencies and signal polarization, and it has been normalized (z-score normalization) to ease its visualization. Note that the blue-dashed line shows the exact thickness at which both signals present a minimum.

Figure 15. Modeled IPT using the retrieved snow and ice thickness overlaid on top of the signals retrieved by the PYCARO-2 instrument in the MOSAiC campaign during December 2019. Both the median curves and the modeled interference patterns are represented in black.

5. Conclusions

This work has presented the GNSS-R theory and techniques required to estimate the snow and the ice thickness of an ice floe. Multifrequency and dual-polarization GNSS-R can be used to determine both snow and ice thickness by means of the IPT. A peculiarity is that when nonlinear least-square minimization is applied to both RHCP and LHCP signals, it shows a periodicity in the error function, leading to multiple solutions. These uncertainties can be solved by comparing the error functions from different bands. Snow thickness can be easily estimated from this process, as the periodicity of the error with the snow thickness has the same order than the observation range (i.e., a period every 10 cm, observation range ~40 cm). However, for the ice thickness case, the periodicity of the error with the ice thickness is one order of magnitude smaller than the observation range (i.e., a period every 20 cm, observation range ~2 m), which leads to multiple solutions for ice thickness that, from the data available for PYCARO-2, cannot be resolved without a priori information. It has been seen that the snow thickness can be more easily retrieved, as it is the one inducing larger changes in the LHCP signal, which led to centimeter errors as compared to the in-situ measurements. However, this is not the case for the ice thickness, where ambiguities may appear, and a priori information is required. Furthermore, it has been seen that the error in the ice thickness retrieval for the second example differs 10 cm from the in-situ measurements.

The interference pattern produced by the four-layer model exhibits some sensitivity to ice thickness variations, especially at the lower frequency band. This means that signals do reflect off the bottom of the ice and reach the air media with still enough power to induce an interference with the direct signal. This behavior supports the hypothesis in [6], from which measurements of the sea ice draft could be possible with GNSS-R. However, there is a wide range of elevation angles where the snow thickness is dominant in the reflection (i.e., between 30° and 42.5° of elevation angle, the ice thickness cannot be measured). In this range of angles, the reflection takes place mostly in the snow-ice interface, while for other elevation angles, the interference pattern is produced in the air-snow, snow-ice, and ice-water interfaces.

From the results presented, it is clearly seen that the snow layer on top of the sea-ice is dominant, while the ice thickness seems to have a smaller impact to GNSS reflections. This phenomena can be also translated to other sensors, where the snow layer shall be properly modeled and taken into account to, for instance, provide an ice thickness measure.

The MOSAiC campaign ended on 30 September 2020, and data from different sensors will be released in the future. For this reason, the acquired data processed through the described algorithm can be used for validation of other MOSAiC instruments such as microwave radiometers. Furthermore, the combination of GNSS-R measurements and measurements from other instruments may increase the accuracy of the GNSS-R retrieved snow and ice thickness, thus helping to remove the ambiguities for different ice thicknesses.

Author Contributions: Conceptualization, J.F.M.-M., A.P., and A.C.; methodology, J.F.M.-M., A.P., A.C., and S.R.; software, J.F.M.-M.; validation, J.F.M.-M., A.C., J.S., and P.I.; formal analysis, J.F.M.-M. and A.C.; investigation, J.F.M.-M.; resources, A.C., J.S., V.N., P.I., R.T., M.H., S.H., G.S., and M.P.; data curation J.F.M.-M. and A.C.; visualization J.F.M.-M., J.S., and P.I.; supervision A.C.; project administration, A.C., and M.P.; funding acquisition, A.C., M.P., E.C., and S.R.; writing—original draft preparation, J.F.M.-M., A.C.; writing—review and editing, J.F.M.-M., A.P., A.C., S.R., and E.C. All authors have read and agreed to the published version of the manuscript.

Funding: This work was supported by 2017 ESA S3 challenge and Copernicus Masters overall winner award ("FSSCat" project) and ESA project "FSSCat Validation Experiment in MOSAIC" (ESA CN 4000128320/19/NL/FF/ab). This work was also supported by ESA under the PO 5001025474. The PYCARO-2 instrument was developed within the SPOT project: Sensing with Pioneering Opportunistic Techniques grant RTI2018-099008-B-C21/AEI/10.13039/501100011033 and RTI2018-099008-B-C22, and by EU EDRF funds and the Spanish Ministry of Science, Innovation and Universities, and by the Unidad de Excelencia Maria de Maeztu MDM-2016-0600. Data used in this manuscript was produced as part of the international Multidisciplinary drifting Observatory for the Study of the Arctic Climate (MOSAiC) and all their operators and investigators with the following tags: MOSAiC20192020, NSF-1820927, NFR-287871. Project-ID for Polarstern expedition AWI_PS122_00.

Acknowledgments: The authors are thankful to Gunnar Spreen, Oguz Demir, Marcus Huntemann, Julienne Stroeve, Vishnu Nandan, Robert Ricker, Lars Kaleschke, Reza Naderpour, Stefan Hendricks, and Rasmus Tonboe, who installed and operated the GNSS-R instruments during MOSAIC Legs 1 to 3. The authors also want to thank Bernardo Carnicero (from ESA) for his support during the whole development and implementation of the FSSCat mission.

Conflicts of Interest: The authors declare no conflict of interest.

References

1. FSSCat—Towards Federated EO Systems. Available online: https://copernicus-masters.com/winner/ffscat-towards-federated-eo-systems/ (accessed on 29 September 2020).
2. Camps, A.; Golkar, A.; Gutierrez, A.; de Azua, J.R.; Munoz-Martin, J.; Fernandez, L.; Diez, C.; Aguilella, A.; Briatore, S.; Akhtyamov, R.; et al. Fsscat, the 2017 Copernicus Masters' "Esa Sentinel Small Satellite Challenge" Winner: A Federated Polar and Soil Moisture Tandem Mission Based on 6U Cubesats. In Proceedings of the IGARSS 2018—2018 IEEE International Geoscience and Remote Sensing Symposium, Valencia, Spain, 22–27 July 2018.
3. Munoz-Martin, J.F.; Capon, L.F.; de Azua, J.A.R.; Camps, A. The Flexible Microwave Payload-2: A SDR-Based GNSS-Reflectometer and L-Band Radiometer for CubeSats. *IEEE J. Sel. Top. Appl. Earth Obs. Remote Sens.* **2020**, *13*, 1298–1311. [CrossRef]
4. Alonso-Arroyo, A.; Zavorotny, V.; Camps, A. Sea Ice Detection Using U.K. TDS-1 GNSS-R Data. *IEEE Trans. Geosci. Remote Sens.* **2017**, 1–13. [CrossRef]
5. Camps, A. Spatial Resolution in GNSS-R Under Coherent Scattering. *IEEE Geosci. Remote Sens. Lett.* **2019**. [CrossRef]
6. Li, W.; Cardellach, E.; Fabra, F.; Rius, A.; Ribó, S.; Martín-Neira, M. First spaceborne phase altimetry over sea ice using TechDemoSat-1 GNSS-R signals. *Geophys. Res. Lett.* **2017**, *44*, 8369–8376. [CrossRef]
7. Cardellach, E.; Wickert, J.; Baggen, R.; Benito, J.; Camps, A.; Catarino, N.; Chapron, B.; Dielacher, A.; Fabra, F.; Flato, G.; et al. GNSS Transpolar Earth Reflectometry exploriNg System (G-TERN): Mission Concept. *IEEE Access* **2018**, *6*, 13980–14018. [CrossRef]
8. Kaleschke, L.; Tian-Kunze, X.; Maaß, N.; Mäkynen, M.; Drusch, M. Sea ice thickness retrieval from SMOS brightness temperatures during the Arctic freeze-up period. *Geophys. Res. Lett.* **2012**, *39*. [CrossRef]
9. European Space Agency. Eight Years of SMOS Arctic Sea Ice Thickness Level Now Available from SMOS Data Dissemination Portal. Available online: https://earth.esa.int/web/guest/missions/esa-operational-eo-missions/smos/news/-/article/eight-years-data-of-smos-arctic-sea-ice-thickness-level-now-available-from-smos-data-dissemination-portal (accessed on 11 November 2019).
10. Carreno-Luengo, H.; Camps, A. First Dual-Band Multiconstellation GNSS-R Scatterometry Experiment Over Boreal Forests From a Stratospheric Balloon. *IEEE J. Sel. Top. Appl. Earth Obs. Remote Sens.* **2016**, *9*, 4743–4751. [CrossRef]
11. Expedition, M. MOSAiC Expedition Web Page. Available online: https://follow.mosaic-expedition.org/ (accessed on 1 September 2020).
12. Knust, R. Polar Research and Supply Vessel POLARSTERN operated by the Alfred-Wegener-Institute. *J. Large-Scale Res. Facil. JLSRF* **2017**, *3*. [CrossRef]
13. Munoz-Martin, J.F.; Camps, A.; Cardelach, E.; Pastena, M. Circular Polarization GNSS-R Measurements of MOSAiC RS Site during January 2020. PANGAEA. Available online: https://doi.org/10.1594/PANGAEA924837 (accessed on 15 November 2020).
14. Carreno-Luengo, H.; Amézaga, A.; Vidal, D.; Olivé, R.; Martin, J.M.; Camps, A. First Polarimetric GNSS-R Measurements from a Stratospheric Flight over Boreal Forests. *Remote Sens.* **2015**, *7*, 13120–13138. [CrossRef]
15. Carreno-Luengo, H.; Camps, A.; Via, P.; Munoz, J.F.; Cortiella, A.; Vidal, D.; Jané, J.; Catarino, N.; Hagenfeldt, M.; Palomo, P.; et al. 3Cat-2 An Experimental Nanosatellite for GNSS-R Earth Observation: Mission Concept and Analysis. *IEEE J. Sel. Top. Appl. Earth Obs. Remote Sens.* **2016**, *9*, 4540–4551. [CrossRef]
16. Sturm, M.; Holmgren, J. An Automatic Snow Depth Probe for Field Validation Campaigns. *Water Resour. Res.* **2018**, *54*, 9695–9701. [CrossRef]
17. Hunkeler, P.A.; Hendricks, S.; Hoppmann, M.; Paul, S.; Gerdes, R. Towards an estimation of sub-sea-ice platelet-layer volume with multi-frequency electromagnetic induction sounding. *Ann. Glaciol.* **2015**, *56*, 137–146. [CrossRef]

18. Rodriguez-Alvarez, N.; Bosch-Lluis, X.; Camps, A.; Vall-llossera, M.; Valencia, E.; Marchan-Hernandez, J.; Ramos-Perez, I. Soil Moisture Retrieval Using GNSS-R Techniques: Experimental Results Over a Bare Soil Field. *IEEE Trans. Geosci. Remote Sens.* **2009**, *47*, 3616–3624. [CrossRef]

19. Rodriguez-Alvarez, N.; Camps, A.; Vall-llossera, M.; Bosch-Lluis, X.; Monerris, A.; Ramos-Perez, I.; Valencia, E.; Marchan-Hernandez, J.F.; Martinez-Fernandez, J.; Baroncini-Turricchia, G.; et al. Land Geophysical Parameters Retrieval Using the Interference Pattern GNSS-R Technique. *IEEE Trans. Geosci. Remote Sens.* **2011**, *49*, 71–84. [CrossRef]

20. Rodriguez-Alvarez, N.; Bosch-Lluis, X.; Camps, A.; Aguasca, A.; Vall-llossera, M.; Valencia, E.; Ramos-Perez, I.; Park, H. Review of crop growth and soil moisture monitoring from a ground-based instrument implementing the Interference Pattern GNSS-R Technique. *Radio Sci.* **2011**, *46*. [CrossRef]

21. Rodriguez-Alvarez, N.; Bosch-Lluis, X.; Camps, A.; Ramos-Perez, I.; Valencia, E.; Park, H.; Vall-llossera, M. Water level monitoring using the interference pattern GNSS-R technique. In Proceedings of the 2011 IEEE International Geoscience and Remote Sensing Symposium, Vancouver, BC, Canada, 24–29 July 2011; pp. 2334–2337.

22. Alonso-Arroyo, A.; Camps, A.; Park, H.; Pascual, D.; Onrubia, R.; Martin, F. Retrieval of Significant Wave Height and Mean Sea Surface Level Using the GNSS-R Interference Pattern Technique: Results From a Three-Month Field Campaign. *IEEE Trans. Geosci. Remote Sens.* **2015**, *53*, 3198–3209. [CrossRef]

23. Rodriguez-Alvarez, N.; Aguasca, A.; Valencia, E.; Bosch-Lluis, X.; Ramos-Perez, I.; Park, H.; Camps, A.; Vall-llossera, M. Snow monitoring using GNSS-R techniques. In Proceedings of the 2011 IEEE International Geoscience and Remote Sensing Symposium, Vancouver, BC, Canada, 24–29 July 2011; pp. 4375–4378.

24. Cardellach, E.; Fabra, F.; Rius, A.; Pettinato, S.; D'Addio, S. Characterization of dry-snow sub-structure using GNSS reflected signals. *Remote Sens. Environ.* **2012**, *124*, 122–134. [CrossRef]

25. Zavorotny, V.U.; Gleason, S.; Cardellach, E.; Camps, A. Tutorial on Remote Sensing Using GNSS Bistatic Radar of Opportunity. *IEEE Geosci. Remote Sens. Mag.* **2014**, *2*, 8–45. [CrossRef]

26. Kaleschke, L. *SMOS Sea Ice Retrieval Study (SMOSSIce): Final Report*; ESA ESTEC Contract No: 4000202476/10/NL/CT; Zenodo: Cham, Switzerland, 2013. [CrossRef]

27. Vant, M.R.; Ramseier, R.O.; Makios, V. The complex-dielectric constant of sea ice at frequencies in the range 0.1–40 GHz. *J. Appl. Phys.* **1978**, *49*, 1264–1280. [CrossRef]

28. Lérondel, G.; Romestain, R. Fresnel coefficients of a rough interface. *Appl. Phys. Lett.* **1999**, *74*, 2740–2742. [CrossRef]

29. Ulaby, F.; Long, D. *Microwave Radar and Radiometric Remote Sensing*; Artech House: Norwood, MA, USA, 2015.

30. Perovich, D.K.; Grenfell, T.C.; Richter-Menge, J.A.; Light, B.; Tucker, W.B., III; Eicken, H. Thin and thinner: Sea ice mass balance measurements during SHEBA. *J. Geophys. Res. Ocean.* **2003**, *108*. [CrossRef]

Publisher's Note: MDPI stays neutral with regard to jurisdictional claims in published maps and institutional affiliations.

 *remote sensing*

Article

An Aircraft Wetland Inundation Experiment Using GNSS Reflectometry

Stephen T. Lowe [1,*], Clara Chew [2], Jesal Shah [1] and Michael Kilzer [1]

[1] Engineering and Science Division, Jet Propulsion Laboratory, California Institute of Technology, Pasadena, CA 91109, USA; jesal.a.shah@jpl.nasa.gov (J.S.); michael.j.kilzer@jpl.nasa.gov (M.K.)
[2] COSMIC Group, University Corporation for Atmospheric Research, P.O. Box 3000, Boulder, CO 80307-3000, USA; clarac@ucar.edu
* Correspondence: stephen.t.lowe@jpl.nasa.gov; Tel.: +1-818-354-1042

Received: 6 December 2019; Accepted: 24 January 2020; Published: 5 February 2020

Abstract: In early May of 2017, a flight campaign was conducted over Caddo Lake, Texas, to test the ability of Global Navigation Satellite System-Reflectometry (GNSS-R) to detect water underlying vegetation canopies. This paper presents data from that campaign and compares them to Sentinel-1 data collected during the same week. The low-altitude measurement allows for a more detailed assessment of the forward-scattering GNSS-R technique, and at a much higher spatial resolution, than is possible using currently available space-based GNSS-R data. Assumptions about the scattering model are verified, as is the assumption that the surface spot size is approximately the Fresnel zone. The results of this experiment indicate GNSS signals reflected from inundated short, thick vegetation, such as the giant *Salvinia* observed here, results in only a 2.15 dB loss compared to an open water reflection. GNSS reflections off inundated cypress forests show a 9.4 dB loss, but still 4.25 dB above that observed over dry regions. Sentinel-1 data show a 6-dB loss over the inundated giant *Salvinia*, relative to open water, and are insensitive to standing water beneath the cypress forests, as there is no difference between the signal over inundated cypress forests and that over dry land. These results indicate that, at aircraft altitudes, forward-scattered GNSS signals are able to map inundated regions even in the presence of dense overlying vegetation, whether that vegetation consists of short plants or tall trees.

Keywords: GNSS-R; terrestrial hydrology; soil moisture; biomass; remote sensing

---

1. Introduction

Mapping wetland extent and quantifying how the extent changes over time is an important task for both scientific and societal applications. Wetlands are the largest natural emitter of atmospheric methane, a potent greenhouse gas, yet with the widest uncertainty range [1], and feedbacks between wetlands and the global carbon cycle are still poorly understood [1–4]. Understanding wetland dynamics is important for predicting the transmission of mosquito-borne diseases like malaria [5], and quantifying the rapid rate of wetland collapse is also important, as they provide economic and ecologic benefits to surrounding communities [6].

Current techniques used to map wetlands are not able to fully capture the sometimes-rapid changes in wetland extent, high spatial variability in inundated or saturated areas, or successfully map water beneath dense vegetation canopies. For example, in situ observations of wetlands are often sparse and seldom made, as many wetlands are located in areas too remote to map regularly. Satellite remote sensing has been widely used to map wetlands and seasonal changes in wetland extent. Optical techniques (e.g., [7]) are unable to map water underneath vegetation or see the land surface through cloud cover, both of which decrease the temporal frequency of observations and underestimate the total extent of inundated areas. Microwave instruments onboard satellites, either monostatic synthetic aperture radars (SARs) or radiometers, are able to see through clouds and, depending on the wavelength, some amount of vegetation cover. Currently available products deriving inundation extent from microwave instruments have primarily used data from radiometers, with an extremely coarse (>25 km) spatial footprint [8,9]. Data from SARs have also been used extensively to quantify seasonal changes in inundation extent in wetland complexes (e.g., [10]), though the only publicly available data are from C-band, which cannot penetrate dense vegetation cover, and the long temporal repeat cycles of a single SAR instrument inhibits their ability to capture short term changes in inundation dynamics.

A new observational technique, Global Navigation Satellite System Reflectometry (GNSS-R), has shown promise to map wetland inundation, even in the presence of dense, overlying vegetation. In [11], significantly higher power was observed in reflections over inundated rice fields compared to dry land. These results motivated a dedicated flight campaign to observe GNSS reflections over a variety of wetland sites and vegetation types, with a focus on Caddo Lake, located on the Texas-Louisiana border. Caddo Lake was chosen because it is one of the most heavily vegetated bodies of water in North America—large portions of the lake are densely covered with tall cypress forests. The cypress forests stand up to 50 m tall, and the trees are sometimes surrounded by cypress knees, which are small knobby structures that support the cypress trees' stability. Cypress knees stand 20–30 cm above the water surface and prevent the passage of even the smallest watercraft through the forests. In addition to the cypress trees, a distinctive feature of Caddo Lake is the presence of an invasive floating weed, giant *Salvinia* (*Salvinia molesta*). These floating mats can completely obscure the open water and move with the lake currents—infestations can appear and disappear within a matter of weeks.

In addition to [11], there have been additional investigations into the potential of GNSS-R to map wetlands using CYGNSS (Cyclone Global Navigation Satellite System; e.g., [12–14]), a constellation of space-based GNSS-R instruments. However, these studies have been limited by the spatial footprint of the CYGNSS measurement (7 km × 1 km), which is artificially large due to its optimization for ocean surface remote sensing. With a footprint of this size, it is nearly guaranteed that there will be multiple landcover types contributing to surface scattering within each footprint. The aircraft GNSS-R experiment presented here allows a more detailed assessment of GNSS-R's forward-scattering geometry, and at a much higher spatial resolution. In particular, it allows for the quantification of the attenuation of the GNSS-R signal through specific canopy types, with less uncertainty introduced by mixed pixels. Other GNSS-R aircraft flight experiments have been done in the past (e.g., [15,16]), though these experiments were for the purpose of understanding the sensitivity of GNSS-R to soil moisture and typical agricultural crops and not for the purpose of sensing standing water under trees.

This study will present GNSS-R data from the aircraft experiment over Caddo Lake and quantify the attenuation of the signal through the vegetation that typifies the Caddo Lake area to better understand the potential of GNSS-R to map inundation underneath different types of vegetation canopies. In order to put the results in the context of other remote sensing techniques, we will compare the results to those from Sentinel-1, a C-band SAR instrument that is routinely used to map inundation dynamics in wetland areas. Although it is well known that, in theory, L-band data should be able to penetrate further into a dense vegetation canopy and therefore be more sensitive to water underneath vegetation, there have been no quantitative assessments of the increased ability of GNSS-R to sense inundation relative to Sentinel-1 when the data collected by the two instruments are on a similar spatial scale. The comparison with Sentinel-1 will quantitatively show the increase in sensitivity of GNSS-R to inundation for specific wetland types relative to the more well-known and trusted Sentinel-1 data.

2. Experimental Campaign

This section describes the GNSS-R instrument, flight setup, and antenna calibration experiment.

2.1. Aircraft Flights

Keystone Ariel Survey, Inc. (Keystone) provided the aircraft, a Cessna 310, for this experiment, as shown in Figure 1a. Keystone also provided a pilot, while we ran the equipment, verified it was working in flight, and oversaw the data collection and flight plan. Most flight data were collected while flying at between 2.4 km (8000′) and 3 km (10,000′) altitude. The flight over the Caddo Lake region took place on 2 May 2017, and consisted of a flight in the morning, when large North/South (N/S) and East/West grids centered on Caddo Lake were performed. During the afternoon flight, another N/S grid at a lower altitude was performed, followed by multiple passes over the Northwest region of Caddo Lake, which has especially dense vegetation over the water. We separately toured this Northwest region that morning by boat to collect in situ data on inundation and tree density.

(a)

(b)

Figure 1. (**a**) Keystone's Cessna 310 used for all three flights. (**b**): Antenna calibration experiment.

2.2. Flight Equipment

The Cessna 310 had been modified to have a large hole on the underside of the aircraft, which was typically used by Keystone for their commercial optical surveys. For our experiments, Keystone removed their optical camera, and the hole covered with an aluminum plate hosting a 9-element, down-looking, patch array antenna, mounted flush with the airframe. Each patch antenna element provided dual polarization, horizontal and vertical (H/V), at both GPS L1 and L2 frequencies. We flew a custom open-loop recorder, capable of recording up to 40 MHz 1-bit samples from 32 channels. Our front-end electronics had eight antenna inputs, each electronically tunable to two simultaneous local-oscillator (LO) frequencies, resulting in 16 complex (I/Q) sampled data streams, or 32 single-bit sampled streams, matching our recorder's capability. Figure 2 shows our instruments and down-looking antenna. The recorder was monitored in real time during the flight using a laptop computer.

The Cessna had installed an up-looking commercial aircraft GPS antenna, mounted on top of the aircraft, along with a commercial Novatel GPS receiver, for its navigation system. We split the signal going from this antenna to the Novatel to gain access to the direct signal. For our experiments, we used seven of the eight antenna inputs, recording the H and V outputs from three of the down-looking patch antennas, along with the direct signal from the aircraft's up-look antenna, all at both the GPS L1 and L2 frequencies. For the Caddo Lake experiment, we combined the H and V outputs from a 4th patch element to form a Left-Hand Circularly Polarized (LCP) signal, and recorded the GPS L1 and L5 frequencies from that patch element.

Most of our data were collected with a sampling rate of 20 MHz, however, some 40 MHz sampled data were collected during the afternoon flight. Our sampled data was stored onto a disk array of SSD hard drives having a total capacity of 8 TB. This allowed about 15 h of continuous recording using a 20 MHz sampling rate.

In addition to the patch antenna and corresponding recording system, we flew a downward-pointing optical camera fastened to the aluminum antenna mount over a small viewing hole, which took high quality optical images every 2 s. Finally, for aircraft positioning, we used the output from the Novatel receiver Keystone had installed on their plane.

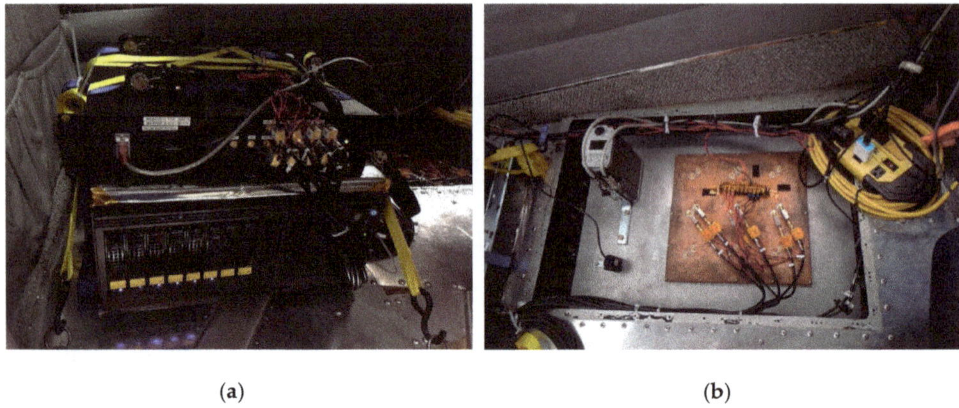

(a)   (b)

Figure 2. (**a**): Sampler and front-end electronics (top) with 8 antenna inputs and low-noise amplifiers (LNAs), and solid state disk (SSD) array (bottom) with 8 TB of storage. (**b**): View of 9-element patch antenna from inside the aircraft, with central three patches' H and V-Pol outputs cabled. IR camera is on upper left of antenna, and optical camera is small black box on lower left of antenna.

2.3. Antenna Calibration

After a preliminary examination of the data, it was clear that the theoretical antenna gain pattern being used for the downward-looking patch elements resulted in geometrically correlated inconsistencies in the resulting surface scattering coefficient. It was assumed that a measurement of the antenna gain pattern would help remove these effects. With limited funding, we chose to perform a simple antenna calibration experiment, differencing the measured gains between our flight antenna and a well-measured geodetic antenna while observing GPS signals as their positions in the sky moved over several hours.

On 21 August 2018, both our 9-element patch array antenna and a well-measured geodetic choke-ring antenna were located near Jet Propulsion Laboratory's (JPL's) antenna range at a site with clear open sky views and few nearby objects to minimize multipath, as shown in Figure 1b. The antennas were placed up-looking, with the patch antenna configured as it was during flight, mounted on its aluminum plate with the same low-noise amplifiers (LNAs) on each antenna input. Data were collected as in the aircraft experiment, but with the aircraft's up-looking geodetic antenna replaced with our choke ring antenna. Although these antennas were placed near each other, we did not observe any mutual coupling. We collected over 7 h of data in two continuous sessions, rotating the patch antenna 180 degrees in the second session for better azimuthal coverage. Since both antennas observed the same GPS transmitters and were located the same distance from them, the ratio of observed power between the antennas would be equal to the ratio of the antenna gains in that direction. By observing over several hours while the geometry of the transmitters changed, and knowing the gain pattern of the geodetic antenna, the patch antenna's relative gain pattern over much of its beam pattern could be measured. The peak measured gain of the geodetic antenna was 6.8 dBIc while each linear polarization of a single patch was measured to have a peak gain of 5.4 dBIc.

3. Data Processing

The primary GNSS-R observable is a delay-Doppler map (DDM), two-dimensional cross-correlations between the received reflected GNSS signal, and a replica of the transmitted signal. Our DDMs were processed in a similar way as in previous studies [11,17,18].

3.1. Software Receiver and Geometric Modeling

The 20 MHz and 40 MHz raw samples recorded from the aircraft's usual GPS antenna on top of the plane were processed through a software GPS receiver. All of the GPS signals in view were detected and tracked with delay and phase locked loops. The models used to track these signals were recorded, along with the correlation results. The main processing pass for GNSS-R used these models derived from the direct signals to correlate with data from the down-looking patch arrays.

All correlations were performed coherently each 20 msec. The correlations were performed while offsetting the data relative to the models by between −50 and +500 samples, or lags. The 550 correlations were computed for five Doppler values ranging from −2 to +2 times 39 Hz, using Fast Fourier Transforms (FFTs) and cyclic convolution for efficient computation. For an aircraft flying horizontally, the Doppler offset is close to zero—the additional offsets were to model slight upward and downward movements of the plane. The early correlations were computed to calibrate the system noise, while the later correlations covered the expected range of delays between the direct and reflected signal reception. The 500-lag range corresponds to a path delay of up to 7330 m, while all data were recorded below about 10,000′ altitude, or a maximum of 6100 m path delay. Due to the significant processing time to create (550 × 5) DDMs at 50 Hz, only three antenna inputs were processed: the direct signal from the plane's up-looking antenna and the H and V signals from one of the three down-looking patches used in the experiment. The three DDMs were written to disk along with ancillary data each 20 msec for all GPS signals in view as the primary software-receiver output.

The aircraft position was obtained by processing the aircraft's Novatel GNSS receiver data through JPL's Gipsy software, a state-of-the-art sub-cm GNSS navigation software used by hundreds of academic and industry partners and dozens of NASA space missions. A geometric model was developed by combining these receiver positions with a surface height model based on SRTM 1″ (30 m) data, and JPLs GPS transmitter positions. The 50-Hz DDMs were combined incoherently into 10-Hz DDMs by summing the power from five consecutive DDMs. This was to increase the signal to power ratio (SNR), average down the measurement noise, and to reduce speckle for incoherent reflections. For each of these resulting DDMs, an iterative procedure was used to compute the corresponding specular point on the surface given the uneven surface topology. The output of this processing step consists of the 10-Hz DDMs' peak power and corresponding delay and Doppler values for the three processed antenna inputs, along with noise estimates and geometric model information, including the specular point location.

The peak power of the DDM is proportional to the surface reflectivity, however, it is also affected by other factors like system noise levels and receiver instrument gain. To mitigate these factors, the signal-to-noise ratio (SNR) was calculated, which is the peak power of the DDM normalized to a noise floor, which is defined as the mean cross-correlation before the leading edge of the reflection in the DDM.

3.2. Antenna Calibration Data Processing

The antenna calibration data were also processed through the software receiver and the signal-to-noise ratio was averaged each 5 s, for both the patch antenna H and V outputs, and the geodetic antenna. In processing the choke-ring antenna data, a simple quadratic polynomial as a function of boresight angle was found to fit the data well, after accounting for an average transmitter power for each satellite, with scatter of the 5-sec residuals under 0.4 dB. This function also agreed with published measurements to better than 0.5 dB. This agreement gave confidence in the derived average transmit power values from each transmitter, which differed from each other by up to 4 dB.

The patch antenna data were processed similarly, however the average transmitter power values were fixed to those found from the choke-ring antenna. Here, a clear azimuthal dependence was seen, and modeled with a simple sin(cos) of the azimuthal angle for the H(V) patch output with amplitude 4.2 dB, times the sin of the inclination angle. With these corrections, a quadratic polynomial fit the data reasonably well with 1.0 dB scatter. The power residuals showed slowly varying correlated trends from each transmitter on the order of 1 dB, presumably from multipath effects. Thus, the polynomial gain models for the patch H and V channels are likely good to better than 1 dB, but there may be 1 dB of systematic effects in the aircraft data.

3.3. Scattering Model

In assessing the observables to be used for inundation detection, two scattering models were considered, both assuming distributed targets (see [19] for a more refined scattering model): one where the signal is absorbed by the surface and retransmitted, and another where the surface is assumed flat and conducting. The first model follows the standard derivation of the radar equation [20], where the received surface-reflected power, P_r, is given by

$$P_r = \frac{P_t G_t}{4\pi R_{ts}^2} \sigma \frac{A_r}{4\pi R_{sr}^2} \tag{1}$$

The RHCP transmitted power, P_t, is amplified by the transmitter's antenna in the direction of the surface specular point with effective gain G_t. The signal then suffers a free-space loss of $4\pi R_{ts}^2$ traveling from the transmitter to the surface, a distance of R_{ts}. The signal's power is absorbed by the surface and retransmitted in the direction of the receiver from an effective cross-sectional area given by σ. The signal suffers another free-space loss of $4\pi R_{sr}^2$ traveling the distance R_{sr} from the surface to the receiver. The signal is then collected by the receiver with an antenna having effective area A_r. This development ignores losses due to the atmosphere and instrumentation, and this equation holds separately for each polarization. Figure 7.1 in [20] shows the geometry explicitly.

The second model assumes the surface approximates a flat conductor, so the signal travels from the transmitter to the receiver's mirror image below the surface a distance $(R_{ts} + R_{sr})$, and suffers an attenuation expressed as the reflectivity Γ_r at the surface. The received power is then given by

$$P_r = \frac{P_t G_t \Gamma_r A_r}{4\pi (R_{ts} + R_{sr})^2} \tag{2}$$

At the aircraft altitudes used in these experiments, R_{sr} is much smaller than R_{ts}, and can be neglected in Equation (2). With that change, the difference in the two scattering models is in how the signal interacts with the surface and travels to the receiver. In the first, the signal power is reradiated from an effective area σ and suffers a free-space loss traveling to the receiver, while in the second model, the signal reflects coherently with surface reflectivity Γ_r. The surface reflectivity will change depending on the surface dielectric constant (e.g., wet vs. dry ground) and the roughness of the surface. Surface reflectivity is also expected to be attenuated by overlying vegetation canopies.

Both models can be expressed as

$$P_r = \frac{P_t G_t G_r \lambda^2}{(4\pi)^2 R_{ts}^2} \alpha \tag{3}$$

where λ is the 0.19 cm GPS L1 wavelength, and α is given by:

$$\alpha = \frac{\sigma}{4\pi R_{sr}^2} \tag{4}$$

for the former, incoherent model and $\alpha = \Gamma_r$ for the latter, coherent model, and G_r is the gain of the down-looking reflection antenna. A model for the expected received power from the direct transmitter-to-receiver signal is given by:

$$P_d = \frac{P_t G_t G_d \lambda^2}{(4\pi)^2 R_{tr}^2} \tag{5}$$

where P_d is the received direct-signal power, and G_d is the gain of the up-looking antenna mounted on top of the aircraft.

3.4. Surface Spot Size

The surface spot size that is sensed by a coherent GNSS-R reflection has been assumed to be given by the size of the Fresnel zone [20], which is typically defined to be the iso-delay ellipse corresponding to a wavelength delay compared to the specular arrival (some authors use 1/2 wavelength). The major axis of an iso-delay ellipse corresponding to a delay (distance) of Δ, is:

$$a = \frac{1}{\cos\theta}\sqrt{\frac{2\Delta R_{ts}R_{sr}}{R_{ts}+R_{sr}}} \cong \frac{\sqrt{2\Delta R_{sr}}}{\cos\theta} \tag{6}$$

while the minor axis is given by the same expression without the cosine term, and θ is the surface incidence angle.

4. Analysis Data Sets

4.1. Caddo Lake Calibrated Data

The received power values from the Caddo Lake aircraft passes, which were taken as the peak values from the 10 Hz DDMs, were divided by the received power from the aircraft's up-looking antenna. Dividing Equation (3) by Equation (5), this ratio is given by:

$$\frac{P_r}{P_d} = \frac{G_r}{G_d}\alpha \tag{7}$$

where R_{tr}/R_{ts} is taken to be one. By computing this ratio, the effects of unknown GPS transmit power and gain are canceled out. Correcting for the direct and reflected antenna gains, the calibrated power P_{cal} results in an estimate for α:

$$P_{cal} \equiv \frac{P_r G_d}{P_d G_r} \quad = \alpha \sim \frac{\sigma}{R_{sr}^2} \text{ (incoherent scattering)} \\ = \alpha \sim \Gamma_r \text{ (coherent scattering)} \tag{8}$$

where $\sim$ denotes proportionality. Note that P_{cal} is not a calibrated power in the usual sense, but an SNR that is proportional to P_r in Equation (2). By mapping spatial changes in P_{cal}, we can thus relate them to spatial changes in the surface reflectivity. Our dataset included GNSS-R reflections from the following PRNs: 3, 7, 8, 9, 11, 14, 16, 22, 23, 26, 27, 28, 30, and 31. Data were obtained from a range of incidence angles (min: 10.7 deg, max: 60 deg, mean: 34.7 deg). No notable relationship between P_{cal} and incidence angle was observed.

4.2. Ancillary Data

4.2.1. Sentinel-1 Data

Sentinel-1 data for 8 May 2017 was obtained from the University of Alaska Fairbanks Vertex data portal, which was the overpass closest in time to the aircraft flight. The Sentinel-1 toolbox (S1TBX) was used to process the data, following the recommended practices set forth by the United Nations Office for Outer Space Affairs [21]. VV-pol amplitude data were calibrated to produce sigma nought values, and these were speckle filtered using a 7×7 Lee filter. Finally, speckle-filtered sigma nought values were geometrically corrected and re-projected using the digital elevation model from the shuttle radar topography mission (SRTM).

4.2.2. Landsat Imagery

Landsat 8 data were obtained over Caddo Lake for 8 May 2017, which was the overpass closest in time to the aircraft flight. The Level 1 geotiff data product was converted into an RGB image using the band specific multiplicative and additive rescaling factors and corrected for the sun elevation angle using values within the metadata. The resulting 30 m image was pan-sharpened to 15 m. The Landsat 8 image identifies the extent and location of floating Giant Salvinia mats in Caddo Lake at the time of the flight experiment.

4.2.3. Landcover Classification Map

Landcover classification data for the Caddo Lake area were obtained from the Caddo Lake Institute, which simplifies the National Wetlands Inventory (NWI) habitat data into seven categories [22]. Here, we use both the NWI definitions as well as the simplified landcover categories to describe the different environments in Caddo Lake. Table 1 shows the wetland classes and respective abbreviations.

Table 1. Landcover classification abbreviations used in this study. General and specific classes and the wetland ID follow conventions found in [22].

Abbreviation	General Class	Specific Class	Wetland ID
OW	open water	open water	L1UBHh
EM1	emergents	emergent reeds, semipermanently flooded	PEM5/FLCh
EM2	emergents	emergent reeds, permanently flooded	PEM5/UBFH
EM3	emergents	emergent reeds, seasonally flooded	PEM5C
CF1	cypress forests	broad-leaved deciduous, scrub-shrub temporarily flooded	PFO/SS1A
CF2	cypress forests	needle-leaved deciduous, scrub-shrub semipermanently flooded	PFO/SS2F
BH1	bottomland hardwoods	broad-leaved deciduous, temporarily flooded	PFO1A
BH2	bottomland hardwoods	broad-leaved deciduous, seasonally flooded	PFO1C
CF3	cypress forests	needle-leaved deciduous, semipermanently flooded	PFO2/1F
CF4	cypress forests	needle-leaved deciduous, seasonally flooded	PFO2/UBCh
CF5	cypress forests	needle-leaved deciduous, semipermanently flooded	PFO2/UBF
CF6	cypress forests	needle-leaved deciduous, intermittently exposed	PFO2/UBG
PFCF1	permanently flooded cypress forests	needle-leaved deciduous, permanently flooded	PFO2/UBH
CF7	cypress forests	needle-leaved deciduous, semipermanently flooded	PFO2F
CF8	cypress forests	needle-leaved deciduous, intermittently exposed	PFO2Gh
CF9	cypress forests	dead forest, permanently flooded	PFO5/UBHh
CF10	cypress forests	deciduous, temporarily flooded	PFO6A
CF11	cypress forests	deciduous, seasonally flooded	PFO6C
SSW1	scrub-shrub wetlands	scrub-shrub broad-leaved deciduous, temporarily flooded	PSS1A
SSW2	scrub-shrub wetlands	scrub-shrub broad-leaved deciduous, seasonally flooded	PSS1C
SSW3	scrub-shrub wetlands	scrub-shrub broad-leaved deciduous, semipermanently flooded	PSS1F
SSW4	scrub-shrub wetlands	scrub-shrub needle-leaved deciduous, permanently flooded	PSS2/UBHh
U	upland	upland	U
EM4	emergents	emergent reeds, evergreen semipermanently flooded	PEM5/AB7F
EM5	emergents	emergent reeds, temporarily flooded	PEM5A
SSW5	scrub-shrub wetlands	scrub-shrub, needle-leaved deciduous intermittently exposed	PSS2/UBG
SSW6	scrub-shrub wetlands	scrub-shrub, broad-leaved deciduous semipermanently flooded	PSS1/EMFh
CF12	cypress forests	broad-leaved deciduous, seasonally flooded	PFO/SS1C
EM6	emergents	emergent persistent semipermanently flooded	PEM1Fh
EM7	emergents	emergent reeds semipermanently flooded	PEM5Fh

5. Results

5.1. Scattering Model

It has been assumed that the received power from GNSS-R signals reflecting from water are coherent and follow the scattering model given by Equation (2) [11,23]. We attempted to verify this by selecting data over the open-water portions of Caddo Lake, which should show a relatively constant calibrated power level, and assessed whether P_{cal} or $P_{cal}R_{sr}^2$ fit the data better. The P_{cal} values should ideally show a constant value, however there are systematic effects at the 1–2 dB level that appear correlated with the plane's direction. Since strong wind currents were encountered during the experiment, these systematics may be related to the plane's unmodeled orientation, and thus, an unmodeled change in the antenna gain. Despite these systematic changes in power, the standard deviation of the open-water data was 2.2 dB for P_{cal} but 2.8 dB for $P_{cal}R_{sr}^2$, indicating the coherent model was favored. There was also one satellite (PRN 16) that was visible in both the 3 km (10,000′) and 2.4 km (8000′)-altitude grids. The difference in average P_{cal} between these grids for that satellite was 1.0 dB while the difference in average $P_{cal}R_{sr}^2$ was 3.8 dB. Again, the coherent scattering model seems favored, and is assumed below.

5.2. Estimate of Surface Spot Size

We performed a data search for ground tracks passing cleanly from water to land, or tracks passing over a river or small lake. Measuring the transition time as the power level rises or falls, and knowing the receiver velocity, we can estimate the surface spot size. We found several dozen examples, all showing approximately the same rise and fall times, and present one example here.

Figures 3 and 4 show an example track over three bodies of water—a small pond, an inundated area, and a river. The track over the pond and river are reasonably clear and can be used to estimate the surface spot size, while the inundated area's extent is uncertain due to overlying vegetation. The total duration of the pond and river crossings are approximately 0.8 s and 1.6 s, and their sizes under the track are about 31 m and 72 m, respectively. With an aircraft velocity of 81 m/s and assuming a simple boxcar (step-function) model for the spot, the spot size is estimated to be 34 m from the pond data and 58 m from the river data. This scene was chosen because it exemplifies the range of sizes deduced for essentially all the clear water transitions and crossings observed. Figure 5 shows a blowup of these crossings with their respective spot sizes indicated as yellow ellipses. The spots shown also account for the 33° incidence angle and the reflection plane being 170° CCW from the velocity vector. Both crossings have vegetation, flat open land, and/or buildings nearby, which can affect the measurements. With R_{sr} being within a few meters of 3130 m over this track, Equation (6) would predict the spot size along the track to be about 41 m, which is in good agreement with the observed values, given the boxcar model approximation and measurement uncertainties.

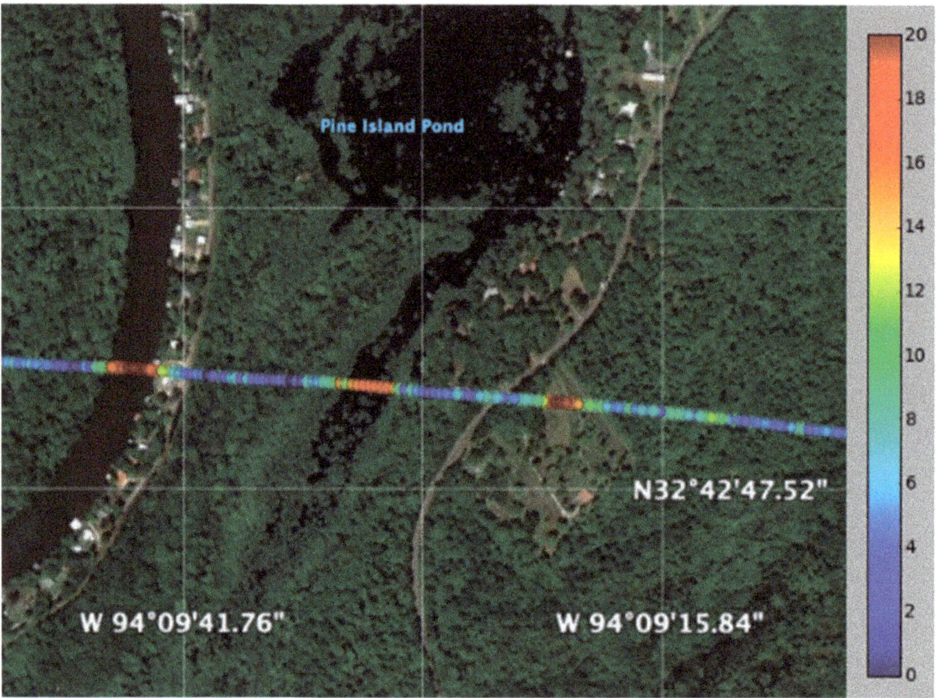

Figure 3. Westward ground track near Caddo Lake of PRN 23 over a Google Earth image from April 2017. The track's calibrated power in dBs is color coded from low to high as blue to red respectively, over a 20-dB range.

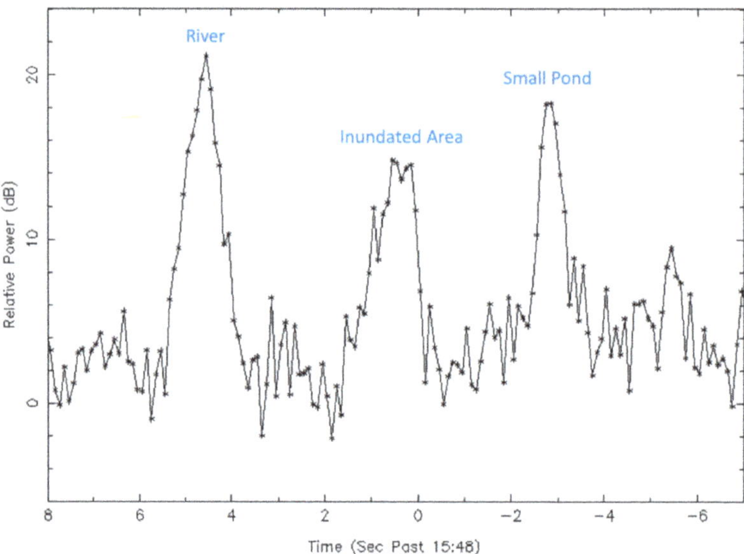

Figure 4. Received power vs. time corresponding to the westward ground track in Figure 3. Note that time increases to the left to align with Figure 3 features. Points are 0.1 s apart, and the aircraft velocity was 81 m/s.

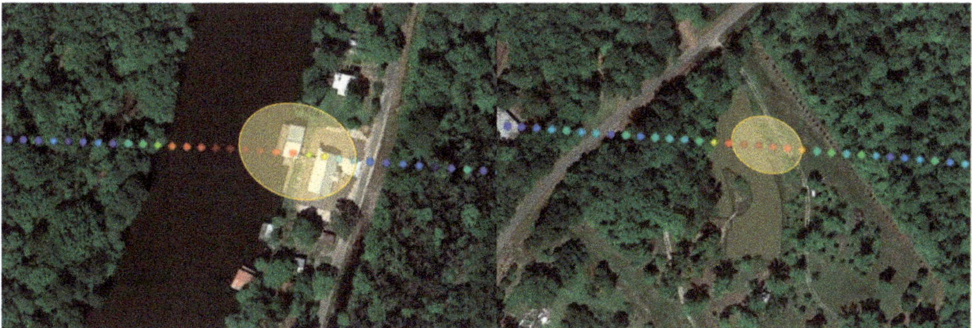

Figure 5. Blowup of the pond (right) and river (left) crossings from Figure 3, with yellow ellipses indicating the approximate orientation and size of the surface spot size, as derived from their respective power measurements vs time.

5.3. Caddo Lake

Figure 6a shows a pansharpened RGB image of the Caddo Lake area. Superimposed yellow lines are the paths the boat took concurrently with the GNSS-R flight. This region, the northwestern corner of Caddo Lake, is the area where inundated cypress forests are abundant. Giant *Salvinia* appears light green in the image shown in 6a.

Figure 6b shows a landcover classification map of Caddo Lake [22]. Abbreviated classes shown in legend are expanded in Table 1. The northwestern region of Caddo Lake contains a variety of intermittently, seasonally, semi-permanently, and permanently inundated cypress forests, scrub-shrub wetlands, and emergent species. The moving transient mats of giant *Salvinia* are not included in the landcover map. At the time of the aircraft experiment, they were prevalent predominantly in the areas denoted as permanently flooded cypress forests (PFCF1) but were also found in the semi-permanently flooded cypress forests (CF7/CF5), which can be seen in the Landsat 8 image in Figure 6a.

Selected photos depicting the conditions of Caddo Lake at the time of the aircraft experiment are shown in Figure 6c–e, and their respective locations are indicated by the colored dots in 6a. Figure 6c,d show typical cypress forests with a mixture of open water and giant *Salvinia* underneath the cypress trees. Figure 6e,f show two locations where giant *Salvinia* mats were particularly thick—the mats in 6f were thick enough to prevent further passage of the boat. Part of this analysis will show the effect of this aquatic vegetation on the ability for both C- and L-band signals to sense the underlying water.

Figure 7 shows GNSS-R data (LHCP transmit, V pol received as well as LHCP transmit, Hpol received) collected during the flight experiment (a,b) and C-band data from Sentinel-1 (c; VV polarization). The Sentinel-1 data were collected on the same day as the Landsat image (Figure 6a), 8 May 2017. We do not expect there to have been significant variation in the inundated areas within and around Caddo Lake in the six days that spanned between the aircraft experiment and the acquisition date of the Sentinel-1 and Landsat data. Here, and in all figures that follow, we present the GNSS-R data as their values relative to their mean value over the upland landcover class (abbreviated U). In other words, we subtract the mean value obtained for observations falling over uplands from all observations. We do this in order to more clearly show the sensitivity of the GNSS-R signal to different wetland types. Overall, GNSS-R data from both V and H pol are very similar to one another, though on average the H pol data showed higher Equation (8)'s P_{cal} by 1.7 dB.

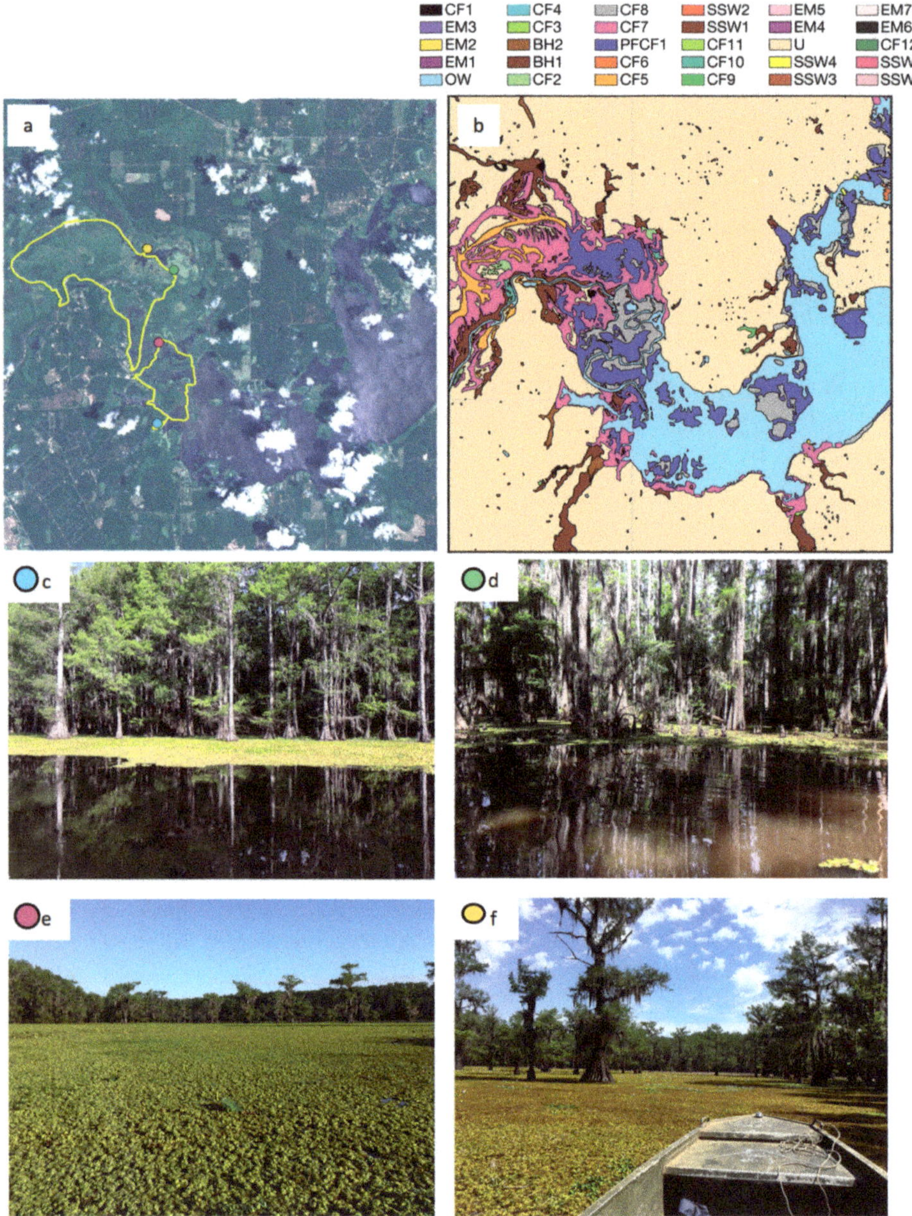

Figure 6. (**a**) Pansharpened Landsat 8 image of Caddo Lake from 8 May 2017. Yellow lines are GPS tracks of the boat. Colored dots are locations of photos in (**c**–**f**). (**b**). Landcover classification map of Caddo Lake [22]. Acronyms are defined in Table 1.

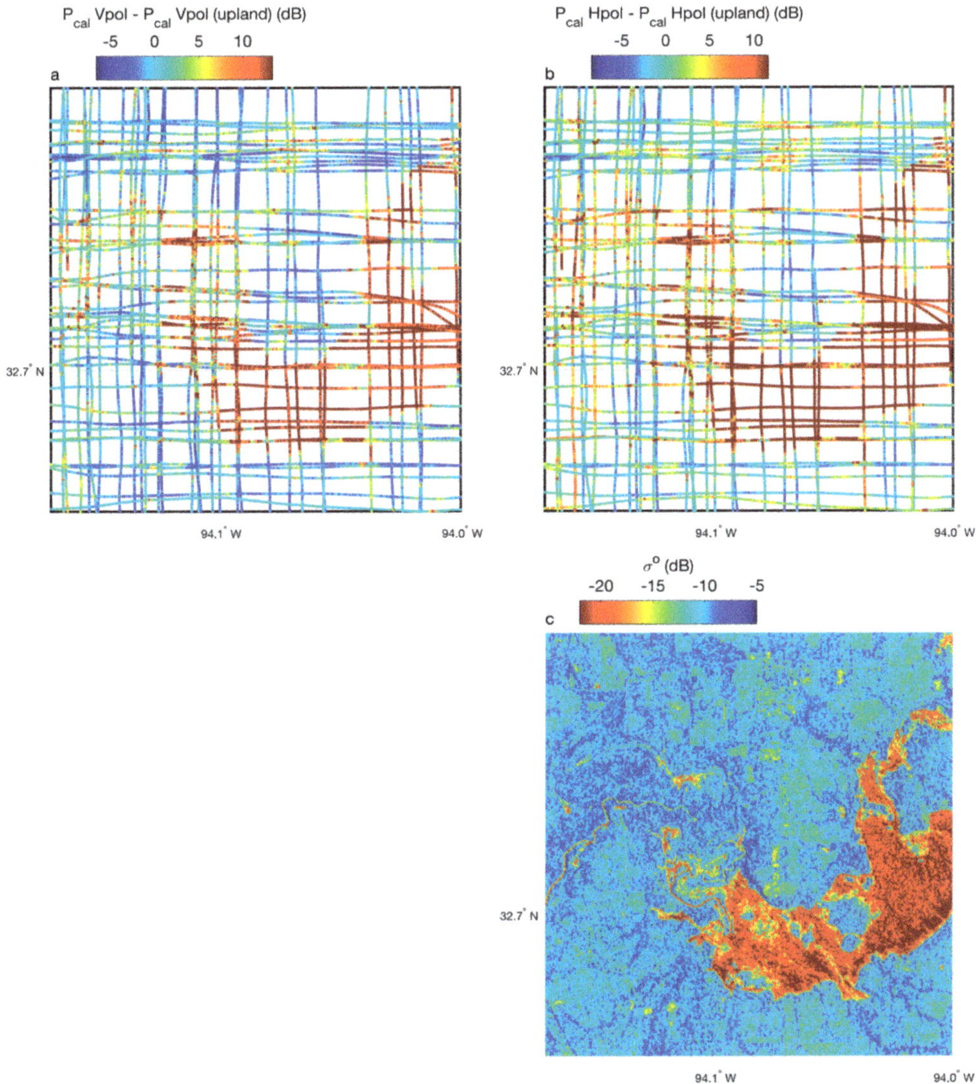

Figure 7. (**a**) Global Navigation Satellite System-Reflectometry (GNSS-R) Equation (8)'s P_{cal} Vpol observations over Caddo Lake, acquired on 2 May 2017. Observations are relative to their mean value over uplands (156.63 dB). (**b**) Same as (**a**), except showing GNSS-R P_{cal} Hpol observations, relative to their mean value over uplands (158.39 dB). (**c**) Sigma nought observations (VV pol) from Sentinel-1, acquired on 8 May 2017.

From Figure 7c, it is clear that the open water towards the eastern side of Caddo Lake is easily distinguishable in the Sentinel-1 data, apparent in the low sigma nought values. As stated above, low sigma nought values can indicate a very flat surface (e.g., smooth water) that results in little to no backscattering. As expected, the GNSS-R data in Figure 7a,b show the converse of this: the smooth open water on the eastern side of Caddo Lake produces high observed P_{cal}.

The northwestern side of Caddo Lake, with both cypress trees and giant *Salvinia* to obscure the water, is a more complex scattering environment. There is no qualitative distinction between the flooded cypress forests in the northwest and the surround upland areas, at least not as obvious of a distinction as that between the open water to the east and nearby uplands. There appears to be higher P_{cal} in the northwestern flooded forests relative to surrounding uplands, but without complete spatial sampling, this cannot be definitively concluded without a quantitative analysis.

In order to quantify the sensitivity of both the GNSS-R aircraft data and that from Sentinel-1 to water obscured by vegetation, we first binned the observations by the landcover classes shown in Figure 6b. When binning the Sentinel-1 observations, we only used the data that corresponded to the locations where the GNSS-R data were collected, such that a direct comparison could be made. We calculated the mean and standard deviation of the sigma nought and P_{cal} observations within each landcover class, and we show this in Figure 8. For this analysis we again show GNSS-R data relative to its mean value over the upland land cover class, and now we show Sentinel-1 data also relative to its own mean value over uplands. Thus, for both the GNSS-R and Sentinel-1 data, the upland class (U) is plotted with a value of 0, and all other classes are relative to this value. Since results for the GNSS-R V- and H-pol data were so similar, the following results are described for only the V-pol data, though the H-pol data may be seen in Figure 8.

First, we focus on the sensitivity of the GNSS-R and Sentinel-1 data to the open water (OW) class, relative to uplands (U). Both show a several dB change between the classes: the mean increase in P_{cal} between uplands and open water was 13.61 dB, and the mean decrease in backscattering was 8.55 dB. In both the GNSS-R and Sentinel-1 data, the landcover class that came closest to open water was the intermittently exposed cypress forests (CF6). This type of landcover is only found in a very small portion of Caddo Lake—it is in the extreme northeastern part of the lake shown in Figure 3b. Although the sample size is small ($n = 64$), it appears that there is no attenuation of the GNSS-R signal through the vegetation canopy, as the observed mean P_{cal} was actually 1.29 dB higher than that observed over open water. This could indicate that the vegetation sheltered the water surface from any slight roughening from the wind, though given the relatively large standard deviation of the observations, the difference between open water and this vegetation class was small. The Sentinel-1 data show some attenuation through the vegetation, as there was a 4.4 dB increase in backscattering in the Sentinel-1 data.

P_{cal} over the permanently flooded cypress forests (PFCF1, $n = 8345$) also showed a high signal, only 2.15 dB lower than P_{cal} over open water. Sentinel-1 data over this landcover class was 6 dB higher than that over open water and 2.5 dB different than the backscattering over uplands. We attribute changes in the GNSS-R signal and Sentinel-1 data mostly to the presence of the giant *Salvinia* mats, and only partially to the presence of cypress trees in this class, as the spacing between the trees in this class is usually at least 50 m, which is more than the spatial resolution of both the GNSS-R and Sentinel-1 data. This indicates that there is some attenuation of the forward-scattered L-band signal through the giant *Salvinia*, but only 2.15 dB. For the backscattered C-band signal, the giant *Salvinia* geometry or biomass is not sufficient to produce a double bounced signal, but the vegetation itself is dense enough to produce significant backscatter, which makes the signal closer to that from upland rather than open water.

Another notable landcover class was the semi-permanently flooded cypress forests (CF5, $n = 1781$), which produced a P_{cal} of 8.3 dB higher than upland, and a sigma nought of 0.14 dB below the upland value. We again attributed this to the extensive presence of giant *Salvinia* in these areas at the time of acquisition.

In the densest cypress forests (CF8, $n = 3816$), we still saw a mean P_{cal} value 4.25 dB above the upland value, which was an indication that at least some of these forests were likely inundated. The sigma nought value was 0.5 dB higher than that for uplands. Despite the fact that parts of these forests were flooded, and the cypress trees having significant geometric components that might at L-band produce double bounce, it appeared that the C-band signals were not able to penetrate these types of canopies. Similar conclusions could be drawn for the other landcover classes in and around Caddo Lake.

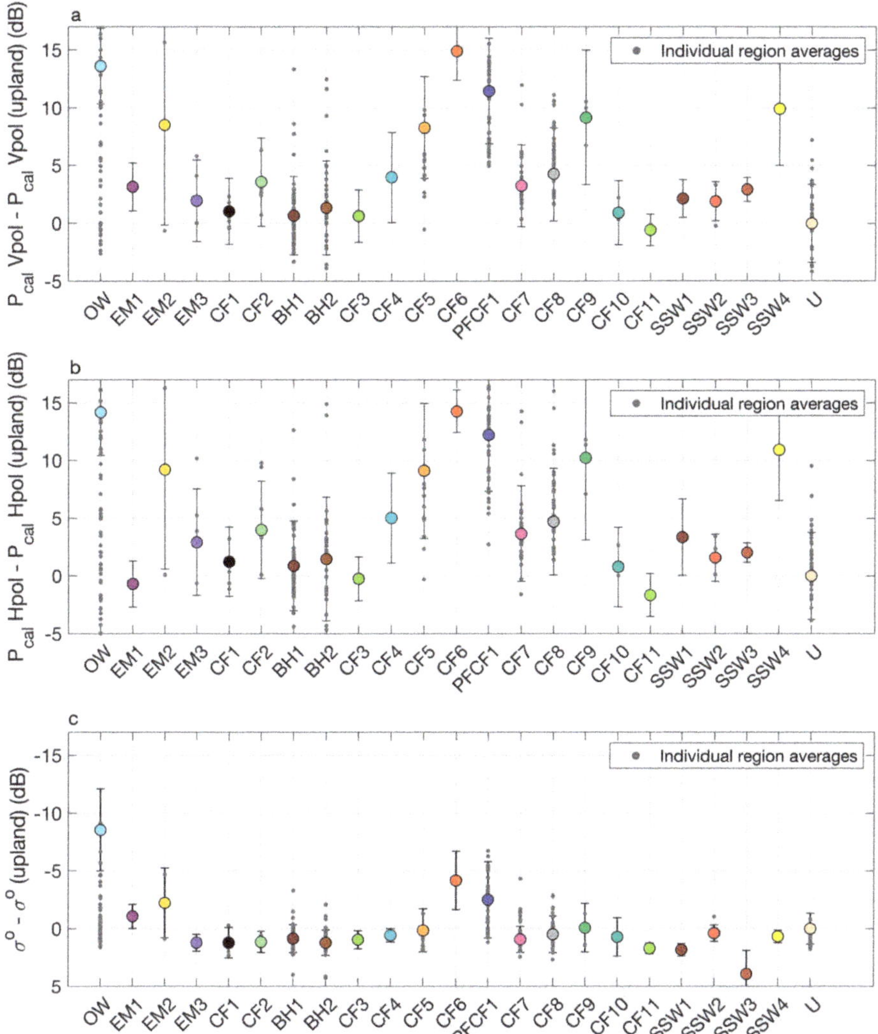

Figure 8. (**a**) Mean GNSS-R P_{cal} Vpol observation, binned by the landcover classes in the Caddo Lake area, defined in Table 1. Error bars are ± one standard deviation. Observations are relative to the mean value over uplands (U), which was 156.63 dB. (**b**) Same as (**a**) except for GNSS-R P_{cal} Hpol observations (mean value over uplands = 158.39 dB). (**c**) Same as (**a**) except for sigma nought observations from Sentinel-1. The mean value over uplands (U) was -10.4 dB. Individual region averages are shown by the grey dots.

The only notable distinction between the GNSS-R P_{cal} H-pol and V-pol data in terms of sensitivities to different landcover classes is the difference in mean P_{cal} over the EM1 landcover class (emergent reeds, semi-permanently flooded). In this case, the V-pol data show a few dB increase relative to the upland landcover class, whereas the H-pol data show a slight decrease relative to uplands. After closer examination, we found this single class to be misidentified, and instead of emergent reeds it should have been classified as uplands.

6. Discussion

Exploring the sensitivity of the GNSS-R signal to different wetland types based on a static landcover map is inherently an imperfect exercise. For example, just because the landcover map classifies an area as 'semi-permanently flooded' does not mean it actually was flooded during the flight experiment. Due to the extremely dense nature of the vegetation in Caddo Lake, it was not possible to conduct an extensive survey of flooded areas. In particular, we found many misclassifications of open water (OW) after comparing the Landsat 8 image and Google Earth images to the landcover classification map. These misclassifications were primarily small ponds to the north of Caddo Lake which had dried up or been converted to uplands in the time since the wetland inventory was conducted in 1983. These misclassifications can be seen in Figure 8, which also contains the mean P_{cal} value for individual landcover regions (i.e., each region classified as a particular landcover class instead of the class as a whole), indicated by the grey dots. Some individual regional averages for the open water (OW) class are quite low compared to the total average—these are the misclassified regions.

We plotted all individual region averages for each landcover class in Figure 8, not just for open water. For the most part, these regional averages cluster together, with individual region averages in the Sentinel-1 data mimicking what is observed in the GNSS-R data. A notable exception to the overall clustering of individual regions is for the EM2 (emergent reeds) class. In this case, there were two small clusters of observations over two different areas classified as EM2 in the southeast portion of Caddo Lake, separated by a larger region classified as CF7 (cypress forests). The cluster of observations closer to open water had a much higher mean P_{cal} V-pol than the cluster of observations adjacent to the upland/dry land. We interpret this as meaning that the cluster of observations in the EM2 region close to open water were probably responding to water underneath the reeds, whereas the reeds closer to the uplands/dry lands might have been dry. By considering these observations as being a part of one group only, a large standard deviation is the result.

7. Conclusions

From this aircraft experiment we have shown the scattering model in Equation (2) was appropriate for inundation and that the spot size in this case was approximately the Fresnel zone. From a comparison of L-band GNSS-R data with C-band Sentinel-1 data, we could conclude the following: The GNSS-R signal was able to penetrate the dense cypress canopies that typify Caddo Lake and other swamps in the southeast United States. There were several dB of attenuation, but the signal was still significantly higher than that over dry land. In addition, the presence of giant *Salvinia* only minimally attenuates the GNSS-R signal by ~2 dB. C-band backscatter data, on the other hand, was significantly attenuated by both of these vegetation types. This study was not able to conclude whether L-band backscatter signals, such as those collected by PALSAR-2, would be able to detect water underneath these types of vegetation, as there were no data publicly available. This study also did not quantify the saturation point for GNSS-R with respect to vegetation, as the GNSS-R signal was able to penetrate the vegetation canopies overflown here. Future GNSS-R aircraft flights, which could fly over denser vegetation canopies, might be able to quantify the saturation point, if there is one. Future studies should also focus on obtaining coincident L-band backscatter data, which could identify the relative strengths of the two data types without wavelength being a confounding factor.

Author Contributions: Conceptualization, S.T.L. and C.C.; methodology, S.T.L. and C.C.; experiment hardware, S.T.L., J.S. and M.K.; antenna calibration, S.T.L., J.S. and M.K.; data-collection software, J.S. and M.K.; experiment campaign, S.T.L., J.S. and C.C; analysis software, S.T.L.; validation, S.T.L and C.C.; formal analysis, S.T.L. and C.C.; resources, S.T.L., C.C., J.S. and M.K.; data curation, S.T.L., J.S. and M.K.; writing—original draft preparation, S.T.L. and C.C.; writing—review and editing, S.T.L. and C.C.; supervision, S.T.L.; project administration, S.T.L.; funding acquisition, S.T.L. All authors have read and agreed to the published version of the manuscript.

Funding: This research was carried out at the Jet Propulsion Laboratory, California Institute of Technology, under a contract with the National Aeronautics and Space Administration (NASA). Support for this work was provided by NASA's Solid Earth Program, JPL's Research and Technology Development Program, and JPL's Advanced Concepts Program. Part of this work was funded by NASA NNH17ZDA001N-THP. © 2020. All rights reserved.

Acknowledgments: We thank Keystone Arial Surveys, Inc. for excellent performance and handling the 2-year flight delay, members of the Caddo Lake Institute for their information on this region, Dr. Roy Darville of East Texas Baptist University for providing a boat tour around Caddo Lake, JPL's Bruce Haines for processing the aircraft Novatel data through Gipsy, and JPL's Larry Young for his help with the antenna calibration experiment.

Conflicts of Interest: The authors declare no conflict of interest. The funders had no role in the design of the study; in the collection, analyses, or interpretation of data; in the writing of the manuscript, or in the decision to publish the results.

References

1. Stocker, T.F.; Qin, D.; Plattner, G.-K.; Tignor, M.; Allen, S.K.; Boschung, J.; Nauels, A.; Xia, Y.; Bex, V.; Midgley, P.M. (Eds.) *IPCC, 2013: Climate Change 2013: The Physical Science Basis. Contributions of Working Group I to the Fifth Assessment Report of the Intergovernmental Panel on Climate Change*; Cambridge University Press: Cambridge, UK; New York, NY, USA, 2013.

2. Melton, J.R.; Wania, R.; Hodson, E.L.; Poulter, B.; Ringeval, B.; Spahni, R.; Bohn, T.; Avis, C.A.; Beerling, D.J.; Chen, G.; et al. Present state of global wetland extent and wetland methane modelling: Conclusions from a model inter-comparison project (WETCHIMP). *Biogeosciences* **2013**, *10*, 753–788. [CrossRef]

3. Ringeval, B.; de Noblet-Ducoudré, N.; Ciais, P.; Bousquet, P.; Prigent, C.; Papa, F.; Rossow, W.B. An attempt to quantify the impact of changes in wetland extent on methane emissions on the seasonal and interannual time scales. *Glob. Biogeochem. Cycles* **2010**, *24*, GB2003. [CrossRef]

4. Shindell, D.T.; Walter, B.P.; Faluvegi, G. Impacts of climate change on methane emissions from wetlands. *Geophys. Res. Lett.* **2004**, *31*, 21. [CrossRef]

5. Catry, T.; Li, Z.; Roux, E.; Herbreteau, V.; Gurgel, H.; Mangeas, M.; Seyler, F.; Dessay, N. Wetlands and Malaria in the Amazon: Guidelines for the Use of Synthetic Aperture Radar Remote-Sensing. *Int. J. Environ. Res. Public Health* **2018**, *15*, 368. [CrossRef] [PubMed]

6. Gopal, B. Future of wetlands in tropical and subtropical Asia, especially in the face of climate change. *Aquat. Sci.* **2013**, *75*, 39–61. [CrossRef]

7. Pekel, J.F.; Cottam, A.; Gorelick, N.; Belward, A.S. High-resolution mapping of global surface water and its long-term changes. *Nature* **2016**, *540*, 418–422. [CrossRef] [PubMed]

8. Schroeder, R.; McDonald, K.; Chapman, B.; Jensen, K.; Podest, E.; Tessler, Z.; Bohn, T.; Zimmermann, R. Development and evaluation of a multi-year fractional surface water data set derived from active/passive microwave remote sensing data. *Remote Sens.* **2015**, *7*, 16688–16732. [CrossRef]

9. Du, J.; Kimbal, J.; Galantowicz, J.; Kim, S.; Chan, S.; Reichle, R.; Jones, L.; Watts, J. Assessing global surface water inundation dynamics using combined satellite information from SMAP, AMSR2 and Landsat. *Remote Sens. Environ.* **2018**, *213*, 1–17. [CrossRef] [PubMed]

10. Martinez, J.M.; le Toan, T. Mapping of flood dynamics and spatial distribution of vegetation in the Amazon floodplain using multitemporal SAR data. *Remote Sens. Environ.* **2007**, *108*, 209–223. [CrossRef]

11. Nghiem, S.V.; Zuffada, C.; Shah, R.; Chew, C.; Lowe, S.; Mannucci, A.; Cardellach, E.; Brakenridge, G.; Geller, G.; Rosenqvist, A. Wetland monitoring with global navigation satellite system reflectometry. *Earth Sp. Sci.* **2017**, *4*, 16–39. [CrossRef] [PubMed]

12. Jensen, K.; McDonald, K.; Podest, E.; Rodríguez-Álvarez, N.; Horna, V.; Steiner, N. Assessing L-band GNSS-Reflectometry and Imaging Radar for Detecting Sub-Canopy Inundation Dynamics in a Tropical Wetlands Complex. *Remote Sens.* **2018**, *10*, 1431. [CrossRef]

13. Rodriguez-Alvarez, N.; Podest, E.; Jensen, K.; McDonald, K. Classifying Inundation in a Tropical Wetlands Complex with GNSS-R. *Remote Sens.* **2019**, *11*, 1053. [CrossRef]

14. Morris, M.; Chew, C.; Reager, J.T.; Shah, R.; Zuffada, C. A novel approach to monitoring wetland dynamics using CYGNSS: Everglades case study. *Remote Sens. Environ.* **2019**, *233*, 111417. [CrossRef]

15. Zribi, M.; Motte, E.; Baghdadi, N.; Baup, F.; Dayau, S.; Fanise, P.; Guyon, D.; Huc, M.; Wigneron, J. Potential Applications of GNSS-R Observations over Agricultural Areas: Results from the GLORI Airborne Campaign. *Remote Sens.* **2018**, *10*, 1245. [CrossRef]

16. Egido, A.; Paloscia, S.; Motte, E.; Guerriero, L.; Pierdicca, N.; Caparrini, M.; Santi, E.; Fontanelli, G.; Floury, N. Airborne GNSS-R polarimetric measurements for soil moisture and above-ground biomass estimation. *IEEE J. Sel. Top. Appl. Earth Obs. Remote Sens.* **2014**, *7*, 1522–1532. [CrossRef]

17. Chew, C.C.; Small, E.E. Soil Moisture Sensing Using Spaceborne GNSS Reflections: Comparison of CYGNSS Reflectivity to SMAP Soil Moisture. *Geophys. Res. Lett.* **2018**, *45*, 4049–4057. [CrossRef]

18. Chew, C.C.; Shah, R.; Zuffada, C.; Hajj, G.; Masters, D.; Mannucci, A.J. Demonstrating soil moisture remote sensing with observations from the UK TechDemoSat-1 satellite mission. *Geophys. Res. Lett.* **2016**, *43*, 3317–3324. [CrossRef]

19. Comite, D.; Ticconi, F.; Dente, L.; Guerriero, L.; Pierdicca, N. Bistatic Coherent Scattering from Rough Soils with Application to GNSS Reflectometry. *IEEE Trans. Geosci. Remote Sens.* **2019**, *58*, 612–625. [CrossRef]

20. Ulaby, F.; Moore, R.; Fung, A. *Microwave Remote Sensing: Active and Passive*; Addison-Wesley Publishing Company: Boston, MA, USA, 1982; p. 02116.

21. Stefanski, J. Step by Step: Recommended Practice Flood Mapping. UN-SPIDER Knowledge Portal. 2015. Available online: http://www.un-spider.org/ (accessed on 2 September 2019).

22. Data Server for Caddo Lake Information. 2017. Available online: http://caddolakedata.us/maps (accessed on 9 September 2019).

23. Camps, A. Spatial Resolution in GNSS-R Under Coherent Scattering. *IEEE Geosci. Remote Sens. Lett.* **2019**, *17*, 1–5. [CrossRef]

Article

Description of the UCAR/CU Soil Moisture Product

Clara Chew [1,*] **and Eric Small** [2]

[1] COSMIC, University Corporation for Atmospheric Research, Boulder, CO 80301, USA
[2] Department of Geological Sciences, University of Colorado, Boulder, CO 80309, USA;
    eric.small@colorado.edu
*   Correspondence: clarac@ucar.edu

Received: 14 April 2020; Accepted: 13 May 2020; Published: 14 May 2020

Abstract: Currently, the ability to use remotely sensed soil moisture to investigate linkages between the water and energy cycles and for use in data assimilation studies is limited to passive microwave data whose temporal revisit time is 2–3 days or active microwave products with a much longer (>10 days) revisit time. This paper describes a dataset that provides soil moisture retrievals, which are gridded to 36 km, for the upper 5 cm of the soil surface at sparsely sampled 6-hour intervals for +/− 38 degrees latitude for 2017–present. Retrievals are derived from the Cyclone Global Navigation Satellite System (CYGNSS) constellation, which uses GNSS-Reflectometry to obtain L-band reflectivity observations over the Earth's surface. The product was developed by calibrating CYGNSS reflectivity observations to soil moisture retrievals from NASA's Soil Moisture Active Passive (SMAP) mission. Retrievals were validated against observations from 171 in-situ soil moisture probes, with a median unbiased root-mean-square error (ubRMSE) of 0.049 cm^3 cm^{-3} (standard deviation = 0.026 cm^3 cm^{-3}) and median correlation coefficient of 0.4 (standard deviation = 0.27). For the same stations, the median ubRMSE between SMAP and in-situ observations was 0.045 cm^3 cm^{-3} (standard deviation = 0.025 cm^3 cm^{-3}) and median correlation coefficient was 0.69 (standard deviation = 0.27). The UCAR/CU Soil Moisture Product is thus complementary to SMAP, albeit with a larger random noise component, providing soil moisture retrievals for applications that require faster revisit times than passive microwave remote sensing currently provides.

Keywords: soil moisture; GNSS-R; CYGNSS

---

1. Introduction

Near-surface (0–5 cm) soil moisture data are used for a variety of applications, such as drought monitoring [1–3], flood forecasting [4,5], initializing climate models [6], numerical weather prediction [7], quantifying the linkages between the surface and atmospheric boundary layers [8–10], optimizing irrigation strategies [11], and predicting infectious disease transmission [12]. All of these applications require soil moisture data at different spatial and temporal scales. Due to its ability to retrieve soil moisture over large parts of the globe, satellite remote sensing is often used to provide these data. Most often, microwave instruments are used due to their ability to see through clouds and penetrate at least some amount of overlying vegetation. However, even the microwave sensor with the fastest revisit time (passive radiometer), is constrained to providing data every 2–3 days [13]. While for many soil moisture applications, such as climate model initialization, a 2–3 day repeat period is more than sufficient, other applications require a higher density of observations. For example, the quantification of soil moisture memory is limited to the time scale of observations [10]. Several ongoing efforts are also using soil moisture as a proxy for other quantities, such as rainfall [14,15] or evaporation [16,17], and having daily or sub-daily data would increase the success of these techniques.

A constellation of microwave instruments, the Cyclone Global Navigation Satellite System (CYGNSS), is currently on orbit over the tropics, and because it is a constellation with numerous satellites, it can provide data more frequently than currently possible using a single instrument alone [18]. Although not designed for soil moisture remote sensing, this paper describes a dataset [19] that has been developed from the CYGNSS constellation that provides soil moisture retrievals for the majority of the tropics (+/− 38 degrees latitude) at daily and sub-daily time steps. Here, we describe the algorithm, its assumptions, and validation of the retrievals using in-situ soil moisture observations. In regions where the algorithm performance is acceptable, the CYGNSS soil moisture retrievals could be used to augment existing soil moisture satellite data for the aforementioned applications that require data at a higher temporal resolution.

1.1. CYGNSS

The CYGNSS mission was launched in December 2016. A NASA Earth Ventures Mission, CYGNSS consists of eight small satellites that orbit the tropics with an inclination of 35 degrees [18]. CYGNSS was designed to retrieve ocean surface wind speed during hurricane intensification events, and, as such, the receivers and software were optimized for ocean surface remote sensing. Each of the CYGNSS satellites carries with it two downward-looking antennas and a Global Navigation Satellite System-Reflectometry (GNSS-R) receiver.

GNSS-R is a form of L-band bistatic radar that utilizes transmitted navigation signals as the signal source. GNSS is an umbrella term that encompasses constellations like the United States' GPS, but also the EU's Galileo, Russia's GLONASS, China's BeiDou, India's IRNSS, and Japan's QZSS. In total, there are over 80 GNSS satellites currently in orbit (32 of which are GPS satellites), with more being planned in the coming years. To date, GNSS-R most commonly utilizes right hand circular polarized (RHCP) L-band signals transmitted from GPS satellites, which upon reflection are predominantly left hand circular polarized (LHCP).

Because L-band signals are commonly used for near-surface (0–5 cm) soil moisture remote sensing (e.g., NASA's Soil Moisture Active Passive (SMAP) or the European Space Agency's Soil Moisture and Ocean Salinity (SMOS) missions [13,20]), there is significant interest in quantifying the sensitivity of spaceborne GNSS-R data to changes in soil moisture. There had been evidence of sensitivity during previous GNSS-R flight campaigns (e.g., [21–25]), though the first analyses investigating the sensitivity of spaceborne GNSS-R data to soil moisture were only done after the launch of TechDemoSat-1 (TDS-1) in 2014. TDS-1 carried the same GNSS-R receiver as is used by CYGNSS, and these preliminary investigations did show that there appeared to be some sensitivity of GNSS-R to soil moisture [26,27]. Due to the fact that TDS-1 only collected GNSS-R data for one out of every eight days, however, developing a full soil moisture retrieval algorithm was not attempted.

The sheer amount of data collected by CYGNSS has allowed for an empirical investigation into the sensitivity of spaceborne GNSS-R data to soil moisture. The dataset that we describe here is the result of a modified version of the methodology presented in [28]. The full dataset is available online [19]. The results presented here are the outcome of just one of several ongoing efforts investigating the potential of CYGNSS to retrieve soil moisture. Several of these efforts are also using SMAP retrievals as a reference data set, though differences arise in assumptions regarding gridding, open water masking, and vegetation considerations. For example, [29] assumed that the vast majority of CYGNSS observations have a coarse (>25 km) spatial resolution and spatially averaged the data under this assumption, whereas in the algorithm described here we assume a 3 km spatial resolution. The authors of [30] also used SMAP data for retrieval algorithm development, though also made considerations for changes in vegetation, and [31] incorporated surface roughness data from IceSat-2, neither of which do we do here. Both [29,32] scale their soil moisture retrievals by knowing a priori the maximum and minimum values of soil moisture for a particular region, which we also do not do here. Several other researchers are exploring machine learning techniques for CYGNSS soil moisture retrieval (e.g., [33,34]). Given the differences in validation data sets and time periods used for validation, it is difficult to directly compare the output of algorithms, though most have shown at least moderate success in retrieving soil moisture from CYGNSS.

1.2. GNSS-R Background

Unlike monostatic radar, which measures backscatter, GNSS-R measures the forward-scattered signal, which has reflected off of the surface of the Earth and back into space. Figure 1 presents a schematic of the signal geometry. A satellite in low Earth orbit, with a GNSS-R receiver onboard, has one or more downward-looking antennas, which record the forward-scattered signals. Information contained within the scattered signals can be related to land surface parameters, such as the surface roughness, dielectric constant of the soil, and vegetation properties.

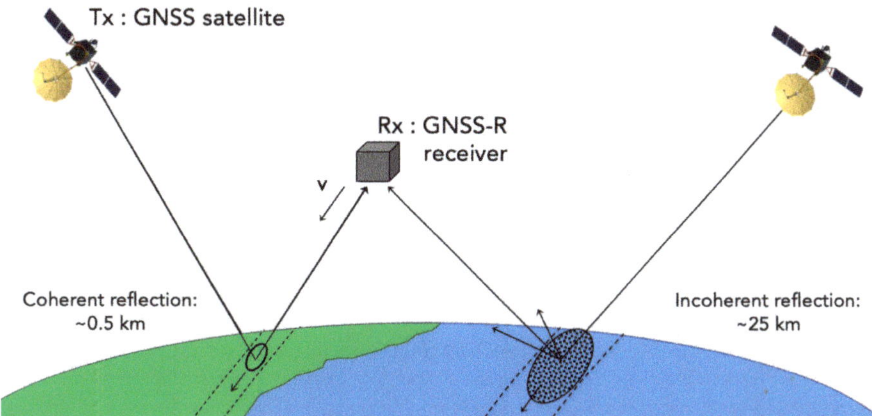

Figure 1. Schematic of the GNSS-R technique. A GNSS satellite transmits (Tx) a signal towards the Earth's surface. Part of this signal reflects in the forward (specular) direction and back into space. A GNSS-R receiver (Rx) onboard a low-Earth-orbiting satellite, with a downward-looking antenna, records this signal. The point on the Earth's surface where the signal reflects depends upon the positions of the transmitting and receiving satellites. The roughness of the surface at the reflection point determines the spatial resolution of the signal, with rougher surfaces producing larger spatial footprints. Nearly always, the receiver integrates the reflected signal over a period of time, which elongates the spatial footprint in the along-track direction.

The point of reflection on the Earth's surface is determined by the positions of the transmitting and receiving satellites. Because these positions are constantly changing, the reflection points are pseudo-randomly distributed on the Earth's surface (see Figure 2a for examples), which is different than traditional remote-sensing techniques, which collect data in repeatable swaths. This means that, for a given point of the Earth's surface, observations could be recorded one hour apart, and then there could be no observations for the next several hours, for example [35]. Observations are recorded at all times of day, again, unlike traditional remote-sensing techniques, which tend to observe a particular location at a particular time of day. The pseudo-random distribution of observations, over time, aggregate such that complete maps of the reflected signal can be made (Figure 2b).

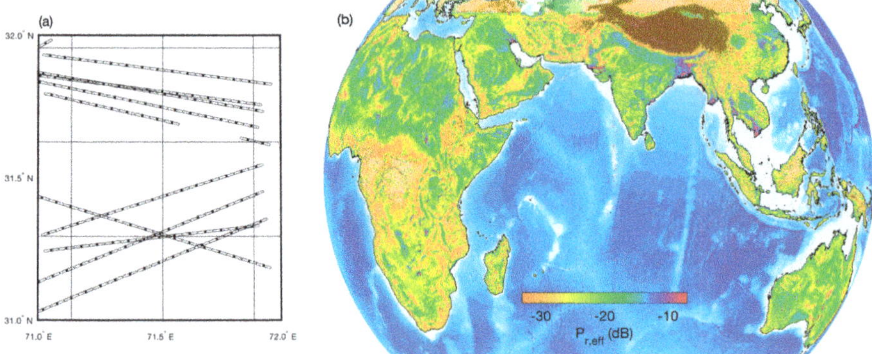

Figure 2. (**a**) Illustration of the pseudo-random surface sampling by CYGNSS (ellipses with dots). Ellipses are approximately 7×0.5 km in size, which is the expected footprint if the surface has little surface or topographic roughness. Dots are the location of the specular reflection points recorded by CYGNSS. This example shows coverage for one typical 24-hr period. Grid cells are the size of the EASE2 36-km grid cells. (**b**) Over time, observations made by CYGNSS completely cover the land surface, producing maps such as this (CYGNSS observations overlaid on DEM). Here, higher values could indicate a wet surface or a relatively flat surface.

The spatial resolution of the reflecting signal depends on the roughness of the surface at and near the reflection point [36]. Here, roughness includes possible contributions from many factors and not just the micro-scale roughness of the soil, such as roughness introduced from vegetation canopies, wind-roughened water, and macro-scale roughness from topography. If the surface is relatively rough, then the reflected signal is incoherent and comes from an area called the "glistening zone," which is on the order of several kilometers (~25 km in the case of the ocean surface), though recent research is beginning to show that where incoherent scattering originates from over the land surface is highly dependent on the local topography and may not be definable with a simple 25-km radius [37]. If the surface is relatively smooth, then the reflected signal is coherent and comes from an area defined by the first Fresnel zone. For a low-Earth-orbiting GNSS-R satellite, this area is on the order of 0.5 km, though this also depends on incidence angle (0.33 km at 0 deg incidence, 1.3 km at 60 deg incidence) [38]. In practical terms, determining whether the reflecting surface is smooth enough to produce a coherent reflection is challenging, as surface roughness is an extremely difficult parameter to measure. Existing measurements show that surface roughness tends to be on the order of 2–3 cm [39–41], though surface roughness likely varies considerably on scales as large as the first Fresnel zone. There are also likely complications from macro-scale surface roughness due to topographic variations that are not well understood. Modelling efforts are beginning to shed light into the role of topography in GNSS-R signal scattering (e.g., [42,43]), with some studies showing that the sensitivity of the GNSS-R signal to soil moisture is relatively unaffected by complex topography [42].

In all likelihood, most signals are probably a combination of incoherent and coherent scattering. In the algorithm presented here, we ignore contributions from incoherent scattering. Assuming that the reflected signal is always coherent may lead to increased uncertainties in final soil moisture retrievals, though currently these uncertainties are not able to be quantified without a better knowledge of the precise conditions that lead to incoherence.

Due to the fact that CYGNSS was designed to be an ocean sensor, where the reflected signal is relatively weak, the processing software integrates the signal over a period of 1 s for each "observation." During that time, the spacecraft has moved approximately 7 km, which means that the smallest along-track spatial resolution possible over land is 7 km, though the across track resolution could still be the theoretical 0.5 km. This results in the spatial footprint having a minimum size of 7×0.5 km, with the signal being smeared out along track (Figure 2a). In mid-2019, the CYGNSS integration time was decreased from 1 to 0.5 s, which means the minimum spatial footprint is currently 3.5 × 0.5 km.

The reflected GNSS signal is recorded by the receiver in the form of what is called a delay-Doppler map (DDM). A DDM is created by cross-correlating the received signal with a locally generated replica that has been modified considering different path delays (resulting from the path distance between the transmitter, reflecting surface, and receiver, as shown in Figure 1) and Doppler shifts (resulting from the relative motions of the transmitter, reflecting surface, and the receiver). Two examples of DDMs are shown in Figure 3. Figure 3a is an example of a DDM recorded by CYGNSS over the land surface, and Figure 3b is an example of a DDM recorded over the ocean surface. The horseshoe shape of the ocean DDM is an indication that the reflection is incoherent and comes from a large, rough area. The absence of a horseshoe shape in Figure 3a indicates that the reflection is mostly coherent, and comes from a smaller, smoother area. The maximum power of each DDM is affected by surface roughness, the dielectric constant of the surface, and the vegetation overlying the surface, which is explained further below.

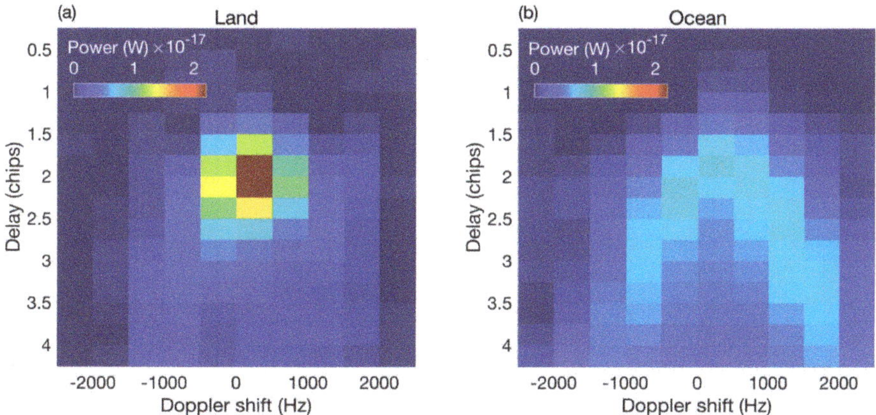

Figure 3. (**a**) DDM recorded over the land surface; (**b**) DDM recorded over the ocean.

DDMs are most commonly used by summarizing them into one metric or observable. The observables that are commonly used for soil moisture estimation are the peak cross-correlation of each DDM or the peak divided by the noise floor (signal-to-noise ratio, SNR). The value of the peak cross-correlation of each DDM is related to surface characteristics at the specular reflection point of the GNSS signal, including the roughness of the surface, the surface dielectric constant, and properties related to any overlying vegetation such as vegetation water content and structure of the canopy. However, the peak of each DDM is also affected by variables unrelated to the reflecting surface, such as antenna gain and range. We describe our procedure to correct for these effects in the next section.

2. Materials and Methods

2.1. Introduction to the Algorithm

The algorithm presented here uses collocated soil moisture retrievals from the Soil Moisture Active Passive (SMAP) mission to calibrate concurrent (same calendar day) CYGNSS observations throughout a calibration period [44]. For a given location, a linear relationship between SMAP soil moisture and CYGNSS surface reflectivity observations is determined, and the relationship is used to transform all CYGNSS observations into soil moisture, even at times when there are no corresponding SMAP data points. Once the calibration is performed, it is applied to data outside the calibration period such that SMAP data are no longer required for ongoing CYGNSS soil moisture retrievals.

Using SMAP data for calibration of course comes with many drawbacks, the major one being that SMAP soil moisture retrievals are not soil moisture observations and have their own error and uncertainties. One must be careful when using CYGNSS data in areas where it is known that SMAP performs poorly. In addition, SMAP's 40 km spatial resolution is likely coarser than that of CYGNSS. Intelligent upscaling of CYGNSS data to the 36-km EASE-2 grid that SMAP uses is necessary. If the resolution of CYGNSS is smaller than 36 km, then this effectively degrades the CYGNSS data and does not utilize it to its full potential. Despite the drawbacks associated with calibrating CYGNSS data with SMAP retrievals, SMAP data are considered to be one of, if not the, most accurate of the existing soil moisture products [45,46].

2.2. Algorithm Description

2.2.1. Derivation of $P_{r,eff}$

The first step in the retrieval algorithm is to calculate the effective surface reflectivity, which is the peak value of each delay-Doppler map (DDM) corrected for gain, range, and incidence angle effects. We call the uncorrected peak value of each DDM P_r.

P_r is affected by surface characteristics, such as the dielectric constant, roughness, and vegetation, as well as the gain of the receiving antenna, the bistatic range, and the power transmitted by each GPS satellite. We correct P_r for antenna gain, range, and GPS transmit power assuming a coherent reflection:

$$P_r = \frac{P^t G^t}{4\pi(R_{ts} + R_{sr})^2} \frac{G^r \lambda^2}{4\pi} \Gamma_{rl} \tag{1}$$

where P^t is the transmitted RHCP power, G^t is the gain of the transmitting antenna, R_{ts} is the distance between the transmitter and the specular reflection point, R_{sr} is the distance between the specular reflection point and the receiver, G^r is the gain of the receiving antenna, λ is the GPS wavelength (0.19 m), and Γ_{rl} is the effective surface reflectivity.

We then solve for Γ_{rl}, which is the term affected by the surface roughness, dielectric constant of the soil, and vegetation by first converting all terms to dB:

$$\Gamma_{rl}[dB] = 10\log P_r - 10\log P^t - 10\log G^t - 10\log G^r + 20\log(R_{ts} + R_{sr}) - 20\log \lambda + 20\log 4\pi, \tag{2}$$

Incidence angle is also expected to affect a coherent reflection, though this effect is only significant when the incidence angle is above 40 or 50 degrees. We correct for incidence angle in a similar way as in [29] by modelling how the effective surface reflectivity should change as a function of incidence angle, using established relationships (e.g., [47]). We call Γ_{rl} observations that have been corrected for incidence angle "$P_{r,eff}$," which stands for the effective surface reflectivity.

2.2.2. Outlier Identification

Because CYGNSS was not optimized for remote sensing of the land surface, we remove observations that are flagged with standard quality measures as well as use empirical quality control that we have found to increase the effectiveness of our algorithm. Standard quality flags that we use are the following: "S-band transmitter powered up," "spacecraft attitude error," "black body DDM," "DDM is a test pattern," "direct signal in DDM," and "low confidence in the GPS EIRP estimate." Although it has been recommended that observations from the GPS Block IIF satellites be removed due to larger variations in GPS transmit power and consequently larger uncertainties in P_r [48], we keep these observations, as removing them reduces data volume by more than 30%.

Any data recorded before December 2017 reflecting from surface elevations greater than 600 m are removed. Prior to this time, the satellites did not record DDMs that contained the full surface reflection coming from these elevations.

We perform additional quality control measures that are not standardized in the analysis of CYGNSS data over oceans. These measures were developed after a detailed examination of outliers in regions where the surface is relatively constant over time (e.g., deserts). We remove any observation where P_r is less than 2 dB above the noise floor (SNR < 2 dB). We remove observations with a receiver antenna gain less than 0 dB, observations with an incidence angle greater than 65 degrees, any data with P_r occurring in a delay bin outside of 7–10 pixels (exclusive), and any observations that do not have a SNR less than or equal to the receiver antenna gain plus 14 dB.

2.2.3. Removal of Open Water Observations

The removal of specular reflection points that are affected by open water is a critical step before retrieving soil moisture. Even small water bodies ~25 m wide can significantly affect $P_{r,eff}$ (e.g., Figure 1 from [28]). Our algorithm uses the Global Surface Water Explorer (GSWE) dataset described in [49], which is a 30-m water mask derived from Landsat data. Because it is derived from optical data, it cannot sense water beneath vegetation, though the L-band CYGNSS data is likely sensitive to some amount of water underneath a vegetation canopy. Additionally, because the GSWE is quasi-static in that it describes open water occurrence or when water occurs seasonally, it is not concurrent with the CYGNSS observations. The open water masking effort is thus imperfect, though we have found that it successfully removes a large amount of CYGNSS observations that are affected by open water.

The current algorithm removes open water using the "seasonality" data product provided by the GSWE. This product indicates how many months out of a year a pixel is inundated (0–12). For our purposes, we make this product binary by considering any value greater than 1 to be flagged as open water, and anything below this not to be open water. This is done because occasionally permanent water bodies are seasonally covered by vegetation, which makes the GSWE represent them as less than 12 (permanent). For each specular reflection point, we find the amount of water within a 7×7 km region surrounding the point. This is a simplification of the actual footprint, but it is computationally more efficient than rotating axes to form actual ellipses, which themselves are simplifications and not well quantified. If the amount of water in the 7×7 km region exceeds 1%, we remove that CYGNSS observation from consideration. Changing these thresholds or the size of the 7×7 km region does change the results, though never uniformly increasing or decreasing error across regions. It is possible that future versions of the soil moisture product will use a different open water masking procedure or incorporate CYGNSS coherence detectors, such as that described in [37], to aid in the detection of small water bodies.

2.2.4. Conversion of $P_{r,eff}$ to Soil Moisture

We now describe how $P_{r,eff}$ is transformed into soil moisture, using SMAP soil moisture retrievals to calibrate CYGNSS observations. Our calibration period was chosen to be 17 March 2017–1 October 2018. This is an extended period beyond what was shown in [28]. In our calibration, we use all SMAP retrievals regardless of SMAP quality flags to allow for the retrieval of soil moisture from CYGNSS for the entirety of CYGNSS' observational area. Of course, users should be cautious to use CYGNSS soil moisture retrievals from regions regularly flagged by SMAP, such as the dense forests of South America and Africa, which are indicated in the CYGNSS quality flags.

The algorithm is based on the assumption that $P_{r,eff}$ is linearly related to SMAP soil moisture. The linear relationship is allowed to vary spatially, but we assume that it does not change over time. For a given location, we calculate the slope of the best-fit linear regression between concurrent (same calendar day) SMAP soil moisture retrievals and CYGNSS $P_{r,eff}$, after having removed the mean of each for the entire time series. Before we can describe this in more detail, however, we must explain what "a given location" means in this context.

In Section 1.2, we described how the smallest theoretical (no roughness) spatial footprint of CYGNSS over land is approximately 7×0.5 km, or 3.5×0.5 km for data recorded after mid-2019. What the actual spatial resolution is over land is a matter of debate within the GNSS-R community, though our own analysis of how $P_{r,eff}$ varies with small landcover or topographic features indicates to us that the effective footprint is likely only a few km, much smaller than SMAP's 40 km resolution [28]. If, for every SMAP observation, the CYGNSS observations completely sampled the 36-km EASE-2 grid cell used by SMAP, then a simple averaging could be used to aggregate CYGNSS observations to match with the SMAP retrieval. However, this is not the case. For every SMAP observation, there will likely be several CYGNSS observations within the grid cell, though not enough to completely sample the grid cell, if the spatial footprint is small (< 10 km). In this case, simple averaging will lead to variations in the day-to-day signal due to differential sampling of landcover types and topography within the SMAP pixel, which could be mistaken for variations in soil moisture.

In order to avoid this, we first grid our $P_{r,eff}$ observations to ~3 × 3 km "subcells," retrieve soil moisture from the subcells, and then aggregate the gridded observations to the 36-km SMAP EASE-2 grid resolution (Figure 4a). This subcell approach minimizes the confounding effects of landcover and topography on $P_{r,eff}$. The number of points per subcell in the calibration period are shown in Figure 5—subcells with less than three observations were not used for calibration.

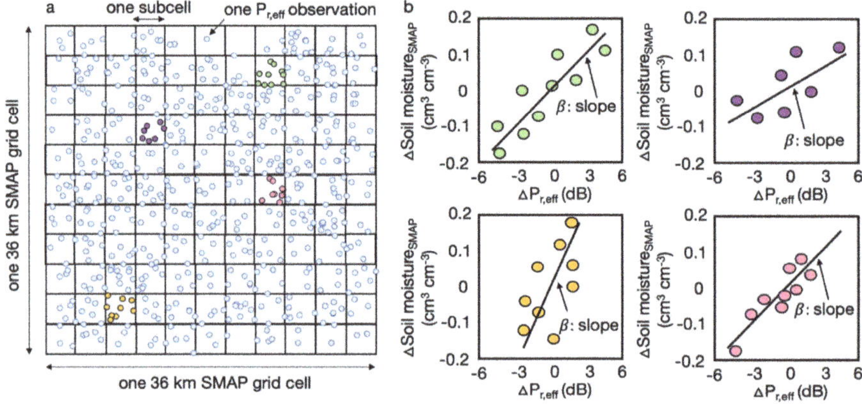

Figure 4. (**a**) Depiction of subcells within each 36-km EASE-2 grid cell, along with individual $P_{r,eff}$ observations (simplified as colored circles) used for calibration. (**b**) Depiction of how β is calculated for subcells, with four different example subcells shown, which correspond to the different colored circles in (**a**). β can be different for each subcell.

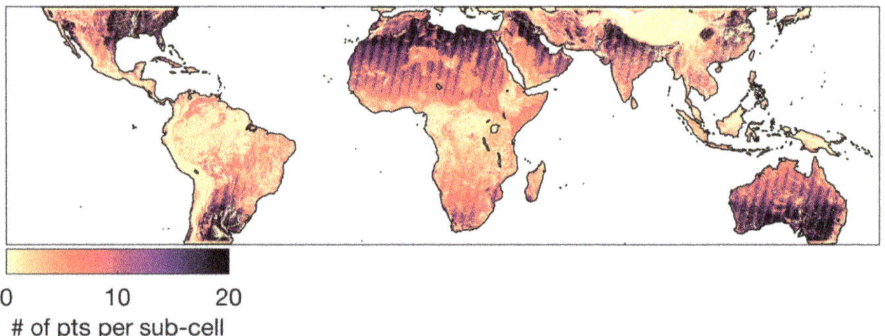

Figure 5. The number of CYGNSS observations for each sub-cell that were used for calibration. Fewer observations are found in higher elevation areas, which did not have data for most of 2017. Observations over open water have already been removed.

Within each subcell, we calculate the linear regression between SMAP soil moisture and $P_{r,eff}$ match-ups (occurring on the same calendar day), after having removed the mean values of both SMAP soil moisture and $P_{r,eff}$ in that subcell. The slope of the linear regression is β, which is conceptualized in Figure 4b. The correlation coefficients for these relationships are shown in Figure 6. β varies spatially (Figure 7). Some of the spatial variability is likely real, and it could be the result of spatial variations in topography or landcover. Some of it, however, is likely the result of the fact that some arid regions do not have enough soil moisture variability throughout the year in order to adequately calculate β. In these regions, β is artificially low because any noise in the CYGNSS observations will significantly affect the linear regression [28].

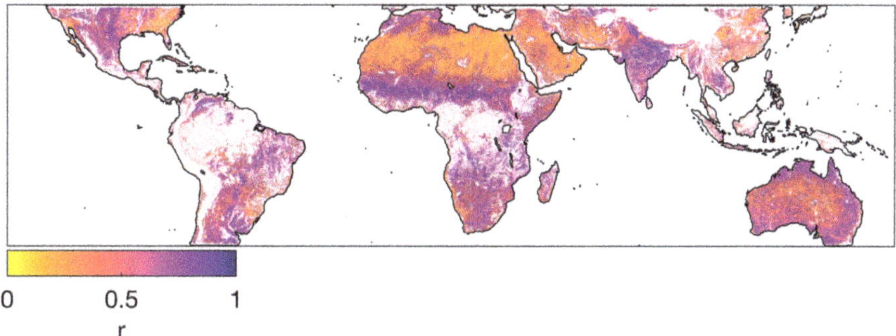

Figure 6. The correlation coefficient between SMAP soil moisture and CYGNSS effective reflectivity observations. Open water points have been removed. In general, there is a higher correlation coefficient when there is high soil moisture variability throughout the year. Areas in white either contained open water and were thus not used in the analysis, or they are regions too vegetated or mountainous to produce reflections over the SNR threshold of 2 dB.

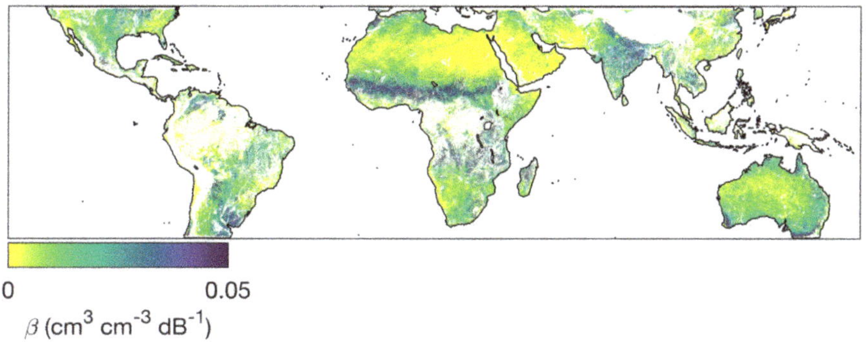

Figure 7. The slope of the linear regression between CYGNSS effective reflectivity observations and SMAP soil moisture (β). This represents the sensitivity of CYGNSS to soil moisture, with lower values indicating a higher sensitivity—though low values are also found in regions where soil moisture does vary significantly. Higher values of β mean that CYGNSS observations are not as sensitive to soil moisture. Imperfect open water masking will cause an apparent insensitivity to soil moisture.

β is then used to estimate soil moisture from CYGNSS for data falling outside the calibration period as well as data within the calibration period when there are no SMAP match-ups (since SMAP has a 2–3 day overpass period) using the following equation:

$$SM_{CYGNSS,\,t} = \beta \times \left(P_{r,eff,\,t} - \overline{P_{r,eff,cal}} \right) + \overline{Soil\ moisture_{SMAP,\,cal}} \tag{3}$$

where $SM_{CYGNSS,\,t}$ is the soil moisture derived from one CYGNSS observation at time, t, $P_{r,eff,\,t}$ is one observation of $P_{r,eff}$ at time, t, $\overline{P_{r,eff,cal}}$ is the mean value of $P_{r,eff}$ for a particular subcell during the calibration period, and $\overline{Soil\ moisture_{SMAP,\,cal}}$ is the mean SMAP soil moisture during the calibration period. The mean values of both SMAP and CYGNSS during the calibration period serve as our reference values, in order to return an absolute value of soil moisture from CYGNSS. Once retrievals are made for individual subcells, the mean of the retrievals for all subcells within each 36-km grid cell is used as the final CYGNSS soil moisture retrieval. Note that not all subcells within one 36-km grid cell will be sampled for each time step, and there could be times when just one subcell is used for the retrieval. The timesteps at which retrievals are averaged is described below.

2.2.5. Daily and Sub-Daily Retrievals

Soil moisture retrievals are currently provided on daily and sub-daily (6 hourly) time steps. For the daily retrievals, we average all CYGNSS soil moisture retrievals within a particular 36-km grid cell that fall within the 24-hour time period. For the sub-daily retrievals, we average all retrievals for a grid cell in 6-hour intervals, which are currently midnight–6 am, 6 am–noon, noon–6 pm, and 6 pm–midnight (UTC). Note that, just because retrievals are provided at 6-hour time steps, it does not mean that there will be values everywhere at every time step. There will be missing values in some grid cells even when aggregated to a 24-hour time step.

2.2.6. Quality Control

There is currently only minimal quality control of the soil moisture retrievals themselves. Soil moisture retrievals that are either less than 0.01 cm^3 cm^{-3} or greater than 0.65 cm^3 cm^{-3} are removed, though users should apply their own additional thresholds when using retrievals in specific regions that may have higher or lower residual or saturated moisture contents.

2.2.7. Soil Moisture Retrieval Uncertainty

Figure 8 shows the unbiased root-mean-square difference (ubRMSD) between CYGNSS and SMAP soil moisture retrievals for the calibration period (18 March 2017–1 October 2018). Semi-transparent regions are those frequently flagged by SMAP as being poor quality. Higher values of ubRMSD tend to cluster in areas that flood seasonally, which indicates imperfect open water masking, or have greater soil moisture variability throughout the year. It is "easier" to have a lower ubRMSD in regions with lower soil moisture variability.

2.2.8. Quality Flags

Static quality flags are provided in a separate file from the soil moisture retrievals. Examining the quality flags is a crucial first step before using the CYGNSS soil moisture retrievals or interpreting the empirical sensitivity of CYGNSS to soil moisture (β). The flags were developed to encourage users to exercise caution when using soil moisture from certain regions, or to be careful in the interpretation of β from these regions. The following criteria were used in the development of the flags:

1. Regions where CYGNSS observations were calibrated to SMAP data where a large portion (>90%) of the SMAP soil moisture retrievals were flagged as "not recommended for retrieval." These data tend to be in regions that are forested, with significant topography, or near coastlines. Although the overall ubRMSE between CYGNSS retrievals and in-situ observations remains largely unchanged for sites located in these regions, there are fewer instances where the ubRMSE is < 0.04 cm^3 cm^{-3}.

2. Regions where CYGNSS was calibrated to SMAP data with a small range of soil moisture values (<0.1 cm^3 cm^{-3}). This indicates a larger uncertainty in β [28]. The ubRMSE between CYGNSS and in-situ observations in these regions is low (0.0395 cm^3 cm^{-3}) due to the fact that there is only small variability in soil moisture. In these regions, because there is a larger uncertainty

in β, we do not want users to make any interpretations about β, or attempt to compare it to modelled sensitivity.

3. Regions where the ubRMSD between CYGNSS and SMAP was large for the calibration period (> 0.08 cm^3 cm^{-3}). The ubRMSE between CYGNSS and in-situ stations with this condition was higher than average (0.0561 cm^3 cm^{-3}). Users are advised to use caution when analysing retrievals from these areas. In-situ stations used for validation are described in the next section.

4. Regions with few observations in the 36 km grid cell for calibration, leading to less certain retrievals outside the calibration period (n < 100). β is also more uncertain in these regions.

5. Regions where $\overline{P_{r,eff,cal}}$ is low (<5 dB). There is a higher likelihood that roughness or vegetation effects dominate in these areas. Soil moisture retrievals from these areas are particularly suspect—the ubRMSE between CYGNSS and in-situ observations located in these regions is 0.07 cm^3 cm^{-3}. We advise users against using CYGNSS soil moisture retrievals at these locations.

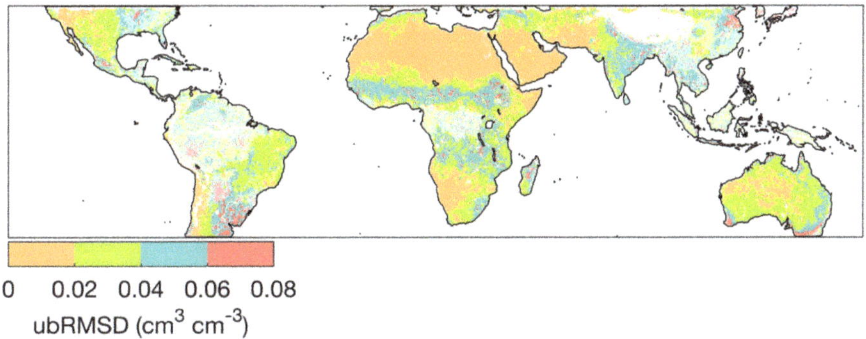

Figure 8. Unbiased root-mean-square difference (ubRMSD) between SMAP and CYGNSS soil moisture retrievals. Regions where SMAP always flags the data as being 'poor quality' are semi-transparent, such as the Amazon, Central Africa, Indonesia, Japan, Southeast Asia, and the majority of the Eastern United States. Higher ubRMSD in regions with "good-quality" SMAP data tend to be found in regions that are seasonally flooded or near coastlines. It is possible that in these areas, the seasonal water influence on CYGNSS effective reflectivity may overwhelm the soil moisture signal. Alternatively, it is also possible that the SMAP brightness temperature observations are actually responding to the increase in flooded area instead of soil moisture.

3. Results

We validated the UCAR/CU CYGNSS Soil Moisture Product at 171 in-situ soil moisture sites from six different networks: COSMOS [50], PBOH2O [51], SCAN [52], SNOTEL [53], USCRN [54], and OzNet [55]. The time period chosen for validation was 2 October 2018–31 December 2019, so as to not overlap with the calibration period. However, not all stations had data for the entire validation time period. In particular, sites within the PBOH2O network (a ground-based GNSS reflectometry network) only contained data through the first six weeks of the validation time period. Note that the majority of the validation sites are located in the United States—just because there may be acceptable agreement between CYGNSS and in-situ observations at these locations, it does not mean that CYGNSS is expected to perform as well in environments extremely disparate from those typical of the United States (e.g., tropical rainforests). Before validation, we removed obviously non-sensical soil moisture data from the in-situ records, for example soil moisture observations below 0 cm^3 cm^{-3} or greater than 1 cm^3 cm^{-3}.

We calculated the median unbiased root-mean-square error (ubRMSE) between daily averaged CYGNSS retrievals and in-situ observations for each individual station (Table A1) as well as aggregated by network (Table 1). Note that here we use ubRMSE, whereas before we used ubRMSD to compare

SMAP and CYGNSS retrievals. Though we calculated them the same way, we use ubRMSE here to indicate that this is a validation exercise, whereas the comparison between SMAP and CYGNSS was only for the purpose of showing where the two products are dissimilar from one another. For context, we also calculated the ubRMSE between SMAP retrievals and in-situ observations for the same time period. These values are also contained in Tables 1 and A1. Overall, the median ubRMSE between CYGNSS and in situ (0.049 cm^3 cm^{-3}) and SMAP and in situ (0.045 cm^3 cm^{-3}) were very similar, given that the standard deviations of each were ~0.025 cm^3 cm^{-3}. Similar ubRMSEs are expected since that CYGNSS was calibrated from SMAP retrievals.

Table 1. Median unbiased root-mean-square error (ubRMSE) and correlation coefficient (r) between CYGNSS and in-situ soil moisture sites, with those from SMAP shown for context. ubRMSE and r values were calculated for the time period between 2 October 2018, and 31 December 2019, unless an individual station did not have data for the full time period. Note that the distribution of r values for SMAP is significantly non-normal, which limits the interpretation of the standard deviation.

| | ubRMSE (cm^3 cm^{-3}) | | | | r | | | |
| | Median | | Standard Deviation | | Median | | Standard Deviation | |
	CYGNSS	SMAP	CYGNSS	SMAP	CYGNSS	SMAP	CYGNSS	SMAP
All (n = 171)	0.049	0.045	0.026	0.025	0.40	0.69	0.27	0.27
COSMOS (n = 11)	0.054	0.040	0.026	0.020	0.39	0.69	0.19	0.22
PBOH2O (n = 46)	0.033	0.024	0.019	0.025	0.14	0.58	0.26	0.36
SCAN (n = 63)	0.051	0.048	0.023	0.021	0.55	0.78	0.20	0.16
SNOTEL (n = 11)	0.089	0.082	0.016	0.020	0.10	0.36	0.18	0.15
USCRN (n = 38)	0.055	0.047	0.024	0.022	0.45	0.71	0.23	0.26
OzNet (n = 2)	0.043	0.058	0.006	0.013	0.68	0.68	0.02	0.02

A map of all in-situ stations used in this validation exercise along with their respective ubRMSE values is shown in Figure 9. Note that there is a wide range of ubRMSE values, depending on the site and network. In particular, SNOTEL sites performed poorly (mean ubRMSE ~= 0.09 cm^3 cm^{-3}). SNOTEL sites tend to be in mountainous areas with surrounding trees. $P_{r,eff}$ in these regions tends to be low, which is an indicator that either topographic roughness or dense vegetation is significantly affecting the reflected signal. We caution users who are interested in using the CYGNSS retrievals in such areas, given the high validation ubRMSE at the in-situ sites. The quality flags described in the last section delineate where these effects are likely to be at play.

Table A1 also includes other validation metrics, such as the correlation coefficient (r) between CYGNSS and in-situ sites and the mean bias between in-situ observations and SMAP retrievals, which by extension is also the bias between in-situ observations and CYGNSS retrievals. The median correlation coefficient across sites is somewhat low (r = 0.4) but with a large standard deviation (0.27). Figure 10a shows a histogram of the correlation coefficient (r) between CYGNSS and in-situ observations, with that from SMAP shown for context. Compared to SMAP, CYGNSS has a wider distribution of correlation coefficients, with a lower median value (Table 1). Similar to what was discussed in relation to the correlation between CYGNSS effective reflectivity and SMAP soil moisture, areas with low or no soil moisture variation during the validation time period often results in a low correlation coefficient, as any effects due to random noise are amplified when there is no soil moisture variation. CYGNSS retrievals are expected to have more noise than SMAP retrievals, given the aforementioned imperfections in signal calibration and lack of full grid cell coverage. Figure 10b shows that the distribution of r values does change significantly when only in-situ sites with moderate or large soil moisture variability are considered, which helps mask the effect of noise. In addition to the correlation coefficient, the median absolute bias between SMAP and in-situ observations, which will also be the bias between CYGNSS and in-situ, is 0.05 cm^3 cm^{-3}. This dry bias is a known issue in the SMAP retrievals [56].

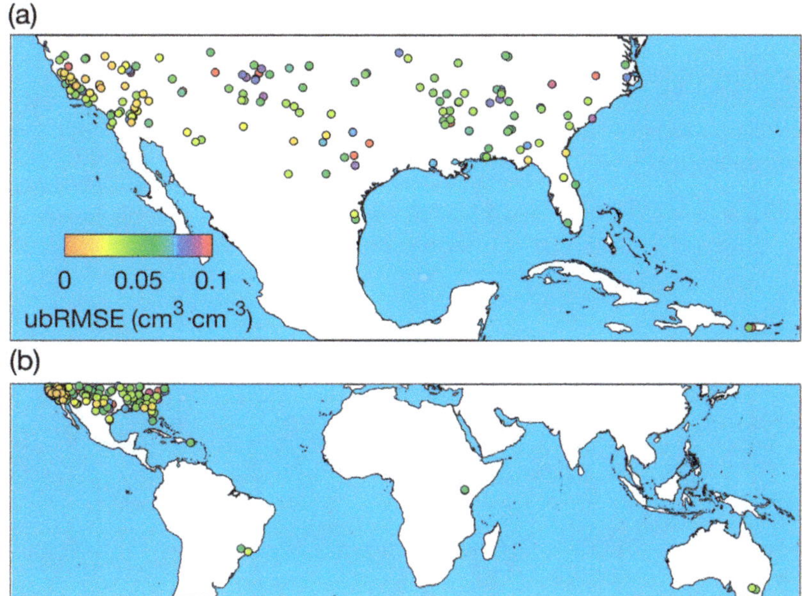

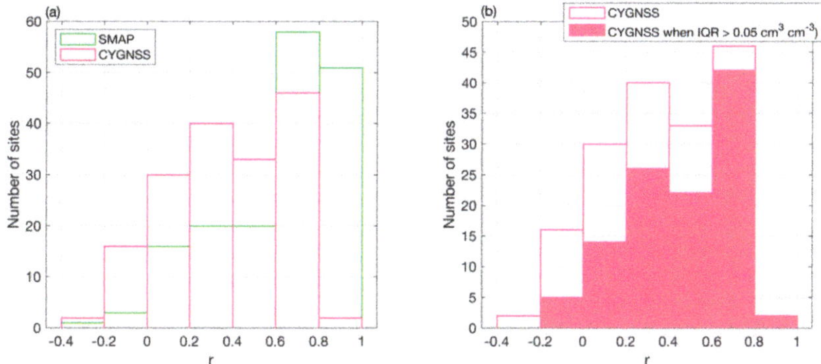

Figure 9. (**a**) Unbiased root-mean-square error (ubRMSE) between individual in-situ soil moisture stations (colored dots) and CYGNSS soil moisture retrievals for the continental United States and Puerto Rico. ubRMSE values from these stations were used to derive the values shown in Table A1. (**b**) Same as (**a**), except also showing global sites.

Figure 10. (**a**) Distributions of the correlation coefficient (r) between CYGNSS and in-situ observations (pink bars) and SMAP and in-situ observations (green bars). (**b**) The distribution of the correlation coefficient between CYGNSS and in-situ observations (open pink bars, same as in (**a**)) and the distribution of the correlation coefficient when only in-situ sites with moderate to high soil moisture variability are considered (closed pink bars).

Validating a satellite remote-sensing product based off of point measurements, as we have done here, is an imperfect exercise, as the individual soil moisture probe observations may not be representative of the larger areal average. The validation effort conducted by SMAP, for example, averaged in-situ observations from many probes spread out over a large area, for each of their validation sites [46]. The sites that comprise OzNet do indeed contain multiple probes at each site, though there are only two sites with data within the calibration time period (see Table A1). In this case, we used the mean of all probe measurements within one site to calculate the ubRMSE. CYGNSS soil moisture retrievals did agree better with the in-situ observations from these sites (ubRMSE = 0.043 cm^3 cm^{-3}) than they did from the point-based observations.

Examples of CYGNSS soil moisture time series that agree well with in-situ observations are shown in Figure 11. These sites show that CYGNSS has the ability to retrieve both low and high soil moisture contents. An example of the increased temporal resolution of daily averaged CYGNSS retrievals with respect to SMAP is shown in Figure 12. In this example, the increased resolution of CYGNSS results in the observation of increased soil moisture due to two precipitation events in late June and early July of 2018, which are missed by SMAP. Table A1 shows how many daily averaged CYGNSS soil moisture retrievals were available during the validation time period, which can be compared to the number of SMAP retrievals for the same time period. On average CYGNSS soil moisture retrievals increase the temporal revisit time by 63.4%.

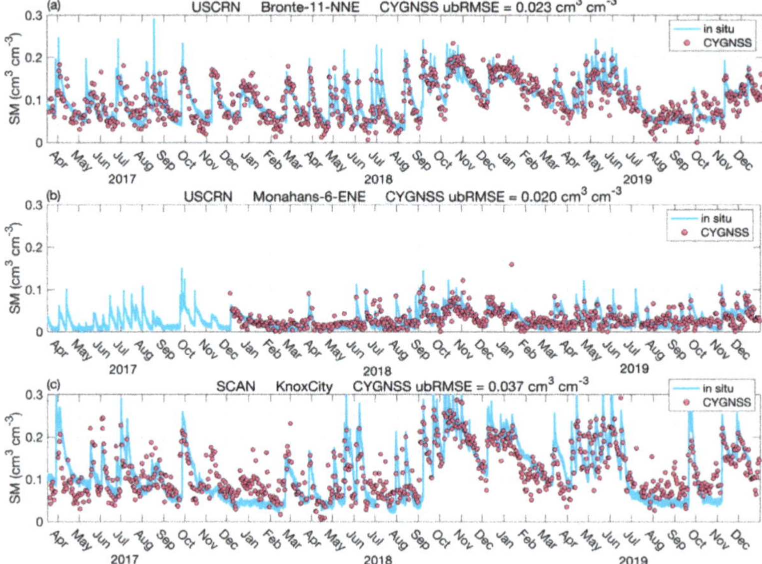

Figure 11. Examples of CYGNSS soil moisture retrievals that agree well with in-situ probes from three sites: USCRN Bronte-11-NNE (**a**), USCRN Monahans-6-ENE (**b**), and SCAN Knox City (**c**). CYGNSS retrievals from USCRN Monahans-6-ENE start in December 2017 because the elevation of the station is greater than 600 m. The values of ubRMSE shown for each station is for the validation time period only (2 October 2018–31 December 2019).

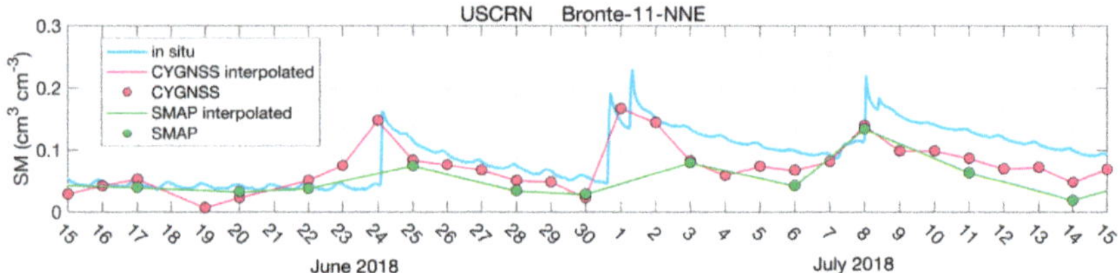

Figure 12. Time series of CYGNSS (pink dots) and SMAP (green dots) soil moisture retrievals and in-situ soil moisture observations (blue line) from the USCRN Bronte-11-NNE station. Pink and green lines interpolate between the CYGNSS and SMAP observations, respectively. The shortened time series shows the increased temporal resolution of the CYGNSS retrievals relative to SMAP.

For each in-situ site, we calculated the number of rain events during the validation period, which we defined as an increase in the daily averaged soil moisture greater than 0.02 cm^3 cm^{-3}. We then calculated the number of the rainy days that had CYGNSS observations and the number that had SMAP observations. From this, we could calculate the percentage of rain events "observed" by both CYGNSS and SMAP and found that CYGNSS observed a median of 87% of the rain events, and SMAP observed a median of 43% (distributions shown in Figure 13). At sites with acceptable CYGNSS performance, CYGNSS would be able to provide more information about soil drying and wetting than SMAP currently is able to.

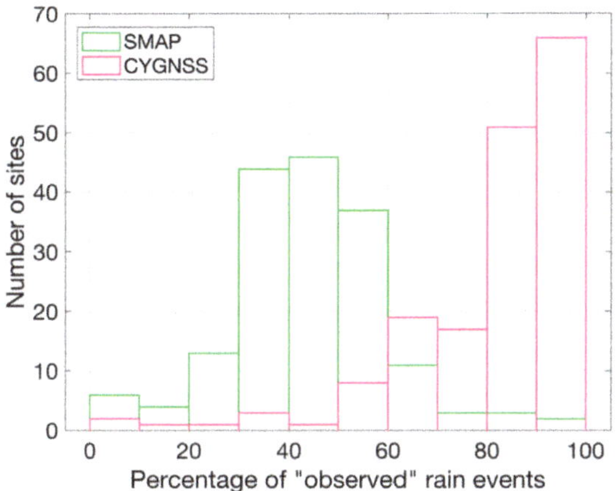

Figure 13. The percentage of rain events during the validation period observed by SMAP (green bars) and CYGNSS (pink bars). A rain event was defined as the daily averaged soil moisture value increasing by more than 0.02 cm^3 cm^{-3} with respect to the previous day's value.

Figure 14 shows three examples of soil moisture time series from CYGNSS and SMAP where CYGNSS does not agree well with in-situ observations. Generally, if SMAP does not perform well at a site, CYGNSS is likely to perform poorly, too. However, there are other sites where SMAP performance is significantly more acceptable than that from CYGNSS (e.g., Figure 14c). Understanding why CYGNSS soil moisture retrievals at some locations do not perform as well as SMAP is one subject of future research, though preliminary analyses indicate that inadequate open water masking could be one contributing factor.

Figure 15a shows a histogram of the ubRMSEs between CYGNSS and in-situ observations (values in Table A1), with that from SMAP shown for context. Although the median ubRMSE for both CYGNSS and SMAP retrievals are lower than 0.05 cm^3 cm^{-3}, it is important to note that ubRMSE statistics are correlated with the variability of soil moisture at a particular site. Figure 15b shows the ubRMSE of CYGNSS and SMAP as a function of the interquartile range (IQR) of soil moisture for each site. As soil moisture variability increases, so does the ubRMSE. This means it may be unreasonable to assume an error of 0.04 cm^3 cm^{-3} in areas that undergo extreme fluctuations in soil moisture throughout the year. For context, the median IQR at validation sites was 0.07 cm^3 cm^{-3}, and the IQR of USCRN station Bronte-11-NNE shown in Figures 11 and 12 is 0.071 cm^3 cm^{-3}.

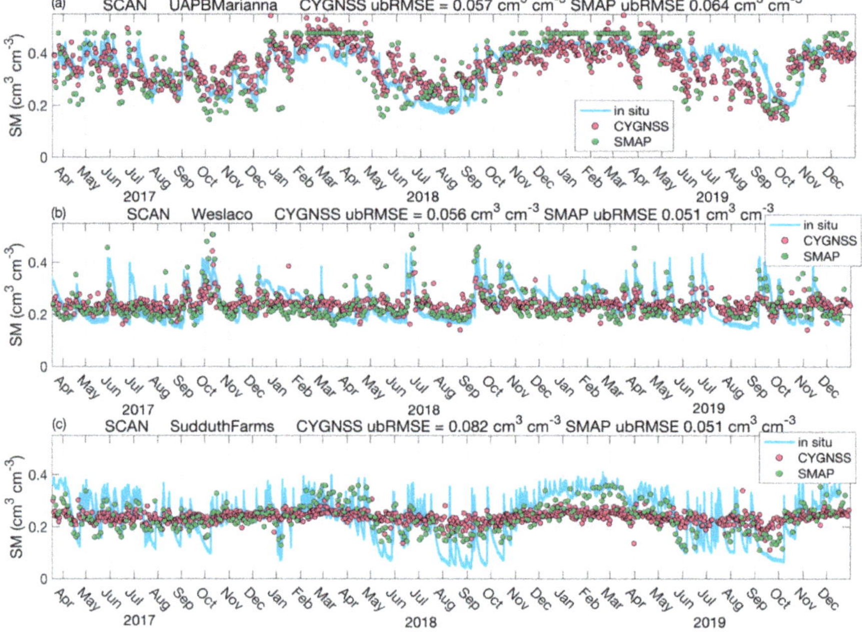

Figure 14. Examples of CYGNSS soil moisture retrievals that do not agree as well with in-situ probe measurements from SCAN UAPB Marianna (**a**), SCAN Weslaco (**b**), and SCAN SudduthFarms (**c**). SMAP data for these locations are shown for reference. The values of ubRMSE shown for each station is for the validation time period only (2 October 2018–31 December 2019).

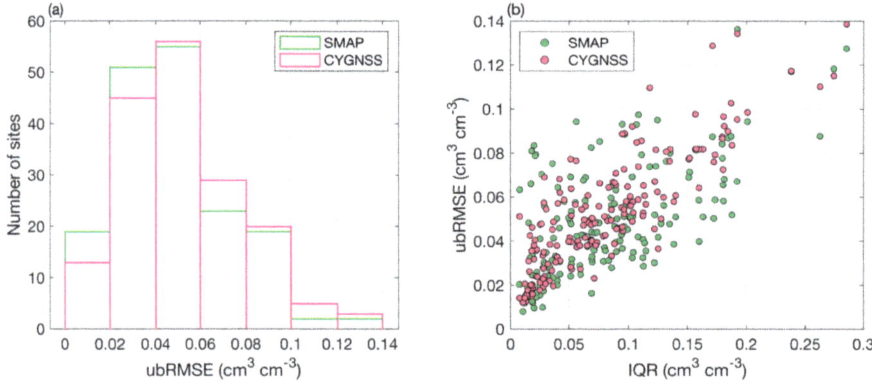

Figure 15. (**a**) Histograms of ubRMSE statistics for CYGNSS and in-situ observations (pink) and SMAP and in-situ observations (green). (**b**) The relationship between the interquartile range (IQR) of soil moisture at individual stations and the ubRMSE between CYGNSS and observations from that station (pink) and SMAP and observations from that station (green).

4. Discussion

As with all remote-sensing techniques, there are limitations to what information retrieval algorithms can provide. Below are some of the more significant limitations of the UCAR/CU retrieval algorithm. We avoid commenting on limitations of CYGNSS in general and instead only address the specific algorithm presented here.

1. Errors in SMAP retrievals will propagate into CYGNSS soil moisture retrievals. Because CYGNSS is calibrated using SMAP, any systemic errors in the SMAP retrievals (particularly, persistent bias) will also be present in CYGNSS retrievals. As discussed above, at validation sites with poor SMAP performance, CYGNSS also performs poorly.

2. As with all empirical approaches, investigation into the "true" sensitivity to soil moisture is difficult. As mentioned above, unless there is enough soil moisture variability, calculation of β is difficult due to noise in the CYGNSS observations. Additionally, if there is variability of $P_{r,eff}$ within a subcell due to, for example, spatial variations in land cover, then β may appear artificially high (i.e., low sensitivity to soil moisture).

3. The relationship between $P_{r,eff}$ and soil moisture may not actually be linear. Although we approximate the relationship as being linear, it may not be—it may appear to be linear either due to noise overwhelming an obvious non-linearity, or it may appear linear because in many regions soil moisture does not often fluctuate between 0.02 cm^3 cm^{-3} and 0.5 cm^3 cm^{-3} or higher, which would be necessary to elucidate significant non-linear relationships. The empirical linear relationships may thus not match those eventually derived from a model and should not be compared.

4. The assumption that the sensitivity of $P_{r,eff}$ to soil moisture does not change over time is likely incorrect. Fluctuations in vegetation water content, particularly in agricultural regions, will likely change β, though we currently ignore that possibility.

5. Aggregating CYGNSS observations to 36 km does not take advantage of the finer spatial resolution. Any advantages CYGNSS might have vis-à-vis providing higher spatial resolution soil moisture retrievals is not permitted by the approach that we use here. Other approaches using either machine learning methods or models could be successful, if provided with accurate, high-resolution ancillary data.

5. Conclusions

This paper described the UCAR/CU CYGNSS Soil Moisture Product [19], which uses spaceborne GNSS-R reflections, calibrated to SMAP soil moisture, to produce soil moisture retrievals. Validation of the product at 171 in-situ soil moisture stations resulted in a median ubRMSE of 0.049 cm^3 cm^{-3}. This product can be used by hydrologists who are interested in soil moisture data at a higher temporal resolution than currently available using other products. As was shown in Figures 12 and 13, CYGNSS can observe changes in soil moisture due to precipitation events that may be too quick for the SMAP overpass period. The quantification of soil moisture memory, observation of moisture conditions giving rise to flooding, and the inverse estimation of precipitation using soil moisture time series all require soil moisture data on short time scales, and a daily soil moisture product may be able to provide more complete information on soil moisture dynamics at needed time scales.

GNSS-R researchers interested in developing independent CYGNSS soil moisture retrievals using model-based approaches may also benefit from using this product, as it can provide an indication of where simple algorithms like this one can successfully retrieve soil moisture and where more complicated approaches may be warranted.

Author Contributions: Conceptualization, C.C. and E.S.; methodology, C.C. and E.S.; software, C.C.; validation, C.C. and E.S.; writing, C.C. and E.S. All authors have read and agreed to the published version of the manuscript.

Funding: This research received no external funding.

Acknowledgments: We would like to thank Jan Weiss and Maggie Sleziak of UCAR for putting the soil moisture product online and the CYGNSS team for providing high quality GNSS-R data.

Conflicts of Interest: The authors declare no conflict of interest.

Appendix A

Table A1. Location information and statistics for each in-situ soil moisture station used for validation. Unbiased root-mean-square errors (ubRMSEs) for CYGNSS are shown as well as those for SMAP, which are shown for context. The correlation coefficient (r) between CYGNSS soil moisture retrievals and in-situ observations is shown. The bias between SMAP retrievals and in-situ observations is the same as for CYGNSS retrievals and in-situ observations, since CYGNSS retrievals are already bias-corrected with respect to SMAP. The number of CYGNSS and SMAP observations used for calculation of ubRMSEs, r, and the bias are also shown.

Network	Station	Latitude (deg)	Longitude (deg)	ubRMSE CYGNSS ($cm^3\ cm^{-3}$)	ubRMSESMAP ($cm^3\ cm^{-3}$)	r CYGNSS	Bias SMAP	# Obs CYGNSS	# Obs SMAP
COSMOS	COSMOS_064	35.19	−102.10	0.054	0.056	0.61	0.16	182	91
COSMOS	COSMOS_101	−22.68	−45.00	0.033	0.022	0.06	−0.08	220	125
COSMOS	COSMOS_023	33.61	−116.45	0.051	0.037	0.17	0.01	289	167
COSMOS	COSMOS_057	29.95	−98.00	0.086	0.079	0.50	0.12	269	105
COSMOS	COSMOS_067	34.26	−89.87	0.042	0.050	0.50	−0.19	301	164
COSMOS	COSMOS_055	0.28	36.87	0.071	0.040	0.50	0.09	152	88
COSMOS	COSMOS_050	0.49	36.87	0.062	0.031	0.50	−0.10	113	60
COSMOS	COSMOS_034	37.07	−119.19	0.129	0.087	0.34	−0.03	98	154
COSMOS	COSMOS_044	−21.62	−47.63	0.057	0.030	0.04	−0.18	90	57
COSMOS	COSMOS_014	36.06	−97.22	0.049	0.038	0.39	−0.07	237	118
COSMOS	COSMOS_033	37.03	−119.26	0.053	0.041	0.31	−0.07	21	12
PBOH2O	bkap	35.29	−116.08	0.016	0.016	0.20	0.01	33	20
PBOH2O	crrs	33.07	−115.74	0.077	0.069	-0.01	−0.10	36	14
PBOH2O	csci	34.17	−119.04	0.052	0.054	0.05	−0.04	33	14
PBOH2O	ctdm	34.52	−118.61	0.034	0.023	−0.14	0.09	26	14
PBOH2O	fgst	34.73	−120.01	0.046	0.034	0.03	0.05	22	19
PBOH2O	glrs	33.27	−115.52	0.032	0.084	−0.04	−0.18	35	15
PBOH2O	gnps	34.31	−114.19	0.022	0.010	0.38	−0.02	35	19
PBOH2O	hunt	35.88	−120.40	0.020	0.013	0.33	−0.02	22	19
PBOH2O	hvys	34.44	−119.19	0.036	0.079	0.00	−0.30	34	19
PBOH2O	imps	34.16	−115.15	0.019	0.021	0.51	0.04	34	19
PBOH2O	masw	35.83	−120.44	0.069	0.056	0.08	0.01	22	19
PBOH2O	ndap	34.77	−114.62	0.021	0.016	0.40	−0.01	32	17

Table A1. *Cont.*

Network	Station	Latitude (deg)	Longitude (deg)	ubRMSE CYGNSS ($cm^3\ cm^{-3}$)	ubRMSESMAP ($cm^3\ cm^{-3}$)	r CYGNSS	Bias SMAP	# Obs CYGNSS	# Obs SMAP
PBOH2O	p035	34.60	−105.18	0.088	0.078	0.69	0.07	38	19
PBOH2O	p038	34.15	−103.41	0.041	0.016	0.70	0.05	38	15
PBOH2O	p039	36.45	−103.15	0.068	0.046	0.54	0.18	32	16
PBOH2O	p070	36.04	−104.70	0.058	0.039	0.79	0.02	40	19
PBOH2O	p094	37.20	−117.70	0.019	0.012	0.27	0.04	21	15
PBOH2O	p107	35.13	−107.88	0.050	0.043	0.40	0.05	30	19
PBOH2O	p123	36.64	−105.91	0.052	0.055	0.52	0.04	30	16
PBOH2O	p250	36.95	−121.27	0.043	0.032	−0.11	−0.05	21	19
PBOH2O	p284	35.93	−120.91	0.030	0.024	−0.04	0.00	33	18
PBOH2O	p288	36.14	−120.88	0.018	0.019	0.26	−0.01	34	19
PBOH2O	p472	32.89	−117.10	0.039	0.027	−0.03	−0.01	34	19
PBOH2O	p474	33.36	−117.25	0.055	0.035	−0.14	−0.01	34	20
PBOH2O	p475	32.67	−117.24	0.040	0.079	−0.11	−0.15	36	19
PBOH2O	p498	32.90	−115.57	0.015	0.025	0.14	−0.08	37	15
PBOH2O	p505	33.42	−115.69	0.048	0.083	0.07	−0.11	27	12
PBOH2O	p508	33.25	−115.43	0.036	0.081	−0.21	−0.19	35	15
PBOH2O	p511	33.89	−115.30	0.021	0.015	0.14	0.03	34	18
PBOH2O	p514	35.01	−120.41	0.036	0.026	0.26	0.00	36	19
PBOH2O	p525	35.43	−120.81	0.066	0.085	0.40	−0.25	25	19
PBOH2O	p530	35.62	−120.48	0.026	0.015	0.15	−0.01	22	19
PBOH2O	p532	35.63	−120.27	0.024	0.015	0.26	−0.01	22	19
PBOH2O	p536	35.28	−120.03	0.024	0.023	0.02	0.04	23	19
PBOH2O	p537	35.32	−119.94	0.013	0.013	0.19	0.01	23	19
PBOH2O	p538	35.53	−120.11	0.033	0.020	0.18	0.03	26	19
PBOH2O	p553	34.84	−118.88	0.022	0.028	0.33	0.04	32	19
PBOH2O	p565	35.74	−119.24	0.014	0.020	−0.02	−0.04	36	19
PBOH2O	p568	35.25	−118.13	0.020	0.013	−0.01	0.02	23	15
PBOH2O	p569	35.38	−118.12	0.018	0.020	0.10	0.02	23	15
PBOH2O	p591	35.15	−118.02	0.012	0.008	0.19	0.00	33	15
PBOH2O	p742	33.50	−116.60	0.035	0.021	−0.11	−0.02	25	20

Table A1. *Cont.*

Network	Station	Latitude (deg)	Longitude (deg)	ubRMSE CYGNSS (cm^3 cm^{-3})	ubRMSESMAP (cm^3 cm^{-3})	r CYGNSS	Bias SMAP	# Obs CYGNSS	# Obs SMAP
PBOH2O	p807	30.49	−98.82	0.056	0.069	0.68	0.01	30	19
PBOH2O	p811	35.15	−118.02	0.014	0.012	0.02	−0.01	33	15
PBOH2O	qcy2	36.16	−121.14	0.024	0.017	−0.31	0.01	31	14
PBOH2O	sdhl	34.26	−116.28	0.038	0.010	0.41	0.01	31	20
SCAN	AAMU-jtg	34.78	−86.55	0.045	0.044	0.69	−0.13	386	202
SCAN	AdamsRanch#1	34.25	−105.42	0.056	0.048	0.29	0.03	365	190
SCAN	AllenFarms	35.07	−86.90	0.065	0.063	0.70	0.02	174	94
SCAN	BraggFarm	34.90	−86.60	0.067	0.065	0.42	−0.01	385	202
SCAN	BroadAcres	32.28	−86.05	0.050	0.030	0.47	0.08	75	41
SCAN	Charkiln	36.37	−115.83	0.077	0.069	0.36	0.05	364	200
SCAN	CochoraRanch	35.12	−119.60	0.051	0.033	0.66	−0.01	313	199
SCAN	DeathValleyJCT	36.33	−116.35	0.027	0.032	0.39	−0.04	285	139
SCAN	DesertCenter	33.80	−115.31	0.023	0.028	0.28	0.00	116	61
SCAN	Dexter	36.78	−89.93	0.045	0.075	0.78	−0.06	389	197
SCAN	Essex	34.67	−115.17	0.042	0.031	0.46	−0.03	257	104
SCAN	FordDryLake	33.65	−115.10	0.028	0.017	0.36	−0.04	288	148
SCAN	FortReno#1	35.55	−98.02	0.060	0.051	0.78	0.08	398	199
SCAN	GoodwinCreekTimber	34.23	−89.90	0.064	0.031	0.62	−0.10	374	201
SCAN	GuilarteForest	18.15	−66.77	0.134	0.136	NaN	−0.16	107	70
SCAN	KnoxCity	33.45	−99.87	0.037	0.042	0.85	−0.04	385	198
SCAN	KoptisFarms	30.52	−87.70	0.046	0.040	0.67	−0.17	378	202
SCAN	Levelland	33.55	−102.37	0.042	0.058	0.28	−0.01	394	200
SCAN	LittleRiver	31.50	−83.55	0.039	0.032	0.18	−0.10	154	83
SCAN	LosLunasPmc	34.77	−106.77	0.046	0.050	0.23	0.05	399	187
SCAN	LovellSummit	36.17	−115.62	0.090	0.087	0.31	0.06	310	200
SCAN	MammothCave	37.18	−86.03	0.065	0.045	0.64	−0.05	298	200
SCAN	MaricaoForest	18.15	−67.00	0.049	0.051	NaN	−0.18	310	151
SCAN	Mayday	32.87	−90.52	0.139	0.128	0.73	0.03	333	201
SCAN	McalisterFarm	35.07	−86.58	0.079	0.059	0.74	0.00	387	202
SCAN	MccrackenMesa	37.45	−109.33	0.061	0.051	0.43	0.08	35	181
SCAN	MonoclineRidge	36.54	−120.55	0.095	0.067	0.64	−0.02	357	199

Table A1. *Cont.*

Network	Station	Latitude (deg)	Longitude (deg)	ubRMSE CYGNSS ($cm^3\ cm^{-3}$)	ubRMSESMAP ($cm^3\ cm^{-3}$)	r CYGNSS	Bias SMAP	# Obs CYGNSS	# Obs SMAP
SCAN	MorrisFarms	32.42	−85.92	0.055	0.043	0.55	−0.03	378	202
SCAN	MtVernon	37.07	−93.88	0.038	0.050	0.52	0.01	297	197
SCAN	NorthIssaquena	33.00	−91.07	0.051	0.063	0.41	−0.03	397	155
SCAN	Onward	32.75	−90.93	0.048	0.041	0.17	−0.08	169	176
SCAN	PeeDee	34.30	−79.73	0.039	0.049	0.63	−0.11	267	133
SCAN	PerdidoRivFarms	31.12	−87.55	0.059	0.042	0.73	−0.04	392	202
SCAN	PineNut	36.57	−115.20	0.058	0.051	0.24	0.00	273	171
SCAN	Riesel	31.48	−96.88	0.110	0.088	0.70	0.03	164	76
SCAN	RiverRoadFarms	31.02	−85.03	0.040	0.044	0.62	−0.13	291	149
SCAN	SanAngelo	31.55	−100.51	0.073	0.066	0.60	0.07	387	152
SCAN	SandHollow	37.10	−113.35	0.031	0.025	0.50	−0.06	258	190
SCAN	SandyRidge	33.67	−90.57	0.048	0.070	0.66	−0.17	371	201
SCAN	Scott	33.62	−91.10	0.039	0.057	0.70	0.00	388	151
SCAN	SellersLake#1	29.10	−81.63	0.031	0.048	0.04	−0.39	351	202
SCAN	Sevilleta	34.35	−106.68	0.040	0.033	0.36	0.00	223	100
SCAN	SilverCity	33.08	−90.52	0.049	0.041	0.68	−0.11	333	201
SCAN	StanleyFarm	34.43	−86.68	0.084	0.052	0.74	0.04	394	202
SCAN	Starkville	33.63	−88.77	0.040	0.030	0.67	0.04	396	200
SCAN	Stephenville	32.25	−98.20	0.076	0.048	0.74	0.02	400	159
SCAN	Stubblefield	34.97	−119.48	0.076	0.050	0.39	0.05	144	113
SCAN	SudduthFarms	34.18	−87.45	0.082	0.051	0.53	−0.06	389	156
SCAN	Tidewater#1	35.87	−76.65	0.078	0.072	−0.05	−0.18	221	125
SCAN	TidewaterArec	36.68	−76.77	0.054	0.043	0.72	0.00	228	129
SCAN	Tuskegee	32.43	−85.75	0.061	0.049	0.43	−0.27	389	163
SCAN	UAPBDewitt	34.28	−91.35	0.061	0.039	0.67	−0.08	256	186
SCAN	UAPBLonokeFarm	34.85	−91.88	0.046	0.037	0.73	−0.02	381	202
SCAN	UAPBMarianna	34.78	−90.82	0.057	0.064	0.65	0.03	384	151
SCAN	UAPBPointRemove	35.22	−92.92	0.040	0.038	0.37	−0.07	241	130
SCAN	Uvalde	29.36	−100.25	0.058	0.052	0.56	0.04	385	202
SCAN	WTARS	34.90	−86.53	0.046	0.030	0.80	0.02	87	44
SCAN	Wakulla#1	30.30	−84.42	0.021	0.030	0.42	−0.43	381	204

Table A1. *Cont.*

Network	Station	Latitude (deg)	Longitude (deg)	ubRMSE CYGNSS ($cm^3\ cm^{-3}$)	ubRMSESMAP ($cm^3\ cm^{-3}$)	r CYGNSS	Bias SMAP	# Obs CYGNSS	# Obs SMAP
SCAN	WalnutGulch#1	31.73	−110.05	0.043	0.036	0.60	0.00	395	147
SCAN	Watkinsville#1	33.88	−83.43	0.082	0.040	0.33	−0.05	274	132
SCAN	Wedowee	33.33	−85.52	0.053	0.035	0.21	−0.22	349	141
SCAN	Weslaco	26.16	−97.96	0.056	0.051	0.34	0.04	335	198
SCAN	YoumansFarm	32.67	−81.20	0.040	0.051	0.53	−0.16	391	201
SNOTEL	BRISTLECONETRAIL	36.32	−115.70	0.117	0.117	0.01	0.09	361	188
SNOTEL	BarM	34.86	−111.61	0.067	0.052	0.50	0.14	310	159
SNOTEL	ElkCabin	35.70	−105.81	0.082	0.074	0.30	−0.03	239	129
SNOTEL	LEECANYON	36.31	−115.68	0.082	0.080	0.08	0.06	361	188
SNOTEL	MormonMountain	34.94	−111.52	0.115	0.118	0.45	0.09	310	159
SNOTEL	NAVAJOWHISKEYCK	36.18	−108.95	0.110	0.074	0.11	0.21	283	109
SNOTEL	PALO	36.41	−105.33	0.085	0.093	0.06	−0.06	333	111
SNOTEL	RAINBOWCANYON	36.25	−115.63	0.099	0.094	0.11	0.04	361	188
SNOTEL	SantaFe	35.77	−105.78	0.089	0.089	0.07	0.03	239	129
SNOTEL	TresRitos	36.13	−105.53	0.098	0.082	−0.03	0.03	122	109
SNOTEL	VacasLocas	36.03	−106.81	0.082	0.081	0.31	0.03	322	159
USCRN	Asheville-13-S	35.42	−82.56	0.092	0.068	0.33	−0.01	361	192
USCRN	Austin-33-NW	30.62	−98.08	0.103	0.088	0.74	0.11	381	147
USCRN	Batesville-8-WNW	35.82	−91.78	0.048	0.044	0.65	−0.07	395	198
USCRN	Blackville-3-W	33.36	−81.33	0.049	0.032	0.62	−0.11	81	45
USCRN	Bowling-Green-21-NNE	37.25	−86.23	0.054	0.051	0.67	−0.04	248	168
USCRN	Bronte-11-NNE	32.04	−100.25	0.023	0.036	0.86	−0.07	394	179
USCRN	Brunswick-23-S	30.81	−81.46	0.020	0.066	0.40	−0.38	336	193
USCRN	Durham-11-W	35.97	−79.09	0.097	0.058	0.45	−0.11	366	195
USCRN	Edinburg-17-NNE	26.53	−98.06	0.030	0.033	0.42	0.00	368	196
USCRN	Elgin-5-S	31.59	−110.51	0.033	0.028	0.72	−0.03	382	145
USCRN	Everglades-City-5-NE	25.90	−81.32	0.059	0.058	0.16	−0.13	188	96
USCRN	Fairhope-3-NE	30.55	−87.88	0.062	0.095	0.13	−0.33	267	141
USCRN	Fallbrook-5-NE	33.44	−117.19	0.065	0.029	−0.03	0.05	355	198
USCRN	Gadsden-19-N	34.29	−85.96	0.057	0.036	0.72	−0.12	384	199
USCRN	Goodwell-2-E	36.60	−101.60	0.060	0.056	0.67	0.08	361	193

Table A1. *Cont.*

Network	Station	Latitude (deg)	Longitude (deg)	ubRMSE CYGNSS (cm^3 cm^{-3})	ubRMSESMAP (cm^3 cm^{-3})	r CYGNSS	Bias SMAP	# Obs CYGNSS	# Obs SMAP
USCRN	Goodwell-2-SE	36.57	−101.61	0.064	0.059	0.66	0.14	361	193
USCRN	Holly-Springs-4-N	34.82	−89.43	0.039	0.044	0.72	0.07	380	197
USCRN	Joplin-24-N	37.43	−94.58	0.081	0.076	0.24	0.04	127	194
USCRN	Lafayette-13-SE	30.09	−91.87	0.074	0.097	0.09	−0.11	351	149
USCRN	Las-Cruces-20-N	32.61	−106.74	0.027	0.031	0.38	0.00	382	147
USCRN	Los-Alamos-13-W	35.86	−106.52	0.087	0.084	0.29	0.04	23	115
USCRN	McClellanville-7-NE	33.15	−79.36	0.089	0.093	−0.04	−0.41	291	149
USCRN	Merced-23-WSW	37.24	−120.88	0.050	0.060	0.46	−0.18	285	155
USCRN	Mercury-3-SSW	36.62	−116.02	0.028	0.024	0.41	−0.03	345	198
USCRN	Monahans-6-ENE	31.62	−102.81	0.020	0.023	0.56	−0.03	382	199
USCRN	Muleshoe-19-S	33.96	−102.77	0.037	0.042	0.57	0.08	383	187
USCRN	Newton-5-ENE	32.34	−89.07	0.067	0.047	0.57	−0.09	392	198
USCRN	Newton-8-W	31.31	−84.47	0.077	0.094	0.09	−0.08	369	197
USCRN	Panther-Junction-2-N	29.35	−103.21	0.038	0.029	0.45	0.01	347	199
USCRN	Socorro-20-N	34.36	−106.89	0.038	0.042	0.22	0.00	368	191
USCRN	Stillwater-2-W	36.12	−97.09	0.092	0.069	0.47	0.09	383	196
USCRN	Stillwater-5-WNW	36.13	−97.11	0.073	0.047	0.49	0.03	379	195
USCRN	Stovepipe-Wells-1-SW	36.60	−117.14	0.016	0.018	0.37	−0.05	333	198
USCRN	Titusville-7-E	28.62	−80.69	0.057	0.058	0.18	−0.29	196	123
USCRN	Tucson-11-W	32.24	−111.17	0.027	0.025	0.66	−0.04	400	196
USCRN	Watkinsville-5-SSE	33.78	−83.39	0.046	0.035	0.34	−0.20	345	180
USCRN	Williams-35-NNW	35.76	−112.34	0.050	0.048	0.66	0.00	362	138
USCRN	Yuma-27-ENE	32.83	−114.19	0.064	0.048	0.21	0.03	123	68
OzNet	Yanco	−34.85	146.12	0.038	0.049	0.66	−0.02	405	200
OzNet	Kyeamba	−35.32	147.53	0.047	0.068	0.69	−0.02	224	128

References

1. Oglesby, R.J.; Erickson, D.J. Soil Moisture and the Persistence of North American Drought. *J. Clim.* **1989**, *2*, 1362–1380. [CrossRef]
2. Dai, A.G. Increasing drought under global warming in observations and models. *Nat. Clim. Chang.* **2013**, *3*, 52–58. [CrossRef]
3. Sheffield, J.; Wood, E.F. Global trends and variability in soil moisture and drought characteristics, 1950-2000, from observation-driven simulations of the terrestrial hydrologic cycle. *J. Clim.* **2008**, *21*, 432–458. [CrossRef]
4. Wanders, N.; Karssenberg, D.; De Roo, A.; De Jong, S.M.; Bierkens, M.F.P. The suitability of remotely sensed soil moisture for improving operational flood forecasting. *Hydrol. Earth Syst. Sci.* **2014**, *18*, 2343–2357. [CrossRef]
5. Komma, J.; Blöschl, G.; Reszler, C. Soil moisture updating by Ensemble Kalman Filtering in real-time flood forecasting. *J. Hydrol.* **2008**, *357*, 228–242. [CrossRef]
6. Bisselink, B.; Van Meijgaard, E.; Dolman, A.J.; De Jeu, R.A.M. Initializing a regional climate model with satellite-derived soil moisture. *J. Geophys. Res. Atmos.* **2011**, *116*, 1–13. [CrossRef]
7. Drusch, M. Initializing numerical weather prediction models with satellite-derived surface soil moisture: Data assimilation experiments with ECMWF's integrated forecast system and the TMI soil moisture data set. *J. Geophys. Res. Atmos.* **2007**, *112*, 1–14. [CrossRef]
8. Hirschi, M.; Seneviratne, S.I.; Alexandrov, V.; Boberg, F.; Boroneant, C.; Christensen, O.B.; Formayer, H.; Orlowsky, B.; Stepanek, P. Observational evidence for soil-moisture impact on hot extremes in southeastern Europe. *Nat. Geosci.* **2011**, *4*, 17–21. [CrossRef]
9. Koster, R.D.; Dirmeyer, P.A.; Guo, Z.; Bonan, G.; Chan, E.; Cox, P.; Gordon, C.T.; Kanae, S.; Kowalczyk, E.; Lawrence, D.; et al. Regions of Strong Coupling Between Soil Moisture and Precipitation. *Science* **2004**, *305*, 1138–1140. [CrossRef]
10. McColl, K.A.; Alemohammad, S.H.; Akbar, R.; Konings, A.G.; Yueh, S.; Entekhabi, D. The global distribution and dynamics of surface soil moisture. *Nat. Geosci.* **2017**, *10*, 100–104. [CrossRef]
11. Brocca, L.; Tarpanelli, A.; Filippucci, P.; Dorigo, W.; Zaussinger, F.; Gruber, A.; Fernandez-Prieto, D. How much water is used for irrigation? A new approach exploiting coarse resolution satellite soil moisture products. *Int. J. Appl. Earth Obs. Geoinf.* **2018**, *73*, 752–766. [CrossRef]
12. Mousam, A.; Maggioni, V.; Delamater, P.L.; Quispe, A.M. Using remote sensing and modeling techniques to investigate the annual parasite incidence of malaria in Loreto, Peru. *Adv. Water Resour.* **2017**, *108*, 423–438. [CrossRef]
13. Entekhabi, D.; Njoku, E.G.; Neill, P.E.O.; Kellogg, K.H.; Crow, W.T.; Edelstein, W.N.; Entin, J.K.; Goodman, S.D.; Jackson, T.J.; Johnson, J.; et al. The Soil Moisture Active Passive (SMAP) Mission. *Proc. IEEE* **2010**, *98*, 704–716. [CrossRef]
14. Brocca, L.; Moramarco, T.; Melone, F.; Wagner, W. A new method for rainfall estimation through soil moisture observations. *Geophys. Res. Lett.* **2013**, *40*, 853–858. [CrossRef]
15. Brocca, L.; Ciabatta, L.; Massari, C.; Moramarco, T.; Hahn, S.; Hasenauer, S.; Kidd, R.; Dorigo, W.; Wagner, W.; Levizzani, V. Soil as a natural rain gauge: Estimating global rainfall from satellite soil moisture data. *J. Geophys. Res. Atmos.* **2014**, *119*, 5128–5141. [CrossRef]
16. Lu, Y.; Steele-Dunne, S.C.; Farhadi, L.; van de Giesen, N. Mapping Surface Heat Fluxes by Assimilating SMAP Soil Moisture and GOES Land Surface Temperature Data. *Water Resour. Res.* **2017**, *53*, 10858–10877. [CrossRef]
17. Zhang, J.; Bai, Y.; Yan, H.; Guo, H.; Yang, S. Linking observation, modelling and satellite-based estimation of global land evapotranspiration. *Big Earth Data* **2020**. [CrossRef]
18. Ruf, C.; Gleason, S.; Jelenak, Z.; Katzberg, S.; Ridley, A.; Rose, R.; Scherrer, J.; Zavorotny, V. The CYGNSS nanosatellite constellation hurricane mission. In Proceedings of the 2012 IEEE International Geoscience and Remote Sensing Symposium, Munich, Germany, 22–27 July 2012; pp. 214–216.
19. Chew, C.; Small, E. The UCAR/CU CYGNSS Soil Moisture Product. Available online: https://data.cosmic.ucar.edu/gnss-r/soilMoisture/cygnss/level3/ (accessed on 1 January 2020).
20. Kerr, Y.H.; Waldteufel, P.; Richaume, P.; Wigneron, J.P.; Ferrazzoli, P.; Mahmoodi, A.; Al Bitar, A.; Cabot, F.; Gruhier, C.; Juglea, S.E.; et al. The SMOS soil moisture retrieval algorithm. *IEEE Trans. Geosci. Remote Sens.* **2012**, *50*, 1384–1403. [CrossRef]

21. Egido, A.; Paloscia, S.; Motte, E.; Guerriero, L.; Pierdicca, N.; Caparrini, M.; Santi, E.; Fontanelli, G.; Floury, N. Airborne GNSS-R polarimetric measurements for soil moisture and above-ground biomass estimation. *IEEE J. Sel. Top. Appl. Earth Obs. Remote Sens.* **2014**, *7*, 1522–1532. [CrossRef]

22. Valencia, E.; Camps, A.; Vall-llossera, M.; Monerris, A.; Bosch-Lluis, X.; Rodriguez-Alvarez, N.; Ramos-Perez, I.; Marchan-Hernandez, J.F.; Martínez-Fernández, J.; Sánchez-Martín, N.; et al. GNSS-R Delay-Doppler Maps over land: Preliminary results of the GRAJO field experiment. *Int. Geosci. Remote Sens. Symp.* **2010**, *29*, 3805–3808. [CrossRef]

23. Alonso-Arroyo, A.; Camps, A.; Monerris, A.; Rudiger, C.; Walker, J.P.; Forte, G.; Pascual, D.; Park, H.; Onrubia, R. The light airborne reflectometer for GNSS-R observations (LARGO) instrument: Initial results from airborne and Rover field campaigns. *Int. Geosci. Remote Sens. Symp.* **2014**, 4054–4057. [CrossRef]

24. Masters, D.; Zavorotny, V.; Katzberg, S.; Emery, W. GPS signal scattering from land for moisture content determination. In Proceedings of the IGARSS 2000 IEEE 2000 International Geoscience and Remote Sensing Symposium. Taking the Pulse of the Planet: The Role of Remote Sensing in Managing the Environment (Cat. No.00CH37120), Honolulu, HI, USA, 24–28 July 2000; Volume 7, pp. 3090–3092. [CrossRef]

25. Zribi, M.; Motte, E.; Baghdadi, N.; Baup, F.; Dayau, S.; Fanise, P.; Guyon, D.; Huc, M.; Wigneron, J.P. Potential Applications of GNSS-R Observations over Agricultural Areas: Results from the GLORI Airborne Campaign. *Remote Sens.* **2018**, *10*, 1245. [CrossRef]

26. Chew, C.; Shah, R.; Zuffada, C.; Hajj, G.; Masters, D.; Mannucci, A.J. Demonstrating soil moisture remote sensing with observations from the UK TechDemoSat-1 satellite mission. *Geophys. Res. Lett.* **2016**, *43*. [CrossRef]

27. Camps, A.; Park, H.; Pablos, M.; Foti, G.; Gommenginger, C.P. Sensitivity of GNSS-R Spaceborne Observations to Soil Moisture and Vegetation. *IEEE J. Sel. Top. Appl. Earth Obs. Remote Sens.* **2016**, *9*, 4730–4742. [CrossRef]

28. Chew, C.; Small, E. Soil Moisture Sensing Using Spaceborne GNSS Reflections: Comparison of CYGNSS Reflectivity to SMAP Soil Moisture. *Geophys. Res. Lett.* **2018**, *45*, 4049–4057. [CrossRef]

29. Al-Khaldi, M.M.; Johnson, J.; O'Brien, A.; Balenzano, A.; Mattia, F. Time-Series Retrieval of Soil Moisture Using CYGNSS. *IEEE Trans. Geosci. Remote Sens.* **2019**, *57*, 4322–4331. [CrossRef]

30. Clarizia, M.P.; Pierdicca, N.; Costantini, F.; Floury, N. Analysis of CYGNSS data for soil moisture retrieval. *IEEE J. Sel. Top. Appl. Earth Obs. Remote Sens.* **2019**, *12*, 2227–2235. [CrossRef]

31. Calabia, A.; Molina, I.; Jin, S. Soil Moisture Content from GNSS Reflectometry Using Dielectric Permittivity from Fresnel Reflection Coefficients. *Remote Sens.* **2020**, *12*, 122. [CrossRef]

32. Kim, H.; Lakshmi, V. Use of Cyclone Global Navigation Satellite System (CyGNSS) Observations for Estimation of Soil Moisture. *Geophys. Res. Lett.* **2018**, *45*, 8272–8282. [CrossRef]

33. Eroglu, O.; Kurum, M.; Boyd, D.; Gurbuz, A.C. High spatio-temporal resolution cygnss soil moisture estimates using artificial neural networks. *Remote Sens.* **2019**, *11*, 2272. [CrossRef]

34. Senyurek, V.; Lei, F.; Boyd, D.; Kurum, M.; Gurbuz, A.C.; Moorhead, R. Machine Learning-Based CYGNSS Soil Moisture Estimates over ISMN sites in CONUS. *Remote Sens.* **2020**, *12*, 1168. [CrossRef]

35. Park, J.; Johnson, J.T.; Yi, Y.; O'Brien, A. Using "Rapid Revisit" CYGNSS Wind Speed Measurements to Detect Convective Activity. *IEEE J. Sel. Top. Appl. Earth Obs. Remote Sens.* **2019**, *12*, 98–106. [CrossRef]

36. Comite, D.; Ticconi, F.; Dente, L.; Guerriero, L.; Pierdicca, N. Bistatic Coherent Scattering From Rough Soils With Application to GNSS Reflectometry. *IEEE Trans. Geosci. Remote Sens.* **2019**, *58*, 612–625. [CrossRef]

37. Gleason, S.; O'Brien, A.; Russel, A.; Al-Khaldi, M.M.; Johnson, J.T. Geolocation, Calibration and Surface Resolution of CYGNSS GNSS-R Land Observations. *Remote Sens.* **2020**, *12*, 1317. [CrossRef]

38. Katzberg, S.J.; Garrison, J.L. Utilizing GPS to Determine Ionospheric Delay over the Ocean. *NASA Tech. Memo. TM-4750* **1996**, 1–16.

39. Hornbuckle, B.; Walker, V.; Eichinger, B.; Wallace, V.; Yildirim, E. Soil surface roughness observed during SMAPVEX16-IA and its potential consequences for SMOS and SMAP. In Proceedings of the International Geoscience and Remote Sensing Symposium (IGARSS)2, Fort Worth, TX, USA, 23–28 July 2017.

40. Snapir, B.; Hobbs, S.; Waine, T.W. Roughness measurements over an agricultural soil surface with Structure from Motion. *ISPRS J. Photogramm. Remote Sens.* **2014**, *96*, 210–223. [CrossRef]

41. Thomsen, L.M.; Baartman, J.E.M.; Barneveld, R.J.; Starkloff, T.; Stolte, J. Soil surface roughness: Comparing old and new measuring methods and application in a soil erosion model. *SOIL* **2015**, *1*, 399–410. [CrossRef]

42. Dente, L.; Guerriero, L.; Comite, D.; Pierdicca, N. Space-Borne GNSS-R Signal Over a Complex Topography: Modeling and Validation. *IEEE J. Sel. Top. Appl. Earth Obs. Remote Sens.* **2020**, *13*, 1218–1233. [CrossRef]

43. Campbell, J.D.; Melebari, A.; Moghaddam, M. Modeling the effects of topography on delay-Doppler maps. *IEEE J. Sel. Top. Appl. Earth Obs. Remote Sens.* **2020**. [CrossRef]

44. O'Neill, P.E.; Chan, S.; Njoku, E.G.; Jackson, T.; Bindlish, R. *SMAP L3 Radiometer Global Daily 36 km EASE-Grid Soil Moisture, Version 5*; National Snow and Ice Data Center: Boulder, CO, USA, 2018.

45. Burgin, M.S.; Colliander, A.; Njoku, E.G.; Chan, S.K.; Cabot, F.; Kerr, Y.H.; Bindlish, R.; Jackson, T.J.; Entekhabi, D.; Yueh, S.H. A Comparative Study of the SMAP Passive Soil Moisture Product with Existing Satellite-Based Soil Moisture Products. *IEEE Trans. Geosci. Remote Sens.* **2017**, *55*, 2959–2971. [CrossRef]

46. Colliander, A.; Jackson, T.J.; Bindlish, R.; Chan, S.; Das, N.; Kim, S.B.; Cosh, M.H.; Dunbar, R.S.; Dang, L.; Pashaian, L.; et al. Validation of SMAP surface soil moisture products with core validation sites. *Remote Sens. Environ.* **2017**, *191*, 215–231. [CrossRef]

47. Fuks, I.M. Wave diffraction by a rough boundary of an arbitrary plane-layered medium. *IEEE Trans. Antennas Propag.* **2001**, *49*, 630–639. [CrossRef]

48. Voosen, P. Satellites see hurricane winds despite military signal tweaks. *Science* **2019**, *364*, 1019. [CrossRef] [PubMed]

49. Pekel, J.-F.; Cottam, A.; Gorelick, N.; Belward, A.S. High-resolution mapping of global surface water and its long-term changes. *Nature* **2016**, *540*, 418–422. [CrossRef] [PubMed]

50. Zreda, M.; Desilets, D.; Ferre, T.P.A.; Scott, R.L. Measuring soil moisture content non-invasively at intermediate spatial scale using cosmic-ray neutrons. *Geophys. Res. Lett.* **2008**, *35*. [CrossRef]

51. Small, E.E.; Larson, K.M.; Chew, C.C.; Dong, J.; Ochsner, T.E. Validation of GPS-IR Soil Moisture Retrievals: Comparison of Different Algorithms to Remove Vegetation Effects. *IEEE J. Sel. Top. Appl. Earth Obs. Remote Sens.* **2016**, 1–12. [CrossRef]

52. Schaefer, G.L.; Cosh, M.H.; Jackson, T.J. The USDA Natural Resources Conservation Service Soil Climate Analysis Network (SCAN). *J. Atmos. Ocean. Technol.* **2007**, 2073–2077. [CrossRef]

53. Schaefer, G.L.; Paetzold, R.R. SNOTEL (SNOwpack TELemetry) And SCAN (Soil Climate Analysis Network). In Proceedings of the Automated Weather Stations for Applications in Agriculture and Water Resources Management: Current Use and Future Perspectives, Lincoln, NB, USA, 6–10 March 2000.

54. Diamond, H.J.; Karl, T.R.; Palecki, M.A.; Baker, C.B.; Bell, J.E.; Leeper, R.D.; Easterling, D.R.; Lawrimore, J.H.; Meyers, T.P.; Helfert, M.R.; et al. U.S. Climate Reference Network after one decade of operations: Status and assessment. *Bull. Am. Meteorol. Soc.* **2013**, *94*, 489–498. [CrossRef]

55. Smith, A.B.; Walker, J.P.; Western, A.W.; Young, R.I.; Ellett, K.M.; Pipunic, R.C.; Grayson, R.B.; Siriwidena, L.; Chiew, F.H.S.; Richter, H. The Murrumbidgee Soil Moisture Monitoring Network Data Set. *Water Resour. Res.* **2012**, *48*. [CrossRef]

56. Chen, Q.; Zeng, J.; Cui, C.; Li, Z.; Chen, K.S.; Bai, X.; Xu, J. Soil Moisture Retrieval from SMAP: A Validation and Error Analysis Study Using Ground-Based Observations over the Little Washita Watershed. *IEEE Trans. Geosci. Remote Sens.* **2018**, *56*, 1398–1408. [CrossRef]

 *remote sensing*

Article

The Polarimetric Sensitivity of SMAP-Reflectometry Signals to Crop Growth in the U.S. Corn Belt

Nereida Rodriguez-Alvarez [1,*], **Sidharth Misra** [2] **and Mary Morris** [2]

[1] Planetary Radar Radio Science Systems Group, Jet Propulsion Laboratory, California Institute of Technology, Pasadena, CA 91109, USA

[2] Microwave Instrument Science Group, Jet Propulsion Laboratory, California Institute of Technology, Pasadena, CA 91109, USA; sidharth.misra@jpl.nasa.gov (S.M.); mary.g.morris@jpl.nasa.gov (M.M.)

* Correspondence: nereida.rodriguez.alvarez@jpl.nasa.gov; Tel.: +1-818-393-7507

Received: 20 February 2020; Accepted: 19 March 2020; Published: 21 March 2020

Abstract: Crop growth is an important parameter to monitor in order to obtain accurate remotely sensed estimates of soil moisture, as well as assessments of crop health, productivity, and quality commonly used in the agricultural industry. The Soil Moisture Active Passive (SMAP) mission has been collecting Global Positioning System (GPS) signals as they reflect off the Earth's surface since August 2015. The L-band dual-polarization reflection measurements enable studies of the evolution of geophysical parameters during seasonal transitions. In this paper, we examine the sensitivity of SMAP-reflectometry signals to agricultural crop growth related characteristics: crop type, vegetation water content (VWC), crop height, and vegetation opacity (VOP). The study presented here focuses on the United States "Corn Belt," where an extensive area is planted every year with mostly corn, soybean, and wheat. We explore the potential to generate regularly an alternate source of crop growth information independent of the data currently used in the soil moisture (SM) products developed with the SMAP mission. Our analysis explores the variability of the polarimetric ratio (PR), computed from the peak signals at V- and H-polarization, during the United States Corn Belt crop growing season in 2017. The approach facilitates the understanding of the evolution of the observed surfaces from bare soil to peak growth and the maturation of the crops until harvesting. We investigate the impact of SM on PR for low roughness scenes with low variability and considering each crop type independently. We analyze the sensitivity of PR to the selected crop height, VWC, VOP, and Normalized Differential Vegetation Index (NDVI) reference datasets. Finally, we discuss a possible path towards a retrieval algorithm based on Global Navigation Satellite System-Reflectometry (GNSS-R) measurements that could be used in combination with passive SMAP soil moisture algorithms to correct simultaneously for the VWC and SM effects on the electromagnetic signals.

Keywords: GNSS-R; SMAP-R; VWC; vegetation opacity; crop type; crop height; soil moisture; crop health; crop productivity; agriculture

---

1. Introduction

Sustaining and enhancing the economical production of crops continues to be an important focus of agricultural research. Information and knowledge on crop vegetative growth and crop reproductive development is vital to agricultural producers with the goal of more efficient production of high-quality crops. Vegetative growth of crops is defined as the accumulation of dry matter, which is the weight of the crops including all its constituents, excluding water. For example, the vegetative stage of corn begins when the seedling emerges and continues until tasseling, when the reproductive stage begins. During the vegetative stage, leaves develop and grow, the stalk forms, and reproductive structures (ear and tassel) begin to form. The reproductive development of crops is related to the crop transition into the reproductive phase. Reproductive development stages begin with fertilization of the floret (pollination) and end when the grain reaches its maximum dry weight. This later stage is called physiological maturity. To ensure high productivity of the crops, there are certain requirements that need to be met at the crop growth and development stages. By understanding how corn grows and develops, producers can more confidently assess crop damage, estimate if it will recover, and apply herbicides and other crop treatments at the best time. Information on soil conditions and crop growth related parameters, such as the vegetation water content (VWC) at every stage, are the main factors that can help understand the final crop's production.

Missions like the Soil Moisture Active Passive (SMAP) [1] and the Soil Moisture Ocean Salinity (SMOS) [2] are currently providing information on the soil moisture (SM) conditions at scales of 9 km $\times$ 9 km [3] in the case of SMAP and 25 km $\times$ 25 km [4] in the case of SMOS. Note that for the SMAP product, event it is posted to the 9km $\times$ 9km grid, and the representative area of each grid cell is 36 km. In [5], it was found that the performance of the enhanced 9 km SMAP SM product was equivalent to that of the standard 33 km SMAP SM product, attaining a retrieval uncertainty below 0.040 m^3/m^3 unbiased root-mean-squared error and a correlation coefficient above 0.800. Soil moisture estimation remains a challenge due to the lack of validated VWC products at a global scale at a temporal resolution that matches the speed of the crop growth. VWC impacts the electromagnetic signals through volume scattering due to the leaves, branches, trunks, and attenuation due to the dielectric constant of the vegetation. The transmissivity, computed as $e^{(-VOP/cos(\theta))}$, describes the amount of soil emission passing through the vegetation layer. The transmissivity depends on both the incidence angle θ and the vegetation opacity (VOP), computed as $VOP = b_p VWC$, where b_p is a parameter describing the vegetation type [6]. Note that VOP and vegetation optical depth (VOD) are synonymous. Therefore, VWC impacts the microwave emission received by SMAP such that under the same soil moisture conditions, differences in VWC result in different microwave emissions from the same surface. In trying to estimate soil moisture, the lack of VWC information results in an uncertainty of the soil moisture retrievals. The availability of high spatial and temporal resolution global estimates of VWC would therefore improve SM estimates. In particular, the SMAP mission has developed a VWC dataset used in generating SMAP science data products, based on Moderate Resolution Imaging Spectroradiometer (MODIS) [7] data. The drawback of the MODIS-based product is that it relies on Normalized Differential Vegetation Index (NDVI) as a proxy to VWC; NDVI loses sensitivity to VWC as the water content increases [8]. In addition, the SMAP VWC products are based on 10-day NDVI climatology of 10 years (2000 to 2010), and this results in SM errors when vegetated areas undergo dynamic changes, as is sometimes the case within agricultural landscapes. Other approaches include large-scale mapping of VWC [9], but the validity is limited by the data used. Even so, VWC datasets are not updated routinely. For agricultural areas characterized by very dynamic landscapes in terms of SM, VWC, VOP, and vegetation height, a 10-day independent VWC product would have a great impact on the soil moisture retrievals obtained by satellite missions as SMAP. Limited in situ measurements are obtained, but those are usually sparse and cover only limited crop types. In order to understand the impact of using products based on climatology, it is important to understand first the dynamics of the vegetated landscapes: forest of any type, tundra, meadows, and in general, the majority of natural spaces not used for the purpose of agriculture are less dynamic vegetated landscapes, and therefore, their VWC signature changes less drastically in a seasonal regime. For most natural spaces, the low-frequency nature of their VWC variations produces a high correlation between VWC seasonal climatology and real-time VWC [10]. Equally, when considering VOP (dependent on VWC) information, agricultural areas will show a bigger discrepancy between seasonal climatology and daily estimates as compared to natural spaces. Another descriptor of the vegetation is the vegetation height, i.e., the thickness of the vegetation layer. The vegetation height for agricultural landscapes is very dynamic during the growing season and may change in scales of a few days. Therefore, some vegetated landscapes are well represented with a static/seasonal information, while agricultural areas need finer temporal scales.

In August 2015, the SMAP mission started collecting GPS signals as they scatter off the Earth's surface [11–15]. These signals are measured by the SMAP radar receiver, whose band pass filter center frequency was switched to the GPS L2C band (1227.6 MHz). These bistatic radar L-band signals are measured at V-polarization and H-polarization simultaneously. Since bistatic radar and radiometer measurements are impacted by vegetation cover differently [16,17], the combination of both active (SMAP-R) and passive (SMAP radiometer) measurements could act to further constrain vegetation effects for the SM retrieval. SMAP-R based vegetation parameters are independent of any NDVI product and could improve SM estimation performance. There have been previous studies employing left hand and right hand circularly polarized GNSS-R measurements to compute a polarimetric ratio, such as [18,19] and a previous work using SMAP-R data [12].

This manuscript presents a sensitivity analysis of the SMAP-R signals to VWC, VOP, crop height, and type, including corn and soybean. The impact of SM on the sensitivity to the crop growth parameters is also assessed. The analysis is limited to an area characterized by low roughness: the United States Corn Belt during the 2017 crop season. Section 2 describes the SMAP-R measurements and presents the main observable used in this study, the polarimetric ratio (PR), as well as provides a discussion on the spatial resolution of SMAP-R measurements. Section 3 explains the methodology and analyzes each one of the datasets involved in the process. In Section 4, we investigate the variability of the PR through the crop season and the sensitivity of the PR to crop growth parameters. Section 5 discusses a path towards a retrieval algorithm based on Global Navigation Satellite System-Reflectometry (GNSS-R) and on improvements and requirements for future research. Section 6 states the final conclusions of this study.

2. SMAP-R Measurements Description

The GPS signals reflected off the Earth's surface are collected at the SMAP radar receiver in the form of in-phase and quadrature (I/Q) samples. The I/Q samples are publicly available at the NASA Earthdata website [20]. Using a modified version of the SMAP-R processor used in [11,12], data are filtered for those geometries where there is potential to capture a specular point, i.e., within the -3 dB beam width of the SMAP antenna pointed to 40°, which provides a range of incidence angles between 37.3° and 42.7°. This selection ensures that the measured specular points are not degraded by the decay of SMAP antenna gain away from −3dB. Those selected I/Q samples are post-processed into delay-Doppler maps (DDM) [21,22], with a 5 ms coherent time and 25 ms incoherent time (five incoherent accumulation). A DDM is defined as the delay and Doppler power distribution of the GPS signal scattered over the Earth's surface. Since the SMAP antenna rotates, consecutive DDMs integrated at 25 ms are spaced approximately 25 km apart. The DDMs are then calibrated to account for GPS transmitter power and GPS antenna gain at the reflection angle and also for the filtering effect of the SMAP high gain antenna, whose small footprint partially observes the scattering surface. The next subsection provides more details on the calibration performed to the DDM and the observables used in the assessment of VWC estimates.

2.1. SMAP-R Calibration

To calibrate the SMAP-R DDMs, we applied the modified equation presented in [15]. The equation in [12] followed the CYclone Global Navigation Satellite System (CYGNSS) [23–25] mission calibration procedure presented in [26], but adds the filtering effect of the SMAP antenna pattern. The equation takes the form in eqn. (1):

$$\sigma_0 = \frac{(4\pi)^3 Y(\tau, f_d)\ R^2_{rx_{sp}} R^2_{tx_{sp}}}{T_i^2 P_{tx} G_{tx} \lambda^2 G_{rx_{sp}} \overline{B}(\tau,\ f_d)} \tag{1}$$

where:

- $Y(\tau, f_d)$ is the DDM received power distribution at each delay τ and Doppler f_d bin.

- T_i is the coherent integration time. $T_i = 25$ ms for SMAP-R data, instead of the typical value of 1000 ms used in other GNSS-R missions.
- P_{tx} is the GPS transmitted power.
- G_{tx} is the GPS transmitter antenna gain.
- λ is the GPS signal wavelength (at GPS-L2C is 24.42 cm).
- $R_{rx_{sp}}$ corresponds to the distance from the transmitter specular point.
- $R_{tx_{sp}}$ corresponds to the distance from the receiver to the specular point.
- $G_{rx_{sp}}$ is the SMAP antenna gain value at the specular point incidence angle θ_i. It is computed from eqn. (2), where we assume the gain presents a linear decay within the -3 dB beam width (2.7 deg.) that goes from its maximum value ($G_{\mathrm{max}\_dB} = 36$ dBi @ 40 deg. incidence angle pointing) to 33 dB.

$$G_{rx}(\theta_i) = G_{\mathrm{max}\_dB} - \frac{3\,\mathrm{dB}}{2.7^\circ} * abs\left(40^\circ - \theta_i^\circ\right)\,[\mathrm{dB}] \tag{2}$$

- $\overline{B}(\tau,\,f_d)$ is the filtered effective surface scattering area. It is computed following the methodology described in [15], where the size of the area corresponding to each τ and f_d bin filtered by the SMAP antenna is normalized to the size of the same τ and f_d bin considering an omnidirectional antenna.

The calibration methodology developed in [15] and implemented in this study considers two corrections:

- Calibration of the direct power information ($P_{tx}G_{tx}$). $P_{tx}G_{tx}$ is calibrated by collocating CYGNSS information for the same day of measurements.
- Calibration of the SMAP antenna filtering effect ($\overline{B}(\tau,\,f_d)$).

SMAP-R measurements have a non-uniform spatial resolution. The variability on the spatial resolution is linked to the nature of the SMAP-R signals, since the spatial resolution depends on the characteristics of the scattering area. A rougher surface enlarges the scattering area, enlarging the spatial resolution. The presence of vegetation also enlarges the spatial resolution as equally, it results in the scattering coming from a more extensive area. To compute the approximate spatial resolution of each measurement, we applied the methodology also explained in [15]. According to the results in [15], a threshold of 70% was selected, and the delay gathering the 70% of the power was transformed to an ellipse of constant delay on the surface (iso-delay line) [21]. The size of the scattering area was set to the semi-major axis of the computed ellipse to represent the scattering area of that particular measurement. Figure 1 shows the variability on the scattering area sizes computed for two extreme scenarios: (1) bare soil, characterized by low VWC, low height, and low VOP; and (2) crop seasonal growth peak, characterized by high VWC, high height, and high VOP. The two extreme scenarios were selected as the reference states (RS1 and RS2, respectively) of our dataset and will be further explained in Section 3.

The spatial resolutions observed in the measurements involved in this study were between 5 km and 24 km, depending on the crop type and growth stage that characterize the scattering surface. The mean of the bare soil conditions was ~12 km, while for the same area, after crop grew, the observed spatial resolution increased to a mean of ~17 km. Our conclusions agreed with the results presented in [15], where an expected spatial resolution between 5 km to 26 km was acceptable for none to high vegetation, although roughness conditions and the surface topography had an impact on the final spatial resolution. For the purpose of this study, analyzing the sensitivity, and in order to easily match the spatial resolution of the reference products used in this study (SMAP SM and VWC), we used a simplistic strategy of gridding the SMAP-R data to a 9×9 km where all specular points falling within each grid cell were considered to contribute only to that grid cell and were averaged together. Future research focused on generating products from this dataset will include the appropriate spatial resolution gridding.

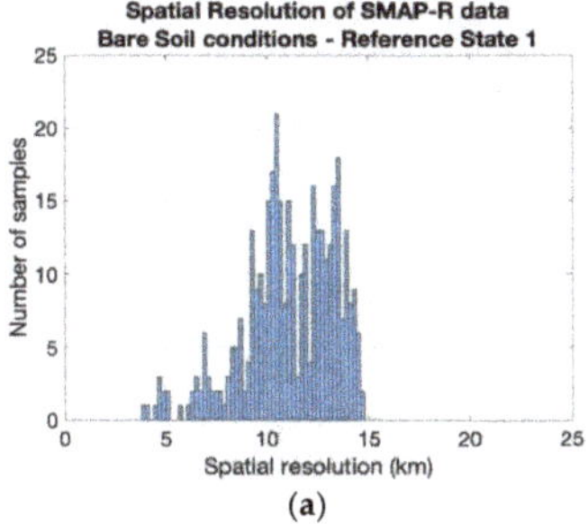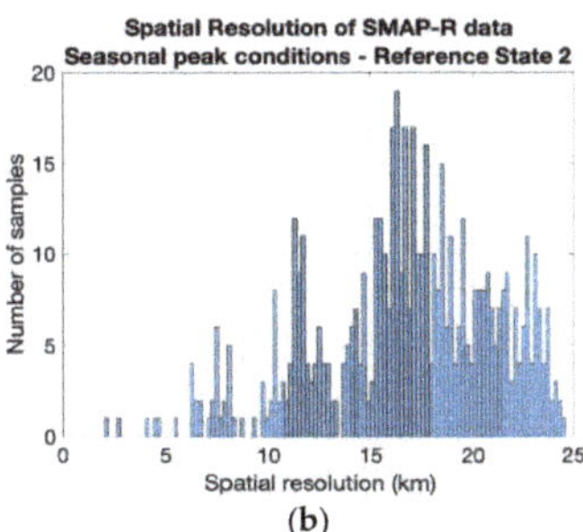

Figure 1. Computed scattering area for the measurements during the two reference states: (**a**) low VWC, low height, and low vegetation opacity (VOP) (reference state RS1) and (**b**) high VWC, high height, and high VOP (reference state RS2).

2.2. The Observable: Polarimetric Ratio

We used an observable built from the DDM power information, in particular the power of the peak. Due to the complexity of the filtering effect of the SMAP antenna, we decided not to include any observable based on the shape of the DDMs. Therefore, the corrections described in Section 2.1 were only applied to the τ and f_d bin that corresponded to the peak of the DDM. The observable used in this study was the polarimetric ratio (PR). PR was computed as in [27] using the formula:

$$PR\left(\tau_{peak}, f_{d\_peak}\right) = \frac{Y_V\left(\tau_{peak}, f_{d\_peak}\right) - Y_H\left(\tau_{peak}, f_{d\_peak}\right)}{Y_V\left(\tau_{peak}, f_{d\_peak}\right) + Y_H\left(\tau_{peak}, f_{d\_peak}\right)} \tag{3}$$

where:

- τ_{peak} is the delay at the bin of the DDM peak.
- f_{d_peak} is the frequency Doppler at the bin of the DDM peak.
- $Y_H\left(\tau_{peak}, f_{d\_peak}\right)$ is the DDM received power at H-polarization measured by SMAP-R at $\left(\tau_{peak}, f_{d\_peak}\right)$.
- $Y_V\left(\tau_{peak}, f_{d\_peak}\right)$ is the DDM received power at V-polarization measured by SMAP-R at $\left(\tau_{peak}, f_{d\_peak}\right)$.

The sensitivity of SMAP-R signals to crop growth will therefore be evaluated through the variability of the PR observable to changes on crop growth parameters (VWC, VOP, NDVI, and crop height), as well as SM conditions.

3. Methodology

In order to analyze the sensitivity of SMAP-R signals to the crop growth parameters, we selected data collected from the United States (U.S.) Corn Belt, an extensive agricultural area that is planted with primarily corn, soybean, and wheat every year. Figure 2 shows information on the U.S. Corn Belt.

Figure 2a shows the selected site with information from the U.S. Department of Agriculture (USDA) National Agricultural Statistics Service (NASS) database [28] showing the amount of planted corn acres per county. The number of acres planted varies year-to-year as can be seen in Figure 2b. SMAP-R data allowed for the temporal series analysis of the U.S. Corn Belt seasonal changes from 2016 to 2019. The maps shown are available at [28]. Figure 2c shows a sketch of the growth stages of a corn plant. Figure 2d to Figure 2f show images of different stages of the corn, bare soil, plant growing, and harvesting, respectively.

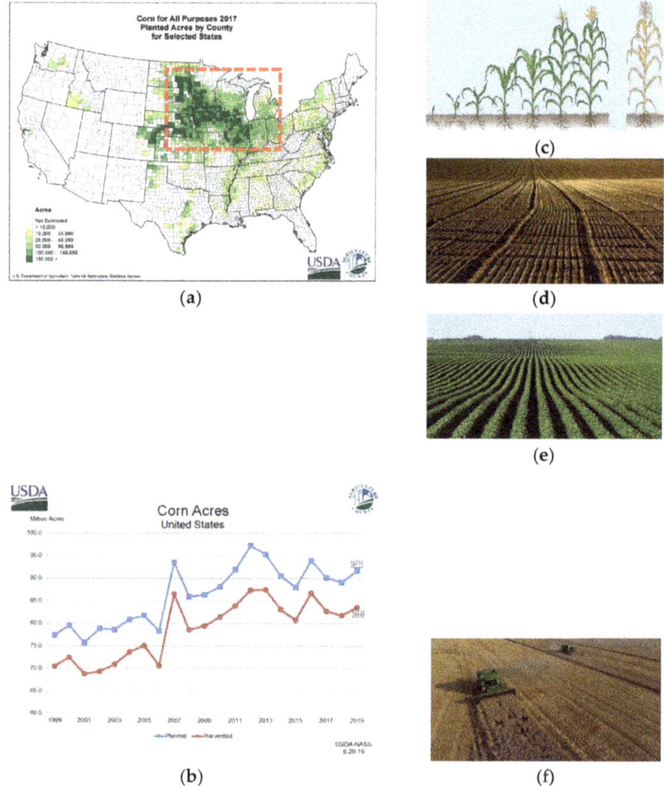

Figure 2. Site selected: (**a**) U.S. Department of Agriculture (USDA) National Agricultural Statistics Service (NASS) [28] map showing the corn for all purposes planted in 2017, the red box indicating the specific area under study; (**b**) graph showing the total number of acres planted in consecutive years (2017 and 2018 show similar numbers; 2019 shows an increase); (**c**) stages of growth of a corn plant ([29], University of Illinois Extension program); and three images illustrating the agricultural landscape at different moments of the season; (**d**) terrain preparation for plantation (bare soil); (**e**) Vegetative-stage soybean planted and growing (high VWC, increasing height, and VOP); (**f**) soybean at the end of the season (low VWC, maximum height, medium VOP) being harvested (a layer of dried harvesting leftovers, low VWC, low VOP).

The sensitivity analysis of SMAP-R measurements was therefore explored over the U.S. Corn Belt. We used the observable, PR, described in Section 2.2. The methodology implemented is illustrated in Figure 3.

Summarizing Figure 3, there are three main steps:

- SMAP-R data preparation: computing PR and gridding the data.
- Use of ancillary datasets: gathering information on roughness, SM, and crop type using it to filter and bin the PR observable.
- Sensitivity analysis: Analyze PR observable against reference datasets; i.e., VWC, VOP, NDVI, and crop height.

The next subsections provide detailed information on the key elements shown in Figure 3: the two-state analysis strategy, the ancillary datasets employed to characterize the scattering area and the resulting grouped the data, as well as the reference datasets used to explore the sensitivity of the

SMAP-R signals to the crop growth parameters (VWC, VOP, NDVI, crop height). Sensitivity analysis are provided in Section 4.

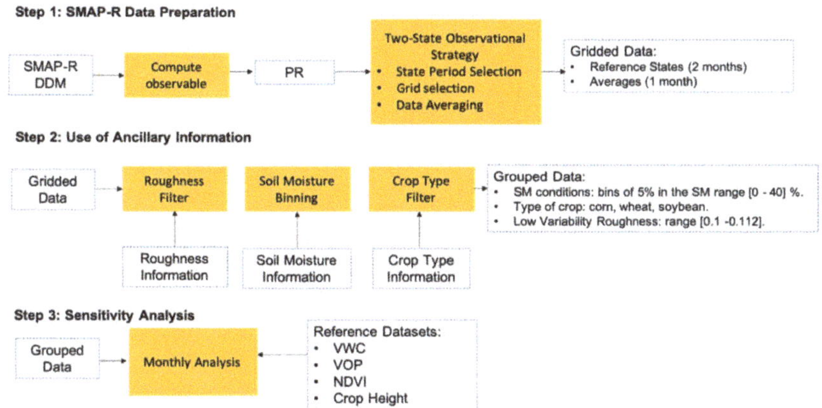

Figure 3. Block diagram of the methodology implemented to assess the sensitivity of SMAP-R signals to crop growth parameters. The methodology is divided into three steps: SMAP-R data preparation, ancillary information use, and sensitivity analysis of SMAP-R signals to VWC, VOP, NDVI, and crop height. All datasets are gridded to the same 9 km × 9 km resolution. DDM, delay-Doppler maps; PR, polarimetric ratio; SM, soil moisture.

3.1. Two-State Analysis Strategy

SMAP-R measurements have low sampling and poor coverage, but those characteristics can be overcome by implementing analysis strategies as using measurements over long periods of stability in two extremes as the references and analyzing the variability of the measurements in transitional periods. We refer to this as a two-state analysis strategy. We analyzed the mean variability of the observable PR, described in Section 2.2, at these two states. Data between states could then be analyzed at shorter times referencing them to the two extreme states and discarding samples that are out of the limits marked by the two reference states.

In this study, we averaged the information from two months, March and April 2017, as one reference state describing the bare soil conditions (no VWC, 0 VOP, 0 cm crop height: reference state RS1). We used July and August 2017 as the other reference state describing the season peak of the crops (high VWC, high VOP, high height: reference state RS2). The transitional periods from especially April to July 2017 (growing period with increasing VWC, VOP, and height) and August to October 2017 (drying period with decreasing VWC, low height variability, and harvesting) would therefore be referenced to the two reference states described. Data that did not fall within the PR ranges identified for the reference states RS1 and RS2 were discarded. Figure 4 shows the PR observable for the two reference states.

PR plots were computed using a fixed grid of 9 × 9 km to match current official SMAP products. Both Y_H and Y_V were then computed and used to derive the PR values, as described in eqn. (1). Note that PR values, computed from the forward scattered GPS signal after it interacted with the vegetation, differed from those expected in the radiometry field. PR values were averaged together using a drop in the box approach, i.e., measurements whose specular points fell within each grid cell were averaged together (regardless of their true spatial resolution). Figure 4 shows the heterogeneity of the PR observable; i.e., it shows the variability within the scene due to differences in the crop type and growth stage during the selected period, as well as the variability of the soil moisture conditions. For bare soil and constant roughness conditions (Figure 4a), the variability was due primarily to soil moisture. During the peak season (Figure 4b), VWC, VOP, crop height, crop type, and soil moisture played an important role. From Figure 4, a general drop in the levels of the PR observable can be observed. In Section 4, we provide a deeper analysis of the SMAP-R signal sensitivity to the different geophysical parameters that compose the scene.

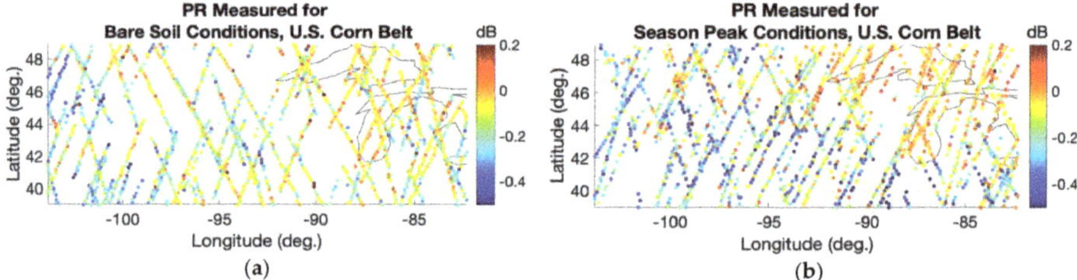

Figure 4. Two-state analysis strategy: (**a**) corresponds to the PR values for bare soil state, and (**b**) corresponds to the PR values for the peak season state.

3.2. Ancillary Datasets

Below, we describe the ancillary information used to filter and bin our data as a part of the process outlined in Figure 3: crop type, surface roughness, and monthly soil moisture variability.

The USDA NASS CropScape–Cropland Data [30] provides information on the type and quantity of crops per county in the U.S. and is freely available at [30]. We used the information in the dataset to assign the predominant crop type to a pixel of 9 km × 9 km, matching SMAP spatial resolution, for the year 2017. The map is shown in Figure 5.

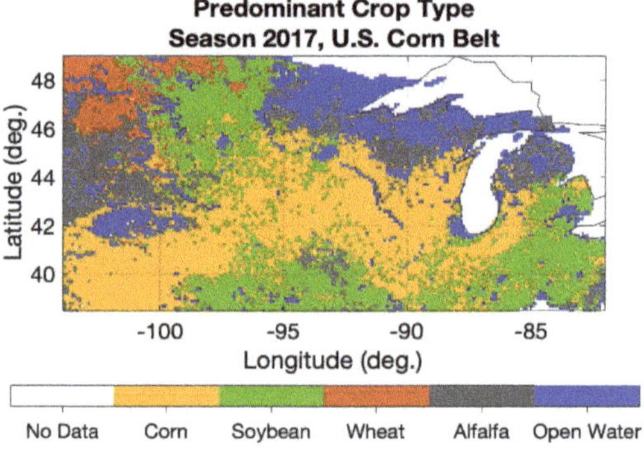

Figure 5. Crop type map developed from the USDA NASS CropScape–Cropland Data database showing the predominant corn type in pixels of 9 × 9 km, matching SMAP official product spatial resolution [30].

Following the methodology described in Figure 3, the crop type information was used to analyze the SMAP-R PR observable for the different crops separately avoiding the mix due to the different scattering properties characteristic of each crop type and distinguishing effects of the different growth stages of each crop type.

Roughness had a big impact on the variability of the strength of the SMAP-R signals, as well as the variability of the spatial resolution of the measurements. The United States Corn Belt crop area presented a low surface roughness with small variability. SMAP's roughness ancillary product [31,32], stored in the SMAP SM official product, corroborated this statement (see Figure 6).

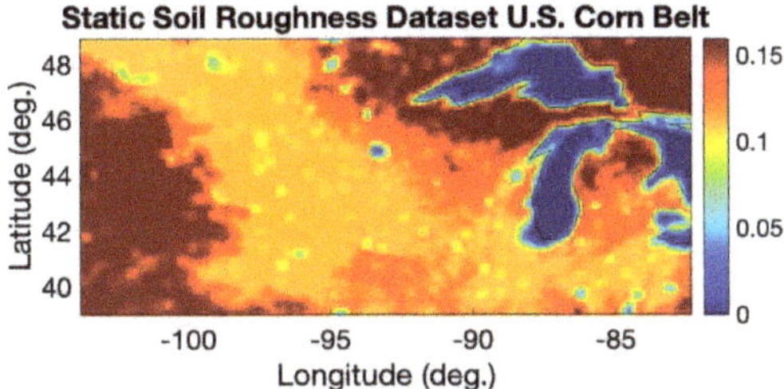

Figure 6. Roughness information: SMAP ancillary product for surface roughness (static product).

The SMAP roughness ancillary dataset is a static product. Even where the roughness as low (orange areas) as compared to surrounding rougher (red) areas, the assumption that the roughness was static was imperfect. We recognized that roughness varied throughout the year in the U.S. Corn Belt, primarily due to tillage [33]. Tillage may occur after harvesting the previous season and before the first snowfall—around November—or after the snow melts in spring—around March. In addition, this tillage was spatially variable, as it depended on farm management practices. Additionally, rainfall smooths the soil surface [33] until the crops grow and can shelter the surface. Therefore, both the tillage and the rain smoothing affect the roughness of the area dynamically rather than statically—as was assumed in this work. This assumption was one source of error in our analysis of the relationships between observations and geophysical parameters explored in this study: SM, VWC, VOP, and vegetation height. Given limited information on tilling and other dynamic variables, we utilized the static product stored in the official SMAP SM product and acknowledged that this may limit the robustness of the analysis herein.

Much of the central region shown in Figure 6 had a roughness value of around ~0.1 with little variability in the surrounding agricultural areas. The roughness ancillary product is a static product that has been modeled as the surface reflectivity of a rough surface (r_{p_rough}) computed as the surface reflectivity of a smooth surface (r_{p_smooth}) multiplied by a factor dependent on a parameter (h) linearly dependent on the root-mean-squared surface height [34,35] and the incidence angle (θ) as: $r_{p\_rough} = r_{p\_smooth}e^{-hcos^x(\theta)}$, with $x = 0, 1$ or 2; see [36]. Fresnel equations were used to compute r_{p_smooth} at each polarization; see [36]. For SMAP, h values were obtained from a land cover-driven lookup table, available in [37]. The roughness values corresponded to unitless values that were indicative of bare soil roughness within SMAP 9 km grid cell (0 min, 1 max). For this study, we selected those areas in the map that had a low roughness with low variability (orange area in Figure 6). Using the roughness information, we generated a mask where roughness value remained between the range [0.1–0.112], shown in Figure 7a, and we used this map to filter the crop information (Figure 7b).

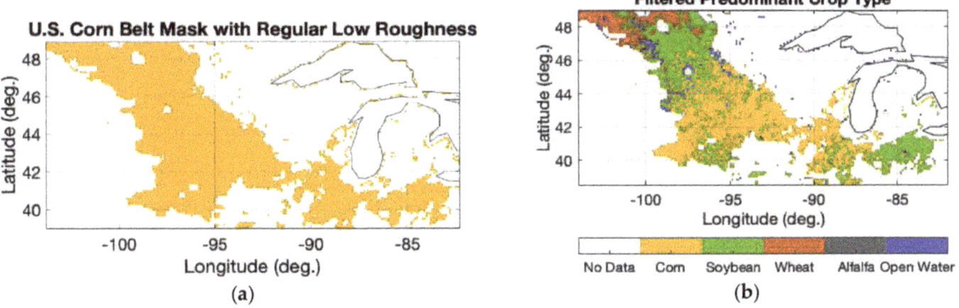

Figure 7. Roughness information: (**a**) roughness mask built from low variability low roughness areas in Figure 6a and (**b**) masked predominant crop type information.

After using the roughness mask, there were only three types of crop to include potentially in our study: great extensions of corn and soybean and some areas in the northern part of wheat. The next important parameter to control in the area under study was the SM, whose variability would have an impact on the reflected signal strength, and therefore on the PR. To isolate the impact of SM on the sensitivity to crop growth parameters, we used the SMAP SM official product at 9 × 9 km [3]. Figure 8 shows the mean and standard deviation of the SM on a monthly basis for areas represented by each crop type. In addition, in Appendix A, we included the monthly averaged SMAP SM maps from April 2017 to November 2017. SM values are expressed in units of % throughout the manuscript. The % corresponds to the volumetric soil water content computed from the volume of water (cm^3) over the volume of soil (cm^3) expressed in percentage.

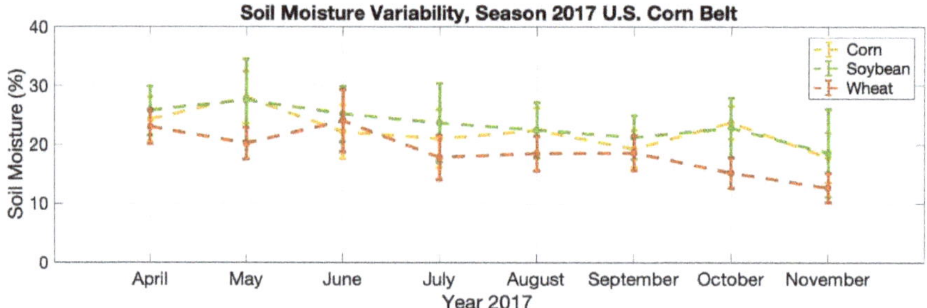

Figure 8. Monthly soil moisture information for the 2017 season in the U.S. Corn Belt.

The monthly means of SM observed by the SMAP mission in our area under study showed high variability. During May and June, SMAP soil moisture retrievals showed an increase in soil moisture, most probably due to rain events since most part of the area under study was rain-fed. Rain benefits the growth and health of the crops. During the months of July, August, and September, SM variations seemed to decrease linearly in the mean. During October, variations in soil moisture were also observed for most of the fields and were also probably due to rain events. We used these data in order to understand what the impact of SM variation on the sensitivity of SMAP-R signals to the crop growth parameters was. In order to do this and following the scheme presented in Figure 3, we initially binned together areas with SM values in the range of 0% to 40% in steps of 5%, providing a total of eight levels of SM, and then analyzed the sampling population of binned PR values to discard those statistically poorly represented. We defined those statistically poorly represented SM ranges with a number of samples below five for most of the PR bins. This analysis is presented in Section 4.

3.3. Reference Datasets

Following the scheme in Figure 3, in order to study the sensitivity of SMAP-R signals to the different crop growth parameters, we selected a number of reference datasets: the VWC from the SMAP ancillary dataset, the VOP from the SMAP ancillary dataset, and the crop height that was estimated from information on typical growth values for the different crop types.

The NDVI reference dataset was obtained from the NDVI product from the USDA NASS VegScape–Vegetation Condition Explorer [38], where MODIS measurements were employed to derive the NDVI measurements. Appendix A shows the NDVI maps for the whole crop season (April to November, 2017). Figure 9 shows the mean and standard deviation of the NDVI on a monthly basis for areas represented by each crop type.

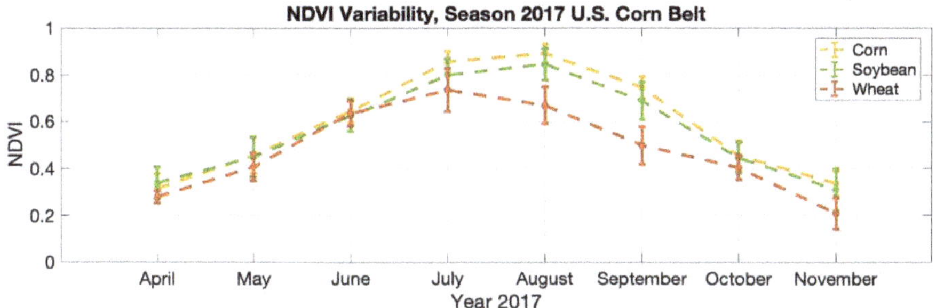

Figure 9. Monthly crop NDVI information for the 2017 season in the U.S. Corn Belt.

The NDVI provides a numerical indicator used to assess whether the observed surface contains live green vegetation or not. As can be seen in Figure 9, NDVI showed the highest values during July and August 2019 for both corn and soybean crop types. In the case of wheat, the peak of the season happened in June.

The selected VWC and VOP reference datasets were those stored as ancillary products in the SMAP SM official product [3]. VWC and VOP products used in this study were therefore derived from 10 day NDVI climatology rather than obtained from actual 2017 data. We included monthly averaged VWC and VOP maps in Appendix A. Because VWC and VOP were derived from NDVI climatology, there would be a correlation between all three products; even the selected NDVI reference was derived from actual measurements. Some differences were expected because VWC estimations considered foliage and stem components adjusted for land cover types using the MODIS International Geosphere Biosphere Programme (IGBP) classification scheme [39], and therefore, the analysis of the VWC could add information that the NDVI dataset was missing. Equally, since VOP was computed from VWC estimations, the two products would be correlated and would provide similar information. Because VOP was computed as the VWC multiplied by a b-factor and this b-factor was dependent on the type of vegetation and the measurement frequency, the VOP dataset may bring relevant information that both NDVI and VWC were missing. Figure 10a,b respectively show the mean and standard deviation of the VWC and VOP datasets in a monthly basis for areas represented by crop type.

As can be seen from Figure 10 and as expected, both VWC and VOP were very similar products that showed a high dependency on NDVI (Figure 9). It is important to note that a comparison against VWC and VOP based on real measurements, rather than climatology, would have been more accurate. By using a climatological dataset, we would observe larger errors due to the real variability of the fields not captured by the selected reference datasets. In [40], the authors compared SMAP VOP (based on climatology) and SMOS VOP (based on real data) in the South Fork Network in Iowa day-to-day for three seasons (in 2015, 2016, and 2017), and the general trends of both products were very similar. This work intends to demonstrate that SMAP-R contains information related to crops and does not support a geophysical model function derived from the selected VOP dataset. The average month-to-month VOP changes were well represented by the selected reference VOP and therefore could be used to determine SMAP-R sensitivity to crop growth. In the next sections, we will analyze the dependence of SMAP-R signals on NDVI, VWC, and VOP.

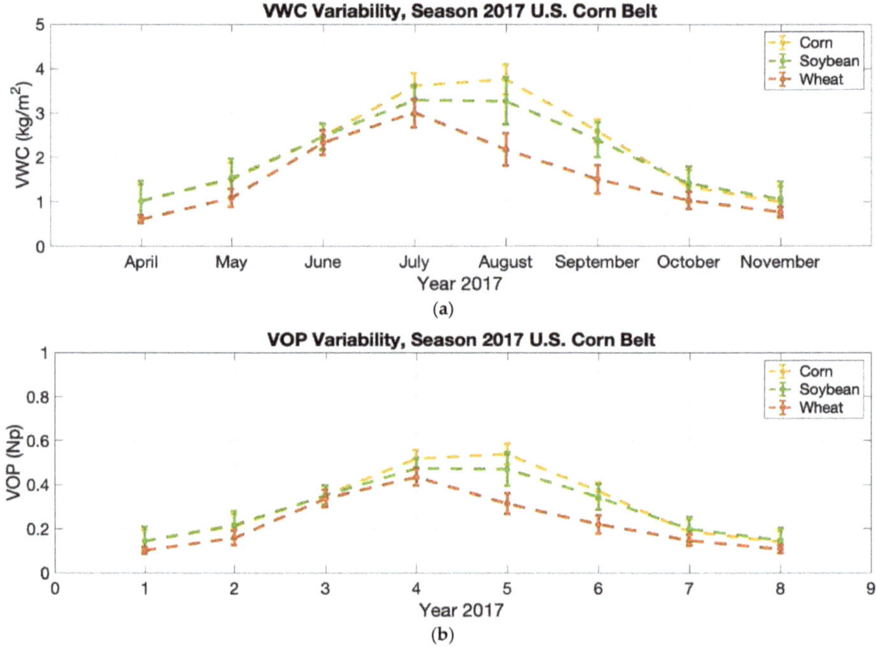

Figure 10. Monthly crop condition information for the 2017 season in the U.S. Corn Belt: (**a**) SMAP ancillary product for VWC and (**b**) SMAP ancillary product for VOP.

Each crop had a different associated growth rate. The crop height was therefore a function of the crop type, which was represented by a particular growth speed and final height. In order to understand if the thickness of the vegetation layer affected the SMAP-R signals, it was important to add the crop height to our analysis. We were not able to find a monthly product of the vegetation height over the U.S. Corn Belt during 2017, and for this reason, we developed an algorithm to estimate the crop height evolution. The generated product may not be accurate at a daily rate, but it proved sufficient information for monthly mean analysis. Characterizing agricultural fields is a difficult task since there is a lack of seasonal ground-truth, such as yearly updated information on the specific variety of crops of each field or the planting date. In addition, there are also many aspects that can be different not only year-to-year, but also field-to-field, and even plant-to-plant. We made assumptions on the height of the plants based on a few field studies, but we recognized these assumptions may not always be robust. The algorithm was mainly based on standard knowledge for the growth stages of a specific crop type and the initial time at which the crops started to grow. The initial growth time was obtained from the NDVI dataset. For the crop growth (height vs. days from emergence), we employed information available from previous studies and various resources ([41–43]). The values provided from those studies were used as the representative means of the fields for each crop type. Although not accurate to specific plants, the overall growth was expected to be robust enough for testing the monthly sensitivity of SMAP-R to crop height. Since there were only three types of crops within our study area, we limited the information to corn, soybean, and wheat. The information is shown in Tables 1–3.

Table 1. Corn height based on growth stage. Based on the growth observed reported in [41] for the ground-truth of corn field. Vx = Vegetative stages, Rx = Reproductive stages.

Growth Stage	Height (cm)	Days
Emergence (VE)	0	0
1 leaves (V1)	5	7
5 leaves (V5)	14	14
8 leaves (V8)	48	28
12 leaves (V12)	89	42
16 leaves (V16)	125	56
18/21 leaves (V18/V21) Tassel and silk (VT)	250	66
Reproductive stages (R1-R5)	260–275	70–120
Maturity (R6)	275	126

Table 2. Wheat height based on growth stage. Based on the growth observed reported in [42] for the ground-truth of wheat field.

Growth Stage	Height (cm)	Days
Emergence (1)	0	0
Tillering (2)	3	7
Tillers formed (3)	6	25
Tillers erect (4)	9	50
Tillers strong (5)	12	60
First node (6)	16	70
Second node (7)	21	75
Last leaf (8)	27	85
Ligule visible (9)	35	90
Boot, head swollen (10)	44	98
Heading (10.1)	47	105
Flowering (10.5)	51	110
Ripening (11)	55	125

Table 3. Soybean height based on growth stage. From VE to V5, soybean moves to a new V stage every 6 days and grows about 1.9 cm/day. From R1 to R6, soybean moves to a new stage every 4 days and grows about 4 cm per day [43]. Vx = Vegetative stages, Rx = Reproductive stages.

Growth Stage	Height (cm)	Days
Emergence (VE)	0	0
Unrolled unifoliate (VC)	3	7
1 trifoliate developed (V1)	8.65	13
2 trifoliate developed (V2)	18.18	19
3 trifoliate developed (V3)	27.71	25
4 trifoliate developed (V4)	37.23	31
5/6 trifoliate developed (V5/V6)	46.76	37
Flowering (R1-R2) Pod development (R3-R4) Seed filling (R5-R6)	53.11–129	41–57
Maturity (R7-R8)	131	61–100

The methodology used to generate crop height information is shown in Figure 11.

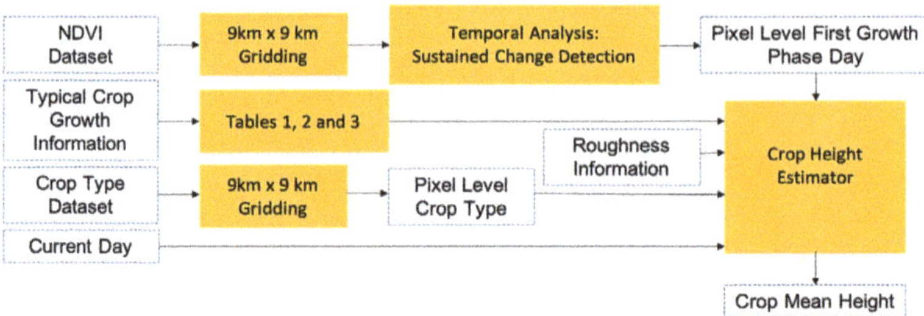

Figure 11. Block diagram of the methodology implemented to estimate crop height monthly means.

The information in Table 1 to Table 3 was used in combination with crop type information in Figure 7b to identify which area of the surface corresponded to each crop type. Using NDVI change information at a weekly rate, we determined the first phase of vegetation presence for each 9 × 9 km pixel and assigned an estimated initial growth day. Following the methodology in Figure 11, we used the 9 × 9 km grid initial day map, the crop type information, and the information in Table 1 to Table 3 to create the crop height maps showing the estimated mean height of the plants at a monthly rate. Those are included in Appendix B. Figure 12 shows the mean and standard deviation of the crop height on a monthly basis for areas represented by each crop type.

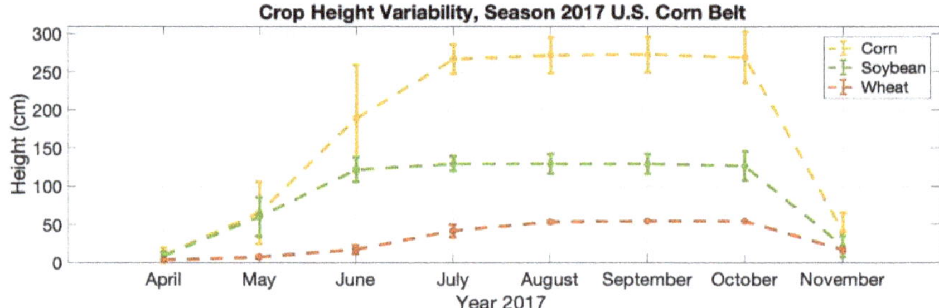

Figure 12. Crop height information for the 2017 season in the U.S. Corn Belt. Note that the monthly mean of the estimated crop height is already filtered using roughness information to the area of low variability, low roughness area shown in Figure 4b.

The resulting crop mean height information represents an estimated height of the predominant crops within each grid cell every month, from April to November 2017. Corn presented in general a higher standard deviation. The higher standard deviation occurred in June and could be explained through the height variability of corn plants growing from 50 cm to 250 cm in a matter of two months. This period of high variability was directly linked to the initial day of growth at the pixel level found through NDVI measurements using the methodology in Figure 11. Differences in the start day would cause differences in growth, and because the plants grew ~150 cm in a short period of time, this translated into dispersion. As mentioned at the beginning of this subsection, the validation of the height maps was not feasible since we were not able to find ground-truth data for the U.S. Corn Belt during the 2017 growing season. Nevertheless, since we were doing monthly averages of the typical crop height values at 9 km × 9 km, we believe the approach was accurate for our purposes and that the crop growth observed for all plants was reasonable. We wanted to observe if there was an overall correlation between monthly averages of crop height and monthly averages of peak SNR; even though there was a bias on the crop height or an incorrect final height assumption, correlation should still show the sensitivity of SMAP-R signals to crop height. Since there was no validation of the height maps, absolute relationships between the two variables would not be possible.

4. Sensitivity Analysis

This section shows the results obtained following the methodology outlined in Figure 3. A sensitivity analysis was performed using the SMAP-R gridded data, the roughness information, soil moisture information, crop type information, and the reference datasets VWC, VOP, NDVI, and crop height, following Steps 1, 2, and 3, as described in Figure 3. As a first step, we applied the roughness mask (Figure 7b) to the PR observable, and we binned the data using the soil moisture information for each crop type. Therefore, the resulting roughness-filtered data were binned into eight SM ranges, from 0 to 40% in steps of 5%, and separated into different crops: corn, soybean, and wheat (the only crop types present within our roughness mask). After grouping the data, we selected SM ranges with enough samples to be statistically representative. We then analyzed the monthly variations of the different observables and correlated those variations to crop growth parameters: VWC, VOP, NDVI, and crop height.

4.1. PR Sample Distribution

First, the PR sample distribution for five different SM ranges and for each crop type is shown in Figure 13.

As was shown in Figure 5, there was only a small area of wheat on the north-east part of the area under study. Figure 13 proves that the crop type wheat was not well represented for any of the SM ranges. We excluded wheat when analyzing the data, which stayed below five samples for all PR bins and all SM ranges. Corn and soybean had a good representation of samples for SM ranges [15–20]%, [20–25]%, and [25–30]%. Ranges [10–15]% and [30–35]% showed a reduced number of samples, and the results would show a higher uncertainty over those ranges. There were no samples below SM = 10% or over SM = 35%. Consequently, we performed the sensitivity analysis of the SMAP-R PR observable to corn and soybean crop types within the SM ranges [10–15]%, [15–20]%, [20–25]%, [25–30]%, and [30–35]%.

4.2. Variability of the PR during the Crop Season

In order to analyze the PR observable on a month-to-month basis, we performed a pre-calculation of the minimum and maximum values expected for bare soil (reference state RS1) and high peak season conditions (reference state RS2), using a two-month average. We observed the PR mean transitioning from RS1 state to RS2 state and the as it dries and gets harvested transitions back to RS1 state. We then analyzed the monthly mean and standard deviation of the PR observable through the season, discarding values that were showed to be outliers based on RS1 and RS2 (~3% of the data). Figure 14 shows the temporal analysis of PR, for two of the SM ranges.

Figure 14 shows that PR showed a clear signature that decreased as crops grew and returned to initial levels after harvesting (October-November). For each plot in Figure 14, SM variation was less than or equal to 5%, roughness variation was minimal, and crop type is shown with different colors. Note that the two selected SM ranges showed similar seasonal behavior, but the PR dynamic range of the higher SM range was reduced. Figure 14a shows a PR drop for corn crop type. This drop could be explained by the dynamic roughness effects since corn is usually tilled post-harvest, causing an increase in roughness. Soybean is occasionally left on the surface with no tilling. An increase in the surface roughness, especially after harvesting, could impact the PR together with SM variations. To investigate this effect, Figure 15 shows the dependence of PR on SM. We selected April and November as representative of bare soil condition and harvested fields, respectively, and June and August as representative of growing season with two levels of growth, intermediate and peak.

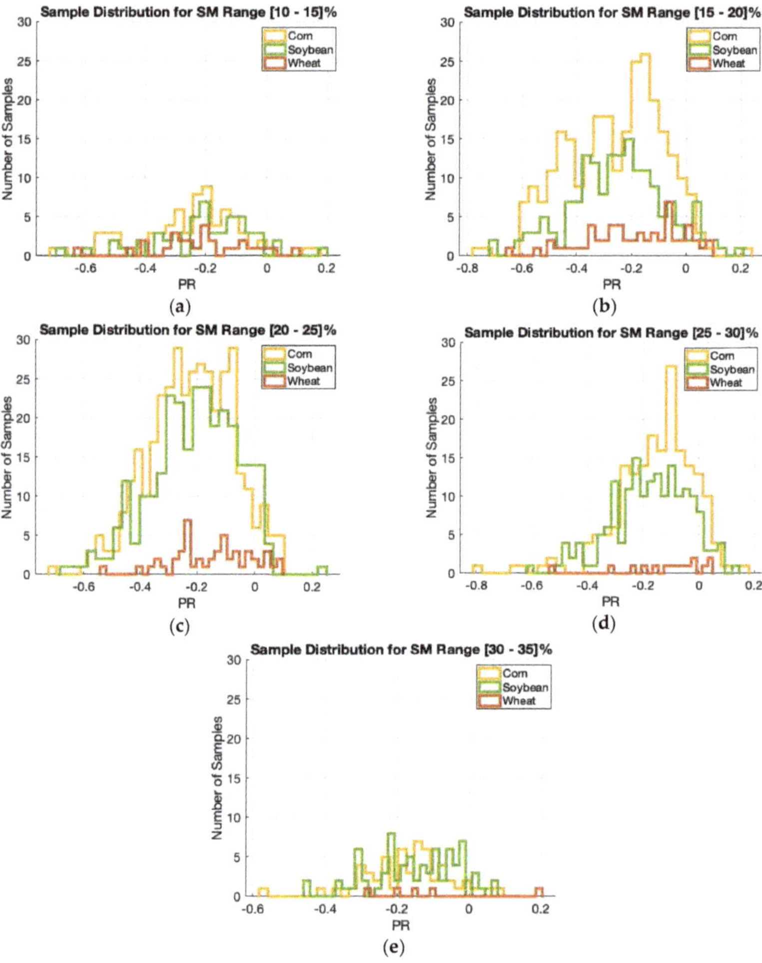

Figure 13. Sample distribution for SM ranges of (**a**) [10–15]%, (**b**) [15–20]%, (**c**) [20–25]%, (**d**) [25–30]%, and (**e**) [30–35]%. Roughness mask applied to ensure low variability of low roughness values.

Under bare soil conditions, as April and November plots, the relationship between the polarimetric ratio and the soil moisture showed a linear behavior. For the November plot (Figure 15d), the surface was not actually bare soil as it was at the beginning of the season. After harvesting, the fields contained residues and stalks that were left on the field. Those left-overs may retain moisture and act as a homogenous layer of higher SM content and low surface roughness. This would explain the higher PR values of November (Figure 15d) with respect to April (Figure 15a). The range of PR under bare soil conditions was contained between −0.05 and −0.25. As the crops grew, the PR dynamic range was increased. Figure 15b shows the mean values for June, during the growing phase with a height variability of [90–260] cm for corn, a height variability of [63–131] cm for soybean, a VWC variability of [2.2–2.5] kg/m2, a VOP variability of [0.3–0.4], and an NDVI variability of [0.55–0.6]. Figure 15c shows the mean values for August, during the growing phase with a height of 274 cm for corn and height of 131 cm for soybean, a VWC variability of [2.8–4] kg/m2, a VOP variability of [0.4–0.6], and an NDVI variability of [0.75–0.9].

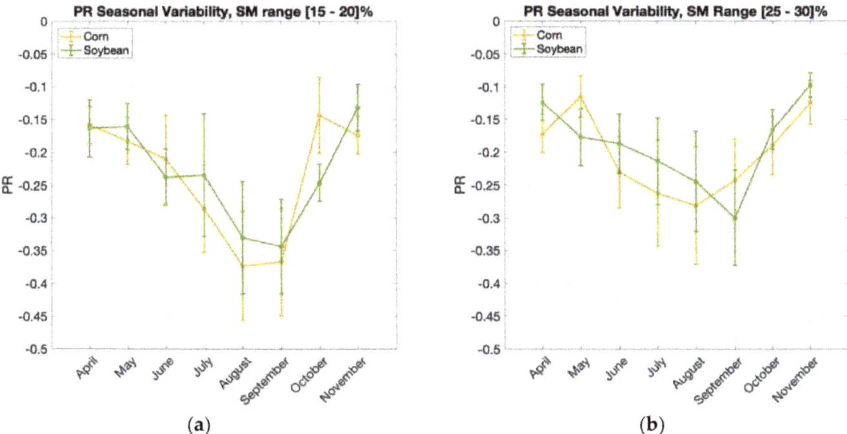

Figure 14. PR seasonal variations (mean and std) for the SM range of (**a**) [15–20]% and (**b**) [25–30]%. Roughness mask applied to ensure the low variability of low roughness values.

Comparing the mean PR values of Figure 15a to Figure 15c, for SM = 10%, PR decreased from −0.25 to −0.3 and −0.45 as the crops grew. For SM = 25%, PR decreased from −0.18 to −0.25 and −0.28 as the crops grew. Finally, for SM = 40%, PR decreased from −0.08 to −0.15 and −0.16 as the crops grew. The PR dependence on both SM and crop growth stage seemed to saturate as SM increased, but it showed sensitivity even under season peak conditions. In summary, Figure 15 showed that both SM and crop growth related parameters had an impact on the PR. Next, we individually analyzed the dependence of PR on VWC, VOP, NDVI, and crop height for the different ranges of SM.

4.3. Sensitivity of the PR to the Crop Growth Parameters

In this subsection, we analyze the PR observable sensitivity to the different crop growth parameters to better understand the effect of those parameters in the SMAP-R signal. PR describes the degree of the de-polarization of the signal and was therefore expected to describe crop growth.

PR and Crop Height Dependency

The PR observable could be affected by the thickness of the vegetation layer, by its water content, or by a combination of both. First, we investigated the direct dependency of the PR observable on crop height alone. Figure 16 shows the mean variability of SMAP-R as the crops increased in height.

Figure 16 shows the PR observable analyzed for corn and soybean data in the range SM = [15–20]%. The PR had an overall linear decreasing behavior with the increase of crop height, but Figure 16 shows information that helped discard the dependency of the PR observable on crop height alone. If it was only dependent on crop height, both crops should display same slope and PR values for the same height. On the contrary, the slopes were different, indicating that the dielectric constant value of the layers had an impact on the PR value. In addition, towards the last stage of growth, we could see that the mean PR increased back to initial levels, even if the crops were at their maximum height. The sudden increase in PR denoted that, as the crops reached their maximum height and started to dry, the drying of the crop plant was the main factor. Corn exhibited higher VWC for most of the season, peaking around the reproductive stages (R2/R3) [44]. Soybean did not exhibit a clear peak of VWC, but it experienced a smooth transition between increasing VWC and the start of VWC decrease in the reproductive stage R1, always below the VWC of corn plants [40]. These differences in VWC along the season could explain the differences in the slopes for the two crops observed in Figure 16. Figure 16 indicates that the main drivers in the PR observable values were likely to be the VWC and/or the VOP of the layer.

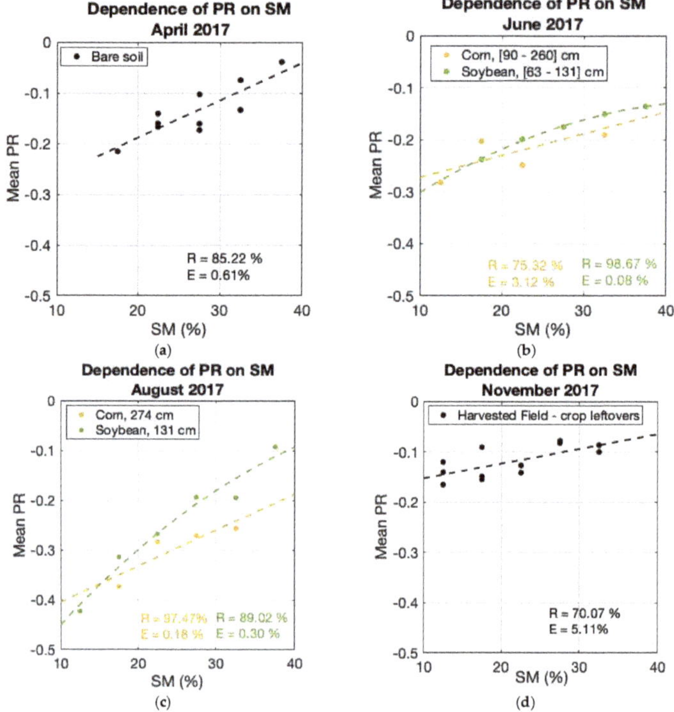

Figure 15. Analysis of the dependence of PR on SM for (**a**) April (mostly bare soil, planting season), (**b**) June (growing stage, variability of heights for the same crop type), (**c**) August (season peak, plants at maximum height), and (**d**) November (after harvesting, the field contains residues and stalks from both corn and soybean left on the field after harvesting).

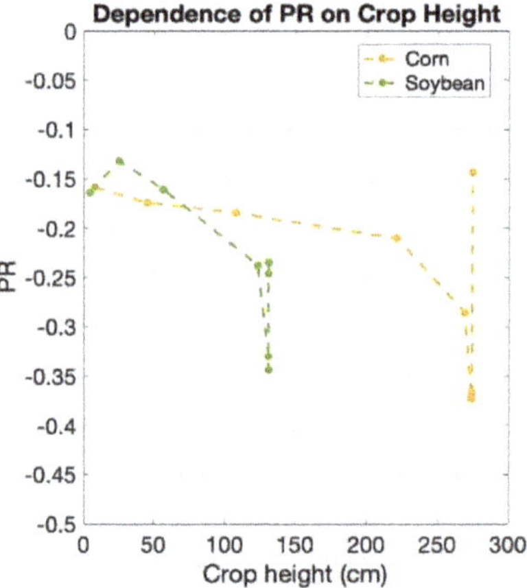

Figure 16. PR dependency on crop height. Crop height information for the 2017 season in the U.S. Corn Belt obtained from the methodology explained through Figure 11. SMAP-R data analyzed contain corn and soybean for the SM range of [15–20]%.

PR and VWC Dependency

Second, we investigated the direct dependency of PR observable on VWC alone. Since VWC was derived from a 10-day NDVI climatology, we would observe errors associated with the discrepancies between climatology and actual values. Figure 17 shows the mean variability of the SMAP-R PR observable as the crops increased in VWC during the first crop growth stages and then decreased in VWC during the last stages of the crop life. The VWC cycle is shown in Figure 10a.

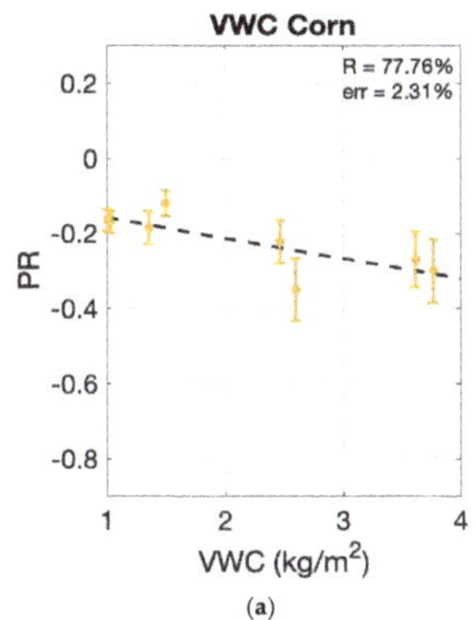

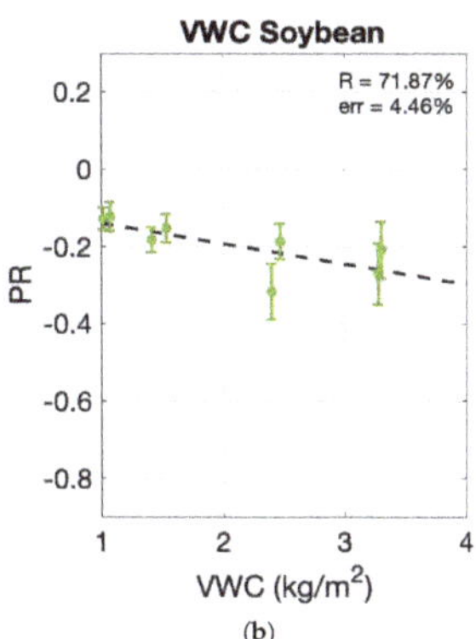

Figure 17. PR dependency on VWC. VWC information for the 2017 season in the U.S. Corn Belt obtained from ancillary products stored in the SMAP SM official product [3]. SMAP-R data analyzed for (**a**) corn and (**b**) soybean.

Figure 17 shows a linear decrease of the PR observable as the VWC increased. The range of VWC for the data observed was between 1 kg/m^2 and 4 kg/m^2, typical of the crops analyzed. We observed a consistent sensitivity for both crop types, with a PR variation of −0.0546 per kg/m^2 of VWC for the corn data and a PR variation of −0.0527 per kg/m^2 of VWC for the soybean data. Figure 17 includes all SM measurements, and as is shown in Figure 15, SM had a relevant impact on the dynamic range of the PR and therefore was impacting the standard deviation of the average PR values in Figure 17. The estimated uncertainty for the corn data was 30% in the case of corn and 29.28% in the case of soybean. In order to reduce the uncertainty due to SM, we followed the approach in Figure 3, binning the data into SM bins of 5%. Table 4 provides the PR sensitivity to changes in VWC and the uncertainty if we were to use a linear approach to estimate VWC from PR. We provide results for each SM range for each crop type.

Table 4. PR sensitivity to VWC, with the uncertainty of the estimations for corn and soybean crop type. The green dashed box highlights the SM ranges with a statistically significant number of samples. R is the correlation [45]; E is the probability of R being random [45]; S is the sensitivity; and U is the uncertainty. The VWC maximum range is [0–4] kg/m^2.

Soil Moisture Range	Corn				Soybean			
	R (%)	E (%)	S (kg/m²)$^{-1}$	U (%)	R (%)	E (%)	S (kg/m²)$^{-1}$	U (%)
[10–15] %	54.33	34.40	-0.0748	18.3	49.20	39.9	-0.0756	21.3
[15–20] %	62.35	9.86	-0.0541	9.93	74.05	3.56	-0.0505	11
[20–25] %	86.68	0.53	-0.0469	10.14	75.25	3.12	-0.0488	8.75
[25–30] %	65.09	8.04	-0.0531	9.5	55.49	15.34	-0.0483	12.54
[30–35] %	58.15	15.22	-0.0574	12.32	61.07	12.34	-0.0468	10.84

Note that ranges with low sampling had worse results. The green dashed box highlights the most statistically significant results. As shown in Table 4, by binning the data, the VWC uncertainty was generally reduced from ~30% to ~9%. The sensitivity S was obtained as the slope of the linear approximation; the black dashed line in Figure 17a,b. The uncertainty U was computed as the mean standard deviation in a retrieval based on the linear approximation, divided by the maximum range of VWC.

PR and VOP Dependency

Third, we investigated the direct dependency of the PR observable to VOP. As in the case of VWC, since VOP was derived from a 10 day NDVI climatology, we would observe errors associated with the discrepancies between climatology and actual values. Similar to the VWC analysis, Figure 18 shows the mean variability of SMAP-R PR observable as the crops increased in VOP during the first crop growth stages and then decreased in VOP during the last stages of the crop life. The VWC cycle is shown in Figure 10b.

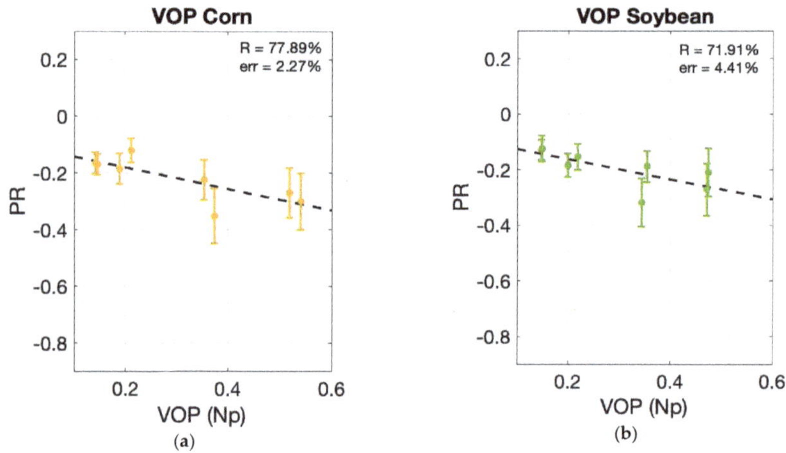

Figure 18. PR dependency on VOP. VOP information for the 2017 season in the U.S. Corn Belt obtained from ancillary products within the SMAP SM official product [3]. SMAP-R data analyzed for (**a**) corn and (**b**) soybean.

Figure 18 shows a linear decrease of the PR observable as the VOP increased. The range of VOP for the data observed was between 0.1 and 0.6 Nep.er (Np) We observed a consistent sensitivity for both crop types, with a PR variation of −0.3771 per unit of VOP for the corn data and a PR variation of −0.3630 per unit of VOP for the soybean data. Figure 18 includes all SM measurements, and as was shown in Figure 15, SM had a relevant impact on the dynamic range of the PR and therefore was impacting the standard deviation of the averaged PR values in Figure 17. The estimated uncertainty for the corn data was 17.44% in the case of corn and 17.01% in the case of soybean. As we did with the VWC, to reduce the uncertainty due to SM, we followed the approach in Figure 3, binning the data into SM bins of 5%. Table 5 provides the PR sensitivity to changes in VOP and the uncertainty if we were to use a linear approach to estimate VOP from PR. We provide results for each SM range for each crop type.

The same as Table 4, ranges with low sampling had worse results, and the green dashed box highlights the most statistically significant results. As shown in Table 5, by binning the data, the VOP uncertainty was generally reduced from ~17% to ~6%. The sensitivity S was obtained as the slope of the linear approximation; the black dashed line in Figure 18a,b. The uncertainty U was computed as the mean standard deviation in a retrieval based on the linear approximation, divided by the maximum range of VOP.

Table 5. PR sensitivity to VOP, with the uncertainty of the estimations for corn and soybean crop type. The green dashed box highlights the SM ranges with a statistically significant number of samples R is the correlation [45]; E is the probability of R being random [45]; S is the sensitivity; and U is the uncertainty. The VOP maximum range is [0 to 1] Np.

Soil Moisture Range	Corn				Soybean			
	R (%)	E (%)	S (Np)⁻¹	U (%)	R (%)	E (%)	S (Np)⁻¹	U (%)
[10–15] %	53.93	34.82	−0.5161	10.26	48.05	41.26	−0.4952	9.44
[15–20] %	62.95	9.44	−0.3808	5.65	73.67	3.71	−0.3533	6.36
[20–25] %	86.84	0.52	−0.3239	5.87	75.30	3.11	−0.3397	5.03
[25–30] %	62.87	9.5	−0.3477	6.80	57.69	13.43	−0.3431	7.06
[30–35] %	52.75	17.16	−0.3830	8.48	57.07	13.87	−0.3173	14.12

PR and NDVI Dependency

Finally, we investigated the direct dependency of PR observable on NDVI. In this case, the NDVI reference product was obtained from actual weekly MODIS data rather than climatology. Similar to the VWC and VOP analysis, Figure 19 shows the mean variability of the SMAP-R PR observable as the crops increased in NDVI during the first crop growth stages and then decreased in NDVI during the last stages of the crop life. The NDVI cycle is shown in Figure 9.

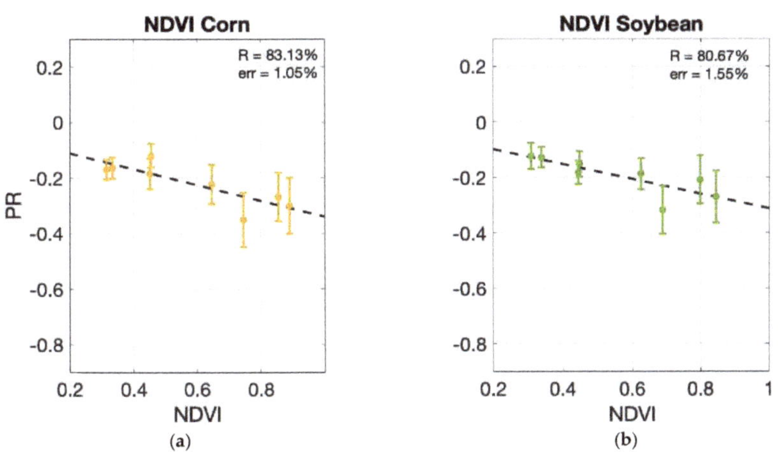

Figure 19. PR dependency on NDVI. NDVI information for the 2017 season in the U.S. Corn Belt obtained from [34]. SMAP-R data analyzed for (**a**) corn and (**b**) soybean.

Figure 19 shows a linear decrease of the PR observable as the NDVI increased. The range of NDVI for the data observed was between 0.2 and 0.9. We observed a consistent sensitivity for both crop types, with a PR variation of −0.2840 per unit of NDVI for the corn data and a PR variation of −0.2668 per unit of VOP for the soybean data. Figure 18 includes all SM measurements, and as was shown in Figure 15, SM had a relevant impact on the dynamic range of the PR and therefore was impacting the standard deviation of the averaged PR values in Figure 17. The estimated uncertainty for the corn data was 0.2315 in the case of corn and 0.2313 in the case of soybean. As we did with the VWC and VOP, in order to reduce the uncertainty due to SM, we followed the approach in Figure 3, binning the data into SM bins of 5%. Table 6 provides the PR sensitivity to changes in NDVI and the uncertainty if we were to use a linear approach to estimate NDVI from PR. We provide results for each SM range for each crop type.

Table 6. PR sensitivity and NDVI uncertainty of the estimations for corn and soybean crop type. The green dashed box highlights the SM ranges with a statistically significant number of samples. R is the correlation [45]; E is the probability of R being random [45]; S is the sensitivity; and U is the uncertainty. The NDVI maximum range is [0 to 1].

Soil Moisture Range	Corn				Soybean			
	R (%)	E (%)	S	U (%)	R (%)	E (%)	S	U (%)
[10–15] %	51.78	17.15	−0.3757	16.6	60.60	10.42	−0.3863	15.19
[15–20] %	64.69	8.3	−0.2836	7.58	81.15	1.45	−0.2600	8.65
[20–25] %	89.20	0.29	−0.2358	8.06	81.93	1.28	−0.2484	6.88
[25–30] %	55.92	14.96	−0.2154	9.36	65.99	7.5	−0.2661	9.10
[30–35] %	51.86	17.94	−0.2453	13.24	55.49	15.32	−0.2091	10.85

Similar to Tables 4 and 5, ranges with low sampling had worse results, and the green dashed box highlights the most statistically significant results. As shown in Table 6, by binning the data, the NDVI uncertainty was generally reduced from ~23% to ~8.5%. The sensitivity S was obtained as the slope of the linear approximation; the black dashed line in Figure 19a,b. The uncertainty U was computed as the mean standard deviation in a retrieval based on the linear approximation, divided by the maximum range of NDVI.

Conclusions on Sensitivity to Crop Growth Parameters

The assessment of the PR observable sensitivity indicated that all VWC, VOP, and NDVI datasets had a similar impact. This result was expected since VWC and VOP were both derived from a 10-day NDVI climatology, which was expected to be correlated with NDVI data. Even though the correlation between them was high, the analysis showed small sensitivity differences; VOP exhibited the lowest uncertainty for all crop types. The PR observable and VOP showed a linear dependency that produced a 6% uncertainty if we were to apply the linear relationship to estimate the VOP. Our analysis indicated that SMAP-R data could be used as a proxy to obtain independent VOP estimations. Overall, improved VOP estimates could produce better SM estimations. Uncertainty related to VWC was ~9%, i.e., 0.36 kg/m^2.

5. Discussion

Given our need to use CYGNSS data for calibration purposes, there was currently a total of 2.5 years of SMAP-R data overlapping with CYGNSS data that could be used to develop a GNSSR-based SM/VWC/VOP retrieval algorithm. The explored dependency of the SMAP-R PR observable on SM and vegetation descriptors (crop height, VWC, VOP, and NDVI) supported the potential of polarimetric GNSS-R signals to be combined with radiometric measurements to produce more accurate SM and VWC/VOP estimates. In order to develop a GNSS-R-based retrieval algorithm, it was important to have multi-year measurements over a controlled area where there was availability of in situ and independent information regarding SM and vegetation parameters. Furthermore, it was important that the variability of the SM and the variability of the different vegetation parameters provided statistically representative population of samples for all their values within their possible ranges.

SMAP calibration and validation (Cal/Val) activities were intended to verify and improve the performance of the science algorithms, as well as validate the accuracies of the science data products. The Cal/Val sites, given their overall range of vegetation and seasonal variability, would be an ideal target to develop a GNSSR-based retrieval algorithm. The algorithm would use both SMAP official product SM retrievals (or SMAP radiometric data directly) and GNSS-R polarimetric measurements to reduce the error on the estimations of both SM and vegetation related parameters. Since both the radiometer and the GNSS-R receiver were sensitive to both the SM and vegetation parameters, we would be able to resolve the two variables through a system of equations that was no longer ill posed. Furthermore, having in situ data of VWC/VOP would help us build an algorithm that did not rely on NDVI measurements, which would benefit the retrieval algorithm since NDVI saturated, losing sensitivity to VWC as the water content increased [7]. In addition, using an alternate reference product that relied on real measurements rather than climatology would allow for the development of a geophysical model function that would link polarimetric GNSS-R measurements to VWC or VOP estimates. If GNSS-R measurements could be used as a source of independent information to produce better estimates, both SMAP SM and their ancillary VWC/VOP final products would result in more accurate estimates.

The true spatial resolution of SMAP-R signals was not a fixed number. The spatial resolution varied measurement-to-measurement given the surface characteristics. Surfaces such as rivers, lakes, wetlands, and sea ice produce a highly coherent scattering of the GPS signals reflecting from them. Surfaces such as the ocean or forests produce highly incoherent scattering. In our study, the scattered GPS signals that SMAP-R observed were reflected off crop fields. Therefore, the spatial resolution, which was calculated here based on the methodology developed in [15], varied as the crops grew. The mean scattering area for the bare soil conditions was ~12 km. For the same area, post-growth, the observed area increased to a mean of ~17 km. In our study, we implemented a drop in the box approach, where all specular points within each grid cell were averaged together, and future research will include the true spatial resolution information in the analysis.

Future research will utilize SMAP Cal/Val sites for in situ measurements of the relevant parameters (SM and vegetation parameters). Additionally, we plan on selecting locations strategically, ensuring we have a dataset spanning many years, with large geophysical variability. Given a large, diverse dataset, it will be possible to develop an algorithm based on an ensemble of data that are statistically representative. In addition, having validated crop height measurements facilitates a methodology to link the size of the scattering area derived from SMAP-R to the incoherency related to the height and type of the vegetation. Furthermore, future research will include the practical implementation of a retrieval algorithm based on GNSS-R measurements combined with other sensors and ancillary information, following the results in this paper.

6. Conclusions

This manuscript analyzed the polarimetric sensitivity of SMAP-R signals to crop growth descriptors under different soil moisture conditions. The calibration methods described in [15] were applied, achieving a calibrated dataset over the U.S. Corn Belt, where the sensitivity analysis of the SMAP-R signals was performed. The sensitivity analysis performed in Section 4.2, considering SMAP-R PR observable dependency on SM and Section 4.3 considering SMAP-R PR observable dependency on crop growth parameters, can be summarized as follows:

- The SMAP-R PR observable showed a dependency on SM, regardless of the crop growth stage (Figure 15).

  - For bare soil conditions, we observed a linear behavior of the SMAP-R PR observable (Figure 15a,d).
  - The different crop growth stages had an impact on the dynamic range of the SMAP-R PR observable.

    - As crops grew, the SMAP-R PR observable showed a logarithmic behavior, with more sensitivity to crop growth parameters at low SM values (Figure 15b,c, Table 4 to Table 6).

- During the peak season (maximum height, VWC, VOP, and NDVI), we observed the maximum dynamic range (Figure 15c).

- The SMAP-R PR observable showed a degree of correlation with crop height, but it was not the main driver:

  - As the crops reached their maximum height and started to dry, the crop height did not have any impact.
  - Different crops did not display the same slope and PR values for the same crop height.

- The SMAP-R PR observable showed a linear dependency on crop growth parameters:

  - PR decayed at a mean rate of -0.054 per kg/m^2 of VWC. Applying an SM binning of the data, a 9% uncertainty was obtained (Section 4.3), i.e., 0.36 kg/m^2.
  - PR decayed at a mean rate of -0.37 per unit of VOP. Applying an SM binning of the data, a 6% uncertainty was obtained (Section 4.3).
  - PR decayed at a mean rate of -0.23 per unit of NDVI. Applying an SM binning of the data, an 8.5% uncertainty was obtained (Section 4.3).

As previously discussed, this work has implications for SM retrieval algorithm work in the future. GNSS-R polarimetric measurements could be used synergistically with passive radiometric observations for improved estimates of soil moisture under dynamic vegetation. Bistatic radar measurements from the GPS constellation complement those from radiometers by providing an independent source of vegetation and soil moisture information. Both SMAP and SMAP-R measurements are dependent on crop growth stage and SM. Therefore, a combined radiometer-reflection-based retrieval will constrain and improve estimates of vegetation parameters and SM.

Author Contributions: Conceptualization, N.R.-A., S.M., and M.M..; methodology, N.R.-A.; software, N.R.-A.; validation, N.R.-A; formal analysis, N.R.-A.; investigation, N.R.-A.; visualization, N.R.-A.; writing, original draft preparation, N.R.-A.; writing, review and editing, N.R.-A., S.M., and M.M.; funding acquisition, S.M. All authors read and agreed to the published version of the manuscript.

Funding: This research was carried out at the Jet Propulsion Laboratory, California Institute of Technology, under a contract with the National Aeronautics and Space Administration. In particular, this research was supported by the R&A Hydrology & Weather from the Soil Moisture Active Passive (SMAP) project. © 2020. All rights reserved. California Institute of Technology. Government sponsorship acknowledged.

Acknowledgments: Authors would like to thank Stephen Lowe and Stephan Esterhuizen, from the Jet Propulsion Laboratory, for the earlier developments of the SMAP delay-Doppler map processor.

Conflicts of Interest: The authors declare no conflict of interest.

Appendix A

This Appendix shows the monthly maps for the different products used within our study: SM, NDVI, VWC, and VOP. Note that NDVI was obtained from actual 2017 MODIS data as opposed to VWC and VOP datasets, which were derived from 10 day NDVI climatology. These products are provided in Equal-Area Scalable Earth (EASE) grid.

SM Maps

Generated maps of the SMAP Enhanced L3 Radiometer Global Daily 9 km EASE-Grid Soil Moisture Version 2 product [3] for the area under study during the 2017 season.

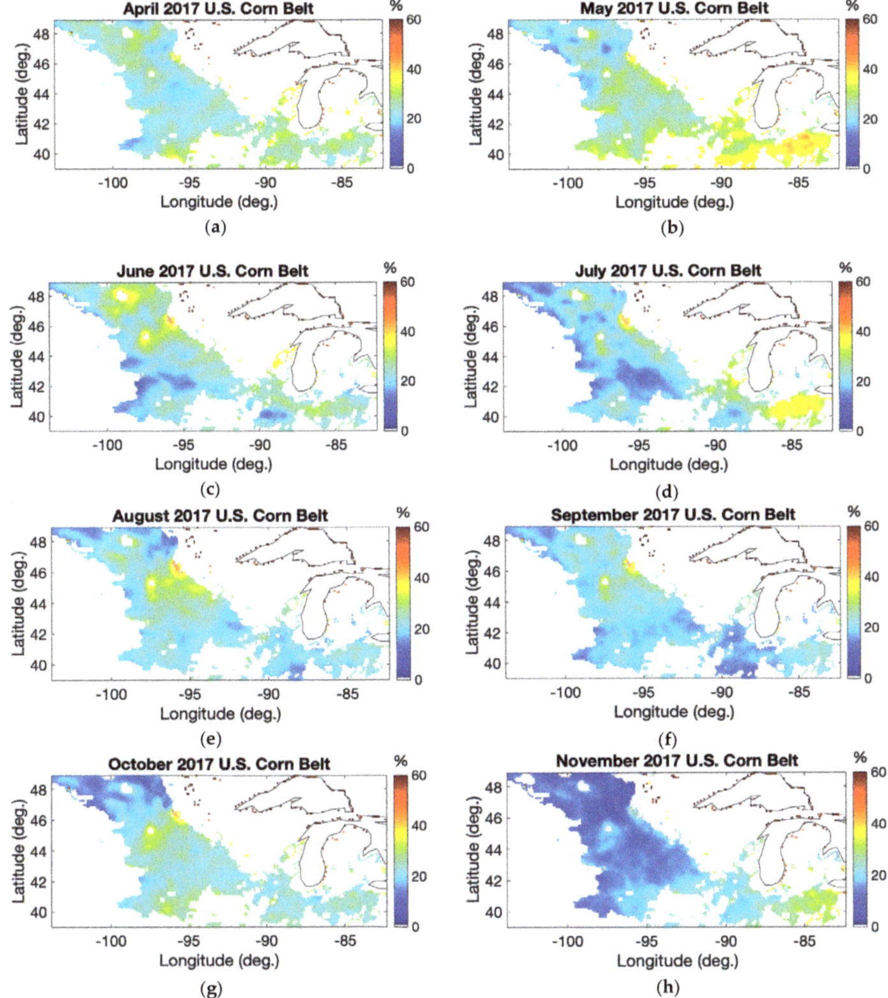

Figure A1. Soil moisture maps from the SMAP Enhanced L3 Radiometer Global Daily 9 km EASE-Grid Soil Moisture Version 2 [3]: (**a**) to (**h**) corresponds to April to November 2017.

NDVI Maps

The USDA NASS VegScape [38]–Vegetation Condition Explorer offers information on the NDVI at a bi-weekly rate. Data are freely available at https://nassgeodata.gmu.edu/VegScape/. The crop NDVI information corresponds to the 2017 production and was gridded to match the 9 km × 9 km SMAP spatial resolution, assigning the monthly mean NDVI value to each grid cell for a period. Figure 9 shows the NDVI maps used in this study [38].

The Surface Reflectance Daily L2G Global 250m of MODIS is selected as the input for generating normalized difference vegetation index (NDVI). The L2G product has two bands–Band1 (620–670 nm) and Band2 (841–876 nm) that represent respectively red and near-infrared band. Therefore, the equation for computing daily NDVI is as follows:

$$\text{NDVI} = \frac{\text{Band2} - \text{Band1}}{\text{Band2} + \text{Band1}} \tag{A1}$$

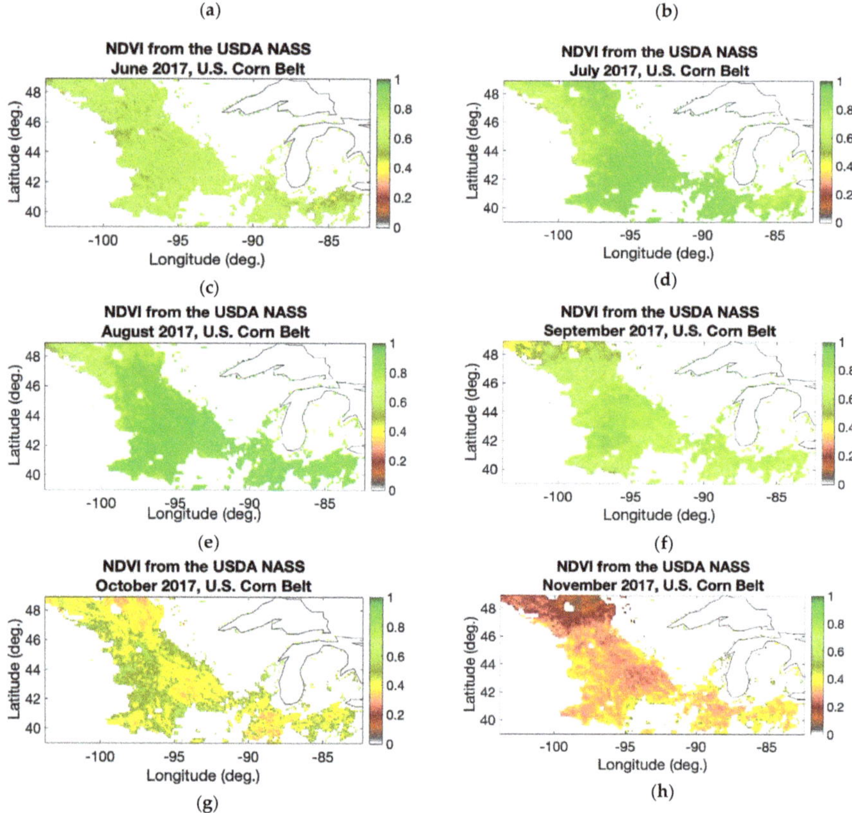

Figure A2. NDVI map developed from the USDA NASS VegScape–Vegetation Condition Explorer database showing the monthly mean NDVI values gridded to 9 km × 9 km, matching SMAP official product spatial resolution, from April to November 2017, respectively from (**a**) to (**h**) [38]. Note: These data were obtained from NDVI from actual MODIS data.

VWC Maps

The SMAP mission employed the Terra/MODIS Vegetation Indices (MOD13Q1) [46] Version 6 product and the stem factor values for different MODIS International Geosphere Biosphere Programme (IGBP) land cover types to obtain their VWC ancillary dataset [37]. MOD13Q1 is provided every 16 days at 250 m spatial resolution, and it is available from 2000-02-18 to the present. The equation applied is:

$$VWC = \left(1.9134 \times NDVI^2 - 0.3215 \times NDVI\right) + sf \times \frac{NDVI_{max} - NDVI_{min}}{1 - NDVI_{min}} \tag{A2}$$

where $NDVI$ is the Normalized Difference Vegetation Index, $NDVI_{max}$ is the maximum annual value, $NDVI_{min}$ is the minimum annual value, and sf is the steam factor whose values for different land cover type were shown in [37].

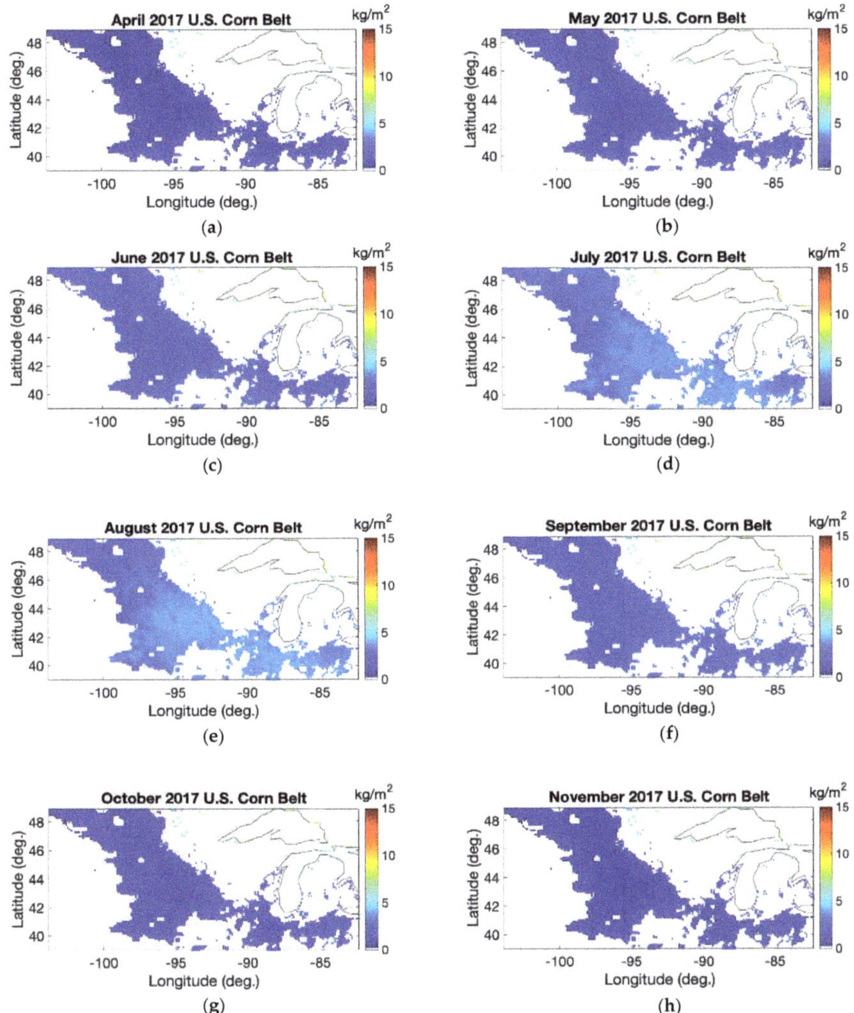

Figure A3. VWC maps from the SMAP ancillary products: (**a**) to (**h**) correspond to April to November 2017. Units of kg/m². Note: This product was derived from 10 day NDVI climatology.

VOP Maps

The SMAP mission also provides information on the VOP, and it is available in the SMAP SM official product [3]. The VOP is defined as the VWC multiplied by a parameter that depends on the frequency and the vegetation type [36], in this case, the crop type. We gathered the VOP as a reference dataset to study the sensitivity of the SMAP-R signals to this crop type dependent dataset.

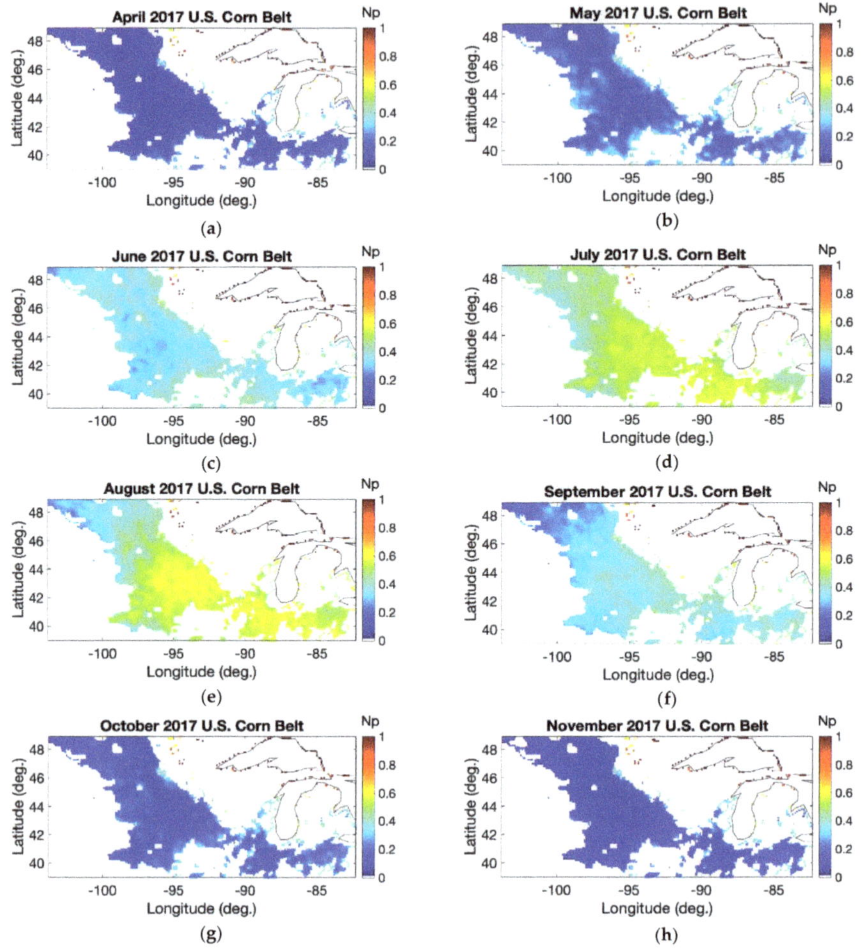

Figure A4. VOP maps from the SMAP ancillary products: (**a**) to (**h**) corresponds to April to November 2017. Units in nepers (Np). Note: This product was derived from 10 day NDVI climatology.

Appendix B

This Appendix shows the monthly maps for crop height maps used within our study. This product was originally from this work.

Crop Height Maps

The algorithm was mainly based on standard knowledge for the growth stages of a specific crop type and the initial time at which the crops started to grow. The initial time was obtained from the NDVI dataset. The methodology is explained through Figure 11.

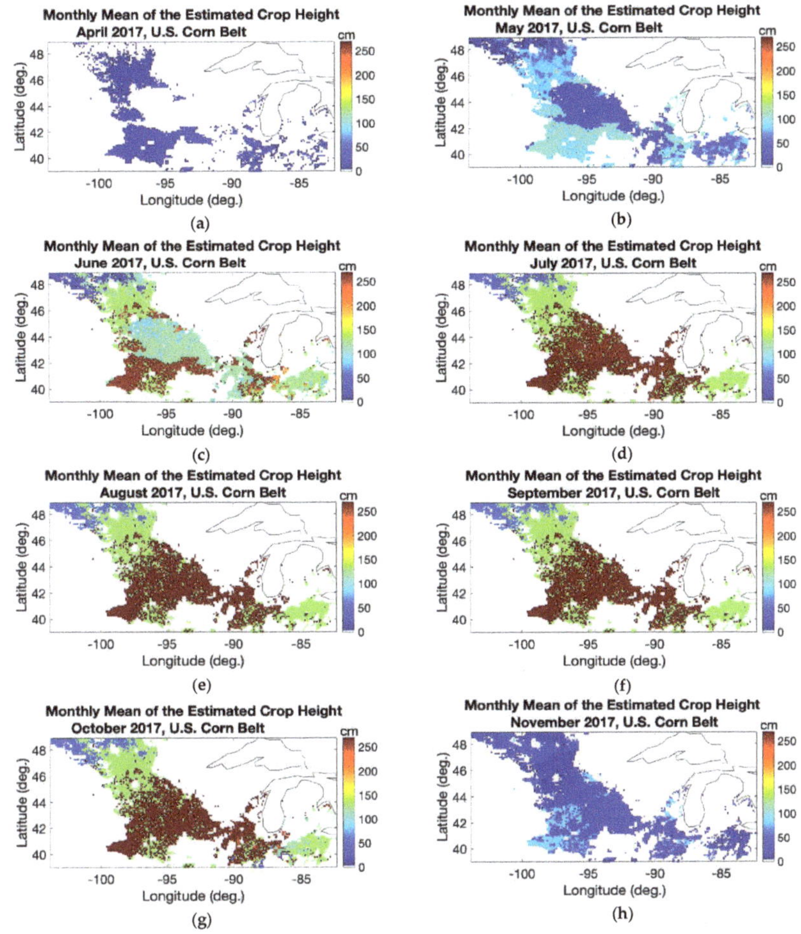

Figure A5. Height map developed from crop type information, NDVI information, and typical growth values for the different crop types: corn, wheat, and soybean from April to November 2017, respectively from (**a**) to (**h**).

References

1. Entekhabi, D. The Soil Moisture Active Passive (SMAP) mission. *Proc. IEEE* **2010**, *98*, 704–716. [CrossRef]
2. Silvestrin, P.M.; Berger, Y.H.; Kerr, J. Font, ESA's Second Earth Explorer Opportunity Mission: The soil Moisture and Ocean salinity Mission—SMOS. *IEEE Geosci. Remote Sens. Newsl.* **2001**, *118*, 11–14.
3. O'Neill, P.E.; Chan, S.; Njoku, E.G.; Jackson, T.; Bindlish, R. SMAP Enhanced L3 Radiometer Global Daily 9 km EASE-Grid Soil Moisture, Version 2. Available online: https://nsidc.org/data/SPL3SMP_E/versions/3 (accessed on 20 March 2020).
4. Centre Aval de Traitement des Données SMOS (CATDS). CATDS-PDC L3SM Aggregated—3-Day, 10-Day and Monthly Global Map of Soil Moisture Values from SMOS Satellite. Available online: https://www.catds.fr/Products/Available-products-from-CPDC/Catalogue/Catds-products-from-Sextant#/metadata/b57e0d3d-e6e4-4615-b2ba-6feb7166e0e6 (accessed on 9 March 2020).
5. Chan, S.K. Development and assessment of the SMAP enhanced passive soil moisture product. *Remote Sens. Environ.* **2018**, *204*, 931–941. [CrossRef]

6. Jackson, T.J.; Schmugge, T.J. Vegetation effects on the microwave emission from soils. *Rem. Sens. Environ.* **1991**, *36*, 203–212. [CrossRef]

7. Salomonson, V.V.; Barnes, W.I.; Maymon, P.W.; Montgomery, H.; Ostrow, H. MODIS—Advanced Facility Instrument For Studies Of The Earth As A System. *IEEE Trans. Geosci. Remote Sens.* **1989**, *27*, 145–153. [CrossRef]

8. Cosh, M.H.; White, W.A.; Colliander, A.; Jackson, T.J.; Prueger, J.H.; Hornbuckle, B.K.; Hunt, E.R.; McNairn, H.; Powers, J.; Walker, V.; et al. Estimating vegetation water content during the soil moisture active passive validation experiment in 2016. *J. Appl. Remote Sens.* **2019**, *13*, 2019. [CrossRef]

9. Jackson, T.J.; le Vine, D.M.; Hsu, A.Y.; Oldak, A.; Starks, P.J.; Swift, C.T.; Isham, J.D.; Haken, M. Soil moisture mapping at regional scales using microwave radiometry: The Southern Great Plains hydrology experiment. *IEEE Trans. Geosci. Remote Sens.* **1999**, *37*, 2136–2151. [CrossRef]

10. Dong, J.; Crow, W.T.; Bindlish, R. The error structure of the SMAP single- and dual-channel soil moisture retrievals. *Geophys. Res. Lett.* **2018**, *45*, 758–765. [CrossRef]

11. Chew, C.; Lowe, S.; Parazoo, N.; Esterhuizen, S.; Oveisgharan, S.; Podest, E.; Zuffada, C.; Freedman, A. SMAP radar receiver measures land surface freeze/thaw state through capture of forward-scattered L-band signals. *Remote Sens. Environ.* **2017**, *198*, 333–344. [CrossRef]

12. Carreno-Luengo, H.; Lowe, S.; Zuffada, C.; Esterhuizen, S.; Oveisgharan, S. Spaceborne GNSS-R from the SMAP Mission: First Assessment of Polarimetric Scatterometry over Land and Cryosphere. *Remote Sens.* **2017**, *9*, 362. [CrossRef]

13. Rodriguez-Alvarez, N.; Podest, E. Characterization of the Land Freeze/Thaw State with SMAP-Reflectometry. In Proceedings of the IGARSS 2019–2019 IEEE International Geoscience and Remote Sensing Symposium, Yokohama, Japan, 28 July–2 August 2019; pp. 4024–4027. [CrossRef]

14. Rodriguez-Alvarez, N.; Misra, S.; Morris, M. Sensitivity Analysis of SMAP-Reflectometry (SMAP-R) Signals to Vegetation Water Content. In Proceedings of the IGARSS 2019–2019 IEEE International Geoscience and Remote Sensing Symposium, Yokohama, Japan, 28 July–2 August 2019; pp. 7395–7398. [CrossRef]

15. Rodriguez-Alvarez, N.; Misra, S.; Podest, E.; Morris, M.; Bosch-Lluis, X. The Use of SMAP-Reflectometry in Science Applications: Calibration and Capabilities. *Remote Sens.* **2019**, *11*, 2442. [CrossRef]

16. Njoku, E.; Li, L. Retrieval of land surface parameters using passive microwave measurements at 6-18 GHz. *IEEE Trans. Geosci. Remote Sens.* **1999**, *37*, 79–93. [CrossRef]

17. Guerriero, L.; Pierdicca, N.; Pulvirenti, L.; Ferrazzoli, P. Use of Satellite Radar Bistatic Measurements for Crop Monitoring: A Simulation Study on Corn Fields. *Remote Sens.* **2013**, *5*, 864–890. [CrossRef]

18. Carreno-Luengo, H.; Amèzaga, A.; Vidal, D.; Olivé, R.; Munoz, J.F.; Camps, A. First Polarimetric GNSS-R Measurements from a Stratospheric Flight over Boreal Forests. *Remote Sens.* **2015**, *7*, 13120–13138. [CrossRef]

19. Motte, E. GLORI: A GNSS-R Dual Polarization Airborne Instrument for Land Surface Monitoring. *Sensors* **2016**, *16*, 732. [CrossRef]

20. NASA Earthdata Website. Available online: https://earthdata.nasa.gov (accessed on 18 March 2020).

21. Zavorotny, V.U.; Voronovich, A.G. Scattering of GPS signals from the ocean with wind remote sensing application. *IEEE Trans. Geosci. Remote Sens.* **2000**, *38*, 951–964. [CrossRef]

22. Voronovich, A.G.; Zavorotny, V.U. Bistatic radar equation for signals of opportunity revisited. *IEEE Trans. Geosci. Remote Sens.* **2018**, *56*, 1959–1968. [CrossRef]

23. Ruf, C.S. The CYGNSS nanosatellite constellation hurricane mission. In Proceedings of the 2012 IEEE International Geoscience and Remote Sensing Symposium, Munich, Germany, 22–27 July 2012; pp. 214–216. [CrossRef]

24. Ruf, C.S. CYGNSS: Enabling the Future of Hurricane Prediction Remote Sensing Satellites. *IEEE Geosci. Remote Sens. Mag.* **2013**, *1*, 52–67. [CrossRef]

25. Ruf, C.S.; Atlas, R.; Chang, P.S.; Clarizia, M.P.; Garrison, J.L.; Gleason, S.; Katzberg, S.J.; Jelenak, Z.; Johnson, J.T.; Majumdar, S.J.; et al. New Ocean Winds Satellite Mission to Probe Hurricanes and Tropical Convection. *Bull. Amer. Meteor. Soc.* **2016**, *97*, 385–395. [CrossRef]

26. Ruf, C.S.; Scherrer, J.; Rose, R.; Provost, D. Algorithm Theoretical Basis Document Level 1B DDM Calibration. Available online: https://clasp-research.engin.umich.edu/missions/cygnss/reference/ATBD%20L1B%20DDM%20Calibration%20R1.pdf (accessed on 20 March 2020).

27. Owe, M.; de Jeu, R.; Walker, J. A methodology for surface soil moisture and vegetation optical depth retrieval using the microwave polarization difference index. *IEEE Trans. Geosci. Remote Sens.* **2001**, *39*, 1643–1654. [CrossRef]

28. U.S. Department of Agriculture National Agricultural Statistics Service Database. Available online: https://www.nass.usda.gov/Charts_and_Maps/Crops_County/cr-pl.php (accessed on 20 March 2020).

29. Neeser, C.; Dille, J.; Krishnan, G.; Mortensen, D.; Rawlinson, J.; Martin, A.; Bills, L. WeedSOFT®: A weed management decision support system. *Weed Sci.* **2004**, *52*, 115–122. [CrossRef]

30. U.S. Department of Agriculture National Agricultural Statistics Service Database CropScape—Cropland Data. Available online: https://nassgeodata.gmu.edu/CropScape/ (accessed on 20 March 2020).

31. Colliander, A. Vegetation & Roughness Parameters. SMAP Report, JPL D-53065.

32. Peng, J.; Mohammed, P.; Chaubell, J.; Chan, S.; Kim, S.; Das, N.; Dunbar, S.R.B.; Xu, X. Soil Moisture Active Passive (SMAP) L1-L3 Ancillary Static Data, Version 1. Available online: https://nsidc.org/data/SMAP_L1_L3_ANC_STATIC/versions/1 (accessed on 20 March 2020).

33. Zobeck, T.M.; Onstad, C.A. Tillage and rainfall effects on random roughness: A review. *Soil Till. Res.* **1987**, *9*, 1–20. [CrossRef]

34. Choudhury, B.J.; Schmugge, T.J.; Chang, A.; Newton, R.W. Effect of surface roughness on the microwave emission from soil. *J. Geophys. Res.* **1979**, *84*, 5699–5706. [CrossRef]

35. Wang, J.R. Passive microwave sensing of soil moisture content: The effects of soil bulk density and surface roughness. *Remote Sens. Environ.* **1983**, *13*, 329–344. [CrossRef]

36. SMAP Algorithm Theoretical Basis Document: L2 & L3 Radiometer Soil Moisture (Passive) Products. Available online: https://nsidc.org/sites/nsidc.org/files/technical-references/L2_SM_P_ATBD_v7_Sep2015-po-en.pdf (accessed on 20 March 2020).

37. Chan, S. Vegetation Water Content. JPL D-53061. Available online: https://www.google.com/url?sa=t&rct=j&q=&esrc=s&source=web&cd=2&ved=2ahUKEwiR-eD2n6noAhVUvJ4KHQ6pB2gQFjABegQIAxAB&url=https%3A%2F%2Fsmap.jpl.nasa.gov%2Fsystem%2Finternal_resources%2Fdetails%2Foriginal%2F289_047_veg_water.pdf&usg=AOvVaw1ccuGfDNPM6-rdRqF2_SSj (accessed on 20 March 2020).

38. U.S. Department of Agriculture National Agricultural Statistics Service database VegScape—Vegetation Condition Explorer. Available online: https://nassgeodata.gmu.edu/VegScape/ (accessed on 20 March 2020).

39. Hunt, E.R., Jr.; Piper, S.C.; Nemani, R.; Keeling, C.D.; Otto, R.D.; Running, S.W. Global net carbon exchange and intra-annual atmospheric CO2 concentrations predicted by an ecosystem process model and three-dimensional atmospheric transport model. *Glob. Biogeochem. Cycles* **1996**, *10*, 431–456. [CrossRef]

40. Togliatti, K.T.; Hartman, T.; Walker, V.A.; Arkebaur, T.J.; Suyker, A.E.; VanLoocke, A.; Hornbuckle, B.K. Satellite L–band vegetation optical depth is directly proportional to crop water in the US Corn Belt. *Remote Sens. Environ.* **2019**, *223*, 111. [CrossRef]

41. Berglund, D.R. Corn Growth and Management Quick Guide—A1173. North Dakota State University Extension Service. Available online: https://www.ag.ndsu.edu/pubs/plantsci/rowcrops/a1173.pdf (accessed on 21 March 2020).

42. Wise, K. Managing Wheat by Growth Stage. Purdue Extension, ID-422. Available online: https://www.extension.purdue.edu/extmedia/ID/ID-422.pdf (accessed on 21 March 2020).

43. Berglund, D.R. Soybean Growth and Management Quick Guide—A1174, North Dakota State University. Available online: https://www.ag.ndsu.edu/pubs/plantsci/rowcrops/a1174.pdf (accessed on 21 March 2020).

44. Hornbuckle, B.K.; Patton, J.C.; VanLoocke, A.; Suyker, A.E.; Roby, M.C.; Walker, V.A.; Iyer, E.R.; Herzmann, D.E.; Endacott, E.A. SMOS optical thickness changes in response to the growth and development of crops, crop management, and weather. *Remote Sens. Environ.* **2016**, *180*, 320–333. [CrossRef]

45. Kendall, M.G. *The Advanced Theory of Statistics*, Inference and relationship, 4th ed.; Charles Griffin: London, UK, 1979; Volume 2.

46. Didan, K. Mod13q1 Modis/Terra Vegetation Indices 16-Day L3 Global 250 m SIN Grid V006. Distributed by NASA Eosdis Land Processes DAAC. Available online: https://lpdaac.usgs.gov/products/mod13q1v006/ (accessed on 25 September 2019).

Article

L-Band Vegetation Optical Depth Estimation Using Transmitted GNSS Signals: Application to GNSS-Reflectometry and Positioning

Adriano Camps [1,2,*], Alberto Alonso-Arroyo [1], Hyuk Park [1,2], Raul Onrubia [1], Daniel Pascual [1] and Jorge Querol [3]

[1] CommSensLab-UPC, Department of Signal Theory and Communications, UPC BarcelonaTech, c/Jordi Girona 1-3, 08034 Barcelona, Spain; alberto.alonso.arroyo@tsc.upc.edu (A.A.-A.); park.hyuk@tsc.upc.edu (H.P.); onrubia@tsc.upc.edu (R.O.); daniel.pascual@tsc.upc.edu (D.P.)

[2] Institut d'Estudis Espacials de Catalunya-IEEC/CTE-UPC, Gran Capità, 2-4, Edifici Nexus, despatx 201, 08034 Barcelona, Spain

[3] Interdisciplinary Centre for Security, Reliability and Trust (SnT), University of Luxembourg, 29 Avenue John F. Kennedy, Luxembourg L-1855, Luxembourg; jorge.querol@uni.lu

* Correspondence: camps@tsc.upc.edu

Received: 23 June 2020; Accepted: 21 July 2020; Published: 22 July 2020

Abstract: At L-band (1–2 GHz), and particularly in microwave radiometry (1.413 GHz), vegetation has been traditionally modeled with the τ-ω model. This model has also been used to compensate for vegetation effects in Global Navigation Satellite Systems-Reflectometry (GNSS-R) with modest success. This manuscript presents an analysis of the vegetation impact on GPS L1 C/A (coarse acquisition code) signals in terms of attenuation and depolarization. A dual polarized instrument with commercial off-the-shelf (COTS) GPS receivers as back-ends was installed for more than a year under a beech forest collecting carrier-to-noise (C/N_0) data. These data were compared to different ground-truth datasets (greenness, blueness, and redness indices, sky cover index, rain data, leaf area index or LAI, and normalized difference vegetation index (NDVI)). The highest correlation observed is between C/N_0 and NDVI data, obtaining R^2 coefficients larger than 0.85 independently from the elevation angle, suggesting that for beech forest, NDVI is a good descriptor of signal attenuation at L-band, which is known to be related to the vegetation optical depth (VOD). Depolarization effects were also studied, and were found to be significant at elevation angles as large as ~50°. Data were also fit to a simple τ-ω model to estimate a single scattering albedo parameter (ω) to try to compensate for vegetation scattering effects in soil moisture retrieval algorithms using GNSS-R. It is found that, even including dependence on the elevation angle ($\omega(\theta_e)$), at elevation angles smaller than ~67°, the $\omega(\theta_e)$ model is not related to the NDVI. This limits the range of elevation angles that can be used for soil moisture retrievals using GNSS-R. Finally, errors of the GPS-derived position were computed over time to assess vegetation impact on the accuracy of the positioning.

Keywords: GNSS; vegetation; opacity; albedo; depolarization; propagation; positioning

---

1. Introduction

It is known that the accuracy of the GPS navigation solution under deciduous forests is degraded due to the obstruction by leaves. A degradation in the positioning of approximately 2 mm per percentage (1%) of sky cover blocked has been reported [1]. However, the degradation of the received carrier-to-noise ratio (C/N_0)can also be used to retrieve vegetation properties that can be related to the vegetation water content. The model widely used in passive microwaves at L-band is the τ-ω model [2,3], where the vegetation opacity (τ) and single scattering albedo (ω) are usually assumed to be constant over all elevation angles. This study uses a dual-input GPS receiver connected to a dual-polarization antenna to extend the work conducted in [4] to characterize both the co-polar (RHCP) and cross-polar (LHCP) received powers as a function of the elevation angle, and the vegetation properties, characterized by the NDVI, the LAI, or the greenness, blueness or redness levels, as derived from zenith-looking images. Section 2 presents the methodology: the instrument developed, the field experiment, and the ground-truth data acquired. Section 3 analyzes and discusses the results obtained. Finally, Section 4 summarizes the main results and presents the conclusions of this study.

2. Methodology

2.1. Test Site Description

The field experiment started on August 3rd 2015 at La Fageda d'en Jordà (Girona, Spain), and ended in October, 2016. La Fageda d'en Jordà is a densely populated beech forest. The instrument was located at 42°10′56″ N, 2°29′20″ E, in the north east of Spain (Figure 1). Although data were acquired continuously, the fall season (defoliation), and the spring season (growing) are the most interesting ones to analyze the effect of the leaves. The effect of branches and trunks was analyzed when the leaves had completely fallen and the open sky conditions were nearly met.

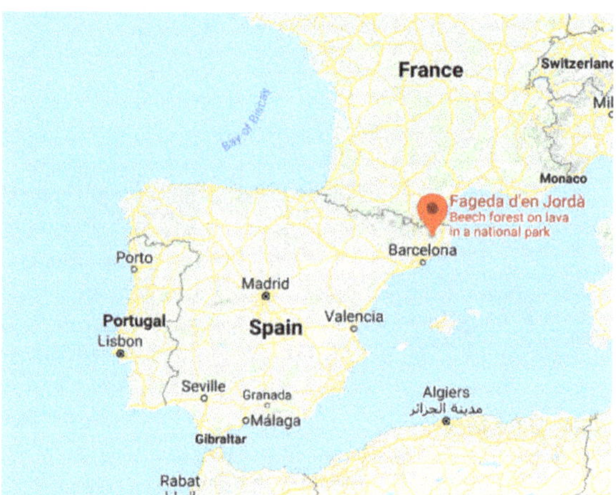

Figure 1. Location of the field experiment at Fageda d'en Jordà, in the North East of Spain.

2.2. The Global Navigation Satellite Systems (GNSS)-Transmissivity Instrument

The instrument developed for this experiment is based on a commercial off-the-shelf (COTS) GNSS receiver with two inputs that are connected to a dual-polarization zenith-looking antenna so as to measure the received GNSS signals at RHCP and LHCP, as they pass through the vegetation layer with different elevation angles. The CN_0 values are discretized at 0.1 dB, which prevents the realistic modelling of the observed data. The forest structure leads to different scattering and attenuation processes; while attenuation increases with the leaves' vegetation water content (VWC), the signal polarization purity degrades with increasing VWC and the presence of branches, etc. Additionally, multiple scattering occurs, which is very difficult to model.

Figure 2a,b show the φ (azimuth) cuts of the co-polar and cross-polar antenna pattern measured at the UPC Anechoic Chamber (https://www.tsc.upc.edu/en/facilities/anechoic-chamber). Including reflections in the anechoic chamber and measurement noise, antenna pattern measurement errors are <0.1 dB in amplitude and <1° up to 50° off-boresight angle. Details are provided in the Appendix A.

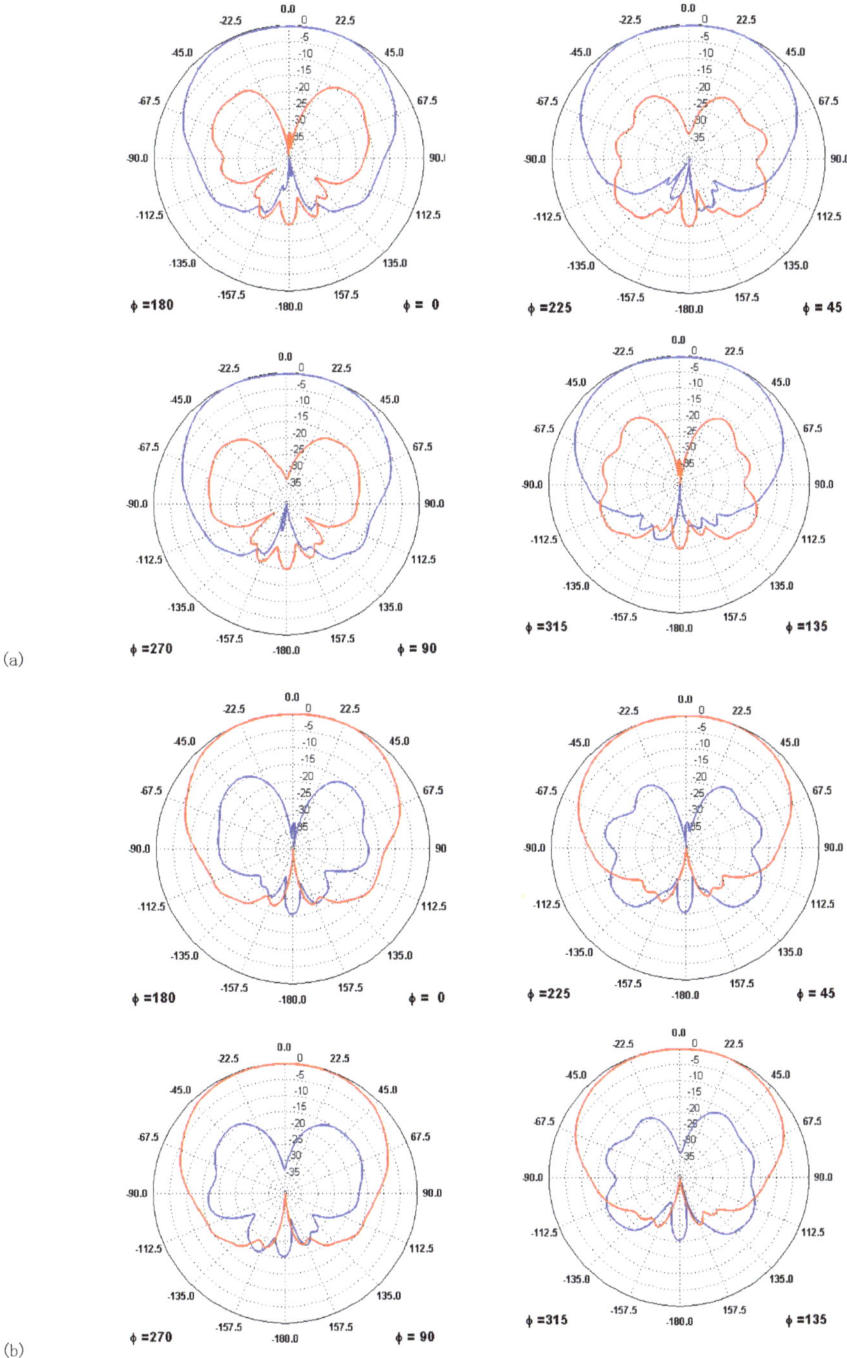

Figure 2. (**a**) Co-polar (RHCP, blue) and cross-polar (LHCP, red) RHCP antenna pattern; (**b**) Co-polar (LHCP, blue) and cross-polar (RHCP, red) RHCP antenna pattern. Cuts at $\varphi = 0°$, $45°$, $90°$, and $135°$, as indicated.

Overall, the error (GNSS receiver discretization and antenna pattern characterization) is much smaller than the data scatter encountered due to multiple-scattering in the vegetation canopy, so that it can be neglected.

The antenna design consists of two linear polarization antennas, connected to a 90° hybrid to compose the RHCP and LHCP, a technique that enlarges the antenna bandwidth.

Figure 3 shows the instrument final assembly installed in the field experiment. During the field experiment, the instrument was covered with a radome to protect it.

Figure 3. Dual-polarization up-looking antenna and receiver (behind the antenna) installed at La Fageda d'en Jordà (Girona, Spain).

In addition to the antenna pattern calibration conducted in the UPC-Antenna Lab Anechoic chamber [5], the antenna pattern was cross-checked by measuring continuously the received signal power in both channels (RHCP and LHCP) over 6 complete days. Figure 4a shows the average measured C/N_0 for different steps of azimuth ($\Delta\varphi = 10°$) and incidence ($\Delta\theta = 5°$) angles after compensating for the antenna patterns for June 7th. White spots occur because that particular day, the GPS constellation did not pass through that angular extent area. The large white area in the top corresponds to the Northern hemisphere, where there are no GPS satellites.

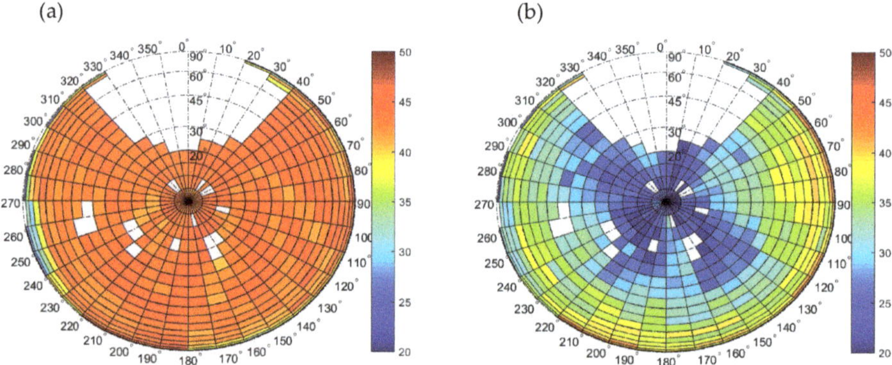

Figure 4. (**a**) RHCP and (**b**) LHCP average received carrier-to-noise ratio (C/N_0) for 7th June 2015 during the preliminary tests after compensating for the antenna pattern effect. Incidence angle in 5° steps, azimuth angle in 10° steps.

Figure 4b is similar to Figure 4a, but for the LHCP channel (including the compensation of the LHCP antenna radiation pattern). Note that only a small part of the LHCP radiation pattern around the boresight (0° to ~30°) has a cross-polar rejection > 20 dB. This information was used to calibrate the observed C/N_0 values, i.e., it is constant from all directions.

2.3. Ground-truth data

Different sources of ground-truth data were gathered for the analysis. Figure 5 shows the antenna zenith view evolution during the fall and spring. Images were acquired with a Canon EOS 50-D oriented towards the North, as the antenna. The camera FOV is 66.5° × 47.3°, which fits the antenna beam. As it can be seen, the dominant vegetation on the test site was beeches, with nearly no understory during the entire field campaign. In the next paragraphs, different parameters are computed from these images for comparison with the received signal power.

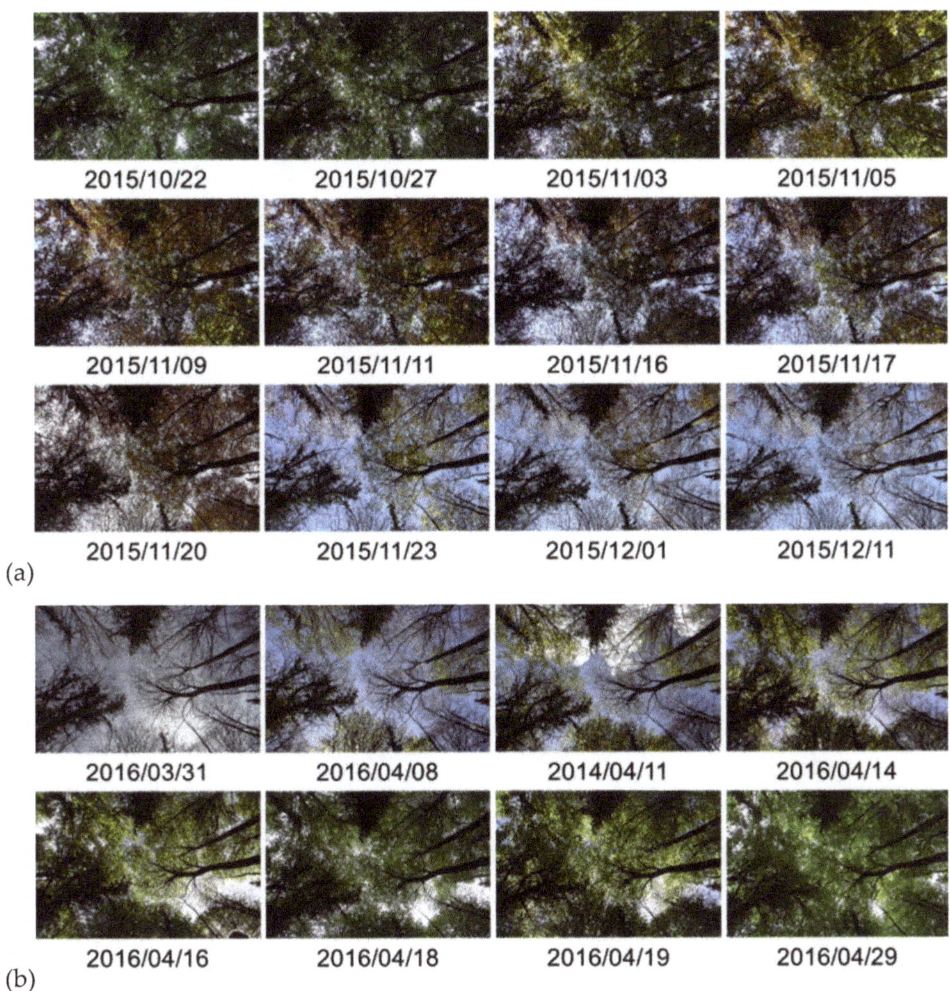

Figure 5. Vegetation observed from a camera located at the instrument's position looking to the zenith during (**a**) the fall season and (**b**) during the spring season.

Additionally, MODIS-derived LAI and NDVI maps from [6] were used. LAI and NDVI maps have a 0.1° resolution, and the revisit time is 8 and 16 days, respectively. Only the information from the nearest pixel was used. The scale of NDVI maps ranges from −0.1 to 0.9, and that of LAI maps is from 0 to 7 m^2/m^2. Finally, rain data from a meteorological station located in Olot were used to analyze the effect of rain on the measurements [7].

3. Results

Figures 6 and 7 show polar plots of the received C/N$_0$ values for the RHCP and LHCP channels, respectively. During the fall seasons, it is clearly seen in the RHCP plot in Figure 6a that the smaller the number of leaves, the smaller the attenuation is. Accordingly, the C/N$_0$ is measured as being higher in December than in October. Leaves also cause depolarization of the waves, but from the two effects (attenuation and scattering), attenuation is dominant since the LHCP received power actually increases after the leaves have fallen, as shown in Figure 7a. Figures 6b and 7b show different examples of the received C/N$_0$ when the leaves are growing in spring. The decrease in the received power shows again that the attenuation effect is dominant for leaves.

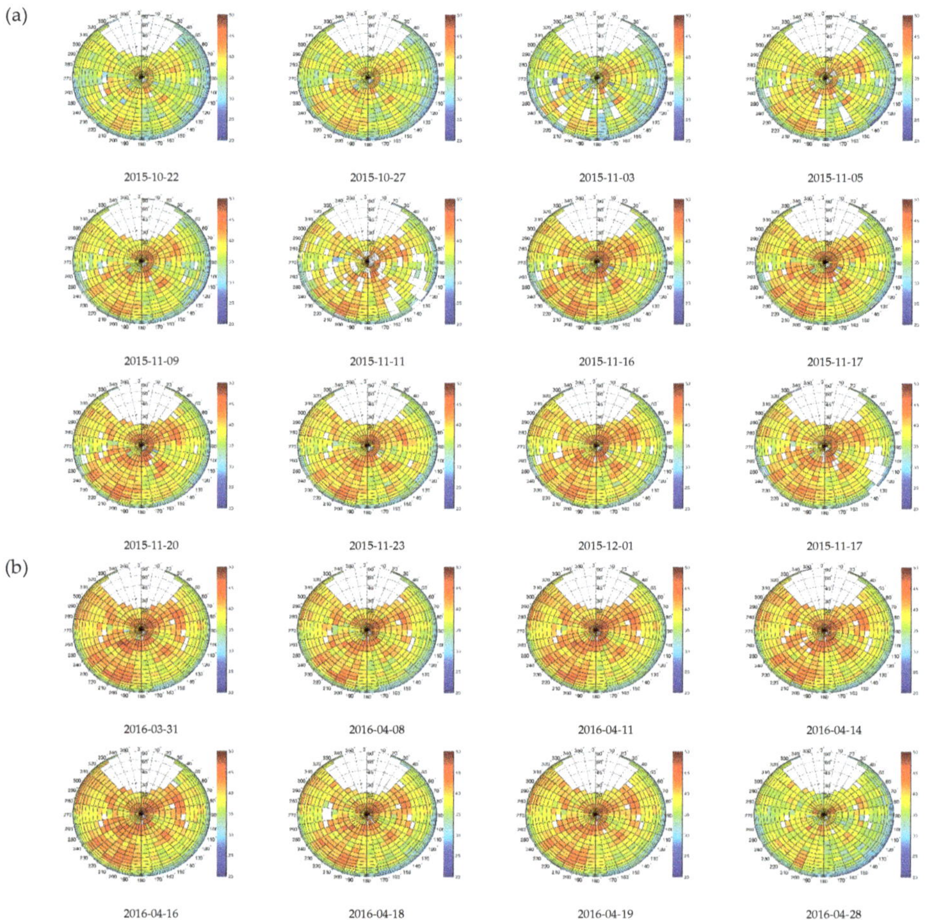

Figure 6. RHCP received C/N$_0$ during the (**a**) fall; and (**b**) spring season for different dates when vegetation pictures were taken. Incidence angle in 5° steps, azimuth angle in 10° steps.

(a)

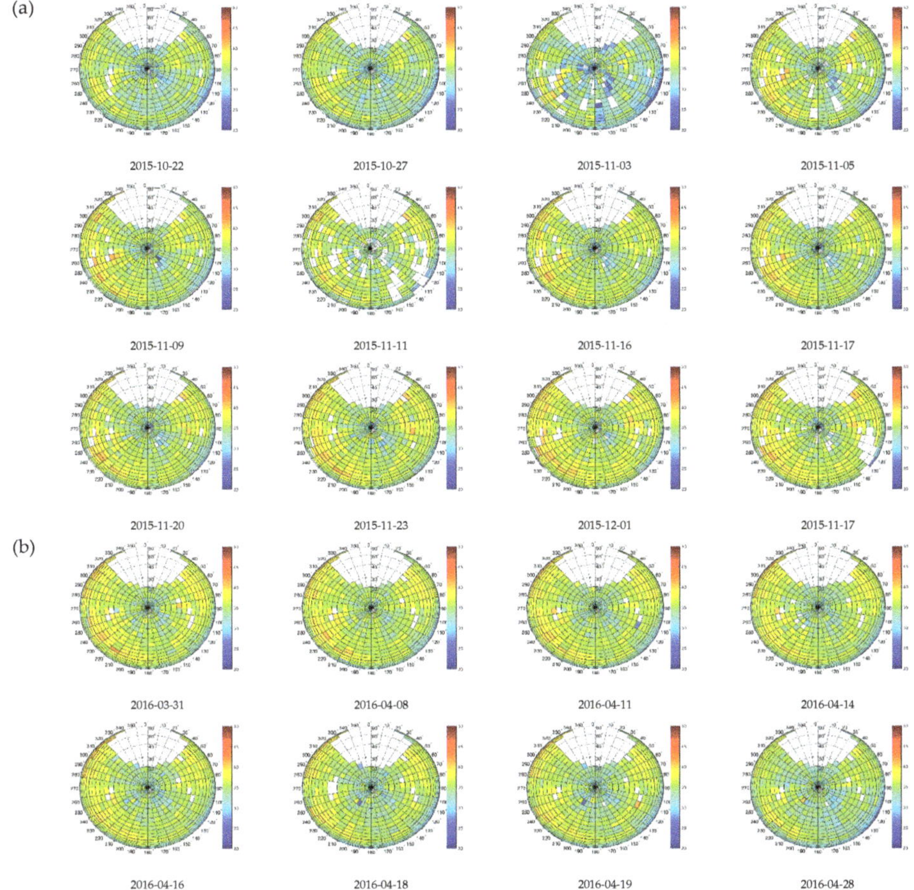

Figure 7. LHCP received C/N$_0$ during (**a**) the fall and (**b**) the spring season for different dates when vegetation pictures were taken. Incidence angle in 5° steps, azimuth angle in 10° steps.

In the following, the received RHCP C/N$_0$ is azimuthally averaged to analyze the evolution with the elevation/incidence angle. Resulting C/N$_0$ curves are analyzed with respect to:

- rain rates, obtained from the regional meteorological station,
- blueness, greenness, redness, and sky cover percentage computed from the RGB and gray scale pictures (Figure 5),
- LAI and NDVI, both computed from MODIS,

To investigate which parameter estimates best the vegetation effects in the signal propagation.

3.1. Rain Effects

Figure 8 shows the azimuthally averaged RHCP and LHCP C/N$_0$ curves for different satellite elevation angles as a function of time, together with the rain events during the field campaign. It can be appreciated that rain events induce a fading on the C/N$_0$ plots, especially in the RHCP channel, due to two main factors: (1) the presence of water drops in the atmosphere, and the water that stays on the leaves' surface, increasing the attenuation induced by the leaves; (2) the fact that after the rain event, trees absorb the water from the soil, increasing the vegetation water content. Note that during the period without leaves (December 2015–April 2016), fading events due to rain are very smaller in depth.

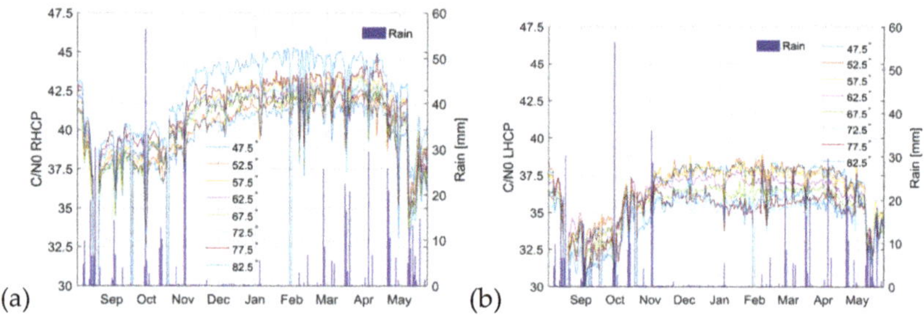

Figure 8. Effect of rain to the azimuthally averaged C/N$_0$ curves: (**a**) RHCP; and (**b**) LHCP. The curves are plotted corresponding to the elevation angles from 47.5° to 82.5° angles in the legend.

3.2. Dependence on the Greenness/Redness/Blueness

The greenness, redness, and blueness can be estimated from the color histograms as:

$$G, R, B = \frac{\rho_{G,R,B}}{\rho_G + \rho_R + \rho_B},$$

(1)

where $\rho_{G,R,B}$ is the amount of G,R,B color bits from the pictures taken with the Canon 50-D. Figure 9a shows the evolution of the greenness estimated from the pictures together with the azimuthally averaged RHCP C/N$_0$ curves. It is expected that the larger the greenness parameter, the larger the amount of leaves, and therefore the larger the attenuation. During the defoliation process (October–December 2015), the R^2 parameter with a linear fit computed between the greenness and the different C/N$_0$ curves is 0.76–0.87; it does not depend on the incidence angle, and the mean slope of the fit is −31 dB/au (au: arbitrary unit). During the leaf growing period (March–April 2016), the R^2 parameter goes down to 0.46–0.66.

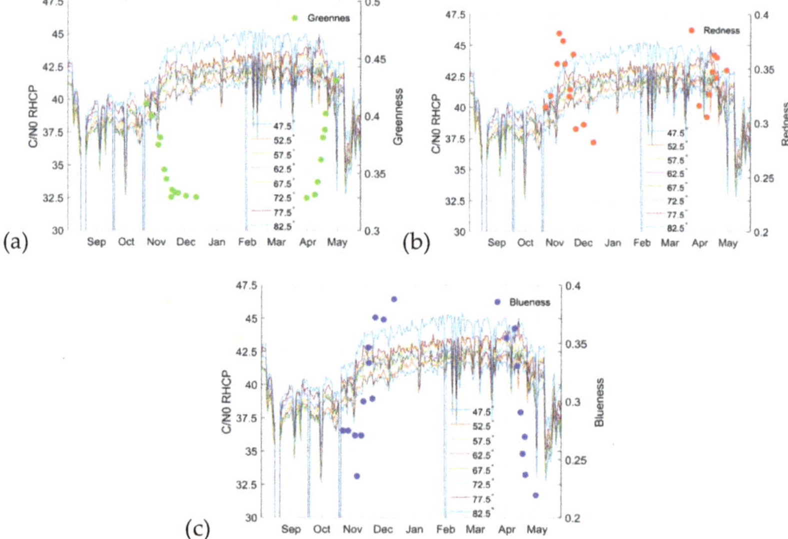

Figure 9. Evolution of (**a**) greenness; (**b**) redness; and (**c**) blueness, and C/N$_0$ curves as a function of time.

Figure 9b shows the RHCP C/N_0 curves together with the redness parameter. The R^2 parameter for both the falling and growing season is below 0.05 for any elevation angle, and it does not depend on the season.

Finally, Figure 9c shows the RHCP C/N_0 curves together with the blueness parameter. As for the greenness parameter, C/N_0 curves and blueness are correlated, but not as correlated as with the greenness. During the defoliation process, R^2 is ~0.46–0.58 with a slope of ~14 dB/au, while during the growing process, R^2 is ~0.25–0.43, with a slope of 10 dB/au. The correlation between curves appears because the amount of blue is related to the amount of sky observed, and therefore the larger the amount of sky observed, the lower the amount of leaves; however, the amount of blue color seems to be a poor vegetation indicator. Apart from that, on a cloudy day, such as 2015/11/20 or 2016/03/31, the sky is white and not blue (see Figure 4), and therefore the blueness is not such a good indicator.

3.3. Dependence on the Sky Cover

Figure 10 shows the RHCP C/N_0 curves together with the fraction of sky covered computed in two different ways. In Figure 10a, it is computed from the gray-scale image (intensity, 0–255). A value threshold of 155 is selected [8,9] to differentiate between the vegetation (vegetation < 155), and the sky (open sky > 155). In Figure 10b, the blue channel of the RGB image is used for sky classification, and a similar threshold is applied.

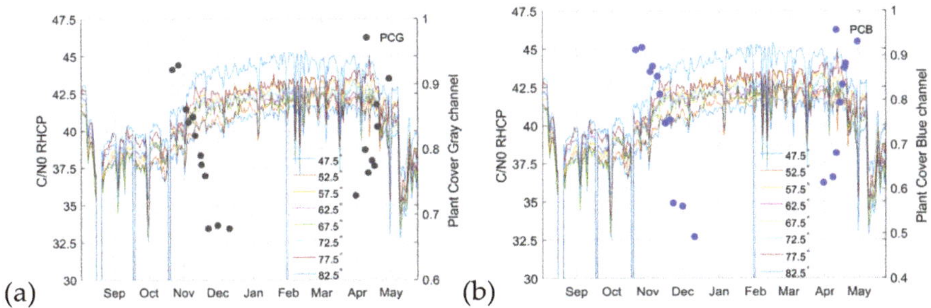

Figure 10. (a) Evolution of the percentage of sky covered and C/N_0 curves as a function of time: (a) using a gray-scale image; and (b) using the blue channel of the RGB image.

Regarding the percentage of sky cover computed from the gray-scale image, the R^2 parameter with a linear fit between the different C/N_0 curves and the percentage of sky cover is 0.6–0.7 for the falling season, whereas it is between 0.67 and 0.82 for the growing season. However, when using the blue channel to estimate the percentage of sky cover, the R^2 parameter is between 0.47 and 0.57 for the fall season, and 0.3 and 0.5 for the spring season. Again, both are independent of the elevation angle.

3.4. Dependence on the LAI

Figure 11 shows the evolution of the LAI parameter and the RHCP C/N_0 curves for different elevation angles. The R^2 coefficient of the regression lines that relate the RHCP C/N_0 to the LAI ranges from 0.50 and 0.62, which is still lower than the greenness parameter. However, a trend can be clearly seen in Figure 11 where the lower the LAI, the larger the C/N_0 observed.

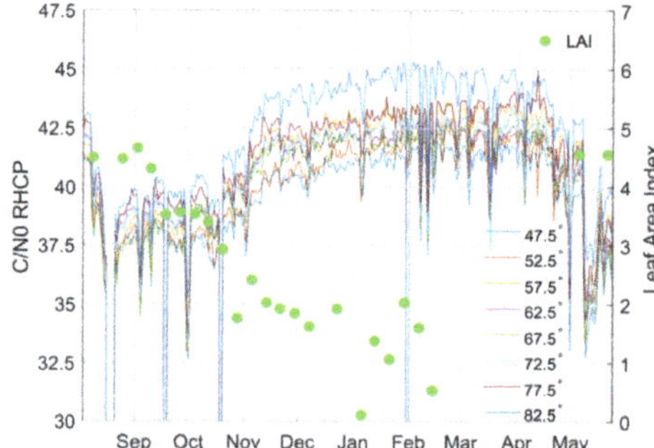

Figure 11. Evolution of LAI and RHCP C/N0.

3.5. Dependence on NDVI

Figure 12 shows the evolution of the NDVI parameter and the RHCP C/N_0 curves for different elevation angles. There is a very high correlation between the received signal power or C/N_0 and the NDVI.

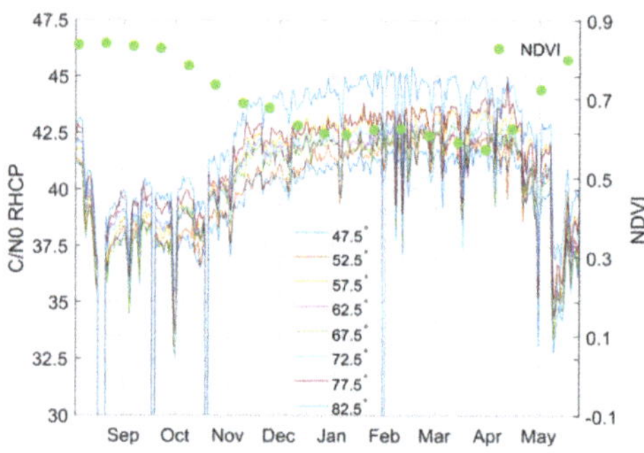

Figure 12. Evolution of normalized difference vegetation index (NDVI) and RHCP C/N_0.

Figure 13 compares the NDVI values against the mean C/N_0 value for different satellite elevation angles. For all satellites and elevation angles the R^2 parameter is between 0.87 and 0.94, and the slope of the fit from −16.9 dB/au to −22.6 dB/au (au denotes an arbitrary unit of the NDVI from 0 to 1). Table 1 (columns 2 to 5) shows the fitting parameters of the regression of the RHCP C/N_0 with respect to NDVI as a function of the elevation angle (Figure 13).

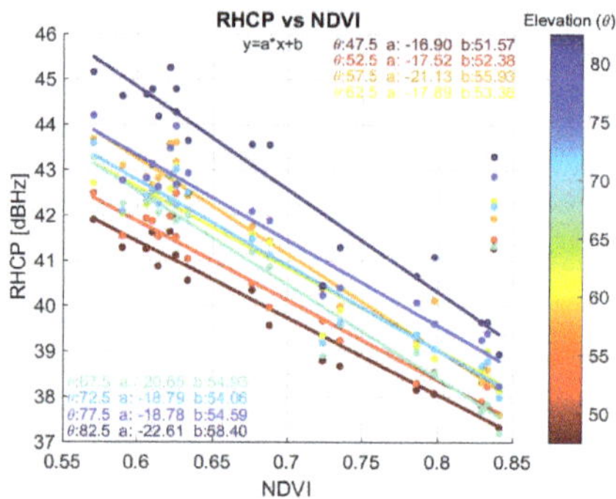

Figure 13. Dependence of the RHCP C/N_0 and the NDVI for different satellite elevation angles, and linear fits.

Table 1. RHCP C/N$_0$ vs. NDVI fitting parameters.

Elevation Angle [deg]	*a* [dB/au]	*b* [dB]	RMSE [dB]	R^2
		RHCP		
47.5	−16.90	51.57	0.55	0.89
52.5	−17.52	52.38	0.49	0.92
57.5	−21.13	55.93	0.61	0.91
62.5	−17.89	53.36	0.65	0.86
67.5	−20.65	54.93	0.48	0.94
72.5	−18.79	54.06	0.62	0.89
77.5	−18.78	54.59	0.66	0.88
82.5	−22.61	58.40	0.71	0.90

($y = a \cdot$ NDVI $+ b$, RMSE: root mean square error, and coefficient of determination R^2).

4. Discussion

From all the previous analysis, it can be concluded that the NDVI is the best descriptor to account for the vegetation attenuation for the forest type of our experiment. This parameter will now be used in our discussion on the signal depolarization through the vegetation.

The dependence between the LHCP C/N$_0$ and the NDVI (Figure 14) is also found to be strongly correlated. Because of the increased attenuation, the larger the NDVI, the lower the received power at LHCP. Table 2 (columns 2 to 5) shows the fitting parameters of the regression of the LHCP C/N$_0$ wrt. NDVI as a function of the elevation angle (Figure 14).

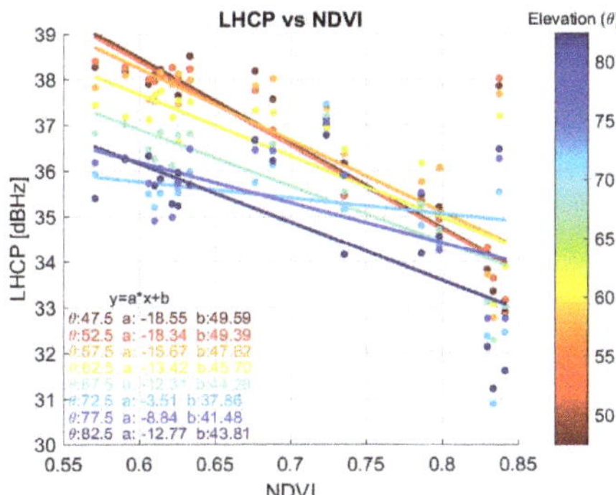

Figure 14. Dependence of the LHCP C/N$_0$ and the NDVI for different satellite elevation angles, and linear fits.

However, it must be noted that:

- The correlation drops at high elevation angles (77.5° and 82.5°) because the path through the vegetation layer is shorter, and scattering effects (responsible for signal depolarization) are less important.
- The slope is (in absolute value) smaller for LHCP than for RHCP, suggesting a combined effect of depolarization that transfers power from the RHCP signal to the LHCP.

The ratio LHCP/RHCP vs NDVI (Figure 13 minus Figure 14 in dB) also exhibits an interesting behavior. At mid-low elevation angles, the dependence on the vegetation is small (small slope) and the independent coefficient *b* is also very small, indicating that the incoming RHCP wave is almost completely depolarized. As the elevation angle increases, the absolute value of the slope (*a*) and the independent coefficient (*b*) both increase (Table 2, columns 6 and 7). These results are in agreement with [10], showing how the polarization ratio of the GNSS-R observables decreased with increasing vegetation, in [10] parametrized with the Leaf Area Index.

Table 2. RHCP and LHCP C/N$_0$, and Co- to Cross-polar Ratio vs. NDVI Fitting Parameters.

Elevation Angle [deg]	LHCP				RHCP to LHCP Ratio			
	a [dB/au]	b [dB]	RMSE [dB]	R^2	a [dB/au]	b [dB]	RMSE [dB]	R^2
47.5	−18.55	49.59	0.80	0.82	1.65	1.98	0.97	0.86
52.5	−18.34	49.39	0.62	0.88	0.82	2.99	0.79	0.73
57.5	−15.67	47.62	1.09	0.65	−5.46	8.31	1.25	0.59
62.5	−13.42	45.70	0.96	0.63	−4.47	7.66	1.16	0.54
67.5	−12.31	44.28	0.98	0.59	−8.34	10.65	1.09	0.55
72.5	−3.51	37.86	0.79	0.76	−15.28	16.20	1.00	0.68
77.5	−8.84	41.48	1.31	0.31	−9.94	13.11	1.47	0.27
82.5	−12.77	43.81	1.31	0.43	−9.90	14.59	1.49	0.39

($y = a \cdot \text{NDVI} + b$, RMSE: root mean square error, and coefficient of determination R^2).

These results also indicate that, as pointed out in [11], a correction based on the compensation of the (two-way) vegetation optical depth is not appropriate and a more refined vegetation model that properly accounts for vegetation scattering must be used in soil moisture retrieval algorithms using GNSS-Reflectometry.

In an attempt to refine the inclusion of vegetation effects, a more refined τ-ω model is proposed, with variable parameters with the elevation angle. By fitting the observed data to a simple τ-ω model, an albedo ($\omega(\theta_e)$) value can be estimated. Figure 15 shows the evolution of the estimated albedo with respect to time for several elevation angles. As it can be appreciated, there is a strong dependence with the elevation angle, and apparently a weak dependence with the NDVI (not shown in this plot, but NDVI varies from 0.6 to 0.9, see Figure 12). Figure 16 shows the regression lines of the albedo with respect to NDVI for different elevation angles, and Table 3 shows the fit parameters of Figure 15. The albedo varies from ~0.1 to 0.2 at the zenith, but up to ~0.35 at 47.5°. Note, however, that only at high elevation angles ($\theta_e \geq 67.5°$) is the single scattering albedo correlated with the NDVI, and at lower elevation angles, the presence of multiple scattering makes the τ-ω model more likely to be invalid.

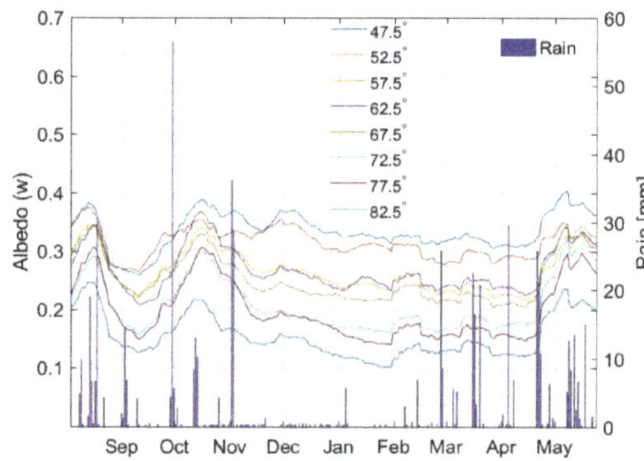

Figure 15. Evolution of the estimated albedo (ω) versus time at different elevation angles.

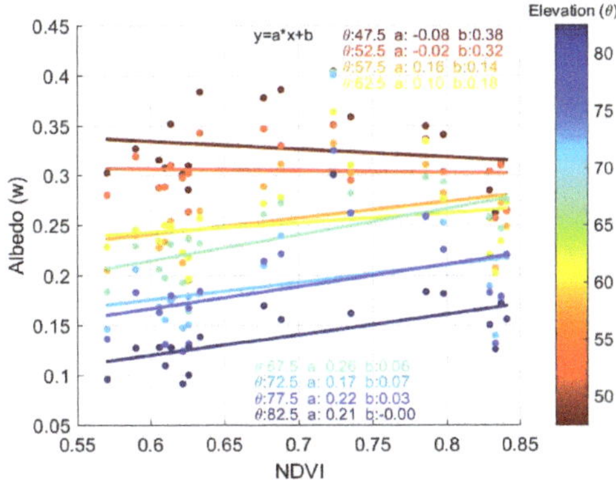

Figure 16. Comparison between the estimated single scattering albedo (ω) and the NDVI for different satellite elevation angles, and linear fits.

Table 3. ω vs. NDVI Fitting Parameters.

Elevation Angle	a [-]	b [-]	RMSE [-]	R^2
47.5	−0.08	0.38	0.05	−0.17
52.5	−0.02	0.32	0.03	−0.20
57.5	0.16	0.14	0.04	0.05
62.5	0.10	0.18	0.04	0.05
67.5 *	0.26	0.06	0.04	0.52
72.5 *	0.17	0.07	0.05	0.46
77.5	0.22	0.03	0.05	0.19
82.5 *	0.21	0.00	0.03	0.66

Note: Only elevation angles marked with an "*" are significant ($R^2 > 0.4$).

As a side parameter, Figure 17 shows the evolution of the 6-h position error (stem plot), the weekly root man squared errors (circles), and the median errors (diamonds) in north-south (Y-component, in red) and east-west (X-component, in blue). In general, the error increases over time, from August 2015 to the end of May 2016. In general, the increase in spring can be due to extra attenuation due to the water content in the leaves (Figure 12). However, there is a sudden NDVI increase in mid-April that translates into a smaller positioning error. Additionally, the smaller rmse and monotonic increase in the late summer and early fall cannot be explained by the NDVI (Figure 12); as leaves get drier and finally fall, the attenuation also decreases, and so the NDVI slowly decreases in October until mid-November. Here, the only plausible explanation encountered would be the increased scattering in the tree branches that creates a multipath, which is then reduced as leaves appear and attenuate the signal, but also the multiple-scattering (multi-path) as well. This empirical result turns out to be in very good agreement with Figure 4, left, where Zimbelman et al. [12], predicted a 10 m rmse error for 20 m tall trees, as the ones shown in Figure 5.

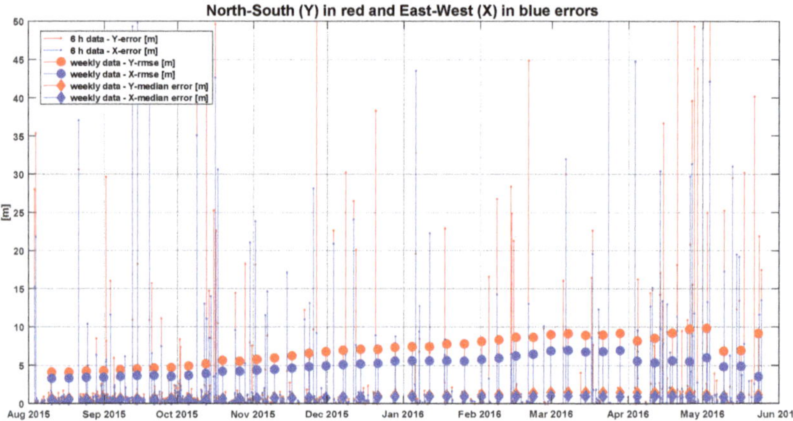

Figure 17. Evolution of the 6-h position error (stem plot), the weekly root man squared errors (circles), and the median errors (diamonds) in north-south (Y-component, in red) and east-west (X-component, in blue). Six instances with errors larger than 50 m rmse were found.

5. Summary and Conclusions

A one year long field experiment was conducted between 8/2015 and 10/2016 at La Fageda d'en Jordà forest in the north east of Spain, to assess the vegetation impact on the propagation of GNSS signals, as this is a critical correction for the accurate soil moisture retrieval using GNSS-Reflectometry.

The correlation of the vegetation co-polar (RHCP) attenuation has been evaluated against different vegetation descriptors, such as the rain, the greenness, blueness and redness indices, the sky cover, the LAI, and the NDVI. It has been found that among all of them, the correlation with the NDVI shows the highest R^2 parameter (>0.85), with sensitivities ranging from −17 dB/au to −23 dB/au. This indicates that at L-band, auxiliary NDVI data can be used as a descriptor for beech forest vegetation attenuation in GNSS-R soil moisture retrievals. Alternatively, L-band multi-angular attenuation measurements can be used to infer the vegetation water content, which is related to the vegetation optical depth (VOD).

The correlation of the vegetation cross-polar (LHCP) attenuation with the NDVI has also been evaluated, finding lower values (~0.55), but still significant. The LHCP signal is ~9 dB to ~3 dB below the RHCP signal around zenith and elevation angles of 82.5° and 47.5°, respectively. This indicates that the lower the elevation angle, but even as high as 47.5°, the more important the multiple scattering effects are, and so the signal depolarization.

Trying to find an equivalent "single scattering albedo" (ω) dependent on the elevation angle, that could be used in a τ-ω model, it was found that it varies from ~0.1 to 0.2 at the zenith, and increases up to ~0.35 at 47.5°. However, only at high elevation angles ($\theta_e \geq 67.5°$), the estimated albedo is significant, and can be related to the NDVI. At lower elevation angles, signal depolarization and multiple scattering effects must be taken into account to properly model vegetation effects in GNSS-Reflectometry, for this type of forest, and probably for other types of dense vegetation as well. This limitation of the model is what nowadays limits the range of elevation angles that can be used for soil moisture retrievals using GNSS-R, as shown in [13].

Finally, the evolution of the rmse positioning error is shown wrt time, which exhibits an increase from 3–4 to 6–10 m from fall to late spring, as shown in [12].

Author Contributions: Conceptualization, A.C. and A.A.-A.; methodology, A.C.; software, A.A.-A. and R.O.; validation, A.A.-A.; formal analysis, A.A.-A.; investigation, A.C. and A.A.-A.; resources, A.C., A.A.-A., H.P. and J.Q., R.O. and D.P.; data curation, A.A.-A.; writing—original draft preparation, A.C.; writing—review and editing, ALL; visualization, A.A.-A.; supervision, A.C.; project administration, A.C. and H.P.; funding acquisition, A.C. All authors have read and agreed to the published version of the manuscript.

Funding: This work has been funded by the Spanish MCIU and EU ERDF project (RTI2018-099008-B-C21) "Sensing with pioneering opportunistic techniques" and grant to "CommSensLab-UPC" Excellence Research Unit Maria de Maeztu (MINECO grant MDM-2016-600).

Acknowledgments: The authors would like to express their gratitude to S. Blanch for the antenna pattern measurements, to Emili Bassols Isamat (webassol@gencat.cat) of the La Garrotxa Volcanic Zone Natural Park and Lluís Serrat (info@cisteller.com) for their support to the execution of the field experiment, and to Emili Bassols Isamat for taking the pictures used to compute the greenness, redness, blueness, and sky cover.

Conflicts of Interest: The authors declare no conflict of interest.

Appendix A

Antenna pattern measurements were conducted in the UPC anechoic chamber. The following figures were produced by Prof. S. Blanch (private communication) during the characterization of the L-band patch antennas of a scale model of the SMOS instrument for the European Space Agency (ESA). At L1, the reflection coefficient in the absorbers in the walls is about −35 dB and the Signal-to-Noise Ratio (SNR) about 45 dB. The following figures show the measurement setup (Figure A1), and the computed antenna pattern error (amplitude and phase) associated to reflections, thermal noise, and to both (Figure A2). As it can be noted, the maximum error up to 50° off-boresight angle is less than 0.1 dB and 1°.

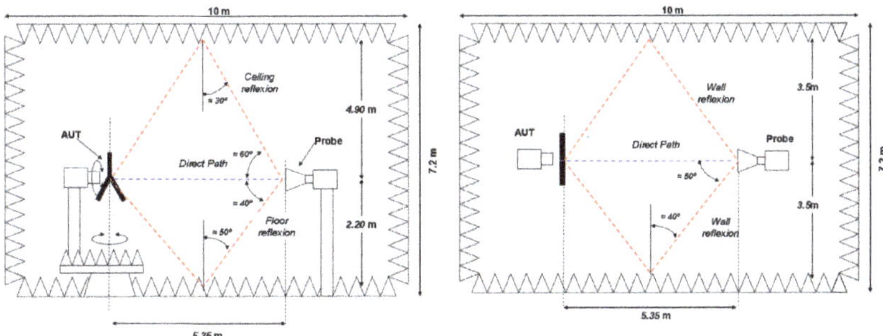

Figure A1. Antenna pattern measurement test setup in UPC anechoic chamber (https://www.tsc.upc.edu/en/facilities/anechoic-chamber).

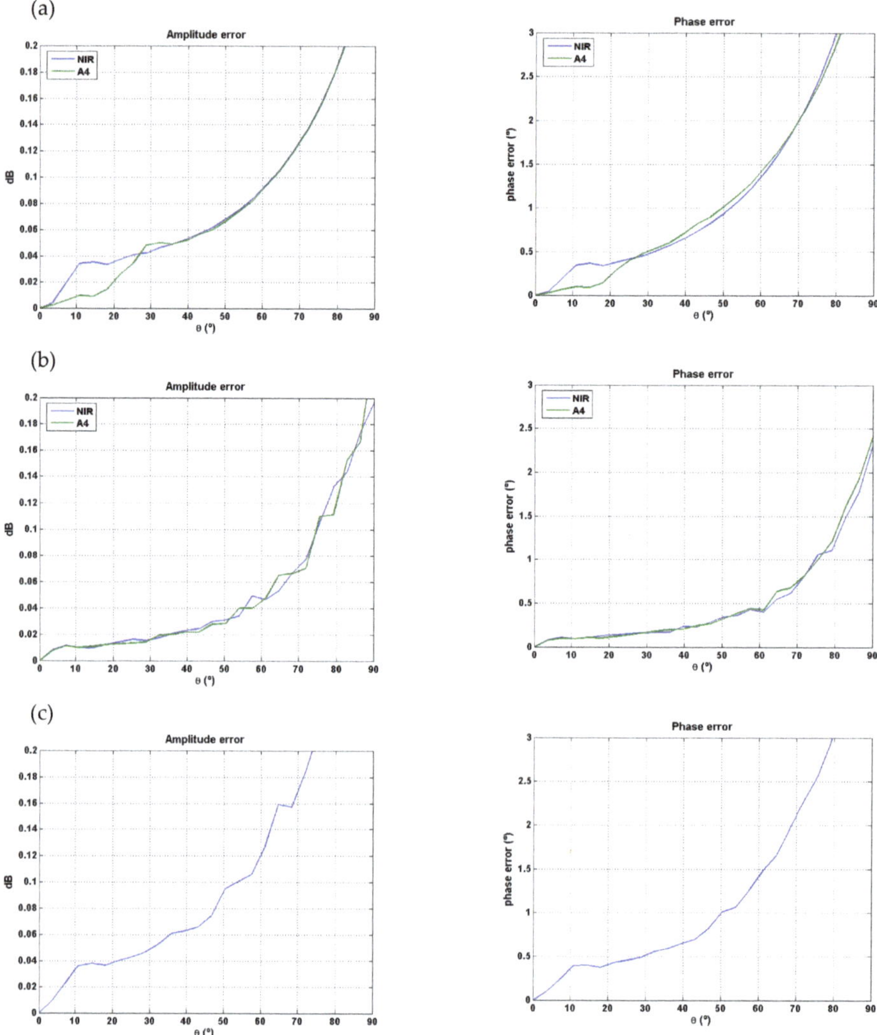

Figure A2. Antenna pattern measurement errors for two antennas (green and blue plots, left: amplitude, right: phase) associated to: (**a**) wall reflections; (**b**) thermal noise; and (**c**) both wall reflections and thermal noise.

References

1. Meyer, T.H.; Bean, J.E.; Ferguson, C.R.; Naismith, J.M. The Effect of Broadleaf Canopies on Survey-grade Horizontal GPS/GLONASS Measurements. *Surv. L. Inf. Sci.* **2002**, *62*, 215–224.
2. Wigneron, J.-P.; Chanzy, A.; Calvet, J.-C.; Bruguier, N. A simple algorithm to retrieve soil moisture and vegetation biomass using passive microwave measurements over crop fields. *Remote Sens. Environ.* **1995**, *51*, 331–341. [CrossRef]
3. Wigneron, J.-P.; Parde, M.; Waldteufel, P.; Chanzy, A.; Kerr, Y.; Schmidl, S.; Skou, N. Characterizing the Dependence of Vegetation Model Parameters on Crop Structure, Incidence Angle, and Polarization at L-Band. *IEEE Trans. Geosci. Remote Sens.* **2004**, *42*, 416–425. [CrossRef]

4.	Rodriguez-Alvarez, N.; Bosch-Lluis, X.; Camps, A.; Ramos-Perez, I.; Valencia, E.; Park, H.; Vall-llossera, M. Vegetation Water Content Estimation Using GNSS Measurements. *IEEE Geosci. Remote Sens. Lett.* **2012**, *9*, 282–286. [CrossRef]

5.	AntennaLab. Anechoic Chamber. Available online: http://www.tsc.upc.edu/antennalab/ (accessed on 22 July 2020).

6.	NASA. LAI and NDVI Maps. Available online: http://neo.sci.gsfc.nasa.gov/ (accessed on 22 July 2020).

7.	MeteOlot. Available online: www.meteolot.com (accessed on 22 July 2020).

8.	Zhang, Y.; Chen, J.M.; Miller, J.R. Determining digital hemispherical photograph exposure for leaf area index estimation. *Agric. For. Meteorol.* **2005**, *133*, 166–181. [CrossRef]

9.	Goodenough, A.E.; Goodenough, A.S. Development of a Rapid and Precise Method of Digital Image Analysis to Quantify Canopy Density and Structural Complexity. *ISRN Ecol.* **2012**, *2012*, 1–11. [CrossRef]

10.	Zribi, M.; Motte, E.; Baghdadi, N.; Baup, F.; Dayau, S.; Fanise, P.; Guyon, D.; Huc, M.; Wigneron, J.P. Potential Applications of GNSS-R Observatons over Agricultural Areas: Results from the GLORI Airborne Campaign. *Remote Sens.* **2018**, *10*, 1245. [CrossRef]

11.	Camps, A.; Vall·llossera, M.; Park, H.; Portal, G.; Rossato, L. Sensitivity of TDS-1 GNSS-Reflectivity to Soil Moisture: Global and Regional Differences and Impact of Different Spatial Scales. *Remote Sens.* **2018**, *10*, 1856. [CrossRef]

12.	Zimbelman, E.G.; Keefe, R.F. Real-time positioning in logging: Effects of forest stand characteristics, topography, and line-of-sight obstructions on GNSS-RF transponder accuracy and radio signal propagation. *PLoS ONE* **2018**, *13*, e0191017. [CrossRef] [PubMed]

13.	Camps, A.; Park, H.; Castellví, J.; Corbera, J.; Ascaso, E. Single-Pass Soil Moisture Retrievals Using GNSS-R: Lessons Learned. *Remote Sens.* **2020**, *12*, 2064. [CrossRef]

remote sensing

Article

Single-Pass Soil Moisture Retrieval Using GNSS-R at L1 and L5 Bands: Results from Airborne Experiment

Joan Francesc Munoz-Martin [1,*], Raul Onrubia [1], Daniel Pascual [1], Hyuk Park [1], Miriam Pablos [2,3], Adriano Camps [1], Christoph Rüdiger [4], Jeffrey Walker [4] and Alessandra Monerris [5]

1   CommSensLab—UPC, Universitat Politècnica de Catalunya—BarcelonaTech, and IEEC/CTE-UPC, 08034 Barcelona, Spain; onrubia@tsc.upc.edu (R.O.); daniel.pascual@upc.edu (D.P.); park.hyuk@tsc.upc.edu (H.P.); camps@tsc.upc.edu (A.C.)

2   Physical and Technological Oceanography Group, Consejo Superior de Investigaciones Científicas (ICM-CSIC), Centre of Excellence Severo Ochoa, Institut de Ciències del Mar, Passeig Marítim de la Barceloneta 37-49, 08003 Barcelona, Spain; mpablos@icm.csic.es

3   Barcelona Expert Center on Remote Sensing (BEC), Passeig Marítim de la Barceloneta 37-49, 08003 Barcelona, Spain

4   Department of Civil Engineering, Monash University, Clayton, VIC 3800, Australia; chris.rudiger@monash.edu (C.R.); jeff.walker@monash.edu (J.W.)

5   Department of Infrastructure Engineering, The University of Melbourne, Parkville, VIC 3010, Australia; alessandra.monerris@gmail.com

*   Correspondence: joan.francesc@tsc.upc.edu; Tel.: +34-626-253-955

Abstract: Global Navigation Satellite System—Reflectometry (GNSS-R) has already proven its potential for retrieving a number of geophysical parameters, including soil moisture. However, single-pass GNSS-R soil moisture retrieval is still a challenge. This study presents a comparison of two different data sets acquired with the Microwave Interferometer Reflectometer (MIR), an airborne-based dual-band (L1/E1 and L5/E5a), multiconstellation (GPS and Galileo) GNSS-R instrument with two 19-element antenna arrays with four electronically steered beams each. The instrument was flown twice over the OzNet soil moisture monitoring network in southern New South Wales (Australia): the first flight was performed after a long period without rain, and the second one just after a rain event. In this work, the impact of surface roughness and vegetation attenuation in the reflectivity of the GNSS-R signal is assessed at both L1 and L5 bands. The work analyzes the reflectivity at different integration times, and finally, an artificial neural network is used to retrieve soil moisture from the reflectivity values. The algorithm is trained and compared to a 20-m resolution downscaled soil moisture estimate derived from SMOS soil moisture, Sentinel-2 normalized difference vegetation index (NDVI) data, and ECMWF Land Surface Temperature.

Keywords: GNSS-R; dual-band; airborne; soil moisture; surface roughness; vegetation

Citation: Munoz-Martin, J.F.; Onrubia, R.; Pascual, D.; Park, H.; Pablos, M.; Camps, A.; Rüdiger, C.; Walker, J.; Monerris, A. Single-Pass Soil Moisture Retrieval using GNSS-R at L1 and L5 bands: Results from Airborne Experiment. *Remote Sens.* **2021**, *13*, 797. https://doi.org/10.3390/rs13040797

Academic Editor: Ferdinando Nunziata

Received: 11 January 2021
Accepted: 17 February 2021
Published: 22 February 2021

Publisher's Note: MDPI stays neutral with regard to jurisdictional claims in published maps and institutional affiliations.

1. Introduction

Soil is a natural reservoir of water, making it the main supply store for plants to live. The surface storage is mainly depleted by the natural process of evaporation, percolation to lower layers in the soil, water uptake by plants, etc. Conversely, moderate-to-high surface soil moisture (SM) values increase flood [1] risks and affect soil erosion [2] by wind and rain. Consequently, monitoring the soil moisture content of this near-surface layer of soil is crucial for sustainable irrigation of crop fields (smart irrigation), forest fire risk prediction, assessment of vegetation senescence, and to have a better knowledge of the water cycle, which plays a key role in the climate feedback loops [3].

Soil moisture can be measured using in-situ probes, or by means of remote sensing techniques, for which several approaches have been shown to have the ability to retrieve surface soil moisture states at different spatiotemporal scales. Using L-band microwave radiometry, the European Space Agency (ESA) Soil Moisture and Ocean Salinity (SMOS)

115

mission [4] and the National Aeronautics and Space Administration (NASA) Soil Moisture Active Passive (SMAP) mission [5] are providing soil moisture maps at a native resolution of ∼55 km [6] and 36 km [7], respectively. Thermal Infrared spectrometers have also been used to estimate soil moisture, although with lower accuracy than microwave radiometers [8]. Moreover, radar scatterometers (e.g., [9]) and Synthetic Aperture Radars (SAR) (e.g., [10–12]) at L- and C-bands have been used to derive soil moisture indexes. More recently, Global Navigation Satellite System—Reflectometry (GNSS-R) has been proven to allow for the retrieval of soil moisture from ground [13,14], airborne [15–17], and spaceborne [18–24] configurations.

GNSS-R offers the promise of an enhanced spatial resolution when compared to microwave radiometry sensors. The spatial resolution of GNSS-R receivers is mostly linked to the size of the first Fresnel zone [25], where for an airborne receiver at an altitude of ∼1000 m the spatial resolution ranges from 7.2 m/cos(θ_i) to 16.6 m/cos(θ_i) [26] and for a spaceborne receiver it scales with the squared root of the height, i.e., at 500 km height it is 22.4 times larger. Such high resolution from space has been analyzed in [27], where the GNSS-R signal collected by CyGNSS allowed the detection of riverbeds of a width of 200–250 m. However, current methods to retrieve SM from GNSS-R space-borne data are not providing such resolution. As discussed in [28], few reflections (less than 16%) contain a noticeable coherent component, thus showing that incoherent scattering is dominant over land. Under these conditions, the spatial resolution is degraded, as many contributions coming from the entire glistening zone are collected by the receiving antenna (i.e., 25–37 km as shown in [29]). Because of that, many algorithms to retrieve SM using GNSS-R data require averaging the GNSS-R observable to the SMAP native resolution of 36 km [23,30] or by applying spatial and temporal averaging (i.e., 22 km resolution in [31]). Other works have shown an enhanced spatial resolution as compared to the previous ones. As CyGNSS data is now tagging reflections containing a large coherent component [32], new algorithms are being developed providing large spatial resolutions up to 2 km [33] or 3 km [34] just including coherent reflections retrieved by CyGNSS. It is worth mentioning that the use of coherent reflections and an Artificial Neural Network (ANN) based algorithm is key to provide a high-resolution SM product [34].

Machine learning algorithms, and in particular ANNs, are now the latest approach to retrieve soil moisture from a wide range of remote sensing techniques. For instance, the SMOS soil moisture product assimilated by the European Centre for Medium-range Forecast (ECMWF) is obtained using an ANN [35]. However, ANNs challenge is to correctly train the algorithm: in this process, the key is not only to select the correct target but also the amount of ancillary data inputs used for the algorithm. Known ANN implementations are summarized in Table 1 from [30], and they require a large spatial averaging and the use of ancillary data due to the incoherent scattering that is produced over land. In most of the cases discussed in [30], the ancillary data used is either the normalized difference vegetation index (NDVI), the SMAP vegetation optical depth (VOD), or a combination of the soil texture and topography data.

Current reflectivity models (Equation (1) [36]) compensate for the vegetation attenuation and the surface roughness according to:

$$\Gamma(\theta) = R(\theta)^2 \gamma^2 exp(-4k^2 \sigma_h^2 cos^2(\theta)), \tag{1}$$

where Γ is the reflectivity, θ stands for the local incidence angle, R corresponds to the amplitude of the Fresnel reflection coefficient, γ is the transmissivity—which accounts for the vegetation attenuation—and it is modeled by the VOD or the NDVI as a proxy, k is the wavenumber (i.e., $\frac{2\pi}{\lambda}$) and σ_h is the surface root-mean square (RMS) height.

If any of the terms included in Equation (1) is not estimated accurately, the soil moisture cannot be properly retrieved. This is shown experimentally in [17,37]: at L-band, soil moisture produces a change on the retrieved reflectivity of up to 17 dB for a range between 0 and 0.45 m^3/m^3, while under coherent scattering conditions, the surface roughness effect by itself may reduce the reflectivity up to 18 dB for a local RMS surface height variation of 0–4 cm [37] or up to 11 dB when a Kirchhoff Approximation simulator is applied to the multielevation surface [38]. Furthermore, vegetation may reduce the reflectivity by up to 11 dB for a VOD variation of 0–0.6. Therefore, it is critical to adequately estimate and include both parameters into the retrieval algorithm. While the vegetation impact can be, in principle, more easily corrected by means of the NDVI or the VOD, this is not the case for the surface roughness, as the effective surface RMS height is extremely complicated to be accurately modeled or retrieved. Previous studies from an airborne platform [17] showed that, even when estimating the surface roughness using laser profilers on ground, the reflectivity could not be properly corrected for, as the reflection of the GNSS signal actually takes place in the near soil surface, whose depth varies depending on the soil moisture content of the reflection area, and the signal wavelength [39]. Therefore, the effective RMS height also depends on the soil moisture content [40].

This study analyzes the effects of the surface roughness variations using GNSS-R data from the Microwave Interferometer Reflectometer (MIR) instrument [41,42], acquired during two flights across an extensively used agricultural area New South Wales, Australia. MIR is a state-of-the-art GNSS-R instrument designed to work as an interferometric GNSS-Reflectometer [43], but raw data was also stored at 32 MS/s at 1 bit. These data have been postprocessed following the conventional GNSS-R (cGNSS-R) technique at different integration times, as presented in [44,45] for the MIR flights over the ocean. This work shows the very first results of the MIR instrument over land, and it is organized as follows: Section 2 describes the data set used and the ancillary data used to calibrate/validate the instrument performance. Section 3 analyzes the statistics of the reflectivity of the different MIR flights over Yanco, where the surface roughness effect is evaluated at different integration times; its effect on the soil moisture retrieval is also evaluated. Finally, Section 4 shows the results of an ANN algorithm applied to the MIR data to retrieve soil moisture.

2. Data Description

MIR flew over the Yanco-designated portion covered by OzNet [46], New South Wales, Australia, on 1 May 2018, and on 18 June 2018 (see Figure 1). The two flights covered the same area, but the first flight was conducted after a long period without rain, under very dry soil conditions, while the second flight was conducted the day after a rainy day, in which the water content of the soil was substantially higher. These two flights are called herein as "Dry" and "Wet" flights, respectively.

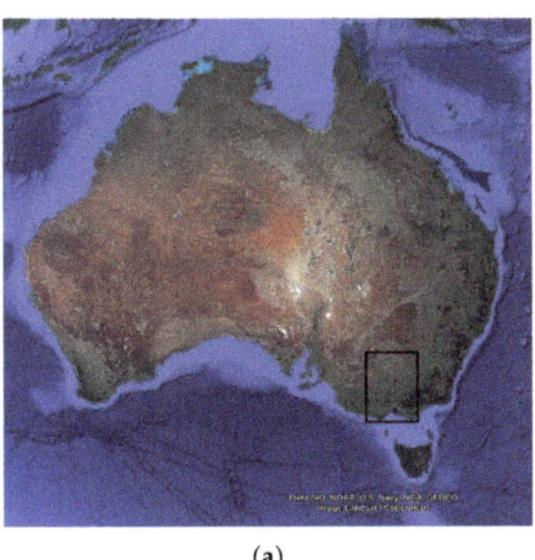

(a)   (b)

Figure 1. (**a**) Highlighted in black is the NSW area of Australia where the flight was conducted, and (**b**) definition of the Yanco areas A (34°43′S, 146°05′E) and B (34°59′S, 146°18′E).

2.1. Ground-Truth and Ancillary Data

MIR covered Yanco areas A and B (see Figure 1) in the two flights. As shown in Figure 2, the OzNet soil moisture of both days is significantly different. In the area covered by the plane track, the ground stations showed an average SM $\sim$0.05 m^3/m^3, and $\sim$0.27 m^3/m^3, for the dry and wet flights, respectively. However, considering only a limited set of in situ measurements is not enough to compare it with the GNSS-R measurements. For that reason, other remote sensing products have been used as reference data.

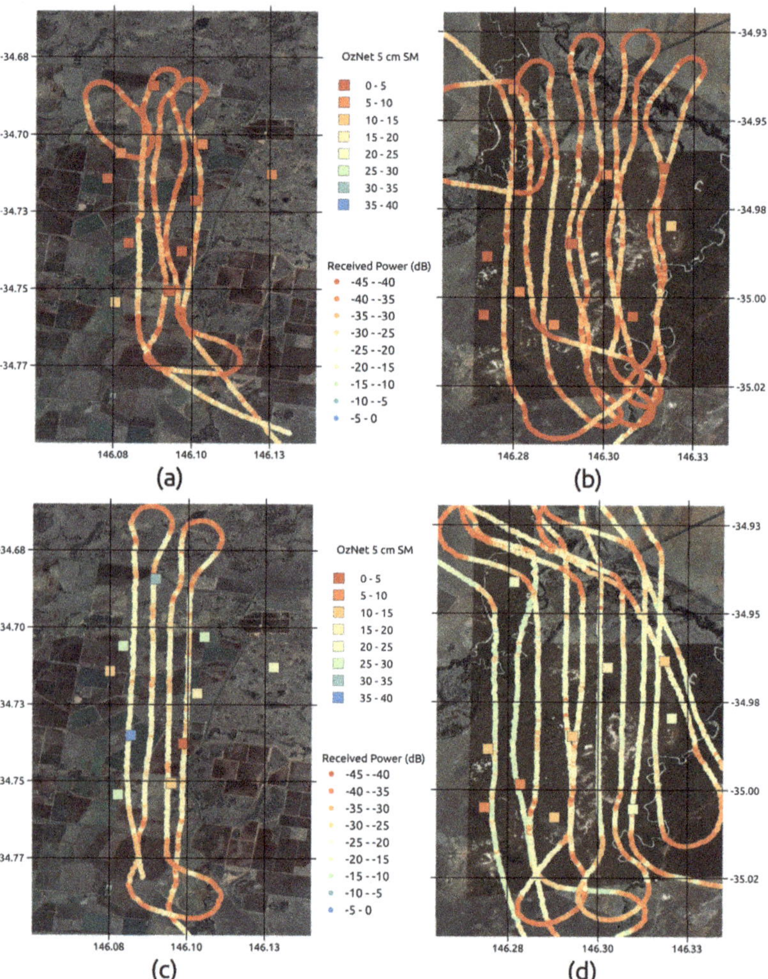

Figure 2. The tracks denoted by the *Received Power* color scale represent the received power (in arbitrary units) by the Microwave Interferometer Reflectometer (MIR) instrument, and the coloured squares are the in-situ OzNet soil moisture (m^3/m^3) sensors at 5 cm depth. (**a**) Shows the received power and in-situ measurement during *Dry* flight at (**a**) Site A, and (**b**) Site B; and during *Wet* flight at (**c**) Site A, and (**d**) Site B. Note: anomalous low reflectivity values for large banking angles during the turns are not used in the study.

Two ancillary SM products are key for the correct interpretation of the data in this work. The first ones are the SMOS and SMAP soil moisture products. However, as the resolution is too large for our data set ($\sim$55 and 36 km), a pixel downscaling [47,48] of the coarse resolution SMOS L3 soil moisture product has been enhanced down to 20 m, using SMOS brightness temperatures, ECMWF land surface temperature, and visible and near infra-red (VNIR) data from Sentinel-2. This technique has been validated with different soil moisture networks [49]. Moreover, the downscaled data have been validated against the OzNet SM network at both "Dry" and "Wet" flights, showing a bias of 0.01 m^3/m^3, with a root-mean-square deviation (RMSD) of 0.032 m^3/m^3 for the "Dry" flight, and a bias of 0.02 m^3/m^3, with a RMSD of 0.038 m^3/m^3 for the "Wet" flight. The RMSD for both maps is lower than the SMOS accuracy at native resolution (0.04 m^3/m^3), and it is lower than the RMSD presented in other studies [50]. Therefore, the presented downscaled product is suitable as ground-truth data.

As it can be seen in Figure 3 and 4, the NDVI from Sentinel-2 presents some differences between the *Dry* and *Wet* flights, and the downscaled SMOS soil moisture product also presents large differences between both flights. However, the important values for this study are the ones collocated with the GNSS specular reflection points. For that reason, both the NDVI and the SM maps are interpolated to each of the specular points of the MIR instrument, and the histograms of both measurements for the two flights are shown in Figure 5.

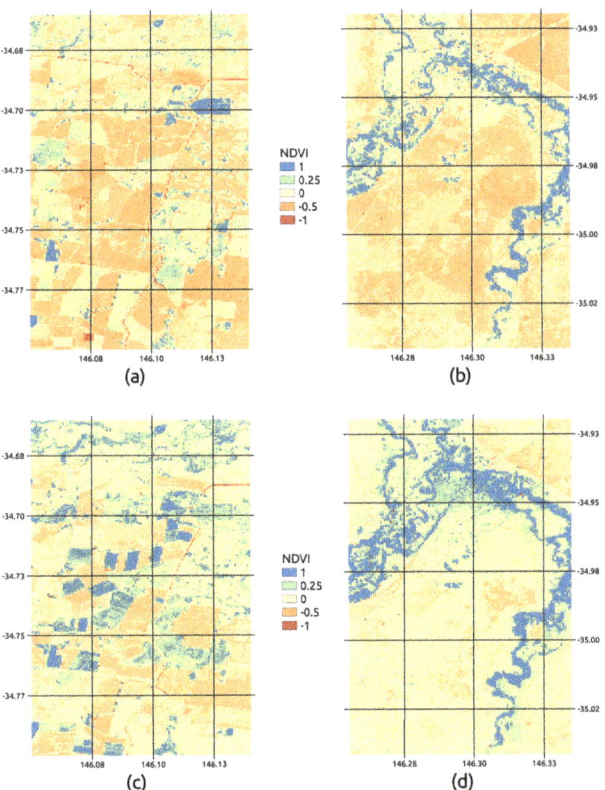

Figure 3. Normalized difference vegetation index (NDVI) retrieved by Sentinel-2. The selected data sets contain Copernicus Sentinel data corresponding to 2018-05-01 (Dry) and 2018-06-16 (Wet) from the Sentinel Hub [51]. *Dry* flight at (**a**) Site A, and (**b**) Site B; and during *Wet* flight at (**c**) Site A, and (**d**) Site B. Note that negative NDVI values in (**a,c**) correspond to the Coleambally Canal.

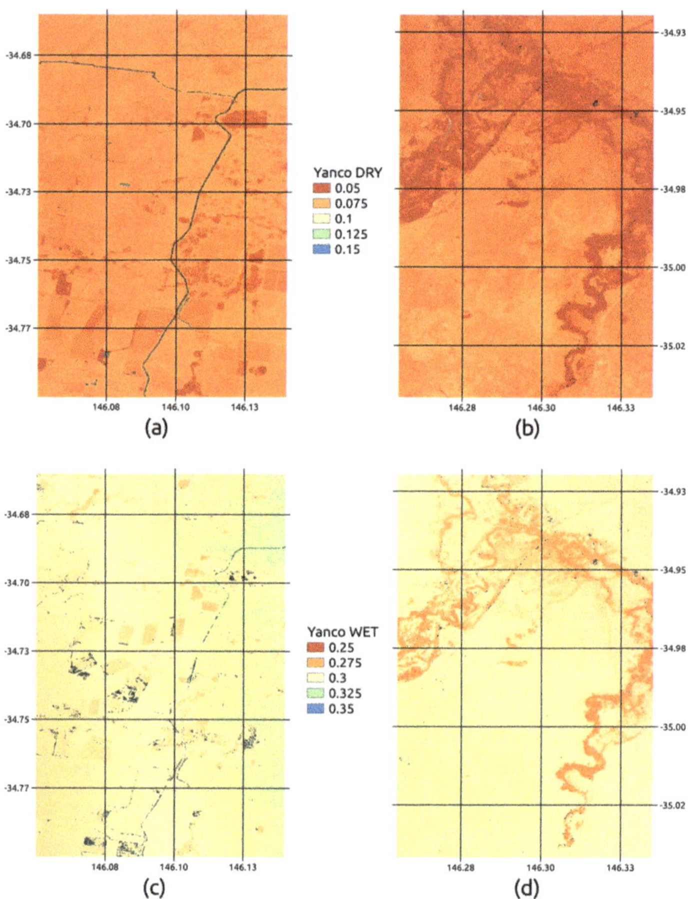

Figure 4. SM retrieved by the combination of SMOS SM and downscaled using Sentinel-2 NDVI. *Dry* flight at (**a**) Site A, and (**b**) Site B; and during *Wet* flight at (**c**) Site A, and (**d**) Site B. Credits: Barcelona Expert Center [52]. Note that the color scale is different in each plot to maximize the contrast and ease its visualization.

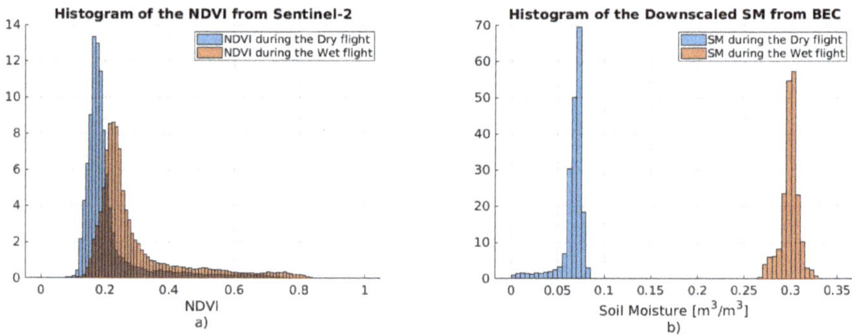

Figure 5. Histograms of the collocated Sentinel-2 NDVI (**a**) and downscaled SM (**b**) values over the MIR specular points for both the Dry and Wet flights.

2.2. GNSS-R Data

The MIR uplooking and downlooking antenna arrays are composed of 19-element dual-band patches. Each array creates 2 beams at L1/E1 and 2 beams at L5/E5a using an automatic beam-steering algorithm that compensates for the aircraft attitudes. Therefore, MIR is able to track up to four different GNSS satellites simultaneously with the uplooking antenna and four specular points with the downlooking antenna (two for each band). The instrument automatically tracks different satellites when the plane has a stable attitude in the Local-Vertical Local-Horizontal frame. In the case that a given satellite goes out of the antenna pattern, the algorithm automatically changes to a new one. Moreover, the instrument notifies whenever the plane is not having a stable attitude (i.e., during turns or maneuvers), and therefore flagged data need to be filtered out. The plane flew at an altitude of $h \sim 500$ m at an average speed of $v \simeq 75$ m/s. From this altitude, the size of the Fresnel zone l_{F_r} is (Equation (2) [43,53]):

$$l_{Fr} = \frac{\sqrt{\lambda R_r}}{\cos(\theta_{inc})},$$
$$R_r = \frac{h}{\cos(\theta_{inc})}, \tag{2}$$

where λ is the signal wavelength, ($\lambda = 19$ cm at L1/E1 and $\lambda = 25$ cm at L5/E5a), and θ_{inc} denotes the incidence angle of the GNSS signal. Thus, for an incidence angle $\theta_{inc} = 0°$ (i.e., smaller Fresnel zone), $l_{Fr_{L1}} = 9.75$ m at L1/E1, and $l_{Fr_{L5}} = 10.75$ m at L5/E5a.

Based on the flight characteristics, the maximum integration time to prevent blurring of the first Fresnel zone is bounded by $T_{int} < \frac{l_{Fr_{L1}}}{v} \simeq 130$ ms. For this study, and in order to have enough oversampling, $T_{int} = 20$ ms has been selected. In that case, and depending on the local incidence angle, the number of samples overlapped in the same Fresnel zone varies from 6 to 11 samples, which corresponds to $\theta_{inc} = 0°$ and to $\theta_{inc} = 45°$, respectively.

Because of the flight height and speed, the delay and Doppler spreads are negligible, and therefore the selected GNSS-R observable is the reflectivity as shown in Equation (1). Applying the Pseudo-Random Noise injection technique conceived in [54], the MIR instrument is calibrated as described in [55], including the antenna pattern compensation as in [56]. Furthermore, the reflectivity is calculated assuming only the peak of the waveform minus the noise according to Equation (3):

$$\Gamma = \frac{P_{ref} - P_{N_{ref}}}{P_{dir} - P_{N_{dir}}} \times \frac{G_{up}(\theta_{dir}, \phi_{dir})}{G_{dn}(\theta_{ref}, \phi_{ref})}, \tag{3}$$

where P is the amplitude of the waveform at its maximum, P_N is the noise power of the waveform, computed as the average of the delay bins before the peak of the Delay-Doppler Map, G is the gain of either the uplooking or the downlooking antenna, θ and ϕ are the look angles to either the transmitting satellite ("dir"), and to the specular reflection point ("ref"). In this case, the GNSS satellite antenna pattern is not taken into account, neither the additional path losses, as the increased distance is negligible (500 m) with respect to the already traveled distance (>20,000 km).

The reflectivity is absolutely-calibrated using different water bodies that the MIR instrument crossed during both flights. As shown in Figure 6, the reflectivity over the selected water bodies is ~ -2.1 dB at incidence angle $\sim 20°$. Figure 7 shows the histogram, normalized as a Probability Density Function (PDF), of the reflectivity. As it can be seen, mean reflectivity values have a difference between *Dry* and *Wet* flights of ~ 7–8 dB. Considering Figure 21 from [37], the difference in received power is expected to be ~ 4 dB for a difference of ~ 0.22 m^3/m^3. However, Ref. [37] does not take into account the effect of the penetration depth of the GNSS reflection signal in different soil moisture conditions. In this work, the surface roughness plays a significant role in this averaging operation, as the average of a magnitude affected by an attenuation is a biased estimator of the actual magnitude. Additionally, as previously discussed, this roughness also depends on the soil moisture content [39], and therefore an analytic solution to that problem is very complex.

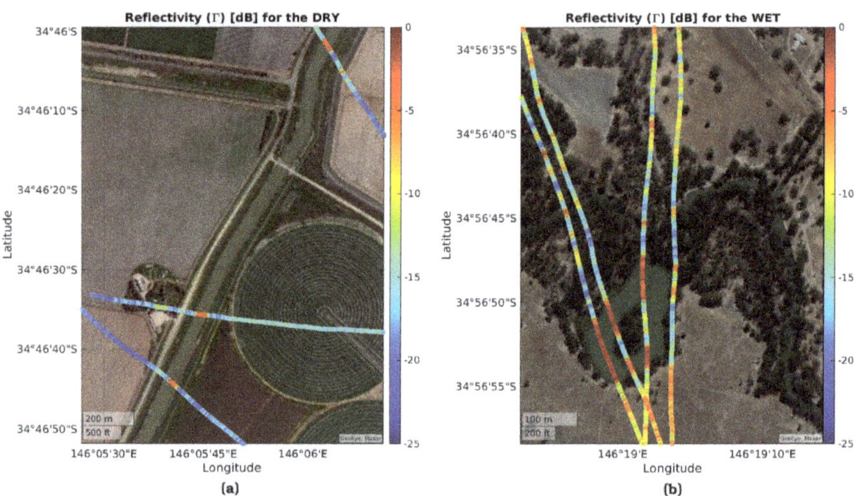

Figure 6. MIR reflectivity for both (**a**) Dry and (**b**) Wet flights (dB). Average Reflectivity over the selected water bodies is ~ -2.1 dB at an incidence angle $\sim 20°$. The resultant reflectivity has been evaluated against a saline water model [57], assuming a very low salinity (psu = 1 ppm) and a temperature of 15 °C.

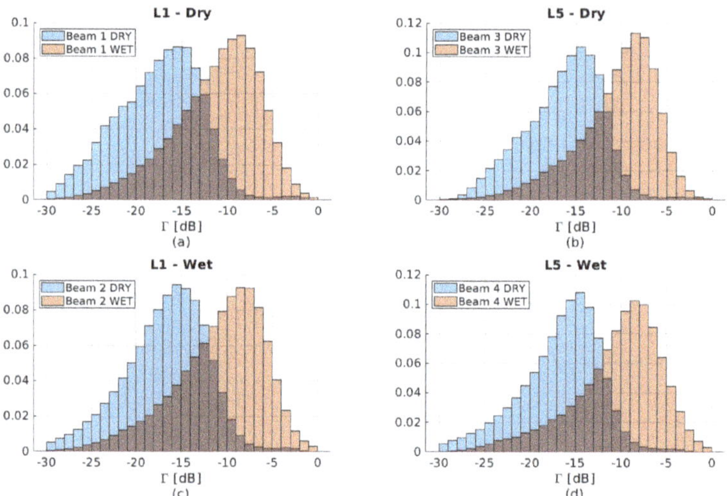

Figure 7. MIR reflectivity Probability Density Function (PDF) for both (**a**,**b**) Dry and (**c**,**d**) Wet flights. The Y-axis is the normalized counts values of the PDF, and the average reflectivity for the four beams at the *Dry* flight are: -17.6 dB, -16.8 dB, -16.3 dB, and -16.7 dB, with a standard deviation of ~ 4.7 dB; and for the *Wet* flight are: -11.2 dB, -10.8 dB, -10.5 dB, and -10.4 dB, with a standard deviation of ~ 4.8 dB. Note that, the tracking algorithm selects those satellites whose reflection has a very small incidence, and the *average* incidence angle for the four beams is below $\sim 30°$.

3. Soil Moisture Retrieval Using GNSS-R

As shown in Figure 23 from [37], and discussed in Section 1, herein a difference of just 4 cm in the local RMS height may produce a 18 dB drop in the power of the reflected GNSS signal. Furthermore, as the first Fresnel zone of the reflection is very small, the local surface roughness also has a high variability between overlapped nearby reflections.

3.1. Reflectivity Statistics Using Different Integration Times

The discussion of the previous section has shown that performing the average of the reflectivities (in linear units, and then taking its logarithm) of a certain area can reduce the effect of terrain inhomogeneity, and reduces the Speckle noise. In this section, the effect of increasing the integration time is addressed from a statistical point of view. In this case, consecutive incoherently-integrated reflectivity samples at T_{int} = 20 ms are averaged (linearly, using a moving average filter, and then taking the logarithm of the resulting magnitude) up to 5 s. Different histograms for Beams 1 (L1) and 3 (L5) for both *Dry* and *Wet* flights are shown in Figure 8a,b, respectively for L1 and L5 cases. As it can be seen, as the effective integration time increases, the average value remains constant, but the standard deviation of the reflectivity decreases.

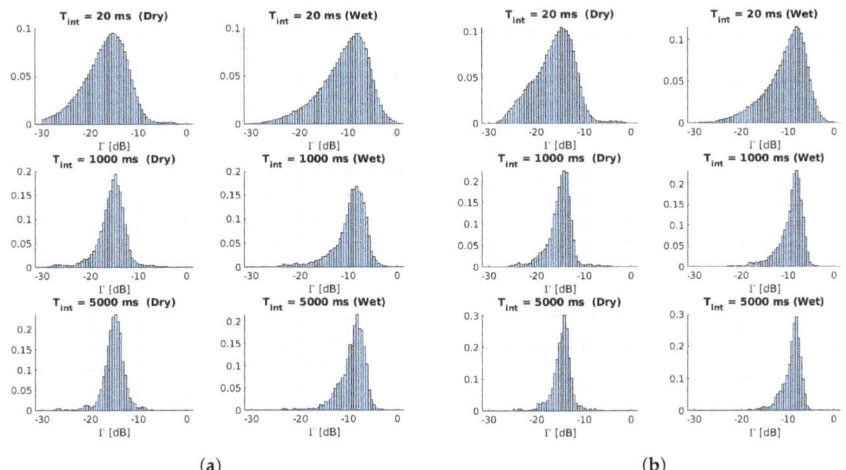

Figure 8. Normalized histogram PDFs of the MIR reflectivities retrieved at different effective integration times at (**a**) L1 and (**b**) L5.

In addition, Figure 9 shows a selected track over a certain region where the local surface roughness of the crop field changes. As it can be seen, for shorter averaging times, a larger local variability of the signal is encountered, capturing small variations in the terrain, as in the middle of the image (outlined with a black box), where the signal crosses a small irrigation channel. However, as the effective integration time increases, small variabilities of the terrain (i.e., surface roughness) are averaged, and therefore the attenuation variability caused by the surface roughness is diluted, providing a negative bias in the reflectivity, which varies as a function of the integration time.

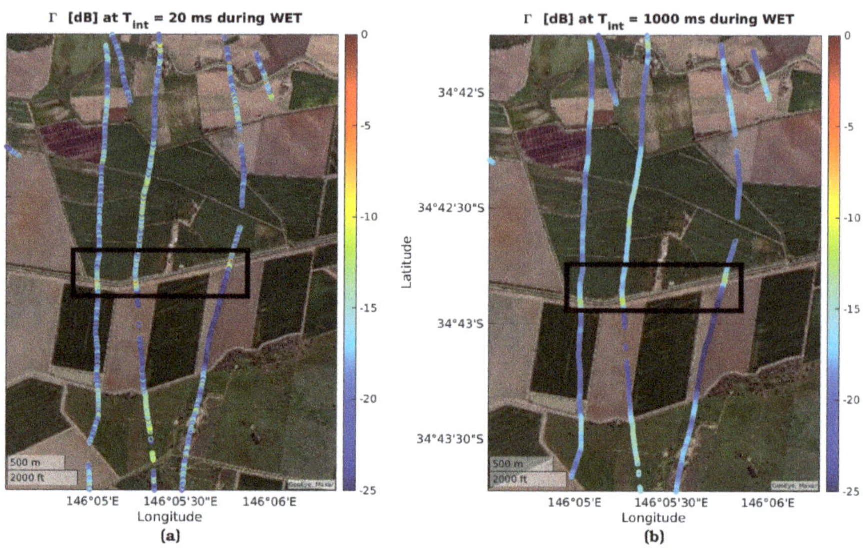

Figure 9. Geolocated MIR reflectivities in Yanco site A during *Wet* flight at different T_{int}: (**a**) 20 ms and (**b**) 1000 ms.

3.2. Surface Roughness Effect in Soil Moisture Retrievals

Since at high SM the reflectivity tends to saturate (Figure 21 of [37]), its sensitivity to SM decreases. In this case, if the scattering occurs close to the surface, the roughness effects on reflectivity, and consequently SM, are more noticeable. This is observed in Figure 9, in which the large variability in the reflectivity is mostly produced by local surface roughness variations. When the incoherent integration time is very short (i.e., 20 ms), the terrain contribution is not smoothed, and therefore the power changes caused by local roughness variations are highlighted.

In order to analyze in detail the effect of the surface roughness, the Wet flight over Yanco site B has been selected. The NDVI of the area is very low and almost constant, and the attenuation due to vegetation opacity can be neglected. The influence of the soil surface roughness on SM can be studied by isolating its corresponding term in Equation (1) and assuming a saturated reflectivity due to very high soil moisture values. Therefore, the surface RMS roughness can be estimated from Equation (4), as follows:

$$\sigma_h = \sqrt{\frac{(-ln(10^{\frac{\Gamma[dB]+5}{10}})}{(4k^2 \times \cos(\theta_{inc})^2))}}.\tag{4}$$

The Fresnel reflection coefficient of Equation (1) is set to a constant value of -5 dB for a saturated SM value. As it is seen in [37], the GNSS-R reflectivity gets saturated for SM values larger than $\sim$0.25–0.3 m^3/m^3. Moreover, as it can be seen in Figure 7c from [21], these SM values produce a range of reflectivities from -6 to -4 dB. Thus, for the sake of simplicity, a reflectivity value of -5 dB has been selected for this study. Note that, this range of reflectivities is assuming a VOD < 0.2. As it is shown in Figure 2a [58], an L-band VOD < 0.2 is provided for NDVI values < 0.4.

Figure 10a shows the results of applying Equation (4) to the Yanco site B during the *Wet* flight, after removing all reflectivities from water bodies or with an NDVI > 0.4, so as to assume negligible attenuation due to vegetation. To perform this study, the selected beams at L1 and L5 are from the same GNSS spacecraft so that the specular reflection point is the same. In this case, for moist soils the penetration depth at both bands is limited to $\sim$2 cm [39]. As it can be seen, the estimated roughness at both bands is very similar, with an average value of ∼1.23 cm and a standard deviation of ∼0.68 cm at L1/E1, and an average value of ∼1.37 cm, and a standard deviation of ∼0.59 cm at L5/E5a, which is consistent with a slightly larger penetration depth at L5/E5a, and a reflection accuracy over a slightly flatter interface. In this case, the impact of an average roughness of 1.3 cm produces a degradation in the reflectivity of ∼1.4 dB, while a surface roughness of 2 cm produces a degradation up to 4 dB.

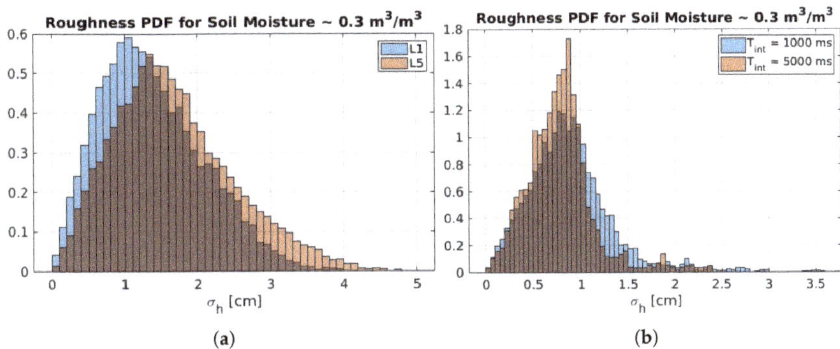

Figure 10. Estimated Surface Roughness normalized PDF in Site B during *Wet* flight seen by L1 and L5 signals, assuming a flat reflectivity $\Gamma = -5$ dB, and using (**a**) $T_{int} = 20$ ms, and (**b**) at $T_{int} = 1000$ ms and 5000 ms only for the L1 case.

Finally, the roughness at L1 depending on the integration time is displayed in Figure 10b. As the effective integration time increases, the estimated surface roughness decreases to values lower than 1 cm. Averaging up to 1000 ms produces an average roughness of 0.88 cm, with a standard deviation of 0.45 cm, and averaging up to 5000 ms produces an average roughness of 0.78 cm and a standard deviation of 0.35 cm. Increasing the effective integration time reduces the surface roughness variability, but this introduces a bias in the estimation of the reflectivity.

4. Soil Moisture Retrieval Algorithms

Surface roughness is the variable limiting the accuracy of SM retrievals using GNSS-R data [17]. As introduced in Section 1, algorithms based on ANNs have proven to be very powerful tools to detect and solve nonlinear problems by minimizing the error of the algorithm output with respect to a known target. Despite that, up to now, most of the algorithms have shown a dependence on the ancillary data (e.g., [30]). In our proposed approach, the use of ancillary data has been reduced by adding statistical metrics of the reflectivity itself to the algorithm input, such as the standard deviation of the reflectivity. However, those statistics are computed from data collected at different time instants, and there is not yet an approach for single-pass retrieval. Despite that, this points out an interesting question: is there a relationship between the standard deviation of the reflectivity and the surface roughness?

4.1. Surface Roughness and Reflectivity Standard Deviation

To address this question, some examples are provided in both *Dry* and *Wet* flights over Yanco site B. The geolocated averaged Γ and the moving standard deviation (movstd) of Γ are computed and presented in Figures 11 and 12.

Both figures present the Γ, averaged to 5000 ms, the movstd(Γ) for the same integration interval. The movstd is calculated in linear units over N 20 ms integrated reflectivity measurements (i.e., for a $T_{int} = 5000$ ms, $N = 250$), and then converted into dB units. Finally, the Noise-to-Signal Ratio (NSR), computed as movstd(Γ) minus Γ, is presented in the third column. The color axis is evenly defined for the two flights and each of the three parameters.

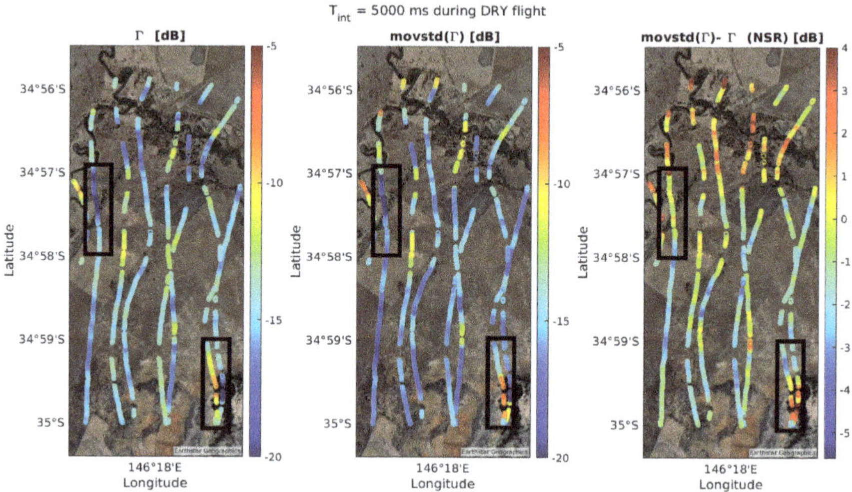

Figure 11. Geolocated Γ, movstd(Γ), and Noise-to-signal ratio (NSR), defined as movstd(Γ) $-\ \Gamma$, at $T_{int} = 5000$ ms at L1 C/A for the *Dry* flight. Black boxes identify areas with reflectivity drops due to vegetated areas and an increase of the surface roughness.

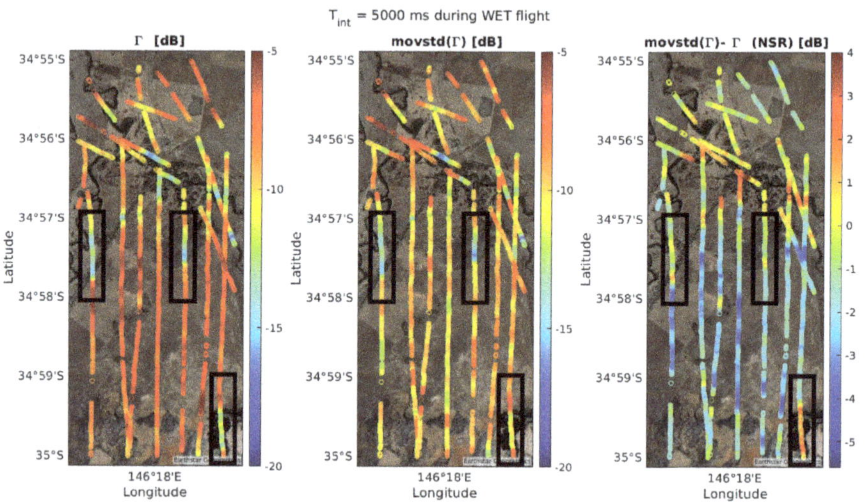

Figure 12. Same as Figure 11 but for the *Wet* flight.

As it can be seen in Figure 11, the two highlighted areas present a large decrease of the average reflectivity down to ~ -20 dB. The moving standard deviation is also affected, and the computed NSR in both areas increases. In this case, the increase in this NSR term can be linked to a rougher area, as in the bottom one. In this case, the reflectivity drop is linked to the loss of coherency caused by a vegetated area. However, it is important to remark that the *average* NSR is large: -0.78 dB. As it can be seen, lower NSR values correspond to areas with a larger reflectivity, and a lower reflectivity variation, which can be linked to smoother surfaces.

Moving to the *Wet* flight (Figure 12), the two previously highlighted areas are also shown, plus a third one in the middle of the map. First of all, it is important to remark that the average reflectivity is larger due to the higher soil moisture content of the flight. Moreover, the NSR is slightly lower than the *Dry* flight: -1.13 dB, and there are converging to the Speckle Noise NSR limit: -5.6 dB (p. 608 from [59]). This noise is the effect of the environmental conditions of the reflection scenario (i.e., surface facets, geometry, etc.), and it is a multiplicative noise that, among others, can be reduced by low-pass filtering, applying averages, or using neural networks [60]. The two selected sites of the *Dry* and *Wet* flights present a similar NSR, $\sim$1 dB for the top left one, and $\sim$2–3 dB for the one at the bottom right. Therefore, even though the soil moisture content is different, and the reflectivity value of the *Dry* flight is $\sim$7 dB lower than the *Wet* flight, the NSR is in the same order; therefore, the terrain inhomogeneity is similar in both cases.

Finally, Figure 13 presents a scatter density plot of the reflectivity computed at 20 ms, compared to the NSR computed at 5000 ms, for Dry and Wet cases, and at L1 and L5 bands. It is important to remark in the regime where the Speckle Noise is dominant (i.e., NSR close to -5.6 dB), the reflectivity tends to its average value, whilst for larger NSR values, the reflectivity displays a much larger variability due to terrain inhomogeneity and surface roughness.

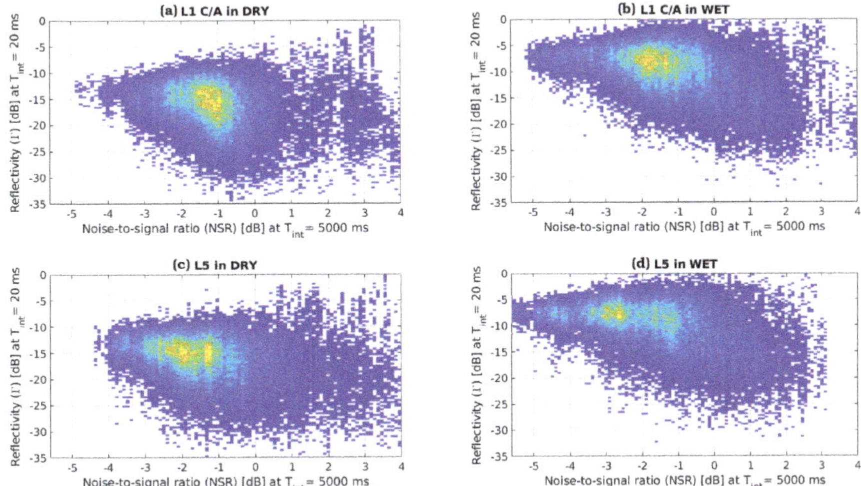

Figure 13. Scatter density plot of the Noise-to-Signal Ratio (NSR) computed at T_{int} = 5000 ms with respect to the reflectivity values at T_{int} = 20 ms at L1 for (**a**) *Dry* and (**b**) *Wet* flights, and at L5 for (**c**) *Dry* and (**d**) *Wet* flights.

The differences between the *Dry* and *Wet* flights are noticeable. In the *Dry* flight, the contribution from the surface roughness is dominant, and the NSR does not reach the -5.6 dB Speckle Noise SNR limit. On the contrary for the *Wet* flight, and especially at L5, the NSR presents a larger number of points with NSR lower than -4 dB, with some of the realizations in the -5.6 dB of the Speckle Noise SNR limit. The differences in NSR in both flights are linked to the surface roughness variation depending on the moisture content, and therefore the penetration depth of the incidence wave. The higher NSR of the *Dry* flight indicates a rougher terrain than the *Wet* flight. Furthermore, the flights at L5 also present a lower NSR than the ones at L1, due to the change in the penetration depth of the L5 signal, and the difference in the autocorrelation function width.

4.2. Artificial Neural Network

Due to the nonlinear behavior of Γ and σ_h in relation to SM, the soil moisture retrieval process does not have an analytical closed-form solution. However, the use of machine learning algorithms, and neural networks in particular, is a growing technique broadly used to solve nonlinear problems. In this case, selecting the proper inputs is crucial to accurately retrieve soil moisture. It has been shown in the previous section that the NSR is related in some way to the surface roughness, but as this parameter is computed from the subtraction of movstd(Γ) and Γ, both parameters have been used separately in an ANN algorithm, therefore letting the network use both parameters independently. To do that, the following ANNs are proposed using the following four cases:

1. Γ, movstd(Γ) as a proxy for σ_h, NDVI, and θ_{inc},
2. Γ, NDVI, and θ_{inc},
3. Γ, movstd(Γ) as a proxy for σ_h, and θ_{inc},
4. Γ, and θ_{inc}.

Furthermore, different integration times ($T_{int} = 0.1, 1, 2$, and 5 s) are used, and the target output for all cases is the collocated and downscaled to 20 m SM data from SMOS/Sentinel-2 described in Section 2. Finally, the input data is split in two: one network is deployed for L1 and another for L5. Both networks are based on a three hidden-layer feed-forward network with 6 neurons each, and each data set (L1 and L5) is randomly divided into two parts: training and test. The training set is 20% of all the data available for both *Dry* and *Wet* flights, for the two Yanco A and B locations. The test set is the resulting 80%. This training set is the one used to train the network, randomly divided again in 70%, 15%, 15%, for training, validation, and testing of the network, respectively. In order to avoid neural network overfitting, early stopping and pruning techniques are applied to the trained network [61]. Finally, the results are applied to the overall data set. Note that the entire data set (i.e., both Yanco A and B areas, and all the possible NDVI values) are used in the algorithm to cover a wider reflectivity and terrain variability.

4.3. Results

Due to the environmental constraints of the two flights, the SM values of the *Dry* flight and the SM values of the *Wet* flight are quite distant. It is important to remark that the *Dry* flight was conducted after $\sim$1 month without rain and most of the area is not irrigated. Moreover, during the *Dry* flight SM values from 0.05 m^3/m^3 up to 0.1 m^3/m^3 are found. In contrast, the *Wet* flight was conducted after a strong rain event, and therefore most of the soil is very moist, in this case, values from $\sim$0.26 m^3/m^3 up to 0.33 m^3/m^3 are found.

Figures 14 and 15 show the scatter density plot of the GNSS-R L1 SM output from the ANN, with respect to the SMOS/Sentinel-2 downscaled SM. The color axis is the density of points in logarithmic units. Analyzing Figure 14, it can be seen that the information provided by the NDVI increases the R parameter in any case. Moreover, the standard deviation of the error decreases, even without using the movstd(Γ), but the best case is when this parameter is used. For short effective integration times, the R and the standard deviation of the error are slightly worse than the case for longer integration times. As the averaging increases, the estimation of the attenuation due to both the surface roughness and the vegetation are averaged, and the use of the movstd(Γ) increases the correlation between the target and the ANN output. Furthermore, by looking at the shape of the output of the ANN, it can be seen the misclassification produced by the algorithm, where points falling into the *Wet* area, are *classified* as low soil moisture. This is mostly due to the effect of the reflectivity reduction due to surface roughness. In this case, by increasing the effective integration time (Figure 14 right, $T_{int} = 5$ s), the attenuation is *smoothed* and the algorithm shows a better behavior, showing the lowest error. Finally, it is important to remark that, in the case where the reflectivity is used alone, the ANN output is the least accurate one, clearly indicating that additional information is required to retrieve soil moisture.

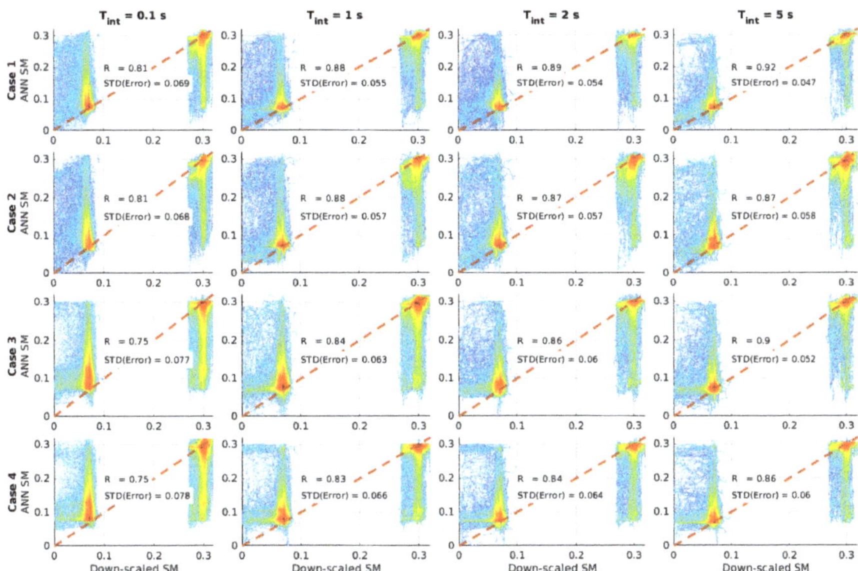

Figure 14. ANN estimated SM vs. SMOS/Sentinel-2 downscaled SM at L1. Columns from left to right increasing T_{int} for 0.1, 1, 2, and 5 s. Row from top to bottom, ANN cases 1 to 4.

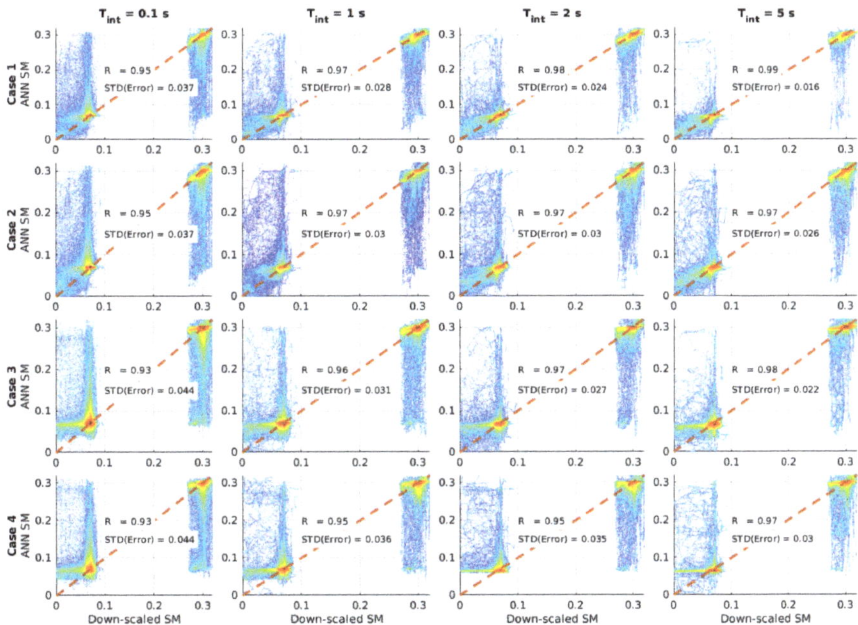

Figure 15. As for Figure 14 but for L5.

Moving to the L5 case (Figure 15), it can be seen that both the R and the standard deviation of the error are clearly better than in the L1 case. In any of the selected cases, the standard deviation of the error is ~2–3 times lower than in the L1 case. Furthermore, even with a small averaging (i.e., $T_{int} = 0.1$ s), the dispersion is smaller than at L1. Note that, these results are consistent with Figure 13, where the NSR at L5 was very close to the Speckle Noise limit, indicating that the L5 signal is less affected by surface roughness variations, especially during the *Wet* flight.

As the effective integration time increases, the ANN output shows lower error and a higher correlation coefficient with respect to the SMOS/Sentinel-2 downscaled soil moisture. Note that errors are drastically reduced for the case with larger averaging, where the standard deviation of the error is almost negligible (i.e., 0.016 m^3/m^3), showing a very low dispersion in both *Dry* and *Wet* flights.

4.4. Discussion

The presented results show a significant difference between L1 and L5 bands, with a standard deviation of the error at L1 being three times larger than at L5, despite the lower antenna directivity (D_{L_1} = 21 dB, D_{L_5} = 18 dB). This is due to a longer wavelength and a larger penetration depth, and by design, a much narrower autocorrelation function (30 m in space) at L5, which translates into a higher resolution. In this case, the peak of the L5 waveform contains contributions from a smaller glistening zone, increasing the coherency of the received signal. On the contrary, the L1 signal has a much larger autocorrelation function (300 m in space), and therefore contributions from a larger glistening zone are added in the L1 waveform, producing larger fluctuations than at L5.

Aside from the difference in bands, there is also a large variability depending on the selected integration time. In order to illustrate it, the same neural network is now deployed for T_{int} = 0.1, 0.5, 1, 2, 3, 4, and 5 s. Results are shown in Figure 16.

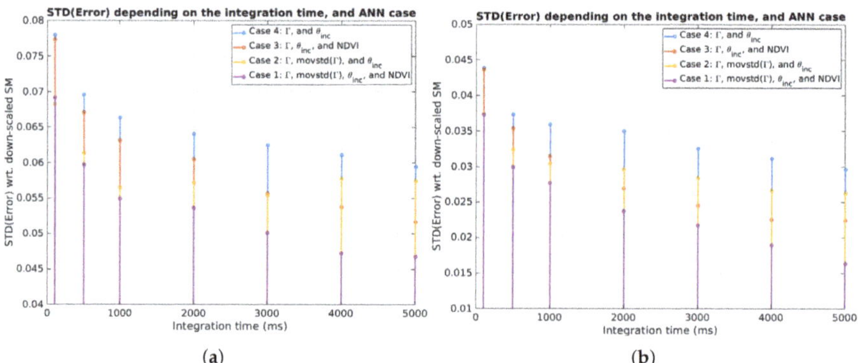

Figure 16. Standard deviation of the error of the ANN with respect to Sentinel-2/SMOS downscaled soil moisture for the four different cases for different integration times, at (**a**) L1 and (**b**) L5.

Comparing cases 2 and 3 for large integration times, case 3 (i.e., using movstd(Γ)) introduces larger errors than case 2 (i.e., using NDVI), which is not happening for lower integration times (i.e., 2000 ms at L1 and 1000 ms at L5). However, the combined use of the movstd(Γ) parameter together with the NDVI provides the lowest error for all integration times. In this case, as the vegetated areas are quite small (see Figure 3), a very large integration time produces errors, as a T_{int} = 5000 ms is equal to a specular point movement of ~375 m. However, the lower the integration time, the larger the surface roughness effect. On the contrary, for the smallest integration time, 100 ms, where the plane movement is equal to the size of the first Fresnel zone, case 3 provides the same error as case 1, and a much smaller error than cases 3 and 4. Thanks to the oversampling in the along-track direction, several measurements of speckle noise and surface roughness are included in the recovery algorithm by means of the movstd(Γ) term. However, for longer integration periods, e.g., T_{int} = 2000 ms, the signal covers up to ~15 Fresnel zones (i.e., v_{plane} = 75 m/s, and l_{F_r} ~10 m). The terrain inhomogeneity while crossing these areas together with the presence of different vegetated areas induces errors in the recovery algorithm, which can be only corrected using NDVI.

It is clear that there is a clear trade-off between spatial resolution and radiometric resolution, where it is not possible to achieve a good radiometric resolution and low root-mean-square error of the retrieved parameter and good spatial resolution at the same time. As shown in Figure 13 from [30], the standard deviation of their spaceborne GNSS-R SM retrieval algorithm is decreased to ± 0.1 m^3/m^3 when more than 25 averages are performed, and the smaller the number of averages, the larger the error, and hence the worse the radiometric resolution.

As shown in this study, when increasing the integration time, and therefore lowering the spatial resolution, the standard deviation of the error decreases, but effects due to terrain changes in the along-track direction of the GNSS-R measurement induce errors that need to be corrected for using NDVI measurements. On the contrary, reducing the averaging leads to a much higher resolution, and the NDVI term is not providing additional information to the ANN algorithm. However, the standard deviation of the error at such low integration times is larger.

Just to remark that the radiometric and the spatial resolutions cannot be maximized at the same time and there will always be a trade-off (if no ancillary data is used) between the required SM error and the spatial resolution of the GNSS-R-derived SM product.

5. Conclusions

This study has analyzed the data collected by the MIR instrument during two flights ("Dry" and "Wet") over the OzNet Yanco sites in New South Wales, Australia, during May–June 2018. The effect of increasing the averaging and its impact on the surface roughness estimation are addressed, showing that the effective integration time has to be increased to 5 s to neglect surface roughness effects. A statistical parameter based on the moving standard deviation over N samples of the reflectivity (movstd(Γ)) has been presented as a proxy of the surface roughness effects when the averaging (N) is large enough. Finally, an ANN-based algorithm has been presented for different combinations of auxiliary data and reflectivity averages, for L1 and L5 cases. In both cases, the use of the movstd(Γ) parameter reduces the error of the retrieved SM to 0.047 m^3/m^3 and 0.016 m^3/m^3 at L1 and L5 respectively, for a T_{int} = 5000 ms. Furthermore, the L5 signal shows a larger correlation coefficient with the expected SM output than the L1 signal because of the higher penetration depth and the narrower autocorrelation function.

Author Contributions: Conceptualization, J.F.M.-M. and A.C.; methodology, J.F.M.-M. and A.C.; software, J.F.M.-M. and H.P.; validation, M.P. and A.C.; formal analysis, J.F.M.-M., H.P. and A.C.; investigation, J.F.M.-M. and A.C.; resources, R.O., D.P., M.P., A.C., C.R., J.W. and A.M.; data curation, J.F.M.-M. and A.C.; visualization, J.F.M.-M.; supervision, H.P. and A.C.; project administration, A.C.; funding acquisition, A.C.; writing—original draft preparation, J.F.M.-M. and A.C.; writing—review and editing, J.F.M.-M., R.O., D.P., H.P., M.P., A.C., C.R., J.W. and A.M. All authors have read and agreed to the published version of the manuscript.

Funding: This work has been (partially) sponsored by project SPOT: Sensing with Pioneering Opportunistic Techniques grant RTI2018-099008-B-C21/AEI/10.13039/501100011033, by the Unidad de Excelencia Maria de Maeztu MDM-2016-0600, by the Spanish Ministry of Science and Innovation through the project ESP2017-89463-C3-1-R, by the Centro de Excelencia Severo Ochoa (CEX2019-000928-S), by the CSIC Plataforma Temática Interdisciplinar de Teledetección (PTI-Teledetect), and by the grant for recruitment of early-stage research staff FI-DGR 2018 of the AGAUR—Generalitat de Catalunya (FEDER), Spain.

Institutional Review Board Statement: Not applicable.

Informed Consent Statement: Not applicable.

Data Availability Statement: Restrictions apply to the availability of these data. Data was obtained from UPC and BEC and are available from the authors with the permission of M.P. and A.C.

Acknowledgments: The authors would like to thank Passive Remote Sensing Lab previous members who contributed in MIR and also Adrian Perez who helped create the MIR database used for data postprocessing. The authors would also like to thank Gerard Portal for his initial survey to retrieve the Sentinel-2 data.

Conflicts of Interest: The authors declare no conflict of interest.

References

1. Wasko, C.; Nathan, R. Influence of changes in rainfall and soil moisture on trends in flooding. *J. Hydrol.* **2019**, *575*, 432–441. [CrossRef]
2. Wei, L.; Zhang, B.; Wang, M. Effects of antecedent soil moisture on runoff and soil erosion in alley cropping systems. *Agric. Water Manag.* **2007**, *94*, 54–62. [CrossRef]
3. Miralles, D.G.; Gentine, P.; Seneviratne, S.I.; Teuling, A.J. Land-atmospheric feedbacks during droughts and heatwaves: State of the science and current challenges. *Ann. N. Y. Acad. Sci.* **2018**, *1436*, 19–35. [CrossRef] [PubMed]
4. Kerr, Y.H.; Waldteufel, P.; Wigneron, J.P.; Delwart, S.; Cabot, F.; Boutin, J.; Escorihuela, M.J.; Font, J.; Reul, N.; Gruhier, C.; et al. The SMOS Mission: New Tool for Monitoring Key Elements ofthe Global Water Cycle. *Proc. IEEE* **2010**, *98*, 666–687. [CrossRef]
5. Entekhabi, D.; Njoku, E.G.; ONeill, P.E.; Kellogg, K.H.; Crow, W.T.; Edelstein, W.N.; Entin, J.K.; Goodman, S.D.; Jackson, T.J.; Johnson, J.; et al. The Soil Moisture Active Passive (SMAP) Mission. *Proc. IEEE* **2010**, *98*, 704–716. [CrossRef]
6. Array Systems Computing Inc. ATBD for SMOS Level 2 Soil Moisture Processor Development Continuation Project. Available online: https://earth.esa.int/documents/10174/1854519/SMOS_L2_SM_ATBD (accessed on 9 November 2020).
7. O'Neill, P.; Bindlish, R.; Chan, S.; Chaubell, J.; Njoku, E.; Jackson, T. ATBD for Level 2 & 3 Soil Moisture (Passive)Data Products. Available online: https://smap.jpl.nasa.gov/system/internal_resources/details/original/484_L2_SM_P_ATBD_rev_F_final_Aug2020.pdf (accessed on 9 November 2020).
8. Jackson, R.D. Soil Moisture Inferences from Thermal-Infrared Measurements of Vegetation Temperatures. *IEEE Trans. Geosci. Remote Sens.* **1982**, *GE-20*, 282–286. [CrossRef]
9. Naeimi, V.; Scipal, K.; Bartalis, Z.; Hasenauer, S.; Wagner, W. An Improved Soil Moisture Retrieval Algorithm for ERS and METOP Scatterometer Observations. *IEEE Trans. Geosci. Remote Sens.* **2009**, *47*, 1999–2013. [CrossRef]
10. Srivastava, H.; Patel, P.; Sharma, Y.; Navalgund, R. Large-Area Soil Moisture Estimation Using Multi-Incidence-Angle RADARSAT-1 SAR Data. *IEEE Trans. Geosci. Remote Sens.* **2009**, *47*, 2528–2535. [CrossRef]
11. Paloscia, S.; Pettinato, S.; Santi, E.; Notarnicola, C.; Pasolli, L.; Reppucci, A. Soil moisture mapping using Sentinel-1 images: Algorithm and preliminary validation. *Remote Sens. Environ.* **2013**, *134*, 234–248. [CrossRef]
12. Gorrab, A.; Zribi, M.; Baghdadi, N.; Mougenot, B.; Fanise, P.; Chabaane, Z. Retrieval of Both Soil Moisture and Texture Using TerraSAR-X Images. *Remote Sens.* **2015**, *7*, 10098–10116. [CrossRef]
13. Rodriguez-Alvarez, N.; Bosch-Lluis, X.; Camps, A.; Aguasca, A.; Vall-llossera, M.; Valencia, E.; Ramos-Perez, I.; Park, H. Review of crop growth and soil moisture monitoring from a ground-based instrument implementing the Interference Pattern GNSS-R Technique. *Radio Sci.* **2011**, *46*. [CrossRef]
14. Yin, C.; Lopez-Baeza, E.; Martin-Neira, M.; Fernandez-Moran, R.; Yang, L.; Navarro-Camba, E.A.; Egido, A.; Mollfulleda, A.; Li, W.; Cao, Y.; et al. Intercomparison of Soil Moisture Retrieved from GNSS-R and from Passive L-Band Radiometry at the Valencia Anchor Station. *Sensors* **2019**, *19*, 1900. [CrossRef]
15. Alonso-Arroyo, A.; Camps, A.; Monerris, A.; Rudiger, C.; Walker, J.P.; Forte, G.; Pascual, D.; Park, H.; Onrubia, R. The light airborne reflectometer for GNSS-R observations (LARGO) instrument: Initial results from airborne and Rover field campaigns. In Proceedings of the 2014 IEEE Geoscience and Remote Sensing Symposium, Quebec City, QC, Canada, 13–18 July 2014; pp. 4054–4057. [CrossRef]
16. Egido, A.; Paloscia, S.; Motte, E.; Guerriero, L.; Pierdicca, N.; Caparrini, M.; Santi, E.; Fontanelli, G.; Floury, N. Airborne GNSS-R Polarimetric Measurements for Soil Moisture and Above-Ground Biomass Estimation. *IEEE J. Sel. Top. Appl. Earth Obs. Remote Sens.* **2014**, *7*, 1522–1532. [CrossRef]
17. Camps, A.; Park, H.; Castellví, J.; Corbera, J.; Ascaso, E. Single-Pass Soil Moisture Retrievals Using GNSS-R: Lessons Learned. *Remote Sens.* **2020**, *12*, 2064. [CrossRef]
18. Camps, A.; Park, H.; Pablos, M.; Foti, G.; Gommenginger, C.P.; Liu, P.; Judge, J. Sensitivity of GNSS-R Spaceborne Observations to Soil Moisture and Vegetation. *IEEE J. Sel. Top. Appl. Earth Obs. Remote Sens.* **2016**, *9*, 4730–4742. [CrossRef]
19. Camps, A.; Vall·llossera, M.; Park, H.; Portal, G.; Rossato, L. Sensitivity of TDS-1 GNSS-R Reflectivity to Soil Moisture: Global and Regional Differences and Impact of Different Spatial Scales. *Remote Sens.* **2018**, *10*, 1856. [CrossRef]
20. Chew, C.C.; Small, E.E. Soil Moisture Sensing Using Spaceborne GNSS Reflections: Comparison of CYGNSS Reflectivity to SMAP Soil Moisture. *Geophys. Res. Lett.* **2018**, *45*, 4049–4057. [CrossRef]
21. Calabia, A.; Molina, I.; Jin, S. Soil Moisture Content from GNSS Reflectometry Using Dielectric Permittivity from Fresnel Reflection Coefficients. *Remote Sens.* **2020**, *12*, 122. [CrossRef]
22. Edokossi, K.; Calabia, A.; Jin, S.; Molina, I. GNSS-Reflectometry and Remote Sensing of Soil Moisture: A Review of Measurement Techniques, Methods, and Applications. *Remote Sens.* **2020**, *12*, 614. [CrossRef]
23. Clarizia, M.P.; Pierdicca, N.; Costantini, F.; Floury, N. Analysis of CYGNSS Data for Soil Moisture Retrieval. *IEEE J. Sel. Top. Appl. Earth Obs. Remote Sens.* **2019**, *12*, 2227–2235. [CrossRef]

24. Chew, C.; Small, E. Description of the UCAR/CU Soil Moisture Product. *Remote Sens.* **2020**, *12*, 1558. [CrossRef]

25. Camps, A. Spatial Resolution in GNSS-R Under Coherent Scattering. *IEEE Geosci. Remote Sens. Lett.* **2019**, [CrossRef]

26. Camps, A.; Munoz-Martin, J.F. Analytical Computation of the Spatial Resolution in GNSS-R and Experimental Validation at L1 and L5. *Remote Sens.* **2020**, *12*, 3910. [CrossRef]

27. Gleason, S.; O'Brien, A.; Russel, A.; Al-Khaldi, M.M.; Johnson, J.T. Geolocation, Calibration and Surface Resolution of CYGNSS GNSS-R Land Observations. *Remote Sens.* **2020**, *12*, 1317. [CrossRef]

28. Wang, Y.; Morton, Y.J. Coherent GNSS Reflection Signal Processing for High-Precision and High-Resolution Spaceborne Applications. *IEEE Trans. Geosci. Remote Sens.* **2020**, *59*, 831–842. [CrossRef]

29. Clarizia, M.P.; Ruf, C.S. On the Spatial Resolution of GNSS Reflectometry. *IEEE Geosci. Remote Sens. Lett.* **2016**, *13*, 1064–1068. [CrossRef]

30. Yan, Q.; Huang, W.; Jin, S.; Jia, Y. Pan-tropical soil moisture mapping based on a three-layer model from CYGNSS GNSS-R data. *Remote Sens. Environ.* **2020**, *247*, 111944. [CrossRef]

31. Al-Khaldi, M.M.; Johnson, J.T.; O'Brien, A.J.; Balenzano, A.; Mattia, F. Time-Series Retrieval of Soil Moisture Using CYGNSS. *IEEE Trans. Geosci. Remote Sens.* **2019**, *57*, 4322–4331. [CrossRef]

32. Al-Khaldi, M.M.; Johnson, J.T.; Gleason, S.; Loria, E.; O'Brien, A.J.; Yi, Y. An Algorithm for Detecting Coherence in Cyclone Global Navigation Satellite System Mission Level-1 Delay-Doppler Maps. *IEEE Trans. Geosci. Remote Sens.* **2020**, 1–10. [CrossRef]

33. Yan, Q.; Gong, S.; Jin, S.; Huang, W.; Zhang, C. Near Real-Time Soil Moisture in China Retrieved From CyGNSS Reflectivity. *IEEE Geosci. Remote Sens. Lett.* **2020**, 1–5. [CrossRef]

34. Senyurek, V.; Lei, F.; Boyd, D.; Kurum, M.; Gurbuz, A.C.; Moorhead, R. Machine Learning-Based CYGNSS Soil Moisture Estimates over ISMN sites in CONUS. *Remote Sens.* **2020**, *12*, 1168. [CrossRef]

35. Rodriguez-Fernandez, N.J.; Aires, F.; Richaume, P.; Kerr, Y.H.; Prigent, C.; Kolassa, J.; Cabot, F.; Jimenez, C.; Mahmoodi, A.; Drusch, M. Soil Moisture Retrieval Using Neural Networks: Application to SMOS. *IEEE Trans. Geosci. Remote Sens.* **2015**, *53*, 5991–6007. [CrossRef]

36. Choudhury, B.J.; Schmugge, T.J.; Chang, A.; Newton, R.W. Effect of surface roughness on the microwave emission from soils. *J. Geophys. Res.* **1979**, *84*, 5699. [CrossRef]

37. Park, H.; Camps, A.; Castellvi, J.; Muro, J. Generic Performance Simulator of Spaceborne GNSS-Reflectometer for Land Applications. *IEEE J. Sel. Top. Appl. Earth Obs. Remote Sens.* **2020**, *13*, 3179–3191. [CrossRef]

38. Zhu, J.; Tsang, L.; Xu, H.; Gu, W. A Patch Model Based on Numerical Solutions of Maxwell Equations for GNSS-R Land Applications. In Proceedings of the IGARSS 2019-2019 IEEE International Geoscience and Remote Sensing Symposium, Yokohama, Japan, 28 July–2 August 2019. [CrossRef]

39. Li, F.; Peng, X.; Chen, X.; Liu, M.; Xu, L. Analysis of Key Issues on GNSS-R Soil Moisture Retrieval Based on Different Antenna Patterns. *Sensors* **2018**, *18*, 2498. [CrossRef]

40. Parrens, M.; Wigneron, J.P.; Richaume, P.; Mialon, A.; Bitar, A.A.; Fernandez-Moran, R.; Al-Yaari, A.; Kerr, Y.H. Global-scale surface roughness effects at L-band as estimated from SMOS observations. *Remote Sens. Environ.* **2016**, *181*, 122–136. [CrossRef]

41. Onrubia, R.; Pascual, D.; Camps, A.; Alonso-Arroyo, A.; Park, H. The Microwave Interferometric Reflectometer. Part I: Front-end and beamforming description. In Proceedings of the International Geoscience and Remote Sensing Symposium (IGARSS), Quebec City, QC, Canada, 13–18 July 2014; pp. 4046–4049. [CrossRef]

42. Pascual, D.; Onrubia, R.; Alonso-Arroyo, A.; Park, H.; Camps, A. The microwave interferometric reflectometer. Part II: Back-end and processor descriptions. In Proceedings of the International Geoscience and Remote Sensing Symposium (IGARSS), Quebec City, QC, Canada, 13–18 July 2014; pp. 3782–3785. [CrossRef]

43. Zavorotny, V.U.; Gleason, S.; Cardellach, E.; Camps, A. Tutorial on Remote Sensing Using GNSS Bistatic Radar of Opportunity. *IEEE Geosci. Remote Sens. Mag.* **2014**, *2*, 8–45. [CrossRef]

44. Munoz-Martin, J.F.; Onrubia, R.; Pascual, D.; Park, H.; Camps, A.; Rüdiger, C.; Walker, J.; Monerris, A. Untangling the Incoherent and Coherent Scattering Components in GNSS-R and Novel Applications. *Remote Sens.* **2020**, *12*, 1208. [CrossRef]

45. Munoz-Martin, J.F.; Onrubia, R.; Pascual, D.; Park, H.; Camps, A.; Rüdiger, C.; Walker, J.; Monerris, A. Experimental Evidence of Swell Signatures in Airborne L5/E5a GNSS-Reflectometry. *Remote Sens.* **2020**, *12*, 1759. [CrossRef]

46. Smith, A.B.; Walker, J.P.; Western, A.W.; Young, R.I.; Ellett, K.M.; Pipunic, R.C.; Grayson, R.B.; Siriwardena, L.; Chiew, F.H.S.; Richter, H. The Murrumbidgee soil moisture monitoring network data set. *Water Resour. Res.* **2012**, *48*. [CrossRef]

47. Piles, M.; Camps, A.; Vall-llossera, M.; Corbella, I.; Panciera, R.; Rüdiger, C.; Kerr, Y.H.; Walker, J. Downscaling SMOS-Derived Soil Moisture Using MODIS Visible/Infrared Data. *IEEE Trans. Geosci. Remote Sens.* **2011**, *49*, 3156–3166. [CrossRef]

48. Portal, G.; Vall-llossera, M.; Piles, M.; Camps, A.; Chaparro, D.; Pablos, M.; Rossato, L. A Spatially Consistent Downscaling Approach for SMOS Using an Adaptive Moving Window. *IEEE J. Sel. Top. Appl. Earth Obs. Remote Sens.* **2018**, *11*, 1883–1894. [CrossRef]

49. Pablos, M.; Piles, M.; Gonzalez-Haro, C. BEC SMOS Land Products Description. Available online: http://bec.icm.csic.es/doc/BEC-SMOS-0003-PD-Land.pdf (accessed on 22 December 2020).

50. Hajj, M.E.; Baghdadi, N.; Zribi, M.; Rodríguez-Fernández, N.; Wigneron, J.; Al-Yaari, A.; Bitar, A.A.; Albergel, C.; Calvet, J.C. Evaluation of SMOS, SMAP, ASCAT and Sentinel-1 Soil Moisture Products at Sites in Southwestern France. *Remote Sens.* **2018**, *10*, 569. [CrossRef]

51. Singergise Ltd. Sentinel Hub. Available online: https://www.sentinel-hub.com/ (accessed on 5 November 2020).

52. Center, B.E. Barcelona Expert Center Webpage. Available online: http://bec.icm.csic.es/ (accessed on 22 December 2020).
53. Giacobazzi, J. 50 - Line of sight radio systems. In *Telecommunications Engineer's Reference Book*; Mazda, F., Ed.; Butterworth-Heinemann: Oxford, UK, 1993; pp. 50-1–50-17. [CrossRef]
54. Ramos-Pérez, I.; Bosch-Lluis, X.; Camps, A.; Rodriguez-Alvarez, N.; Marchán-Hernandez, J.F.; Valencia-Domènech, E.; Vernich, C.; de la Rosa, S.; Pantoja, S. Correction: Ramos-Pérez, I. *et al.* Calibration of Correlation Radiometers Using Pseudo-Random Noise Signals. *Sensors* 2009, *9*, 6131–6149. *Sensors* **2009**, *9*, 7430. [CrossRef]
55. Pascual, D.; Onrubia, R.; Querol, J.; Park, H.; Camps, A. Calibration of GNSS-R receivers with PRN signal injection: Methodology and validation with the microwave interferometric reflectometer (MIR). In Proceedings of the 2017 IEEE International Geoscience and Remote Sensing Symposium (IGARSS), Fort Worth, TX, USA, 23–28 July 2017; pp. 5022–5025. [CrossRef]
56. Onrubia, R. Advanced GNSS-R Instruments for Altimetry and Scatterometry. Ph.D. Thesis, Universitat Politècnica de Catalunya, Barcelona, Spain, 2020.
57. Kaleschke, L. SMOS Sea ice retrieval study (SMOSSIce): Final report. ESA ESTEC Contract No: 4000202476/10/NL/CT. *STSE SMOSIce Final. Rep.* **2013**. [CrossRef]
58. Grant, J.; Wigneron, J.P.; Jeu, R.D.; Lawrence, H.; Mialon, A.; Richaume, P.; Bitar, A.A.; Drusch, M.; van Marle, M.; Kerr, Y. Comparison of SMOS and AMSR-E vegetation optical depth to four MODIS-based vegetation indices. *Remote Sens. Environ.* **2016**, *172*, 87–100. [CrossRef]
59. Ulaby, F.; Moore, R.; Fung, A. *Microwave Remote Sensing: Active and Passive*; Number v. 3 in Artech House Remote Sensing Library; Addison-Wesley Publishing Company, Advanced Book Program/World Science Division: Boston, MA, USA, 1981.
60. Maity, A.; Pattanaik, A.; Sagnika, S.; Pani, S. A Comparative Study on Approaches to Speckle Noise Reduction in Images. In Proceedings of the 2015 International Conference on Computational Intelligence and Networks, Odisha, India, 12–13 January 2015; pp. 148–155. [CrossRef]
61. Ying, X. An Overview of Overfitting and its Solutions. *J. Phys. Conf. Ser.* **2019**, *1168*, 022022. [CrossRef]

Article

Evaluations of Machine Learning-Based CYGNSS Soil Moisture Estimates against SMAP Observations

Volkan Senyurek [1], Fangni Lei [1], Dylan Boyd [2], Ali Cafer Gurbuz [2], Mehmet Kurum [2,*], Robert Moorhead [1,2]

[1]   Geosystems Research Institute, Mississippi State University, Mississippi State, MS 39759, USA; volkan@gri.msstate.edu (V.S.); fangni@gri.msstate.edu (F.L.); rjm@gri.msstate.edu (R.M.)
[2]   Department of Electrical and Computer Engineering, Mississippi State University, Mississippi State, MS 39759, USA; db1950@msstate.edu (D.B.); gurbuz@ece.msstate.edu (A.C.G.)
*   Correspondence: kurum@ece.msstate.edu

Received: 8 September 2020; Accepted: 21 October 2020; Published: 25 October 2020

Abstract: This paper presents a machine learning (ML) framework to derive a quasi-global soil moisture (SM) product by direct use of the Cyclone Global Navigation Satellite System (CYGNSS)'s high spatio-temporal resolution observations over the tropics (within $\pm 38°$ latitudes) at L-band. The learning model is trained by using in-situ SM data from the International Soil Moisture Network (ISMN) sites and various space-borne ancillary data. The approach produces daily SM retrievals that are gridded to 3 km and 9 km within the CYGNSS spatial coverage. The performance of the model is independently evaluated at various temporal scales (daily, 3-day, weekly, and monthly) against Soil Moisture Active Passive (SMAP) mission's enhanced SM products at a resolution of 9 km $\times$ 9 km. The mean unbiased root-mean-square difference (ubRMSD) between concurrent (same calendar day) CYGNSS and SMAP SM retrievals for about three years (from 2017 to 2019) is 0.044 cm^3 cm^{-3} with a correlation coefficient of 0.66 over SMAP recommended grids. The performance gradually improves with temporal averaging and degrades over regions regularly flagged by SMAP such as dense forest, high topography, and coastlines. Furthermore, CYGNSS and SMAP retrievals are evaluated against 170 ISMN in-situ observations that result in mean unbiased root-mean-square errors (ubRMSE) of 0.055 cm^3 cm^{-3} and 0.054 cm^3 cm^{-3}, respectively, and a higher correlation coefficient with CYGNSS retrievals. It is important to note that the proposed approach is trained over limited in-situ observations and is independent of SMAP observations in its training. The retrieval performance indicates current applicability and future growth potential of GNSS-R-based, directly measured spaceborne SM products that can provide improved spatio-temporal resolution than currently available datasets.

Keywords: GNSS-reflectometry; random forest; CYGNSS; soil moisture retrieval; ISMN; SMAP

1. Introduction

Soil moisture (SM) is an essential hydroecological variable that plays an important role in water dynamics, evapotranspiration [1], the energy and carbon flows between the surface and the atmosphere [2], vegetation states [3], climatic conditions [4], and many hydrological and agricultural processes. Microwave remote sensing at L-band (1–2 GHz) is an established method for providing accurate SM retrievals on a global basis because the primary physical property that affects the microwave measurement is directly dependent on the amount of water present in the top 5 cm of soil. The European Space Agency's Soil Moisture and Ocean Salinity (SMOS, launched in late 2009) [5] and the National Aeronautics and Space Administration (NASA)'s Soil Moisture Active Passive (SMAP, launched in early 2015) [6] are two passive microwave spaceborne missions that are currently operating at L-band. Both provide global SM retrievals every 2–3 days with a spatial resolution on the order of several tens of kilometers. While this product is critical for many large-scale climate studies, a higher spatio-temporal resolution SM product is needed to advance applications in hydrometeorology, atmospheric research, and water resource management at regional and local scales [7–9].

There is a growing interest within the hydrology community to utilize spaceborne Global Navigation Satellite System (GNSS) Reflectometry (GNSS-R) observations in SM retrievals due to its relatively high spatial footprint (on the order of a few kilometers) over smooth Earth surface with frequent observation (subdaily or daily) capabilities [10]. The GNSS-R approach re-purposes existing GNSS infrastructure for remote sensing by processing the forward-scattered signal that is reflected off the surface of the Earth [11]. The exploitation of this approach has been particularly accelerated with the availability of large and diverse datasets acquired from recent spaceborne GNSS-R observatories such as the United Kingdom's TechDemoSat-1 (TDS-1, launched in mid-2014 and retired in mid-2019) [12] and NASA's Cyclone Global Navigation Satellite System (CYGNSS, launched in late 2016) [13]. While TDS-1's GNSS-R measurements featured a relatively low revisit time due to the satellites intention for evaluating multiple payloads simultaneously, the dedicated research community surrounding TDS-1 has greatly improved the literature's understanding of spaceborne GNSS-R responses to ocean winds [14,15], ice sheets [16,17], and land geophysical parameter features [18–20] from space due to TDS-1's global coverage created by its polar orbiting configuration. On the other hand, CYGNSS features a high-temporal resolution over a smaller spatial extent in order to optimize its measurements for ocean studies. To achieve this, CYGNSS uses eight small satellites that orbit the tropics (within ±38° latitudes). The considerable amount of CYGNSS land observation data also measured in this region has greatly contributed to the development of new SM retrieval approaches from spaceborne GNSS-R [21–29].

The GNSS-R approach comes with many unique challenges for SM retrieval such as (1) the variable footprint due to the relative contributions of coherent and incoherent scattering mechanisms, (2) non-repeating ground tracks due to quasi-random transmitter/receiver configurations, (3) sensitivity to fine-scale surface features (e.g., topography, water bodies), and (4) other non-geophysical factors (e.g., variation/uncertainty of the GNSS transmitter power). Spatial and/or temporal averaging are often employed in retrieval algorithms to overcome most of the aforementioned challenges. The majority of the previous GNSS-R-based SM inversion studies also calibrate their models and test their performance with SMAP or point-scale in-situ observations. However, it is difficult to directly compare the CYGNSS-based SM data products from these studies against each other because each paper assumes critically different inversion methodologies due to differences in (1) assumptions regarding gridding, open water masking, and surface conditions, (2) ancillary data requirements, (3) validation and reference data sets, (4) time spans, (5) models, and (6) spatial coverage. Despite these differences, most approaches have shown a moderate performance in retrieving SM from CYGNSS as summarised in Appendix A (see Figure A1). The purpose of this study is to achieve a daily 9-km gridded SM product within the CYGNSS spatial coverage (±38° latitudes) and evaluate the predictions against Soil Moisture Active Passive (SMAP) mission's enhanced SM products at a resolution of 9 km × 9 km.

In this paper, we apply a machine learning (ML) framework that is similar to those previously developed by [25,26] to perform SM predictions at a high spatio-temporal resolution. The framework uses in-situ SM data sets provided by the International Soil Moisture Network (ISMN) to train CYGNSS observations that fall within approximately 9 km × 9 km grid centering each ISMN site. Several physics-aware features from both CYGNSS observables and other space-borne ancillary data are incorporated into the model to contend with the nonlinear relationship between the CYGNSS signals and SM. While Eroglu et al. [25] established artificial neural network (ANN) retrieval over limited ISMN sites as a proof-of-concept using a ten-fold validation method, Senyurek et al. [26] generalized the framework by utilizing a larger and more diverse dataset and developing ANN, random forest (RF), and SVM approaches with 5-fold, site-independent, and year-based cross-validation strategies. This paper extends these ML-based studies to global scales by including more reference stations and applying an independent validation strategy. The learning model utilizes in-situ SM data sets provided by the 170 ISMN sites over a nearly 3-year period as the reference. Most importantly, the model's performance is evaluated by using SMAP observations for 9-km gridded resolution at various temporal scales within the CYGNSS coverage.

The most distinct aspect of this study with respect to the previous studies, the framework does not use SMAP data in the training stage but uses it for independent validation only. In other words, the CYGNSS-based SM estimates are independent of SMAP observations, so we make predictions anywhere within CYGNSS coverage as long as it meets the quality control criteria of our approach (see details in our preceding publication [26]). However, the accuracy of our approach over denser vegetation (e.g., the Amazon) is expected to be degraded due to the limited vegetation penetration of L-band and lack of training over forested regions. Additionally, SMAP retrievals exist but their accuracy is not guaranteed over such regions. Therefore, this study often divides its comparison into the two groups "all grids" and "SMAP recommended grids". Both SMAP and CYGNSS results are compared against each other within the CYGNSS coverage and at the same in-situ SM observations at ISMN sites. Temporal and regional studies are also demonstrated over selected areas (the US, India, Africa, and Australia) across the globe. The proposed approach achieves a performance better than 0.05 cm^3 cm^{-3} mean unbiased root-mean-square difference (ubRMSD) at a 9-km spatial resolution within the CYGNSS coverage and with a similar mean unbiased root-mean-square errors (ubRMSE) performance over ISMN sites.

The rest of the paper is organized as follows: Section 2 summarizes the CYGNSS and SMAP data, and ancillary data needed for the ML algorithm. Section 3 describes the training and evaluation methodology for the proposed ML framework. In Section 4, the results of CYGNSS SM retrievals and performance analysis are provided. This includes a study on effects of land cover on the performance of the proposed approach as well as the temporal and spatial illustrations on SM predictions of the proposed approach over selected regions across the globe. Section 5 discusses findings, challenges, and implications for future work. Finally, Section 6 summarizes our conclusions.

2. Datasets

In order to effectively develop an ML-based retrieval algorithm for surface SM from CYGNSS observations, different land surface products must be leveraged to indicate the underlying surface conditions. The following subsections introduce the input feature selection for the retrieval algorithm as well as the physical relationships related to the GNSS-R observables and SM content. The data quality control and spatial-temporal aggregation processes used by this framework to ensure consistency for accurate SM products are given. Here, pertinent details are provided and additional details can be found in [26].

2.1. Cyclone Global Navigation Satellite System

The CYGNSS data used for this study is the Level-1 (L1) version 2.1 product, available at the NASA Physical Oceanography Distributed Active Archive Center (PO.DAAC, https://podaac.jpl.nasa.gov/). The CYGNSS mission consists of eight micro-satellites that record the reflected Global Positioning System (GPS) signals via a four-channel GNSS-R bistatic radar receiver. These micro-satellites primarily orbit the tropics, limiting their spatial coverage to latitudes $\pm 38°$.

In L1 data, the delay Doppler map (DDM) is processed to obtain the bistatic radar cross section (BRCS) and is represented as a 17 delay by 11 Doppler array. Under most circumstances, the surface reflections of GNSS signals are the combinations of coherent and incoherent reflections. Meanwhile, the spatial footprints for varying reflections are dependent on the bistatic geometry among the transmitter, reflecting surface, and receiver. Theoretically, the CYGNSS received DDM signals are impacted by the instrumental specifications and land surface characteristics. After correcting for the GPS transmitting power, antenna gain, signal travel distance, and wavelength, the derived reflectivity can be representative of the soil water content level and land surface roughness and vegetation conditions. Following [21,22,24–29], the CYGNSS reflectivity is calculated assuming a coherent reflection over land. This reflectivity derived using the coherent reflection assumption has demonstrated a satisfactory performance for estimating surface SM in various previous studies. In addition, from the BRCS, the reflectivity delay waveform can be calculated for the Doppler domain between the peak delay bin m and $m + 3$, which is then used for deriving the trailing edge slope (TES) [30]. TES is one of the indexes indicating the surface coherence/incoherence reflection, providing supplementary information for the ML-based retrieval model [25,26]. The instrumental data given in CYGNSS's L1 product along with other metadata provide key information for filtering CYGNSS observations with larger uncertainties. The standard filtering criteria used in [26,27,30] are also applied here. Additional constraints for the CYGNSS data are applied and explained in Section 3.1. The CYGNSS data is spatially aggregated into EASE-Grid 2.0 3-km cells and is temporally averaged for daily values. For the ML process, three features derived from CYGNSS are used, i.e., specular point incidence angle, TES, and the reflectivity [30].

2.2. International Soil Moisture Network

Here, daily averaged in-situ SM data of 170 sites selected from the International Soil Moisture Network (ISMN) are used as ground truth information for the training of the ML model. The spatial distribution of in-situ sites used in this study are shown in Figure 1. The ISMN provides a global in-situ SM database with uniform data format and pre-processing quality flags [31] over the world. While there are some sites in Asia, Australia, and Europe, most ISMN sites within CYGNSS's spatial coverage are in the Continental United States (CONUS). A similar distribution of sites can be found for the validation of other global retrievals such as SMAP's SM products over core validation sites [32] and CYGNSS SM estimates over ISMN sites [27]. Nonetheless, the ISMN sites across CONUS provide a good coverage of heterogeneous land surface conditions. It is important to note that a good agreement between CYGNSS and in situ observations is expected to be achieved as in [26], which does not mean that CYGNSS's SM predictions perform well in regions extremely disparate from those typical of the United States (e.g., tropical rain forests).

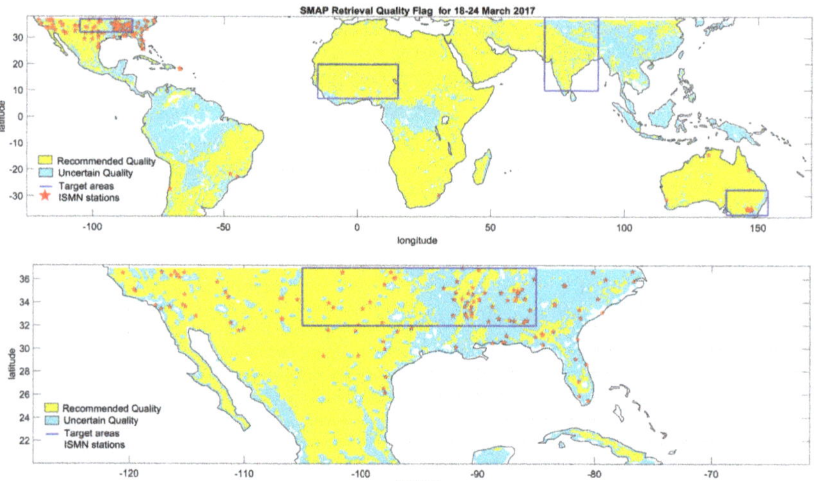

Figure 1. Soil Moisture Active Passive (SMAP) Retrieval Quality Flag (RQF) indicates whether soil moisture (SM) retrieval is recommended (green) or not recommended (cyan) over the specified location. Quality uncertain quality is influenced by factors such as water body fraction, coastal proximity, urban area, precipitation, terrain slope, and high vegetation water content. Red pentagrams show locations of International Soil Moisture Network (ISMN) stations. Blue boxes show target areas that are investigated in Section 4.3.

Detailed information about the ISMN is reported in [33,34]. The ISMN dataset is publicly accessible at http://ismn.geo.tuwien.ac.at. Here, the hourly SM data from ISMN is masked using the provided quality flag (identified as 'good' with 'G') and then averaged to daily values. From this ISMN data, the surface SM at a 5-cm depth is employed which is consistent with the penetration depth of L-band microwave signals.

2.3. Ancillary Data

Geophysical parameters such as vegetation density, topography, surface roughness, and soil texture also play important roles in securing accurate SM predictions in addition to CYGNSS observables. We use the five following ancillary datasets as input features to the learning model: (i) Normalized Difference Vegetation Index (NDVI), (ii) Vegetation Water Content (VWC), (iii) elevation, (iv) soil clay ratio, and (v) soil silt ratio.

The 16-day composite Normalized Difference Vegetation Index (NDVI) from Moderate Resolution Imaging Spectroradiometer (MODIS) data of MYD13A1 is utilized for characterizing vegetation conditions. The original 500-m resolution of NDVI data is spatially averaged to 3 km. The MYD13A1 dataset can be acquired from the NASA Land Processes Distributed Active Archive Center (https://lpdaac.usgs.gov/products/myd13a1v006/). Vegetation Water Content (VWC), used in the model, is derived from NDVI and Land Cover Type (MCD12Q1) product through the same lookup table approach as the SMAP VWC product [35]. The 1-km resolution Digital Elevation Model GTOPO30 product from the United States Geological Survey Earth Resources Observation and Science archive is used to provide surface elevation information. Similarly, the elevation data is spatially averaged from 1 km to 3 km. The Global Gridded Soil Information (SoilGrids) [36] is used to obtain soil clay and silt ratios. In the SoilGrids product, soil profiles are discretized into different layers and only data from the top layer (surface to 5-cm in depth) is used for consistency with the penetration depth of L-band signals. The product is available at 250 m and is spatially aggregated onto 3 km for this study. The presence of surface inland water body is identified by utilizing a 30-m Global Surface Water Dataset from the Joint Research Centre (GSW-JRC) [37]. The water percent is determined by calculating the percentage of 30-m grids within each 3-km grid indicating the presence of either permanent or seasonal water, and this value is used during the retrieval algorithm's quality control phase.

2.4. SMAP Radiometer Soil Moisture Data

The CYGNSS-based SM retrieval framework is independently validated by using the SMAP Enhanced L3 Radiometer Global Daily 9-km EASE-Grid SM product. SMAP collects brightness temperature data via an L-band microwave radiometer and provides SM estimations at a 5-cm depth for each half orbit. Although the SMAP radiometer soil moisture product is generated at 36 km, a 9-km enhanced grid product is available by using Backus–Gilbert optimal interpolation techniques [38]. The data is freely available through the National Snow and Ice Data Center (NSIDC) at https://nsidc.org/data/SPL3SMP_E/versions/3. Retrieval Quality Flag (RQF) information is provided in the product as well. The first bit of the RQF is a summary flag indicating whether the SM retrieval is recommended or not. SM retrievals can have an uncertain quality for several reasons that includes water body fraction, coastal proximity, urban area, precipitation, slope, vegetation water content, etc. Figure 1 shows an example of SMAP RQF for the period of 18–24 March 2017. In the following validation of the proposed method, performance metrics are calculated separately for regions with only the recommended quality and the whole study area.

3. Methodology

The SM product is a complex non-linear function of CYGNSS's measured reflectivity, system geometry, as well as environmental variables such as topography, vegetation, and soil properties. Our approach to a model with such complex and unknown relationships is to use a data-driven ML technique with physics-based features. Since ML techniques are known to be universal function approximators, the goal of our framework is to map the unknown relationships between available datasets to SM values. Our previous approach [26] has shown that a supervised ML technique has remarkable SM estimation performance when trained and tested over the ISMN sites in CONUS. In this study, we extend our previous results and evaluate our ML-based CYGNSS SM estimates against SMAP observations within the CYGNSS coverage. In Section 3.1, the training methodology of the ML model for SM estimation is described. An evaluation and application of the learned model is detailed in Section 3.2.

3.1. Training of Random Forest Model

We consider a supervised learning problem where the ML model maps a set of input features to the SM value, which is the final output label. A supervised training of an ML model requires a labeled dataset where the input features and the corresponding outputs, i.e., labels, are known. Such a training set is constructed using 170 ISMN sites over the globe as described in Section 2. Under this setting and following the feature selection results of our previous study [26], a set of 8 features from a 9-km $\times$ 9-km representative area centered around ISMN sites is used as the inputs to our ML model. These 8 features include observables from CYGNSS (i.e., reflectivity, incidence angle, and TES), MODIS (i.e., NDVI and VWC), SoilGrids soil texture database (i.e., soil clay and silt ratios), and GTOPO30 DEM data (i.e., elevation). The set of these input features is passed through a series of quality checks and additional masks before they are utilized in ML model's training. (1) CYGNSS data with reflection power higher than -5 dB and lower than -30 dB, and specular point incident angle higher than $65°$ are filtered out. (2) CYGNSS observations are removed if more than 2% of the 3-km $\times$ 3-km area centered around the specular point is covered with permanent or seasonal water. The specular reflection from water bodies is typically much stronger than land reflections and, hence, including data samples near open water areas creates incorrect training data. (3) CYGNSS observations from urban areas are removed since reflections from these regions are not reliable for SM estimation. (4) ISMN sites above an altitude of 2000 m are not utilized in training since CYGNSS specular point for high altitudes are not reliable. (5) CYGNSS observations before December 2017 from a surface elevation above 600 m have been masked out due to the altitude limitation of CYGNSS L1 for the specified time frame. The remaining CYGNSS observations in the 9-km $\times$ 9-km area around ISMN sites along with the other features calculated from the same region constitute the input training data. As the labels, for these input feature sets, daily SM measurements from the corresponding ISMN sites are used. At this point, it is important to note that, although ISMN sites provide the SM value at their specific positions, we assume that the ISMN SM observation is a representative SM value for the entire 9-km $\times$ 9-km area around the ISMN site [26].

After the application of quality checks and data masking, from approximately 81,000 initial samples for the period of March 2017–May 2019, the remaining total of approximately 59,000 sets of input features and ISMN SM value data samples have been utilized as the training dataset for the ML model. Figure 2 illustrates the simplified training procedure. For the ML model, a RF regression structure is utilized. Our previous study [26] compared different ML models for the problem of CYGNSS-based SM estimation and showed that RF models provided enhanced SM estimation performance compared to other tested ML techniques. The developed RF model in this study contains 100 trees with a maximum split size of 60 for each tree. The RF model is trained using a least-squares boosting (LSBoost) ensemble strategy with a learning rate of 0.1. The required computations are carried out using the machine learning toolbox of MATLAB R2019b software over a machine with Intel(R) Xeon(R) CPU E5-2643 and 128 GB memory.

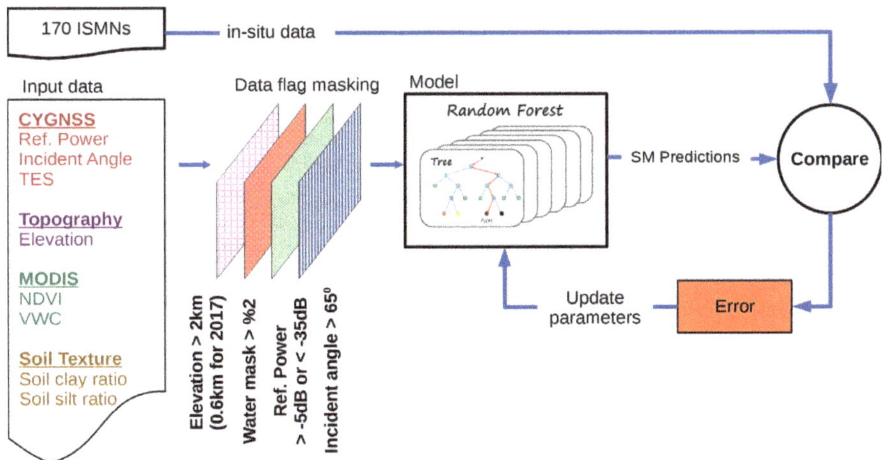

Figure 2. Simplified random forest (RF) model training process.

3.2. Quasi-Global Application and Evaluation of the Model

The learned RF model from Section 3.1 is applied within the CYGNSS coverage, and its performance is evaluated against the SMAP SM predictions. To achieve a proper evaluation, first the whole region bound between latitudes $\pm 38°$ is divided into EASE-Grid 2.0 3-km $\times$ 3-km grids creating a matrix of 2967 by 11,568 for the study region. For each grid cell, the input features for the ML model is acquired. This means that for each observation day an 8×1 feature vector is created for each of the 3-km $\times$ 3-km grids. The same CYGNSS, water body, elevation quality checks, and masking procedures that were utilized in training are also applied to the input data. The input features are fed into the learned RF model that outputs a SM prediction. In this way the learned RF model, which used the training dataset only from ISMN locations, generates SM predictions at unseen locations. To evaluate the performance of the SM predictions from the RF model, its predictions are compared against SMAP observations. However, the proposed ML-based model generates daily SM predictions at 3-km $\times$ 3-km resolutions. The SMAP SM product has a temporal resolution of 2–3 days at both 9-km $\times$ 9-km and 36-km $\times$ 36-km spatial resolutions. To properly compare ML-based SM predictions with SMAP, both datasets are spatially and temporally averaged to directly comparable resolutions. First, SM outputs of both SMAP and the proposed ML approach are temporally averaged over k consecutive days (i.e., $k =$ daily, 3-days, 1-week, and 1-month), creating a k-day average SM product for both SMAP and the proposed approach. The daily CYGNSS data set refers the data averaged over the same calendar day that is concurrent with SMAP. In addition, to achieve the same spatial resolution, the 3-km resolution of the proposed SM estimation results are spatially averaged to SMAP spatial resolution levels. The general outline of the aforementioned steps for a k-days, 9-km $\times$ 9-km SM product is illustrated in Figure 3.

After achieving the same spatial and temporal resolutions, the CYGNSS ML model predictions are compared with both SMAP SM products and in-situ measurements. Several performance metrics including the root-mean-square error (RMSE), unbiased RMSE (ubRMSE), bias, and correlation coefficient (R) are calculated. The root-mean-square difference (RMSD) term is preferred for CYGNSS ML and SMAP SM product comparison because the reference SMAP SM values may contain errors and cannot be considered as the "true" SM values [39] whereas RMSE is used for in-situ evaluation since in-situ measurements are ground truth information for SM. Both RMSD and RMSE metrics are calculated in the same way. The next section provides performance results and detailed analysis on the proposed CYGNSS-ML based SM products.

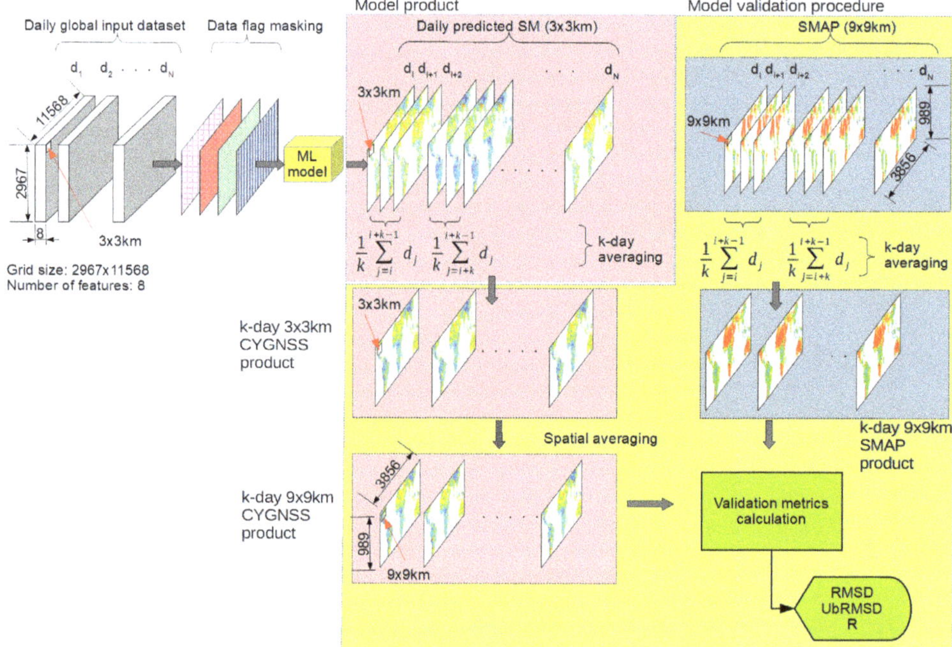

Figure 3. Application of the proposed machine learning (ML) model and k-days evaluation process.

4. Results

In this section, we evaluate the performance of the proposed CYGNSS ML-based SM retrieval against SMAP observations within the CYGNSS coverage. For this evaluation, the results are organized in order to discuss important factors that impact SM inversion quality such as spatial heterogeneity and temporal variation. First, the overall performance using the proposed SM predictions are shown in Section 4.1, along with a detailed analysis of the land cover effects. Next, both SMAP and the CYGNSS ML-based results are compared against in-situ SM observations at ISMN sites in Section 4.2. Spatial and temporal analysis on SM predictions of the proposed approach are detailed in Section 4.3, with examples extracted from different regions of the world.

4.1. Quasi-Global Performance Results of the Proposed ML-Based SM Retrieval

This study's ML-based SM retrieval approach is designed to generate daily SM predictions for 3 km × 3 km grids using CYGNSS observations. In order to compare and evaluate these SM predictions with the SMAP observations, both SM products are sampled to the same spatial and temporal resolutions through gridding onto 9-km cells over k-days periods as detailed in Section 3. SM products are compared over a period of nearly three years (18 March 2017–31 December 2019). Figure 4 shows the ubRMSD between 3-day averaged SMAP and CYGNSS SM retrievals for each cell. For any grid cell with less than 30 collocated SM predictions in total over the 3-day averaging period, its performance metrics are not calculated or shown in the map to limit errors caused by a small number of samples. The mean ubRMSD for all measurements over the entire period is found to be 0.049 cm^3 cm^{-3} with a standard deviation of 0.025 cm^3 cm^{-3}. This result illustrates the potential and applicability of CYGNSS observations for high spatially and temporally-resolved global SM products. Even though the proposed approach uses no SMAP observations in its training, the retrieval still achieves a performance of less than 0.05 cm^3 cm^{-3} ubRMSD when compared to SMAP SM measurements independently. Regions that are generally flagged by SMAP as being 'poor quality' such as the Amazon, central Africa, and seasonal flooding regions (e.g., Southeast Asia) are also shown to have relatively worse ubRMSD values. This result is not surprising since the performance of CYGNSS SM estimates is also expected to degrade over those regions since they are not well represented by ISMN site networks that are used in the model training. On the other hand, most of the remaining regions show a high agreement with SMAP observations with lower ubRMSD results.

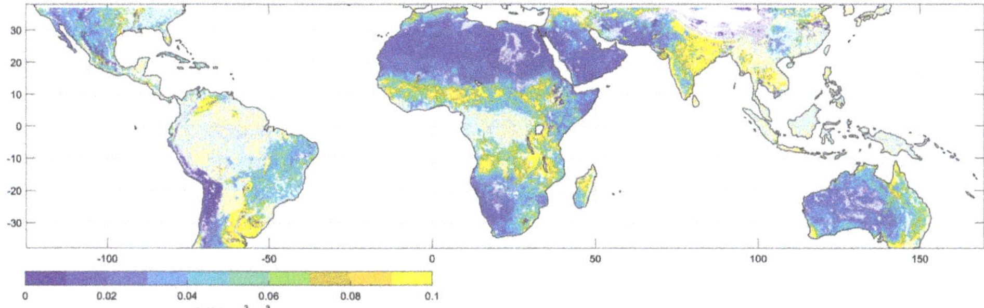

Figure 4. Unbiased root-mean-square difference (ubRMSD) between Cyclone Global Navigation Satellite System (CYGNSS)-based model prediction and SMAP retrievals. Semi-transparent regions are those flagged by SMAP as having uncertain SM retrieval quality.

Figure 5 shows the bias between SMAP and CYGNSS-based SM retrievals for each grid within the CYGNSS coverage. The mean values of SM estimations of SMAP and CYGNSS are close for most of the globe except for the Amazon and central Africa regions where SMAP SM retrievals are also mostly flagged with uncertain quality. Over such regions (with mostly dense vegetation), CYGNSS estimates are often drier than those made by SMAP, leading to negative bias values. This bias could be attributed to uncertainties in SMAP's SM estimates over flagged regions as well as the limited training dataset used in the ML algorithm. The ISMN sites are not equally distributed around the world, and the training dataset does not include data samples from regions with such a high bias (see cyan color around equatorial regions in Figure 1). Additionally, the training dataset constructed from ISMN observations does not include enough samples of high SM conditions and therefore the learned model trained with such data samples can underestimate SM for relatively high SM levels.

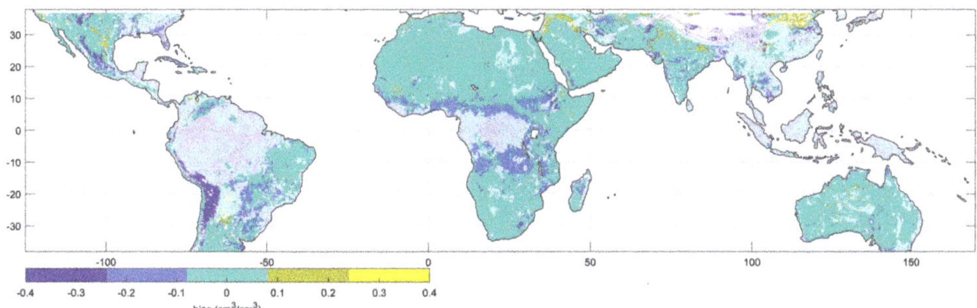

Figure 5. Bias between CYGNSS-based model prediction and SMAP retrievals. Semi-transparent regions are those flagged by SMAP as having an uncertain SM retrieval quality.

Another important metric is the correlation coefficient (R) which can be used to describe how CYGNSS-based ML predictions are spatially correlated with SMAP observations. Figure 6 shows the R values between SMAP and CYGNSS predictions as a function of time. It can be seen that the CYGNSS SM retrieval is highly correlated with SMAP observations with the coefficient changing between 0.6 and 0.8 over time. Although there are small fluctuations and periodicity in R values as a function of time, where the fall season has slightly higher correlation compared to spring, the correlation is mostly steady during the whole evaluation period. Moreover, when calculated only over SMAP RQF recommended cells, the correlation is further increased following a similar trend in time.

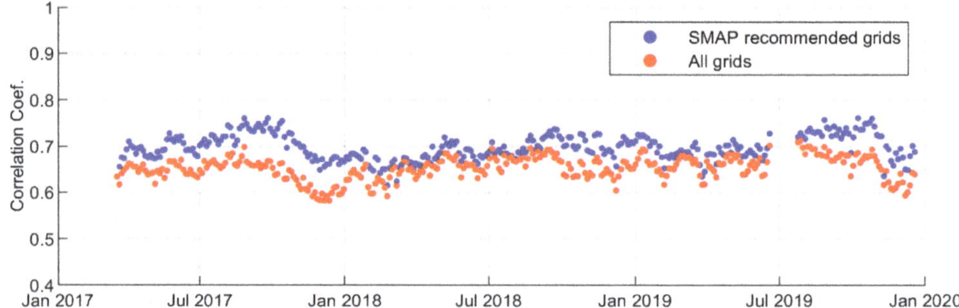

Figure 6. Temporal evaluation of the spatial correlation between CYGNSS-based model prediction and SMAP retrievals for all grids (red circles) and SMAP recommended grids (blue circles).

Figure 7 shows the SM estimation performance of the proposed ML model over different land cover types. The ubRMSD, bias and the number of data samples for each land cover type are provided, including croplands, shrubland, barren area, forest types, and savanna. All land cover types except for ice, snow, wetland, and urban are examined. It is clear that SM estimations over shrublands and barren have comparably lower ubRMSD of 0.032 cm^3 cm^{-3} and 0.022 cm^3 cm^{-3} respectively, while other land type classes have ubRMSD varying between 0.05 cm^3 cm^{-3} and 0.07 cm^3 cm^{-3}. Forest types and woody land covers generally show negative biases indicating that the ML model provides drier SM levels for these land cover classes while the biases in grass, barren, shrublands, and croplands are close to zero.

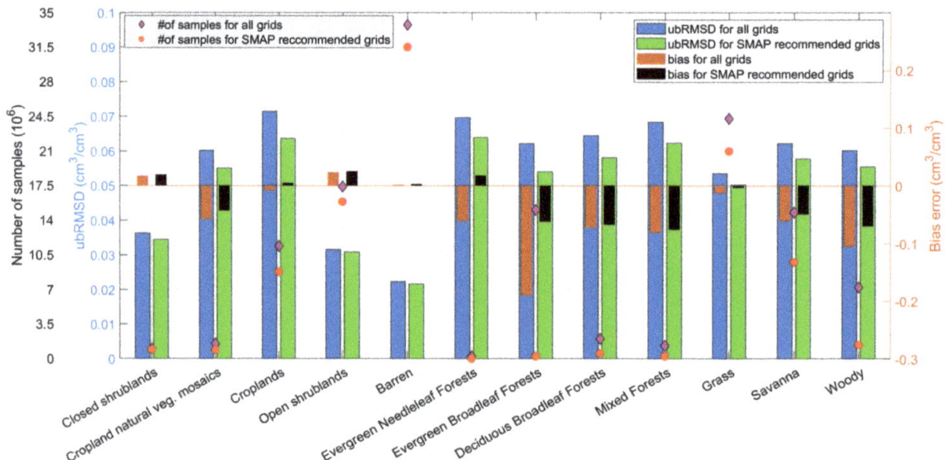

Figure 7. Evaluation of the model for different land cover types. The left axis is for ubRMSD and the right axis is for bias error.

Table 1 provides the overall comparison (i.e., RMSD, ubRMSD, and R metrics) between CYGNSS and SMAP SM predictions over the entire CYGNSS ground track and SMAP's recommended grid points at 9-km and 36-km spatial resolutions for concurrent, 3-day, 1-week, and 1-month temporal averaging. The mean ubRMSD and R over SMAP recommended 9-km grids between concurrent

CYGNSS and SMAP SM retrievals are 0.044 cm^3 cm^{-3} and 0.66, respectively. This suggests that an enhanced quasi-global CYGNSS-based SM product in space and time with a relatively high performance can be generated given that the ML model is trained independently with the inclusion of any SMAP SM data. The metrics show that averaging over lower temporal frequencies (from 3-day to monthly) generally leads to decreased ubRMSD and increased R between CYGNSS and SMAP predictions while the RMSD remains mostly unaffected since the bias error does not change with spatial and temporal averaging. The RMSD, ubRMSD and R values improve when the metrics are calculated exclusively over the SMAP recommended grids. In particular, RMSD values are significantly lowered over recommended grids.

Table 1. Overall performance evaluation metrics computed between CYGNSS and SMAP SM predictions at different spatial and temporal averaging levels for about 3 years (from March 2017 to December 2019).

	Spatial Resolution	Time Resolution	# of Samples	RMSD cm^3 cm^{-3}	Mean ubRMSD (std.) cm^3 cm^{-3}	R-Value
for all grids	9 km × 9 km	concurrent	1.64 × 10^8	0.11	0.054 (±0.027)	0.63
		3-day	1.07 × 10^8	0.11	0.049 (±0.025)	0.63
		week	7.13 × 10^7	0.11	0.044 (±0.024)	0.63
		month	2.24 × 10^7	0.11	0.034 (±0.023)	0.64
	36 km × 36 km	concurrent	2.95 × 10^7	0.11	0.050 (±0.024)	0.65
		3-day	1.66 × 10^7	0.12	0.045 (±0.024)	0.65
		week	7.82 × 10^6	0.12	0.040 (±0.023)	0.66
		month	1.95 × 10^6	0.12	0.032 (±0.022)	0.66
for SMAP recommended grids	9 km × 9 km	concurrent	1.16 × 10^8	0.066	0.044 (±0.021)	0.66
		3-day	9.14 × 10^7	0.066	0.041 (±0.022)	0.66
		week	5.99 × 10^7	0.065	0.038 (±0.021)	0.68
		month	1.87 × 10^7	0.065	0.031 (±0.021)	0.71
	36 km × 36 km	concurrent	2.12 × 10^7	0.070	0.045 (±0.021)	0.69
		3-day	1.16 × 10^7	0.071	0.042 (±0.022)	0.70
		week	5.51 × 10^6	0.071	0.037 (±0.022)	0.72
		month	1.39 × 10^6	0.070	0.030 (±0.022)	0.74

4.2. Performance Evaluation of SMAP and ML Model at ISMN Sites

The previous section has demonstrated the performance of the ML model, which is trained using the SM measurements from ISMN sites, and evaluated against the SMAP SM retrievals within the CYGNSS coverage. However, SMAP SM observations are also remotely sensed estimates and have their own error uncertainties. In this section, we compare both SMAP and CYGNSS SM predictions against the in-situ SM measurements at 170 ISMN stations. The SM observations are averaged over three days for comparison purposes. Table 2 presents the performance of both SMAP- and CYGNSS-based predictions against in-situ measurements over the 18 March 2017–31 December 2019 period. The mean ubRMSE for the SMAP observations is found to be 0.054 cm^3 cm^{-3} with a correlation coefficient (R) value of 0.59. The SMAP RMSE over the same dataset is 0.112 cm^3 cm^{-3}, much higher than the ubRMSE value indicating a bias in SMAP observations. This is expected since SMAP SM products are calibrated against the SMAP core validation site, which are defined as sites that have multiple calibrated and representative soil moisture measurement locations within a SMAP pixel [32]. The CYGNSS-based ML estimations are also compared with the same ISMN data. Since the developed ML model is trained with the ISMN SM measurements, we provide both the training- and 5-fold-cross-validation-based test performance of the ML model. The proposed CYGNSS-based ML model provides a 0.052 cm^3 cm^{-3} ubRMSE with a 0.83 correlation coefficient. While the ubRMSE performance of SMAP and CYGNSS estimates are similar, the ML model achieves a higher correlation and much lower RMSE. It should be noted that SMAP provides SM estimates at a larger spatial scale than the ISMN's point-scale in-situ measurements. Such bias levels are expected due to the spatial representativeness error. In addition, the ML model is trained with the ISMN observations and is expected to perform better on data with similar statistics. Nonetheless, the comparison of CYGNSS SM estimates against those by SMAP within the CYGNSS coverage and ISMN sites locally show a similar ubRMSE performance to SMAP's SM estimates against ISMN sites. This indicates that the quasi-global SM prediction performance of the proposed CYGNSS-based ML model performs with a similar accuracy compared to SMAP, even though no SMAP data is used in the ML model training.

Table 2. Performance at in-situ stations.

In-Situ vs.	RMSE	Mean ubRMSE $cm^3\,cm^{-3}$	Median ubRMSE $cm^3\,cm^{-3}$	R
CYGNSS (training)	0.067	0.052	0.049	0.83
CYGNSS (5-fold)	0.072	0.055	0.051	0.80
SMAP	0.112	0.054	0.052	0.59

Figure 8 provides examples from individual sites for a temporal comparison of the SMAP's and CYGNSS's SM estimates against in-situ measurements. It is evident that both SMAP and CYGNSS SM estimates follow the in-situ observations closely but the SMAP estimates have consistent biases with respect to in-situ observations (particularly in Figure 8c). Again, a low bias of the CYGNSS SM estimates is expected since the ML model learns the site characteristics during the training phase. In summary, CYGNSS SM estimation shows some levels of variability from site to site but generally close to in-situ measurements with low RMSE, high correlation, and negligible bias. In order to illustrate the advantage of CYGNSS's temporal coverage, a shorter duration of detailed SM estimation results at the Knoxcity site is provided in Figure 9. While SMAP provides an estimate every 2–3 days on average, CYGNSS can provide SM predictions almost daily depending on the quality factors and quasi-random transmitter/receiver configurations. This allows CYGNSS to capture rapid changes in SM, such as precipitation events, more frequently than SMAP. For example, two rapid increases in SM between 15–23 May 2018, is missed by SMAP while CYGNSS estimates are able to capture both events due to the increased temporal resolution of CYGNSS relative to SMAP. The more frequent CYGNSS observations in combination with the proposed ML-based prediction model can provide an enhanced SM product. However, the number of CYGNSS observations is also strongly determined by its spatial representing footprint and regridding scheme. For instance, in order to obtain daily CYGNSS estimates, observations need to be spatially aggregated over a larger area (36 × 36 km) as shown in Figure 9.

4.3. Spatial and Temporal Analysis of CYGNSS Soil Moisture Retrievals

Here, four target regions are selected, i.e., Midwest US, India, the Sahara, and Australia, to provide a regional analysis of the temporal and spatial performance of CYGNSS in comparison to SMAP. These regions are chosen to be representative of different vegetation and climate characteristics. Information on each region with specific location coordinates, dominant land cover, and climate types are summarized in Table 3 and the extent of the target area are shown in Figure 1. In order to spatially compare CYGNSS and SMAP SM estimates, the results are averaged over a duration of one month. Both CYGNSS and SMAP 9-km SM maps are accompanied with the land cover map of the same region. In addition, sub-regions (approximately 200 km × 200 km) are selected to illustrate the differences and similarities between CYGNSS and SMAP estimates at all combinations of available resolutions (from 3 km, 9 km, to 36 km). For each sub-target area, the average CYGNSS and SMAP SM values are provided as a function of time from March 2017 to December 2019.

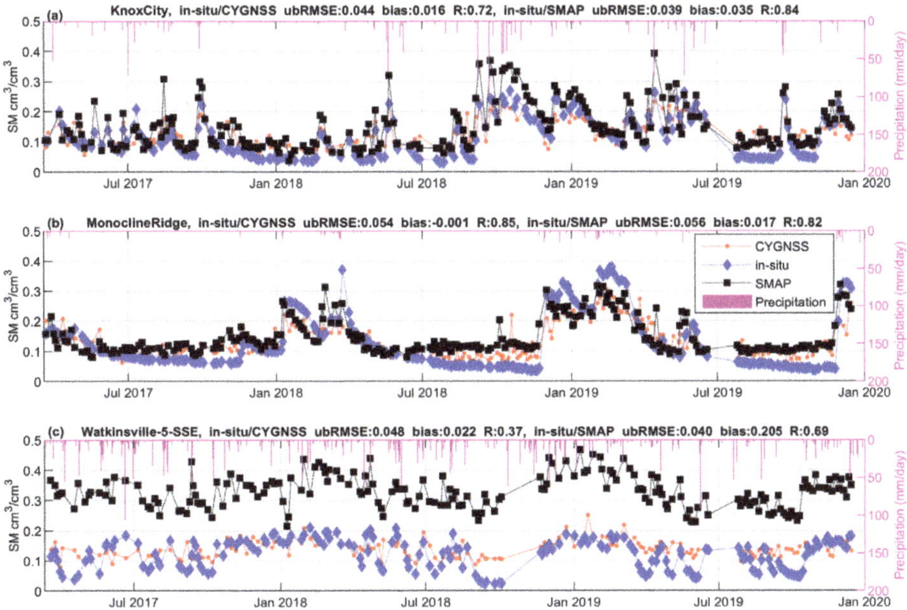

Figure 8. Time series examples of 3-day averaged CYGNSS and SMAP SM predictions that show a moderate performance: (**a**) KnoxCity, (**b**) MonoclineRidge, and (**c**) Watkinsville-5-SSE. The summary table of results from all sites are provided for the same time period in supplementary files.

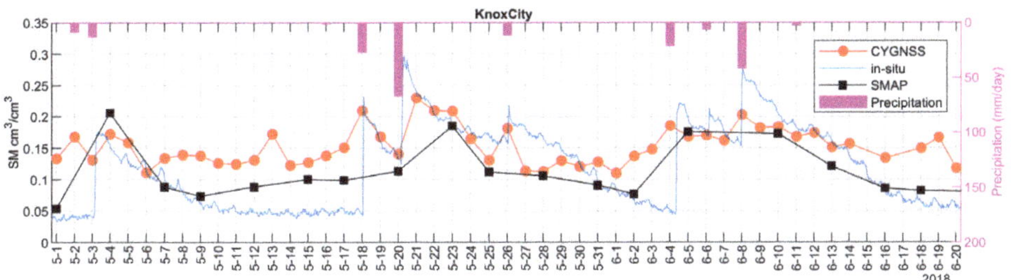

Figure 9. Shortened time series of CYGNS and SMAP soil moisture retrievals and in-situ measurements from the KnoxCity station.

Table 3. Information on selected regions for spatial and temporal analysis (see Figure 1 for locations of these regions on the map).

Region	Location	Dominant Land Cover	Climate Zone
Midwest US	85W:105W–32N:37N	Grass	Cfa
India	70E:90E–10N:20N	Croplands	Aw
Sahara	−15W:15 E–7N:20N	Grass	BWh, Aw
Australia	138E:153 E–27S:37S	Open shrublad	BWh

Figure 10 shows the results over the Midwest US for January 2018. This region has heterogeneous land cover conditions (grass, crop, and woody) and is within a warm temperate fully humid climate zone. From Figure 10b,c, a good spatial correspondence between CYGNSS and SMAP predictions is seen at 9 km while CYGNSS estimates are generally drier. In particular, SM from CYGNSS are underestimated near the Mississippi River and other water bodies. This can be due to the fact that SMAP's native resolution is about 40 km. SMAP observations can blend small water bodies with land leading to higher SM values while the CYGNSS SM algorithm does not attempt to retrieve SM when the water fraction within the 3-km grid of the CYGNSS observation exceeds 2% (see white areas in Figure 10(a4)). A somparison of SM estimates at various resolutions is shown in Figure 10a to illustrate the CYGNSS SM algorithm's ability to resolve small details. The 3-km gridded CYGNSS estimates are visually sharper and provide more detailed features than its coarser resolution (i.e., 9-km and 36-km) results. For the same sub-region, a time-series example for the entire data set is shown in Figure 10e. The CYGNSS-based SM values show good agreement with SMAP retrievals for the entire period. Although the dynamic range of CYGNSS ML-based estimates is slightly less than SMAP, there is a high correlation (0.90) between the two time series.

Figure 11 shows SM estimates for the month of Oct 2018 in the India region, which dominantly features croplands with equatorial-winter-dry climate zone. A high spatial correlation can be seen from the 9-km SMAP and CYGNSS SM maps in Figure 11a,b. CYGNSS tends to underestimate SM in some coastal and woody areas when compared to SMAP. However, these regions are also flagged by SMAP's RQF as having high uncertainty for SM retrieval. The results in the selected sub-region are presented in Figure 11d for SM estimates at fine and coarse resolutions. Note that the Tibet regions in the top-right corner of the CYGNSS SM images are excluded due to high elevation. For other regions, due to the seasonal standing water on the soil surface, SMAP results are wetter compared to the CYGNSS estimates. This is visually apparent in the temporal trend given in Figure 11e where abrupt increases in SM appear during monsoon seasons. Outside of these monsoon events which occur roughly from June to September, a very high correlation is found between the time-series SM values from the proposed CYGNSS and SMAP retrievals. It should be noted that the training dataset does not have any samples from the monsoon areas, since there are no ISMN sites in this area. In addition, SMAP retrievals are not recommended for these regions in the monsoon season. The correlation coefficient of the two time series is calculated to be 0.83. The result shows the proposed model is able to follow dry and rainy seasons.

The selected region in the Sahara shown in Figure 12 captures a transition from barren to grass and savanna land covers in arid and equatorial climate zones. The results are given for the month of October in 2018. While both SMAP and CYGNSS estimates follow a similar spatial pattern, CYGNSS SM tends to underestimate SM over woody and savanna regions when compared to SMAP. In parallel to the previous examples, CYGNSS is drier around small water bodies compared to SMAP estimates as seen in close up images in Figure 12a. The temporal SM change of the sub-region for both CYGNSS and SMAP SM retrievals can be seen in Figure 12e.

The last test region is Australia which is dominantly open-shrublands in central regions and grass near coastal areas. The 9-km CYGNSS and SMAP SM maps are presented for the month of October in 2018 in Figure 13a,b for a spatial comparison while the close-up images are given in Figure 13c to check the details in finer resolution SM predictions. The temporal comparison of the sub-set area of Australia is shown in Figure 13e. Similar to previous observations, here CYGNSS's high spatial resolution capability is able to resolve details over moderately heterogeneous regions. It is clearly seen that there is a large offset between SMAP and CYGNSS estimates for wet seasons and SMAP's flagged grids. CYGNSS's ability to distinguish small features at fine resolutions can potentially contribute to the quality of SM estimation.

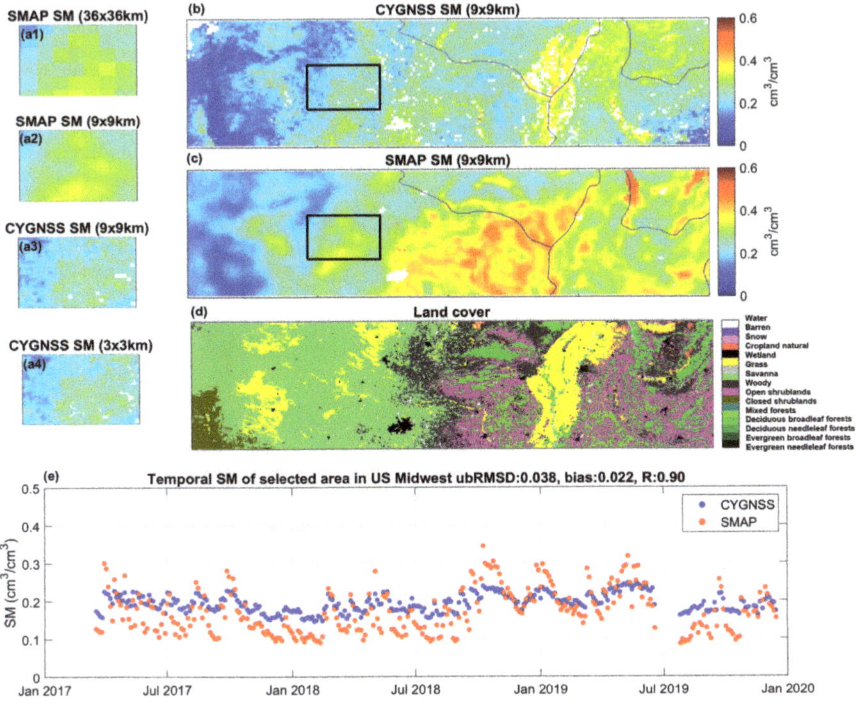

Figure 10. Averaged SM predictions for Midwest US during Jan 2018. (**a**) Detailed images of selected area for SMAP 9-km and 36-km and CYGNSS 3-km and 9-km, (**b**) CYGNSS 9-km, (**c**) SMAP 9-km, (**d**) land cover map, and (**e**) a time-series example from selected area.

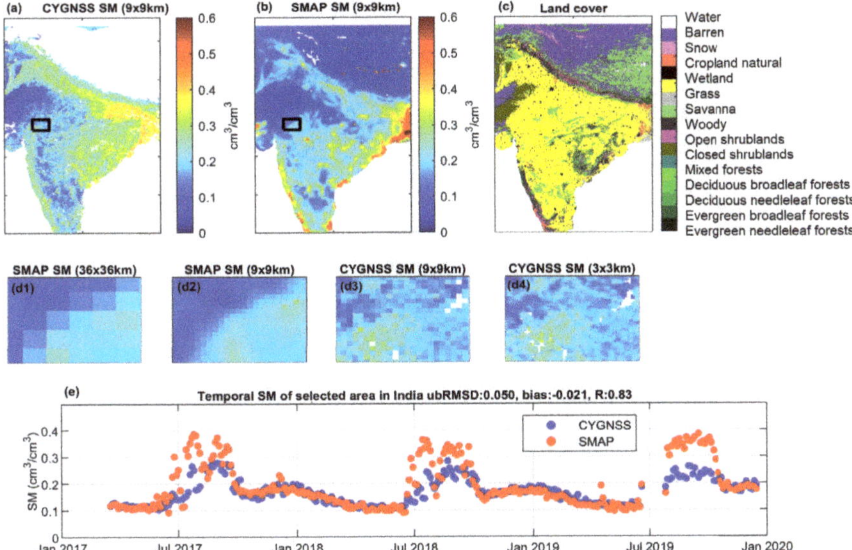

Figure 11. Averaged SM predictions for India during Oct 2018. (**a**) CYGNSS 9-km, (**b**) SMAP 9-km, (**c**) land cover map, (**d**) detailed images of selected area for SMAP 9-km and 36-km and CYGNSS 3-km and 9-km, and (**e**) a time-series example from selected area.

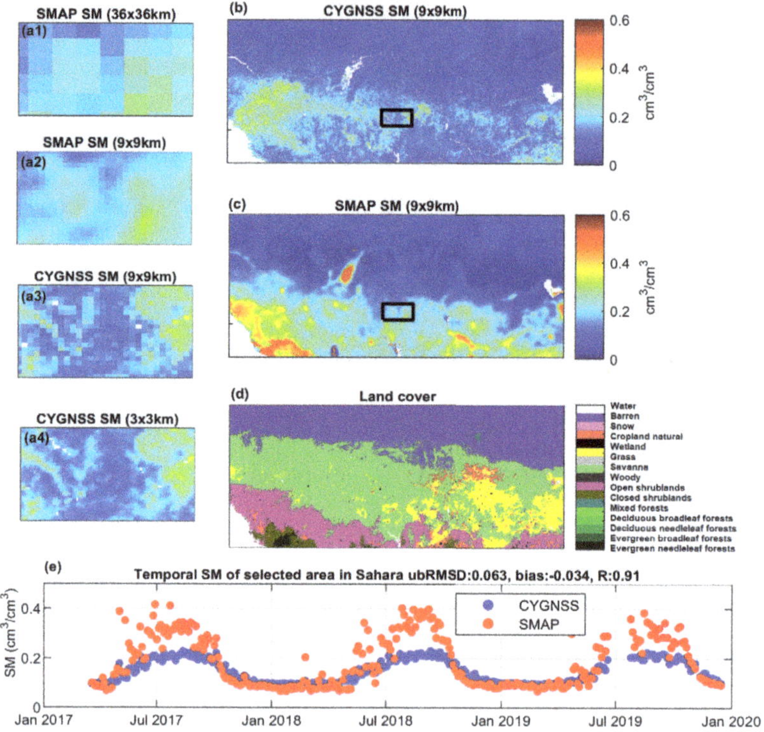

Figure 12. Averaged SM predictions for Sahara during Oct 2018. (**a**) Detailed images of selected area for SMAP 9-km and 36-km and CYGNSS 3-km and 9-km, (**b**) CYGNSS 9-km, (**c**) SMAP 9-km, (**d**) land cover map, and (**e**) a time-series example from selected area.

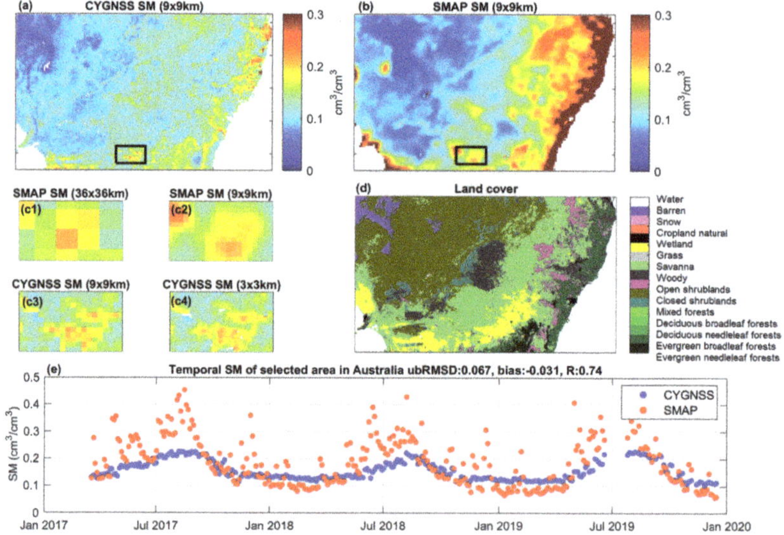

Figure 13. Averaged SM predictions for Australia during Oct 2018. (**a**) CYGNSS 9-km, (**b**) SMAP 9-km, (**c**) detailed images of selected area for SMAP 9-km and 36-km and CYGNSS 3-km and 9-km, (**d**) land cover map, and (**e**) a time-series example from selected area.

5. Discussion

In this paper, we evaluated the performance of a CYGNSS ML-based SM product over the globe using SMAP radiometer SM data. The ML model is trained using measurements from in-situ SM stations that are primarily located in the continental U.S. Despite the lack of a large number of global in-situ SM stations, the SM estimates perform quite well when compared to SMAP even though the model is trained without any SMAP SM information. In general, the model estimation agrees well with SMAP measurements with an average ubRMSD of 0.049 and 0.041 $cm^3\ cm^{-3}$ for the whole globe and the areas encompassed by SMAP's recommended quality flags, respectively.

The results show that the performance of the proposed model estimations is degenerated with increasing canopy cover, especially for forested areas. This is expected since the presence of a vegetation canopy reduces the dynamic range of observed reflectivity at L-band [40,41]. In addition, with increasing canopy affects, the model shows a dry bias compared to SMAP SM estimates. Part of the reason for this mismatch is that the model is trained on a small, point-scale dataset which does not feature sufficient samples from areas that can be classified as equatorial, high canopy, or other conditions such as seasonal snow and monsoon. In addition, SMAP's large native resolution (about 40 km) blends small water bodies that makes the SMAP results contain higher SM values while the higher-resolution CYGNSS SM algorithm excludes small water bodies within the 3 km grid for the CYGNSS observations. The bias is particularly evident during the wet season (e.g., monson) where there exists seasonal surface standing water.

Overall, the proposed model based on CYGNSS and ancillary inputs generates promising results with a comparable accuracy to the reference SMAP data and in-situ measurements. The most challenging components of the study is the limited samples provided in the training dataset. The model is trained using 170 in-situ station records, and this corresponds to an approximate land surface area of $10^3\ km^2$. However, the total surface area of the earth that the model is tested on is roughly $1.67 \times 10^8\ km^2$. It is expected that a CYGNSS-based ML model trained with a larger sample of in-situ measurements over the world can provide a more generalized solution for SM estimation with high spatial resolution and short revisit time.

Here, we implemented a single ML model to predict SM values for all locations within CYGNSS's ground track. As our results suggest, model estimates can vary by location as influenced by different scattering structures within CYGNSS's footprint. The use of such a single ML model can lead to significant biases caused by location-specific factors such as unique SM profiles, terrain, and weather conditions. Instead of a single ML model, future studies can implement multiple ML models that are trained over different geological areas in order to obtain better SM predictions for different terrain. Furthermore, we developed a model without using any seasonal data or location information. Such information plays a significant role in the SM profile structure, and the inclusion of these sources will likely increase the prediction accuracy for surface SM given sufficient training datasets with appreciable spatial variability.

6. Conclusions

In this paper, a ML model designed for surface SM estimation by using CYGNSS measurements in conjunction with ancillary data is presented for surface SM estimation. The model was trained exclusively with 170 in-situ measurements from ISMN sites within CYGNSS's spatial coverage. The model was capable of producing daily SM estimates with a 3-km spatial resolution without requiring any space-borne SM data products such as SMAP. The SM estimates were spatially averaged to 9 km for comparison with SMAP's enhanced, 9-km product within the CYGNSS coverage.

The model evaluation for about 3 years showed that the CYGNSS-based ML model yielded a ubRMSE of 0.049 $cm^3\ cm^{-3}$ when compared to all available SMAP SM values. After excluding all grid values that did no meet SMAP's recommended quality control factors, the ubRMSE was further reduced to 0.041 $cm^3\ cm^{-3}$. Despite being trained independently of SMAP data, it was shown that the ML model obtained high correlation with SMAP data with an approximate correlation coefficient of 0.7 within the CYGNSS coverage. Furthermore, by analyzing for various land cover types, SMAP and the ML model still showed high agreements in areas of light vegetation with ubRMSE below 0.03 cm^3 cm^{-3}. For areas with dense vegetation such as forests and croplands, higher ubRMSE values of roughly 0.07 cm^3 cm^{-3} were observed. In general, the ubRMSE between the two models could be reduced with spatio-temporal averaging as expected.

To further understand the accuracy of the model, a comparison between the CYGNSS-based ML model, SMAP, and in-situ data from the ISMN sites was performed. We note that the ML model was trained using ISMN samples. While this comparison involved three datasets of vastly different spatial extent (i.e., point-scale datasets vs. 3-km CYGNSS-based SM and 9-km SMAP-based SM), the performance showed that both SMAP and the ML model have similar ubRMSE against the ISMN data. SMAP data, however, showed a much larger bias in its dataset that was likely caused by SMAP's coarse native resolutions and uncorrected small water bodies within the footprint. Additionally, it was shown that the ML method's higher temporal resolution allowed it to capture abrupt changes in SM due to rain events.

In addition, all data was temporally averaged for a month-long period at selected regions. Overall, it could be seen that both SMAP and the ML model featured high spatial correspondence for all examined areas across the world. In general, the CYGNSS-based ML method was capable of producing higher resolution data with notable land features, but a reduced dynamic range was apparent when compared to SMAP's datasets. Temporally, some smaller SM values were found for CYGNSS when compared to SMAP. This could be partially attributed to the limited samples in the ML model, as well as the quality control factors of the CYGNSS-based method which excludes grids with significant water bodies at 3-km resolution.

In summary, it can be seen that GNSS-R-based, physics-guided ML techniques could provide an excellent, complimentary dataset for global SM estimation with significant research potential for improved data products. Despite the limited spatial variability of the training dataset, the model showed high agreement with independent SM estimates over the entire CYGNSS coverage.

Supplementary Materials: The following are available online at http://www.mdpi.com/2072-4292/12/21/3503/s1.

Author Contributions: Conceptualization: M.K. and A.C.G.; Methodology: M.K. and A.C.G.; Software: V.S. and F.L.; Validation: V.S. and F.L.; Formal analysis: V.S. and F.L.; Investigation: V.S., F.L., D.B., M.K., and A.C.G.; Resources: M.K. and A.C.G.; Data curation: F.L.; Writing–original draft preparation: V.S., F.L., M.K., D.B., and A.C.G.; Writing–review and editing: V.S., F.L., M.K., D.B., A.C.G., and R.M.; Visualization: V.S., F.L., and D.B.; Supervision: M.K. and A.C.G.; Project administration: M.K. and A.C.G.; Funding acquisition: M.K., A.C.G., and R.M. All authors have read and agreed to the published version of the manuscript.

Funding: This research was funded by USDA Agricultural Research Service(USDA-ARS), Award NACA 58-6064-9-007.

Conflicts of Interest: The authors declare no conflict of interest.

Appendix A. Retrieval Comparisons

Source	CYGNSS Time Span	Spatial Coverage	CYGNSS observables	Reference SM	Validation	Ancilary data	Model	Overall Daily Performance	Spatial Resolutions
Chew and Small (2018) [21]	1 year (March 2017 to March 2018)	Quasi-Global	Reflecvtivity	SMAP	SMAP	Water fraction	Multiple Linear Regression	median ubRMSD = 0.045 cm3/cm3, R=?	36 km X 36 km
Kim and Lakshmi (2018) [22]	1 year (March 2017 to March 2018)	Regional (CONUS)	Reflecvtivity	SMAP	SMAP	?	Linear Regression	RMSE =?, R=0.68/0.77	9 km X 9 km
Al-Khaldi et al. (2019) [23]	6-month (December 2017–May 2018)	Quasi-Global	BRCS	SMAP	SMAP	SMAP VWC	Physical, time-series	ubRMSE=0.038 cm3/cm3, R=0.82	25 km X 25 km
Clarizia et al. (2019) [24]	5 months (April 2017 to August 2017)	Quasi-Global	Reflecvtivity	SMAP	SMAP, SMOS	SMAP h-paramer and vegetation opacity	Physical, Trilinear Regression	RMSE = 0.07 cm3/cm3, R=?	36 km X 36 km
Eroglu et al. (2019) [25]	2 years (2017, 2018)	Regional (CONUS)	Reflecvtivity, TES, LES, incidence angle	18 ISMN Sites	18 ISMN Sites	Elevation, Slope, NDVI,VWC, h-parameter	ANN	RMSE =0.054 cm3/cm3, R=0.90	9 km X 9 km
Senyurek et al. (2020) [26]	3 years (2017-2019)	Regional (CONUS)	Reflecvtivity, TES, incidence angle	106 ISMN Sites	106 ISMN Sites	Elevation, NDVI,VWC, Soil Texture	Random Forest, ANN, SVN	RMSE = 0.052 cm3/cm3, R=0.89	9 km X 9 km
Chew and Small (2020) [27]	3 years (2017-2019)	Quasi-Global	Reflecvtivity	SMAP	SMAP and 171 ISMN Sites	Water fraction	Multiple Linear Regression	median ubRMSE = 0.049 cm3/cm3, median R=0.40	36 km X 36 km
Yang et. al. (2020) [28]	1 year (May 2017-April 2018)	Regional (China)	Reflecvtivity	SMAP, 588 insitu networks	SMAP, 588 insitu networks	Elevation, Slope, NDVI,VWC, h-parameter, precipitation	ANN	ubRMSD=0.062, cm3/cm3, R=0.79 (vs SMAP) ubRMSD=0.053, cm3/cm3, R=0.72 (vs in-situ)	36 km X 36 km
Yan et. al. (2020) [29]	1 year (2018)	Quasi-Global	Reflecvtivity, and DDM statistics	SMAP	SMAP	SMAP vegetation opacity	Linear Regression	RMSE = 0.07 cm3/cm3, R=0.80	36 km X 36 km
This Study	3 years (2017-2019)	Quasi-Global	Reflecvtivity, TES, incidence angle	ISMN sites	SMAP and 170 ISMN sites	Elevation, NDVI,VWC, Soil Texture	Random Forest	mean ubRMSD = 0.044 cm3/cm3, R=0.66	9 km X 9 km

Figure A1. A list of recent CYGNSS-based SM retrieval approaches that mostly differ in (1) assumptions regarding gridding, open water masking, and surface conditions, (2) ancillary data requirements, (3) validation and reference data sets, (4) time spans, (5) models and (6) spatial coverage. The list is given in chronological order.

References

1. Jung, M.; Reichstein, M.; Ciais, P.; Seneviratne, S.I.; Sheffield, J.; Goulden, M.L.; Bonan, G.; Cescatti, A.; Chen, J.; De Jeu, R.; et al. Recent decline in the global land evapotranspiration trend due to limited moisture supply. *Nature* **2010**, *467*, 951–954. [CrossRef] [PubMed]
2. Entekhabi, D.; Rodriguez-Iturbe, I.; Castelli, F. Mutual interaction of soil moisture state and atmospheric processes. *J. Hydrol.* **1996**, *184*, 3–17. [CrossRef]
3. Korres, W.; Reichenau, T.; Schneider, K. Patterns and scaling properties of surface soil moisture in an agricultural landscape: An ecohydrological modeling study. *J. Hydrol.* **2013**, *498*, 89–102. [CrossRef]
4. Seneviratne, S.I.; Corti, T.; Davin, E.L.; Hirschi, M.; Jaeger, E.B.; Lehner, I.; Orlowsky, B.; Teuling, A.J. Investigating soil moisture–climate interactions in a changing climate: A review. *Earth-Sci. Rev.* **2010**, *99*, 125–161. [CrossRef]
5. Kerr, Y.H.; Waldteufel, P.; Wigneron, J.P.; Delwart, S.; Cabot, F.; Boutin, J.; Escorihuela, M.J.; Font, J.; Reul, N.; Gruhier, C.; et al. The SMOS mission: New tool for monitoring key elements ofthe global water cycle. *Proc. IEEE* **2010**, *98*, 666–687. [CrossRef]
6. Entekhabi, D.; Njoku, E.G.; ONeill, P.E.; Kellogg, K.H.; Crow, W.T.; Edelstein, W.N.; Entin, J.K.; Goodman, S.D.; Jackson, T.J.; Johnson, J.; et al. The soil moisture active passive (SMAP) mission. *Proc. IEEE* **2010**, *98*, 704–716. [CrossRef]
7. Brocca, L.; Ciabatta, L.; Massari, C.; Moramarco, T.; Hahn, S.; Hasenauer, S.; Kidd, R.; Dorigo, W.; Wagner, W.; Levizzani, V. Soil as a natural rain gauge: Estimating global rainfall from satellite soil moisture data. *J. Geophys. Res. Atmos.* **2014**, *119*, 5128–5141. [CrossRef]
8. Lu, Y.; Steele-Dunne, S.C.; Farhadi, L.; van de Giesen, N. Mapping surface heat fluxes by assimilating SMAP soil moisture and GOES land surface temperature data. *Water Resour. Res.* **2017**, *53*, 10858–10877. [CrossRef]
9. Zhang, J.; Bai, Y.; Yan, H.; Guo, H.; Yang, S.; Wang, J. Linking observation, modelling and satellite-based estimation of global land evapotranspiration. *Big Earth Data* **2020**, *4*, 94–127. [CrossRef]
10. Zavorotny, V.U.; Gleason, S.; Cardellach, E.; Camps, A. Tutorial on remote sensing using GNSS bistatic radar of opportunity. *IEEE Geosci. Remote Sens. Mag.* **2014**, *2*, 8–45. [CrossRef]
11. Martin-Neira, M. A passive reflectometry and interferometry system (PARIS): Application to ocean altimetry. *ESA J.* **1993**, *17*, 331–355.
12. Unwin, M.; Jales, P.; Tye, J.; Gommenginger, C.; Foti, G.; Rosello, J. Spaceborne GNSS-reflectometry on TechDemoSat-1: Early mission operations and exploitation. *IEEE J. Sel. Top. Appl. Earth Obs. Remote Sens.* **2016**, *9*, 4525–4539. [CrossRef]
13. Ruf, C.S.; Chew, C.; Lang, T.; Morris, M.G.; Nave, K.; Ridley, A.; Balasubramaniam, R. A new paradigm in earth environmental monitoring with the CYGNSS small satellite constellation. *Sci. Rep.* **2018**, *8*, 8782. [CrossRef]
14. Addabbo, P.; Giangregorio, G.; Galdi, C.; di Bisceglie, M. Simulation of TechDemoSat-1 delay-Doppler maps for GPS ocean reflectometry. *IEEE J. Sel. Top. Appl. Earth Obs. Remote Sens.* **2017**, *10*, 4256–4268. [CrossRef]
15. Wang, F.; Yang, D.; Zhang, B.; Li, W. Waveform-based spaceborne GNSS-R wind speed observation: Demonstration and analysis using UK TechDemoSat-1 data. *Adv. Space Res.* **2018**, *61*, 1573–1587. [CrossRef]
16. Rius, A.; Cardellach, E.; Fabra, F.; Li, W.; Ribó, S.; Hernández-Pajares, M. Feasibility of GNSS-R ice sheet altimetry in Greenland using TDS-1. *Remote Sens.* **2017**, *9*, 742. [CrossRef]
17. Li, W.; Cardellach, E.; Fabra, F.; Ribó, S.; Rius, A. Measuring Greenland Ice Sheet Melt Using Spaceborne GNSS Reflectometry From TechDemoSat-1. *Geophys. Res. Lett.* **2020**, *47*, e2019GL086477. [CrossRef]
18. Comite, D.; Cenci, L.; Colliander, A.; Pierdicca, N. Monitoring Freeze-Thaw State by means of GNSS Reflectometry: An Analysis of TechDemoSat-1 Data. *IEEE J. Sel. Top. Appl. Earth Obs. Remote Sens.* **2020**, *13*, 2996–3005. [CrossRef]
19. Camps, A.; Park, H.; Pablos, M.; Foti, G.; Gommenginger, C.P.; Liu, P.W.; Judge, J. Sensitivity of GNSS-R spaceborne observations to soil moisture and vegetation. *IEEE J. Sel. Top. Appl. Earth Obs. Remote Sens.* **2016**, *9*, 4730–4742. [CrossRef]

20. Chew, C.; Shah, R.; Zuffada, C.; Hajj, G.; Masters, D.; Mannucci, A.J. Demonstrating soil moisture remote sensing with observations from the UK TechDemoSat-1 satellite mission. *Geophys. Res. Lett.* **2016**, *43*, 3317–3324. [CrossRef]

21. Chew, C.; Small, E. Soil moisture sensing using spaceborne GNSS reflections: Comparison of CYGNSS reflectivity to SMAP soil moisture. *Geophys. Res. Lett.* **2018**, *45*, 4049–4057. [CrossRef]

22. Kim, H.; Lakshmi, V. Use of Cyclone Global Navigation Satellite System (CYGNSS) observations for estimation of soil moisture. *Geophys. Res. Lett.* **2018**, *45*, 8272–8282. [CrossRef]

23. Al-Khaldi, M.M.; Johnson, J.T.; O'Brien, A.J.; Balenzano, A.; Mattia, F. Time-Series Retrieval of Soil Moisture Using CYGNSS. *IEEE Trans. Geosci. Remote Sens.* **2019**, *57*, 4322–4331. [CrossRef]

24. Clarizia, M.P.; Pierdicca, N.; Costantini, F.; Floury, N. Analysis of CYGNSS Data for Soil Moisture Retrieval. *IEEE J. Sel. Top. Appl. Earth Obs. Remote Sens.* **2019**, *12*, 2227–2235. [CrossRef]

25. Eroglu, O.; Kurum, M.; Boyd, D.; Gurbuz, A.C. High Spatio-Temporal Resolution CYGNSS Soil Moisture Estimates Using Artificial Neural Networks. *Remote Sens.* **2019**, *11*, 2272. [CrossRef]

26. Senyurek, V.; Lei, F.; Boyd, D.; Kurum, M.; Gurbuz, A.C.; Moorhead, R. Machine Learning-Based CYGNSS Soil Moisture Estimates over ISMN sites in CONUS. *Remote Sens.* **2020**, *12*, 1168. [CrossRef]

27. Chew, C.; Small, E. Description of the UCAR/CU Soil Moisture Product. *Remote Sens.* **2020**, *12*, 1558. [CrossRef]

28. Yang, T.; Wan, W.; Sun, Z.; Liu, B.; Li, S.; Chen, X. Comprehensive Evaluation of Using TechDemoSat-1 and CYGNSS Data to Estimate Soil Moisture over Mainland China. *Remote Sens.* **2020**, *12*, 1699. [CrossRef]

29. Yan, Q.; Huang, W.; Jin, S.; Jia, Y. Pan-tropical soil moisture mapping based on a three-layer model from CYGNSS GNSS-R data. *Remote Sens. Environ.* **2020**, *247*, 111944. [CrossRef]

30. Rodriguez-Alvarez, N.; Podest, E.; Jensen, K.; McDonald, K.C. Classifying Inundation in a Tropical Wetlands Complex with GNSS-R. *Remote Sens.* **2019**, *11*, 1053. [CrossRef]

31. Dorigo, W.A.; Xaver, A.; Vreugdenhil, M.; Gruber, A.; Hegyiova, A.; Sanchis-Dufau, A.D.; Zamojski, D.; Cordes, C.; Wagner, W.; Drusch, M. Global automated quality control of in situ soil moisture data from the International Soil Moisture Network. *Vadose Zone J.* **2013**, *12*. [CrossRef]

32. Colliander, A.; Jackson, T.J.; Bindlish, R.; Chan, S.; Das, N.; Kim, S.; Cosh, M.; Dunbar, R.; Dang, L.; Pashaian, L.; et al. Validation of SMAP surface soil moisture products with core validation sites. *Remote Sens. Environ.* **2017**, *191*, 215–231. [CrossRef]

33. Dorigo, W.A.; Wagner, W.; Hohensinn, R.; Hahn, S.; Paulik, C.; Xaver, A.; Gruber, A.; Drusch, M.; Mecklenburg, S.; Oevelen, P.V.; et al. The International Soil Moisture Network: A data hosting facility for global in situ soil moisture measurements. *Hydrol. Earth Syst. Sci.* **2011**, *15*, 1675–1698. [CrossRef]

34. Gruber, A.; Dorigo, W.A.; Zwieback, S.; Xaver, A.; Wagner, W. Characterizing coarse-scale representativeness of in situ soil moisture measurements from the International Soil Moisture Network. *Vadose Zone J.* **2013**, *12*. [CrossRef]

35. O'Neill, P.E.; Njoku, E.G.; Jackson, T.J.; Chan, S.; Bindlish, R. *SMAP Algorithm Theoretical Basis Document: Level 2 & 3 Soil Moisture (Passive) Data Products*; Jet Propulsion Laboratory, California Institute of Technology: Pasadena, CA, USA, 2015.

36. Hengl, T.; de Jesus, J.; Heuvelink, G.B.; Gonzalez, M.R.; Kilibarda, M.; Blagotić, A.; Shangguan, W.; Wright, M.N.; Geng, X.; Bauer-Marschallinger, B.; et al. SoilGrids250m: Global gridded soil information based on machine learning. *PLoS ONE* **2017**, *12*, e0169748. [CrossRef]

37. Pekel, J.F.; Cottam, A.; Gorelick, N.; Belward, A.S. High-resolution mapping of global surface water and its long-term changes. *Nature* **2016**, *540*, 418. [CrossRef]

38. Chan, S.; Bindlish, R.; O'Neill, P.; Jackson, T.; Njoku, E.; Dunbar, S.; Chaubell, J.; Piepmeier, J.; Yueh, S.; Entekhabi, D.; et al. Development and assessment of the SMAP enhanced passive soil moisture product. *Remote Sens. Environ.* **2018**, *204*, 931–941. [CrossRef]

39. Fernandez-Moran, R.; Al-Yaari, A.; Mialon, A.; Mahmoodi, A.; Al Bitar, A.; De Lannoy, G.; Rodriguez-Fernandez, N.; Lopez-Baeza, E.; Kerr, Y.; Wigneron, J.P. SMOS-IC: An alternative SMOS soil moisture and vegetation optical depth product. *Remote Sens.* **2017**, *9*, 457. [CrossRef]

40. Zribi, M.; Guyon, D.; Motte, E.; Dayau, S.; Wigneron, J.P.; Baghdadi, N.; Pierdicca, N. Performance of GNSS-R GLORI data for biomass estimation over the Landes forest. *Int. J. Appl. Earth Obs. Geoinf.* **2019**, *74*, 150–158. [CrossRef]

41. Zribi, M.; Motte, E.; Baghdadi, N.; Baup, F.; Dayau, S.; Fanise, P.; Guyon, D.; Huc, M.; Wigneron, J.P. Potential applications of GNSS-R observations over agricultural areas: Results from the GLORI airborne campaign. *Remote Sens.* **2018**, *10*, 1245. [CrossRef]

Publisher's Note: MDPI stays neutral with regard to jurisdictional claims in published maps and institutional affiliations.

 *remote sensing*

Article

Machine Learning-Based CYGNSS Soil Moisture Estimates over ISMN sites in CONUS

Volkan Senyurek [1],*, Fangni Lei [1], Dylan Boyd [2], Mehmet Kurum [2], Ali Cafer Gurbuz [2] and Robert Moorhead [1]

[1] Geosystems Research Institute, Mississippi State University, Mississippi State, MS 39759, USA; fangni@gri.msstate.edu (F.L.); rjm@gri.msstate.edu (R.M.)
[2] Department of Electrical and Computer Engineering, Mississippi State University, Mississippi State, MS 39759, USA; db1950@msstate.edu (D.B.); kurum@ece.msstate.edu (M.K.); gurbuz@ece.msstate.edu (A.C.G.)
* Correspondence: volkan@gri.msstate.edu

Received: 11 March 2020; Accepted: 3 April 2020; Published: 5 April 2020

Abstract: Soil moisture (SM) derived from satellite-based remote sensing measurements plays a vital role for understanding Earth's land and near-surface atmosphere interactions. Bistatic Global Navigation Satellite System (GNSS) Reflectometry (GNSS-R) has emerged in recent years as a new domain of microwave remote sensing with great potential for SM retrievals, particularly at high spatio-temporal resolutions. In this work, a machine learning (ML)-based framework is presented for obtaining SM data products over the International Soil Moisture Network (ISMN) sites in the Continental United States (CONUS) by leveraging spaceborne GNSS-R observations provided by NASA's Cyclone GNSS (CYGNSS) constellation alongside remotely sensed geophysical data products. Three widely-used ML approaches—artificial neural network (ANN), random forest (RF), and support vector machine (SVM)—are compared and analyzed for the SM retrieval through utilizing multiple validation strategies. Specifically, using a 5-fold cross-validation method, overall RMSE values of 0.052, 0.061, and 0.065 cm^3/cm^3 are achieved for the RF, ANN, and SVM techniques, respectively. In addition, both a site-independent and a year-based validation techniques demonstrate satisfactory accuracy of the proposed ML model, suggesting that this SM approach can be generalized in space and time domains. Moreover, the achieved accuracy can be further improved when the model is trained and tested over individual SM networks as opposed to combining all available SM networks. Additionally, factors including soil type and land cover are analyzed with respect to their impacts on the accuracy of SM retrievals. Overall, the results demonstrated here indicate that the proposed technique can confidently provide SM estimates over lightly-vegetated areas with vegetation water content (VWC) less than 5 kg/m^2 and relatively low spatial heterogeneity.

Keywords: artificial neural networks; random forest; SVM; CYGNSS; soil moisture retrieval; ISMN; machine learning

---

1. Introduction

Soil moisture (SM) is a critical variable for many Earth science models with applications for hydrology, meteorology, crop forecasting, and Earth thermodynamics [1]. The European Space Agency (ESA)'s Soil Moisture and Ocean Salinity (SMOS) and the National Aeronautics and Space Administration (NASA)'s Soil Moisture Active Passive (SMAP) missions are two microwave remote sensing satellite systems dedicated for global SM retrievals [2,3]. They provide critical, global SM measurements between 25–50 km spatial resolution with ±0.04 cm^3/cm^3 volumetric SM accuracy every 2–3 days. The need for global SM products below 9 km spatial resolution and at a temporal resolution of 3 days or less is still present within the hydrology community for applications in hydrometeorology, atmospheric research, and water resource management at microscale and mesoscale resolutions [4,5].

Global Navigation Satellite System (GNSS) Reflectometry (GNSS-R) is a growing area in microwave remote sensing with great potential for providing cost-efficient, high-resolution SM estimations. The GNSS-R technique uses a measured GNSS signal reflected from a scattering surface to determine geophysical information of the examined surface by cross-correlating the reflected signal with either a received direct signal or a GNSS signal replica [6]. It has been established as an effective tool for monitoring ocean surface roughness and wind vectors from airborne and spaceborne platforms [7–9]. Research is on-going for the use of GNSS-R for other vital remote sensing parameters such as altimetry [10], sea ice monitoring [11], biomass estimation [12], wetland classification [13], and SM estimation [14–18].

The Cyclone GNSS (CYGNSS) mission is NASA's most recent spaceborne GNSS-R application, orbiting over tropics (within ±38° latitudes) [19]. CYGNSS, although designed for the estimation of ocean wind vectors, has shown particular sensitivity to variations in SM and high correlation with SMAP SM data products [14–17]. By means of bistatic radar measurements using eight, four-channel micro-satellites, CYGNSS records reflected signals scattered from Earth's land surface during the 95-minute orbital period of each satellite. Under a coherency assumption, it is capable of performing sub-daily observations at a relatively high spatial resolution (on the order of a few kilometers). However, CYGNSS SM retrieval approaches must contend with many more unknown variables within the measurement scene and the received signal when compared to traditional microwave remote sensing. First, a GNSS signal is scattered from the land surface and is received by the CYGNSS receiver, and this received signal is a composite of coherent and incoherent signals. Second, the received signal emanates from quasi-random locations (i.e., non-repeating ground tracks). Third, the soil contribution can be suppressed by a combination of effects from vegetation, topography, surface roughness, soil type, and water bodies under bistatic geometry. Furthermore, CYGNSS data products have been through a number of updates that consider non-geophysical factors such as a variation/uncertainty of the GNSS transmitter power [20,21] as well as the receiver antenna pattern corrections [22]. In order to determine a SM data product, a model is needed which correctly determines the aforementioned effects based on the complex scattering environment within CYGNSS's contributing area. Given the sensitivity to fine-scale surface features and the pseudorandom sampling of CYGNSS, direct application of physical models in the SM retrieval process from CYGNSS measurements at high spatio-temporal resolutions would be much more challenging than SMAP and SMOS which have exact repeat-pass tracks on a regular interval and relatively coarse resolution. Because of these complications, previous studies have employed spatio-temporal averaging in their retrieval algorithms to successfully eliminate measurement uncertainties associated with the measurement configuration and complexity of scattering contributions [15–17]. This, however, sacrifices CYGNSS's capability to provide high spatio-temporal SM datasets.

A SM retrieval approach that considers complex land surface characteristics is needed to obtain accurate, high spatio-temporal SM information. In order to effectively address the problem for the uncertainty of coherent/incoherent assumptions as well as the nonlinear relationships among SM, CYGNSS observations, and geophysical input data at high spatio-temporal resolution, we have recently implemented a machine learning (ML) framework to retrieve SM from CYGNSS measurements [18]. The approach was realized as a proof-of-concept to obtain a SM data product over limited (a total of 18) International Soil Moisture Network (ISMN) sites without the need for a model that explicitly requires (1) incoherent signal detection and modeling, (2) modeling of unpredictable scattering mechanisms within vegetation, or (3) spatio-temporal averaging. The primary data resources included CYGNSS Level 1 data, NASA's Moderate Resolution Imaging Spectrometer (MODIS) which provides normalized difference vegetation index (NDVI) information, 90 m spatial-resolution digital elevation model data, and *in-situ* SM data from ISMN sites. The area surrounding the sites were chosen for low vegetation and surface roughness regions in order to establish the artificial neural network (ANN) retrieval algorithm's basis from its simplest spaceborne case using non-simulated datasets. The technique achieved 0.0544 cm^3/cm^3 retrieval accuracy with Pearson correlation coefficient of 0.9009 for 2017 and 2018 CYGNSS observations.

This paper extends our previous study [18] over larger and more diverse datasets to gain further insights into the use of ML-based CYGNSS SM retrievals at high spatio-temporal scales. The present study utilizes *in-situ* data from over 100 ISMN sites that exist below 38° latitudes in the Continental United State (CONUS) from years 2017 to 2019. The ML-based retrieval model has been evaluated through three different validation strategies, i.e., 5-fold, site-independent, and year-based cross-validation methods. Furthermore, the results are evaluated across different land cover types and soil textures as well as the SM site network types on the SM prediction performance. The effects of ancillary inputs on SM predictions are also compared.

The rest of the paper is organized as follows: Section 2 describes the CYGNSS L1 and auxiliary land surface parameters, and preprocessing of combined dataset including associated quality filters used before the application of ML approaches. Section 3 provides a detailed explanation of the ML framework as well as the cross-validation approaches. Section 4 illustrates the effectiveness of the approaches across different land cover types, soil textures, and the SM site network types. The effects of primary ancillary inputs on the performance are also compared. Section 5 discusses findings, challenges, and implications for extending the ML-based SM retrieval methodology to a global coverage. Finally, Section 6 summarizes our conclusions.

2. Datasets

In order to effectively develop an ML-based retrieval algorithm for surface SM from CYGNSS observations, several datasets must be leveraged. The following subsections detail the input selection for the retrieval algorithm as well as each input's physical relationship to SM and GNSS-R sensitivity. The methods of quality control and multi-resolution dataset combinations are discussed in order to ensure consistent, accurate SM products.

2.1. Cyclone Global Navigation Satellite System

The CYGNSS mission is a constellation of eight micro-satellite observatories, each of which is carrying a four-channel GNSS-R bistatic radar receivers to record the reflected Global Positioning System (GPS) signals. Despite the fact that the constellation primarily orbits the tropics, limiting the spatial coverage to latitudes ±38°, it acquires a considerable amount of land observation data that provides opportunities to exercise SM retrieval approaches.

To retrieve SM over land, the CYGNSS Level-1 (L1) version 2.1 product is used and accessed from the Physical Oceanography Distributed Active Archive Center (PO.DAAC, https://podaac.jpl.nasa.gov/). The key observable from CYGNSS L1 data is the Delay-Doppler map (DDM) which represents the received surface power over a range of time delays and Doppler frequencies (bin-by-bin) for each observed specular reflection point [23]. DDMs are processed in L1 to account for non-surface related terms through inverting CYGNSS' forward-scattering model and obtaining the surface's bistatic radar cross section (BRCS) and the effective scattering area. The bin-by-bin BRCS is provided as an 11 × 17 array of DDM in L1 data. Additionally, the geometric and instrumental variables are included to provide detailed acquisition information for each specular point with factors such as incidence angle as well as the distances between the GPS transmitter, CYGNSS receiver, and the specular point.

Using the observables provided in L1 data, the surface reflectivity can be estimated via several approaches with different coherence and incoherence assumptions [13,16,18]. Assuming that the observed GNSS-R signal is dominated by coherent reflections, the approach of [13] is selected for calculating reflectivity. Namely, the BRCS (variable *brcs* in CYGNSS L1) (σ_{RL}) and the range terms are used to calculate the reflectivity ($\Gamma_{RL}(\theta_i)$) as

$$\Gamma_{RL}(\theta_i) = \frac{\sigma_{RL}(r_{st} + r_{sr})^2}{(4\pi)r_{st}^2 r_{sr}^2} \tag{1}$$

where r_{st} and r_{sr} denote the distances between the specular reflection point and the GNSS transmitter (*tx_to_sp_range* in L1) and the GNSS-R receiver (*rx_to_sp_range* in L1), respectively. The peak value of the *brcs* DDM is used with the coherency assumption. Furthermore, additional CYGNSS observables such as trailing edge slope (TES) and leading edge slope (LES) are computed from the reflectivity delay waveform which is the integration of BRCS within the Doppler domain. Following [13,24], TES and LES are calculated from the reflectivity delay waveform values at the delay bins m (peak delay bin) to $m+3$ and m to $m-3$, respectively. Both TES and LES are indicators related to the conditions of coherent or incoherent scattering and provide supplementary information in addition to the CYGNSS reflectivity. Therefore, the derived reflectivity is combined with TES, LES, and incidence angle (*sp_inc_angle* in L1) as input layer features from CYGNSS data for SM retrieval in the ML framework. The CYGNSS data used in this work span from March 2017 to December 2019.

2.2. International Soil Moisture Network

The aforementioned CYGNSS observables need to be accompanied with several auxiliary land surface parameters to describe the interaction of received signals with the land surface. To construct the nonlinear relationship between CYGNSS observations and these parameters including SM information through ML approaches, labeled *in-situ* SM measurements are needed. The *in-situ* SM data available at the ISMN sites [25] are used as the reference data and are assumed to be representative over a 9 km × 9 km grid cell [18]. The ISMN has assembled over 50 operational and experimental SM networks worldwide, providing a global *in-situ* SM database with uniform data format and pre-processing quality flags [26]. While there are some sites in Asia, Australia and Europe, sites that provide both temporally and spatially collocated observations with regards to CYGNSS data are mainly in North America. In this study, we consider all available sites over CONUS within the CYGNSS spatial coverage (shown in Figure 1). Detailed information about the ISMN is reported in [25,27]. The ISMN dataset is publicly accessible (http://ismn.geo.tuwien.ac.at).

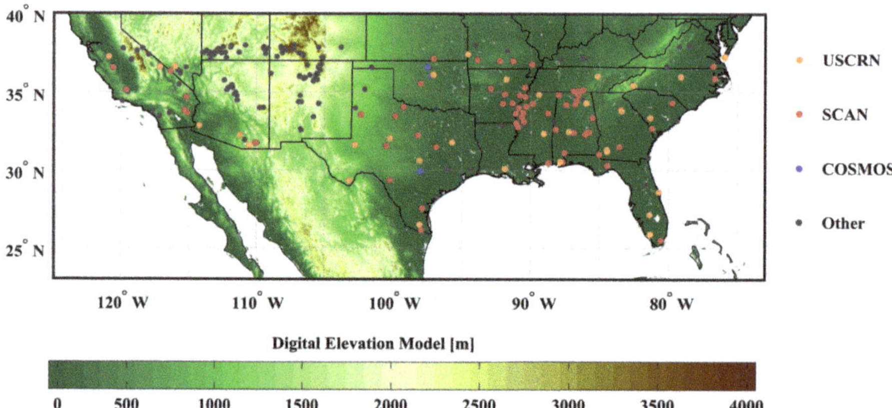

Figure 1. The spatial distribution map of the International Soil Moisture Network (ISMN) sites. Sites shown in gray are excluded based on the masking criteria described in Section 2.4.

The hourly SM data from ISMN are masked using the provided quality flag (identified as 'good' with 'G') and then averaged to daily values. The surface SM at 5 cm depth is used which is consistent with the penetration depth of L-band microwave signals. In total, there are more than 200 ISMN sites with ground-based observations between March 2017 and December 2019 within the CYGNSS spatial coverage in CONUS. These sites are mainly from the Soil Climate Analysis Network (SCAN), U.S. Climate Reference Network (USCRN), and the COsmic-ray Soil Moisture Observing System (COSMOS). Since many effects including complex topography and surface open water can significantly affect CYGNSS received observations, a detailed masking criteria is applied to all ISMN sites as described in Section 2.4. Quality control masks and the number of ISMN for each utilized network is provided in Table 1.

2.3. Ancillary Data

Received GNSS-R signals are affected by various land surface characteristics such as land cover, topography, water bodies, and soil texture in addition to the SM value as stated previously. To account for the impact of land surface characteristics, various time-varying or static physical remote sensing-based land surface parameters are utilized as features in the ML framework. The spatial resolution of CYGNSS observations (that dictates the spatial extent of the ancillary data) is linked to the nature of the scattering surface. It is determined by the first Fresnel zone (several kilometers) in the case of specular (coherent) scattering, and by the glistening zone (several tens of kilometers) in the case of the non-specular (diffuse) scattering. In this study, a coherent reflection is assumed, which is valid when the contributing area is relatively flat and smooth ground with no or light vegetation cover. The "effective size" of the first Fresnel zone is highly variable, and depends on the ranges from specular point to the transmitter and the receiver, specular reflection angle, and operating frequency. It also gets smeared along the track depending on the coherent integration time of the receiver. For most applications, the CYGNSS data is usually gridded to regular grid cells with a fixed resolution under the coherent assumption [13,15]. In our previous work [18], we considered an approximate 4 km × 4 km grid cell that centered the specular point to generate the mean terrain characteristics. Here, the spatial resolution of CYGNSS data is assumed to be 3 km × 3 km which leads to the optimal results from several tests under different spatial resolution assumptions in our pre-test experiments. Therefore, the auxiliary land surface characteristics are spatially aggregated from their native resolutions to 3 km. For CYGNSS specular points located within the 9 km grid centered at each ISMN site, the reference labels are assumed to be the same for one particular day. Each specular point has its own auxiliary data from a 3 km grid centered at the specular point, including physical parameters of DEM elevation, slope, NDVI, and soil texture.

The 500 m resolution, 16-day composite NDVI from MODIS (MYD13A1) is utilized for characterizing the spatio-temporal variations of vegetation conditions. To be consistent with the assumed spatial resolution of CYGNSS data, MODIS NDVI data has been spatially averaged from its original 500 m to 3 km for grids centered at specular points. The MYD13A1 dataset can be acquired from the NASA Land Processes Distributed Active Archive Center (https://lpdaac.usgs.gov/products/myd13a1v006/).

A dominant land cover type map at 500 m resolution is also obtained from the MODIS yearly Land Cover Type (MCD12Q1) product in 2018. This product provides the dominant land cover type for each grid cell. Six classification schemes are included, and the primary International Geosphere Biosphere Programme (IGBP) land cover scheme is selected for further analysis. IGBP contains 17 land cover classes, including water, forest, shrublands, grasslands, cultivated land, wetlands, artificial surfaces, permanent snow and ice, and bareland. For each 3 km grid, the most frequent land cover type is determined. In addition, the land cover information is used to calculate Vegetation Water Content (VWC) and surface roughness (H-value) parameters using the same lookup tables as in the SMAP product [28]. Specifically, both VWC and H-values are obtained using the weighted averages of the lookup table values with weights determined by the percentages of corresponding land cover types within each 3 km grid cell.

The topographic information, known to greatly affect the reflectivity of GNSS-R signals [29], is derived from the 1 km Digital Elevation Model GTOPO30 product. This can be obtained from the United States Geological Survey Earth Resources Observation and Science archive (doi:

/10.5066/F7DF6PQS). Elevation and slope are calculated and aggregated from 1 km to 3 km. Similar to other datasets, the spatial regridding for topography is conducted for each 3 km grid centered at the specular point and averages of elevation and slope are used to represent the underlying topographic complexities.

The presence of surface inland water body is identified by utilizing a 30 m Global Surface Water Dataset from the Joint Research Centre (GSW-JRC) [30]. Particularly, the seasonality map in 2018 provides the intra-annual behavior of surface water and the number of months where permanent or seasonal surface water were present. These 30 m seasonality maps are aggregated to 3 km for representing the percentages of surface inland water within the CYGNSS observed reflection area. The water percent is determined by calculating the percentage of 30 m grids within each 3 km grid indicating the presence of either permanent or seasonal water, and this value is used during the retrieval algorithm's quality control phase.

The Global Gridded Soil Information (SoilGrids) [31] is used to represent soil texture that controls hydraulic properties such as water retention and capillary action within the profile. In SoilGrids, the soil profile is vertically discretized to seven layers with a maximum depth of 2 m. For each layer, the soil is classified into 12 standard soil texture classes based on the sand, clay, and silt proportions. Here, the 5 cm depth is used for consistency with the penetration depth of L-band signals. The product is available at 250 m and spatially aggregated to 3 km for each specular point. Sand, clay, and silt proportions are spatially averaged, and the dominant soil texture class is determined by the percentages of the 12 soil texture classes.

2.4. Quality Control Mechanisms

In total, there are over 160,000 CYGNSS observations available from March 2017 to December 2019 over ISMN sites in CONUS. However, critical screening for the quality of CYGNSS observations and underlying land surface conditions is needed before conducting SM retrieval. Several filtering criteria for quality control are applied to CYGNSS observations, ancillary data, and *in-situ* SM measurements.

CYGNSS metadata contains many quality control flags for both the raw instrument data and processing information. Thus, observations with relatively low quality can be easily removed by quality flags that indicate a potentially unreliable GNSS-R observation. In the study, we use the specific flags (S-band powered up, large spacecraft attitude error, black-body DDM, DDM test pattern, low confidence GPS EIRP estimate) and methods reported in [13,15]. Additionally, CYGNSS observations with a negative antenna gain are filtered from the input dataset. Observations with an incidence angle higher than 65° [14] are removed to avoid the inclusion of noisy DDMs. Also, observations with a DDM peak value outside of 5 to 11 delay bins are removed from the dataset to avoid the inclusion of high-altitude measurements [13].

Open water near the specular point is a critical error source for SM retrieval products. The power of a forward-scattered signal emanating from a water's surface is generally several orders of magnitude higher than a signal scattered from soil due to the very strong coherency over water surfaces [32]. The SM retrieval near water bodies, thus, becomes infeasible if the surface water is sufficiently large within the CYGNSS footprint. Hence, a CYGNSS observation is excluded if more than two percent of the 3 km grid centered on a specular point is covered with permanent or seasonal water.

In addition, CYGNSS observations that fall over forested regions with VWC > 5 kg/m^2 (dense vegetation canopy) [33] and with a dominant urban land cover type [34] are filtered and removed using the land cover type data described in Section 2.3. Furthermore, CYGNSS observations, which corresponds to a total of 84 ISMN sites in the CONUS, are also masked out due to a limited elevation algorithm at the first stage of CYGNSS mission.

Table 1 summarizes some information about the raw dataset, quality mechanisms, and their ratios on the raw dataset and filtered dataset statistics. From 2017 to 2019, there are 234 sites over CONUS reporting SM data from COSMOS, SCAN, and USCRN networks. The concurrent number of data samples from CYGNSS and ISMN is 160, 400. After applying specified quality control criteria, total of 106 sites with 33, 553 concurrent CYGNSS observations for SM retrieval is achieved.

Table 1. The ISMN data statistics before and after quality control flags.

	Initial		Quality Control Mechanisms and the Ratio on the Raw Dataset	%	Final	
	# of Sites	# of Data			# of Sites	# of Data
COSMOS	14	7381	CYGNSS quality flags	27	5	1923
SCAN	104	80, 446	Incidence angle > 65°	3	68	22, 951
USCRN	53	39, 928	Rx_gain < 0	27	33	8679
2017	225	39, 888	Peak power delay row bin	20	89	7580
2018	219	47, 304	Water land percent > 2%	16.6	99	9485
2019	222	73, 208	Elevation > 600 m (for 2017)	11	100	16, 488
Overall	234	160, 400	Urban areas	0.9	106	33, 553
			VWC > 5 kg/m^2	12.5		

3. Methodology

The accuracy of SM estimation retrieved from GNSS-R observations is dependent on a proper modeling of the complex and nonlinear relationship between the forward-scattered signal and environmental variables such as system geometry, surface roughness, topography, and soil properties. It is highly complex to model all such interactions with high fidelity. Instead, our methodology uses data-driven ML approaches with physics-based features that have direct, physical relations to SM estimation. Since ML algorithms are excellent function approximators and have a remarkable capability in modeling complex and nonlinear relationships, ML is a well-suited approach for the CYGNSS-based SM retrieval problem. All available ISMN sites over CONUS are utilized to conduct extensive analysis on both the ML approaches and their performance dependence on utilized physical features. The overall SM retrieval algorithm used in this study is depicted in Figure 2 and briefly summarized in section 3.3.

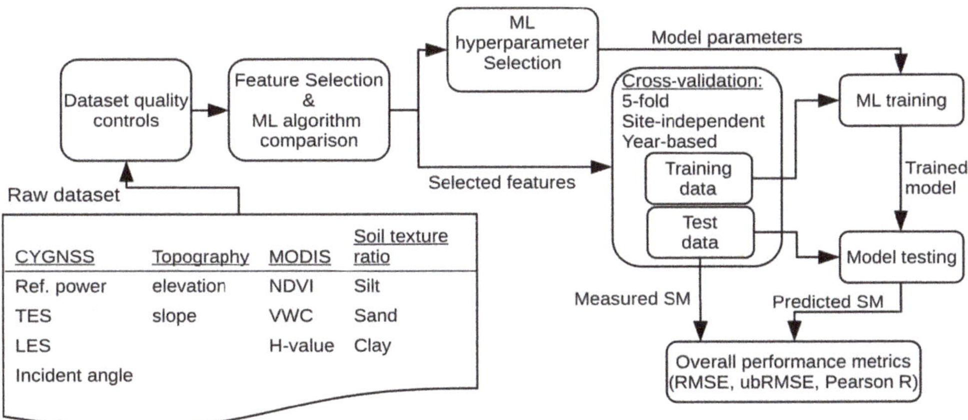

Figure 2. Flowchart of the proposed soil moisture (SM) retrieval algorithm.

In this study, a supervised learning problem is considered that maps the inputs, which are physical features related to land scattering characteristics, to the labels, which are the measured SM values at ISMN sites. CYGNSS and the ancillary dataset that form the input features for the proposed supervised learning approach are presented in Section 2 along with the preprocessing and quality control mechanisms. The physical features for CYGNSS-based SM retrieval are separated into four main groups: CYGNSS observables, topography, MODIS, and soil texture features. All of the 12 physical features in these groups are listed and described in Table 2. To interpret the effect of different types of ancillary data, different ML models are trained and validated by excluding one ancillary input group at a time. Therefore, four different schemes of input feature groups are constructed as below and discussed in Section 4.1.

Scheme 1: default scheme with {*all ancillary inputs*} + CYGNSS.

Scheme 2: {*Soil + MODIS*} + CYGNSS.

Scheme 3: {*Soil + Topography*} + CYGNSS.

Scheme 4: {*Topography + MODIS*} + CYGNSS.

In the following subsections, we describe in detail the trade-offs in use of different ML-algorithms, determination of optimum hyperparameters of the algorithms, selecting most relevant features as well as strategies to evaluate the performance of the framework.

Table 2. Physical features considered for the machine learning (ML)-based SM retrieval model.

Input Group	Feature Name	Description
CYGNSS	Reflectivity	Reflectivity calculated via [13]
	TES	Slope of the trailing edge of the reflectivity
	LES	Leading edge slope of the reflectivity
	SP incidence angle	Incidence angle of specular point
Topography	Elevation	Mean elevation for each specular point 3-km grid
	Slope	Mean Slope for each specular point 3-km grid
MODIS	NDVI	Mean normalized difference vegetation index
	VWC	Mean vegetation water content
	H-value	Dominant land cover type based roughness parameter
Soil texture	Soil clay ratio	Mean clay proportion for each specular point 3-km grid
	Soil silt ratio	Mean silt proportion for each specular point 3-km grid
	Soil sand ratio	Mean sand proportion for each specular point 3-km grid

3.1. Machine Learning Algorithm and Feature Selection

The selection of a ML algorithm and its hyperparameters that are most suitable for the SM retrieval problem is a critical decision and has a significant impact on the performance of SM prediction. We compare ANN, Random Forest (RF), and Support Vector Machine (SVM) approaches, which are popularly used for supervised regression problems.

ANN is one of the most common ML algorithms for nonparametric and nonlinear classification or regression problems [35]. ANNs are networks formed by interconnections between neurons with nonlinear activations. Each neuron is a model that receives a linearly weighted combination of outputs from previous layers and outputs a result passing that linear combination through its nonlinear activation function. A feed-forward ANN of a multi-layer perceptron structure is usually composed of an input layer, one or more hidden layers, and an output layer. The total number of layers and number of nodes used in each hidden layer determines the total number of weights that must be learned through minimizing the total cost between ANN predictions and the actual measured outputs in the training data.

SVM is a supervised nonparametric learning method [36]. The basic idea of the SVM is to find hyperplanes that separates training samples into a predefined number of classes. SVM can also be applied to nonlinearly separable problems by using a kernel function [37]. The accuracy of SVM depends on the hyperparameters of the error penalty parameter and the parameters of kernel functions. These two critical parameters need to be optimized if a radial basis function (RBF) is selected.

RF is an ensemble ML algorithm that forms multiple decision trees. Each decision tree contains a root node, internal nodes, and leaf nodes. In the forest, each decision tree makes its prediction, and then the mean prediction of the trees is calculated as the output of the RF for a regression problem [38].

The important hyperparameter defining the RF performance is the number of decision trees utilized in the RF algorithm.

Hyperparameters are model parameters specified before the learning process. Typical hyperparameters include the number of hidden layers and neurons in ANN, the regularization coefficients and parameters in the kernel function of SVM, maximum split size, and the size of the ensemble tree in RF algorithms, etc. The selection of hyperparameters mainly determines the ML model and thus has a critical impact on SM prediction performance. Optimal hyperparameters should be used to prevent overfitting and underfitting of the ML technique to the training dataset. The optimal operating points of each ML model are obtained by utilizing a grid search method for their hyperparameters as presented in Section 4.1.

Initially, 12 features from CYGNSS, MODIS, soil texture and topography feature groups are used (Table 2). However, using too many features or too complex models may lead to an overfitting with the ML model. A large feature set can contain noisy features or cross-correlated features that might lead to marginal improvements or even deteriorations in final performance. Moreover, using too many features will increase the computational cost. Thus, it is highly essential to select a subset of relevant features. In this work, the sequential feature selection, forward or backward, algorithms are used to choose the most relevant feature subset. The sequential feature selection is widely used for its simplicity and speed in many applications [39,40]. Forward sequential feature selection is an iterative technique that selects at each iteration the subset of features that minimizes the defined cost function. Starting with the best-performing single feature and sequentially adding the next best feature, the iteration continues until a certain stopping criterion is satisfied or all features are used. In the backward feature selection method, the algorithm removes one less significant feature at a time. This process is repeated until there is no feature to be removed or when a stopping criterion is reached. The feature selection results are described in Section 4.1.

3.2. Performance Metrics and Evaluation

Results are evaluated in terms of the root-mean-square error (RMSE), unbiased RMSE (ubRMSE), bias, and correlation coefficient (R) between the predicted SM values and the measured SM from *in-situ* sites. To evaluate how well the performance of the proposed method could be generalized, three different cross-validation techniques are deployed: (1) 5-fold, (2) site-independent, and (3) year-based.

For a 5-fold cross-validation, the data set is first split into 5 folds, then 4 folds are used as the training set, and the remaining fold is used as the testing set. The final evaluation result is the averaged result of each fold. To evaluate the capability of the prediction model to be generalized, a site-independent cross-validation approach is also tested. In this approach, data for a single SM site is used as test dataset while data for all other SM sites are used in training. By this way, the prediction performance of ML for a totally unseen site can be analyzed. The RMSE, ubRMSE, bias, and R values are computed separately for each site under this validation case. For a year-based cross-validation method, the proposed model is first trained with data from 2017 and 2018 and then tested on data from the year 2019. This year-based validation process is repeated for each combination of the three observation years. As seen in Table 1, the number of observations of each year is different. Hence, it should be noticed that the contribution of years to the training model would not be similar.

3.3. Machine Learning Framework Summary

Figure 1 encapsulates the following, overall workflow of the ML-based SM product. The raw dataset which contains CYGNSS-based reflectivity and geometry, MODIS-based vegetation information, GTOPO30-based topography information, and SoilGrid-based soil texture data are compiled into a single dataset for analysis. These individual datasets are described in detail within Sections 2.1 and 2.3. Data that contain potentially unreliable information are filtered out using the dataset quality control processes described in Section 2.4. The optimal hyperparameters for ML methods are chosen as discussed in Section 3.1. Redundant input features are then eliminated using the feature selection process described in Section 3.1. Using a portion of the ISMN data described in Section 2.2 as reference labels, the selected ML method is trained with the filtered input dataset. The remaining portion of ISMN data is considered a testing dataset and is used for model testing. The metrics for this model testing are defined in Section 3.2. Finally, this entire process is performed for different input dataset schemes as defined in Section 3. All analyses and model development processes are performed using the machine learning toolbox of MATLAB R2019b software.

4. Results

In this section, the SM retrieval results from varying ML-based approaches are presented from four perspectives. In Section 4.1, the performance of different ML algorithms and input features are first explored. With the selected ML technique and input features, the overall performance of the ML model for SM retrieval is evaluated in Section 4.2 through three cross-validation strategies (as described in Section 3.2). The spatial distribution of the ML-based SM retrieval performance is also presented in this part. Section 4.3 analyzes the effect of different land cover types and soil texture conditions on the SM prediction performance via a 5-fold cross validation method. In addition, the impact of different *in-situ* SM networks on the performance is examined. In Section 4.4, two representative ISMN sites are selected and their performance are demonstrated in the temporal domain.

4.1. Examination of Different Machine Learning Algorithms and Input Features

As stated in Section 3.1, hyperparameters and, hence, the ML-based model itself require careful selection in order to prevent overfitting or underfitting. Here, we first analyze the ML algorithm performance with a varying set of model parameters with 5-fold cross-validation. The selected grid search ranges for hyperparameters are the number of trees (from 10 to 1250 with a 10-step interval) and maximum split size (from 1 to 250 with a 5-step interval) for RF, hidden neuron size (from 5 to 100 with a 5-step interval) and layer size (from 1 to 3) for ANN, kernel scale (from 2^{-6} to 2^6) and penalty parameter (from 10^{-1} to 10^3) for SVM. During the grid search process, the model complexity is determined in terms of total weight number for ANN and total nodes number for RF. In ANN, the number of weights is a function of the number of features, the number of layers, and the number of neurons. For an RF model, the total number of nodes is the sum of the number of nodes of each decision tree. For SVM, the number of features and kernel scale are the main parameters that affect the model complexity.

Figure 3 shows the validation curves for each evaluated ML approach. The training and validation RMSE curves are shown as a function of the varying model parameters obtained from the grid search process for each ML algorithm. The black circle as shown on each ML approach's validation curve indicates the optimal model order and hyperparameter selection that generates the minimum RMSE on the testing dataset. It is clear and expected that the RMSE of the training dataset generally decreases with increasing model complexity for all compared ML algorithms. However, exclusively minimizing the RMSE value for the training dataset can produce overfitting if the RMSE of the testing dataset is not considered during hyperparameter selection. For ANN, a minimum RMSE value of 0.061 cm^3/cm^3 on the testing dataset is obtained with three layers and 25 hidden neurons for each hidden layer (Figure 3a). For RF the minimum testing dataset RMSE of 0.052 cm^3/cm^3 can be obtained with 120 maximum split size and 200 trees (Figure 3b). Similarly, the optimal penalty parameter and kernel scale of SVM are obtained as one and two, respectively, with a minimum RMSE of 0.065 cm^3/cm^3 (Figure 3c). Comparing the three ML algorithms, RF delivers the smallest RMSE value on the testing dataset. Therefore, the RF is chosen as the ML algorithm for SM retrieval in this work and is used in subsequent analysis.

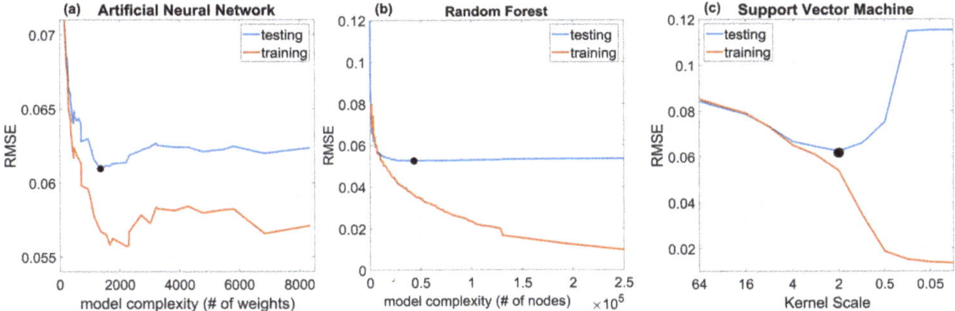

Figure 3. Root-mean-square-error (RMSE) of the training and testing data as a function of the model complexity for three different ML algorithms: (**a**) Artificial Neural Network (ANN), (**b**) Random Forest (RF), and (**c**) Support Vector Machine (SVM).

To understand the impact of the four raw input datasets depicted in Table 2 on the total RMSE, four schemes with different combinations of input feature groups (specified in Section 3) are analyzed using the RF regression model and a 5-fold cross-validation method. Note that the initial 12 features are separated into four groups and are fully included in the Scheme 1 as a benchmark for evaluating the effect of each independent ancillary feature group. Figure 4 shows predicted SM values compared against *in-situ* observations for four different schemes. For the case with all 12 features as input data, a minimum ubRMSE of 0.052 cm^3/cm^3 and a maximum R of 0.894 [-] is obtained via a 5-fold cross-validation method (Figure 4a). For cases where one of the three ancillary feature groups is excluded from the input data, e.g., the topography information is excluded in Figure 4b, the ubRMSE/R values are changed from 0.052 cm^3/cm^3/0.894 [-] to 0.055 cm^3/cm^3/0.879 [-] when comparing the ML model predicted SM and *in-situ* data. Likewise, the removal of either MODIS (Figure 4c) or soil information (Figure 4d) leads to degraded model performance . Particularly, soil texture features are identified as the most influential ancillary input for the SM prediction with a net ubRMSE increase of 0.006 cm^3/cm^3 in Scheme 4. Both MODIS features (i.e., NDVI, VWC, and H-value) and topography features (elevation and slope) are critical for predicting SM as indicated in Figure 4b,c. In combination, the MODIS, topography, and soil texture feature groups provide complementary information of underlying land surface conditions with respect to CYGNSS observables and therefore are necessary for accurate SM retrieval in the ML modeling process.

However, within each feature group, there exists repetitive and cross-correlated information that can slow down the ML modeling process and introduce irrelevant noises. To reduce the input feature size without reducing the retrieval accuracy, the sequential feature selection is conducted. Figure 5 shows the forward and backward sequential feature selection results via a 5-fold cross-validation method. For either forward or backward feature selection, a minimum RMSE of 0.052 cm^3/cm^3 is achieved with eight input features. Further inclusion of new features, i.e., soil sand ratio, H-value, slope, and LES, does not introduce any significant improvements to the regression performance (Figure 5a,b). Thus, an optimal feature number of 8 is determined for the rest of this study. The eight most relevant input features used for SM retrieval are elevation, soil silt and clay ratios, NDVI, VWC, reflectivity, TES, and SP incidence angle (see Table 2 for full descriptions).

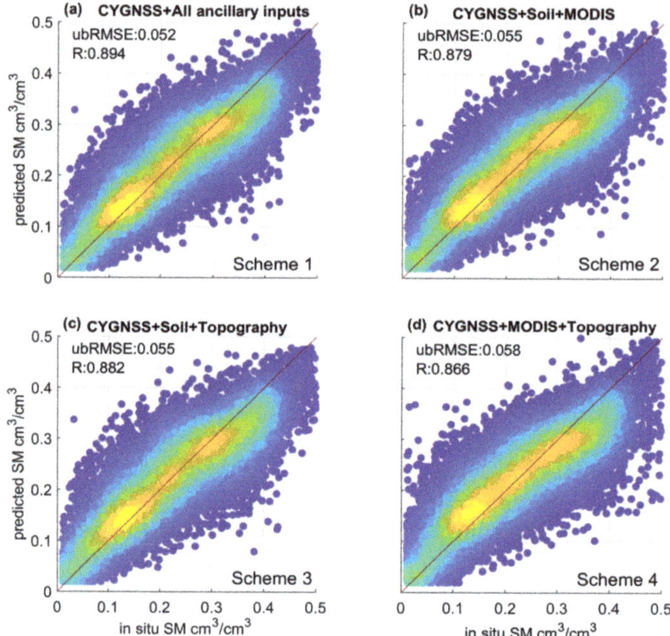

Figure 4. Scatter plots of the predicted SM estimates versus *in-situ* SM observations for ML models with different input feature groups. The input features are described in Table 2. Color of scatter points indicates the density of data points where yellow means a dense data sampling.

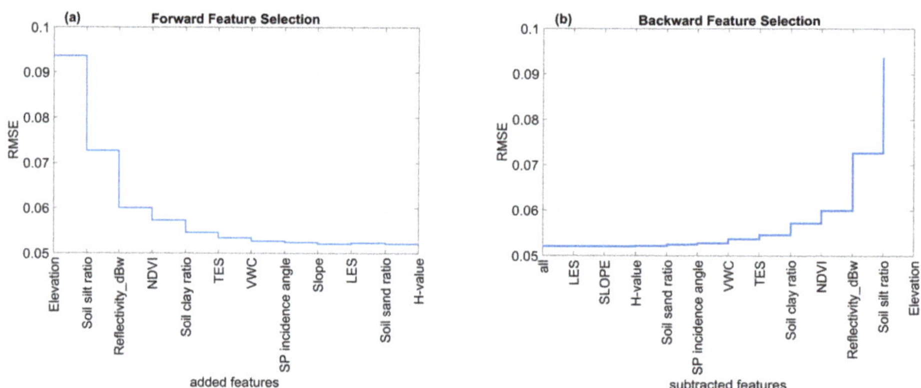

Figure 5. The RMSE values of the sequential feature selection through (**a**) forward and (**b**) backward selection sequences.

4.2. Overall Performance of the Machine Learning Retrieval Model

Figure 6 and Table 3 show the overall SM prediction performance derived via the RF regression model with eight input features. The RF method in a 5-fold cross-validation approach reaches an overall ubRMSE of 0.052 cm^3/cm^3 and a R value of 0.894 [-] over the whole dataset. For all 106 ISMN sites, the mean ubRMSE is obtained as 0.047 cm^3/cm^3 with a standard deviation of 0.016 cm^3/cm^3 whereas the best and poorest scores are 0.09 and 0.085 cm^3/cm^3, respectively. The averaged absolute biases across sites is 0.011 cm^3/cm^3 with a standard deviation of 0.013 cm^3/cm^3 (Table 3). Note that different ISMN sites can show distinct SM climatology, leading to the prediction biases through the ML model with incomplete sampling space. For the 5-fold cross-validation method, 80% of the whole dataset is sampled randomly for use as a training dataset and the ML-based prediction model is tested on the remaining 20% of the data. The biases suggest that the ML model is dependent on the representativeness of the sampling set.

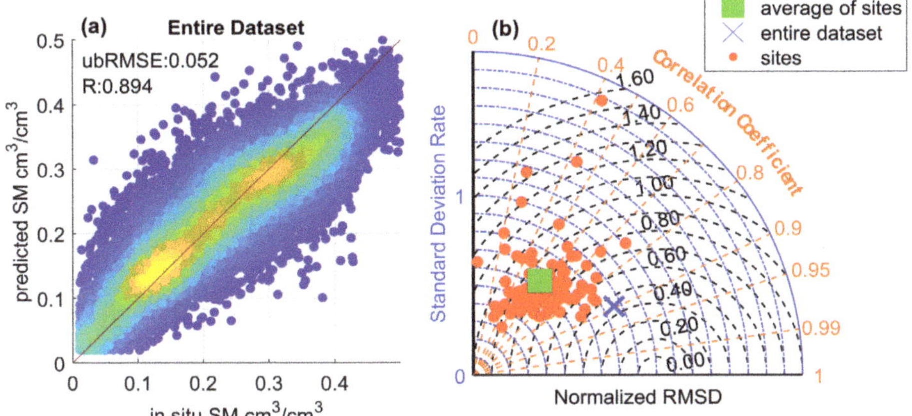

Figure 6. (**a**) The scatter plot of the predicted SM versus *in-situ* SM through a RF-based model with 8 selected input features and (**b**) the Taylor diagram for all sites .

Table 3. Overall and site-averaged performance metrics for different cross-validation methods. Numbers in parentheses are the standard deviation (std.) of metrics across different sites.

Validation Method	Overall Performance			Average of Sites			
	RMSE	ubRMSE	R	RMSE (std.)	bias (std.)	ubRMSE (std.)	R value (std.)
5fold	0.0523	0.0523	0.89	0.050 (±0.017)	0.011 (±0.013)	0.047 (±0.016)	0.56 (±0.20)
Site independent	0.0883	0.0883	0.64	0.084 (±0.037)	0.056 (±0.044)	0.054 (±0.021)	0.42 (±0.20)
year based (2019)	0.0639	0.0639	0.84	0.06 (±0.023)	0.027 (±0.025)	0.05 (±0.018)	0.49 (±0.27)
year based (2018)	0.0586	0.0584	0.86	0.055 (±0.02)	0.024 (±0.019)	0.047 (±0.016)	0.43 (±0.30)
year based (2017)	0.0602	0.0599	0.84	0.058 (±0.022)	0.027 (±0.024)	0.048 (±0.016)	0.40 (±0.24)

Figure 7 shows the spatial distribution and variations of the RF-based SM retrieval model across the CONUS sites. A satisfactory performance is achieved with ubRMSE smaller than 0.045 cm^3/cm^3 (Figure 7a) and R larger than 0.7 [-] (Figure 7b) for the majority of sites. Sites with small R values generally correspond to those with relatively few concurrent *in-situ* observations and CYGNSS data (the number of concurrent samples is less than 100) as shown in Figure 7b. Overall, the 5-fold cross-validation results indicate that the RF-based SM retrieval model is capable of generating satisfactory SM estimates.

In addition to the 5-fold cross-validation, a site-independent cross-validation, or the "leave-one-subject-out" method, is applied which depicts the most challenging strategy to determine how well the proposed method generalizes for new sites' observations that are totally unseen during the training process. For this purpose, the RF model is trained over 105 sites and tested on all observations at the unseen site. This validation procedure is conducted independently for 106 sites. The overall performance statistics across 106 sites in the site-independent cross-validation are provided in Table 3. The mean ubRMSE of all sites is relatively low with a value of 0.054 cm^3/cm^3 indicating that the RF-based SM prediction model can predict the temporal variations of SM for new site or unseen regions. However, relatively large RMSE (0.084 cm^3/cm^3) and mean absolute bias (0.056 cm^3/cm^3) across sites suggest that the ML-based prediction model is less capable of dealing with systematic bias issues. As noted above, different sites can have distinct climatology that can be difficult for the ML model to capture if no *a priori* information is provided. In this site-independent cross-validation method, this phenomenon is further exaggerated since no information will be available for the learning process over the unseen site. Increasing the spatial coverage of the training dataset with more complete characterization of various land surface conditions can potentially improve the performance of ML-based SM retrieval model. However, with limited *in-situ* sites, increasing the spatial representativeness of the training data will require a global satellite-based SM data which is beyond the scope of this work.

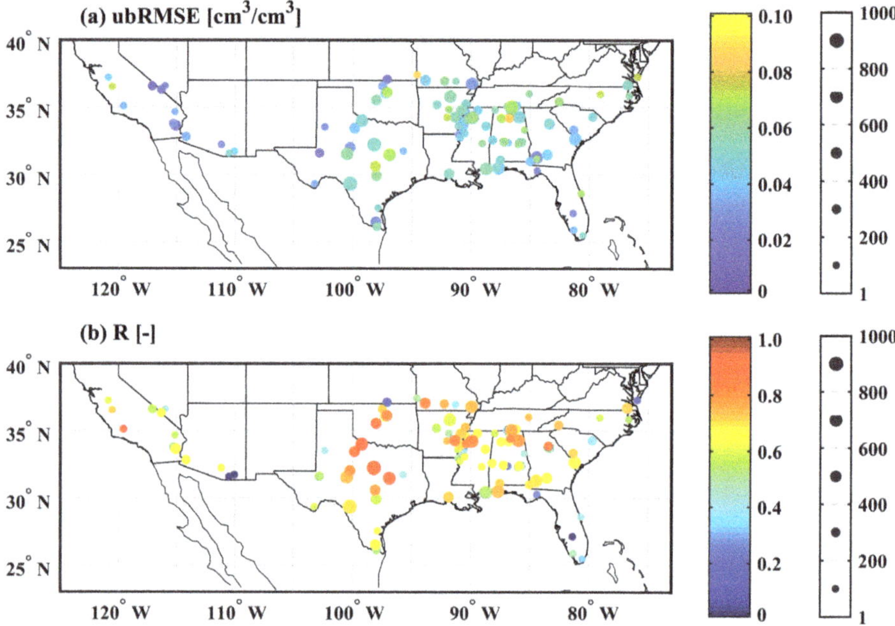

Figure 7. Spatial distribution maps of the (**a**) unbiased RMSE (ubRMSE) and (**b**) correlation coefficient (R) for all sites over contiguous U.S. The sizes of the filled circles are scaled as a function of the number of observations.

Moreover, in order to test the ML-based model for predicting SM estimation under yearly temporal variations, the model is first trained on the data from 2017 and 2018 and then tested on the data from 2019. This cross-validation process is referred as year-based validation and has been repeated for 2017 and 2018. The performance metrics are shown in Table 3 for each testing year. The ubRMSE values for 2017, 2018, and 2019 in the year-based validation are 0.048, 0.047, and 0.050 cm^3/cm^3, respectively. The relatively low ubRMSE values suggest that the ML-based prediction model can be applied to new observations. In the year-based validation, each site provides partial time series data for the training, and the corresponding ML-based model contains a certain amount of the site-specific information for SM retrieval. The low absolute bias error and RMSE, as compared to the site-independent validation, further indicates the importance of *a priori* information on the prediction capability of the ML-based model. The smaller RMSE scores for 2017 and 2018 testing years can be primarily traced back to the number of observation. For year 2019, the ML model is trained with a relatively small dataset and tested on a large dataset.

4.3. Effect of Underlying Land Surface Conditions

To evaluate the overall prediction performance of the ML-based SM retrieval model, it is also important to quantify the impact of different land surface conditions since factors such as soil texture are known to be critical parameters that affect both GNSS-R measurements and retrieval performance. In Figure 8, the SM predictions are compared to *in-situ* SM under 12 main soil texture classes. As demonstrated, the predicted and observed SM estimates are generally well aligned with the 1:1 line. For clay and clay loam classes, the observed SM changes from a minimum of 0.01 cm^3/cm^3 to a maximum of 0.5 cm^3/cm^3 with large variations. The predicted SM estimates have ubRMSE values greater than 0.06 cm^3/cm^3 as compared to ISMN observations. Particularly for clay (Figure 8a), the SM data are concentrated with either high or low values. Moreover, the sampling size of clay is relatively small which further impedes the ML process. On the contrary, the SM is consistently high/low for silty clay loam/sandy clay loam, leading to smaller ubRMSE of 0.042/0.036 cm^3/cm^3 and higher R of 0.884/0.928 [-]. For loam, silt loam, and sandy loam, the SM observations are more evenly distributed as shown in Figure 8e,f,g. The ubRMSE values for these three soil texture classes are around 0.050 cm^3/cm^3 and R values are generally high. As shown in Figure 8h,i, the observed SM values are generally below 0.20 cm^3/cm^3 and thus relatively small ubRMSE values (0.028 and 0.036 cm^3/cm^3) are as expected.

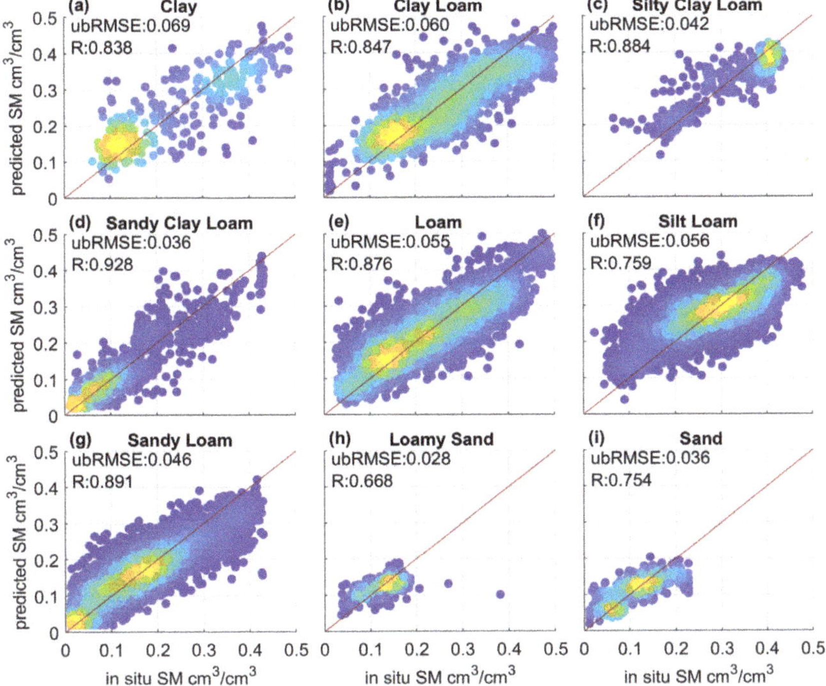

Figure 8. Scatter plots of SM retrievals for different types of soil texture.

Furthermore, the ML-based model prediction capabilities for different land cover types are analyzed and shown in Figure 9. In total, there are eight primary land cover types that are examined. Regions with open shrublands (Figure 9a) and barren (Figure 9h) land cover are generally associated with relatively dry soil at the ISMN sites. Thus, the ubRMSE values are relatively small (0.031 and 0.021 cm^3/cm^3) and correlations tend to be comparatively low (0.661 and 0.749 [-]). For other land cover types, i.e., woody (Figure 9b), savanna (Figure 9c), grass (Figure 9d), croplands (Figure 9f), the observed SM varies from a minimum of 0.01 cm^3/cm^3 to a maximum of 0.50 cm^3/cm^3. In addition, the sampling sizes for these four land cover types are comparably large which benefit a ML-based model to capture the empirical relationship between input features and reference data. The achieved ubRMSE and R values are around 0.050 cm^3/cm^3 and over 0.85 [-] respectively. The relatively small sampling size may be the main reason for a low R (0.786 [-]) over cropland with natural vegetation mosaics land cover type (Figure 9g).

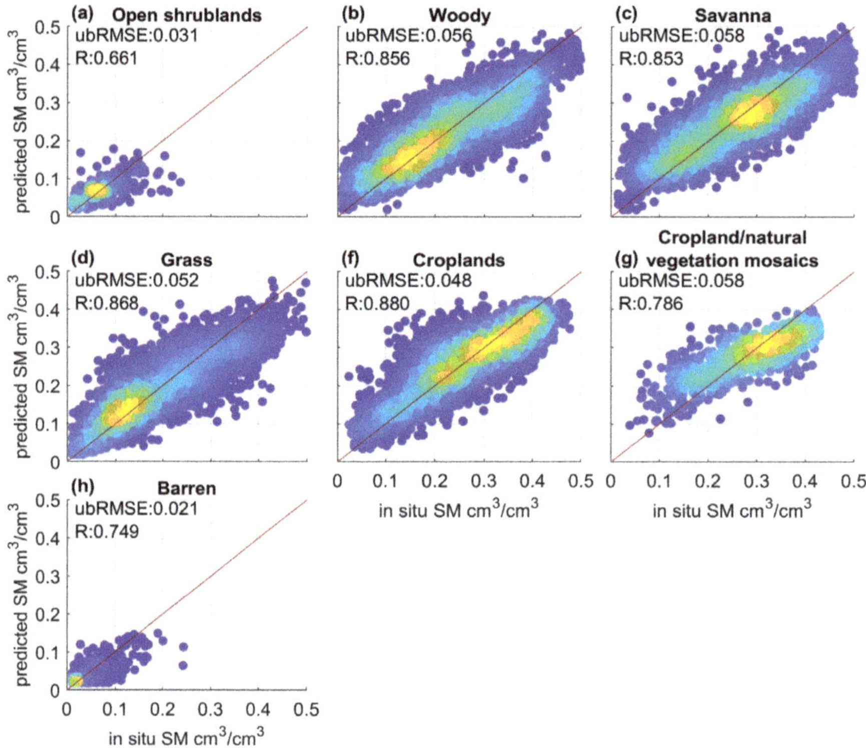

Figure 9. Scatter plots of SM retrievals for different types of vegetation land cover.

When synergistically considering the land cover type and soil texture class, the prediction performance of the ML-based model is shown in Figure 10. As described above, the ubRMSE values are generally small for open shrublands, barren, loamy sand, and sand types. For cases where the dominant soil texture is sandy loam, the ubRMSE scores for different land cover types are mostly below 0.05 cm^3/cm^3 except for cropland/natural vegetation mosaics. More importantly, the predicted SM estimates generally have consistently high accuracy (with ubRMSE less than 0.05 cm^3/cm^3) for croplands under different soil texture classes. The accurate soil water monitoring over croplands is important, and hence, the prediction capability demonstrated here suggests that the ML-based retrieval model can be utilized for agricultural soil water monitoring using the CYGNSS data.

The preceding analysis makes use of the entire ISMN dataset using the RF algorithm with eight selected input features. It is important to note that the *in-situ* SM observations are collected using different ground-based sensors with distinct set-up environments for the three examined observation networks. It is advisable to investigate the impact of the different *in-situ* SM networks on the performance of the learned ML model. To this end, three different RF-based regression models are learned with datasets from the three *in-situ* SM networks separately, i.e., COSMOS, SCAN, USCRN. The 5-fold cross-validation results are shown in Figure 11 and Table 4 for each network-based ML model. By training and testing the ML model with SM observations that are separated by SM network, we find that the SM retrieval accuracy is enhanced such that each network-based ML model reaches lower ubRMSE (an overall performance of 0.049 cm^3/cm^3, Table 4) than the ubRMSE (0.052 cm^3/cm^3, Figure 6) of combining all SM networks. Hence learning three different ML models specific to each SM network performs better than learning a single model for all SM networks. This indicates that the representativeness of *in-situ* data and underlying land surface conditions of the different SM observation networks affect the performance of the ML-based SM inversion model. Despite having less collocated data for the training process, the ubRMSE of predicted SM estimates is 0.043 cm^3/cm^3 for COSMOS. Note that the COSMOS instruments are cosmic-ray water probes that provide SM estimates at the spatial resolutions of dozens to hundreds of meters [41]. Compared to the point-scale SM measurements obtained from SCAN and USCRN, the COSMOS can be more representative for large-scale soil water conditions which benefits the ML-based model for satellite-based SM retrieval. The predicted SM estimates are slightly underestimated for wet conditions when SCAN sites are separately considered in Figure 11b. Thus, the ubRMSE/R values are respectively a bit lower/higher for USCRN as compared to SCAN. Nevertheless, results demonstrated here suggest that the accuracy and representativeness of the reference data are also important for the prediction capability of the ML-based model.

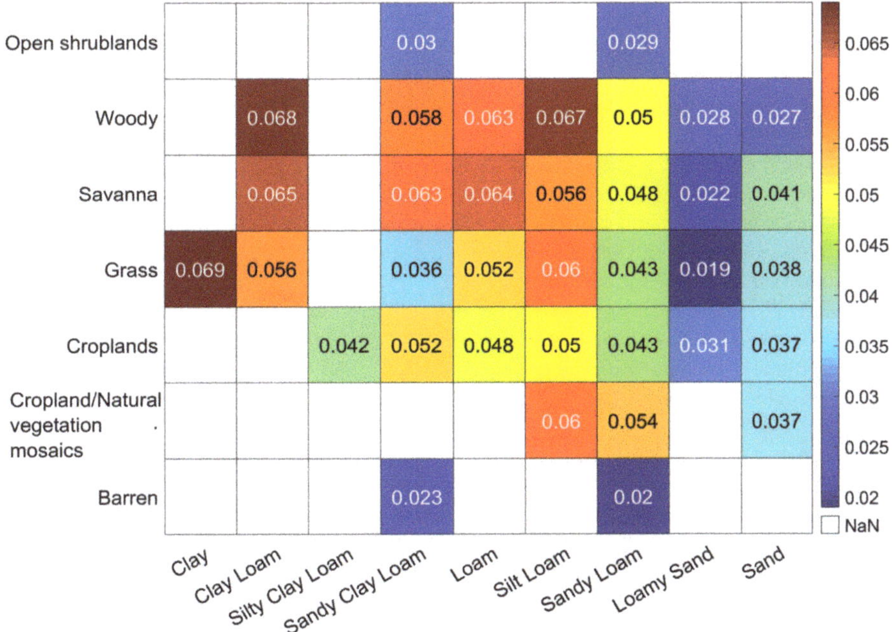

Figure 10. SM prediction performance (ubRMSE value) comparison for varying soil texture and land cover types.

Table 4. Overall and averaged performance metrics for multiple ML models based on SM networks. Numbers in parentheses are the standard deviations of metrics across different network-based models.

	RMSE	Bias	ubRMSE	R
Overall	0.049	-1×10^{-4}	0.049	0.9
Average of sites	0.048($\pm$0.016)	0.0085($\pm$0.01)	0.046($\pm$0.015)	0.58($\pm$0.18)

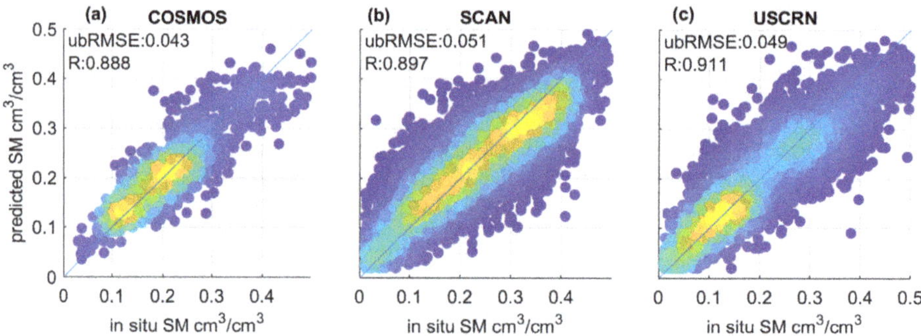

Figure 11. Scatter plots and performance metrics of multiple ML models for each network type: (**a**) the COsmic-ray Soil Moisture Observing System (COSMOS), (**b**) the Soil Climate Analysis Network (SCAN), and (**c**) U.S. Climate Reference Network (USCRN).

4.4. Temporal Variations of Predicted Soil Moisture Retrievals

In addition to evaluating the overall performance metrics, it is important to understand the ML-based model's capability for capturing SM temporal variations. Here, two representative sites are selected and demonstrated in Figures 12 and 13. For both sites, the predicted SM estimates closely follow the temporal trend of the SM observations and correctly capture the precipitation events and the drydown process (Figures 12a and 13a). It is interesting that the CYGNSS reflectivity estimates have a generally good correlation with SM and NDVI for the site with grass land cover type (Figure 12b). For the cropland site shown in Figure 13, NDVI is high (NDVI > 0.7 [-]) and the soil is relatively dry (SM < 0.3 cm³/cm³) for the growing season (from May to September). The CYGNSS reflectivity well captures the soil water condition instead of the vegetation information. Generally, the predicted and observed SM estimates are align with the 1:1 line (Figures 12c and 13c), and the empirical cumulative distribution function (CDF) lines (Figures 12d and 13d) further validate the high accuracy of predicted SM estimates.

As demonstrated previously in Section 4.2, there are several sites with low R values. When examining the land surface conditions of these sites, it is clearly seen that these sites are highly heterogeneous with mixed grass, crops, savanna, surface water and occasionally urban land cover types. The high heterogeneity not only can decrease the representativeness of *in-situ* SM observations, but also it can lower the signal-to-noise ratio of CYGNSS observables leading to more problematic input features and reference labels for the ML process. Nevertheless, the relatively low accuracy of these few sites does not contradict the overall high performance of RF-based retrieval model.

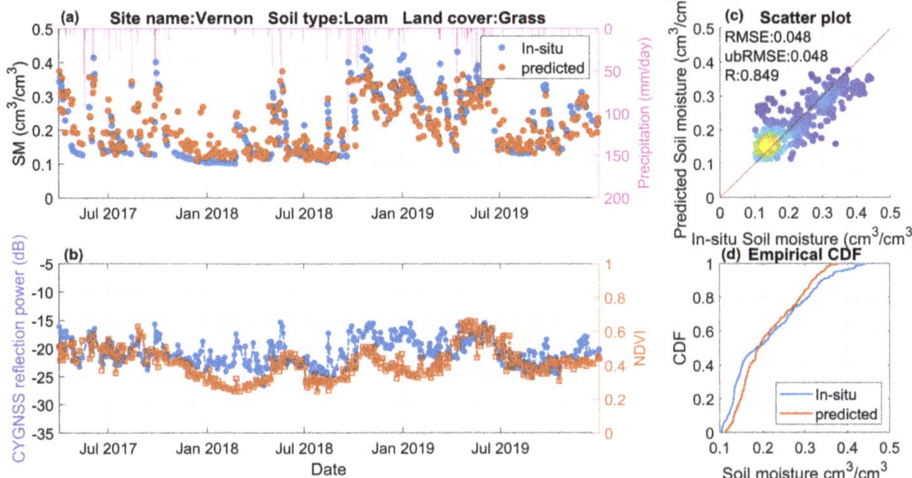

Figure 12. Comparison of (**a**) the *in-situ* observed and predicted SM time series and (**b**) CYGNSS reflectivity and normalized difference vegetation index (NDVI) time series for site Vernon. (**c**) The scatter plot and (**d**) cumulative density function (CDF) between predictions and measurements.

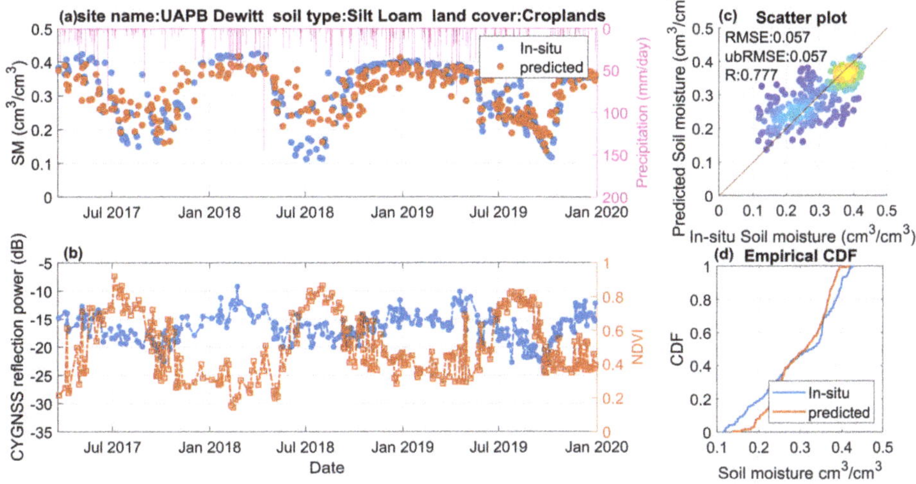

Figure 13. Same as Figure 12 but for site Uapb Dewitt.

5. Discussion

There is a growing interest within the hydrology community to utilize spaceborne GNSS-R observations in SM retrievals. This trend has been particularly accelerated with the availability of recent spaceborne GNSS-R observatories such as TDS-1 and CYGNSS. The allure for using such techniques resides in GNSS-R's relatively high spatial footprint over smooth Earth surface with frequent observation capabilities. This potential can open new applications in hydrometeorology, atmospheric research, and water resource management at microscale and mesoscale resolutions. The goal of this paper's research is to exploit CYGNSS data at high spatio-temporal resolution by taking advantage of recent developments in machine learning algorithms that are excellent function approximators and have a remarkable capability in learning complex and nonlinear relationships.

The choice of ML approach particularly stems from the challenges of CYGNSS's pseudorandomly sampled measurements and sensitivity to fine-scale surface features which are challenging to manage at high spatiotemporal resolutions within a physics-based modeling framework. However, effective utilization of an ML algorithm for SM retrievals requires well prepared data which are labeled and include reliable, physics-based ancillary input features in training phase.

The large number of ISMN sites over CONUS provides an opportunity to extensively exploit ML approaches. Our analysis with ISMN sites demonstrates the potentiality of the ML-algorithms in SM retrievals over various underlying land surface conditions such as soil textures and land covers at high spatio-temporal resolutions. Particularly, the performance over croplands and sandy loam soil provide promising results with higher accuracy. The achieved accuracy is further improved when the ML-model is trained and tested over individual SM networks as opposed to combining all available SM networks. In addition, the generalized methodology is investigated in both space and time using site-independent and year-based cross-validation. The ML-based model is able to capture the temporal variation with variable biases, and the results indicate the importance of *a priori* information on the prediction capability of the ML-based model. While soil texture features are identified as the most influential ancillary input for the SM prediction, both H-parameter and slope are determined as two least significant features in our ML-model. This result is somewhat surprising from a physics-based perspective since the small-scale roughness and topography are two important factors that can alter the relative contributions of the coherent/incoherent energy observed in CYGNSS measurements. This result could be attributed to the locations of ISMN sites which do not show significant variations within and across individual SM network types in either slope or H-value as shown in Figure 14. This indicates that the sites are located on relatively flat surfaces and that coherent reflections are dominant. However, even if those grid cells are perfectly flat (although somewhat tilted), naturally, a small-scale surface roughness of the order of the wavelength is always present. Spatial and temporal changes of the decoherence due to surface roughness would interfere with the analogous changes in the SM. Future studies perhaps could leverage physics-based modeling frameworks by guiding the ML models with simulations to quantify potential source of errors in the retrievals in the absence of alternative ancillary data.

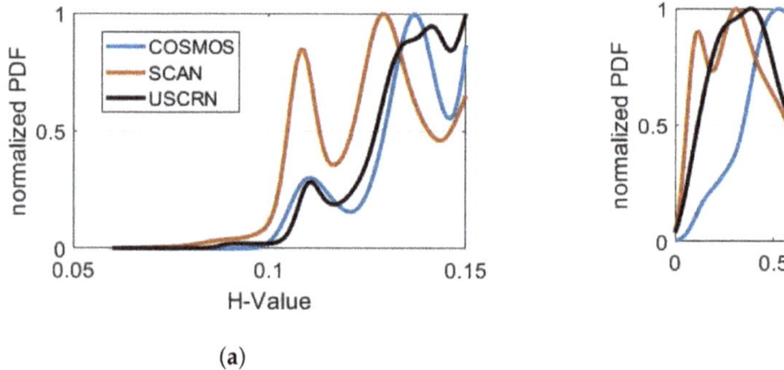

Figure 14. Distribution of (a) small scale roughness (H-value) and (b) topographic slope input features for each SM network.

The correlation coefficient between the *in-situ* data and spaceborne SM retrieval depends on the number of measurements and the dynamic range of the data. The overall performance of the proposed algorithm should be evaluated over data from all sites which provide a wider dynamic range and higher number of observations. As it is also shown in Table 3, correlation coefficient is over 0.8 in all cases where the method is tested for more than a single site. In terms of averaging for each site, both the number of measurements and the dynamic range of each site are lower leading to comparably lower correlation values. Also for small number of measurements, the effects of outliers and the uncertainty of correlation can be higher. It is definitely a goal to obtain higher correlation for each site in future studies. This could be done through site-based learning approaches or developing a model for a group of highly similar sites.

The proposed methodology is potentially limited to similar terrains for which there exist *in-situ* data. Direct application of this paper's ML-model to Earth surfaces beyond CONUS requires further study since the land conditions at ISMN sites are not expected to be representative of the majority of the land scenes crossed by the CYGNSS flight tracks. However, the earth surfaces could be grouped into similar land types by soil texture, topography, and land cover. Perhaps several ML-models for each group could be investigated. In addition, reliable metrics are needed for the ML-based models to learn heterogeneous and mixed scenes which do not necessarily lead to strong coherent reflections. Future work will be needed to fully utilize CYGNSS data for a quasi-global SM data product. This can, perhaps, be achieved by using the SMAP-based SM data as the reference [42]. The key difference between the ISMN and SMAP as the reference label will be the spatial scale. The mismatch of spatial scale representativeness and land surface heterogeneity effects will need further investigation.

6. Conclusions

In this work, an ML-based framework has been presented for estimating SM using the CYGNSS observations over ISMN sites in CONUS. Three widely-used ML algorithms have been tested and validated, among which the RF is observed to be the optimal ML inversion method for this study. A feature selection process reduces the algorithm complexity with a refined input feature set. Several key features are identified, including CYGNSS reflectivity, TES, incidence angle, NDVI, VWC, terrain elevation, and the soil's silt and clay proportions. Using RF as the utilized ML algorithm and with selected input features, an overall ubRMSE of 0.052 cm^3/cm^3 is achieved via the 5-fold cross validation strategy. More importantly, sufficient accuracy can be obtained via the site-independent (ubRMSE of 0.054 cm^3/cm^3) and year-based (ubRMSE less than 0.050 cm^3/cm^3) validation methods, suggesting that the proposed ML-based SM retrieval model can be generalized in space and time with promising confidence. Additionally, the ML inversion performance can be further improved when the training process is separately conducted for different SM observation networks. Although the scale of *in-situ* SM data from different networks varies, the results demonstrated here indicate that a proper consideration of the spatial scales of CYGNSS observations, soil moisture reference data, and ancillary land surface conditions is important for accurately retrieving SM estimates. Meanwhile, the ML-based model performance is analyzed with respect to the land cover and soil texture conditions. Particularly, soil texture features are identified as the most influential ancillary input for the SM prediction. Overall, the ML model predicted SM estimates have high accuracy for croplands (with ubRMSE less than 0.05 cm^3/cm^3), indicating that the ML-based SM retrieval framework can be applied for agricultural soil water monitoring.

Author Contributions: Conceptualization: M.K. and A.C.G.; Methodology: M.K. and A.C.G.; Software: V.S. and F.L.; Validation: V.S. and F.L.; Formal analysis: V.S and F.L.; Investigation: V.S., F.L., D.B., M.K., and A.C.G.; Resources: M.K. and A.C.G.; Data curation: F.L.; Writing–original draft preparation: V.S., F.L., M.K., D.B., and A.C.G.; Writing–review and editing: V.S., F.L., M.K., D.B., A.C.G., and R.M.; Visualization: V.S., F.L., and D.B.; Supervision: M.K. and A.C.G.; Project administration: M.K. and A.C.G.; Funding acquisition: M.K., A.C.G., and R.M. All authors have read and agreed to the published version of the manuscript.

Funding: This research was funded by USDA Agricultural Research Service(USDA-ARS), Award NACA 58-6064-9-007.

Conflicts of Interest: The authors declare no conflict of interest.

References

1. Vereecken, H.; Huisman, J.; Bogena, H.; Vanderborght, J.; Vrugt, J.; Hopmans, J. On the value of soil moisture measurements in vadose zone hydrology: A review. *Water Resour. Res.* **2008**, *44*. doi:10.1029/2008WR006829.
2. Kerr, Y.H.; Al-Yaari, A.; Rodriguez-Fernandez, N.; Parrens, M.; Molero, B.; Leroux, D.; Bircher, S.; Mahmoodi, A.; Mialon, A.; Richaume, P.; et al. Overview of SMOS performance in terms of global soil moisture monitoring after six years in operation. *Remote Sens. Environ.* **2016**, *180*, 40–63.
3. Colliander, A.; Jackson, T.J.; Bindlish, R.; Chan, S.; Das, N.; Kim, S.; Cosh, M.; Dunbar, R.; Dang, L.; Pashaian, L.; et al. Validation of SMAP surface soil moisture products with core validation sites. *Remote Sens. Environ.* **2017**, *191*, 215–231.
4. Brocca, L.; Ciabatta, L.; Massari, C.; Camici, S.; Tarpanelli, A. Soil Moisture for Hydrological Applications: Open Questions and New Opportunities. *Water* **2017**, *9*, 140. doi:10.3390/w9020140.
5. Santanello, J.A.; Lawston, P.; Kumar, S.; Dennis, E. Understanding the Impacts of Soil Moisture Initial Conditions on NWP in the Context of Land-Atmosphere Coupling. *J. Hydrometeorol.* **2019**, *20*, 793–819. doi:10.1175/JHM-D-18-0186.1.
6. Zavorotny, V.U.; Gleason, S.; Cardellach, E.; Camps, A. Tutorial on remote sensing using GNSS bistatic radar of opportunity. *IEEE Geosci. Remote Sens. Mag.* **2014**, *2*, 8–45.
7. Komjathy, A.; Armatys, M.; Masters, D.; Axelrad, P.; Zavorotny, V.; Katzberg, S. Retrieval of ocean surface wind speed and wind direction using reflected GPS signals. *J. Atmos. Ocean. Technol.* **2004**, *21*, 515–526.
8. Valencia, E.; Zavorotny, V.U.; Akos, D.M.; Camps, A. Using DDM asymmetry metrics for wind direction retrieval from GPS ocean-scattered signals in airborne experiments. *IEEE Trans. Geosci. Remote Sens.* **2013**, *52*, 3924–3936.
9. Guan, D.; Park, H.; Camps, A.; Wang, Y.; Onrubia, R.; Querol, J.; Pascual, D. Wind direction signatures in GNSS-R observables from space. *Remote Sens.* **2018**, *10*, 198.
10. Li, W.; Cardellach, E.; Fabra, F.; Ribó, S.; Rius, A. Assessment of Spaceborne GNSS-R Ocean Altimetry Performance Using CYGNSS Mission Raw Data. *IEEE Trans. Geosci. Remote Sens.* **2019**, *58*, 238–250.
11. Rodriguez-Alvarez, N.; Holt, B.; Jaruwatanadilok, S.; Podest, E.; Cavanaugh, K.C. An Arctic sea ice multi-step classification based on GNSS-R data from the TDS-1 mission. *Remote Sens. Environ.* **2019**, *230*, 111202.
12. Santi, E.; Paloscia, S.; Pettinato, S.; Fontanelli, G.; Clarizia, M.; Guerriero, L.; Pierdicca, N. Forest Biomass Estimate on Local and Global Scales Through GNSS Reflectometry Techniques. In Proceedings of the IGARSS 2019—2019 IEEE International Geoscience and Remote Sensing Symposium, Yokohama, Japan, 28 July–2 August 2019; pp. 8680–8683.
13. Rodriguez-Alvarez, N.; Podest, E.; Jensen, K.; McDonald, K.C. Classifying Inundation in a Tropical Wetlands Complex with GNSS-R. *Remote Sens.* **2019**, *11*, 1053.
14. Al-Khaldi, M.M.; Johnson, J.T.; O'Brien, A.J.; Balenzano, A.; Mattia, F. Time-Series Retrieval of Soil Moisture Using CYGNSS. *IEEE Trans. Geosci. Remote Sens.* **2019**, *57*, 4322–4331.
15. Chew, C.; Small, E. Soil moisture sensing using spaceborne GNSS reflections: Comparison of CYGNSS reflectivity to SMAP soil moisture. *Geophys. Res. Lett.* **2018**, *45*, 4049–4057.
16. Clarizia, M.P.; Pierdicca, N.; Costantini, F.; Floury, N. Analysis of CYGNSS Data for Soil Moisture Retrieval. *IEEE J. Sel. Top. Appl. Earth Obs. Remote Sens.* **2019**, *12*, 2227–2235.
17. Kim, H.; Lakshmi, V. Use of Cyclone Global Navigation Satellite System (CYGNSS) observations for estimation of soil moisture. *Geophys. Res. Lett.* **2018**, *45*, 8272–8282.
18. Eroglu, O.; Kurum, M.; Boyd, D.; Gurbuz, A.C. High Spatio-Temporal Resolution CYGNSS Soil Moisture Estimates Using Artificial Neural Networks. *Remote Sens.* **2019**, *11*, 2272.
19. Ruf, C.; Asharaf, S.; Balasubramaniam, R.; Gleason, S.; Lang, T.; McKague, D.; Twigg, D.; Waliser, D. In-orbit performance of the constellation of CYGNSS hurricane satellites. *Bull. Am. Meteorol. Soc.* **2019**, *100*, 2009–2023.
20. Ruf, C.S.; Gleason, S.; McKague, D.S. Assessment of CYGNSS wind speed retrieval uncertainty. *IEEE J. Sel. Top. Appl. Earth Obs. Remote Sens.* **2018**, *12*, 87–97.
21. Wang, T.; Ruf, C.S.; Block, B.; McKague, D.S.; Gleason, S. Design and performance of a GPS constellation power monitor system for improved CYGNSS L1B calibration. *IEEE J. Sel. Top. Appl. Earth Obs. Remote Sens.* **2018**, *12*, 26–36.

22. McKague, D.S.; Ruf, C.S. On-orbit trending of CYGNSS data. In Proceedings of the IGARSS 2019—2019 IEEE International Geoscience and Remote Sensing Symposium, Yokohama, Japan, 28 July–2 August 2019; pp. 8722–8724.

23. Gleason, S.; Ruf, C.S.; O'Brien, A.J.; McKague, D.S. The CYGNSS Level 1 calibration algorithm and error analysis based on on-orbit measurements. *IEEE J. Sel. Top. Appl. Earth Obs. Remote Sens.* **2018**, *12*, 37–49.

24. Carreno-Luengo, H.; Lowe, S.; Zuffada, C.; Esterhuizen, S.; Oveisgharan, S. Spaceborne GNSS-R from the SMAP mission: First assessment of polarimetric scatterometry over land and Cryosphere. *Remote Sens.* **2017**, *9*, 362.

25. Dorigo, W.A.; Wagner, W.; Hohensinn, R.; Hahn, S.; Paulik, C.; Xaver, A.; Gruber, A.; Drusch, M.; Mecklenburg, S.; Oevelen, P.V.; et al. The International Soil Moisture Network: A data hosting facility for global in situ soil moisture measurements. *Hydrol. Earth Syst. Sci.* **2011**, *15*, 1675–1698.

26. Dorigo, W.A.; Xaver, A.; Vreugdenhil, M.; Gruber, A.; Hegyiova, A.; Sanchis-Dufau, A.D.; Zamojski, D.; Cordes, C.; Wagner, W.; Drusch, M. Global automated quality control of in situ soil moisture data from the International Soil Moisture Network. *Vadose Zone J.* **2013**, *12*. doi:10.2136/vzj2012.0097.

27. Gruber, A.; Dorigo, W.A.; Zwieback, S.; Xaver, A.; Wagner, W. Characterizing coarse-scale representativeness of in situ soil moisture measurements from the International Soil Moisture Network. *Vadose Zone J.* **2013**, *12*. doi:10.2136/vzj2012.0170.

28. O'Neill, P.E.; Njoku, E.G.; Jackson, T.J.; Chan, S.; Bindlish, R. *SMAP Algorithm Theoretical Basis Document: Level 2 & 3 Soil Moisture (Passive) Data Products*; Jet Propulsion Laboratory, California Institute of Technology: Pasadena, CA, USA, 2015; p. JPL D-66480.

29. Carreno-Luengo, H.; Luzi, G.; Crosetto, M. Impact of the elevation angle on CYGNSS GNSS-R bistatic reflectivity as a function of effective surface roughness over land surfaces. *Remote Sens.* **2018**, *10*, 1749.

30. Pekel, J.F.; Cottam, A.; Gorelick, N.; Belward, A.S. High-resolution mapping of global surface water and its long-term changes. *Nature* **2016**, *540*, 418.

31. Hengl, T.; de Jesus, J.; Heuvelink, G.B.; Gonzalez, M.R.; Kilibarda, M.; Blagotić, A.; Shangguan, W.; Wright, M.N.; Geng, X.; Bauer-Marschallinger, B.; et al. SoilGrids250m: Global gridded soil information based on machine learning. *PLoS ONE* **2017**, *12*, e0169748.

32. Balakhder, A.M.; Al-Khaldi, M.M.; Johnson, J.T. On the coherency of ocean and land surface specular scattering for GNSS-R and signals of opportunity systems. *IEEE Trans. Geosci. Remote Sens.* **2019**, *57*, 10426–10436.

33. O'Neill, P.; Chan, S.; Njoku, E.; Jackson, T.; Bindlish, R. *SMAP Enhanced L3 Radiometer Global Daily 9 km EASE-Grid Soil Moisture*; Version 1. [SPL3SMP _ E]; NASA National Snow and Ice Data Center Distributed Active Archive Center: Boulder, CO, USA, 2016.

34. Konings, A.G.; Entekhabi, D.; Chan, S.K.; Njoku, E.G. Effect of Radiative Transfer Uncertainty on L-Band Radiometric Soil Moisture Retrieval. *IEEE Trans. Geosci. Remote Sens.* **2011**, *49*, 2686–2698. doi:10.1109/TGRS.2011.2105495.

35. Wasserman, P.D. *Neural Computing: Theory and Practice*; Van Nostrand Reinhold Co.: New York, NY, USA, 1989.

36. Cortes, C.; Vapnik, V. Support Vector Networks. *Mach. Learn.* **1995**, *20*, 273–295.

37. Drucker, H.; Burges, C.J.; Kaufman, L.; Smola, A.J.; Vapnik, V. Support vector regression machines. In *Advances in Neural Information Processing Systems*; MIT Press: Cambridge, MA, USA, 1997; pp. 155–161.

38. Cutler, A.; Cutler, D.R.; Stevens, J.R. Random forests. In *Ensemble Machine Learning*; Springer: Berlin, Germany, 2012; pp. 157–175.

39. Senyurek, V.; Imtiaz, M.; Belsare, P.; Tiffany, S.; Sazonov, E. Cigarette Smoking Detection with An Inertial Sensor and A Smart Lighter. *Sensors* **2019**, *19*, 570.

40. Marcano-Cedeno, A.; Quintanilla-Domínguez, J.; Cortina-Januchs, M.; Andina, D. Feature selection using sequential forward selection and classification applying artificial metaplasticity neural network. In Proceedings of the IECON 2010—36th Annual Conference on IEEE Industrial Electronics Society, Glendale, AZ, USA, 7–10 November 2010; pp. 2845–2850.

41. Zreda, M.; Desilets, D.; Ferré, T.; Scott, R. Measuring soil moisture content non-invasively at intermediate spatial scale using cosmic-ray neutrons. *Geophys. Res. Lett.* **2008**, *35*, L21402.
42. Fangni, L.; Senyurek, V.; Boyd, D.; Kurum, M.; Gurbuz, A.; Moorhead, R. Machine-Learning based retrieval of Soil Moisture at High Spatio-temporal Scales Using CYGNSS and SMAP Observations. In Proceedings of the IGARSS 2020 IEEE International Geoscience and Remote Sensing Symposium, Waikoloa, HI, USA, 19–24 July 2020.

Article

Computation Approach for Quantitative Dielectric Constant from Time Sequential Data Observed by CYGNSS Satellites

Junchan Lee [1,2,*], Sunil Bisnath [1], Regina S.K. Lee [1] and Narin Gavili Kilane [1]

[1] Department of Earth and Space Science and Engineering, Lassonde School of Engineering, York University, Toronto, ON M3J 1P3, Canada; sbisnath@yorku.ca (S.B.); reginal@yorku.ca (R.S.K.L.); narin123@yorku.ca (N.G.K.)
[2] Satellite Technology Research Center, Korea Advanced Institute of Science and Technology, Daejeon 34141, Korea
* Correspondence: ljunchan@yorku.ca

Abstract: This paper describes a computation method for obtaining dielectric constant using Global Navigation Satellite System reflectometry (GNSS-R) products. Dielectric constant is a crucial component in the soil moisture retrieval process using reflected GNSS signals. The reflectivity for circular polarized signals is combined with the dielectric constant equation that is used for radiometer observations. Data from the Cyclone Global Navigation Satellite System (CYGNSS) mission, an eight-nanosatellite constellation for GNSS-R, are used for computing dielectric constant. Data from the Soil Moisture Active Passive (SMAP) mission are used to measure the soil moisture through its radiometer, and they are considered as a reference to confirm the accuracy of the new dielectric constant calculation method. The analyzed locations have been chosen that correspond to sites used for the calibration and validation of the SMAP soil moisture product using in-situ measurement data. The retrieved results, especially in the case of a specular point around Yanco, Australia, show that the estimated results track closely to the soil moisture results, and the Root Mean Square Error (RMSE) in the estimated dielectric constant is approximately 5.73. Similar results can be obtained when the specular point is located near the Texas Soil Moisture Network (TxSON), USA. These results indicate that the analysis procedure is well-defined, and it lays the foundation for obtaining quantitative soil moisture content using the GNSS reflectometry results. Future work will include applying the computation product to determine the characteristics that will allow for the separation of coherent and incoherent signals in delay Doppler maps, as well as to develop local soil moisture models.

Keywords: GNSS-R; CYGNSS; SMAP; dielectric constant; soil moisture

Citation: Lee, J.; Bisnath, S.; Lee, R.S.K.; Kilane, N.G. Computation Approach for Quantitative Dielectric Constant from Time Sequential Data Observed by CYGNSS Satellites. *Remote Sens.* **2021**, *13*, 2032. https://doi.org/10.3390/rs13112032

Academic Editors: Nereida Rodriguez-Alvarez and Mary Morris

Received: 9 April 2021
Accepted: 17 May 2021
Published: 21 May 2021

1. Introduction

Soil moisture research has been recognized as an important subject, as the amount of water that is stored in soil is a key parameter in understanding the hydrological and geophysical processes in the Earth's climate. Soil moisture has significant effects on various biological subjects, such as agriculture, ecology, wildlife, public health, and climate change [1]. There have been numerous attempts at large-scale soil moisture observations using various techniques aboard satellites; for example, the microwave imager [2], multi-channel microwave radiometer [3–5], as well as radar techniques [6,7] have been used for obtaining soil moisture contents. Aside from these numerous endeavors for detecting soil moisture, it is still a challenging problem to retrieve accurate and calibrated soil moisture from electromagnetic waves.

Global Navigation Satellite System reflectometry (GNSS-R) observes freely available, ubiquitously reflected GNSS signals over the surface of the Earth to estimate the geophysical characteristics. Since Martin-Neira first proposed the method in 1993 during the ocean altimetry project Passive Reflectometry and Interferometry System [8], GNSS-R has been used to measure particular geophysical characteristics of the Earth's surface properties.

For instance, inferring ocean roughness and ocean winds were examined in [9]. Over the last two decades, [10], and other similar research has examined sea ice coverage [11], soil moisture [12–18], and potentially mean sea slope and topography [19–21]. In addition, the compact size, low mass, and power consumption are major advantages of a GNSS-R receiver when compared with other remote sensing instruments, such as Synthetic Aperture Radar (SAR). Diverse research has implemented GNSS-R receivers on multiple platforms, from ground to airborne missions, and has estimated or inferred various geophysical characteristics of the Earth's surface [22,23]. The NASA Cyclone Global Navigation Satellite System (CYGNSS) is the recent satellite mission that has popularized GNSS-R instruments, which is designed to measure wind speed over Earth's oceans and predict hurricanes using the GPS signals received by the Delay Doppler Map Instrument (DDMI) aboard each of the eight nanosatellites [24].

Soil moisture retrieval by GNSS-R remains a challenging problem in formulating the mathematical model and understanding other surface source effects on GNSS reflected signal changes. There has been significant research contributing to obtaining qualitative comparisons between the soil moisture content from reference data and GNSS-R reflectivity, which is the ratio of direct signal power from the zenith direction antenna and reflected signal power measured from the nadir direction antenna. For example, [12] demonstrated the positive linear correlation between soil moisture data from the SMAP satellite and reflectivity from CYGNSS satellites [12], and [25] evaluated the CYGNSS-derived relative (zenith versus nadir antenna) signal-to-noise ratio (SNR) to estimate soil moisture over a moderately vegetated region [26]. In addition, they attempted to retrieve soil moisture using UK's TechdemoSat-1 (TDS-1) and CYGNSS based on a back-propagation artificial neural network (BP-ANN) [27]. The referred approaches offer the valuable perspective of a direct relationship between soil moisture and GNSS-R relative SNR; however, understanding the physical relationship between dielectric constant and soil moisture constitutes an important part of soil related research.

Amongst the parameters that are known to be related to soil moisture, such as surface soil temperature, composition of soil, and salinity, it is known that the relative dielectric constant presents a strong dependence on soil moisture contents, especially at microwave frequency ranges. Hallikainen presented an empirical polynomial model for dielectric constant as a function of volumetric water, clay, and sand contents based on five soil types, and a wide range of moisture conditions from 1.4 to 18 GHz; this is the range of microwave frequencies [28]. The Dobson model, which is also called the semi-empirical model [29], was developed on the basis of the same dielectric measurements as the Hallikainen model, where the observation frequency was extended down to 0.3 GHz. The more recent Mironov model is based on the refractive dielectric mixing model. It was developed from dielectric measurements from 15 soil types, covering a wide range of moisture and frequency conditions at the temperature of 20 °C. In contrast to the Dobson model, the Mironov model employs the spectra that are explicitly related to the surface of water either bound to soil or free to the soil [26]. However, the direct relationship between dielectric constant and soil moisture remains unknown.

This paper presents a straightforward approach for computing the quantitative value of dielectric constant using GNSS-R data that were measured by the CYGNSS satellites. This work is part of on-going research at the GNSS Laboratory at York University, Toronto, Canada, being funded by the Canadian Space Agency (CSA) with a GNSS-R soil moisture research grant. Reflectivity, which defined as the ratio of SNR from direct and reflected signals from GNSS satellites, is converted to the Fresnel coefficient to estimate dielectric constants. In order to implement this approach, low incidence angles of reflected signals have been chosen because it is known that the magnitude of horizontal and vertical Fresnel coefficients is identical at low incident angles of less than 60° [30,31]. It is also known that scattering features turn into incoherent scattering when the incident angle is close to normal direction of the reflection surface [21]. The dielectric constant calculated at different locations has showed a positive correlation with the radiometer data from the

SMAP satellite, which is used to validate the presented analysis by means of conversion from soil moisture to dielectric constant using the semi-empirical model for simplicity. It is noticed that the semi-empirical model delivers soil moisture content, including unexpected errors when the remote sensing data are applied for soil moisture calculation, as the semi-empirical model was based on in-situ measurements by microwave.

Section 2 provides brief descriptions of the CYGNSS and SMAP missions. In Section 3, the developed methodology for calculating dielectric constant is described in detail, followed by the results of calculations with the method shown in Section 4. Further validation results for dielectric constant by comparing SMAP data are provided in Section 5. Finally, conclusions and future work are summarized in Section 6.

2. Site Study and Database

The relation between the reflected GNSS signal and surface roughness is trivial knowledge in electromagnetic scattering. Regardless of the complexity in electromagnetic scattering theory, considerable research has contributed to determining soil moisture from GNSS-R, as summarized in the previous section. This research is different from the past research, as it contributes to the computational process of soil moisture with the derived dielectric constant, rather than reflectivity, because the dielectric constant plays a crucial role in understanding the physical characteristics of soil moisture. Accordingly, only flat surfaces were considered in the analysis in order to make the computation process for dielectric constant efficient. The sites used in the analysis were selected for the purposes of calibrating and validating SMAP soil moisture data. Descriptions regarding the sites and instruments used in analysis are discussed in below.

2.1. Site Study

An analysis of the locations for GNSS-R reflection specular point comparisons is provided in [32], which describes the core validation sites to use in assessing the soil moisture retrieval algorithm using SMAP. Thirty-four candidate sites for core validation were presented in [32], and these sites provided well-calibrated, in-situ soil moisture measurements within SMAP grid pixels (the structure of the pixels is called EASE grid, and it is presented in Section 3). Out of the 34 candidate sites, 18 sites satisfied all resolutions that SMAP provided in public, such as 36-km, 9-km, and 3-km for down-scaling. The data from the core validation sites were quality controlled by applying a site-specific spatial scaling function and comparing with SMAP-derived soil moisture to acquire precise soil moisture values. Therefore, by using the core validation sites, the derived result in this study can be guaranteed through the qualified references.

Figure 1 represents the International Geosphere-Biosphere Programme (IGBP) land classification, which arranged the type of Earth's surface into 16 classes of land, including water. The IGBP is created by using Moderate Resolution Imaging Spectroradiometer (MODIS) data and it is included in SMAP metadata with equivalent resolution to soil moisture. Among the 16 classes of the land, the land cover classification of "grasslands" is used in the analysis process. As an initial study for calculating quantitative dielectric constant using GNSS-R, this research has only considered the reflection occurring on flat surfaces with grasslands. Some researchers have reported the results of the relationship between land classification and the qualities of the reflected signals from GNSS satellites [33–35]. In relation to this study, the quantitative effect of land classification to soil moisture will be the subject of a future study.

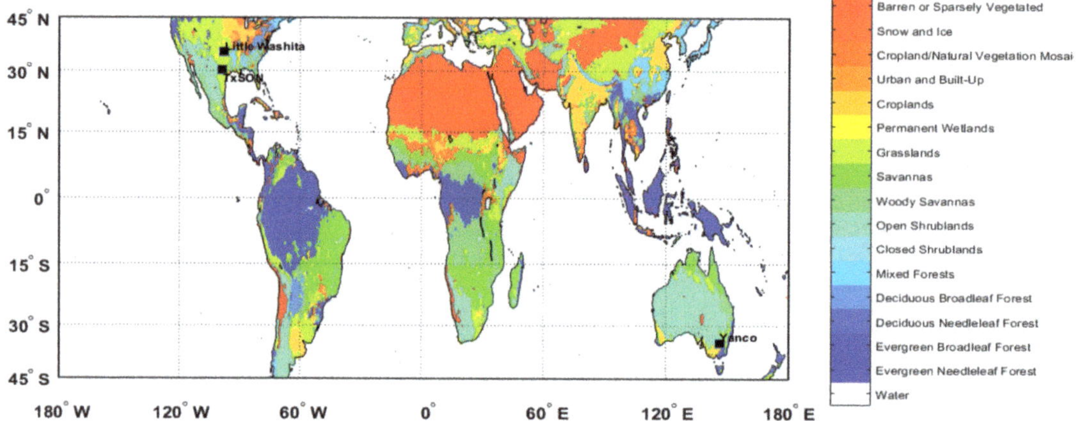

Figure 1. Global image of International Geosphere–Biosphere Programme (IGBP) Land Classification, including water, categorized by using Moderate Resolution Imaging Spectroradiometer (MODIS) data.

2.2. Data

As described, there have been numerous remote sensing satellite missions for measuring and monitoring soil moisture around the world. This paper focuses on data sets derived from two recent missions: SMAP and CYGNSS. The SMAP satellite was launched on 31 January 2015 with the objective of producing global soil moisture mapping and landscape freeze/thaw state via the L-band radar and radiometer employed [36]. The radiometer observes microwave emissions from the Earth's surface through a 6 m antenna, measuring both vertical and horizon polarization. The observation results are provided by four levels of data product with various special resolutions being derived by the scientific objective for each level and available to access through the Alaska Satellite Facility (ASF) and the National Snow and Ice Data Center (NSIDC) website [37]. This research uses Level 3 version 6 soil moisture data with 36 km × 36 km spatial resolution being drawn on same distance of geophysical latitude and longitude, called the Equal-Area Scalable Grid (EASE-Grid) [38]. Additionally, a quality flag indicating the condition of the data accompanying SMAP metadata is used to acquire qualified and validated dielectric constant. The detailed process can be found in Section 4.

CYGNSS performing GNSS-R technology consists of eight nanosatellites for observing wind speed over the Earth's tropical oceans to understand and predict the characteristics of hurricanes. The satellites were launched on 15 December 2016, orbiting an equatorial region with an inclination of 35°, 514 km perigee, and 536 km apogee. Each satellite's orbit limits the instrument surface coverage to a latitude range of ±38°. The observations from each satellite have been made available by the Physical Oceanography Distributed Active Archive Center (PODAAC) (https://podaac.jpl.nasa.gov (accessed on 15 May 2021)) [39] since March 2017.

Each CYGNSS satellite contains a Delay Doppler Mapping Instrument (DDMI), which measures both the direct GPS signals and the GPS signals that are reflected from Earth's surface. The DDMI can process up to four reflections simultaneously. The antenna on each satellite is capable of sensing the Right Hand Circular Polarization (RHCP) signals in the zenith direction and the Left Hand Circular Polarization (LHCP) signals in the nadir direction for the GPS L1 frequency. The observed signals are compressed and downloaded to the ground station, and then a post-processing procedure is applied to generate scientific observables satisfying the mission objective.

There are three accessible data levels published for the community. Level 1 (L1), version 2.1 data are used for the quantitative dielectric constant analysis in this study. Each observation file contains a number of observables and metadata, including Delay Doppler Maps (DDM—scattered power from specular point), as well as geometry information arranged by GPS satellite as transmitter, receiver, and specular power stored in daily NetCDF files. Detailed specifications and processing schemes can be found in [40]. The DDM is the fundamental product that is related to scattering physics affected by scattering geometry and conditions; this information is essential in progressing the analysis of GNSS-R data. Figure 2 provides a DDM example using CYGNSS data from 1 January 2019. CYGNSS data have been used for a variety of geophysical research, because the provided metadata include reflection geometry. The GNSS-R bistatic configurability differs from other active/passive remote sensing instruments, allowing for high surface spatial resolution. The first Fresnel zone or first glistening zone is regarded as the reflection zone for obtaining spatial resolution for a GNSS-R instrument. For example, in [41], the author evaluates that the spatial resolution of a CYGNSS signal, indicating the first Fresnel reflection zone, is approximately 0.65×0.85 km^2. However, it is still challenging to specify a definitive reflection region to produce authentic spatial resolution.

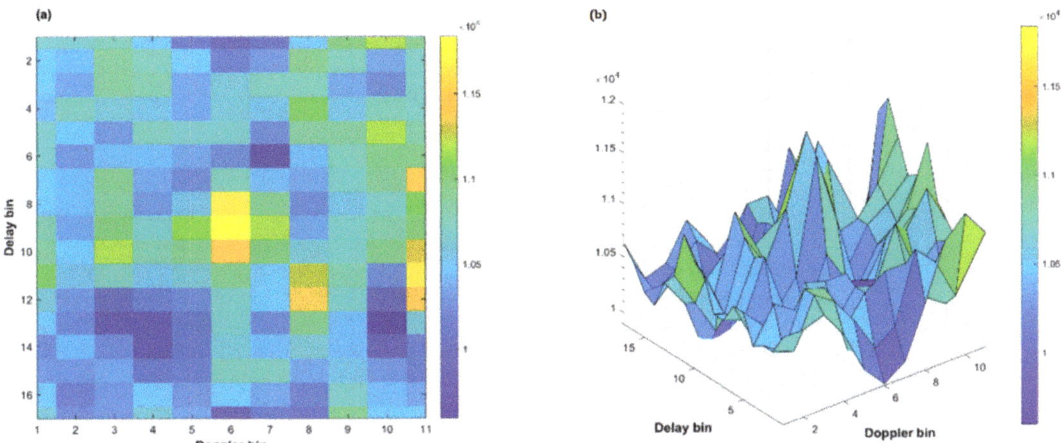

Figure 2. Example of the Delay Doppler Map (DDM) in (**a**) 2D plot and (**b**) 3D plot using observation results from v 2.1 Level 1 CYGNSS data measured by satellite 1 on 1 Jan 2019 in the case of specular point on land. X and Y axis in both plots denote the number of pixels in Doppler bin and delay bin, respectively.

3. Methodology

This section presents the equations used in the data process as well as grid information to draw a 2D global image in analysis. The equations are composed of a reflection equation from electromagnetic theory combined with coherent radar equation. Computation results are drawn as 2D images using EASE grid v2.0, which have same distance in the latitudinal and longitudinal directions.

3.1. Analysis Method

According to conditions, such as surface roughness, incident angle, and vegetation type, GNSS signals experience two types of reflections: coherent scattering with identical incident and reflection angles; and, incoherent scattering, which has randomly distributed reflection angles. The scattering type can be determined by the maximum value and shape and magnitude of values around the maximum value in the DDM, as shown in Figure 2 and described in [13]. Theoretically, coherent scattering corresponds to a sharp peak in the DDM, because it has less obstacles for disrupting the original signal. An incoherent signal can be shown broadening around the maximum value as compared to signals from coherent scattering. Because a GNSS satellite broadcasts at the microwave frequency, both of the scattering signals are measured simultaneously by a nadir antenna.

The measured signal power, P_{tot}, from the nadir antenna can simply be expressed as the summation of coherent scattering P_{coh} and incoherent scattering P_{incoh}, as in Equation 1.

$$P_{tot} = P_{coh} + P_{incoh} \tag{1}$$

In recent research, it is reported that the presence of an inland water body is associated with coherent reflection over land according to the spaceborne motivated observation and the extracting method developed for coherent scattering signals from an observed DDM by applying the Woodward ambiguity function [42]. Nevertheless, a restricted incident angle approach is used in this analysis to make the analysis process less complicated. As mentioned previously, higher incident angles have large opportunities for observing coherent scattering signals, given the concept that a small incident angle will increase the signal-to-noise ratio at the receiver and be applicable for neglecting coherent scattered signals from GNSS satellites. It was shown by [43] that the incident angles of less than $35°$ are able to represent the coherent scattering; thus, incident angles less than $20°$ from CYGNSS data are applied in this computation. The incident angle of $20°$ presenting the most correlated results has been selected to apply the computation progress in this analysis.

Equation (2) describes the coherent scattering equation, which is used to calculate the reflectivity Γ using the bi-static geometry of GNSS-R with coherent scattered signal power from the DDM [44].

$$\Gamma_{lr}(\theta) = \frac{(4\pi)^2 (P_{DDM} - N)(R_r + R_t)^2}{\lambda^2 \, G_r \, G_t \, P_t} \tag{2}$$

Where P_{DDM} is the peak value of DDM scattered power that is calculated from the raw count in CYGNSS data; N denotes the noise floor; R_t and R_r are the distant of transmitter and receiver from specular point, respectively; $G_t \, P_t$ is the transmitter equivalent isotopically radiated power (EIRP); and, G_r represents the receiver antenna gain in the direction of the specular point. In addition, the subscript 'lr' on the left-hand side expresses the polarization of incident signals and reflected signals, and r and t on the right-hand side indicate the reflected signals and direct signals, respectively. All of the variables are obtained from the v2.1 Level 1 CYGNSS product, as denoted.

The Fresnel coefficient, $\mathcal{R}$, is another important parameter in this analysis is that describes the reflection and transmission of electromagnetic radiation when an incident on an interface between different media occurs. The Fresnel coefficient connects the observable reflectivity and surface parameter, such as roughness and vegetations, and it is employed to convert the circular polarization properties of GNSS signals into linear polarization waves, which are generally adopted in electromagnetic reflection theory.

Equation (3) expresses the relation between the reflectivity and Fresnel coefficient as a function of wave number k, surface roughness σ, and elevation angle Θ. The surface roughness is derived from the Shuttle Radar Topography Mission (SRTM) Digital Elevation Model (DEM) [45], and the wave number and incident angle are obtained from Level 1 CYGNSS data.

$$\Gamma = |\mathcal{R}|^2 \exp((2k\sigma \sin \Theta)^2) \tag{3}$$

It is noted that the current electromagnetic scattering theory explains the natural phenomena with linear polarization waves; however, this circumstance exposes the necessity of expanding scattering theory in terms of circular polarization waves to infer accurate geophysical properties while using reflected GNSS signals. The Fresnel coefficient of circularly polarized waves being expressed as a linear combination of horizontal, h, and vertical, v was discussed by [46]. Fresnel coefficients use the transformation matrix shown in Equation (4):

$$\begin{bmatrix} \mathcal{R}_{ll} & \mathcal{R}_{lr} \\ \mathcal{R}_{rl} & \mathcal{R}_{rr} \end{bmatrix} = \frac{1}{2} \begin{bmatrix} \mathcal{R}_{hh} + \mathcal{R}_{vv} & \mathcal{R}_{hh} - \mathcal{R}_{vv} \\ \mathcal{R}_{hh} - \mathcal{R}_{vv} & \mathcal{R}_{hh} + \mathcal{R}_{vv} \end{bmatrix} \tag{4}$$

The subscripts on both sides of Equation (4) represent the change in the wave polarization from the latter to the former subscript. For example, rl represents that the left-hand circular polarization wave changed to right-hand circular polarization. The linear polarized Fresnel coefficient of the horizontal $\mathcal{R}_{hh}$ and vertical $\mathcal{R}_{vv}$ direction can be expressed as Equation (5) [47]:

$$
\begin{aligned}
\mathcal{R}_{hh}(\theta) &= \frac{\cos\theta - \sqrt{\varepsilon_r - \sin^2\theta}}{\cos\theta + \sqrt{\varepsilon_r - \sin^2\theta}} \\
\mathcal{R}_{vv}(\theta) &= \frac{\varepsilon_r\cos\theta - \sqrt{\varepsilon_r - \sin^2\theta}}{\varepsilon_r\cos\theta\ \sqrt{\varepsilon_r - \sin^2\theta}}
\end{aligned}
\tag{5}
$$

where θ is the signal incident angle and ε_r denotes the dielectric constant, which implies an electrical property of insulating material—a dielectric. The constant is the main parameter that is used to infer soil moisture using GNSS-R. With the inverted Fresnel equation for horizontal polarization, the dielectric constant can be simply expressed as Equation (6) in terms of the surface reflectivity and incident angle, as derived in [48]:

$$
\varepsilon = \sin^2\theta + \left[\cos\theta\left(\frac{-1-\sqrt{\Gamma}}{\sqrt{\Gamma}-1}\right)\right]^2
\tag{6}
$$

For CYGNSS data, the dielectric constant is computed using Equation (6) and it accumulated on the EASE grid. The detail process will be addressed in the following section.

3.2. EASE Grid 2.0

Spatial mismatch is the most common problem when the CYGNSS product is compared with SMAP satellite data. When compared to 2D images with the 36 km $\times$ 36 km spatial resolution from the SMAP product, CYGNSS data are provided with a list of tables containing post-processed results that are a sequence of observation results and metadata regarding orbit and satellite information. We applied the averaging method that was used in [12] to overcome the spatial resolution mismatch in comparing CYGNSS and SMAP reflectivity and dielectric constant. For this reason, CYGNSS reflectivity data on each EASE are located in the same grid of SMAP data. The EASE grid is defined as a global scale of gridding with the use of an equal area projection. Because the resolution of the EASE grid is varies depending on usage, the SMAP data provide soil moisture and other variables with varying spatial resolution, such as 3 km, 9 km, or 36 km. V2.0 denotes the updated method for reducing the spatial distortion when the projection is applied.

Figure 3 illustrates the orbit of CYGNSS satellites when one of eight satellites passes above the site TxSON in the United States. The white straight lines depict the EASE grid border and red dots represent the specular points that are measured by a CYGNSS satellite. The selected locations for analysis can be converted from geographic coordinates into the EASE grid using the conversion function shown in [49]. Subsequently, 2D images can be drawn by removing the average from the reflectivities belonging to same EASE grid of SMAP. When the average is applied, the time interval for the average is one day, and the data from eight satellites are used simultaneously without distinction. It is assumed that every satellite has identical measurement properties when they are measured by the reflected signals over the land. The assumption is caused by the fact that a specialized calibration process was not created for the measurement data from individual satellites. The calibration method for land-reflected GNSS signals will be specified by future work, and the soil moisture and dielectric constant values that are derived from GNSS-R observations will become more accurate.

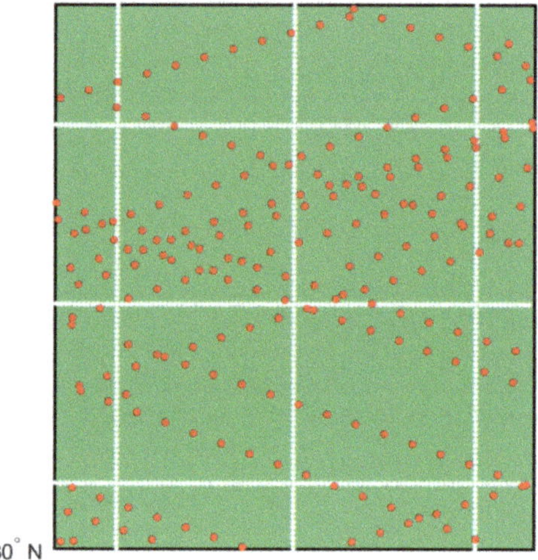

Figure 3. Specular points (red dots) occurred on EASE grid when eight CYGNSS satellite were passing above the TxSON region during Mar 2017. The straight white lines denote an EASE grid, which is the grid system having equal distance in both the latitude and longitude directions.

The averaged 2D global image of reflectivity using CYGNSS data can be obtained every day; then, by iterating the same process with another date of observation, a global number reflectivity is acquired. Figure 4 shows the global reflectivity map that was calculated with CYGNSS data using Equations (2) through (6) on the EASE grid v2.0 and accumulated observation data of satellite 1 from 18 March 2017 to 28 February 2018. Despite the low spatial resolution that is given by the 36 km × 36 km EASE grid, this method simplifies the comparison between CYGNSS and SMAP data. Additionally, it is a straightforward approach for applying the same process to other locations by changing the target pixel of analysis. As noted earlier, the data with incident angle greater than 20° have been removed in CYGNSS data. Iterated results at diverse locales are given in the following sections. The data correlation time series and seasonal variations will be examined in order to determine the time related properties.

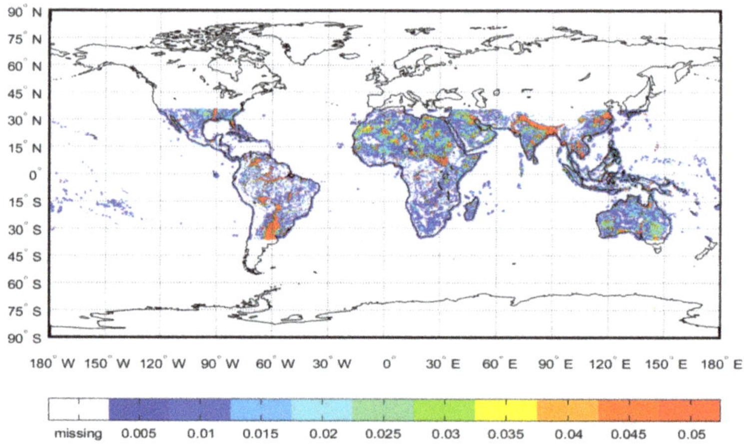

Figure 4. Global image of reflectivity drawn on EASE grid 2.0 using CYGNSS data that were measured by satellite 1 from 18 March 2017 to 28 February 2018.

4. Results

Following the analysis approach that is described in the previous section, the daily variation of dielectric constant from CYGNSS has been examined and compared with the SMAP data. The calculation and comparison processes are applied to three locations in the grass field category. The characteristics of dielectric constant along with the change of the season are also explored by comparing with the reference SMAP datasets.

4.1. Daily Time Variation of Dielectric Constant

A typical feature of the derived dielectric constant using CYGNSS data along with daytime interval is investigated at distinct locations that were selected from the list of core validation sites for SMAP soil moisture. Figure 5 shows the time series comparisons and correlation scatter plots between CYGNSS and SMAP data when specular points occur at Yanco, Australia. The methods for drawing a 2D global image with EASE grid are applied to CYGNSS data, as explained in Section 3. Subsequently, the search process is applied to find the location of the pixel containing the geographical position of Yanco. Because the global soil moisture map from SMAP has been built on an EASE grid, the location of the pixel in CYGNSS 2D global map is identical to the SMAP global image. Analysis data observed on the desired specular point are extracted from the specific pixel in the 2D global map using CYGNSS and SMAP data. By repeating the described process to the entire satellite dataset, a daily time series figure can be obtained, as seen in Figures 5–7. Approximately three years of CYGNSS and SMAP data were examined, spanning from March 2017 to April 2020. In Figure 5a,b, blue dots represent the dielectric constant and reflectivity from CYGNSS satellites and red dots indicate the soil moisture and dielectric constant that originated from the SMAP satellite. Figure 5a demonstrates the results of the dielectric constant derived from Equations (2) through (6) from CYGNSS data as compared with the dielectric constant calculated from the semi-empirical model that was applied to SMAP data. Figure 5b presents the comparison plot between reflectivity from Equation (2) with CYGNSS data and SMAP soil moisture data.

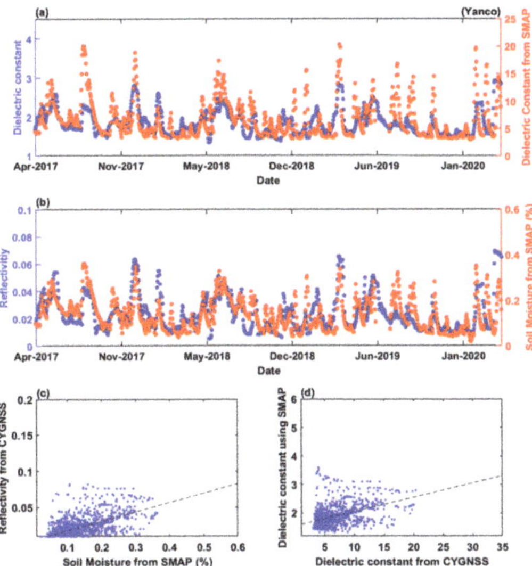

Figure 5. (**a**) Time series plot of dielectric constant at measurement site Yanco, Australia using CYGNSS and SMAP; (**b**) time series plot of reflectivity and soil moisture at same location with reflectivity from CYGNSS and soil moisture from SMAP using (**a**); (**c**) scattering plot of CYGNSS-derived reflectivity and SMAP using (**a**); and, (**d**) scattering plot of CYGNSS-derived dielectric constant and SMAP dielectric constant using results of (**b**).

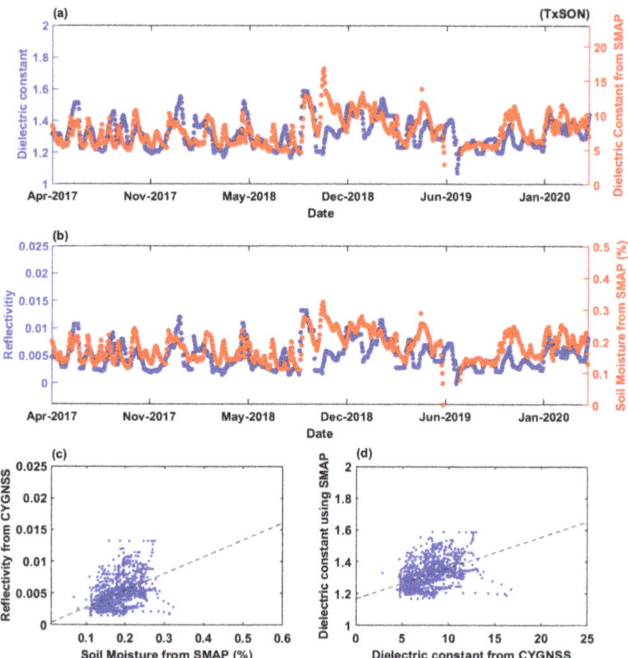

Figure 6. (**a**) Time series plot of dielectric constant at measurement site Texas Soil Moisture Network, U.S using CYGNSS and SMAP; (**b**) time series plot of reflectivity and soil moisture at same location with reflectivity from CYGNSS and soil moisture from SMAP using (**a**); (**c**) scattering plot of CYGNSS-derived reflectivity and SMAP using (**a**); and, (**d**) the scattering plot of CYGNSS-derived dielectric constant and SMAP dielectric constant using results of (**b**).

The CYGNSS-based estimation has the tendency of clearly following a SMAP-based dielectric constant and soil moisture, as seen in both Figure 5a,b. The computed dielectric constant and reflectivity results are also within the expected values: the dielectric constant is larger than 1 and the reflectivity is between 0 and 1, with no shifting of the data axis or normalization method applied to the CYGNSS data. These results are encouraging, as, at 6.6 times less the cost, the CYGNSS mission can provide soil moisture estimates that are comparable to those of the SMAP mission. Additionally, these correlations support quantitative values of soil moisture from CYGNSS satellites, because the dielectric constant, which plays an important role to calculate soil moisture, exists in a reasonable range when the GNSS-R data are applied. The difference of peak values between CYGNSS and SMAP can be observed in both Figure 5a,b. The peak values of dielectric constant presented in Figure 5a are shown to be distinct, and they are considered to have relations to weather condition, season, surface roughness, and, mostly, the calibration process. A more detailed comparison between the dielectric constant from CYGNSS and SMAP with the GNSS-R specified soil moisture model is an important part of future research.

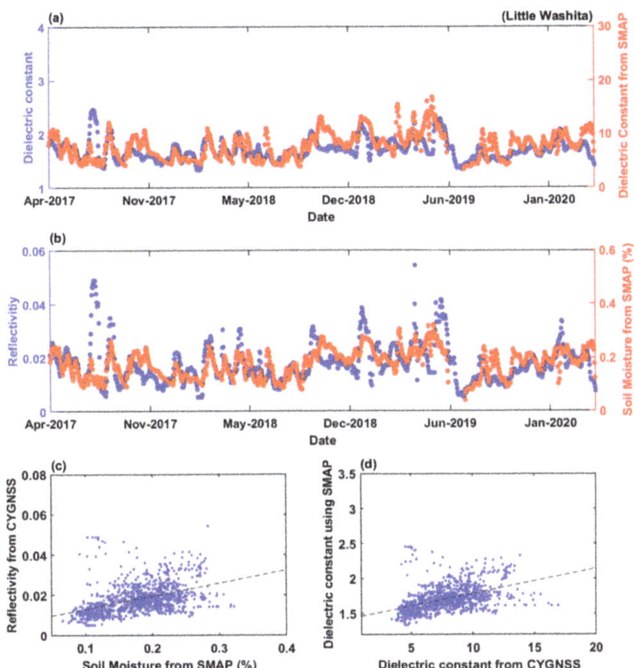

Figure 7. (**a**) Time series plot of dielectric constant at measurement site at Little Washita in United State using CYGNSS and SMAP; (**b**) time series plot of reflectivity and soil moisture at the same location with reflectivity from CYGNSS and soil moisture from SMAP using (**a**); (**c**) scattering plot of CYGNSS-derived reflectivity and SMAP using (**a**); and, (**d**) scattering plot of CYGNSS-derived dielectric constant and SMAP dielectric constant using results of (**b**).

Figure 5c,d demonstrate the statistical results by plotting the scattering features using the results presented in Figure 5a,b: Figure 5c shows the scattering plot using the reflectivity from CYGNSS and soil moisture from SMAP; Figure 5d is used to determine the correlation between the dielectric constant from CYGNSS and SMAP. The Root Mean Square Error (RMSE) and unbiased RMSE (ubRMSE) are used to capture the degree of correspondence between the estimated values and reference values [36], as well as to verify the developed method in the correlation plots. Figure 5a presents an RMSE of 5.73, while Figure 5b depicts a much lower RMSE of 0.13. The RMSE value using reflectivity and soil moisture is reasonable, since it is comparable to the results shown in previous research [12], and the RMSE of the dielectric constant has not been investigated in the published GNSS-R research. These results indicate that: (i) the dielectric constant obtained from GNSS-R using the devised method has greater equivalence than reflectivity as a function of soil moisture from the SMAP results; (ii) the calculated dielectric constant is estimated well with CYGNSS data; and, (iii) it is possible to quantitatively produce a dielectric constant using GNSS-R measurements.

Figures 6 and 7 illustrate additional comparison results between CYGNSS and SMAP satellite data that were measured at two other locations with substantially different environments. The observation results from TxSON, Texas, and Little Washita, Oklahoma in the U.S. have been investigated, and the equivalent analysis was performed as in Figure 5. As seen in Figures 6 and 7, the computed dielectric constant and reflectivity results using CYGNSS data have comparable results with the dielectric constant and soil moisture from SMAP. The RMSE from the dielectric constant results is 6.65 at TxSON and 6.40 at Little Washita, which show comparable, but minimally distinct results from Yanco. This implies that computing the dielectric constant using GNSS-R is required for considering regional characteristics, such as roughness, topography, season, weather, etc. Seasonal variation is investigated in the following section and the topographic effects are analyzed by [50]; yet, the more combined analysis of the regional characteristics is necessary for obtaining precise dielectric constant and soil moisture using GNSS-R.

Table 1 lists additional results of confirmation metric, including unbiased RMSE for the three sets of analyses. The ubRMSE values are also calculated to remove the possible biases incurred in either mean or amplitude fluctuations in the retrieval process [36], and the data for calculating RMSE for the first and ubRMSE for the third row in the table are used in the computation. The second row represents the RMSE calculation and the fourth row depicts the ubRMSE computation that were used for data when Figure 5, Figure 6, and Figure 7b are drawn. The results indicate that the dielectric constants observed at the three different locations are similar to the SMAP results, as confirmed by the statistical analysis, and, when biases are reduced, the GNSS-R estimated dielectric constant is even more similar to the observation results from SMAP. The Pearson correlation coefficient is also calculated and evaluated for the three analysis locations. Two correlation coefficients show positive correlation, and they are comparable with each other, similar to RMSE and ubRMSE, as seen in Table 1. The results indicate that the correlation coefficient results also support the compatibility of the computation approach to compute the dielectric constant using the GNSS reflectometry instrument. On the other hand, it will also arise in future analysis when considering the regional characteristics.

Table 1. The RMSE, ubRMSE, and correlation coefficient results from the scattering plots for measurements at Yanco TxSON and Little Washita (L.W.).

	Yanco	TxSON	L.W.
RMSE (a)	5.73	6.74	6.40
RMSE (b)	0.13	0.18	0.17
ubRMSE (a)	3.31	2.08	2.32
ubRMSE (b)	0.07	0.04	0.05
C.C. (a)	0.4495	0.4680	0.4714
C.C. (b)	0.4116	0.4687	0.4494

Therefore, it can be concluded that the developed method for calculating the dielectric constant using CYGNSS data produces a consistent dielectric constant that corresponds to satellite-based references, especially when the specular point occurs on flat and grass field areas. Future work will determine the dielectric constant that is measured at locations other than core validation sites, and distinct land classifications will be explored. Investigations of other locations or vegetation will also be instructive in finding out the characteristics of the dielectric constant or soil moisture according to topography, as per [50].

4.2. Seasonal Variation at Various Locations

The seasonal variations of the derived dielectric constant are examined in this section, which is necessary for speculating how adequately the dielectric constants are estimated by using reflected GNSS signals. This analysis is used to understand the effect of the environmental change on dielectric constant, as well as vegetation effects. Figure 8 through 10 illustrate the scattering plots separated by the four seasons. The dielectric constants used in analysis are calculated from CYGNSS and SMAP exactly as per the previous section. Each dataset is separated into four seasons: March Equinox, June Solstice, September Equinox, and December Solstice while following the described sequence: (1) determine the Solstice and Equinox dates in each year; (2) gather data from 45 days prior to 45 days after each Solstice/Equinox; (3) apply the data collection process to other years of data; (4) merge the yearly separated data to one variable according to the Solstice and Equinox event; and, finally, (5) draw correlation plots in terms of each season. The entire separation process is repeated for the same core SMAP validation sites: Yanco, TxSON, and Little Washita.

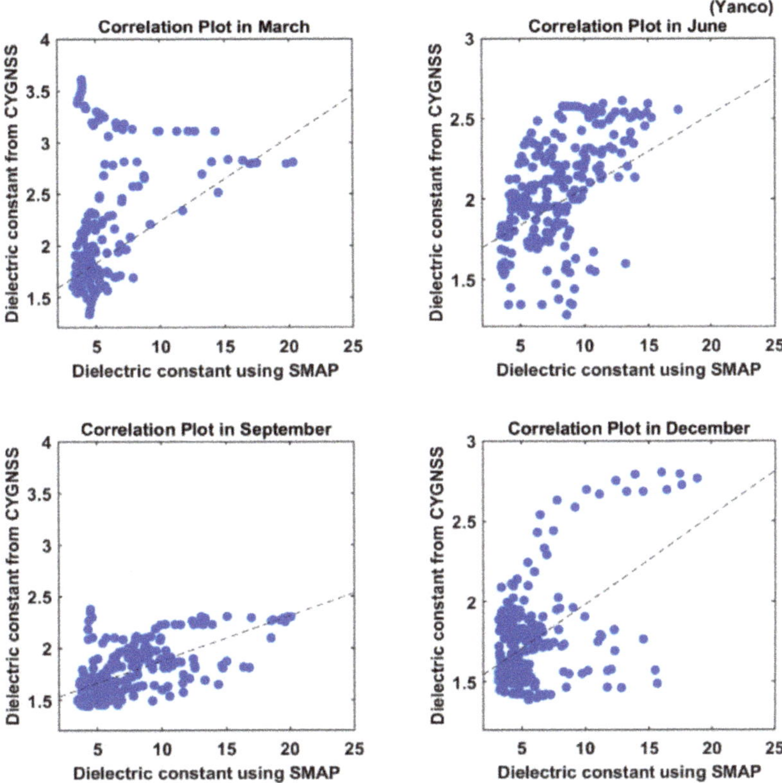

Figure 8. Seasonal scattering plots centered on the spring and fall equinoxes and summer and winter solstices using the dielectric constant estimated from CYGNSS with the new technique and measured by SMAP at Yanco, Australia.

Additionally, it is noted that SMAP Level 3 products are not available from mid-June 2019 to the end of July 2019. On the evening of 19 June 2019, the SMAP satellite turned into safe mode, which is a state of shutting down all instruments and, therefore, data collection was disrupted. After several recovery attempts, SMAP returned to science mode and resumed science data collection on 23 July 2019 [51]. With this absence of observations, the analysis results in June for every seasonal plot will be statistically insufficient and, therefore, cause errors.

The dielectric constant that is determined from CYGNSS is correlated with dielectric constant from SMAP in every season, as shown in Figure 8 through 10. RMSE has been calculated for every season in order to evaluate the variance of the derived dielectric constant when compared with this reference. Table 2 lists the results of RMSE, ubRMSE, and correlation coefficient statistics using the dielectric constant measured at Yanco, Tx-SON, and Little Washita. The equivalent data used to draw Figures 8–10 are adapted to calculate Table 2.

Table 2. RMSE, ubRMSE, and correlation coefficient results from the seasonal scattering plots for measurements at Yanco, TxSON, and Little Washita (L.W.).

	RMSE Yanco	RMSE TxSON	RMSE L.W.
March	4.32	6.43	6.23
June	6.33	6.17	6.78
September	6.57	6.20	5.72
December	4.65	7.69	6.56
	ubRMSE Yanco	ubRMSE TxSON	ubRMSE L.W.
March	2.82	1.52	2.23
June	2.77	1.75	2.69
September	3.29	2.16	2.44
December	2.77	2.12	2.02
	C.C. Yanco	C.C. TxSON	C.C. L.W.
March	0.4904	0.5010	0.5538
June	0.5944	0.4955	0.5759
September	0.6465	0.3875	0.6279
December	0.5492	0.5217	0.6738

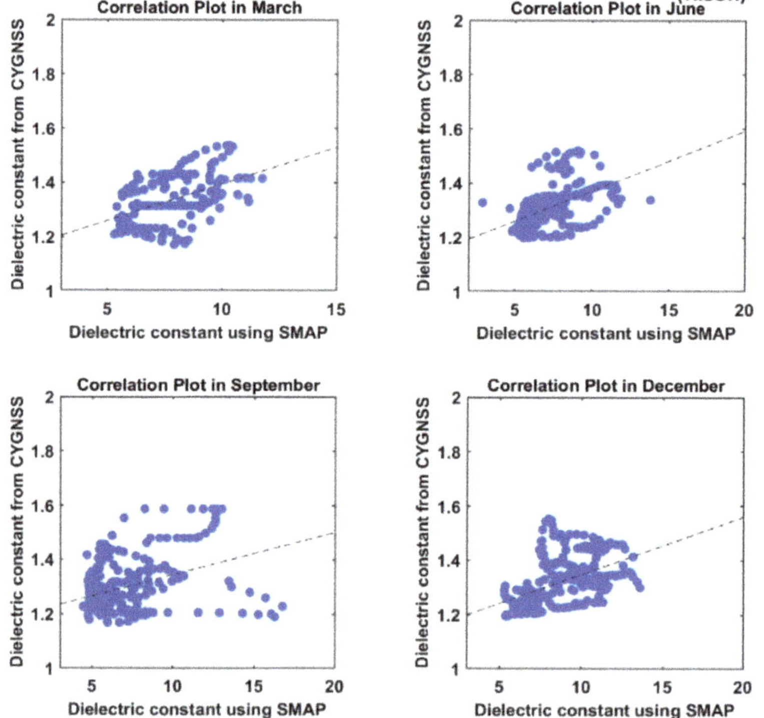

Figure 9. Seasonal scattering plots centered on the spring and fall equinoxes and summer and win solstices using dielectric constant estimated from CYGNSS with the new technique and measured by SMAP at TxSON, Texas in U.S.

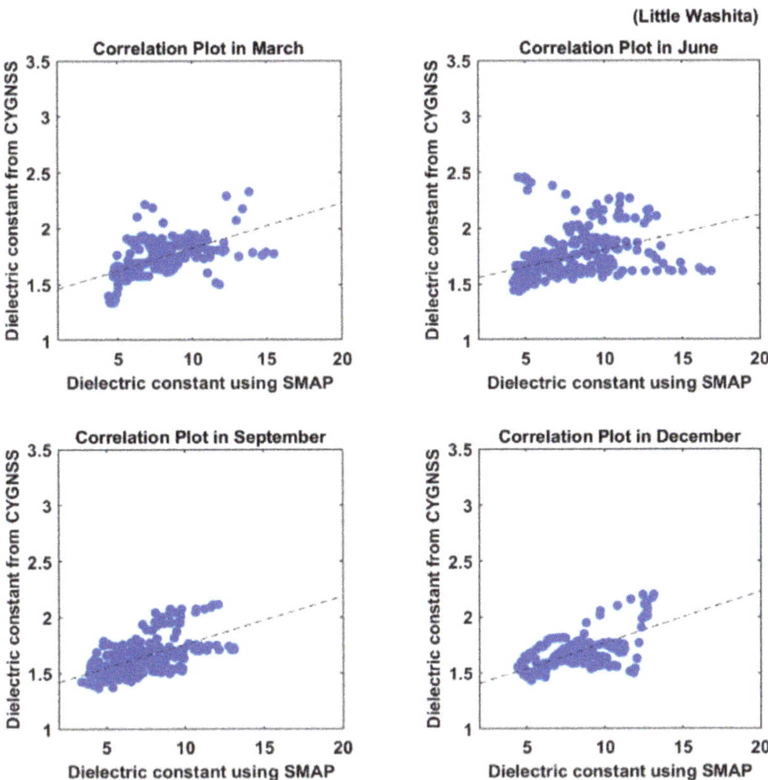

Figure 10. Seasonal scattering plots centered on the spring and fall equinoxes and summer and winter solstices using dielectric constant estimated from CYGNSS with the new technique and measured by SMAP at Little Washita, Oklahoma in U.S.

The results shown in Table 2 indicate that each confirmation parameter is comparable with the value calculated using the entire duration of data. In addition, the confirmation results show that the relevant variation depends on the season, but the factors causing seasonal variation require identification by further investigation. Even though the eight CYGNSS satellites cover the tropical region of the Earth, the specular points are sparsely distributed on the EASE grid, because a single observation using reflected GPS L1 signals will span approximately four months to cover the entire Earth. Thus, by adding more satellites or implementing a multi-GNSS constellation receiver, observing not only GPS L1, but also signal transmitted from GLONASS, Galileo, and Beidou, may allow for reliable determination of seasonal variation.

5. Discussion

Dielectric constant, which is a crucial indicator for soil moisture retrieval, has been estimated using the products from a GNSS-R instrument, specifically from the CYGNSS satellite constellation mission. In this study, sequential and seasonal variations of dielectric constant have been evaluated using the method, as described in [43]. The dielectric constant has been validated by the comparisons of the reference data to the SMAP satellite-derived soil moisture, and it has been statistically quantified by low values of confirmation metrics. A few subjects are addressed in this section in order to identify the parameters that influence the determination of dielectric constant from GNSS-R and soil moisture. Each topic is required to be analyzed further in order to anticipate the quantitative soil moisture content determination using GNSS-R data.

1. Vegetation type or land classification: the classification of land can be of concern due to the parameters that affect the calculation of dielectric constant when the reflectivity is transformed to the Fresnel reflection coefficient (Equation (3)). Therefore, an investigation, including vegetation type, is essential for improving the qualities of soil moisture or dielectric constant retrieval. Over the past few decades, there has been an attempt to classify land type by observing topography using satellite instruments, such as the International Geosphere–Biosphere Programme (IGBP) Land Classification. With comprehensive analysis regarding reflections over distinct land classifications, the results from GNSS-R will be similar to the reference instrument. NDVI or Freeze/thaw state will also be related to the dielectric constant, as referred to in [52]. Elaborated analysis for figuring out the characteristics of derived dielectric constant will be done in future research.

2. Spatial resolution: in this research, the EASE grid v2.0 with 36 km × 36 km of resolution is applied to build two-dimensional images and to acquire information from assigned pixels according to specified locations. From the perspective of global images, the 36 km resolution is good enough to produce usable imagery. The land surface and other classifications, such as urban, cropland, or grassland, were included in the designated pixel. In other words, two specular points in the same pixel are possible for experiencing the reflection on different land classifications. Because the averaging of data over the same pixel has been implemented, the resulting dielectric constant estimates must contain an error. Increasing spatial resolution and investigating land classification is required to remove errors. However, the specific approach is demanding, because high spatial resolution can cause statistical errors for short periods of observations.

3. Soil moisture model: a soil moisture model was applied to convert soil moisture from SMAP to the dielectric constant. In this conversion, the semi-empirical model is used [29]; although, the semi-empirical model was developed based on the in-situ measurements of microwave frequency waves rather than a remote sensing technique. Because the semi-empirical soil moisture model is not specialized to GNSS-R, it consistently contains unexpected errors in conversion. Even though it was established to deal with remote sensing observation results, it still required particular soil moisture to produce coincident results while using GNSS-R.

6. Conclusions

A procedure for calculating dielectric constant using GNSS-R has been tested with data from the CYGNSS satellites. Version 2.1 CYGNSS data were used for the computation method, and Level 3 SMAP soil moisture data with 36 km × 36 km were used as a reference to validate the calculation results. A restriction of incident angle was applied in order to extract coherent scattering components from combined signals that were captured by the LHCP antenna on each CYGNSS satellite. A multi-step mathematical approach has been implemented for approximately three years of data with the objective of synthesizing calculation process to acquire dielectric constant. Reflectivity, which is the ratio of GNSS signal captured by the zenith direction antenna and nadir direction antenna, from the CYGNSS data is converted to Fresnel reflection coefficient with surface roughness information from SRTM. The Fresnel reflection coefficient was applied through the use of an equation developed to derive dielectric constant from radiometer observations. The spatial resolution of CYGNSS data was adjusted on the EASE grid v2.0 coordinates for drawing a 2D global image. The derived dielectric constant has been validated with soil moisture, and the dielectric constant was converted by the semi-empirical soil moisture model. The dielectric constant that was retrieved from CYGNSS data shows values ranging between 1 to 3; whereas, the dielectric constant retrieved from SMAP data shows values that ranged from approximately 5 to 20. Both of the dynamic ranges have different values with a range from 2 to 16 for the dielectric constant retrieved from reflected GPS signals, as presented by [16]. The dynamic range difference is potentially caused by the simplified model that is used to understand the link between land geophysical parameters, because the dielectric constant is dependent on different soil characteristics (soil moisture, soil salinity content, soil type, and/or temperature).

In addition, it is understood that the dielectric constant should not be influenced by changing the observation instruments as well as retrieval models. Nevertheless, selecting the correct mathematical model is important in data analysis, especially in GNSS reflectometry, during the calculation of the dielectric constant or soil moisture because an established retrieval model is based on specific experimental results. This research applied a mathematical model that was developed for handling radiometer data and used to calculate the dielectric constant. This retrieval model is useful for estimating dielectric constant, because it only requires the reflectivity results from the instruments, though it is inevitable to contain errors in the dynamic range, because the model is built to implement radiometer data. On the contrary, [16] used the Wang and Schmugge empirical model in computing the dielectric constant, which requires other parameters, such as frequency and temperature. Therefore, [16] represented the accurate dynamic range of the dielectric constant, even though this research produced significant results emphasizing a practical usage of GNSS Reflectometry instrument by only using the GNSS-R results and generating the dynamic range within the accurate dielectric constant range. In addition, the restriction of the incident angle in this analysis may affect the dielectric constant results, so needs to be further investigated.

Yanco, TxSON, and Little Washita were used in the analysis, which are core validation sites for verifying SMAP soil moisture data using in-situ measurement data. These reference sites were selected for the characteristics of flat and grass field in order to ignore the effect of roughness in the reflected GNSS signals and concentrate on the computation process for dielectric constant. The daily time series plots from these reference sites indicate that the analysis process can be regarded as consistent with other research, because the dielectric constant from CYGNSS corresponds to the dielectric constant that is converted from SMAP soil moisture using the semi-empirical soil moisture model. The confirmation metrics, RMSE, ubRMSE, and Pearson correlation coefficient were used to validate the dielectric constant results. It is reported that the RMSE and ubRMSE values using dielectric constants from CYGNSS and SMAP are 5.71 and 3.31, respectively, and the Pearson correlation coefficient is approximately 0.4 in the case of analysis on Yanco. The other two locations have comparable results in the confirmation metrics to the results at Yanco. It can be concluded that the multi-step mathematical approach is well established, and it is required to include vegetation and topographical analysis to acquire a precise dielectric constant from GNSS-R. A seasonal analysis was also performed, and shows the annual variation of the derived dielectric constant by comparing it to the SMAP data. The confirmation metrics for seasonal analysis are comparable to the time series results, but there was no correlated tendency among the different analysis sites. This result means that it is necessary to determine the parameters influencing the seasonal variation in soil moisture in future analysis. The types of vegetation, surface roughness, and spatial resolution are considered as variables for improving the accuracy of soil moisture retrieval using GNSS-R instruments. Using a distinct soil moisture model that is derived from GNSS-R observation rather than in-situ measurements is also required for acquiring a precise soil moisture model.

7. Future Work

In regard to future research in this field of study, the main concern in soil moisture retrieval will be to obtain quantitative soil moisture estimates from dielectric constant values and advanced research to determine more accurate dielectric constant. Researching the effect of land classification types on the dielectric constant can be achieved using analysis that is similar to that used in this study. In addition, the separation of coherent and incoherent scattering components in delayed Doppler maps is a challenging domain of research due to the combination of electromagnetic scattering theory and data analysis. Regardless of the difficulty, the separation of scattering components will contribute to not only GNSS reflectometry, but also electromagnetic theory. Moreover, the results of this work will maximize the benefit of a GNSS-R instrument by investigating the effect of incident angle on scattering. Observing RHCP reflected GNSS signals in the nadir direction will also require accurate dielectric constant values.

The GNSS Lab at York University, Canada, is developing a Field Programmable Gate Array (FPGA)-based GNSS-R receiver, which is funded by the Canadian Space Agency. Its general objective is to teach students to develop their knowledge of space. Its scientific objective is to estimate the soil moisture content onboard various platforms and it is scheduled to be flown on an unmanned aerial vehicle (UAV) and an aircraft. With these experimental experiences, the GNSS Lab will be designing and developing a receiver for use on a nanosatellite in low Earth orbit. At present, a prototype GNSS-R receiver has been built using commercial off-the-shelf parts. Outdoor testing will be conducted over longer periods, and across various soil types, vegetation, and topography. The results from this GNSS-R receiver will contribute to organizing the calibration method for the instrument on land surfaces and local soil moisture modelling. It is then expected that this process will provide quantitative soil moisture data, not only for local regions, but also global places, and will induce positive interactions with various fields, such as climate research and the agricultural industry.

Author Contributions: Conceptualization, J.L.; Data curation, N.G.K.; Investigation, J.L.; Methodology, J.L.; Project administration, S.B.; Resources, S.B.; Supervision, S.B. and R.S.K.L.; Validation, J.L. and N.G.K.; Visualization, J.L.; Writing—original draft, J.L.; Writing—review & editing, S.B., R.S.K.L. and N.G.K. All authors have read and agreed to the published version of the manuscript.

Funding: This research was funded by Canadian Space Agency (CSA), grant number 18FAYORA09.

Data Availability Statement: The data presented in this study are openly available in PO.DAAC, reference number [39].

Acknowledgments: This work is supported by the Canadian Space Agency's Flights and Fieldwork for the Advancement of Science and Technology (FAST) funding program.

Conflicts of Interest: The authors declare no conflict of interest.

References

1. Lakshmi, V. Remote Sensing of Soil Moisture. *ISRN Soil. Sci.* **2013**, *2013*, 1–33. [CrossRef]
2. Hollinger, J.P.; Peirce, J.L.; Poe, G.A. SSM/I Instrument Evaluation. *IEEE Trans. Geosci. Remote Sens.* **1990**, *28*, 781–790. [CrossRef]
3. Paloscia, S.; Macelloni, G.; Santi, E.; Koike, T. A Multifrequency Algorithm for the Retrieval of Soil Moisture on a Large Scale Using Microwave Data from SMMR and SSM/I Satellites. *IEEE Trans. Geosci. Remote Sens.* **2019**, *39*, 1655–1661. [CrossRef]
4. Guha, A.; Lakshmi, V. Sensitivity, Spatial Heterogeneity, and Scaling of C-Band Microwave Brightness Temperatures for Land Hydrology Studies. *IEEE Trans. Geosci. Remote Sens.* **2002**, *40*, 2626–2635. [CrossRef]
5. Guha, A.; Lakshmi, V. Use of the Scanning Multichannel Microwave Radiometer (SMMR) to Retrieve Soil Moisture and Surface Temperature over the Central United States. *IEEE Trans. Geosci. Remote Sens.* **2004**, *42*, 1482–1494. [CrossRef]
6. Ulaby, F.T.; Dubois, P.C.; van Zyl, J. Radar Mapping of Surface Soil Moisture. *J. Hydrol.* **1996**, *184*, 57–84. [CrossRef]
7. Schmugge, T.J.; Kustas, W.P.; Ritchie, J.C.; Jackson, T.J.; Rango, A. Remote Sensing in Hydrology. *Adv. Water Resour.* **2002**, *25*, 1367–1385. [CrossRef]
8. Martin-Neira, M. A Passive Reflectometry and Interferometry System (PARIS): Applications to Ocean Altimetry. *ESA J.* **1993**, *17*, 331–355.
9. Garrison, J.L.; Komjathy, A.; Zavorotny, V.U.; Katzberg, S.J. Wind Speed Measurement Using Forward Scattered GPS Signals. *IEEE Trans. Geosci. Remote Sens.* **2002**, *40*, 50–65. [CrossRef]
10. Zavorotny, V.U.; Voronovich, A.G. Scattering of GPS Signals from the Ocean with Wind Remote Sensing Application. *IEEE Trans. Geosci. Remote Sens.* **2000**, *38*, 951–964. [CrossRef]
11. Cardellach, E.; Fabra, F.; Rius, A.; Pettinato, S.; D'Addio, S. Characterization of Dry-Snow Sub-Structure Using GNSS Reflected Signals. *Remote Sens. Environ.* **2012**, *124*, 122–134. [CrossRef]
12. Chew, C.C.; Small, E.E. Soil Moisture Sensing Using Spaceborne GNSS Reflections: Comparison of CYGNSS Reflectivity to SMAP Soil Moisture. *Geophys. Res. Lett.* **2018**, *45*, 4049–4057. [CrossRef]

13. Chew, C.; Shah, R.; Zuffada, C.; Hajj, G.; Masters, D.; Mannucci, A.J. Demonstrating Soil Moisture Remote Sensing with Observations from the UK TechDemoSat-1 Satellite Mission. *Geophys. Res. Lett.* **2016**, *43*, 3317–3324. [CrossRef]
14. Camps, A.; Park, H.; Pablos, M.; Foti, G.; Gommenginger, C.P.; Liu, P.-W.; Judge, J. Sensitivity of GNSS-R Spaceborne Observations to Soil Moisture and Vegetation. *IEEE J. Sel. Top. Appl.* **2016**, *9*, 4730–4742. [CrossRef]
15. Egido, A.; Paloscia, S.; Motte, E.; Guerriero, L.; Pierdicca, N.; Caparrini, M.; Santi, E.; Fontanelli, G.; Floury, N. Airborne GNSS-R Polarimetric Measurements for Soil Moisture and Above-Ground Biomass Estimation. *IEEE J. Sel. Top. Appl.* **2014**, *7*, 1522–1532. [CrossRef]
16. Katzberg, S.J.; Torres, O.; Grant, M.S.; Masters, D. Utilizing Calibrated GPS Reflected Signals to Estimate Soil Reflectivity and Dielectric Constant: Results from SMEX02. *Remote Sens. Environ.* **2006**, *100*, 17–28. [CrossRef]
17. Rodriguez-Alvarez, N.; Bosch-Lluis, X.; Camps, A.; Vall-llossera, M.; Valencia, E.; Marchan-Hernandez, J.F.; Ramos-Perez, I. Soil Moisture Retrieval Using GNSS-R Techniques: Experimental Results Over a Bare Soil Field. *IEEE Trans. Geosci. Remote Sens.* **2009**, *47*, 3616–3624. [CrossRef]
18. Small, E.E.; Larson, K.M.; Chew, C.C.; Dong, J.; Ochsner, T.E. Validation of GPS-IR Soil Moisture Retrievals: Comparison of Different Algorithms to Remove Vegetation Effects. *IEEE J. Sel. Top. Appl.* **2016**, *9*, 4759–4770. [CrossRef]
19. Clarizia, M.P.; Ruf, C.S. On the Spatial Resolution of GNSS Reflectometry. *IEEE Geosci. Remote Sens. Lett.* **2016**, *13*, 1064–1068. [CrossRef]
20. Cardellach, E.; Rius, A.; Martin-Neira, M.; Fabra, F.; Nogues-Correig, O.; Ribo, S.; Kainulainen, J.; Camps, A.; D'Addio, S. Consolidating the Precision of Interferometric GNSS-R Ocean Altimetry Using Airborne Experimental Data. *IEEE Trans. Geosci. Remote Sens.* **2014**, *52*, 4992–5004. [CrossRef]
21. Cardellach, E.; Li, W.; Rius, A.; Semmling, M.; Wickert, J.; Zus, F.; Ruf, C.S.; Buontempo, C. First Precise Spaceborne Sea Surface Altimetry with GNSS Reflected Signals. *IEEE J. Sel. Top. Appl.* **2019**, *13*, 102–112. [CrossRef]
22. Zavorotny, V.U.; Voronovich, A.G. Bistatic GPS Signal Reflections at Various Polarizations from Rough Land Surface with Moisture Content. In Proceedings of the IEEE 2000 International Geoscience and Remote Sensing Symposium—Taking the Pulse of the Planet: The Role of Remote Sensing in Managing the Environment, Honolulu, HI, USA, 24–28 July 2000; Volume 7, pp. 2852–2854. [CrossRef]
23. Rodriguez-Alvarez, N.; Camps, A.; Vall-Llossera, M.; Bosch-Lluis, X.; Monerris, A.; Ramos-Perez, I.; Valencia, E.; Marchan-Hernandez, J.F.; Martinez-Fernandez, J.; Baroncini-Turricchia, G.; et al. Land Geophysical Parameters Retrieval Using the Interference Pattern GNSS-R Technique. *IEEE Trans. Geosci. Remote Sens.* **2011**, *49*, 71–84. [CrossRef]
24. Ruf, C.S.; Atlas, R.; Chang, P.S.; Clarizia, M.P.; Garrison, J.L.; Gleason, S.; Katzberg, S.J.; Jelenak, Z.; Johnson, J.T.; Majumdar, S.J.; et al. New Ocean Winds Satellite Mission to Probe Hurricanes and Tropical Convection. *Bull. Am. Meteorol. Soc.* **2016**, *97*, 385–395. [CrossRef]
25. Kim, H.; Lakshmi, V. Use of Cyclone Global Navigation Satellite System (CyGNSS) Observations for Estimation of Soil Moisture. *Geophys. Res. Lett.* **2018**, *45*, 8272–8282. [CrossRef]
26. Guo, P.; Shi, J.; Gao, B.; Wan, H. Evaluation of Errors Induced by Soil Dielectric Models for Soil Moisture Retrieval at L-Band. In Proceedings of the 2016 IEEE International Geoscience and Remote Sensing Symposium (IGARSS), Beijing, China, 10–15 July 2016; pp. 1679–1682. [CrossRef]
27. Yang, T.; Wan, W.; Sun, Z.; Liu, B.; Li, S.; Chen, X. Comprehensive Evaluation of Using TechDemoSat-1 and CYGNSS Data to Estimate Soil Moisture over Mainland China. *Remote Sens.* **2020**, *12*, 1699. [CrossRef]
28. Hallikainen, M.; Ulaby, F.; Dobson, M.; El-rayes, M.; Wu, L. Microwave Dielectric Behavior of Wet Soil-Part 1: Empirical Models and Experimental Observations. *IEEE Trans. Geosci. Remote Sens.* **1985**, *GE-23*, 25–34. [CrossRef]
29. Dobson, M.C.; Ulaby, F.T.; Hallikainen, M.T.; El-Rayes, M.A. Microwave Dielectric Behavior of Wet Soil-Part II: Dielectric Mixing Models. *IEEE Trans. Geosci. Remote Sens.* **1985**, *GE-23*, 35–46. [CrossRef]
30. Jia, Y.; Savi, P.; Canone, D.; Notarpietro, R. Estimation of Surface Characteristics Using GNSS LH-Reflected Signals: Land Versus Water. *IEEE J. Sel. Top. Appl.* **2016**, *9*, 4752–4758. [CrossRef]
31. Egido, A.; Ruffini, G.; Caparrini, M.; Martin, C.; Farres, E.; Banque, X. Soil moisture monitorization using gnss reflected signals. *arXiv* **2008**, arXiv:0805.1881.
32. Colliander, A.; Jackson, T.J.; Bindlish, R.; Chan, S.; Das, N.; Kim, S.B.; Cosh, M.H.; Dunbar, R.S.; Dang, L.; Pashaian, L.; et al. Validation of SMAP Surface Soil Moisture Products with Core Validation Sites. *Remote Sens. Environ.* **2017**, *191*, 215–231. [CrossRef]
33. Pierdicca, N.; Guerriero, L.; Giusto, R.; Brogioni, M.; Egido, A. SAVERS: A Simulator of GNSS Reflections from Bare and Vegetated Soils. *IEEE Trans. Geosci. Remote Sens.* **2014**, *52*, 6542–6554. [CrossRef]
34. Small, E.E.; Larson, K.M.; Braun, J.J. Sensing Vegetation Growth with Reflected GPS Signals. *Geophys. Res. Lett.* **2010**, *37*, L12401. [CrossRef]
35. Hu, C.; Benson, C.; Park, H.; Camps, A.; Qiao, L.; Rizos, C. Detecting Targets above the Earth's Surface Using GNSS-R Delay Doppler Maps: Results from TDS-1. *Remote Sens.* **2019**, *11*, 2327. [CrossRef]
36. Entekhabi, D.; Reichle, R.H.; Koster, R.D.; Crow, W.T. Performance Metrics for Soil Moisture Retrievals and Application Requirements. *J. Hydrometeorol.* **2010**, *11*, 832–840. [CrossRef]
37. SMAP L3 Radiometer Global Daily 36 Km EASE-Grid Soil Moisture, Version 6. Available online: https://nsidc.org/data/SPL3SMP/versions/6 (accessed on 15 May 2021).

38. O'Neill, P.E.; Chan, S.; Njoku, E.G.; Jackson, T.; Bindlish, R.; Chaubell, J. SMAP L3 Radiometer Global Daily 36 Km EASE-Grid Soil Moisture, Version 6. Available online: https://doi.org/10.5067/EVYDQ32FNWTH (accessed on 15 August 2019).
39. PO.DAAC. Available online: https://podaac.jpl.nasa.gov/ (accessed on 14 May 2018).
40. Ruf, C.; Chang, P.S.; Clarizia, M.-P.; Gleason, S.; Jelenak, Z.; Majumdar, S.; Morris, M.; Murray, J.; Musko, S.; Posselt, D.; et al. *CYGNSS Handbook Cyclone Global Navigation Satellite System: Deriving Surface Wind Speeds in Tropical Cyclones*; National Aeronautics and Space Administration: Ann Arbor, MI, USA, 2016; ISBN 978-1-60785-380-0.
41. Chew, C.; Lowe, S.; Parazoo, N.; Esterhuizen, S.; Oveisgharan, S.; Podest, E.; Zuffada, C.; Freedman, A. SMAP Radar Receiver Measures Land Surface Freeze/Thaw State through Capture of Forward-Scattered L-Band Signals. *Remote Sens. Environ.* **2017**, *198*, 333–344. [CrossRef]
42. Al-Khaldi, M.M.; Johnson, J.T.; O'Brien, A.J.; Balenzano, A.; Mattia, F. Time-Series Retrieval of Soil Moisture Using CYGNSS. *IEEE Trans. Geosci. Remote Sens.* **2019**, *57*, 4322–4331. [CrossRef]
43. Calabia, A.; Molina, I.; Jin, S. Soil Moisture Content from GNSS Reflectometry Using Dielectric Permittivity from Fresnel Reflection Coefficients. *Remote Sens.* **2020**, *12*, 122. [CrossRef]
44. Clarizia, M.P.; Pierdicca, N.; Costantini, F.; Floury, N. Analysis of CYGNSS Data for Soil Moisture Retrieval. *IEEE J. Sel. Top. Appl.* **2019**, *12*, 2227–2235. [CrossRef]
45. Jarvis, A.; Reuter, H.I.; Nelson, A.; Guevara, E. Hole-Filled Seamless SRTM Data V4. Available online: http://srtm.csi.cgiar.org/ (accessed on 5 May 2021).
46. Stutzman, W.L. *Polarization in Electromagnetic Systems*; Artech House: Norwood, MA, USA, 1993.
47. Griffiths, D.J. *Introduction to Electrodynamics*; Cambridge University Press: Cambridge, UK, 2017.
48. Jackson, T.J.; Hurkmans, R.; Hsu, A.; Cosh, M.H. Soil Moisture Algorithm Validation Using Data from the Advanced Microwave Scanning Radiometer (AMSR-E) in Mongolia. *Ital. J. Remote Sens.* **2004**, *30*, 23–32.
49. Brodzik, M.J.; Billingsley, B.; Haran, T.; Raup, B.; Savoie, M.H. EASE-Grid 2.0: Incremental but Significant Improvements for Earth-Gridded Data Sets. *ISPRS Int. Geo-inf.* **2012**, *1*, 32–45. [CrossRef]
50. Dente, L.; Guerriero, L.; Comite, D.; Pierdicca, N. Space-Borne GNSS-R Signal Over a Complex Topography: Modeling and Validation. *IEEE J. Sel. Top. Appl.* **2019**, *13*, 1218–1233. [CrossRef]
51. National Snow & Ice Data Center. Update: SMAP in Safe Mode. Available online: https://nsidc.org/the-drift/data-update/update-smap-in-safe-mode/ (accessed on 27 June 2019).
52. Comite, D.; Cenci, L.; Colliander, A.; Pierdicca, N. Monitoring Freeze-Thaw State by Means of GNSS Reflectometry: An Analysis of TechDemoSat-1 Data. *IEEE J. Sel. Top. Appl.* **2019**, *13*, 2996–3005. [CrossRef]

 *remote sensing*

Article

Single-Pass Soil Moisture Retrievals Using GNSS-R: Lessons Learned

Adriano Camps [1,2,*], **Hyuk Park** [1,2,3], **Jordi Castellví** [1,4], **Jordi Corbera** [4] and **Emili Ascaso** [4]

[1] CommSensLab-UPC, Department of Signal Theory and Communications, UPC BarcelonaTech, c/Jordi Girona 1-3, 08034 Barcelona, Spain; park.hyuk@tsc.upc.edu (H.P.); jordi.castellvi@tsc.upc.edu (J.C.)
[2] Institut d'Estudis Espacials de Catalunya-IEEC/CTE-UPC, Gran Capità, 2-4, Edifici Nexus, despatx 201, 08034 Barcelona, Spain
[3] Department of Physics, UPC BarcelonaTech, 1-3 c/Jordi Girona, 08034 Barcelona, Spain
[4] Institut Cartogràfic i Geològic de Catalunya, Parc de Montjuïc, 08038 Barcelona, Spain; jordi.corbera@icgc.cat (J.C.); Emili.Ascaso@icgc.cat (E.A.)
* Correspondence: camps@tsc.upc.edu

Received: 5 May 2020; Accepted: 22 June 2020; Published: 26 June 2020

Abstract: In this paper, an algorithm to retrieve surface soil moisture from GNSS-R (Global Navigaton Satellite System Reflectometry) observations is presented. Surface roughness and vegetation effects are found to be the most critical ones to be corrected. On one side, the NASA SMAP (Soil Moisture Active and Passive) correction for vegetation opacity (multiplied by two to account for the descending and ascending passes) seems too high. Surface roughness effects cannot be compensated using in situ measurements, as they are not representative. An ad hoc correction for surface roughness, including the dependence with the incidence angle, and the actual reflectivity value is needed. With this correction, reasonable surface soil moisture values are obtained up to approximately a 30° incidence angle, beyond which the GNSS-R retrieved surface soil moisture spreads significantly.

Keywords: GNSS-R; reflectivity; surface roughness; vegetation; soil moisture

---

1. Introduction

The first evidence of soil moisture (SM) signatures in GPS reflected signals date back to 1996 [1]. Exploiting this technique, accurate soil moisture retrievals were successfully conducted using ground-based instruments using the so-called "interference pattern technique" (IPT) in which the direct and reflected signals are collected simultaneously by the receiving antenna, creating fringes in the received power as navigation satellites elevation varies over time. This technique can be applied using geodetic zenith-looking right-hand circularly polarized (RHCP) antennas [2], and the reflections are collected at low elevation angles for which the reflected signal is also RHCP; or it can be applied using linearly polarized (vertical, or vertical and horizontal) antennas pointing to the horizon [3,4] which allows collecting reflected signals over a much wider range of elevation angles. The latter technique also allows inferring vegetation height, and with two linear polarizations, surface roughness effects can be corrected. However, none of these techniques can be applied to airborne or spaceborne instruments.

Retrieving soil moisture from airborne and spaceborne instruments can only be performed using the reflected signal power, which requires on the instrument side the compensation of platform attitude impact on the antenna pattern, receiver's noise … and on the target side the compensation of surface roughness (small scale as opposed to topography or large scale), and vegetation effects (both attenuation and possibly scattering). So far, two main approaches have been presented in the literature: using a change detection approach by looking to a time-series of GNSS-R data, or attempting a single-pass retrieval.

The time-series approach [5–7] has been able to produce operationally a soil moisture product using NASA CYGNSS (Cyclone GNSS) mission data [8]. The main advantages of the time-series approach are that assuming surface roughness and vegetation effects are constant within a grid cell (36 km in [8]), and the change in the observed reflectivity is only due to the change in the soil moisture. The main drawbacks are the difficulty to really achieve collocated GNSS-R observations over time as in the case of land, GNSS reflections exhibit a very high spatial resolution (several hundreds of meters even from space, since the spatial resolution is given by the size of the first Fresnel zone [9]), the implicit assumption of the temporal coherence of data (i.e., surface roughness or vegetation can change), and the strong non-linearities of the sensitivity of GNSS-R observables to soil moisture, which is very noticeable at low soil moisture values (Figure 1). To overcome this, in [8], an ad hoc calibration had to be performed as "the product was developed by calibrating CYGNSS reflectivity observations to soil moisture retrievals from NASA's Soil Moisture Active Passive (SMAP) mission".

On the other hand, single pass retrievals [10–14] could provide a faster revisit time, as SM maps could be generated in just one pass, provided that the instrument is accurately calibrated, and suitable ancillary data are used to compensate for surface roughness and vegetation effects without tuning the algorithm for each pixel. In [10,11], the sensitivity to soil moisture was investigated, together with a first attempt to compensate for vegetation effects using the Normalized Differential Vegetation Index (NDVI). Surface topography, spatial scales, and some subsurface "artifacts" were already reported using spaceborne data from the UK TechDemoSat-1 (TDS-1) data. Using airborne data, in [12], the sensitivity to soil moisture was studied, as well as vegetation effects, which were parameterized in terms of the Leaf Area Index (LAI) and a single scattering albedo. More recently [13], IceSat-2 data have been proposed to correct for surface roughness effects and NASA SMAP data to correct for surface roughness and vegetation optical depth (VOD) effects in the Fresnel reflection coefficients at incidence angles smaller than 35°. Even, more recently, Machine Learning techniques have been applied [14].

Despite the encouraging results of these previous works, to the authors' knowledge, the correction of surface roughness and vegetation effects are not yet properly understood so as to perform an accurate soil moisture retrieval. In this work, the goodness of these corrections are analyzed using Sentinel-2 satellite data; GNSS-R, L-band microwave radiometry, VNIR (visible and near-infrared) hyperspectral and TIR (thermal infrared) airborne data, as well as in situ surface roughness, soil type, and soil moisture samples aiming at performing single-pass surface soil moisture (SM) retrievals using GNSS-R observations.

In order to address the above questions, an incremental approach was followed. First, based on our previous experience, very different sensitivities to soil moisture were encountered, from as high as 38 dB/100% for bare dry soils [10,15] to a more moderate approximately 9 dB/100% at global scale [11]. This large variability has to be understood first in order to properly model the sensitivity to soil moisture in the retrieval algorithm.

As it can be appreciated in Figure 1, computed using a comprehensive GNSS-R simulator [16], assuming a flat soil surface and no vegetation, the variation of the peak of the DDM (Delay and Doppler Map) with respect to SM is highly non-linear. Depending on the range of soil moisture values encountered in the field experiments, different slopes can be found if a linear approximation is assumed to compute the sensitivity. For example, in the range 0 to 0.15 m^3/m^3, the sensitivity is as high as 43 dB/100%, while in the range 0.15 to 0.45 m^3/m^3, the sensitivity is much lower 8.6 dB/100%, and if the whole dynamic range is considered 0 to 0.45 m^3/m^3, the sensitivity is 17 dB/100%.

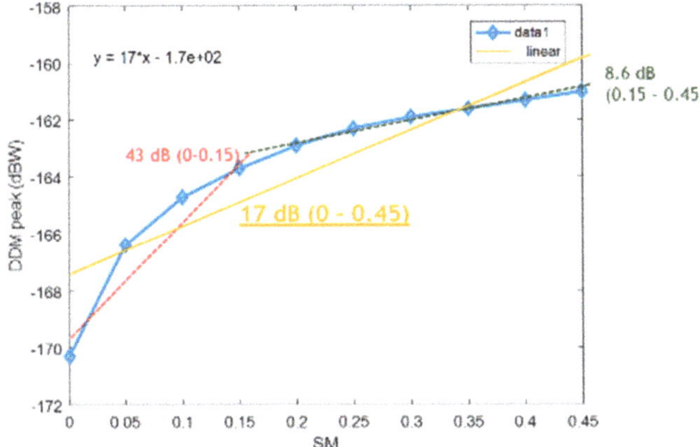

Figure 1. Simulated DDM peak values with respect to the soil moisture (SM) [m^3/m^3].

Actually, these values demonstrate that a linearized (in dB) dependence $\Delta\Gamma/\Delta$SM is not valid for the whole range of SM values, and it cannot be used in SM retrieval algorithms. In addition, time-series algorithms based on the detection of reflectivity changes must also include this non-linearity as depending on the actual SM value, a given reflectivity change will correspond to different soil moisture change.

Second, the retrieval algorithm from GNSS-R observations (airborne and/or spaceborne) proceeds as follows:

1. From the peak of the DDM (direct and reflected signals), and assuming that the scattered signal has a dominant coherent component, the calibrated reflection coefficient is computed taking into account the antenna pattern gains in the directions where the signals are collected, and the distances from the transmitter to the receiver, and from the transmitter to the specular reflection point plus from the specular reflection point to the receiver.

2. Then, following an approach similar to the Single Channel Algorithm in SMAP [17], or in [10,12], but neglecting the single scattering albedo ($\omega = 0$), the computed reflectivity is compensated for surface roughness (h) and vegetation effects (τ: vegetation optical depth, or VOD), in order to derive the flat surface reflectivity:

$$\Gamma_{bare\ [dB]} = \Gamma_{[dB]} + 10log(e^{h \cdot cos^n(\theta_i)}) + 10\log(e^{2 \cdot \tau/cos\,(\theta_i)}), \tag{1}$$

where θ_i is the incidence angle, and n is an empirical factor to account for the angular dependence of the surface roughness. In theory [18,19], it should be $n = 2$. However, in SMAP, the value used is $n = 0$. This effect is discussed afterwards. The factor "2" in the second exponential term of the VOD accounts for the two-way attenuation through the vegetation canopy. Considering the exponential dependence with h, this term is sometimes expressed in dB. Note that at circular polarization (LHCP: left hand circular polarization), at a constant frequency ($f = 1575.42$ MHz), the "flat surface reflectivity" ($\Gamma_{[dB]}$) has a marginal variation with the incidence angle up to approximately 30° to 40°, and the largest variability is linked to the dielectric constant.

3. Then, considering the variation of the SM in the dielectric constant model for soil (among other variables, such as the clay fraction and physical temperature ...), the soil moisture value can be in principle estimated through a minimization process.

The volumetric SM retrieval process is sketched in Figure 2.

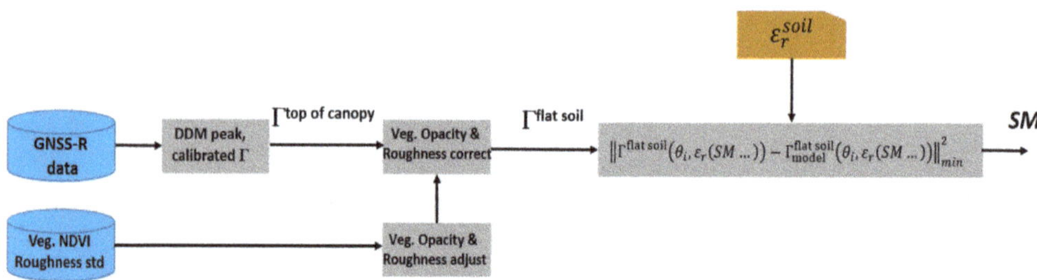

Figure 2. Volumetric SM retrieval algorithm workflow.

At this stage, it is important to remind that in SMAP, there are three main soil moisture retrieval algorithms using microwave radiometer data: the Single Channel Algorithm (SCA), which uses the brightness temperature observations at either horizontal or vertical polarizations, and a vegetation climatology to retrieve soil moisture; and the Dual Channel Algorithm (DCA), which uses the brightness temperature observations at both horizontal and vertical polarizations, while inferring vegetation parameters as well. SMAP also uses static (climatologic) ancillary data such as soil texture and land cover class. The SCA using the brightness temperature at vertical polarization is the current baseline algorithm for SMAP level 2 soil moisture products. A recent study [20] shows that "while the DCA has its lowest unbiased root man squared error (ubRMSE) and highest R^2 when ω is non-zero, the SCA have their lowest ubRMSE and highest R^2 when $\omega = 0$, and the dry bias of all algorithms increases as ω increases. Errors in soil texture are not significant, but soil surface roughness should not be static and have a higher overall value."

When in point 2 above, it is stated that an algorithm similar to the SMAP SCA is used (with $\omega = 0$ as found in [20]), the authors refer to the functional relationships used to account for surface roughness and vegetation effects (Equation (1)), but not to the use of SMAP climatology and ancillary data. Actually, in situ, aiborne and satellite data were acquired to parameterize the vegetation and surface roughness effects, as described in Section 2.

The rest of this work is organized as follows: Section 2 describes an airborne field experiment that was conducted to assess the SM retrieval capabilities of GNSS-R. Section 3 shows the results of the SM retrieval algorithm performance and the limitations of the surface roughness and vegetation corrections. In Section 4, additional considerations for SM retival approaches are discussed. Finally, Section 5 summarizes the main conclusions of this study.

2. Methodology: Field Experiment Description

In 2017, a first airborne flight was conducted in the Tremp area (42.2°N, 0.89°E) of Lleida, Spain. The area was selected because of the availability of soil moisture stations for validation purposes. However, topography effects and the strong wind gusts affected the aircraft trajectory and attitude and did not allow for conducting the SM retrievals successfully. A second airborne campaign was conducted on October 22nd, 2018, in a flatter agricultural region around Balaguer (41.9°N, 0.80°E), Lleida, Spain (Figure 3). In situ and satellite data were acquired to complement the airborne data and allow the estimation of the surface roughness and vegetation optical depth, necessary to perform the vegetation corrections. The acquired data and the sensors are presented below.

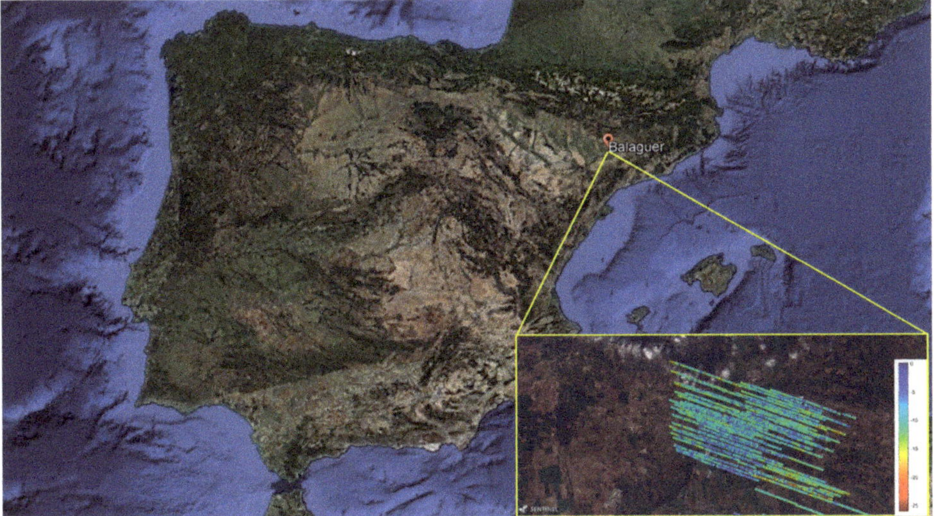

Figure 3. Location of the experiment area in the north east of the Iberian peninsula, near Balaguer (41.9°N, 0.80°E), Lleida, Spain. Zoom shows the collected GNSS-R data strips in [dB].

2.1. Airborne Instrumentation and Configuration

The plane used was a CESSNA Caravan (Figure 4a) from the Institut Cartogràfic i Geològic de Catalunya, with an autonomy of 4 h. The flight height was 1200 m. The instrumentation used was the following:

- AISA [21]: VNIR (406–993 nm) pushbroom hyperspectral sensor with a maximum of 254 bands and a spectral width of 2.4 nm (Figure 4b). The sensor was configured with 126 bands, combining together every two bands leading to an average spectral width of approximately 4.8 nm.
- TASI [22]: TIR (8050–11450 nm) hyperspectral with 32 bands (Figure 4b), and an average spectral width of 109.5 nm.
- ARIEL [23]: L-band radiometer with central frequency at 1.413 GHz and bandwidth of 20 MHz (Figure 4c), and a radiometric accuracy of 0.7 K at 1 Hz. ARIEL was designed by the Universitat Politècnica de Catalunya and is commercialized by BALAMIS [24].
- CORTO (COmpact Reflectometer for Terrain Observations) is a miniature version of the LARGO [25] reflectometer at L1 GPS bands developed by the Universitat Politècnica de Catalunya. It performs the correlations between the direct and the reflected GPS signals and outputs reflectivity.

Figure 4. Experiment instrumentation: (**a**) CESSNA Caravan plane, (**b**) AISA (hyperspectral VNIR) and TASI (TIR), and (**c**) ARIEL L-band radiometer.

The APPLANIX [26] Inertial Navigation System (INS) on-board is a precise GNSS receiver + IMU (Inertial Measurement Unit) System that was used to geo-locate the data acquired and to perform the geometric correction on the images.

In order to receive the GPS signal for the GNSS-R instrument, there are two antennas: an RHCP up-looking GPS antenna already installed on the aircraft for navigation purposes and a LHCP down-looking one that covered the L1/L2 GPS bands. To avoid picking the direct signal through a surface wave propagating around the plane fuselage, a choke ring antenna with a carefully designed mounting was employed. The antenna radiation pattern in that configuration was measured in the UPC anechoic chamber [27] using a dummy of the bottom part of the fuselage.

2.2. Vegetation Optical Depth (VOD) Estimation from Normalized Differential Vegetation Index (NDVI)

VOD is needed to compensate for the vegetation effects (Equation (1)). At the L-band, and taking into account that it is an agricultural area, with some trees, but no dense forests, the SMAP approach can be in principle used [10,12,13]. High-resolution (0.8 m) NDVI was computed from AISA data. The Sentinel-2 L1C NVDI product corresponding to Octobr 23rd, 2018 was also downloaded from the Sentinel Hub [28] (Figure 5). The VOD (τ) can be computed as:

$$\tau = b{\cdot}VWC, \tag{2}$$

where b is a proportionality constant that depends on the vegetation structure and the frequency, and VWC is the vegetation water content, which can be calculated using the following relationship:

$$VWC = b{\cdot}1.9134{\cdot}NDVI^2 - 0.3215{\cdot}NDVI + \frac{StemFactor(NDVI - 0.1)}{0.9}, \tag{3}$$

where the NDVI is given by the ratio of the difference and the sum of the reflectivities in the near infrared (NIR) and the red:

$$NDVI = \frac{\rho_{NIR} - \rho_{Red}}{\rho_{NIR} + \rho_{Red}}. \tag{4}$$

The Stem Factor in Equation (3) is the product of the average vegetation height of a land cover class and the ratio of sapwood area to leaf area [17]. Sample values of b, *StemFactor*, and h are given in Table 3 of [17] for the land use of each observation according to the cadaster, and these are confirmed by in situ observations.

At this stage, it is important to highlight two potential error sources in these parameterizations:

- that the value of the b parameter directly determined at the satellite scale "is nearly identical to what was proposed for crops in the ESA SMOS (Soil Moisture and Ocean Salinity mission) algorithm, but half as large as what is currently used by SMAP" [29], and
- the current values of the h parameter may be too smooth (low) for crop regions, as reported in [20] for the South Fork SMAP Core Validation Site in the Corn Belt state of Iowa.

AISA *NDVI* was calculated using a narrowband (central wavelength) and a wideband approach, with a parametric adjustment to Sentinel-2 spectral sensitivity with respect to the wavelength.

Figure 6 shows the difference between Sentinel 2 *NDVI* and AISA *NDVI* computed either with the narrowband or wideband approaches. As it can be appreciated, the narrowband approach introduces larger errors (overestimation/underestimation) when compared to Sentinel 2 *NDVI*, while results with the wideband approach match much better Sentinel-2 results. As a result of the higher spatial resolution, the AISA *NDVI* using wideband data is used in Equations (3) and (4) to try to compensate for the vegetation effects in Equation (1). The bottom panel in Figure 6 shows the TASI thermal image in [°C]. The calibrated physical land surface temperature (LST) is lower in the bottom right corner due to the presence of clouds.

Figure 5. Normalized Difference Vegetation Index (NDVI) from Sentinel 2 L1C for October 23rd, 2018. Yellow rectangle corresponds to the second quadrant of zoom region of Figure 3. North West corner: 41°47′33″N, 0°50′4″E; South East corner: 41°46′56″N, 0°51′57″E, distance = 2.9 km.

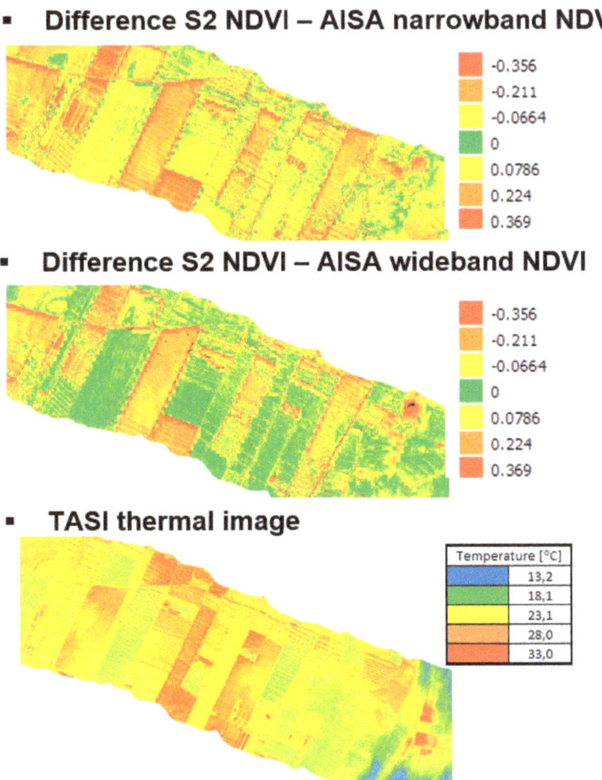

Figure 6. Example showing the difference between Sentinel-2 NDVI and AISA narrowband NDVI (top), and AISA wideband NDVI (middle panel). Bottom panel shows the TASI thermal image in [°C]. Region corresponds to yellow rectangle in Figure 5.

2.3. Soil Surface Roughness and Soil Moisture

A field campaign to validate the land use and to perform the surface roughness, soil sampling, and in situ soil moisture measurements was performed the same day as the airborne campaign. Before the experiment, the Hydrogeology and Soils Unit of the Catalan Institute of Cartography and Geology (ICGC) studied the region and selected 5 different areas representative of the different agricultural activities encountered. The locations of the 29 sampling points are indicated in Figure 7.

Figure 7. Locations where the soil samples, in situ soil moisture measurements, and surface roughness profiles acquired (refer to Figures 3 and 5).

Soil surface roughness was measured with a laser profiler based on the Leica Disto Pro4a laser distance meter, with millimetric precision, in 10 mm steps. Soil Moisture samples were acquired with a Theta Probe sensor, acquiring at least 3 samples per "sampling point" and computing the median value. Then, soil samples were analyzed in the laboratories of the IRTA (Instituto de Investigación y Tecnología Agroalimentaria)/Lleida University facilities and used for calibration of the Theta Probe sensor.

The largest problem with the "effective" surface soil roughness is its difficult estimation, as it depends on numerous additional factors such as the soil composition, the agricultural activity, the state of the crops, etc. [30]. Moreover, this seasonal dependency only correlates with similar fields.

In the aircraft, the ARIEL L-band radiometer was flown to provide a complete 0.8 m resolution SM map (Figure 8) using a downscaling algorithm that combines TASI and AISA data [31,32]. This algorithm has been validated [33] extensively, and in this experiment, the root mean square error (rmse) and bias were approximately 0.06 m^3/m^3, and −0.01 m^3/m^3 with respect to the in situ soil moisture measurements. As a qualitative validation, Figure 9 shows the Sentinel 2 Normalized Difference Water Index (NDWI) for October 23rd, 2018. Similarly to the NDVI, the NDWI index uses the green and near infrared bands (bands 3 and 8), and because of the strong absorbability and low radiation in the range from visible to infrared wavelengths, it is most appropriate for water body mapping, to assess the surface soil moisture and vegetation hydric stress. Figure 10 shows the GNSS-Reflectivities overlaid with the Sentinel-2 L1C NDWI.

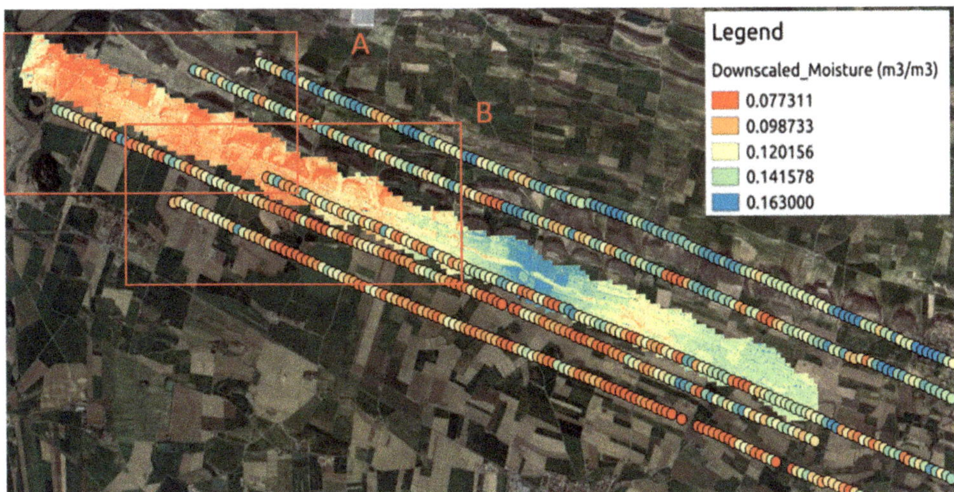

Figure 8. ARIEL-derived surface soil moisture downscaled to 0.8 m with the AISA NDVI and TASI (thermal data, land surface temperature (LST)), and some sample CORTO GNSS-R derived surface soil moisture (discussion on Soil Moisture retrieval in Section 3).

Figure 9. Sentinel-2 L1C Normalized Difference Water Index (NDWI) for October 23rd, 2018. Red rectangles correspond to regions in Figure 8.

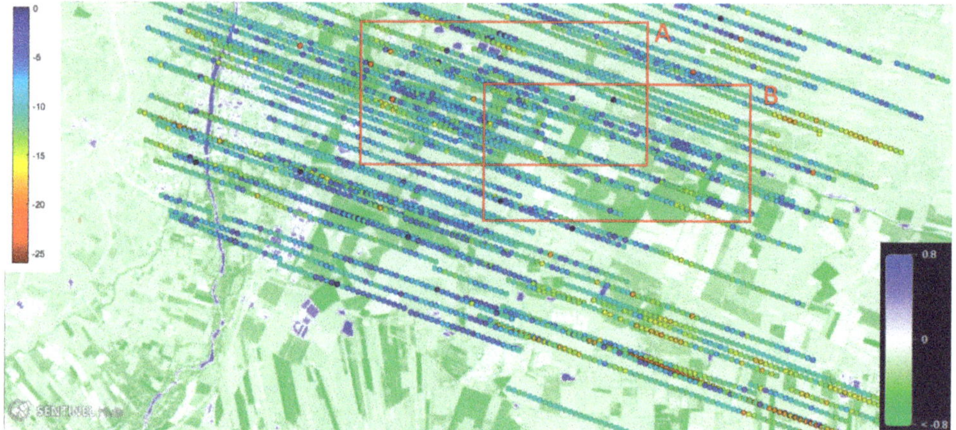

Figure 10. Sentinel-2 L1C Normalized Difference Water Index (NDWI) for October 23rd, 2018, and CORTO GNSS-R reflectivities. Red rectangles correspond to regions in Figure 8.

3. GNSS-R Soil Moisture Retrieval: Results

In this section, the resulting SM maps are assessed for different corrections applied. The results are shown over an ortho-photo of the Region Of Interest (ROI). There are two layers of data: ARIEL downscaled SM (0.8 m spatial resolution), and overlapped samples of the GNSS-R retrieved SM at the same resolution. The ARIEL SM retrievals are used to validate the estimations of GNSS-R SM retrievals.

3.1. In Situ Surface Roughness Correction

In this first attempt, the measured surface roughness with the laser profiler was used to compute the roughness parameter required in Equations (1) and (11) of [18] from the soil surface roughness (σ):

$$h \;=\; 4\,k^2\sigma^2. \tag{5}$$

Due to the limited accuracy and coarse sampling of the soil surface roughness, the results show a large disparity between the ARIEL and the GNSS-R SM retrievals. As it can be appreciated in Figure 11a, for a single track of GNSS-R reflections over a relatively flat area (no topography), the retrieved SM values are highly variable with the incidence angle (see alternating SM retrievals between approximately 0.14 and 0.27), and these are quite different than the ARIEL downscaled SM ~0.21, although on average, it is relatively close (e.g., 0.21 ≈ (0.14 + 0.27)/2). Apart from the different scales (0.8 m versus 10 m), note the high visual correlation between Figure 11a, ARIEL downscaled soil moisture, and Figure 11b Sentinel 2 L1C NDWI.

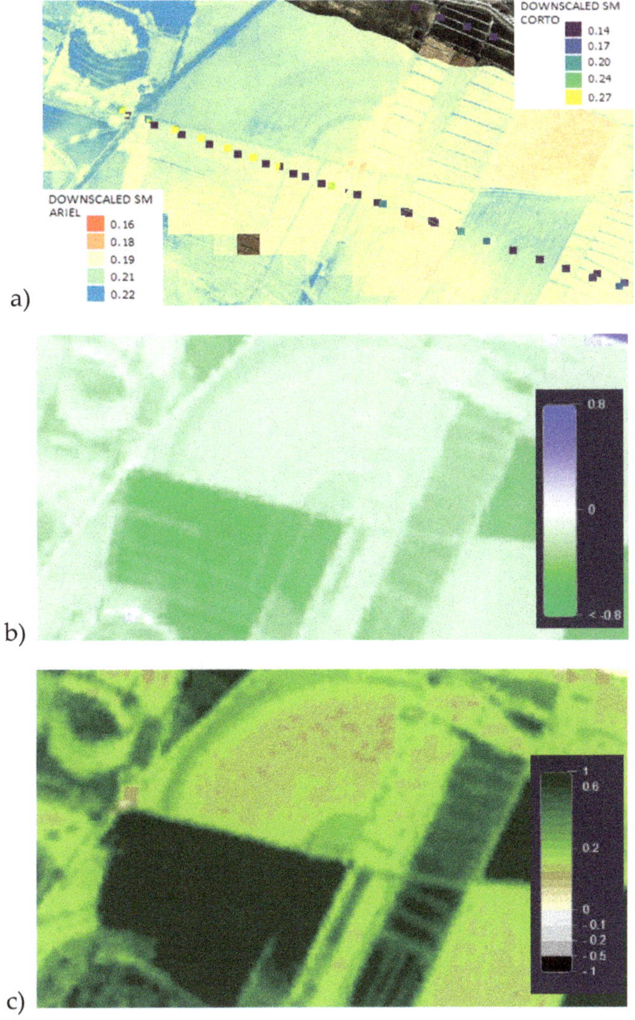

Figure 11. (**a**) Zoom of downscaled ARIEL SM map overlaid with CORTO GNSS-R SM measurements at the Balaguer airborne experiment site, incidence angles between 15° and 30° for October 22nd, 2018; and Sentinel-2 L1C (**b**) Normalized Difference Water Index (NDWI) and (**c**) NDVI for October 23rd, 2018. Region corresponds to left half of rectangle A in Figures 8–10.

3.2. Ad Hoc Soil Surface Roughness Correction

The lack of success in this first SM retrieval approach with the in situ measured soil surface roughness led us to consider that the surface roughness correction applied was not appropriate, either in the value of the h parameter, and/or the power of the cosine function ($n = 0$, 1, or 2) in Equation (1). As discussed in page 35 of [17], the $\cos^2\theta$ term is sometimes changed to $\cos\theta$ or even dropped completely to avoid overcorrecting for roughness.

To derive the "*correct*" surface roughness correction, the same procedure was applied, but adding an *ad hoc* surface roughness correction term h, as in $e^{h \cdot \cos^n(\theta_i)}$, so that the retrieved GNSS-R SM values match as close as possible the ARIEL-derived ones. The step-by-step procedure is described below:

- Compensation of vegetation effects (Equations (1)–(4)).
- Compensation of surface roughness effects (as in Equation (1) and (2): $10 \cdot log(e^{h \cdot cos^n(\theta i)}$), for $n = 0$, 1, and 2.
- Apply SM retrieval algorithm by minimizing the difference between the computed flat bare soil reflectivity using ARIEL-derived SM, and GNSS-R calibrated reflectivity ($\Gamma_{[dB]}$) plus the *ad hoc* surface roughness compensation term (in [dB]).

It was found that for $n = 0$, the scatter in the *ad hoc* surface roughness corrections was the smallest one obtained. For $n = 2$, the corrections were more exaggerated, indicating they were artificially high. Figure 12 shows the map of the computed surface roughness parameters (h). Apparently, there is no evident correlation neither with the different fields nor with the soil moisture values, although some of the largest values occur in the plots with higher humidity, but not all. In addition, as expected, by construction, the matched histograms of the SM values retrieved by downscaling ARIEL's measurements and from CORTO's reflectivity with the ad hoc surface roughness correction are basically the same (Figure 13), except for numerical round off errors in the surface roughness correction (0.1 dB) during the optimization process (third bullet in the above list, or the right box in the diagram in Figure 2).

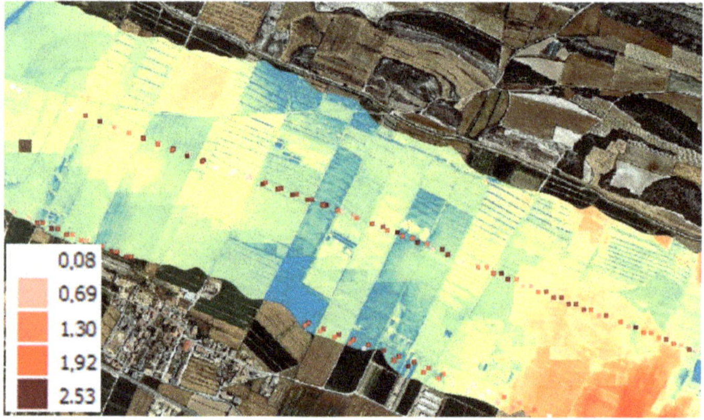

Figure 12. Roughness parameter (square dots) [-] overlaid over the SM [m³/m³] map in the Region of Interest (ROI) (area corresponds to rectangle A in Figures 7–9).

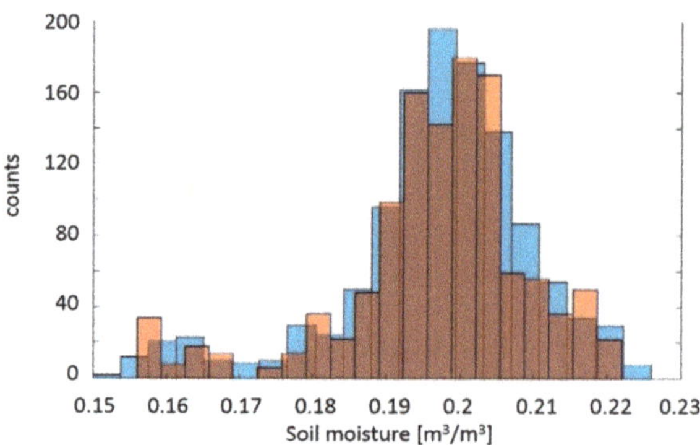

Figure 13. Matched histograms of downscaled ARIEL SM values (blue), and CORTO GNSS-R-derived values (brown).

Figure 14 shows the estimated values of *h* (as in e^{-h}). A more detailed analysis of the *ad hoc* surface roughness correction reveals that it increases for lower reflectivity values. This apparently has no physical meaning, and it is difficult to interpret as an instrumental error. It was also considered that it could be an imperfect estimation of the noise floor of the DDM background, which is needed in the calibration, but this does not explain why, for the same reflectivity values, the correction varies with the incidence angle, increasing for the larger ones. Additionally, negative values in *h* (roughness correction larger than 0 dB at low reflectivities) do not have a physical meaning, but they led us to think that actually the compensation of the vegetation effects as in Equations (1)–(4) was too high, as already shown in Figure 13 of [11]. This effect is examined in more detail in the next section.

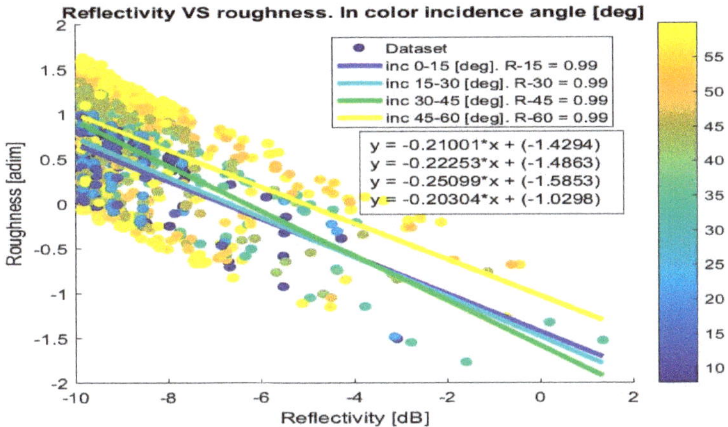

Figure 14. Reflectivity [dB] and ad hoc surface roughness compensation for different incidence angles.

3.3. Vegetation and Roughness Compensation for Different Incidence Angles

Figure 15 shows the scatter plot of the vegetation compensation [dB] (Equation (1)) with respect to the incidence angle. The compensation of vegetation effects increases with incidence angle, as it has a functional dependence with $1/cos(\theta_i)$ (see the variation of peak values wrt. incidence angle). The compensation of the roughness effects has also a dependence on the incidence angle, but it can be modeled neither with $n = 1$, nor with $n = 2$.

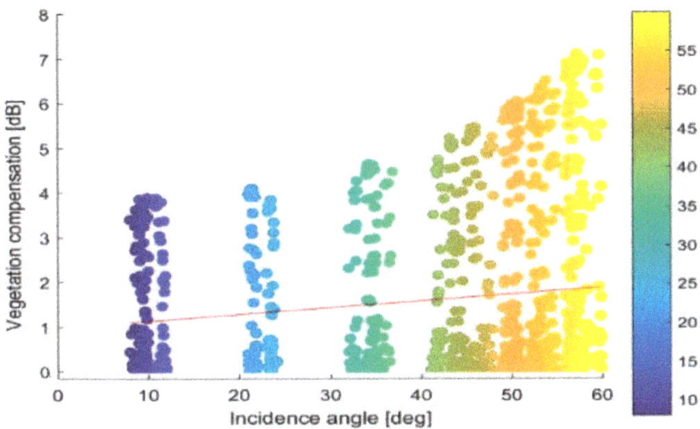

Figure 15. Vegetation compensation [dB] (Equation (1)) for different incidence angles.

Since the compensation of vegetation effects is fixed and depends on the VOD (which is calculated using the NDVI) and the incidence angle, an overcompensation of the vegetation attenuation forces the algorithm to reduce the compensation of the surface roughness component, and that is why the algorithm even finds negative surface roughness coefficients. This effect is more important for large incidence angles, and it suggests that the vegetation compensation is too high at large incidence angles, which is in agreement with [13], who found that corrections were only applicable up to approximately a 35° incidence angle.

Other sources of variability are:

- high reflectivity values encountered in vegetated areas (trees, higher soil moisture), but from the processing point of view, require a too high attenuation compensation, leading to "corrected" flat bare reflectivity values larger than one (0 dB) in Figure 14. This is illustrated in Figure 16 by the dark blue dot over a forested area in the middle of the image, pointed by the triangle, and legend with the parameters.
- there is likely a subsurface variability of the soil moisture that is detected by GNSS-R, as there are regions in the fields where the reflectivity suddenly increases (see lines of light blue or green values in Figure 10, with some dark blue—high reflectivity—observations in the middle), while all other parameters are nearly constant.

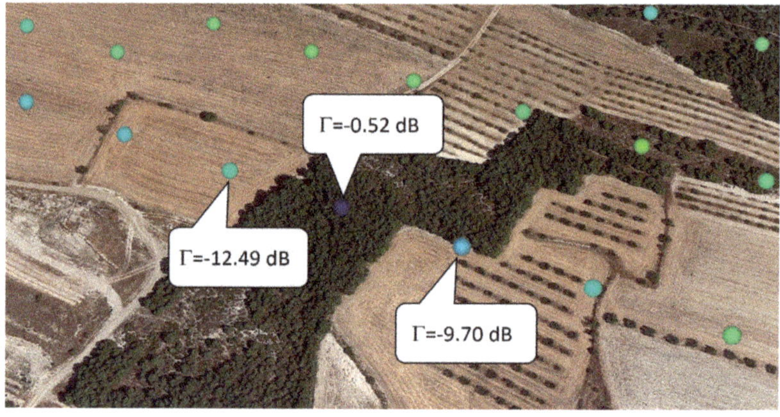

Figure 16. Sample high GNSS-R reflectivity (calibrated for instrumental effects) over the forested area that is overcompensated for vegetation attenuation effects, or areas in which there is likely a variation in the subsurface soil moisture.

3.4. Empirical Roughness Compensation

Not being satisfied with the result that accurate single-pass soil moisture retrievals were only feasible if a pixel-based ad hoc correction of the surface roughness was used, an empirical correction (1st order polynomial) for all pixels within a given range of incidence angles is attempted. The procedure is the same as in the pixel-by-pixel approach described before, but now computing the minimum root mean squared error (RMSE) for all pixels simultaneously. Figure 17 shows the histograms of the ARIEL-retrieved SM values, and the GNSS-R retrieved ones using a generic empirical 1st order polynomial correction for all pixels within a given range of incidence angles. As it can be appreciated, except for a few outliers, the matching is quite reasonable up to approximately a 30° incidence angle, and beyond that, the retrieved histograms widen and GNSS-R derived SM values tend to widen and scatter significantly (sensitivity to SM decreases, while the effect of roughness is still non-negligible). This result is also in agreement with [13] where corrections were only applicable up to an approximately 35° incidence angle.

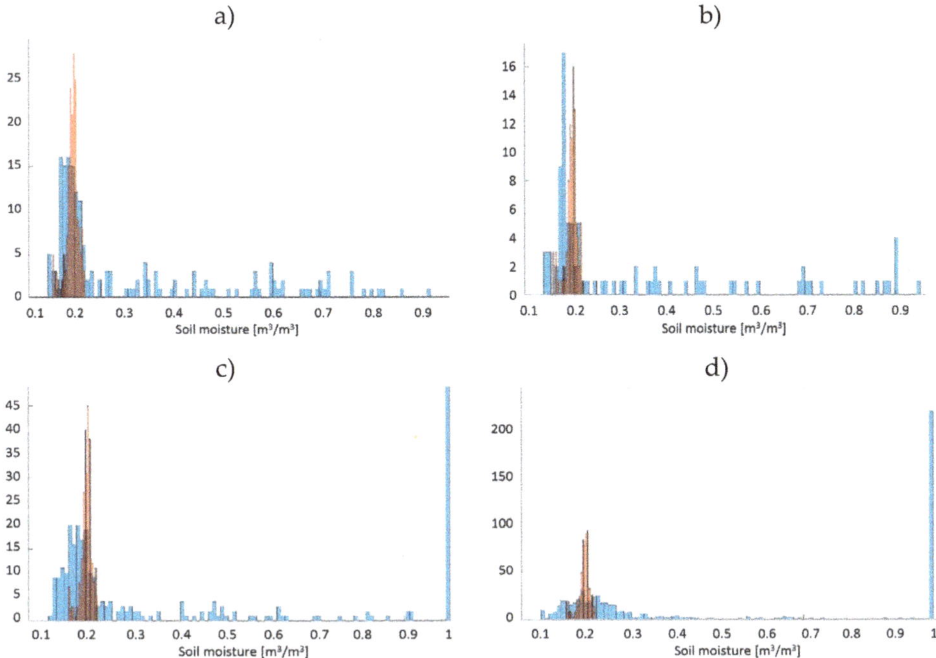

Figure 17. Histograms of ARIEL (brown) and GNSS-R (blue) retrieved SM using surface roughness correction as a function of the incidence angles: (**a**) from 0° to 15°, (**b**) from 15° to 30°, (**c**) from 30° to 45°, and (**d**) from 45° to 60°.

Figure 18 shows the soil moisture maps for two regions (left and right column, corresponding to regions A and B, respectively in Figures 8–10) during the flight for all ranges of incidence angles in steps of 15°. Squares correspond to GNSS-R-derived soil moisture using the generic angular-dependent roughness correction.

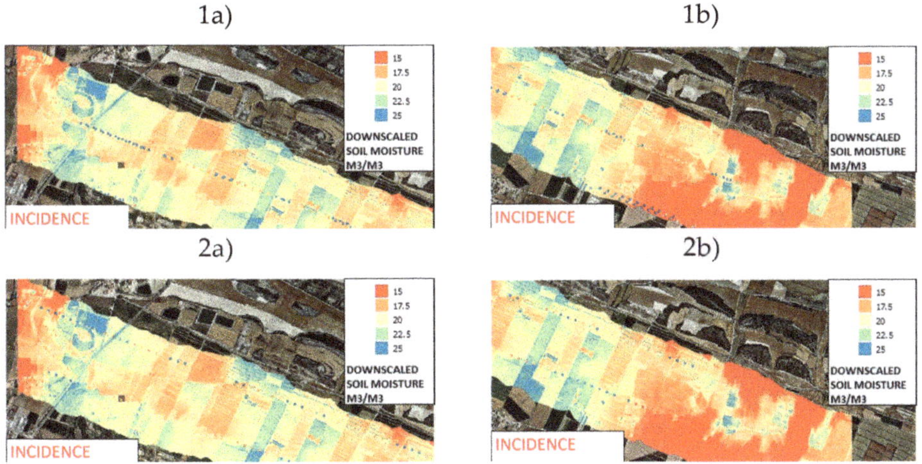

Figure 18. *Cont.*

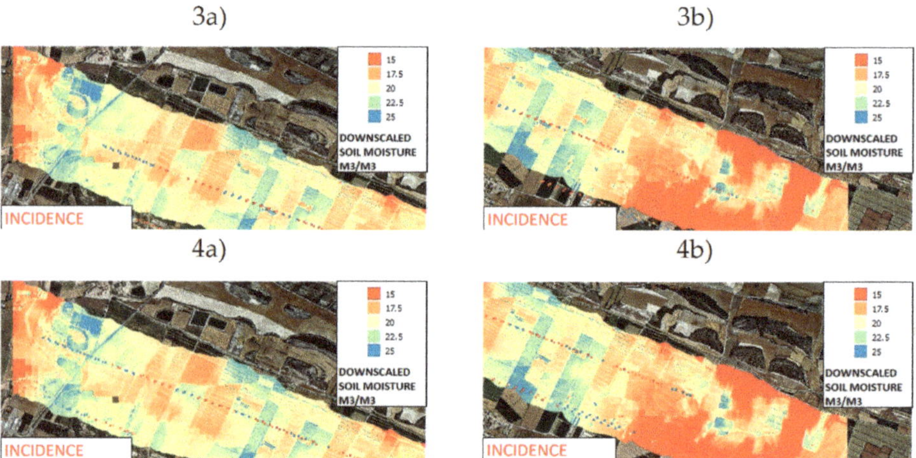

Figure 18. ARIEL and GNSS-R retrieved SM, as a function of the incidence angle in steps of 15°: 1) 0°–15°, 2) 15°–30°, 3) 30°–45°, and 4) 45°–60°. Figures (**a**) and (**b**) correspond to rectangles A and B in Figures 8–10. Note that after a regional calibration of roughness effects, values are much more similar (<5%), and that errors are larger in the 30°–45°, and 45°–60° ranges (larger color disagreement).

Finally, Figure 19 presents the downscaled ARIEL and GNSS-R-retrieved SM overlaid maps for all incidence angles included, showing a reasonable agreement, which degrades for increasing incidence angles.

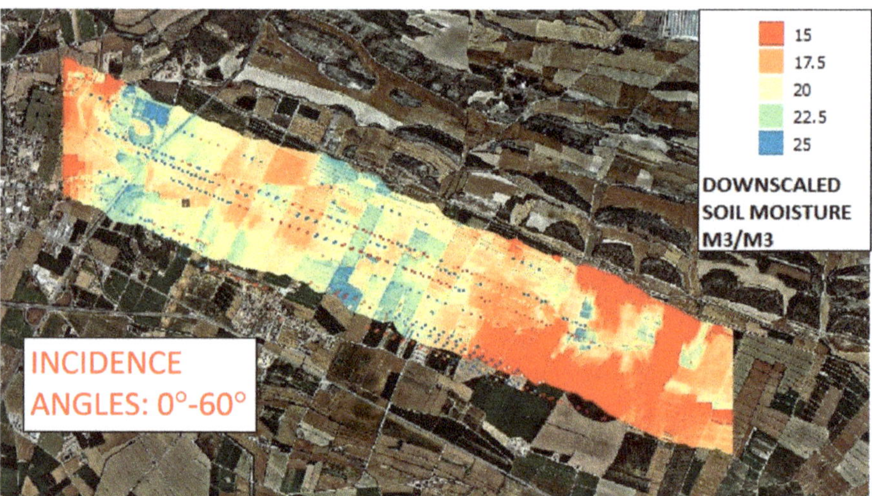

Figure 19. ARIEL downscaled and CORTO GNSS-R retrieved SM: all incidence angles included.

4. Discussion

In the previous sections, it was shown that the sensitivity of the calibrated reflectivity to Soil Moisture is non-linear and varies with the Soil Moisture value (Figure 1) [16]. Linearized retrieval approaches must be cautious in order not to introduce biases and other artifacts. Soil moisture retrieval algorithms require that vegetation and surface roughness effects must be properly compensated first.

Current models adopted from L-band microwave radiometry (1.413 GHz) overcorrect vegetation effects [11], which leads to an overcorrection of the surface roughness to try to compensate for the vegetation correction. Therefore, techniques to better predict the vegetation corrections are needed, as current models adopted from L-band microwave radiometry (1.413 GHz) using a VOD estimate from the *NDVI* [17] seem to overestimate it, which leads to an overcompensation of the surface roughness effects to counteract them. A potential source of this discrepancy may be in the values of the *b* parameter, which are too high in crop regions, as recently reported in [29].

In situ measurements of the surface roughness are not suitable to predict the surface roughness parameter to be used in the compensation, as different roughness values are obtained with different techniques [34], i.e., a laser profiler with respect to mechanical profiler, and that there is an impact of the profiler length and number of independent measurements. It is also possible that the reflection is taking place effectively in the water table underground, behaving then as a much "flatter" surface than the air–soil interface. Volume scattering effects in the soil cannot be neglected, either. The use of a constant value of "*h*" (e.g., as in SMAP) is not suitable either, even if it is different for each field. In addition, it has been recently reported that the current values of the *h* parameter may be too smooth (low) for crop regions [20]. Generic surface roughness empirical corrections as a function of the range of incidence angles are also possible, leading to acceptable results: the root mean squared error between downscaled ARIEL and GNSS_R SM map is < 5%, while today's UCAR/CU soil moisture product developed by calibrating CYGNSS reflectivity observations to soil moisture retrievals from NASA SMAP mission exhibits a median unbiased root-mean-square error (ubRMSE) of 0.049 m^3/m^3 [8].

5. Conclusions

Recent results [13] claim that IceSat-2 surface roughness data can be used to correct for these effects at a global scale. However, the authors of this study believe that further improvements using auxiliary data from other sensors are unlikely, as it is very difficult to match all the observations in time and space, and that they can vary over time, and techniques to infer the effective soil surface roughness parameter, if possible from the data themselves, are needed. A clear example is the estimation of the surface roughness from the GNSS-R observables [35].

Further research is also needed to explore the simultaneous use of several GNSS bands and other GNSS-R observables with variable coherent and incoherent integration times, and to include the non-linear dependence of the VOD with the above ground biomass (ABG), as shown in Supplementary Figure 4 from [36,37]. Vegetation scattering and wave depolarization have to be accounted for as well.

In the meantime, while these corrections are not improved for single pass retrievals, time-series approaches based on temporal changes of the reflectivity (e.g., [5–8]) can be applied, provided that the non-linearities of the dependence of the GNSS-R observables and the SM are properly taken into account, and self-consistency of the retrieved products is demonstrated.

The most relevant result of this study is that despite all the above limitations in the surface roughness and vegetation effects corrections, single-pass GNSS-R soil moisture values seem feasible with a reasonable error (<5%) for incidence angles up to 30°, provided that an incidence angle dependent soil surface roughness correction is applied.

Author Contributions: Conceptualization, A.C.; methodology, A.C.; software, A.C. and J.C. (Jordi Castellví); validation, A.C. and J.C. (Jordi Castellví); formal analysis, A.C.; investigation, A.C., H.P.; resources, A.C., H.P., J.C. (Jordi Corbera); data curation, J.C. (Jordi Castellví), E.A.; writing—original draft preparation, A.C.; writing—review and editing, H.P., J.C. (Jordi Corbera); visualization, A.C., J.C. (Jordi Castellví), H.P.; supervision, A.C.; project administration, A.C., J.C. (Jordi Corbera); funding acquisition, A.C., J.C. (Jordi Corbera). All authors have read and agreed to the published version of the manuscript.

Funding: This work has been funded by the Spanish MCIU and EU ERDF project (RTI2018-099008-B-C21) "Sensing with pioneering opportunistic techniques" and grant to "CommSensLab-UPC" Excellence Research Unit Maria de Maeztu (MINECO grant MDM-2016-600), and by a Doctorat Industrial grant from ICGC.

Acknowledgments: The authors would like to thank the flight department of ICGC for the work carried out to design and implement the new sensors onboard ICGC airplane as well as the operational flight campaign

Conflicts of Interest: The authors declare no conflict of interest.

References

1. Kavak, A.; Xu, G.; Vogel, W.J. GPS multipath fade measurements to determine L-band ground reflectivity properties. In *Proceedings of the 20th NASA Propagation Experiment Meeting (NAPEX 20) Fairbank*; Alaska, Organizer: Jet Propulsion Laboratory: Pasadena, CA, USA, 1996; pp. 257–263.
2. Larson, K.M.; Braun, J.J.; Small, E.E.; Zavorotny, V.U.; Gutmann, E.D.; Bilich, A.L. GPS Multipath and Its Relation to Near-Surface Soil Moisture Content. *IEEE J. Sel. Top. Appl. Earth Obs. Remote Sens.* **2010**, *3*, 91–99. [CrossRef]
3. Rodriguez-Alvarez, N.; Bosch-Lluis, X.; Camps, A.; Vall-llossera, M.; Valencia, E.; Marchan-Hernandez, J.F.; Ramos-Perez, I. Soil Moisture Retrieval Using GNSS-R Techniques: Experimental Results Over a Bare Soil Field. *IEEE Trans. Geosci. Remote Sens.* **2009**, *47*, 3616–3624. [CrossRef]
4. Arroyo, A.A.; Camps, A.; Aguasca, A.; Forte, G.F.; Monerris, A.; Rüdiger, C.; Walker, J.P.; Park, H.; Pascual, D.; Onrubia, R. Dual-Polarization GNSS-R Interference Pattern Technique for Soil Moisture Mapping. *IEEE J. Sel. Top. Appl. Earth Obs. Remote Sens.* **2014**, *7*, 1533–1544. [CrossRef]
5. Chew, C.; Small, E. Soil moisture sensing using spaceborne GNSS reflections: Comparison of CYGNSS reflectivity to SMAP soil moisture. *Geophys. Res. Lett.* **2018**, *45*, 4049–4057. [CrossRef]
6. Clarizia, M.P.; Pierdicca, N.; Costantini, F.; Floury, N. Analysis of CyGNSS Data for Soil Moisture Retrieval. *IEEE J. Sel. Top. Appl. Earth Obs. Remote Sens.* **2019**, *12*, 2227–2235. [CrossRef]
7. Al-Khaldi, M.M.; Johnson, J.T.; O'Brien, A.J.; Balenzano, A.; Mattia, F. Time-Series Retrieval of Soil Moisture Using CYGNSS. *IEEE Trans. Geosci. Remote Sens.* **2019**, *57*, 4322–4331. [CrossRef]
8. Chew, C.; Small, E. Description of the UCAR/CU Soil Moisture Product. *Remote Sens.* **2020**, *12*, 1558. [CrossRef]
9. Camps, A. Spatial Resolution in GNSS-R Under Coherent Scattering. *IEEE Geosci. Remote Sens. Lett.* **2020**, *17*, 32–36. [CrossRef]
10. Camps, A.; Park, H.; Pablos, M.; Foti, G.; Gommenginger, C.P.; Liu, P.; Judge, J. Sensitivity of GNSS-R Spaceborne Observations to Soil Moisture and Vegetation. *IEEE J. Sel. Top. Appl. Earth Obs. Remote Sens.* **2016**, *9*, 4730–4742. [CrossRef]
11. Camps, A.; Vall-llossera, M.; Park, H.; Portal, G.; Rossato, L. Sensitivity of TDS-1 GNSS-R Reflectivity to Soil Moisture: Global and Regional Differences and Impact of Different Spatial Scales. *Remote Sens.* **2018**, *10*, 1856. [CrossRef]
12. Zribi, M.; Motte, E.; Baghdadi, N.; Baup, F.; Dayau, S.; Fanise, P.; Guyon, D.; Huc, M.; Wigneron, J.-P. Potential Applications of GNSS-R observations over agricultural areas: Results from the GLORI airborne campaign. *Remote Sens.* **2018**, *10*, 17. [CrossRef]
13. Calabia, A.; Molina, I.; Jin, S. Soil Moisture Content from GNSS Reflectometry Using Dielectric Permittivity from Fresnel Reflection Coefficients. *Remote Sens.* **2020**, *12*, 122. [CrossRef]
14. Jia, Y.; Jin, S.; Savi, P.; Gao, Y.; Tang, J.; Chen, Y.; Li, W. GNSS-R Soil Moisture Retrieval Based on a XGboost Machine Learning Aided Method: Performance and Validation. *Remote Sens.* **2019**, *11*, 1655. [CrossRef]
15. Camps, A.; Forte, G.; Ramos, I.; Alonso, A.; Martinez, P.; Crespo, L.; Alcayde, A. *2012: Recent Advances in Land Monitoring Using GNSS-R Techniques, 2012 Workshop on Reflectometry Using GNSS and Other Signals of Opportunity (GNSS+R)*; West Lafayette: Indiana, IN, USA, 2012; pp. 1–4. [CrossRef]
16. Park, H.; Camps, A.; Castellvi, J.; Muro, J. Generic Performance Simulator of Spaceborne GNSS-Reflectometer for Land Applications. *IEEE J. Sel. Top. Appl. Earth Obs. Remote Sens.* **2020**, *13*, 3179–3191. [CrossRef]
17. O'Neill, P.; Bindlish, R.; Chan, S.; Njoku, E.; Jackson, T. Soil Moisture Active Passive (SMAP), Algorithm Theoretical Basis Document Level 2 & 3 Soil Moisture (Passive) Data Products. Revision D, June 6, 2018, Technical Note JPL D-66480. Available online: https://smap.jpl.nasa.gov/system/internal_resources/details/original/484_L2_SM_P_ATBD_rev_D_Jun2018.pdf (accessed on 24 May 2020).
18. Choudhury, J.B.; Schmugge, T.J.; Chang, A.; Newton, R.W. Effect of surface roughness on the microwave emission from soil. *J. Geophys. Res.* **1979**, *84*, 5699–5706. [CrossRef]
19. Wang, J.R. Passive microwave sensing of soil moisture content: The effects of soil bulk density and surface roughness. *Remote Sens. Environ.* **1983**, *13*, 329–344. [CrossRef]

20. Walker, A.V.; Hornbuckle, B.K.; Cosh, M.H.; Prueger, J.H. Seasonal Evaluation of SMAP Soil Moisture in the U.S. Corn Belt. *Remote Sens.* **2019**, *11*, 2488. [CrossRef]

21. AISA Technical Data. Available online: https://www.channelsystems.ca/sites/default/files/documents/aisa_products_ver1-2015.pdf (accessed on 22 April 2020).

22. TASI600 Technical Data. Available online: https://itres.com/wp-content/uploads/2019/09/TASI600.pdf (accessed on 22 April 2020).

23. Acevo-Herrera, R.; Aguasca, A.; Bosch-Lluis, X.; Camps, A.; Martínez-Fernández, J.; Sánchez-Martín, N.; Pérez-Gutiérrez, C. Design and First Results of an UAV-Borne L-Band Radiometer for Multiple Monitoring Purposes. *Remote Sens.* **2010**, *2*, 1662–1679. [CrossRef]

24. ARIEL Technical Data. Available online: https://www.balamis.com/products/ (accessed on 22 April 2020).

25. Sánchez, N.; Alonso-Arroyo, A.; Martínez-Fernández, J.; Piles, M.; González-Zamora, A.; Camps, A.; Vall-llosera, M. On the Synergy of Airborne GNSS-R and Landsat 8 for Soil Moisture Estimation. *Remote Sens.* **2015**, *7*, 9954–9974. [CrossRef]

26. Applanix Web Site. Available online: https://www.applanix.com/products/posav.htm (accessed on 24 May 2020).

27. UPC Anechoic Chamber. Available online: https://www.tsc.upc.edu/en/facilities/anechoic-chamber (accessed on 24 May 2020).

28. Sentinel Hub-EO Browser. Available online: https://apps.sentinel-hub.com/eo-browser/ (accessed on 24 May 2020).

29. Togliatti, T.K.; Hartman, T.; Walker, V.A.; Arkebaur, T.J.; Suyker, A.E.; VanLoocke, A.; Hornbuckle, B.K. Satellite L–band vegetation optical depth is directly proportional to crop water in the US Corn Belt. *Remote Sens. Environ.* **2019**, *223*, 111. [CrossRef]

30. Panciera, R.; Walker, J.P.; Merlin, O. Improved Understanding of Soil Surface Roughness Parameterization for L-Band Passive Microwave Soil Moisture Retrieval. *IEEE Geosci. Remote Sens. Lett.* **2009**, *6*, 625–629. [CrossRef]

31. Piles, M.; Camps, A.; Vall·llossera, M.; Corbella, I.; Panciera, R.; Rudiger, C.; Kerr, H.Y.; Walker, J. 2011: Downscaling SMOS-Derived Soil Moisture Using MODIS Visible/Infrared Data. *IEEE Trans. Geosci. Remote Sens.* **2011**, *49*, 3156–3166. [CrossRef]

32. Portal, G.; Vall-llossera, M.; Piles, M.; Camps, A.; Chaparro, D.; Pablos, M.; Rossato, L. 2018: A Spatially Consistent Downscaling Approach for SMOS Using an Adaptive Moving Window. *IEEE J. Sel. Top. Appl. Earth Obs. Remote Sens.* **2018**, *11*, 1883–1894. [CrossRef]

33. Pablos, M.; Piles, M.C. Gonzalez-Haro and BEC TeamBEC SMOS Land Products Description. Technical Note: BEC-SMOS-0003-PD-Land.pdf, Version1.0, Date: 31/07/2019. Available online: http://bec.icm.csic.es/doc/BEC-SMOS-0003-PD-Land.pdf (accessed on 24 May 2020).

34. Mattia, F.; Davidson, M.W.J.; Le Toan, T.; D'Haese, C.M.F.; Verhoest, N.E.C.; Gatti, A.M.; Borgeaud, M. A comparison between soil roughness statistics used in surface scattering models derived from mechanical and laser profilers. *IEEE Trans. Geosci. Remote Sens.* **2003**, *41*, 1659–1671. [CrossRef]

35. Stilla, D.; Zribi, M.; Pierdicca, N.; Baghdadi, N.; Huc, M. Desert Roughness Retrieval Using CYGNSS GNSS-R Data. *Remote Sens.* **2020**, *12*, 743. [CrossRef]

36. Liu, Y.; van Dijk, A.; de Jeu, R.; Canadell, J.G.; McCabe, M.F.; Evans, J.P.; Wang, G. Recent reversal in loss of global terrestrial biomass. *Nat. Clim. Chang.* **2015**, *5*, 470–474. [CrossRef]

37. Supplementary Material for Reference [34]. Available online: https://static-content.springer.com/esm/art%3A10.1038%2Fnclimate2581/MediaObjects/41558_2015_BFnclimate2581_MOESM462_ESM.pdf (accessed on 24 May 2020).

 *remote sensing*

Article

Comparing Winds near Tropical Oceanic Precipitation Systems with and without Lightning

Timothy J. Lang

NASA Marshall Space Flight Center, Huntsville, AL 35812, USA; timothy.j.lang@nasa.gov; Tel.: +1-256-961-7861

Received: 13 November 2020; Accepted: 2 December 2020; Published: 4 December 2020

Abstract: In order to examine how robust updraft strength and ice-based microphysical processes aloft in storms may affect convective outflows near the surface, ocean winds were compared between tropical maritime precipitation systems with and without lightning. The analysis focused on Cyclone Global Navigation Satellite System (CYGNSS) specular point tracks, using straightforward spatiotemporal matching criteria to pair CYGNSS-measured wind speeds with satellite-based precipitation observations, Advanced Scatterometer (ASCAT) wind speeds, and lightning flash data from ground-based and space-based sensors. Based on the results, thunderstorms over the tropical oceans are associated with significantly heavier rain rates (~200% greater) than non-thunderstorms. However, wind speeds near either type of precipitation system do not differ much (~0.5 m s^{-1} or less). Moreover, the sign of the difference depends on the wind instrument used, with CYGNSS suggesting non-thunderstorm winds are slightly stronger, while ASCAT suggests the opposite. These observed wind differences are likely related to lingering uncertainties between CYGNSS and ASCAT measurements in precipitation. However, both CYGNSS and ASCAT observe winds near precipitation (whether lightning-producing or not) to be stronger than background winds by at least 1 m s^{-1}.

Keywords: CYGNSS; ASCAT; ISS LIS; GLM; WWLLN; winds; precipitation; lightning; convection

---

1. Introduction

A common feature of near-surface winds near convection is a short-term (~30 min or less) increase in average wind speed, often coupled with a shift in dominant wind direction. This is caused by evaporatively cooled air within the storm that descends to the surface and then spreads out laterally, and is commonly referred to as a gust front [1]. Behind the gust front, the air is often cooler than the environment. This region is referred to as a cold pool [2–4]. In addition to modulating the evolution of the convection responsible for their genesis, cold pools and their associated gust fronts also play important roles in the initiation and development of new convection [2].

There are important questions regarding how gust front genesis relates to the kinematic and microphysical structure and evolution of convection. For example, there is significant uncertainty about the strength of gust fronts that are produced in convection that is characterized by relatively weaker updrafts and has its precipitation primarily produced by warm-rain microphysical processes, versus gust fronts produced in convection that has relatively stronger updrafts and relatively more of its precipitation produced by ice-based microphysical processes [5]. In the latter case, the precipitation system would be expected to produce lightning, since lightning is the result of charge separation caused by rebounding collisions between ice hydrometeors in the presence of supercooled liquid water [6]. Indeed, many studies have demonstrated that lightning is quantitatively correlated to ice water content, updraft strength, and vertical fluxes of ice hydrometeors [7–10].

Relative to unelectrified convection, electrified storms typically are larger in area, and contain stronger updrafts, larger ice water paths (IWP), and are vertically deeper [9–13]. Thus, it is reasonable to expect that thunderstorms should feature stronger gust fronts compared to storms without lightning. In addition, as lightning flash rate increases, this tends to signal intensification of the convection, to the point that very high flash rate storms are likely to be associated with severe weather, such as damaging winds from intense gust fronts [14,15]. However, due to the relative sparseness of global wind observations, it is difficult to demonstrate this hypothesis conclusively, let alone quantify the actual differences observed between gust fronts associated with thunderstorms and gust fronts associated with storms without lightning.

Over oceans, scatterometers such as the Advanced Scatterometer (ASCAT) are capable of retrieving near-surface vector winds, but these measurements are more uncertain in precipitation [16–18]. The Cyclone Global Navigation Satellite System (CYGNSS) uses Global Positioning System (GPS) reflectometry to retrieve wind speed over the oceans [19]. Because it is L-band, CYGNSS is expected to be less impacted by precipitation, and its ability to retrieve accurate winds in rain has been demonstrated [20]. CYGNSS provides coverage of oceanic winds in the global tropics.

Due to its ability to accurately retrieve winds even in the presence of heavy rain [20], which is common near tropical convection, CYGNSS was chosen as the basis for this study. By combining CYGNSS observations with global precipitation measurements, as well as global/regional lightning datasets, this study will quantify differences in winds near lightning-producing oceanic convection versus convection without lightning. In turn, this study will attempt to determine (in a categorical sense) the impact of relatively stronger updrafts and ice microphysical processes on wind speeds near precipitation systems, as thus by proxy infer potential impacts on cold pool and gust front strength.

2. Materials and Methods

2.1. CYGNSS

The CYGNSS constellation of 8 small satellites provides observations of wind speeds over the tropical and subtropical oceans between approximately ±38° latitude. This study used version 2.1 CYGNSS Level 2 wind data [21]. Version 2.1 data have been validated to meet all CYGNSS level 1 science requirements [20]. However, observations formed by signals from Block IIF GPS satellites are excluded from version 2.1 due to calibration uncertainties. This reduces the available number of wind observations by roughly one third.

CYGNSS wind retrievals are available for specular reflections of GPS signals, and tracks are formed by a series of reflections corresponding to a single GPS satellite and one of the antennas on a single CYGNSS observatory. As noted by [19,22], a single track of specular reflections comprises a fundamental and physically consistent grouping of CYGNSS observations. Therefore, this study's analysis focus was on CYGNSS tracks, and other datasets (e.g., ASCAT, IMERG, lightning) were matched to CYGNSS at the track level.

Version 2.1 Level 2 CYGNSS wind files do not organize CYGNSS observations by track. However, tracks can be identified and grouped via organizing CYGNSS winds by contributing antenna, observatory number, and GPS satellite number (pseudorandom noise or PRN code). However, on a given day there can be multiple tracks with these common values. Therefore, the DBSCAN (Density-Based Spatial Clustering of Applications with Noise) [23] clustering algorithm was applied to these common groups to further decompose to individual tracks. The track-separation algorithm allowed up to 1° difference in latitude/longitude and a time gap of 60 s before a new track was identified from a set of common antenna, observatory, and PRN numbers. The software used to create these tracks was the open source Python Interface to CYGNSS Wind Dataset (PyGNSS) [24].

Each track—including geolocation and timing information, range-corrected gain (RCG), and wind speeds from the Fully Developed Seas (FDS) and Young Seas Limited Fetch (YSLF) retrieval algorithms—was saved to an individual Network Common Data Format (netCDF) file. YSLF wind speeds are generally only valid near tropical cyclones, while FDS wind speeds are accurate in other circumstances [20]. However, it remains useful to study the utility of YSLF winds near more conventional tropical precipitation, so this study analyzed both types of retrievals. Note that the dataset included tropical cyclones, but since these are rare (relative to conventional tropical precipitation systems) they likely did not impact overall results significantly.

Tracks were identified for May 2017 through December 2019. With the full constellation operational, typically 2000–3000 tracks are available per day, and each track can contain an average of 200–300 wind observations over a time period lasting several minutes. CYGNSS samples along an individual track at 1 Hz [20], so based on orbit speed the average spacing along a track is 6 km. This provides roughly 4:1 oversampling relative to the nominal CYGNSS resolution of 25 km. However, data gaps often occur, but, as mentioned previously, the track identification algorithm described above allows for that.

2.2. ASCAT

ASCAT is a C-band scatterometer that measures ocean vector winds. ASCAT consists of a dual-fan-beam radar, which measures along two parallel swaths. Currently, there are three ASCAT instruments in orbit onboard the Meteorlogical Operational (Metop) A, B, and C satellites. The time period of May 2017 through December 2019 was analyzed, so this study only focused on ASCAT-A and -B. For each of these instruments, the 12.5-km coastal-optimized Level 2 data [25,26] were analyzed. Because this study was focused primarily on CYGNSS, only the ASCAT wind speed measurements were used, not wind direction. ASCAT is included in this study in order to examine the sensitivtiy of the results to the choice of wind instrument.

ASCAT winds were matched to CYGNSS tracks in the following way. All CYGNSS tracks that overlapped the time period of a particular ASCAT orbit (approximately 90 min) were examined for that granule. The closest-in-distance ASCAT wind speed observation from either swath was matched to each CYGNSS specular reflection along a track; however, any ASCAT wind measurement more than 20 km away from the track was excluded, even if it was the closest. Only about 25% of CYGNSS tracks have matchups from either ASCAT-A or -B, because the CYGNSS track and ASCAT swaths often did not overlap spatially within the ~90-min orbit window. Moreover, typically any overlap occurred only during a portion of the CYGNSS track's length. In sum, the number of matched samples with ASCAT was <10% of the entire track-based dataset. Matched ASCAT winds were included as a new field in each CYGNSS track file.

2.3. IMERG

IMERG (Integrated Multi-satellitE Retrievals for GPM) is a global precipitation dataset produced as part of the GPM (Global Precipitation Measurement) mission [27]. Precipitation rates are retrieved using intercalibrated passive microwave measurements from a constellation of satellite radiometers in low-Earth orbit (LEO). Geostationary infrared (IR) measurements are used to morph and advect the precipitation in a physically consistent way between LEO passive microwave overpasses.

IMERG data are available every half hour on a global 0.1° grid. Each CYGNSS track was matched to the temporally closest IMERG file (with the caveat that the IMERG time could not be later than the midpoint time of the track), and then the precipitation at the nearest neighbor IMERG gridpoint was matched to each specular reflection within the CYGNSS track. Due to the timing of when the track-based processing occurred, CYGNSS track files prior to 2019 were matched to the version 5 Late IMERG products, while 2019 data were matched to version 6 Final IMERG products. Based on a sensitivity analysis using a subset of data (not shown), these version differences did not significantly impact any of this study's results.

2.4. Lightning

2.4.1. WWLLN

The Worldwide Lightning Location Network (WWLLN) is a ground-based network that detects radio-frequency (RF) emissions from lightning on a global basis [28]. The network is most sensitive to powerful cloud-to-ground (CG) flashes, and thus only detects a small fraction of total lightning [29]. However, it does enable continuous monitoring of thunderstorms over the global oceans, and thus is a useful dataset for comparison with satellite observations of oceanic winds. For this study, the only data used were from 1 August through 31 October of 2018 and 2019 (6 months of data), as they were obtained in support of field campaigns that operated during those periods [30]. However, because of the continuous global observations, this short time period still provided the most matchups of lightning flashes with the CYGNSS tracks, of the three lightning datasets examined in this study.

WWLLN data were matched to CYGNSS tracks in the following manner. Only lightning occurring during a given track's duration (typically several minutes) was considered. Contiguous portions of the track with nonzero IMERG precipitation were automatically identified, and if lightning was identified within 25 km of a specular point in the raining portion of a track, that wind measurement was included in the "winds with lightning" category. The flash rate related to that wind measurement was calculated by totaling all flashes within 25 km of the specular point, and dividing by the duration of the track itself. To limit cross-contamination, matching lightning anywhere along a track led to the exclusion of that track from contributing to the "winds without lightning" category (even if the track featured some raining areas without lightning). For tracks that had no lightning within 25 km anywhere along it, but did have raining portions, winds within the raining portion were added to the "winds without lightning" category. Typically < 10% of the overall raining dataset featured lightning.

The advantage of using WWLLN is that the dataset is global and the temporal coverage is continuous. The disadvantage is that it does not detect all lightning, and thus many wind regions without lightning may in fact be near thunderstorms with undetected lightning. This will tend to cause some overlap between winds associated with lightning, and those without. Figure 1 shows the analysis region for WWLLN, which includes all of the CYGNSS sampling region.

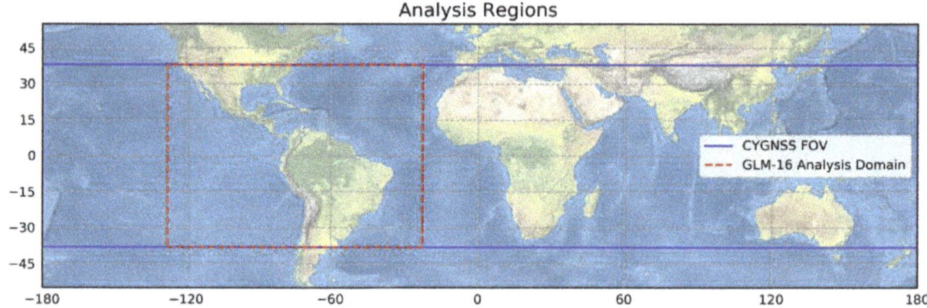

Figure 1. Analysis regions for this study. The observed map is the ISS LIS field of view (FOV), and WWLLN observes the full domain as well. The blue box shows the CYGNSS FOV. The dashed red box is the domain used for GLM analysis in this study. Land regions have no valid CYGNSS winds.

2.4.2. GLM

Observations from the Geostationary Lightning Mapper (GLM) on the Geostationary Operational Environmental Satellite (GOES) 16 were also used. GLM-16 continuously monitors a hemispheric field of view (FOV) centered on 75.2° W, which encompasses North and South America (landmasses where CYGNSS does not provide wind observations), the eastern Pacific Ocean, the western Atlantic Ocean, as well as the Gulf of Mexico and the Caribbean Sea. GLM detects total lightning with a detection efficiency of approximately 70%, and its instantaneous field of view (IFOV) varies between 8–14 km depending on viewing angle [31,32]. Six months from each of 2018 and 2019 (1 July through 31 December), for a total of 12 months of observations, were included in this study.

As noted by [31,32], the GLM-16 FOV varies as a function of latitude. Thus, to ensure continuous monitoring, only CYGNSS tracks occurring (or overlapping) within the box marked by ±38° latitude, and 128 to 22.5 °W longitude, were considered when comparing to GLM-16 (Figure 1). This ensured that these tracks were always within the GLM-16 FOV regardless of latitude. Because of the continuous FOV, GLM-16 lightning was matched to CYGNSS tracks in a similar manner to WWLLN. However, coordinates for flash centroids were used as the basis of comparison. GLM-16 group and event data were ignored in this study.

The advantage of using GLM-16 is that it provides continuous coverage of total lightning, with high detection efficiency. The disadvantage is that the dataset is not global, and thus a longer analysis period was required to obtain a comparable number of samples to the WWLLN dataset, despite the higher lightning detection efficiency. The reason GLM-16 was used in this study was to confirm whether trends observed by the CG-focused WWLLN analysis were also observed when using a total lightning dataset.

2.4.3. ISS LIS

The International Space Station Lightning Imaging Sensor (ISS LIS) is the modified flight spare of the original Tropical Rainfall Measuring Mission (TRMM) LIS [33,34]. This instrument has been in orbit on the ISS since early 2017, and provides routine monitoring of lightning between ±55° latitude. ISS LIS has a detection efficiency of ~60%, and a location accuracy better than its 4-km IFOV [33]. All months of version 1 ISS LIS data from May 2017 through December 2019 were analyzed in this study (32 total months) [35].

Because ISS LIS is a LEO instrument, it does not provide a continuous FOV, and thus instrument viewing time needs to be considered. Lightning matchups were considered differently compared to WWLLN and GLM-16, for which viewing time was not a constraint. Similar to the ASCAT matchup approach, only CYGNSS tracks that intersected the ISS LIS swath during a particular orbit were considered. The ISS LIS swath was defined as the portion of the FOV where viewing time was at least 60 s. ISS LIS viewing time for a given location on the ground is nominally close to 90 s [36], but can be reduced near swath edge, during ISS orbital maneuvers, or when the ISS solar panels pass through the FOV. This viewing time filter thus removes the most obscured portions of the orbit from consideration, although it may potentially cause low-flash-rate storms to appear as having no lightning. Once the properly observed swath was defined, matchups with CYGNSS tracks proceeded similarly to GLM-16, which has a very similar data structure. Only flash centroids, and not groups or events, were considered.

The advantage of ISS LIS is that it provides global coverage (Figure 1) of total lightning with high detection efficiency. The disadvantage is that it does not provide continuous coverage of a particular region. Thus, the number of samples is greatly reduced compared to WWLLN and even GLM-16, despite a significantly longer dataset. The reason ISS LIS was used was to enable a longer sampling period that included additional seasons (namely, January–July), and to confirm whether trends observed using the regional total lightning provided by GLM-16 also manifested globally.

2.5. Synthesis

The synthesized CYGNSS track dataset provided matchups with ASCAT-A and -B within approximately ±90 min (the duration of one Metop orbit), IMERG within 30 min, and three different complementary lightning measurements within the duration of the track itself (typically several minutes). In addition, the analysis was structured relative to the fundamental grouping of CYGNSS wind measurements (a common track of specular points), and thus ensured spatial and temporal continuity.

It is not clear how a different matchup approach could provide comparable spatial and temporal proximity, because the specular reflections provided by CYGNSS are fundamentally sparser in time and space than a swath-based measurement. For example, it is not realistic for CYGNSS to provide contiguous coverage of the winds in a mesoscale convective system (MCS) within the lifetime of said MCS (i.e., a few hours). ASCAT can provide swath coverage of winds (and thus a contiguous snapshot of the wind field), but this approach was not considered because the focus of this study was primarily on CYGNSS due to its ability to retrieve winds in heavy rain.

Figure 2 shows an example of how a track that contained winds near precipitation with lightning appears. The portions of the track marked in magenta (Figure 2b) indicate what was considered in the "winds with lightning" and "precipitation with lightning") categories. A comparable track, but with no lightning anywhere along it, would have the raining portions included in the "winds without lightning" category. The matched ASCAT-A winds agree well with CYGNSS FDS winds in the west swath, but there is significant disagreement in the east, with the largest deviations occurring within the co-observed raining area (Figure 2b). The tendency for ASCAT winds to increase significantly relative to CYGNSS within raining areas was observed in many matched tracks. Additionally, at the east edge of the track, where RCG decreased below 10, CYGNSS wind speeds became noticeably noisier.

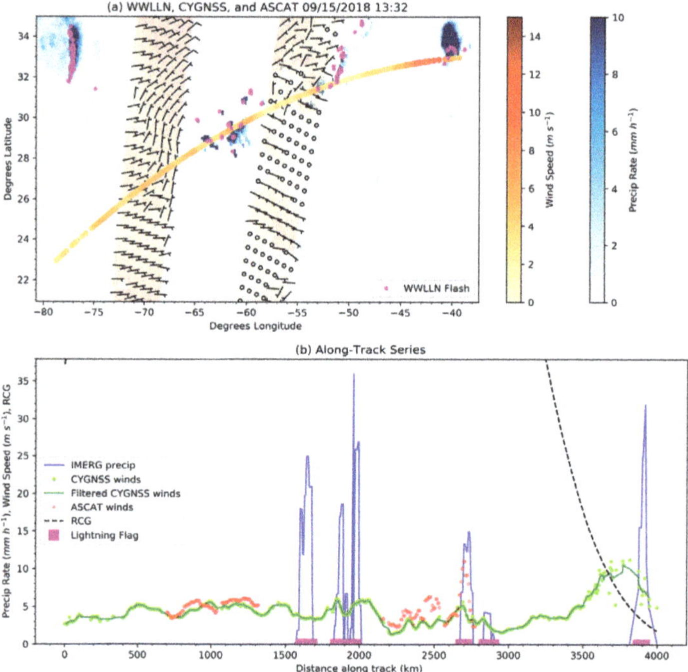

Figure 2. (a) Example of a CYGNSS track (arc of colored wind speeds) on a latitude/longitude map, along with ASCAT-A wind speeds (dual swaths) and wind barbs (black; circles < 5 m s^{-1} speed) for the matching orbit. IMERG precipitation rate is also shown, as are WWLLN flashes during the track (magenta dots). (b) Along-track series of CYGNSS FDS wind speeds (raw and filtered using a 5-point boxcar mean), matched ASCAT-A wind speeds, matched IMERG precipitation rate, and CYGNSS RCG. Shown also are flagged raining features with WWLLN lightning (magenta bars).

3. Results

3.1. Global comparison of CYGNSS and ASCAT

Before delving into the analysis of lightning, it is useful to understand how CYGNSS FDS winds and ASCAT winds compare in a general sense. This will provide valuable context for interpreting the lightning-wind results. For the following analysis, matched tracks for the time period 1 May 2017 through 31 December 2019 were used. Table 1 shows basic statistics for the 32-month period. The comparisons were very similar whether ASCAT-A or -B was used. CYGNSS

FDS winds—the retrieval approach appropriate for non-tropical cyclone wind strengths—had a root mean square difference (RMSD) with ASCAT-A and -B of less than 2 m s^{-1} when matched IMERG rain rates were zero, which was within the expected uncertainty for CYGNSS [20,37]. CYGNSS wind speeds averaged approximately 0.5 m s^{-1} stronger than ASCAT in non-raining circumstances.

Table 1. RMSD, mean offset, and number of matchups for winds outside of rain and winds in rain. Convention is CYGNSS FDS wind speeds minus ASCAT wind speeds (positive offset: CYGNSS winds were stronger).

Variable	ASCAT-A	ASCAT-B
No Rain RMSD (m s^{-1})	1.94	1.93
No Rain Offset (m s^{-1})	+0.52	+0.53
No Rain Matchups	39,763,736	40,099,155
Rain RMSD (m s^{-1})	2.47	2.46
Rain Offset (m s^{-1})	+0.45	+0.47
Rain Matchups	4,238,318	4,275,117

However, when IMERG rain rates were nonzero, the RMSD increased by ~0.5 m s^{-1}, while the mean offset decreased slightly. There was approximately a factor of 10 reduction in the available number of matchups in rain, compared to non-raining matchups.

The magnitudes of the offset and RMSD changes in rain increased as rain rate increased (Figure 3). To develop Figure 3, the raining statistics were reevaluated at increasing thresholds for minimum rain rate. RMSD increased quasi-asymptotically toward ~4 m s^{-1} as rain rate increased toward 30 mm h^{-1} (Figure 3a). Meanwhile, the slight positive offset for CYGNSS wind speeds rapidly became negative (ASCAT wind speeds stronger), reaching approximately −2 m s^{-1} once rain rate reached 30 mm h^{-1} (Figure 3b). Approximately 95% of raining matchups occurred in rain rates below 5 mm h^{-1}, so results above this threshold were increasingly more uncertain as rain rate increased.

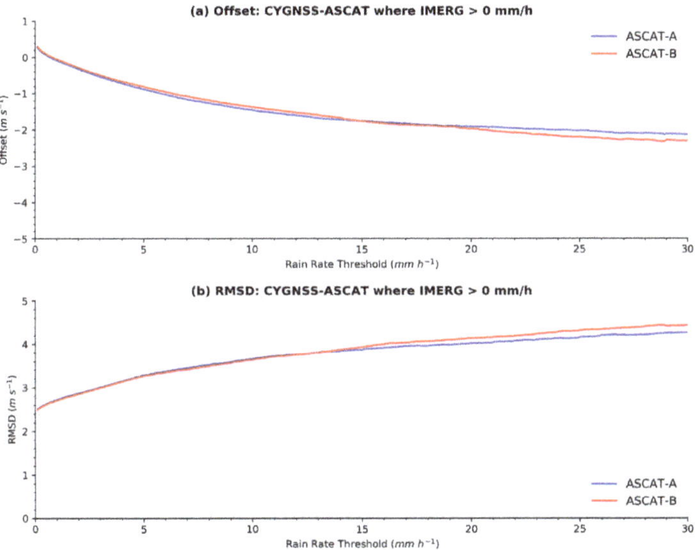

Figure 3. (a) Mean offsets between CYGNSS FDS and ASCAT-A and -B wind speeds as a function of IMERG rain rate, conditioned on the rain rate being nonzero. (b) RMSD between CYGNSS FDS and ASCAT-A and -B wind speeds as a function of IMERG rain rate, conditioned on the rain rate being nonzero. All data were from May 2017 through December 2019.

Figure 4 shows that RMSD and mean offset have, in general, been increasing with time against both ASCAT-A (Figure 4a) and -B (Figure 4b). This was likely due to increasing uncertainty in GPS power calibration, which affects the CYGNSS v2.1 wind retrieval. Newer versions of CYGNSS data are expected to have improved performance, and should remove these trends. In summary, we expect reduced agreement between CYGNSS and ASCAT in rain, and the 2019 matchups should show worse agreement than 2018. In particular, this comparison demonstrates that the ASCAT wind speed enhancement in rainfall (and reduced agreement with CYGNSS), shown in the Figure 2b sample track, was a general trend observed in the overall CYGNSS track dataset.

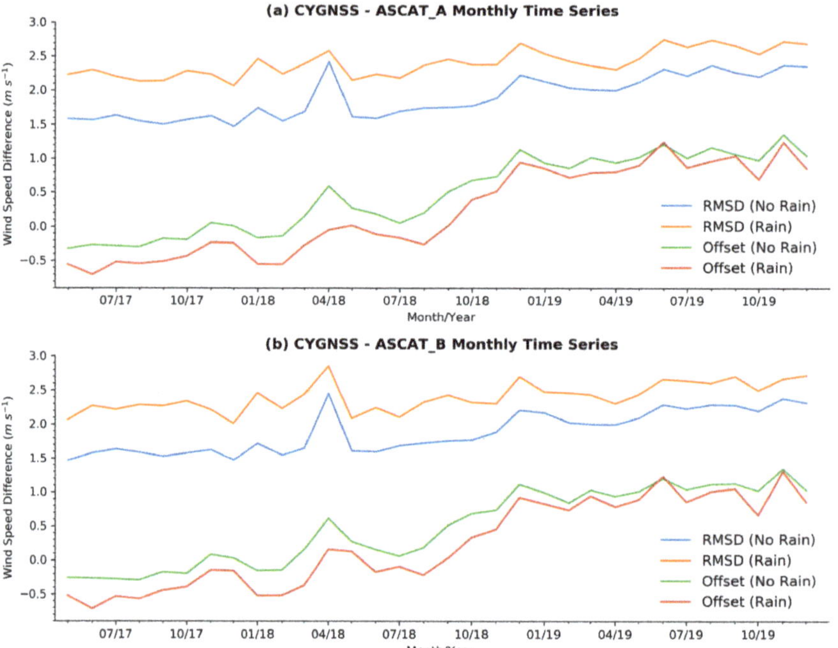

Figure 4. (**a**) Monthly time series of RMSD and mean offsets between CYGNSS FDS and ASCAT-A wind speeds, for zero and nonzero IMERG rain rates. (**b**) Monthly time series of RMSD and mean offsets between CYGNSS FDS and ASCAT-B wind speeds, for zero and nonzero IMERG rain rates. All data are for May 2017 through December 2019.

3.2. Lightning Analysis

3.2.1. Overall Statistics

The statistics for track matchups with WWLLN are shown in Table 2. Recall that the WWLLN period of analysis was August–October during 2018 and 2019. Surprisingly, the CYGNSS FDS winds not associated with lightning had slightly higher speeds (by ~0.2 m s⁻¹) compared to winds associated with lightning. This difference was small but greatly exceeded the 99% confidence intervals for the mean of either category. The difference was also robust if the restrictions were placed, either requiring RCG to exceed 10 or the rain rate to be less than 6 mm h⁻¹ (to control for any possible rain effects). YSLF winds also showed a similar pattern, but with reduced differences between the two categories.

Table 2. Summary statistics for winds and precipitation both associated and not associated with WWLLN lightning.

Variable	Mean	99% Conf. Int. (±)	Samples	RCG > 10 (Mean)	Rain < 6 mm h^{-1} (Mean)
CYGNSS FDS with WWLLN (m s^{-1})	8.44	0.01	1,184,585	8.61	8.34
CYGNSS FDS without WWLLN (m s^{-1})	8.62	0.00	15,227,308	8.85	8.59
CYGNSS YSLF with WWLLN (m s^{-1})	9.14	0.01	1,176,396	8.91	9.00
CYGNSS YSLF without WWLLN (m s^{-1})	9.25	0.00	15,075,239	8.96	9.19
ASCAT with WWLLN (m s^{-1})	8.52	0.03	97,220	N/A	8.14
ASCAT without WWLLN (m s^{-1})	8.12	0.01	1,273,200	N/A	8.05
IMERG with WWLLN (mm h^{-1})	2.95	0.01	1,184,585	N/A	N/A
IMERG without WWLLN (mm h^{-1})	0.89	0.00	15,227,308	N/A	N/A
CYGNSS Std. Dev. with WWLLN (m s^{-1})	1.13	0.01	28,853	N/A	N/A
CYGNSS Std. Dev. without WWLLN (m s^{-1})	1.44	0.00	294,024	N/A	N/A

In a sensitivity study to gauge differences in wind variability in precipitation with and without lightning, the standard deviation of CYGNSS FDS wind speeds in each contiguous region of precipitation along a track (e.g., each region marked by a magenta bar in Figure 2b) was analyzed. The results were similar to the mean wind speeds, with winds in precipitation without lightning about 30% more variable than in precipitation with lightning.

Compared to CYGNSS, an opposite pattern was observed for ASCAT winds, which were stronger when WWLLN lightning was present (by ~0.4 m s^{-1}) than without. However, the magnitude of this difference decreased to only ~0.1 m s^{-1} when controlling for higher rain rates. Note that when ASCAT-A and -B winds were both matched to a particular CYGNSS specular point, all of the analyses in this Section 3.2 defaulted to ASCAT-A. As seen in Section 3.1, A and B are well calibrated against each other so this choice had no significant impact on the results compared to using B (or averaging) instead.

The largest differences between WWLLN lightning and no WWLLN lightning were for IMERG rain rates, which were larger by a factor of 3 on average when WWLLN lightning was present. In summary, the overall trend for CYGNSS was that winds were slightly stronger and more variable when WWLLN lightning was not present, even as precipitation rate increased significantly when lightning occurred. The opposite wind trend was observed in ASCAT winds, but there were fewer samples available, and the ASCAT difference was affected when controlling for precipitation rate.

Since WWLLN is most sensitive to CG lightning, it was possible that some intracloud- (IC-) only thunderstorms were being mistakenly included in the "winds without lightning" categories. Because GLM-16 is sensitive to total (CG + IC) lightning, it should be able to control for that uncertainty. Table 3 shows the equivalent statistics as Table 2, but for GLM lightning. The period of GLM analysis was July–December 2018 and 2019. This lengthened period enabled comparable sample numbers to the WWLLN analysis.

Table 3. Summary statistics for winds and precipitation both associated and not associated with GLM lightning.

Variable	Mean	99% Conf. Int. ($\pm$)	Samples	RCG > 10 (Mean)	Rain < 6 mm h^{-1} (Mean)
CYGNSS FDS with GLM (m s^{-1})	8.35	0.01	937,027	8.50	8.25
CYGNSS FDS without GLM (m s^{-1})	8.72	0.00	19,673,569	8.92	8.69
CYGNSS YSLF with GLM (m s^{-1})	8.84	0.02	929,953	8.58	8.69
CYGNSS YSLF without GLM (m s^{-1})	9.24	0.00	19,534,703	8.97	9.20
ASCAT with GLM (m s^{-1})	8.72	0.03	79,225	N/A	8.37
ASCAT without GLM (m s^{-1})	8.10	0.01	1,752,437	N/A	8.03
IMERG with GLM (mm h^{-1})	2.80	0.01	937,027	N/A	N/A
IMERG without GLM (mm h^{-1})	0.92	0.00	19,673,569	N/A	N/A
CYGNSS Std. Dev. with GLM (m s^{-1})	1.08	0.01	25,070	N/A	N/A
CYGNSS Std. Dev. without GLM (m s^{-1})	1.58	0.00	303,731	N/A	N/A

GLM appeared to confirm that some misidentification of thunderstorms was occurring within the "without lightning" categories. In general, differences between the "with" and "without" categories were magnified, but the same basic trends were observed: CYGNSS winds were stronger and more variable in precipitation without lightning (by ~0.4–0.5 m s^{-1}), while ASCAT winds were stronger in precipitation with lightning by ~0.6 m s^{-1}). However, the CYGNSS differences were robust when controlling for heavy precipitation (> 6 mm h^{-1}), while ASCAT differences between the categories were reduced (to < 0.4 m s^{-1}). The one exception to this intensified pattern was IMERG precipitation rate. This was still significantly stronger in precipitation with lightning than without, but the overall difference was slightly reduced compared to WWLLN (by ~0.15–0.20 mm h^{-1}).

Finally, ISS LIS statistics are shown in Table 4. The ISS LIS analysis covered 32 months during 2017–2019, and also provided global coverage, but, despite this, the number of samples was greatly reduced compared to the WWLLN or GLM-16 analyses. Regardless, the overall trends were very similar—CYGNSS FDS winds were stronger when ISS LIS-detected lightning was not present than when it was, while the opposite was observed for ASCAT. CYGNSS YSLF winds also followed this trend, but the difference was smaller, and it decreased when controlling for RCG but not for precipitation. However, CYGNSS winds appeared to be more slightly more variable when lightning was present. Note, however, that the available number of samples for several of the "with lightning" categories was much smaller than for the other two lightning instruments, so caution is warranted in interpreting these statistics.

Table 4. Summary statistics for winds and precipitation both associated and not associated with ISS LIS lightning.

Variable	Mean	99% Conf. Int. (±)	Samples	RCG > 10 (Mean)	Rain < 6 mm h⁻¹ (Mean)
CYGNSS FDS with LIS (m s^{-1})	8.36	0.02	154,517	8.57	8.22
CYGNSS FDS without LIS (m s^{-1})	8.70	0.01	2,587,259	8.90	8.67
CYGNSS YSLF with LIS (m s^{-1})	9.01	0.04	152,631	8.80	8.82
CYGNSS YSLF without LIS (m s^{-1})	9.19	0.01	2,568,336	8.86	9.15
ASCAT with LIS (m s^{-1})	8.90	0.09	11,664	N/A	8.65
ASCAT without LIS (m s^{-1})	8.30	0.02	235,198	N/A	8.25
IMERG with LIS (mm h^{-1})	3.43	0.04	154,517	N/A	N/A
IMERG without LIS (mm h^{-1})	0.96	0.00	2,587,259	N/A	N/A
CYGNSS Std. Dev. with LIS (m s^{-1})	1.02	0.03	4,394	N/A	N/A
CYGNSS Std. Dev. without LIS (m s^{-1})	0.87	0.01	147,278	N/A	N/A

The overall statistical results, common to all lightning instruments, were the following. CYGNSS FDS winds were stronger when lightning was not present (by ~0.2–0.4 m s^{-1}), and this was largely robust against controls for RCG or precipitation rate. ASCAT winds tended to be stronger when lightning was present (by ~0.4–0.6 m s^{-1}), but this difference was reduced significantly when controlling for heavier precipitation. Meanwhile, the percentage change in precipitation when lightning was present was much larger than the change in winds (~200–270% increase in rain rate when lightning occurred).

3.2.2. Trends versus Precipitation Rates

The general results presented in Tables 2–4 were interrogated more deeply by examining how differences in categories behaved as functions of rain rate and lightning flash rate. For this analysis, rather than examining absolute wind speeds, the mean wind speed along each track was first subtracted from the wind speeds in each raining feature along the same track. This was done to determine the extent to which the results in Tables 2–4 were affected by the influence of larger-scale environmental winds.

Figure 5 shows the results for CYGNSS FDS and ASCAT mean wind differences as functions of IMERG precipitation rate. Whether the precipitation is associated with lightning or not, CYGNSS winds start roughly 1 m s^{-1} above background, indicating that precipitation is associated with stronger winds. For very light rain rates (<1 mm h^{-1}) CYGNSS does see slightly higher wind speed differences when WWLLN lightning is present (Figure 5a), but this trend quickly reverses as rain rate increases and winds without WWLLN lightning obtain and maintain a ~0.5 m s^{-1} advantage for rain rates above ~5 mm h^{-1}. Wind speed differences increase slowly with precipitation rate after ~5 mm h^{-1} as well. For ASCAT winds (Figure 5b), the differences from CYGNSS are significant. Like CYGNSS, ASCAT wind speeds in precipitation start about 1 m s^{-1} above background, and winds with WWLLN lightning are slightly stronger. However, unlike CYGNSS that ~0.5 m s^{-1} advantage for winds with lightning is maintained as rain rate increases until at least 15 mm h^{-1}, when the 99% confidence intervals between the curves start to overlap. Wind speed differences above background also increase more sharply as precipitation rate increases (both with and without WWLLN lightning), compared to CYGNSS.

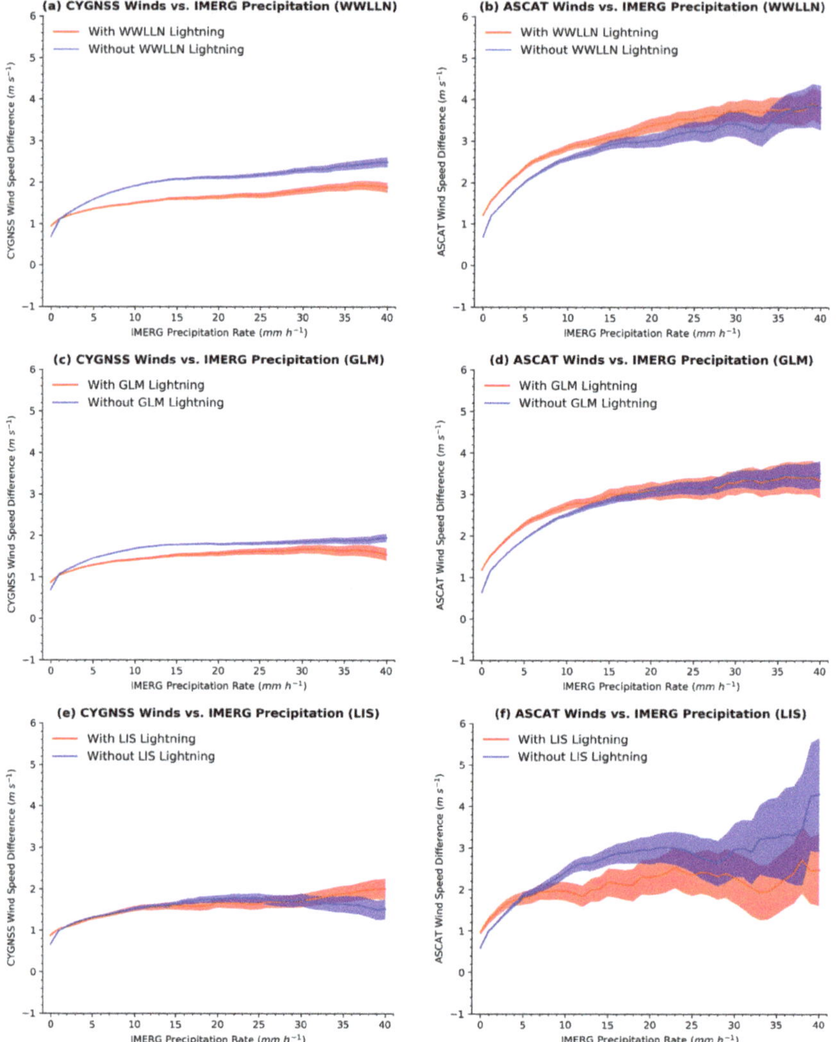

Figure 5. (**a**) CYGNSS FDS wind differences as functions of IMERG precipitation rate, for wind observations associated with WWLLN lightning and those not. (**b**) ASCAT wind differences as functions of IMERG precipitation rate, for wind observations associated with WWLLN lightning and those not. (**c**) CYGNSS FDS wind differences as functions of IMERG precipitation rate, for wind observations associated with GLM lightning and those not. (**d**) ASCAT wind differences as functions of IMERG precipitation rate, for wind observations associated with GLM lightning and those not. (**e**) CYGNSS FDS wind differences as functions of IMERG precipitation rate, for wind observations associated with ISS LIS lightning and those not. (**f**) ASCAT wind differences as functions of IMERG precipitation rate, for wind observations associated with ISS LIS lightning and those not. In all figures, the shaded areas indicate the 99% confidence interval in each 1 mm h^{-1} bin.

The GLM analysis shows a similar pattern. For CYGNSS (Figure 5c), the gap between winds without and without GLM lightning narrows relative to WWLLN, though this is not true for ASCAT (Figure 5d). This narrowing for CYGNSS could reflect more accurate separation between the "with lightning" and "without lightning" categories, since GLM detects total lightning rather than primarily CGs like WWLLN. For ISS LIS and CYGNSS (Figure 5e), the gap between the two wind categories narrows to the point of overlap above ~1 mm h^{-1}. For ASCAT and ISS LIS (Figure 5f), a similar pattern to GLM and WWLLN is seen, but the two curves are closer and start to overlap at 5 mm h^{-1}. It is possible that the reduced samples for ISS LIS may be contributing to the greater volatility in the curves at higher rain rates.

Overall, Figure 5 confirms and expands upon the basic statistics shown in Tables 2–4. For CYGNSS, differences between winds associated with lightning and those not are reduced for total lightning sensors like GLM and ISS LIS. Moreover, the difference between the two wind categories (maximized around +0.5 m s^{-1} for WWLLN, but much smaller for GLM and especially ISS LIS), is significantly smaller than the overall increase in wind speeds above background when precipitation is present (+1–2 m s^{-1}, depending on precipitation rate). Winds associated with lightning are slightly stronger when precipitation rate is low (<1 m s^{-1}), but that changes as precipitation rate increases and winds associated with WWLLN and GLM lightning are slightly weaker than winds not associated with lightning. Similar to Figure 3, ASCAT winds increase more rapidly than CYGNSS as rain rate increases (Figure 5b,d,f). However, for all lightning sensors below at least 5 mm h^{-1} rain rate, ASCAT winds are stronger when lightning is present (by up to 0.5 m s^{-1}). Recall, however, the fewer samples available for ASCAT in the analysis.

3.2.3. Trends versus Flash Rates

Figure 6 shows behavior of CYGNSS and ASCAT wind differences above background as functions of lightning flash rate. For CYGNSS, both FDS and YSLF wind speed differences decrease when WWLLN flash rates increase slightly above 0 min^{-1} (Figure 6a). This confirms the results in Table 2, where wind speeds associated with lightning were less than when lightning was present. However, as flash rate increases above 1 min^{-1}, a quasi-linear increase in wind speed difference is observed, particularly for YSLF winds. However, uncertainty also increases since the number of high-flash-rate samples is low relative to 0–2 min^{-1} flash rates. ASCAT observes a sharper increase in wind speed difference as flash rate increases (Figure 6b); however, when controlling for higher rain rates (>6 mm h^{-1}) the trend is almost flat until about 20 min^{-1}, when the lack of samples begins to lead to dubious results.

Interestingly, CYGNSS wind differences show very little variability relative to GLM flash rates above ~1 min^{-1} (Figure 6a). The initial decrease between 0 and 1 min^{-1} is observed, similar to WWLLN (Figure 6a), but above that range the relationship is essentially flat for both FDS and YSLF winds. Analogous behavior to WWLLN is observed for ASCAT wind differences versus GLM flash rates (Figure 6d). Once again a positive relationship is observed when not controlling for precipitation, but when controlling for heavier precipitation the ASCAT relationship is much flatter, similar to CYGNSS. Note that GLM provided more samples at higher flash rates (>20 min^{-1}) since it is a total flash rate instrument, unlike WWLLN.

ISS LIS shows a weak positive relationship between FDS and YSLF wind differences and flash rate up to about 30 min^{-1}, and the initial decrease in wind difference between 0 and 1 min^{-1} is not as strong compared to WWLLN and GLM (Figure 6e). However, uncertainty in this possible relationship is large enough that it cannot be distinguished from the complete lack of a relationship, based on the spreads of the 99% confidence intervals. ASCAT winds also show a positive relationship with flash rate, even when controlling for precipitation (Figure 6f). However, since ISS LIS provided fewer overall samples it is possible that these relationships are not robust. The 99% confidence interval spread for the ASCAT data is very large.

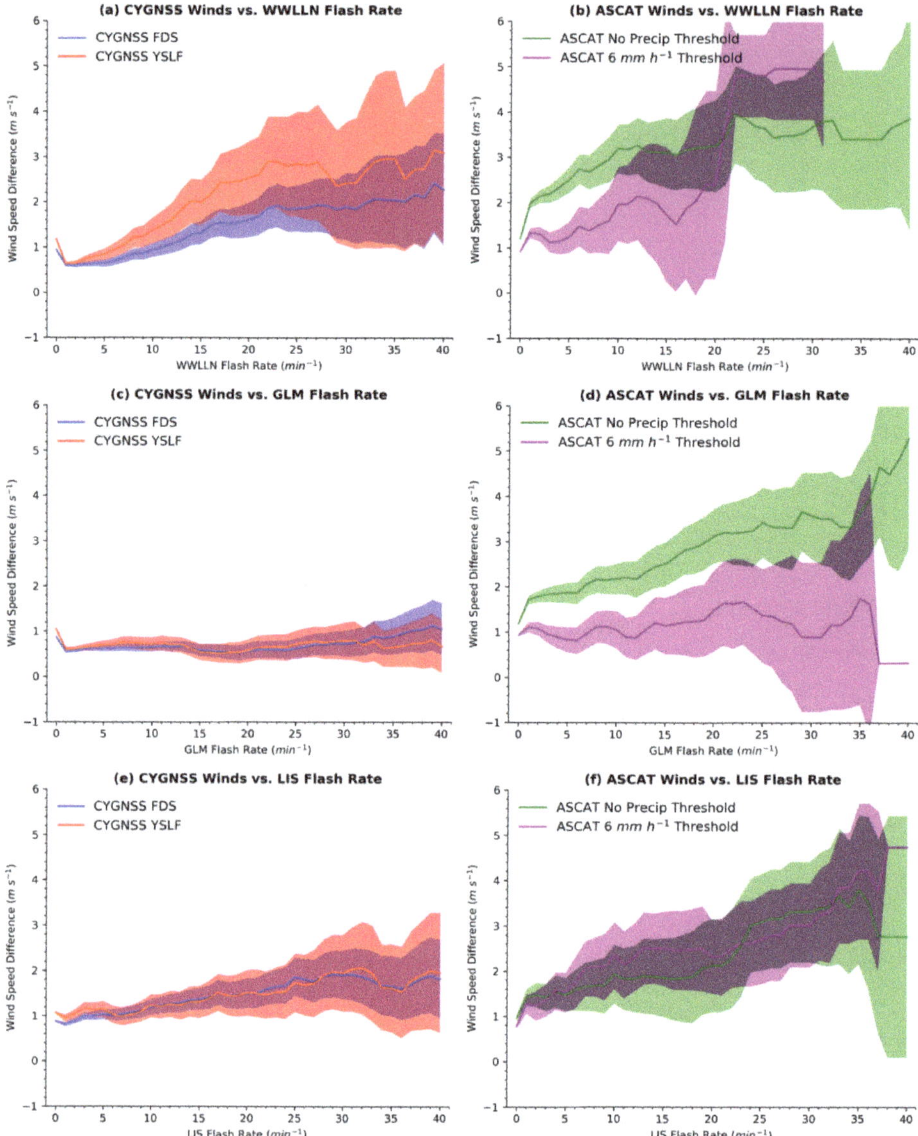

Figure 6. (**a**) CYGNSS FDS and YSLF wind differences as functions of WWLLN flash rate. (**b**) ASCAT wind differences as functions of WWLLN flash rate, for ASCAT observations associated with all precipitation rates and for ASCAT observations with precipitation rate < 6 mm h^{-1}. (**c**) CYGNSS FDS and YSLF wind differences as functions of GLM flash rate. (**d**) ASCAT wind differences as functions of GLM flash rate, for ASCAT observations associated with all precipitation rates and for ASCAT observations with precipitation rate < 6 mm h^{-1}. (**e**) CYGNSS FDS and YSLF wind differences as functions of GLM flash rate. (**f**) ASCAT wind differences as functions of GLM flash rate, for ASCAT observations associated with all precipitation rates and for ASCAT observations with precipitation rate < 6 mm h^{-1}. In all figures, the shaded areas indicate the 99% confidence interval in each 1 min^{-1} bin.

4. Discussion

Focusing first on the existing broad agreement between CYGNSS and matched ASCAT-A and -B wind data near precipitation systems, both types of sensors—despite their very different approaches to wind retrievals—agree that when light precipitation is present wind speeds are approximately 1 m s^{-1} greater than background (Figure 5). This is largely consistent with the independent studies of [3,4]. Additionally, regardless of the analysis approaches used in this study, the mean wind speeds from CYGNSS and matched ASCAT agree to within approximately 0.5 m s^{-1} (Tables 1–4). Finally, and most importantly for this paper, both CYGNSS and ASCAT agree that—if there is a difference in wind speeds near lightning-producing and non-lightning-producing precipitation systems, in the aggregate this effect is small (~0.2–0.6 m s^{-1}), and the overall percentage change in wind speed is much smaller than the observed percentage change in precipitation rate when lightning is occurring (i.e., ~50% change in mean wind speed vs. >200% change in precipitation rate; Tables 2–4). These findings demonstrate the utility of combining distinct satellite-based precipitation and wind datasets to infer wind characteristics near precipitation systems.

Despite the above agreement, there are important differences observed that speak to remaining uncertainties in wind retrievals in and near rainfall. For example, regardless of whether lightning is occurring, ASCAT shows a stronger positive relationship between wind speed and IMERG precipitation rate compared to CYGNSS (Figure 5). Moreover, CYGNSS and ASCAT tend to fundamentally disagree on the sign of the change in wind speed between precipitation systems with and without lightning—ASCAT tends to observe slightly stronger winds when lightning is present, while CYGNSS tends to observe the opposite.

There is substantial evidence (e.g., Tables 1–4, Figures 3–6) that this disagreement stems from fundamental differences in winds measurements between the two sensors in rain. Because CYGNSS is an L-band sensor, it is thought to be potentially less sensitive to rain impacts compared to the C-band ASCAT [20]. The relatively larger sensitivity of the ASCAT results to precipitation rate compared to the CYGNSS results (e.g., Tables 2–4, Figures 5 and 6) supports this inference. Indeed, as [16–18] have shown, ASCAT wind retrievals in rain rates above ~6 mm h^{-1} are likely less accurate. Thus, it makes sense that controlling for precipitation in this study's dataset showed much greater response in the ASCAT dataset. Note also that disagreements between CYGNSS and ASCAT results were reduced when controlling for rain.

However, CYGNSS likely does experience some impact on its wind speed measurements in rainfall, particularly when wind speed is low [38–41]. Moreover, uncertainties in the representativeness of both sensors' geophysical model functions (GMFs) near convection need to be considered. Apart from effects of signal attenuation and capillary waves from rain splash, winds near convection tend to be highly variable and change much faster than the ocean surface waves influenced by background winds can react [42]. Such phenomena will affect the wind retrievals from both types of sensors, and likely causes some of the observed differences since the CYGNSS and ASCAT observations are not matched perfectly in time (approximately a ±45-min window, based on the matching criteria described in Section 2.2).

Tropical cyclones were included in the analysis but were not broken out in the statistics. As argued in Section 2.1, they likely did not significantly impact the results due to their rarity relative to conventional tropical precipitation systems. However, it is possible that tropical cyclones were responsible for a small part of the observed ASCAT/CYGNSS discrepancies. For example, CYGNSS FDS winds are not generally valid in tropical cyclones as their retrieval algorithm generally does not provide wind speeds in excess of 20 m s^{-1}, while the YSLF retrieval algorithm allows for higher wind speeds [20]. Thus, comparing CYGNSS FDS to ASCAT in tropical cyclones would result in a low bias in CYGNSS winds. However, CYGNSS YSLF results were similar to the FDS results, supporting the inference that tropical cyclones were not significantly impacting the overall statistics.

An additional possible influence on the results is the effect of the CYGNSS v2.1 dataset, which was afflicted by variability in calibration, both in time (e.g., Figure 4) as well as between different specular point tracks. This is a known issue that is substantively addressed by track-wise debiasing in the new climate data record (CDR) v1.0 dataset [43]. Select months from this dataset were processed the same way as v2.1 (c.f. Section 2), but the results were not significantly different than for v2.1 (not shown). This is likely because CDR v1.0 does not appear to significantly improve CYGNSS and ASCAT agreement in rainfall [44].

5. Conclusions

In order to examine how robust updraft strength and ice-based microphysical processes aloft in storms may affect convective outflows near the surface, ocean winds were compared between tropical maritime precipitation systems with and without lightning. The analysis focused on CYGNSS specular point tracks, using straightforward spatiotemporal matching criteria to pair CYGNSS-measured wind speeds with IMERG precipitation, ASCAT-A and -B wind speeds, and lightning flash data from WWLLN, GLM-16, and ISS LIS.

Based on the results, thunderstorms over the tropical oceans are clearly associated with significantly heavier rain rates (~200% greater) than non-lightning-producing showers. This is consistent with past work on relationships between lightning and rainfall [45,46]. However, wind speeds near either type of precipitation system do not differ much (~0.5 m s^{-1} or less). Moreover, the sign of the difference depends on the wind instrument used, with CYGNSS suggesting non-thunderstorm winds are (counterintuitively) slightly stronger, while ASCAT suggests the opposite. These observed wind differences are likely related to remaining uncertainties in the comparison of CYGNSS/ASCAT in precipitation (e.g., GMF uncertainties when winds are variable, rain impacts on measurement accuracy, etc.).

If ASCAT observations are considered more accurate in rainfall than CYGNSS, then the hypothesis that thunderstorms are associated with stronger surface outflows (suggesting also stronger cold pools) is supported by the available evidence. However, if this is not the case and CYGNSS winds are more accurate than ASCAT in rainfall, then there is not enough evidence to support the thunderstorm outflow hypothesis (indeed, the opposite argument might be made), which would be unexpected. However, it is possible that the dataset was well-populated with relatively weak, low-flash-rate thunderstorms (this is supported by the greater uncertainty in the Figure 6 curves when flash rates exceeded 1–2 min^{-1}), which indeed may not feature significantly stronger surface winds than non-thunderstorms. In that case, the CYGNSS observations may be consistent with expectations. Another complicating factor to consider is that thunderstorms over the global ocean tend to be larger and more organized than those over land [47,48]. Thus, CYGNSS overpasses may sample stratiform precipitation in lightning-producing systems more often, where winds would be expected to be weaker than near, e.g., the leading-edge gust front. Regardless, both CYGNSS and ASCAT find winds near precipitation (whether lightning-producing or not) to be stronger than background by at least 1 m s^{-1}, consistent with prior studies [3,4].

Future work will focus on a longer period of record, improved CYGNSS wind datasets, and incorporating ASCAT into the analysis in a manner that preserves its swath-based observations. In addition, IMERG data are expected to improve with time. These changes may enable more accurate characterization of wind variability near precipitation systems. Moreover, future improvements to the GLM and ISS LIS datasets are expected to increase their effective detection efficiencies and reduce their false alarm rates [32,33]. This will mitigate remaining uncertainty in whether a precipitation system is a thunderstorm or not.

Funding: This research was funded by the CYGNSS, Ocean Vector Winds Science Team (OVWST), and Earth from ISS programs within the Science Mission Directorate (SMD) of the National Aeronautics and Space Administration (NASA).

Acknowledgments: This research benefitted from fruitful conversations about the various datasets used with the following people: D.M., K.V., S.N., P.G., G.P., C.R., Z.J., S.A., B.R., J.M., X. L., and R.B. The Physical Oceanography Distributed Active Archive Center (PO.DAAC) supplied the CYGNSS and ASCAT data. IMERG data were obtained from Goddard Space Flight Center. University of Washington supplied the WWLLN data. The Global Hydrology Resource Center (GHRC) DAAC supplied ISS LIS and GLM-16 data.

Conflicts of Interest: The author declares no conflict of interest.

References

1. Thorpe, A.J.; Miller, M.J.; Moncrieff, M.W. Two-dimensional convection in non-constant shear: A model of mid-latitude squall lines. *Q. J. R. Meteorol. Soc.* **1982**, *108*, 739–762. [CrossRef]
2. Garg, P.; Nesbitt, S.W.; Lang, T.J.; Priftis, G.; Chronis, T.; Thayer, J.D.; Hence, D.A. Identifying and Characterizing Tropical Oceanic Mesoscale Cold Pools using Spaceborne Scatterometer Winds. *J. Geophys. Res. Atmos.* **2020**, *125*, e2019JD031812. [CrossRef]
3. De Szoeke, S.P.; Skyllingstad, E.D.; Zuidema, P.; Chandra, A. Cold Pools and Their Influence on the Tropical Marine Boundary Layer. *J. Atmos. Sci.* **2017**, *74*, 1149–1168. [CrossRef]
4. Feng, Z.; Hagos, S.; Rowe, A.K.; Burleyson, C.; Martini, M.N.; Szoeke, S.P. Mechanisms of convective cloud organization by cold pools over tropical warm ocean during the AMIE/DYNAMO field campaign. *J. Adv. Model. Earth Syst.* **2015**, *7*, 357–381. [CrossRef]
5. Atkins, N.T.; Wakimoto, R.M. Wet Microburst Activity over the Southeastern United States: Implications for Forecasting. *Weather Forecast.* **1991**, *6*, 470–482. [CrossRef]
6. Saunders, C.P.R.; Keith, W.D.; Mitzeva, R.P. The effect of liquid water on thunderstorm charging. *J. Geophys. Res. Space Phys.* **1991**, *96*, 11007–11017. [CrossRef]
7. Deierling, W.; Petersen, W.A. Total lightning activity as an indicator of updraft characteristics. *J. Geophys. Res. Space Phys.* **2008**, *113*. [CrossRef]
8. Deierling, W.; Petersen, W.A.; Latham, J.; Ellis, S.; Christian, H.J. The relationship between lightning activity and ice fluxes in thunderstorms. *J. Geophys. Res. Space Phys.* **2008**, *113*. [CrossRef]
9. Basarab, B.M.; Rutledge, S.A.; Fuchs, B.R. An improved lightning flash rate parameterization developed from Colorado DC3 thunderstorm data for use in cloud-resolving chemical transport models. *J. Geophys. Res. Atmos.* **2015**, *120*, 9481–9499. [CrossRef]
10. Carey, L.D.; Schultz, E.V.; Schultz, C.J.; Deierling, W.; Petersen, W.A.; Bain, A.L.; Pickering, K.E. An Evaluation of Relationships between Radar-Inferred Kinematic and Microphysical Parameters and Lightning Flash Rates in Alabama Storms. *Atmosphere* **2019**, *10*, 796. [CrossRef]
11. Petersen, W.A.; Christian, H.J.; Rutledge, S.A. TRMM observations of the global relationship between ice water content and lightning. *Geophys. Res. Lett.* **2005**, *32*, 14819. [CrossRef]
12. Price, C.; Rind, D. A simple lightning parameterization for calculating global lightning distributions. *J. Geophys. Res. Space Phys.* **1992**, *97*, 9919–9933. [CrossRef]
13. Lang, T.; Rutledge, S.A. A Framework for the Statistical Analysis of Large Radar and Lightning Datasets: Results from STEPS 2000. *Mon. Weather Rev.* **2011**, *139*, 2536–2551. [CrossRef]
14. Williams, E.; Boldi, B.; Matlin, A.; Weber, M.; Hodanish, S.; Sharp, D.; Goodman, S.; Raghavan, R.; Buechler, D. The behavior of total lightning activity in severe Florida thunderstorms. *Atmos. Res.* **1999**, *51*, 245–265. [CrossRef]
15. Schultz, C.J.; Carey, L.D.; Schultz, E.V.; Blakeslee, R.J. Insight into the Kinematic and Microphysical Processes that Control Lightning Jumps. *Weather Forecast.* **2015**, *30*, 1591–1621. [CrossRef]
16. Lin, W.; Portabella, M.; Stoffelen, A.; Verhoef, A.; Turiel, A. ASCAT Wind Quality Control Near Rain. *IEEE Trans. Geosci. Remote Sens.* **2015**, *53*, 4165–4177. [CrossRef]
17. Portabella, M.; Stoffelen, A.; Lin, W.; Turiel, A.; Verhoef, A.; Verspeek, J.; Ballabrera-Poy, J. Rain Effects on ASCAT-Retrieved Winds: Toward an Improved Quality Control. *IEEE Trans. Geosci. Remote Sens.* **2012**, *50*, 2495–2506. [CrossRef]
18. Kilpatrick, T.; Xie, S.-P. ASCAT observations of downdrafts from mesoscale convective systems. *Geophys. Res. Lett.* **2015**, *42*, 1951–1958. [CrossRef]
19. Ruf, C.; Chew, C.; Lang, T.J.; Morris, M.G.; Nave, K.; Ridley, A.; Balasubramaniam, R. A New Paradigm in Earth Environmental Monitoring with the CYGNSS Small Satellite Constellation. *Sci. Rep.* **2018**, *8*, 1–13. [CrossRef]
20. Ruf, C.S.; Asharaf, S.; Balasubramaniam, R.; Gleason, S.; Lang, T.; McKague, D.; Twigg, D.; Waliser, D. In-Orbit Performance of the Constellation of CYGNSS Hurricane Satellites. *Bull. Am. Meteorol. Soc.* **2019**, *100*, 2009–2023. [CrossRef]

21. CYGNSS. *CYGNSS Level 2 Science Data Record Version 2.1. Ver. 2.1*; PO.DAAC: Pasadena, CA, USA, 2018; Available online: https://doi.org/10.5067/CYGNS-L2X21 (accessed on 1 January 2020).

22. Hoover, K.E.; Mecikalski, J.R.; Lang, T.J.; Li, X.; Castillo, T.J.; Chronis, T. Use of an End-to-End-Simulator to Analyze CYGNSS. *J. Atmos. Ocean. Technol.* **2018**, *35*, 35–55. [CrossRef] [PubMed]

23. Ester, M.; Kriegel, H.P.; Sander, J.; Xu, X. A Density-Based Algorithm for Discovering Clusters in Large Spatial Databases with Noise. In *KDD-96: Proceedings: Proceedings of the 2nd International Conference on Knowledge Discovery and Data Mining, Portland, OR, USA, 2–4 August 1996*; Simoudis, E., Han, J., Fayyad, U.M., Eds.; AAAI Press: Menlo Park, CA, USA, 1996; pp. 226–231.

24. PyGNSS. Python Interface to Cyclone Global Navigation Satellite System (CYGNSS) Wind Dataset. Available online: https://github.com/nasa/PyGNSS (accessed on 1 January 2020).

25. EUMETSAT/OSI SAF. *MetOp-A ASCAT Level 2 Ocean Surface Wind Vectors Optimized for Coastal Ocean: Ver. Operational/Near-Real-Time*; PO.DAAC: Pasadena, CA, USA, 2010. Available online: https://podaac.jpl.nasa.gov/dataset/ASCATA-L2-Coastal (accessed on 1 January 2020).

26. EUMETSAT/OSI SAF. *MetOp-B ASCAT Level 2 Ocean Surface Wind Vectors Optimized for Coastal Ocean: Ver. Operational/Near-Real-Time*; PO.DAAC: Pasadena, CA, USA, 2013. Available online: https://podaac.jpl.nasa.gov/dataset/ASCATA-L2-Coastal (accessed on 1 January 2020).

27. Huffman, G.J.; Bolvin, D.T.; Nelkin, E.J. *Integrated Multi-SatellitE Retrievals for GPM (IMERG) Technical Documentation*; NASA/GSFC Code 612 Tech. Doc.; IMERG Tech Document: Greenbelt, MD, USA, 2015. Available online: http://pmm.nasa.gov/sites/default/files/document_files/IMERG_doc.pdf (accessed on 26 March 2020).

28. Hutchins, M.L.; Holzworth, R.H.; Brundell, J.B.; Rodger, C.J. Relative detection efficiency of the World Wide Lightning Location Network. *Radio Sci.* **2012**, *47*, 6005. [CrossRef]

29. Bürgesser, R.E. Assessment of the World Wide Lightning Location Network (WWLLN) detection efficiency by comparison to the Lightning Imaging Sensor (LIS). *Q. J. R. Meteorol. Soc.* **2017**, *143*, 2809–2817. [CrossRef]

30. Lang, T.J. World-Wide Lightning Location Network. 2020. Available online: https://doi.org/10.5067/Suborbital/CAMP2EX2018/DATA001 (accessed on 1 January 2020).

31. Goodman, S.J.; Blakeslee, R.J.; Koshak, W.J.; Mach, D.; Bailey, J.; Buechler, D.; Carey, L.; Schultz, C.; Bateman, M.; McCaul, E.; et al. The GOES-R Geostationary Lightning Mapper (GLM). *Atmos. Res.* **2013**, *125*, 34–49. [CrossRef]

32. Rudlosky, S.; Goodman, S.; Virts, K.S.; Bruning, E.C. Initial Geostationary Lightning Mapper Observations. *Geophys. Res. Lett.* **2019**, *46*, 1097–1104. [CrossRef]

33. Blakeslee, R.J.; Lang, T.; Koshak, W.J.; Buechler, D.; Gatlin, P.; Mach, D.M.; Stano, G.T.; Virts, K.S.; Walker, T.D.; Cecil, D.J.; et al. Three Years of the Lightning Imaging Sensor Onboard the International Space Station: Expanded Global Coverage and Enhanced Applications. *J. Geophys. Res. Atmos.* **2020**, *125*. [CrossRef]

34. Kummerow, C.; Barnes, W.; Kozu, T.; Shiue, J.; Simpson, J. The Tropical Rainfall Measuring Mission (TRMM) Sensor Package. *J. Atmos. Ocean. Technol.* **1998**, *15*, 809–817. [CrossRef]

35. Blakeslee, R.J. *Non-Quality Controlled Lightning Imaging Sensor (LIS) on International Space Station (ISS) Science Data*; NASA Global Hydrology Resource Center DAAC: Huntsville, AL, USA, 2019. [CrossRef]

36. Mach, D.M.; Christian, H.J.; Blakeslee, R.J.; Boccippio, D.J.; Goodman, S.J.; Boeck, W.L. Performance assessment of the Optical Transient Detector and Lightning Imaging Sensor. *J. Geophys. Res. Space Phys.* **2007**, *112*. [CrossRef]

37. Ruf, C.S.; Gleason, S.; McKague, D.S. Assessment of CYGNSS Wind Speed Retrieval Uncertainty. *IEEE J. Sel. Top. Appl. Earth Obs. Remote Sens.* **2018**, *12*, 87–97. [CrossRef]

38. Balasubramaniam, R.; Ruf, C.S. Improved Calibration of Cygnss Measurements for Downbursts in the Intertropical Convergence Zone. In Proceedings of the IGARSS 2018—2018 IEEE International Geoscience and Remote Sensing Symposium, Valencia, Spain, 22–27 July 2018; pp. 3987–3990.

39. Balasubramaniam, R.; Ruf, C.S. The Impact of Rain on L1 GNSS-R Radar Scattering Cross-Section. In Proceedings of the IGARSS 2019—2019 IEEE International Geoscience and Remote Sensing Symposium, Yokohama, Japan, 28 July–2 August 2019; pp. 7900–7903. [CrossRef]

40. Asgarimehr, M.; Zavorotny, V.U.; Wickert, J.; Reich, S. Can GNSS Reflectometry Detect Precipitation Over Oceans? *Geophys. Res. Lett.* **2018**, *45*, 12585–12592. [CrossRef]

41. Asgarimehr, M.; Wickert, J.; Reich, S. Evaluating Impact of Rain Attenuation on Space-borne GNSS Reflectometry Wind Speeds. *Remote Sens.* **2019**, *11*, 1048. [CrossRef]

42. Priftis, G.; Lang, T.; Chronis, T. Combining ASCAT and NEXRAD Retrieval Analysis to Explore Wind Features of Mesoscale Oceanic Systems. *J. Geophys. Res. Atmos.* **2018**, *123*, 10341–10360. [CrossRef]
43. CYGNSS. *CYGNSS Level 2 Climate Data Record Version 1.0. Ver. 1.0*; PO.DAAC: Pasadena, CA, USA, 2020. Available online: https://doi.org/10.5067/CYGNS-L2C10 (accessed on 22 May 2020).
44. Lang, T.J. *Validation of Satellite-Based Wind Observations during PISTON*; PISTON Science Team Workshop: Fort Collins, CO, USA, 2019. Available online: https://ntrs.nasa.gov/api/citations/20190032972/downloads/20190032972.pdf (accessed on 1 November 2020).
45. Petersen, W.A.; Rutledge, S.A. On the relationship between cloud-to-ground lightning and convective rainfall. *J. Geophys. Res. Space Phys.* **1998**, *103*, 14025–14040. [CrossRef]
46. Tapia, A.; Smith, J.A.; Dixon, M. Estimation of Convective Rainfall from Lightning Observations. *J. Appl. Meteorol.* **1998**, *37*, 1497–1509. [CrossRef]
47. Bang, S.D.; Zipser, E.J. Differences in size spectra of electrified storms over land and ocean. *Geophys. Res. Lett.* **2015**, *42*, 6844–6851. [CrossRef]
48. Bang, S.D.; Zipser, E.J. Seeking reasons for the differences in size spectra of electrified storms over land and ocean. *J. Geophys. Res. Atmos.* **2016**, *121*, 9048–9068. [CrossRef]

Publisher's Note: MDPI stays neutral with regard to jurisdictional claims in published maps and institutional affiliations.

 *remote sensing*

Article

Neural Network Based Quality Control of CYGNSS Wind Retrieval

Rajeswari Balasubramaniam * and Christopher Ruf

Climate and Space Sciences and Engineering, University of Michigan, Ann Arbor, MI 48109, USA; cruf@umich.edu
* Correspondence: rajibala@umich.edu

Received: 26 July 2020; Accepted: 1 September 2020; Published: 3 September 2020

Abstract: Global Navigation Satellite System – Reflectometry (GNSS-R) is a relatively new field in remote sensing that uses reflected GPS signals from the Earth's surface to study the state of the surface geophysical parameters under observation. The CYGNSS is a first of its kind GNSS-R constellation mission launched in December 2016. It aims at providing high quality global scale GNSS-R measurements that can reliably be used for ocean science applications such as the study of ocean wind speed dynamics, tropical cyclone genesis, coupled ocean wave modelling, and assimilation into Numerical Weather Prediction models. To achieve this goal, strong quality control filters are needed to detect and remove outlier measurements. Currently, quality control of CYGNSS data products are based on fixed thresholds on various engineering, instrument, and measurement conditions. In this work we develop a Neural Network based quality control filter for automated outlier detection of CYGNSS retrieved winds. The primary merit of the proposed ML filter is its ability to better account for interactions between the individual engineering, instrument and measurement conditions than can separate thresholded flags for each one. Use of Machine Learning capabilities to capture inherent patterns in the data can create an efficient and effective mechanism to detect and remove outlier measurements. The resulting filter has a probability of outlier detection (PD) >75% and False Alarm Rate (FAR) < 20% for a wind speed range of 5 to 18 m/s. At least 75% of the outliers with wind speed errors of at least 5 m/s are removed while ~100% of the outliers with wind speed errors of at least 10 m/s are removed. This filter significantly improves data quality. The standard deviation of wind speed retrieval error is reduced from 2.6 m/s without the filter to 1.7 m/s with it over a wind speed range of 0 to 25 m/s. The design space for this filter is also analyzed in this work to characterize trade-offs between PD and FAR. Currently the filter performance is applicable only up to moderate wind speeds, as sufficient data is available only in this range to train the filter, as a way forward, more data over time can help expand the usability of this filter to higher wind speed ranges as well.

Keywords: CYGNSS; quality control; Neural Network; outlier detection; wind retrieval; GNSS-R

---

1. Introduction

Global Navigation Satellite System-Reflectometry (GNSS-R) is an emerging trend in space borne ocean remote sensing due to its ability to greatly improve measurement frequency using reflected GNSS signals and also for the simplicity in design and requirements of its receivers. The UK-DMC mission was the first to demonstrate the sensitivity of GNSS signals to ocean winds [1,2], this was followed by the TechDemoSat (TDS) mission [3] and now the Cyclone Global Navigation Satellite System (CYGNSS) mission is actively making continuous measurements across the globe with its constellation of small satellites [4].

The CYGNSS mission consists of 8 small spacecraft deployed in a LEO orbit (~510 km altitude and 35 deg inclination). Each spacecraft has a Delay Doppler Mapping Instrument (DDMI) that can map the signal power of GPS reflections from the Earth's surface onto a range of time delay and Doppler frequency shifts relative to the position and velocity of the GPS transmitters and the CYGNSS spacecraft [5]. These passive measurements, referred to as Delay Doppler Maps (DDMs), enable the individual spacecraft to act as a bistatic scatterometer which estimates the near surface ocean wind speed by measuring the bistatic radar cross-section at the specular reflection point.

The near surface ocean wind speed retrieval by CYGNSS uses empirical Geophysical Model Functions (GMFs) developed for 2 measurement observables—the Normalised Bistatic Radar Cross-Section (NBRCS) and the Leading Edge Slope (LES), derived from the DDMs [5]. Wind speed is then estimated by inverse mapping of these observables to reference winds (NWP models and aircraft measurements) using the GMFs and optimally combining the two estimates using a minimum variance estimate of the wind speed [6].

The error in wind speed retrieval can come from different levels of measurement processing and it is important to identify and eliminate erroneous measurements in order to provide high quality observations for scientific applications. While previous spaceborne GNSS-R missions were primarily focused on technology demonstrations, CYGNSS attempts to provide an operational service to meet its science goals. Hence a high data quality is a top priority.

At the engineering level, the major sources of error can be attributed to errors in estimation of the GPS transmit power, GPS antenna patterns, spacecraft pointing knowledge, and star tracker/science antenna boresight misalignment [7]. At the instrument level, possible errors include science antenna gain error, instrument noise power estimation error, calibration load temperature error, and Digital-Analog quantization error [8]. At the measurement level, errors can occur due to specular point geo-location errors, dependence of observables on other geophysical parameters such as wave age, swell etc., non-unique mapping from wind speed to observable, and error in interpolation in space and time for co-location with reference winds [9].

A number of quality control flags are already in place at different levels of the data processing in order for the science community to pick and choose data with quality requirements specific to their application. At Level 1, some of the major quality flags are star tracker attitude status and one-hertz status flags indicating if Milky Way or the Sun is in the zenith antenna field of view (FOV). The Level 1 quality flags also look out for spacecraft attitude errors, radio frequency interference (RFI), DDM noise floor errors, rapid rate of change of receiver temperature, and telemetry errors [10]. At Level 2, the quality flags look out for retrieval ambiguity, low Range Corrected Gain (RCG) and other data processing errors [11]. Despite such stringent quality flags in place, there remain occasional outlier samples with large discrepancies between the CYGNSS retrieved wind speed and reference validation winds (shown in Figure 1) To improve the data quality of CYGNSS, another layer of quality control is needed which can effectively identify and eliminate these outliers. This is the primary objective of this work.

In this work we develop a Neural Network based quality control filter for CYGNSS Level 2 winds which can effectively identify and remove outliers. We also consider the performance of the CYGNSS retrieved winds before and after this filter is applied to assess its efficacy. The reminder of the paper is structured as follows. Section 2 describes the datasets used. Section 3 explains the details of the proposed quality control filter. In Section 4 the performance of the proposed algorithm is analyzed and the CYGNSS level 2 data performance is assessed before and after the filter. Section 5 discusses the trade-offs in performance and conclusions of this study.

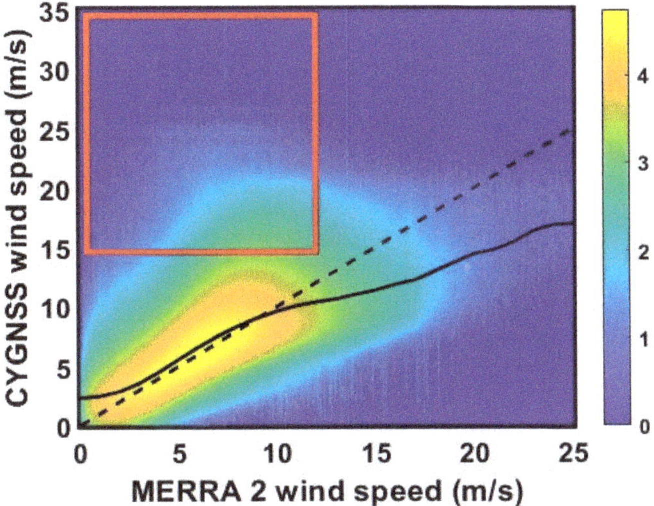

Figure 1. Log density plot of CYGNSS Level 2 retrieved winds matched to MERRA-2 reference winds for the wind speed range 0–25 m/s. The dashed line represents 1:1 agreement between the two winds. The solid line is the average CYGNSS retrieved wind at each MERRA-2 wind speed. A clustering of outliers can be seen near MERRA-2 wind speeds of 0-10 m/s and CYGNSS wind speeds >15 m/s (shown with a red box). One primary objective of the new filter is removal of this cluster.

2. Data Description

The Level 2 CYGNSS winds are minimum variance estimated winds from two observables namely, NBRCS and LES for low-moderate wind speed ranges (0–25 m/s). The CYGNSS retrieved winds are matched to near coincident independent estimates of the ocean surface wind speed referenced to a 10 m height (u_{10}) from the Modern-Era Retrospective Analysis for Research and Applications, version 2 (MERRA-2) [12]. MERRA-2 is a reanalysis product provided by NASA's Global Modelling and Assimilation Office (GMAO). The reference matchup MERRA-2 gridded data product has a spatial resolution of 0.5 deg × 0.625 (lat, lon) and an hourly instantaneous assimilation [13].

Figure 1 shows the density scatter plot of CYGNSS retrieved winds with respect to the MERRA-2 winds. In the figure, the dashed line represents the 1:1 line and the solid line represents the mean retrieved wind speed line, which essentially is the GMF. This plot is generated by dividing the 2-D space into 500 bins or regions. And the matchup winds are assimilated into the nearest bin and finally the log to the base 10 of the number density is taken for better visualization of the density differences. There are several important observations from this plot. Firstly, most of the observations fall along the 1:1 line at lower wind speeds, indicating good retrieval quality. However, a cluster of very high CYGNSS retrieved winds (15–35 m/s) is noticeable at low MERRA-2 winds (5–10 m/s). The improved filtering method developed here targets the removal of these outliers. Secondly, the GMF line and the 1:1 line are very similar up to a MERRA-2 and CYGNSS wind speed of ~10 m/s. Above this range, the GMF line begins to deviate away from the 1:1 line. This inherent bias in the GMF complicates the identification of outliers by the filter algorithm. The purpose of a quality filter is to remove outliers only and not correct for biases in the retrieval. This is another consideration to be accounted for while designing the filter. Finally, the density of samples at high MERRA-2 wind speeds (>20 m/s) is very small relative to the lower wind speed ranges. Therefore trade-off studies must be performed for filter design to balance between efficiency of outlier removal and retaining as many high wind samples as possible. All the above objectives will be addressed in the course of developing the filter.

Over and above the existing quality flags, 13 diagnostic variables are used to distinguish outlier samples from good samples. These diagnostic variables are listed in Table 1. The choice of diagnostic variables is based on previous calibration experience with GNSS-R data and error analyses [7–9]. The diagnostic variables can be categorized into 3 major types—instrument related attributes, measurement geometry related attributes and surface related attributes. In Section 4 these diagnostic variables will be assessed for their individual significance in enabling the filter to distinguish between outliers and good samples.

Table 1. List of diagnostic variables used.

Diagnostic Variable	Description	Type	Symbol/Abbreviation
prn_code	GPS PRN code	Instrument attribute	PRN
antenna	CYGNSS science antenna	Instrument attribute	ant
nst_att_status	Star tracker attitude status	Instrument attribute	nst
sc_roll	Spacecraft roll	Instrument attribute	Roll
zenith_ant_gain	Zenith antenna Gain	Instrument attribute	Z_{gain}
zenith_power	Zenith antenna power	Instrument attribute	Z_{power}
incidence_angle	Incidence angle at SP	Geometry attribute	θ
azimuth	Azimuth angle of SP	Geometry attribute	φ
range_corr_gain	Range Corrected Gain	Geometry attribute	RCG
sp_rx_gain	CYGNSS antenna Gain at SP	Surface attribute	G
ddm_snr	SNR at SP	Surface attribute	SNR
ddm_nbrcs	NBRCS at SP	Surface attribute	DDMA
ddm_les	LES at SP	Surface attribute	LES

SP = Specular Point.

3. Proposed Quality Control Method

Outlier/anomaly detection is an active research field spanning a wide range of applications from manufacturing quality control to astronomical detections. Machine learning techniques are widely used for outlier detection and automation of quality control processes [14]. Despite an emerging trend in the use of machine learning methods for Earth Observation applications, the calibration and validation of satellite measurements are most often handled manually by instrument specialists. Utilizing the capabilities of machine learning tools for calibration and validation activities can help to better understand the behavior of the data.

Outliers can be defined as sample measurements that have a distinct deviation in their properties when compared to the major proportion of the data [15]. Visually, these measurements are regions of low density in the sample space, i.e., have a significantly low number of neighboring points within a threshold distance compared to rest of the sample space. Machine learning tools that are widely used for outlier analysis includes the supervised classification techniques such as Neural Networks (NN), K-Nearest Neighbors (K-NN), Decision trees, Support Vector Machines (SVM) etc. [16]. As noted in Figure 1, there are distinct regions away from the GMF line and the 1:1 line which should be detected and removed. For the CYGNSS quality control filter, supervised training of a Neural Network is used for outlier detection and removal. The details of the quality control filter design are explained in this section.

3.1. Population Definitions

The CYGNSS Level 2 v3.0 data with MERRA-2 wind speed matchups from the year 2018 have a total of ~153 million samples. The sample space is divided into 2 regions. The low wind region consists of all samples with CYGNSS retrieved winds, U_{CYG}, less than or equal to 10 m/s. The high wind region consists of samples with U_{CYG} greater than 10 m/s. In both regions MERRA-2 wind speed, U_M is required to be less than 25 m/s. This division of sample space is due to the behavior of the GMF line relative to the 1:1 line. Below 10 m/s, the GMF line is very similar to the 1:1 line (see Figure 1) and above this wind speed the GMF line begins to underestimate. Therefore, it is appropriate to have two different training datasets, one for each region.

In the low wind region, a good sample satisfies $\left| U_{CYG} - \langle U_{CYG} \rangle |_{U_M} \right| \leq 1$ m/s. and an outlier satisfies $\left| U_{CYG} - \langle U_{CYG} \rangle |_{U_M} \right| > 4$ m/s, where $\langle U_{CYG} \rangle |_{U_M}$ refers to the mean value of the wind speed retrieved by CYGNSS for a given value of the MERRA-2 wind speed. This relationship is described by the solid line in Figure 1. In the high wind region, a good sample is defined by $(U_{CYG} - U_M) \leq 2$ m/s and an outlier as a sample $(U_{CYG} - U_M) > 3$ m/s. The difference in training population definitions at low and high wind speeds is due to the inherent bias in the GMF which can be observed as the deviation of the retrieval mean (solid line) from the 1:1 agreement (dashed line) above 10 m/s in Figure 1. As the wind retrieval is based on the GMF, a bias in the GMF can lead to under/over-estimation of winds despite being a good measurement. To mitigate the effect of GMF-induced bias on the outlier detection capability of the filter, the filter is trained with respect to the GMF. However, as the filter is reliant on the Level 1 diagnostic variables which are independent of GMF, the samples lying near the 1:1 line are also good samples and therefore the modified definition of training data is used at high winds. The training datasets for the 2 different sample spaces are shown in Figure 2. Such conservative training definitions are used to improve the outlier detection capability of the algorithm. Further analysis of the definition of a good or an outlier sample is discussed later in this section. For training, we use ~4 million samples for each wind speed region and ~8 million samples for validation. The performance metrics used to evaluate the outlier detection capability of an algorithm are the Probability of Detection (PD) and False Alarm Rate (FAR). For these metrics, the definition of good and outlier samples are different from the training definitions. The validation definitions are based on the NASA mission requirements on wind retrieval error. Thus, the wind speed differences for a good sample shall be less than 2 m/s from the mean and an outlier is defined as those samples having a difference greater than 5 m/s. Finally, the filter is tested over the total population and its performance is assessed.

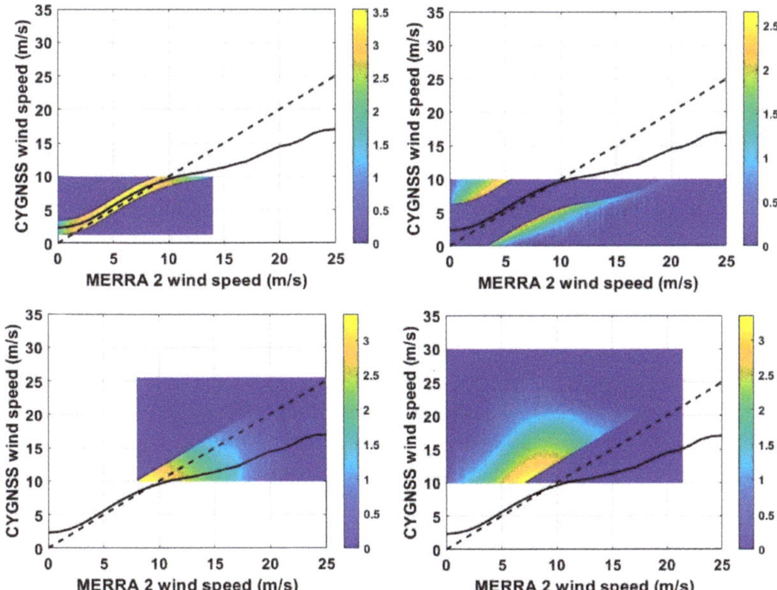

Figure 2. Density plot of CYGNSS Level 2 retrieved winds matched to MERRA-2 reference winds used for training. Top row represents the good (left) and outlier (right) training population for ($0 < U_{CYG} \leq 10$ m/s). The bottom row represents the good (left) and outlier (right) training population for ($U_{CYG} > 10$ m/s).

3.2. Quality Control Process Design

A block diagram representation of the quality control design process is shown in Figure 3. The first stage of the algorithm is feature extraction. The input to this stage is the Fully Developed Seas (FDS) winds over a reference wind speed region of 0–25 m/s. The CYGNSS Level 2 wind retrievals are of two kinds—the FDS and Young Seas Limited Fetch (YSLF) winds. The FDS winds are low to moderate winds (up to 25 m/s) over fully developed waves in the ocean. This forms the major proportion of the total measurements. The YSLF winds are hurricane force winds measured over the tropical cyclones that have varying wave age and fetch conditions. The filter proposed in this work is developed specifically for FDS winds as this dataset encompasses the majority of the measurements and have a well behaved nature relative to its counterpart. The feature extraction stage extracts the different diagnostic variables listed in Table 1 for every sample point. Next is the training stage. One Neural Network (NN) classifier is trained for each of the wind speed space over the individual training datasets described above. The last stage is the validation and testing stage where the skill of the filter is assessed. The performance assessment of this Neural Network filter is discussed in detail in Section 4.

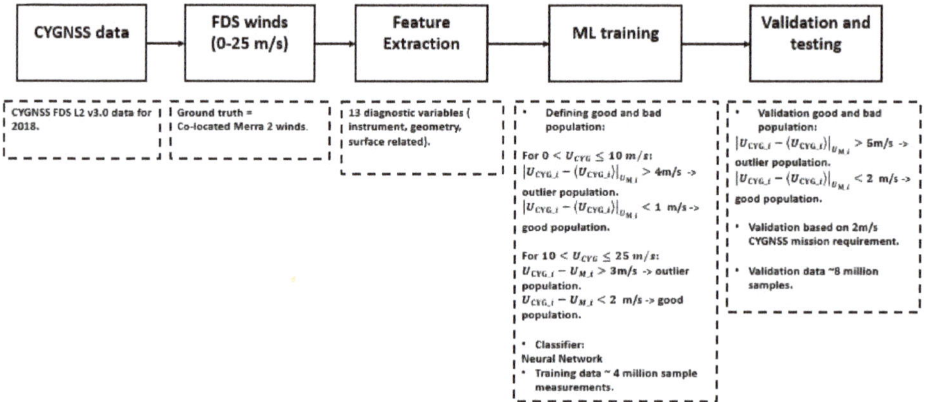

Figure 3. Steps involved in the new quality control algorithm for CYGNSS data. The algorithm has 3 major stages—Feature Extraction, ML training and Validation/testing.

Apart from the Neural Network filter, other standard supervised outlier detection techniques such as Logistic Regression, Decision Trees, Naïve Bayes and K-NN are also considered and their confusion matrices are listed in Table 2 (bottom). In the confusion matrix, the rows represent the true classes, the columns represent predicted classes, and the percentage of samples are mentioned in each of the boxes. Outliers are represented as class '0' and good samples are represented as class '1'. Among the various classifiers experimented with, the K-NN and the NN have a similar performance. In general, NN is preferred over K-NN because of the heavy computational memory requirement of K-NN as compared to the memory requirement for training the NN coefficients. This can be seen in terms of the time requirement for training each of the classifier, shown in Table 2 (top). It can be seen that K-NN requires the most time, followed by the NN.

Table 2. Training time required for individual classifiers (top). Confusion matrices for different classifiers (bottom).

Classifier	Training Time (secs)
Logistic Regression	118.43
Decision tree	78.87
Naïve Bayes	56.39
K-NN	49,608 (13.78 h)
NN	6475 (1.8 h)

	Logistic Regression		Decision Trees		Naïve Bayes		KNN		NN		
0 (True Class)	45.5%	54.5%	27.0%	73.0%	14.1%	85.9%	67.0%	33.0%	58.3%	41.7%	**0**
1 (True Class)	30.2%	69.8%	10.3%	89.7%	8.3%	91.7%	22.3%	77.7%	24.9%	75.1%	**1**
	0	**1**	**0**	**1**	**0**	**1**	**0**	**1**	**0**	**1**	

0 = Outlier
1 = Good sample

Predicted Class

3.3. Neural Network Filter Design

The NN used for this application consists of a single hidden layer with 10 neurons. The input layer consists of 13 neurons, each for one diagnostic variable and the output layer has one neuron that classifies an input sample as an outlier or a good sample. Only one hidden layer is used as it is a sufficient condition to form any bounded/unbounded convex region in the space spanned by the input [17]. The choice of the number of neurons in the hidden layer is decided by experimentation. In general, a feedforward network can have any shape but the commonly used structure is a pyramidal structure with decrease in number of neurons at each layer away from the input. There is practically no upper limit on the number of neurons to be used in this case as the training population is very high (~4 million). So, 3 different neurons counts are experimented here and the performance plot in terms of PD and FAR at different wind speeds is plotted to make a choice on the hidden layer size. The performance plot is shown in Figure 4.

The blue curves represent PD and the red curves represent FAR. It can be noticed that all 3 network sizes have a very similar performance in terms of PD and FAR over the entire wind speed range. For this reason, other performance metrics such a computation time, network complexity and % samples removed as outliers are considered when choosing the optimal structure. In terms of computational time, NN size = 15 is the shortest, followed by NN size =10, and the longest is NN size = 5. This is an expected trend, as simpler networks can take larger time for error convergence. Next, in terms of network complexity, NN size = 5 has the least number of tunable parameters, followed by NN size=10 and the largest being NN size =15. The % of samples removed as outliers by NN size = 5 is ~ 23% of the total data, by NN size =10 is ~ 20% and by NN size = 15 is ~22%. Therefore, after all these considerations, NN size=10 is chosen as the optimal network design for this application.

Thus for purposes of quality control application for CYGNSS, the QC filter has two NNs (NN1 and NN2), each trained for a specific wind speed range (0–10 m/s and 10–25 m/s). The NNs are identical in architecture and contain 2 layers with 10 neurons in the hidden layer, as discussed above. The hidden layer is trained with a sigmoid transfer function and a linear transfer function is used in the output. The optimization algorithm used for this is the widely used Levenberg-Marquardt algorithm.

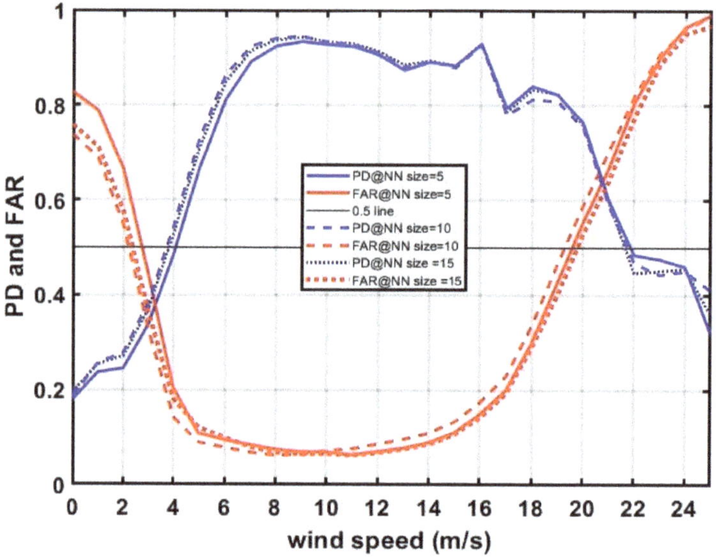

Figure 4. PD and FAR curves for 3 different network sizes (5, 10 and 15).

To evaluate the design space of this filter, the definitions of good and outlier samples are varied and the performance metrics are plotted. Understanding the behavior of the filter for different sample definitions can help users understand how the network handles the outliers and choose an optimum definition based on the application requirements. The family of PD and FAR curves are plotted in Figure 5. The blue curves represent PD and red curves represent FAR. The PD metric is affected by the density of outlier samples and the FAR metric is affected by the density of good samples. Changing the wind speed difference thresholds for good and outlier samples will affect the overall performance of this QC filter. For this study, the wind speed difference from the GMF line for a good sample is varied from 1 m/s to 4 m/s and for an outlier is varied from 3m/s to 7 m/s.

There are many interesting features in Figure 5. Firstly, the FAR curves do not vary much with changes in the definition of the good population but there is a significant jump in PD with changes in the definition of the outlier population. This is due to the relatively small percentage of outliers when compared to the total sample population. Next, the FAR metric has the best performance when the good sample definition is set to 1 m/s and gradually degrades with increase in the difference. However, above a wind speed of ~18 m/s, the trend reverses. This is due to the fact that, at higher wind speeds there is a greater degree of scatter in the data (as seen in Figure 1) resulting in poorer performance in terms of FARs at very stringent definitions of a good sample. Next, as mentioned earlier, the PD metric seems to have a strong jump with change in outlier definition; with the highest PD performance for an outlier definition of >7 m/s for wind speed difference from the GMF line. Again, the trend flips in nature at higher wind speed (>21 m/s), this is again attributed to increased scatter in the data. Finally, it is important to note that the general performance of the filter is not optimal at very low wind speeds (<3 m/s) for any definition of good and outlier sample. Thus, the ideal operating range for this filter is ~5 m/s to 18 m/s; this were most of the samples lie. The choice of the definitions is dependent on the application. For instance, applications that require very high quality control like monitoring long term variations in wind speed data must go for highest PD performance. Applications at higher wind speeds which needs to retain as many higher wind speed samples as possible, must go for lower PD performance. In this work the assessment of wind retrieval performance is used as its definition of good sample a wind speed difference <=2m/s and defines outliers as >5 m/s.

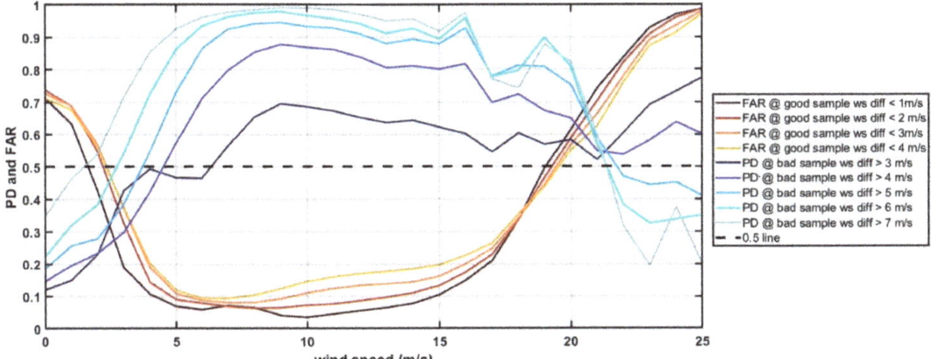

Figure 5. Family of PD and FAR curves for different definitions of good and outlier samples. The dark-light blue curves represent PD and the orange-red curves represent FAR.

4. Results

In this section the performance of the quality controlled CYGNSS wind speed data set is assessed. Two identical Neural Networks, one for each wind speed region discussed in Section 3 are trained. The first NN is applied to CYGNSS winds between 0–8 m/s and the second NN is applied to CYGNSS winds >8 m/s. This slight shift between the training and testing wind speed regimes is to improve the net performance of the filter, as the first NN will be biased towards the lower winds where the highest density of samples are present and the second NN will again be biased towards the lower winds in its range (10–35 m/s). The resulting quality controlled CYGNSS wind speed dataset is shown in Figure 6.

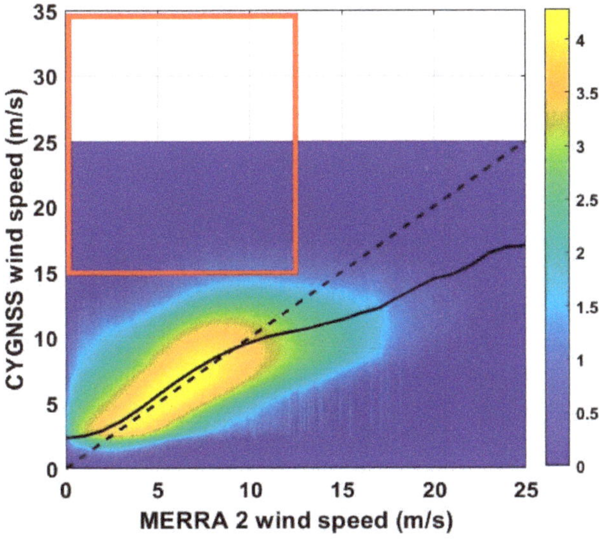

Figure 6. CYGNSS retrieved wind dataset after quality control. The outliers in the red box have been mostly eliminated here relative to Figure 1.

Comparing Figures 1 and 6 demonstrates the effectiveness of the filter. The large cluster of high CYGNSS winds at low MERRA-2 winds has been removed by this filter (compare the red box region between the two figures). Also, the CYGNSS samples are now evenly distributed along the GMF line (solid black line) unlike in the original dataset. Finally, a significant reduction of scatter in the dataset can be observed. The performance of this proposed QC filter is assessed in the following subsections based on the error statistics Mean Difference (MD), Root Mean Squared Difference (RMSD) and variance of data. The test dataset consists of all the sample points (~153 million).

4.1. Algorithm Performance Analysis

To assess the skill of the quality control algorithm, first the validation metrics, PD and FAR, are examined in Figure 7. These metrics are based on the design parameters discussed in the previous section. The optimal range of operation for this filter is ~5 m/s to 17 m/s. In this range the FAR for good samples is consistently <20% and the PD for outliers is >75%. The peak performance is between 6–14 m/s where FAR <10% and PD is >80%. This is also the region of maximum data density as the wind speed distribution has a peak near 7 m/s.

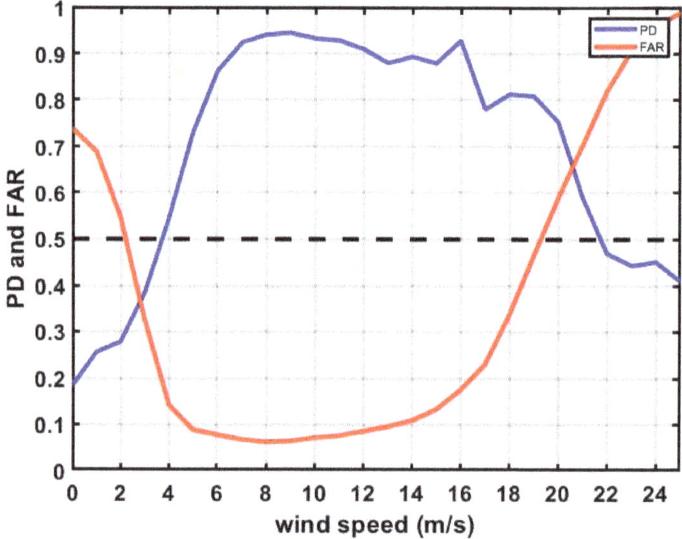

Figure 7. PD and FAR metrics for the CYGNSS test dataset.

Next, the skill of the filter is assessed by looking at the ratio of number of outliers identified by the filter to the total number of outliers for a range of wind speed differences. This is shown in Figure 8. The *x*-axis is the difference between CYGNSS wind speed and the GMF line. As per our validation criteria, we have defined any sample as an outlier if the difference is greater than 5 m/s. The 5 m/s threshold is shown in red. It can be observed that ~70% of the outliers are rightly identified for wind speed difference ~ 5 m/s and the filter eliminates close to ~100% of outliers with wind differences >10 m/s.

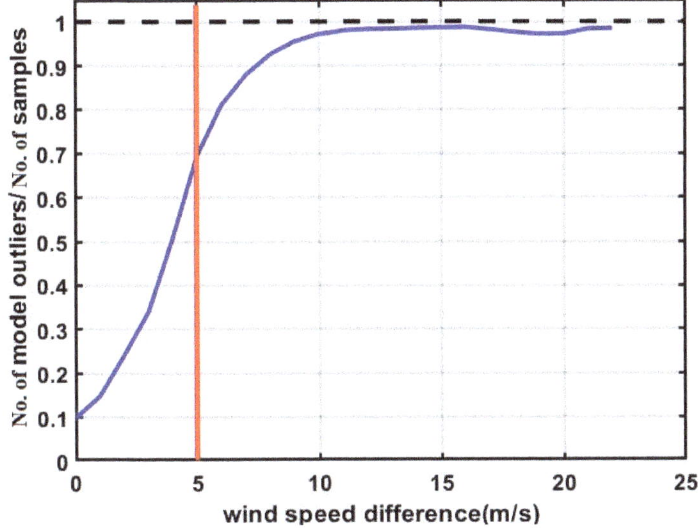

Figure 8. Ratio of outliers rightly identified by the filter to the actual no. of outliers vs. wind speed difference.

To understand Figure 8 better, we look at the distribution of outliers (wind speed difference >= 5 m/s) at different MERRA-2 wind speed bins before and after applying the filter. This data distribution is shown in Figure 9. The red distribution shows the density of outliers in the original dataset and the blue shows the distribution of outliers after applying the filter. Firstly, a very significant decrease in the outlier population can be observed after filtering. The filtered dataset has approximately 4 times less outliers. In the original dataset, most of the outliers are present between 5–10 m/s which is also the peak region for wind speed distribution. In this region the filter has been able to remove a large proportion of the outliers. Next, in the filter design section the low PD and high FAR at high winds region was discussed. Though at first, it may appear as if the filter cannot operate in this wind speed region, the distribution of outliers in this region (plot on the top right) shows that the number of outliers is almost an order of magnitude smaller after the filtering process, indicating that the filter can operate efficiently in this region but the low sample density in the region does not reflect this capability of the filter in the PD and FAR metrics.

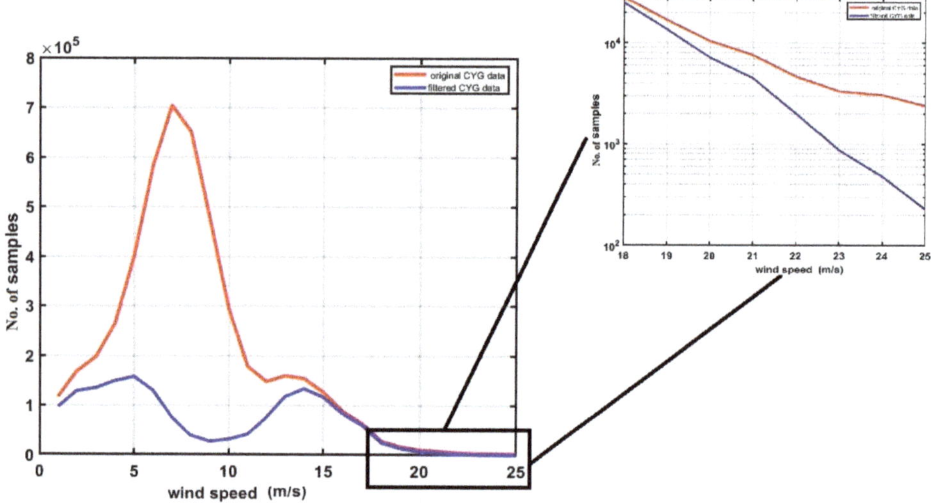

Figure 9. Distribution of outliers at different MERRA-2 wind speed bins before and after QC filter.

Finally, the total wind speed distribution of the dataset before and after applying the filter is plotted in Figure 10. After applying the filter, ~20.5% of the data have been removed by the filter as outliers. From Figure 10 it can be observed that the largest difference in density occurs at high wind speeds (>18 m/s). This is partly due to the high FAR of the filter in this region and partly due to large scatter in the data in this region. A substantial difference in density can also be observed at very low wind speed regions (<3m/s), again owing to the high FAR of the filter in this region.

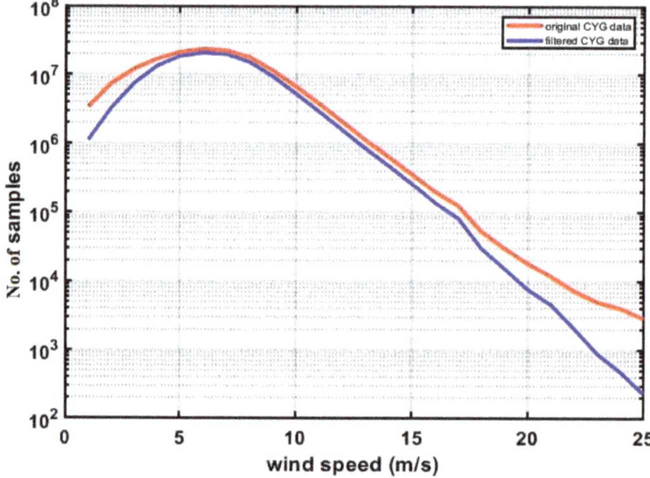

Figure 10. Distribution of CYGNSS retrieved winds before and after QC filter.

4.2. Identifying Dominant Feature Vectors

In this section the importance of each of the diagnostic variable is assessed using the minimum redundancy maximum relevance algorithm. The algorithm minimizes the redundancy of the feature set and maximizes the set with respect to the training data. Pairwise mutual information of the diagnostic variables is used to quantify its redundancy and relevance [18]. Figure 11 shows the score for each of the variable based on its importance in distinguishing outliers from good samples.

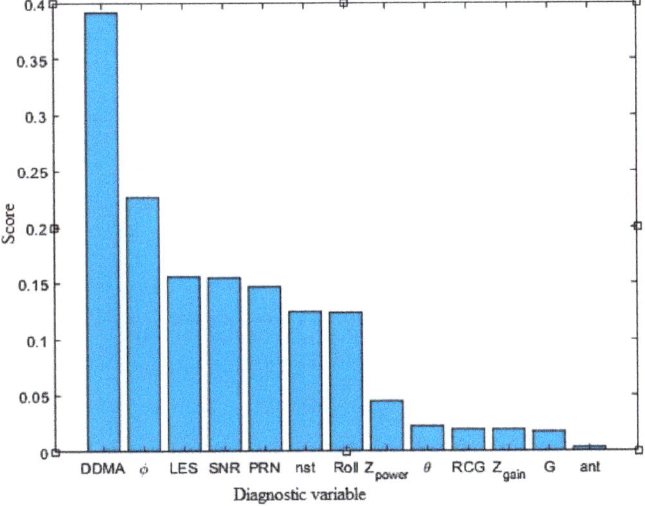

Figure 11. Dominant diagnostic variables in identifying outliers. Variable definitions are provided in Table 1.

The most dominant feature is the DDMA (NBRCS). This is as expected because the wind retrieval by CYGNSS is directly related to the two observables NBRCS and LES. The other dominant features are pre-dominantly instrument related such as azimuth angle, PRN, star tracker attitude status and satellite roll. This suggests that most of the outliers are caused due to improper instrument calibration.

4.3. Wind Retrieval Performance

The CYGNSS wind retrieval is evaluated based on 3 error statistics, namely, the Mean Difference (MD), RMS difference (RMSD) and variance in the data. The MD and RMSD are evaluated with respect to the 1:1 line thus is a superposition of both variance in the data and the intrinsic bias in the GMF. Whereas the variance is a measure of only the degree of scatter in the data. The error statistics are presented in Figures 12 and 13. In Figure 12 the MD and RMSD of the original dataset is shown by solid lines and the filtered dataset is shown by dashed lines. An increase in bias can be observed in the filtered dataset as compared to the original dataset; this is because, after filtering the samples that are identified as good by the filter are aligned closer to the GMF line rather than the 1:1 line. The increase in bias is more dominant above 10 m/s as the GMF line begins to deviate away from the 1:1 line above this wind speed. Figure 13 shows the variance in the data at different wind speed bins. Variance represents the degree of scatter in the data and after applying the filter there is a sharp drop in the scatter. The standard deviation in the filtered dataset is <=2 m/s for a wide range of wind speeds. These error statistics show a significant improvement in the nature of retrieval after the QC filter.

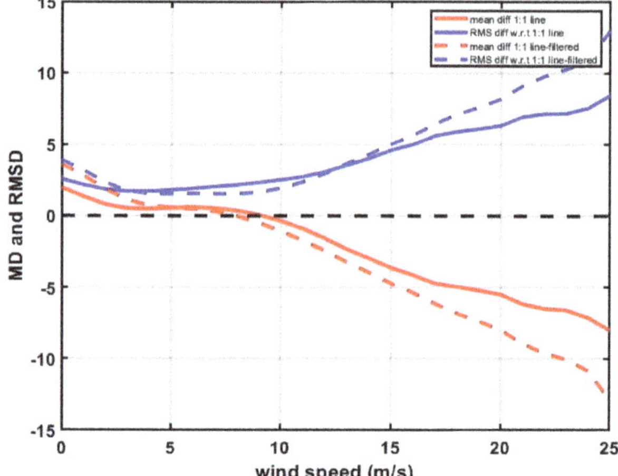

Figure 12. Mean difference and RMS difference statistic on CYGNSS retrieved winds before and after QC filter.

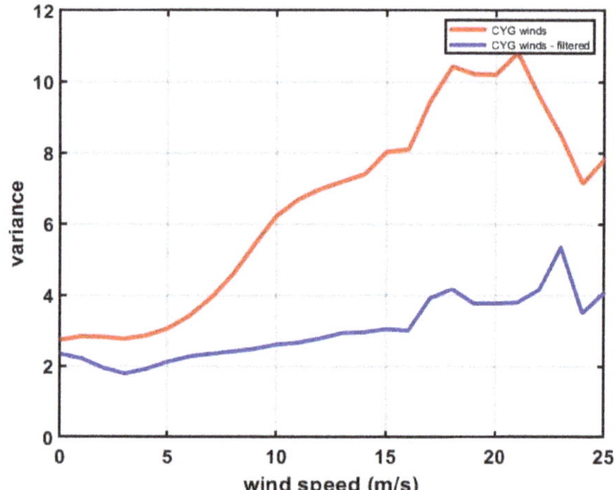

Figure 13. Variance in CYGNSS retrieved winds before and after QC filter.

5. Discussion

The CYGNSS retrieved winds are currently being used for various ocean science applications such as ocean circulation studies, regional and global analysis of ocean winds [19], tropical cyclone studies [20–23], and assimilation into Numerical Weather Prediction (NWP) models. Data reliability plays an important role in aiding such scientific studies. The CYGNSS wind speed data products are of two kinds—the Fully Developed Seas (FDS) wind retrievals and Young Sea Limited Fetch (YSLF) retrievals. Of these two, the FDS winds form the major proportion of the measurements and are therefore used for many scientific applications, especially for assimilation into NWP models. The YSLF data product is for hurricane force winds measured over individual storms, therefore is a substantially smaller set of measurements. The QC mechanism developed in this work is for the CYGNSS FDS winds, in order to reduce errors (in particular, outliers) in the retrieval due to various engineering and measurement related errors.

The primary merit of the proposed ML filter is its ability to better account for interactions between the individual engineering, instrument and measurement conditions than can separate thresholded flags for each one. The current approach upon which we are improving uses individual flags and, despite these existing QC filters, there remains considerable scatter in the data—hinting that individual and independent thresholds is not an effective way of removing the outliers.

The filter proposed here utilizes the capability of ML tools to learn inherent patterns from the training dataset and quickly come up with any convex boundaries separating the outliers from good data. One other advantage of such filters is that, because the system itself is aging with time, and as shown in this work—most of the outliers are due to calibration errors, the new ML-based QC thresholds can be reassessed periodically. In such situations, the ML filters come in handy as their parameters can be tuned easily to respond to any changes.

Assimilating the CYGNSS near surface wind retrievals into NWP models for better forecasting is one of its important uses. In general, NWP models give a weight to meteorological satellite observations based on their error statistics. Thus, reducing errors in the retrieval will help assimilate CYGNSS winds better. Using this filter, the standard deviation of the retrieval is reduced from 2.6 m/s to 1.7 m/s over the wind speed range 0–25 m/s.

At higher wind speed ranges, this filter is too aggressive and removes some valuable high wind measurements. This is due to the fact that high wind data density (> 20 m/s) is very sparse, hence insufficient for the Neural Network to be able to learn significant patterns from it. To address this situation, one possibility is to assimilate more of CYGNSS high wind data in future years, to better train the Neural Network in this region. However, it is also important to consider here that the CYGNSS FDS winds are reliable only up to 25 m/s as they have been developed using NOAA/GDAS ocean surface winds as their reference [24].

The direction of focus of future work will be to develop automated machine learning based QC that can effectively remove outliers at all wind speed ranges. Currently this filter is operable only between 5–18 m/s. The lower sample density at high and very low winds, prevent the QC filter from operating in these regions. One possible solution, as mentioned above, is to wait for more CYGNSS measurements in these wind speed regions before the QC is applied. Using ML based QC for YSLF winds can be complicated by the rapidly varying sea state inside hurricanes. In such cases, a physics based definition of an outlier might be needed. One approach to apply quality control for such data is to observe trends along overlapping tracks within a given spatial boundary around the hurricane.

6. Conclusions

In this work a Neural Network based Quality Control filter for CYGNSS wind retrieval is developed. The inputs to this filter are the 13 diagnostic variables that broadly represent instrument related, measurement geometry related and surface related attributes. Of these diagnostic tools, the surface related attributes (NBRCS, LES, and SNR) and instrument related attributes (azimuth angle, star tracker status, PRN, satellite roll) play a dominant role in distinguishing outliers from good sample population. The Neural Network is trained over two different training datasets at two different CYGNSS wind regimes based on the behavior of the GMF. The operating range of the filter is between 5–18 m/s. Within this range the probability of outlier detection is > 75% and the false alarm rates is <20%. In total ~20.5% of the data is removed as outliers by this filter. Atleast 75% of the outliers with wind speed difference of at least 5 m/s is removed while ~100% of the outliers with wind speed difference of at least 10 m/s is removed. This filter has significantly reduced the scatter in the data. The quality filtered dataset has a standard deviation of <=2 m/s over a wind range of wind speeds. The design space for this filter is also analyzed in this work to identify trade-offs between PD and FAR. The choice of PD and FAR will depend on the application. For example, a low FAR may be especially important for applications in which good spatial and temporal sampling are very important (e.g., to image rapidly changing weather systems) whereas a high PD may be especially important for applications in which the lowest possible uncertainty in wind speed is important (e.g., to detect small trends over long time intervals, such as are associated with global change).

As the next steps in this work, the higher wind regime will be the focus of interest. Strategies to improve the performance of this filter at higher winds while retaining as many samples as possible will be considered. Also, currently this filter is developed only for fully developed seas, in the future the feasibility of extending to young seas will also be studied.

Author Contributions: Conceptualization, R.B. and C.R.; methodology, R.B. and C.R.; software, R.B.; validation, R.B.; formal analysis, R.B.; investigation, R.B. and C.R.; resources, R.B.; data curation, R.B.; writing—original draft preparation, R.B.; writing—review and editing, R.B and C.R..; visualization, R.B.; supervision, C.R.; project administration, C.R.; funding acquisition, C.R. All authors have read and agreed to the published version of the manuscript.

Funding: The work presented was supported in part by NASA Science Mission Directorate contract NNL13AQ00C with the University of Michigan.

Acknowledgments: The MERRA-2 data used in this study/project have been provided by the Global Modeling and Assimilation Office (GMAO) at NASA Goddard Space Flight Center through the online data portal in the Goddard Earth Sciences Data and Information Services Center, 10.5067/3Z173KIE2TPD.

Conflicts of Interest: The authors declare no conflict of interest.

References

1. Clarizia, M.P.; Gommenginger, C.P.; Gleason, S.T.; Srokosz, M.A.; Galdi, C.; Di Bisceglie, M. Analysis of GNSS-R delay-Doppler maps from the UK-DMC satellite over the ocean. *Geophys. Res. Lett.* **2009**, *36*. [CrossRef]

2. Gleason, S. Space-Based GNSS Scatterometry: Ocean Wind Sensing Using an Empirically Calibrated Model. *IEEE Trans. Geosci. Remote. Sens.* **2013**, *51*, 4853–4863. [CrossRef]

3. Unwin, M.; Duncan, S.; Jales, P.; Blunt, P.; Brenchle, M. *Implementing GNSS Reflectometry in Space on the TechDemoSat-1 Mission*; Procedings Institute Navigation: New York, NY, USA, 2014; pp. 1222–1235.

4. Ruf, C.; Chew, C.; Lang, T.J.; Morris, M.; Nave, K.; Ridley, A.; Balasubramaniam, R. A New Paradigm in Earth Environmental Monitoring with the CYGNSS Small Satellite Constellation. *Sci. Rep.* **2018**, *8*, 8782. [CrossRef] [PubMed]

5. Ruf, C.; Chang, P.; Clarizia, M.P.; Gleason, S.; Jelenak, Z.; Murray, J.; Morris, M.; Murray, J.; Musko, S.; Posselt, D.; et al. *CYGNSS Handbook*; United States of America by Michigan Publishing: Ann Arbor, MI, USA, 2016.

6. Clarizia, M.P.; Ruf, C. Wind Speed Retrieval Algorithm for the Cyclone Global Navigation Satellite System (CYGNSS) Mission. *IEEE Trans. Geosci. Remote. Sens.* **2016**, *54*, 4419–4432. [CrossRef]

7. Gleason, S.; Ruf, C.; Clarizia, M.P.; O'Brien, A.J. Calibration and Unwrapping of the Normalized Scattering Cross Section for the Cyclone Global Navigation Satellite System. *IEEE Trans. Geosci. Remote. Sens.* **2016**, *54*, 2495–2509. [CrossRef]

8. Gleason, S.; Ruf, C.; O'Brien, A.J.; McKague, D.S.; OrBrien, A.J. The CYGNSS Level 1 Calibration Algorithm and Error Analysis Based on On-Orbit Measurements. *IEEE J. Sel. Top. Appl. Earth Obs. Remote. Sens.* **2019**, *12*, 37–49. [CrossRef]

9. Ruf, C.; Gleason, S.; McKague, D. Assessment of CYGNSS Wind Speed Retrieval Uncertainty. *IEEE J. Sel. Top. Appl. Earth Obs. Remote. Sens.* **2018**, *12*, 87–97. [CrossRef]

10. Gleason, S. *Level 1B DDM Calibration Algorithm Theoretical Basis Document*; Doc. No: 148–0137; Project Document; CYGNSS: Ann Arbor, MI, USA, 2014.

11. Clarizia, M.P.; Zavarotny, V.; Ruf, C. *Level 2 Wind Speed Retrieval Algorithm Theoretical Basis Document*; Project Document; CYGNSS: Ann Arbor, MI, USA; p. 148-0138.

12. Gelaro, R.; Mccarty, W.; Suárez, M.J.; Todling, R.; Molod, A.; Takacs, L.; Randles, C.A.; Darmenov, A.; Bosilovich, M.G.; Reichle, R.; et al. The Modern-Era Retrospective Analysis for Research and Applications, Version 2 (MERRA-2). *J. Clim.* **2017**, *30*, 5419–5454. [CrossRef] [PubMed]

13. Bosilovich, M.G.; Lucchesi, R.; Suarez, R. *MERRA-2: File Specification*; NASA Goddard Space Flight Center: Greenbelt, MD, USA, 2015.

14. Mehrotra, Kishan, G.; Chilukuri, K.M.; Huang, H. *Anomaly Detection Principles and Algorithms*; Springer International Publishing: New York, NY, USA, 2017.

15. Hawkins, D.M. *Identification of Outliers*; Chapman and Hall: London, UK, 1980; Volume 11.

16. Aggarwal, C.C.; Sathe, S. *Outlier Ensembles: An introduction*; Springer: New York, NY, USA, 2017.

17. Lippmann, R.P. An introduction to computing with neural nets. *IEEE ASSP Mag.* **1987**, *4*, 4–22. [CrossRef]

18. Darbellay, G.; Vajda, I. Estimation of the information by an adaptive partitioning of the observation space. *IEEE Trans. Inf. Theory* **1999**, *45*, 1315–1321. [CrossRef]

19. Leidner, S.M.; Annane, B.; McNoldy, B.; Hoffman, R.; Atlas, R. Variational Analysis of Simulated Ocean Surface Winds from the Cyclone Global Navigation Satellite System (CYGNSS) and Evaluation Using a Regional OSSE. *J. Atmospheric Ocean. Technol.* **2018**, *35*, 1571–1584. [CrossRef]

20. McNoldy, B.; Annane, B.; Majumdar, S.; Delgado, J.; Bucci, L.; Atlas, R.; Brian, M.; Bachir, A.; Sharanya, M.; Javier, D.; et al. Impact of Assimilating CYGNSS Data on Tropical Cyclone Analyses and Forecasts in a Regional OSSE Framework. *Mar. Technol. Soc. J.* **2017**, *51*, 7–15. [CrossRef]

21. Annane, B.; McNoldy, B.; Leidner, S.M.; Hoffman, R.; Atlas, R.; Majumdar, S.J. A Study of the HWRF Analysis and Forecast Impact of Realistically Simulated CYGNSS Observations Assimilated as Scalar Wind Speeds and as VAM Wind Vectors. *Mon. Weather. Rev.* **2018**, *146*, 2221–2236. [CrossRef]

22. Li, X.; Mecikalski, J.R.; Lang, T.J. A Study on Assimilation of CYGNSS Wind Speed Data for Tropical Convection during 2018 January MJO. *Remote. Sens.* **2020**, *12*, 1243. [CrossRef]

23. Mayers, D.; Ruf, C. Tropical Cyclone Center Fix Using CYGNSS Winds. *J. Appl. Meteorol. Clim.* **2019**, *58*, 1993–2003. [CrossRef]

24. Ruf, C.; Balasubramaniam, R. Development of the CYGNSS Geophysical Model Function for Wind Speed. *IEEE J. Sel. Top. Appl. Earth Obs. Remote. Sens.* **2018**, *12*, 66–77. [CrossRef]

Article

Retrieval of Ocean Surface Wind Speed Using Reflected BPSK/BOC Signals

Hao-Yu Wang * and Jyh-Ching Juang

Department of Electrical Engineering, National Cheng Kung University, Tainan City 70101, Taiwan; juang@mail.ncku.edu.tw
* Correspondence: N28044020@gs.ncku.edu.tw

Received: 4 July 2020; Accepted: 19 August 2020; Published: 20 August 2020

Abstract: The Global Navigation Satellite System (GNSS) has become a valuable resource as a remote sensing technique. In the past decade, the use of reflected GNSS signals for sensing the Earth, also known as GNSS reflectometry (GNSS-R), has grown rapidly. On the other hand, with the continuous development of GNSS, multi-frequency multi-modulation signals have been used to enhance not only positioning performance, but also remote sensing applications. It is known that for some constellations, navigation satellites broadcast signals employing BPSK (binary phase-shift keying) modulation and BOC (binary offset carrier) modulation at the same frequency band. This paper proposes a new GNSS-R measurement, called a composite delay-Doppler map (cDDM), by utilizing the received reflected GNSS signals with different modulation techniques for the purpose of retrieving wind speed. The GNSS-R receiver can receive BPSK and BOC signals simultaneously at the same frequency band (e.g., GPS III L1 C/A and L1C or QZSS L1 C/A and L1C) and process the signals to generate GNSS-R measurements. Exploration of the observable features extracted from the composite DDM and the wind speed retrieval algorithm are also provided. The simulation verifies the proposed method under a configuration that is specified for the orbital and instrument specification of the upcoming TRITON mission.

Keywords: GNSS-R; TRITON; wind speed; BPSK; BOC

---

1. Introduction

It has been shown the Global Navigation Satellite System reflectometry (GNSS-R) can be used to observe numerous geophysical parameters above the Earth's surface, including, but not limited to, soil moisture [1–3] and sea ice [4–6]. The capability of GNSS-R for deriving ocean surface wind speed has also been validated in many studies [7–9]. Among all these applications, inversion of sea wind is the most popular subject in GNSS-R and is also crucial to weather forecasting. The algorithms for retrieving wind speed have been developing for about 30 years since 1990. During the initial period, scientists were merely observing variations by processing reflected GPS signals under different sea states. After Zavorotny and Voronovich first proposed the GNSS-R theory in 2000 [10], researchers began to match measured 1-D delay waveforms (DW) or 2-D delay-Doppler maps (DDM) with their local simulated counterparts to estimate ocean wind speed. However, this matching method is time-consuming and requires prior information (i.e., rough wind speed and wind direction). Subsequently, many studies have proposed to extract observables from the DW or DDM and correlate these observables to nearly coincident measurements using other wind sensors, such as buoys. To date, retrieval of wind speed with an empirical model by relating the observables with the collocated wind speed is the most common practice. A summary of the GNSS-R principles and other applications can be found in [11].

The Cyclone GNSS (CYGNSS) mission is a space-based GNSS-R mission, which comprises eight micro-satellites to provide measurements of the sea surface wind field with good spatial and temporal resolution. The evolution of the CYGNSS wind speed retrieval algorithm is briefly reviewed in the following. In 2014, Clarizia et al. [12] developed a preliminary approach to retrieve wind speed using the data recorded by the precursor mission (United Kingdom-Disaster Monitoring Constellation, UK-DMC). They presented a minimum variance estimator to composite five wind speeds estimated from five different observables derived from a small data set of GNSS-R DDMs. Subsequently, Clarizia et al. [13] presented the baseline CYGNSS L2 wind speed retrieval algorithm, with specific characteristics of CYGNSS in-orbit measurements. Instead of using five observables as in the previous study, they used two observations (i.e., DDMA and LES) to develop a weighted wind speed estimator. After CYGNSS was launched on December 15, 2016, Ruf and Balasubramaniam [14] developed a wind speed algorithm using a tremendous amount of on-board measurements, along with two different ocean surface wind speed reference sources. This time, they built independent wind-GMF by considering two different sea states, a fully developed sea (FDS) version and a young sea/limited fetch (YSLF) version. The YSLF GMF is designated for measuring high wind speeds, especially those of tropical cyclones. The overall wind speed retrieval performance of the CYGNSS mission was reported in [15]. Afterward, Park and colleagues, affiliated with National Oceanic and Atmospheric Administration (NOAA), also proposed an improved wind retrieval method in 2019 [16].

Recently, a few studies have been conducted to define new observables for wind speed retrieval. In [17], Gao et al. described the normalized delay waveform (NDW) width as a new observable, along with elevation angle and flight height, to retrieve wind speed. Their wind speed retrieval model has two versions: one is a multiple regression model, and the other is a neural network model. However, both models can only provide the same performance level as the matching method (i.e., comparing the measured delay waveform with a simulated delay waveform). In a study by Wang and co-workers [18], two new observables were proposed based on the variations in the DW distribution. They built a ground-based GNSS-R system to retrieve wind speeds under a gentle wind scenario and a typhoon scenario using the proposed observables by receiving and processing Beidou Geostationary Earth Orbit (GEO) satellite signals. The results demonstrated that optimal wind speed retrieval performance can be obtained by fine-tuning the threshold and coherent time before calculating the proposed observables. On the other hand, Juang et al. proposed a model-based approach that the relationship between the reflected delay waveform and direct delay waveform can be identified as a channel response function [19]. The authors claimed that the proposed method is insensitive to the variation of the transmitter power, and the remote sensing parameters (e.g., ocean wind speed) can be retrieved from the coefficients in the channel model. In addition, the channel response that is established from binary phase-shift keying (BPSK)-modulated signals can be directly applied to binary offset carrier (BOC)-modulated signals. The flight test in their study showed that the proposed method is feasible and potential. However, data retrieval performance of the proposed method has not been verified.

The first GPS III satellite was successfully launched on 23 December 2018 and went into service on 13 January 2020, after a series of rigorous on-orbit operational tests. The remaining nine satellites (IIIA block series) continue to be deployed. It is expected that the last satellite will be launched in the second quarter of 2023. GPS III is constructed to be fully backward compatible with existing GPS systems but with new capabilities related to both military and civilian use, including longer SV lifer, improved accuracy, and improved availability. The GPS III SV will transmit L1 C/A, L1 P(Y), L2 P(Y), the modernized L1M, L2C, and L2M, and new L1C and L5 signals. Among all these signals, the L1C signal, which makes GPS III interoperable with other satellite navigation systems, is a new additional civilian signal. It is believed that the benefits of L1C (including improved accuracy, additional navigation messages, and advanced anti-jamming capability) can provide civilian users with better PVT services throughout the globe. In this regard, it is of interest whether the new L1C signal of the GPS III can be used to enhance GNSS-R performance. In this paper, we proposed a potential solution, described in Section 2, to incorporate this signal in the processing of wind speed retrieval. For a detailed description of the GPS III, including performance requirements, signal and system design, and arrangement of the program, readers are referred to [20].

Thanks to the contribution of TechDemoSat-1 (TDS-1) and CYGNSS, many advances have been made in the GNSS-R. The available data from these missions not only verify the feasibility of GNSS-R but also promote many potential applications in remote sensing. The TRITON satellite, conducted by Taiwan's National Space Organization, will draw on the experience of those missions to carry on a space-based GNSS-R mission to gather ocean surface roughness and wind speed for the purpose of weather research and forecasting. The purpose of this paper is to propose a new GNSS-R measurement by using the signals transmitted under different modulations (i.e., BPSK and BOC) at the same frequency band and source for retrieving ocean wind speeds. The proposed method is based on the difference caused by the reflected signal under different modulations. A benefit of the proposed method is that it may ignore the instrument calibration, which is a crucial pre-processing of scientific data generation, but a challenging procedure during on-board processing. The corresponding wind GMF that is using the derived observable was also developed in this paper. In this paper, the simulation parameters for generating and validating the proposed method are specified under TRITON specifications, including orbital measurement geometry, the nadir antenna gain pattern, and receiver hardware characteristics.

2. Materials and Methods

In this study, a new GNSS-R measurement for wind speed retrieval is proposed that utilizes the signal characteristics of next-generation GNSS under TRITON configurations. This section describes the data simulation under TRITON specifications, including the satellite tracks and reflection tracks. The real weather analysis data, which was used as the ground truth data, is also discussed. An explanation of the software used to simulate GNSS-R measurement is included. The proposed GNSS-R measurement, extracted observables, as well as the wind speed retrieval algorithm, are also described in detail in this section.

2.1. Orbital and Instrumental Specifications of TRITON Mission

The TRITON (initially called FORMOSAT-7 Reflectometry, FS-7R) satellite is a part of the FORMOSAT 7 program developed for a technology demonstration mission by the Taiwanese space agency, National Space Organization (NSPO) and scheduled to be launched in late 2021. Unlike other satellites, which were developed for the purpose of conducting the GNSS radio occultation (GNSS-RO) mission, the TRITON is designed to carry on the GNSS-R experiment. The mission objective of the TRITON is to measure the roughness and wind speed over the ocean surface based on the domestically-developed GNSS-R payload. In this subsection, we introduce the TRITON satellite trajectory and its corresponding reflection events. For more details about the TRITON mission, including the payload design and preliminary test, the reader can refer to [19,21,22].

The TRITON satellite has a sun-synchronous mission orbit at an altitude of 550–650 km. The on-board payload of the TRITON satellite is a GNSS-R receiver tailored for receiving and processing reflected GNSS signals, including GPS, QZSS, and Galileo. The TRITON satellite is equipped with one zenith antenna and one nadir antenna. The zenith antenna is a high gain dual-frequency (L1 + L2) antenna, which is used to receive line-of-sight GNSS signals for GNSS receiver to provide position, velocity, and timing information. The nadir antenna is a left-handed circularly polarized (LHCP) antenna with a gain value of 14.5 dBi at the L1 frequency and 12.7 dBi at the L2 frequency. It should be noted that the RF front-end behind the nadir antenna will sample the incoming signal at 16.368×10^6 samples per second. All these parameters are essential for the following simulations and are summarized in Table 1.

Table 1. Orbital and payload specifications of the TRITON satellite

Parameter	Value
Orbit	Sun-synchronous
Altitude	550–650 km
Period	96 min
Nadir antenna gain	14.5 dBi
Sampling rate	16.368 MHz
Frequency	1575.42 MHz

As mentioned, the TRITON is orbiting the Earth at an inclination angle greater than 24 degrees, and is capable of measuring four simultaneous reflections on the Earth's surface every second. In the following, four GPS satellites with the highest elevation angle are selected to calculate the specular reflection position. Figure 1a shows the 24-h TRITON spatial coverage of ground tracks and the corresponding reflection points. Figure 1b also shows the spatial coverage of reflection points over the ocean only.

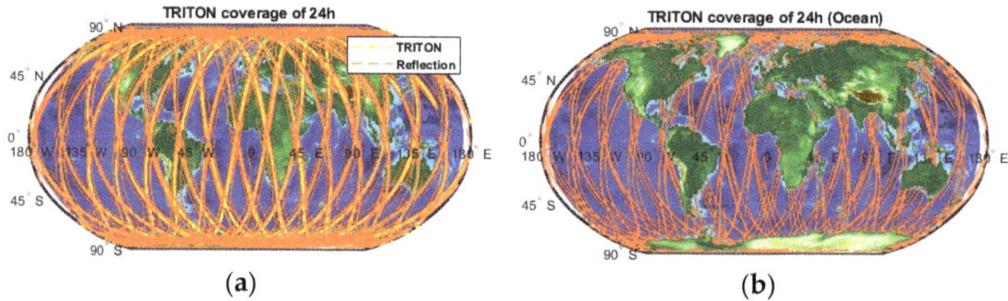

Figure 1. (**a**) Simulated TRITON ground tracks and reflection event coverage concerning GPS satellites for a 24-h period. (**b**) Shown is the reflection point over the ocean only. The broken yellow line indicates the ground tracks of the TRITON satellite, and the broken orange line indicates the reflection points from four GPS satellites with the highest elevation angle.

2.2. The Ground Truth Data: European Centre for Medium-Range Weather Forecasts Product

In the previous section, we simulated the trajectory of the TRITON satellite and its reflection tracks according to given parameters. However, it is also necessary to specify wind speed to generate the GNSS-R measurements, such as the delay-Doppler map (DDM) or the delay-waveform (DW). Therefore, we plan to employ the European Centre for Medium-Range Weather Forecasts (ECMWF) reanalysis product to generate the ground truth wind speed for the purpose of simulating the DDM and DW. To evaluate the retrieval performance of the wind speed of the proposed method, we used the ECMWF reanalysis product as 10-m-referenced ocean surface wind speeds. The reason to use real data, rather than a randomly generated wind speed value, is that the limit and distribution of wind speed values accord with the actual situation, which makes the simulation more realistic. ECMWF is a self-governing organization, and its core mission is to provide weather forecasts services and climate reanalysis products. The climate reanalysis product used here contains the wind speed information at the height of 10 m above the surface of the Earth with a spatial resolution of 0.25° × 0.25°, and temporal resolution of 1 h. Consequently, we used bicubic interpolation to estimate the reference wind speed at the locations and time according to the simulated reflection point of the TRITON satellite. In the following, we used the ECMWF data between 10 February 2020, and 16 February 2020, to produce interpolated 10-m-height ocean surface wind speeds as reference matchups. Figure 2 shows the observed wind speed distribution, which is obtained from the real ECMWF data by interpolation, according to given reflection tracks.

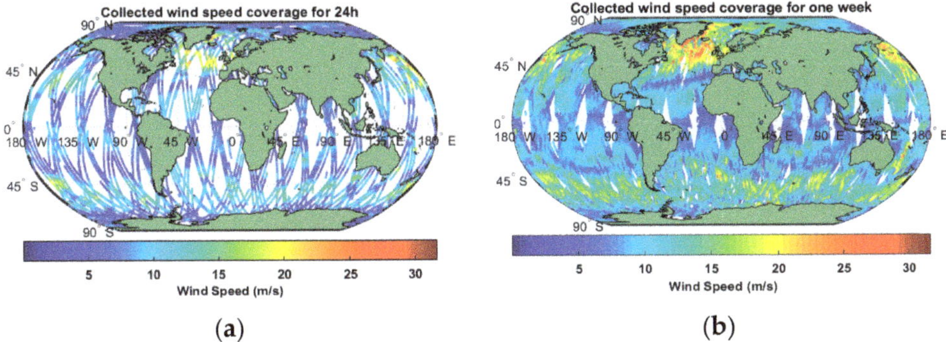

Figure 2. The TRITON reflection observation coverage, respectively, for (**a**) 24 h and (**b**) 7 days represented by wind speed. The color indicates the coincident wind speed obtained by interpolating the European Centre for Medium-Range Weather Forecasts (ECMWF) data.

2.3. The Proposed GNSS-R Measurement: Composite Delay-Doppler Maps

In the GNSS-R, one of the essential products is the delay-Doppler map (DDM). Usually, the DDM will be made in different sizes according to mission designs, such as 20×128 (Doppler bins $\times$ delay bins) for CYGNSS or 64×128 (Doppler bins $\times$ delay bins) for TRITON. In addition, the DDM resolution may also be different. In this paper, it is assumed that a GNSS-R receiver is capable of generating two types of DDMs. One is the DDM generated by processing the reflected L1 C/A signals (BPSK-DDM), and the other is the DDM generated by processing received reflected L1C signals (BOC-DDM). Under these conditions, the proposed GNSS-R measurement can be produced as derived below. As just mentioned, regardless of which size/resolution is specified, any number of pixel values in a DDM can be related to the input signal power as follows [23]:

$$C^M = G\left(P_a + P_r + P_s^M\right) \tag{1}$$

where C represents the DDM values per delay-Doppler bin produced from the GNSS-R receiver in the unit of "counts"; G is the total system gain; P_a is the thermal noise power generated at the antenna, P_r is the thermal noise power generated by the GNSS-R receiver, and P_s^M is the received power of the scattered signal, with respect to L1 C/A ($M = BPSK$) or L1C ($M = BOC$). CYGNSS puts a great deal of effort into removing the effects of system gain and noise components from Equation (1), to retain a clear receiver scatter signal component via the black body calibration algorithm [24,25]. In this paper, we propose a new GNSS-R measurement utilizing the signal characteristics of L1 C/A and L1C without using any calibration methods. It is believed that both the BPSK-DDM and BOC-DDM are suffering from the same noise floor, because the received signals all go through the same instrument path. The DDM noise floor can be obtained by taking the average of a certain specific signal-free area in the DDM and can be expressed as:

$$C_N = G(P_a + P_r) \tag{2}$$

By removing Equation (2) from Equation (1), we obtain:

$$\overline{C}^M = GP_g^M \tag{3}$$

Finally, by taking the decibel computation and subtracting BOC-term from BPSK-term, we obtain

$$\Delta C_{dB} = 10 \cdot \log_{10}\left(P_s^{BPSK}\right) - 10 \cdot \log_{10}\left(P_s^{BOC}\right) \tag{4}$$

To implement the proposed method, we divided the procedure into three parts: (1) produce BPSK-DDM and BOC-DDM for each channel; (2) calculate the noise floor and eliminate it from the

DDM (the noise floor is calculated independently for each channel), and (3) apply the above equations to make the composite delay-Doppler map (cDDM). Figure 3 also shows the processing steps to generate the proposed GNSS measurement. An open-source GNSS-R simulator, WavPy, is used to simulate the proposed DDM measurement [26]. It should be noted that the WavPy currently employs the classic Z-V model [10] to implement the simulation.

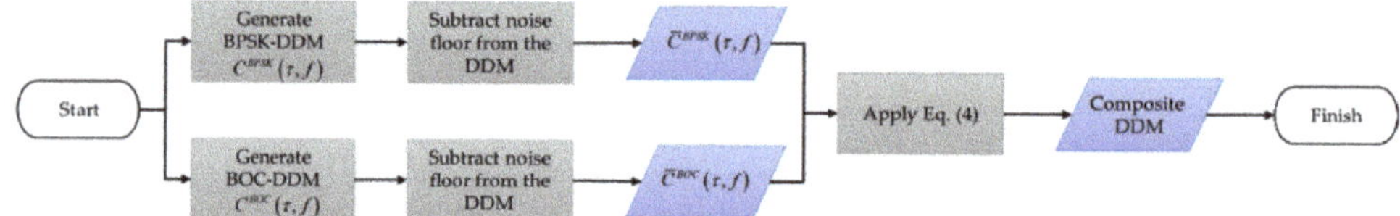

Figure 3. Flow chart of generating composite delay-Doppler map (DDM).

To analyze the relationship between the cDDM and the wind speed, we conducted a simple scenario with the parameters provided in Table 2. Figure 4 shows the simulated BPSK-DDM, BOC-DDM, and their corresponding CDDMs under different wind speeds and incidence angles. Then, we set a constant incidence angle of 30°, and simulate DW under different wind speeds, as shown in Figure 5b. In Figure 5a, we display an example of BPSK-DW and BOC-DW.

It is known that the autocorrelation function of the L1C spreading code has two side lobes, since the L1C spreading code employs the BOC modulation technique. The signal properties are changed due to the noise, fading, and distortion after reflection or scattering. This effect can lead to an extended trailing edge on the BPSK signal and asymmetrical side lobes on the BOC signal, as shown in Figure 5a. In Figure 5b, it can be seen that the area of the waveform around the peak varies with wind speed.

To further analyze the effectiveness of the cDDM in wind speed inversion, we simulate a large amount of data under various conditions, including different wind speeds and receiving geometry, based on realistic parameters. As mentioned in the preceding section, it is assumed that we collected the data generated from TRITON from 10 February 2020, to 16 February 2020. Therefore, a large set of cDDM is simulated according to real configurations, including antenna gain, incidence angle, and sampling rate. The actual ECMWF data during the selected period is used to generate the collocated reference wind speed. Additionally, we also apply some quality control to discard poor-quality simulated cDDM before further processing and analysis. First, measurements with receiving gains of less than 0 dB are excluded. Second, the BPSK-DDM and BOC-DDM, in which peak positions are too far from the theoretical center point, are excluded. Finally, the BPSK-DDM and BOC-DDM, in which the signal-to-noise ratio is less than 3 dB, are excluded. The remaining data set still counts up to 1,300,000 data pairs retained for analysis. We further split the data set into two independent sets: a randomly selected training set, which accounts for 80% of samples, and a remaining test set.

Table 2. Simulation parameters as a preliminary scenario.

Parameter	Value
Incidence angle	10, 20, 30, 40, 50, 60 deg.
Wind speed	3, 5, 10, 15, 20, 30 m/s

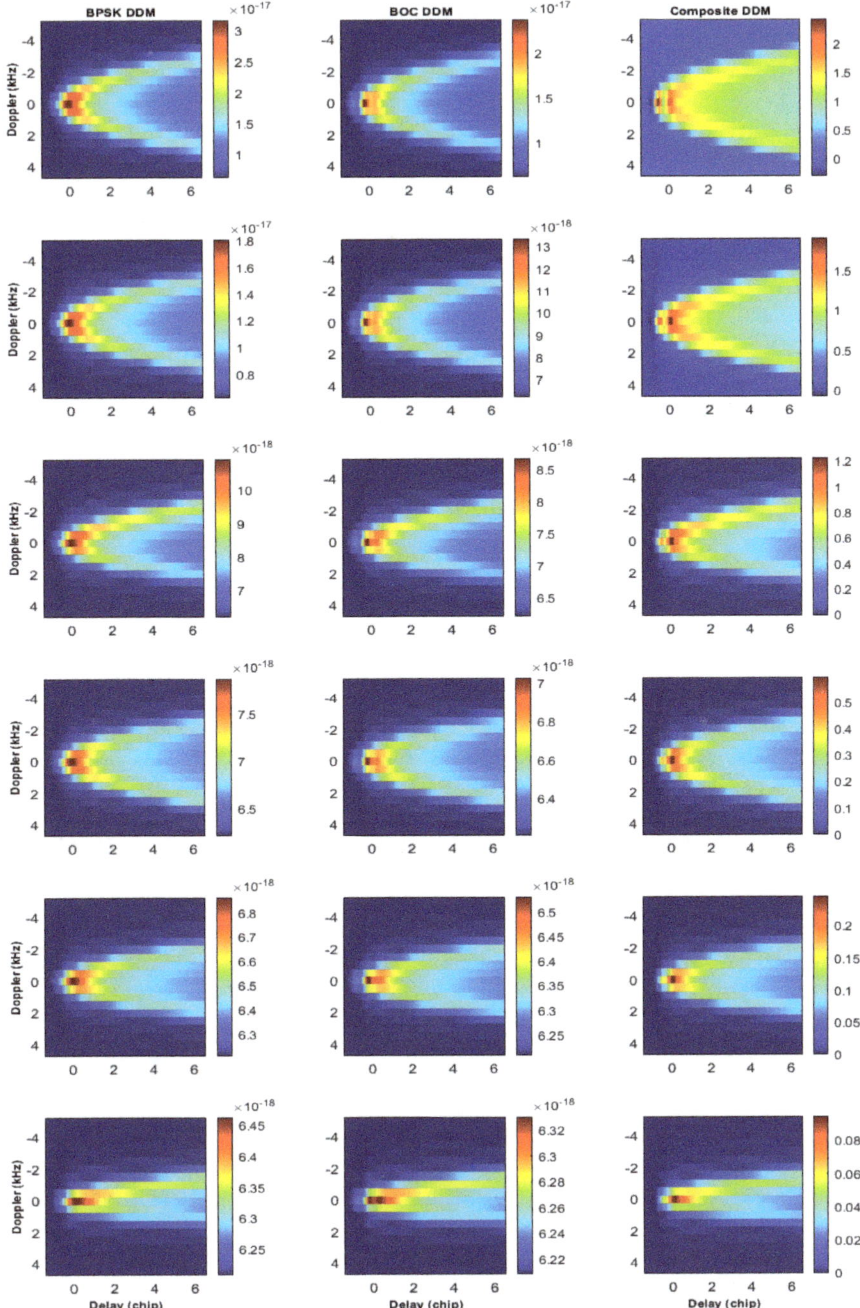

Figure 4. Examples of the binary phase-shift keying (BPSK)-DDM, binary offset carrier (BOC)-DDM, and the composite DDM (cDDM) for different wind speeds and incidence angles. The first, second, and third columns are BPSK-DDM, BOC-DDM, and cDDM, respectively. The specified parameters from the top row to bottom row are, respectively, 10/3, 20/5, 30/10, 40/15, 50/20, and 60/30 (incidence angle/wind speed).

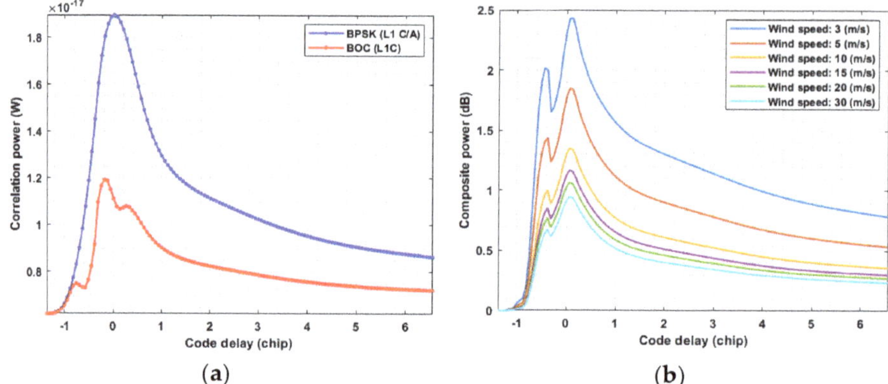

Figure 5. (**a**) Comparison of a BPSK-delay-waveform and a BOC-delay-waveform under a wind speed of 5 m/s. (**b**) Comparison of the composite delay waveform for different wind speeds at an incidence angle of 30°.

2.4. The Observable and Wind Speed Retrieval Algorithm

In principle, regardless of the type of GNSS-R measurement (i.e., delay-Doppler map or delay-waveform), one needs to extract useable observables from the GNSS-R measurement and regress the observable to the collocated wind speeds (or other geophysical parameters). Subsequently, a wind retrieval algorithm can be obtained by deriving a geophysical model function (GMF) that relates the observable to the wind speed, based on the regression parameters. In this study, we plan to extract three observables from the composite delay-Doppler map and deriving their corresponding GMFs. These observables are delay-Doppler map average, the delay-waveform average, and the delay-Doppler map peak, respectively. The following discussion provides a description of the observable and the procedure used to define the wind GMF.

The first observable is the so-called delay-Doppler map average (DDMA), which is a common observable and has been used in many studies on GNSS-R. The concept of the DDMA is the average signal power over a specified region of the DDM around the peak value position. The designer requirement determines the area in which to calculate the DDMA. For example, the Cyclone GNSS (CYGNSS) mission chooses the range of delay as (−0.25 0.25) chips and the range of Doppler as (−1000 1000) Hz, which corresponds to 3 × 5 bins of the DDM, to meet the mission requirement [13]. This selection can provide the retrieved wind speed with a spatial resolution of around 25 km × 25 km. In this paper, we employ the concept of DDMA on the cDDM to produce the composite delay-Doppler map average (cDDMA), which can be expressed as follows:

$$cDDMA = \frac{1}{MN}\sum_{m=1}^{M}\sum_{n=1}^{N}\overline{Y}_{cDDM}(\tau_m, f_n) \tag{5}$$

where $\overline{Y}_{cDDM}$ represents the value of the cDDM obtained from Equation (4); M and N are specified delay and Doppler range, respectively, and τ_m and f_n are the delay and Doppler at indexes m and n, respectively. We inherit the configurations from the CYGNSS; that is, a 3 × 5 cDDMA is calculated from the cDDM and used in the subsequent analysis. The second observable is the composite delay-waveform average (cDWA), which is the average value around a specified range in zero Doppler cDDM calculated using the following equation:

$$cDWA = \frac{1}{M}\sum_{m=1}^{M}\overline{Y}_{cDDM}(\tau_m, f_c) \tag{6}$$

where f_c represents the Doppler index located at the zero Doppler of the cDDM. The last observable is the cDDM peak (cDP), which is defined as the highest value of the cDDM. It should be noted that all the above observables are in units of dB.

Before developing the GMF, one of the issues is the need to correct any effects that may affect the observable. Therefore, we first investigate the relationship between the observable and the nadir antenna gain at the specular point (G_r). Here, we take the DDMA observation as an example. Figure 6 shows the relationship between the $cDDMA_0$, derived from the simulated cDDM using real TRITON settings via WavPy, and the interpolated ECMWF wind speeds. Clearly, the observable has a dependence on G_r.

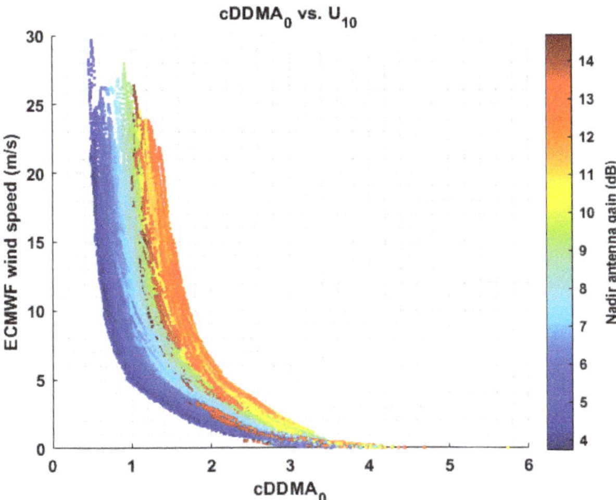

Figure 6. Relationship between composite delay-Doppler map average (cDDMA)$_0$ (dB) and ECMWF U10. The color indicates the nadir antenna gain value at the specular point.

To eliminate the effect of the antenna gain resulting from the observable, we apply the correction method used in [13]. The dependence of cDDMA on G_r is shown in Figure 7a, where the cDDMA, corresponding to different values of G_r, takes the average of different wind speed values. Figure 7a shows a more clear dependence relationship between cDDMA and G_r depicted in Figure 6. It is also shown that the dependence of the cDDMA on G_r does not vary with wind speed, and we can, therefore, developed an empirical correction. Their value has normalized the data sets corresponding to a single wind speed value at $G_r = 0$ dB, and we find that a sum of sines function, given by $f(x) = a_1 sin(b_1 x + c_1) + a_2 sin(b_2 x + c_2)$, where $x = G_r$, can fit the normalized cDDMA data points in that curve. Figure 7b shows the normalized data points and their corresponding best fit polynomial function. The same procedure is applied to the other two observables, cDWA and cDP, since these two observables were also found to have a similar dependency relationship with G_r. Note that the above analyses were all conducted using the training data set. However, we found that the dependence on antenna gain is not entirely directly proportional. That is, there is a decrease in the observable decrease with increase in G_r when G_r is greater than a specific value. As a result, we inspected and observed the dependence of the BPSK-DDM-derived observable and BOC-DDM-derived observable, respectively, on G_r. The results showed that the dependence of the observables on G_r is entirely directly proportional, but they differ from each other. Therefore, the composite combination described in the proceeding section causes the dependence phenomenon shown in Figure 7.

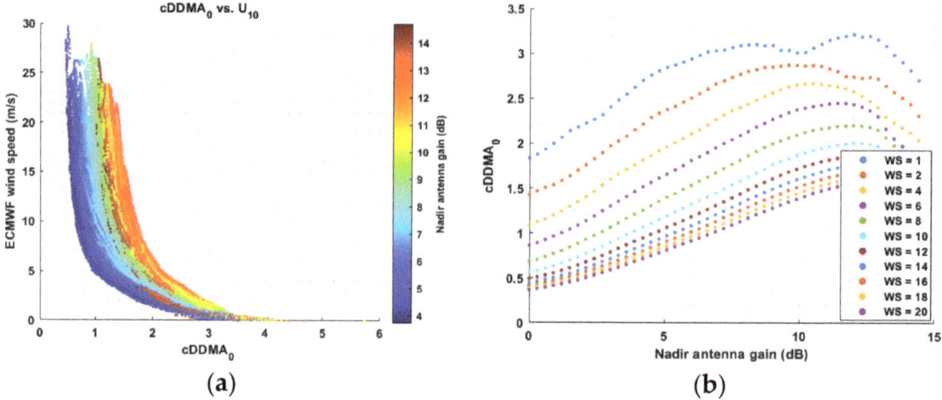

Figure 7. (a) cDDMA$_0$ versus nadir antenna gain at the specular point. The different colors indicate different wind speeds. (b) Normalized cDDMA$_0$ obtained from (a) for all gain values and wind speeds versus the nadir antenna gain, represented by the black dots. The blue curve is a fourth-order polynomial function that fits the data.

We can then obtain the corrected observable with the following expression:

$$O_1 = \frac{O_0}{f(G_r)} \tag{7}$$

where O is the observable (i.e., cDDMA, cDWA, or cDP), and $f(G_r)$ is the polynomial fit as stated above. The cDDMA before and after the G_r correction, represented as scatter density, is shown in Figure 8. The figure manifestly shows that the dependence on the nadir antenna gain value has been eliminated, as compared to Figure 6.

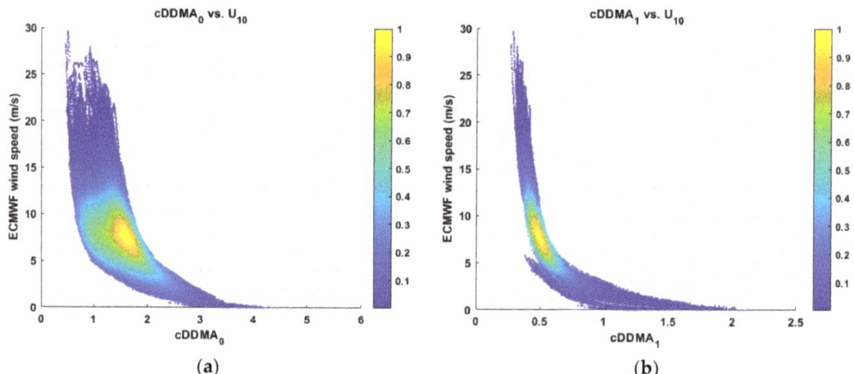

Figure 8. Scatter density plot of the cDDMA versus ECMWF wind speed before nadir gain correction (a) and after nadir gain correction (b).

The final observables used to develop the wind geophysical model function (GMF) are the corrected observables that are obtained from Equation (7). To develop the GMF, we regress the observable against the ECMWF wind speed. In this section, several commonly used fitting models were used to develop the GMF, including power model, exponential model, two-term exponential model, and polynomial model. In addition to the above equations, a dual-form model, which is found to best fit the observed and wind speed, was proposed for developing the GMF. Table 3 provides the mathematical expression for the models described above, which were also used for the subsequent processing.

Table 3. Comparison of fitting models. (In this table, x is the observable, U_{10} is the ECMWF wind speed, and the remaining symbols is the dependent parameters for each model).

Model	Equation	
Power model	$U_{10} = a \times (x - c)^d + b$	(T 3.1)
Exponential model	$U_{10} = b + ae^{-c(x-d)}$	(T 3.2)
Two-term exponential model	$U_{10} = a_1 e^{b_1 x} + a_2 e^{b_2 x}$	(T 3.3)
Polynomial model	$U_{10} = a_3 x^{-3} + a_2 x^{-2} + a_1 x^{-1} + a_0$	(T 3.4)
Exp-power model (dual model)	$U_{10} = \begin{cases} a_l x^{b_l} + c_l, & x < x_{th} \\ a_h e^{b_h x} + c_h, & x \geq x_{th} \end{cases}$	(T 3.5)

Among these five models, the power model, the exponential model, and the two-term exponential model were performed non-linear least squares to find the fitting coefficient. The usage of these three models can be found in [9,27,28], respectively. In a study by Christopher and Rajeswari [14], the wind GMF was developed by dividing the observable into two parts, low wind and high wind, and then regressed each part with a polynomial form. Instead, we did not divide our data, since the behavior of the proposed observable is different from their research. We directly use a three-order polynomial form to regress the observable and the reference wind speed, as expressed in Equation (T 3.4). However, none of the above models, expressed in (T 3.1)–(T 3.4), can perfectly regress the behavior between the observable and the wind speed data. Therefore, we proposed a dual model, which can best fit the relationship between the observable and the wind speed data, to develop the wind GMF, as expressed in Equation (T 3.5). In the dual model, the data set was divided into two portions. The observable below a certain threshold (i.e., O_{th}) is used to determine a_l, b_l, and c_l for the power form. The observable above a certain threshold (i.e., O_{th}) is used to determine a_h, b_h, and c_h for the exponential form. The O_{th} is determined when the least-squares residuals of the two portions are both minimized. Figure 9 shows the fitting results of using different models for training data. In the figure, the black scatter points represent the observable versus ECMWF wind speed, and the fitting curves are represented in different colors for each model. As shown in the figure, the dual model provides the best fit within the whole wind range, and the fitting curve of the exponential model has an obvious bias in low wind speed. However, if we try to reduce such a bias in low wind speed for the exponential model, the fitting result in high wind speed would become worse.

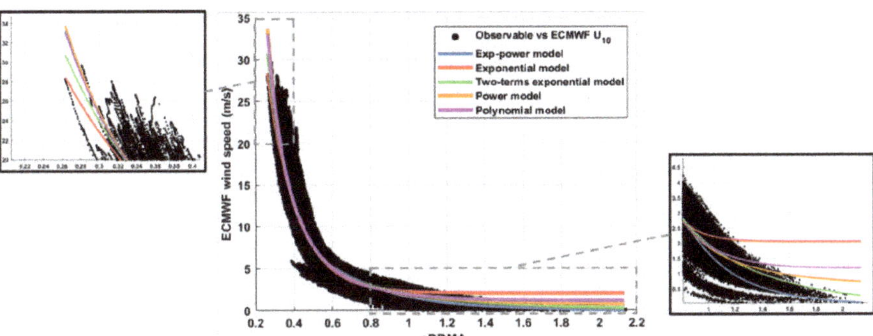

Figure 9. The composite delay-Doppler map average $cDDMA_1$ plotted against the ECMWF wind speed for the training data and the corresponding fit curve for different models.

To clearly compare the performance of wind retrieval between models, we calculate the mean errors and root-mean square (RMS) error using wind speed intervals of width ±1 m/s, as shown in Figure 10. It is shown that the retrieving accuracy and precision between different models are comparable to each other. However, the performance of the dual model is slightly better than other models in low wind speeds, no matter in terms of mean error or RMS error. Thus, in Section 3, we conducted the performance evaluation using the dual model on different observables.

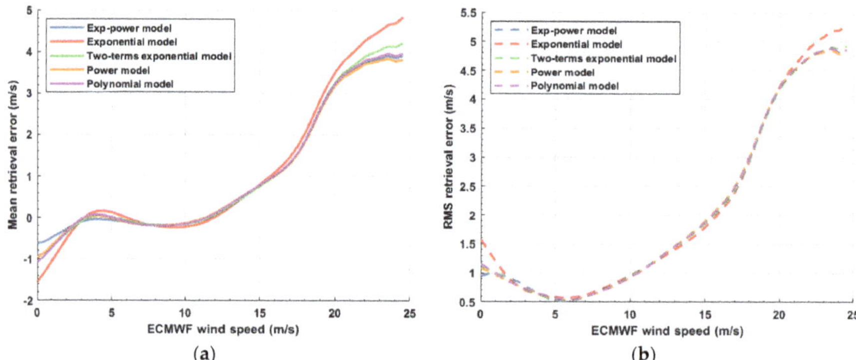

Figure 10. (**a**) Mean error curves and (**b**) root-mean square (RMS) error curves for the retrieved wind speed using different models on test data.

2.5. Summary of the Proposed Method

The complete procedure used in this research can be divided into five parts, as shown in Figure 11. First, we employed the specific orbital parameters of the upcoming TRITON mission to simulate spacecraft ground tracks. Subsequently, the simulated ground tracks were used to calculate reflection points with respect to visible GPS satellite positions. Third, the ground truth for the wind speed data was generated by interpolating realistic ECMWF reanalysis data according to simulated reflection events. Fourth, the proposed GNSS-R measurements, the composite delay-Doppler maps (cDDMs), were generated via an open-source GNSS-R simulator, WavPy, using the observable derived from the previous step. Finally, we developed the independent wind GMF for the different observables. It should be noted that the purpose of this paper was to propose a new GNSS-R measurement that can be used to retrieve the ocean surface wind speed. The differences between the estimated wind speed from different observables and the ground truth wind speed were analyzed to evaluate the feasibility of the proposed method.

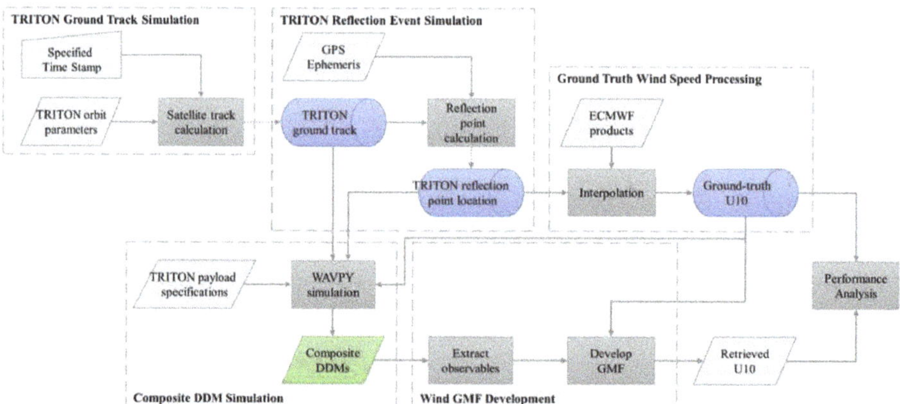

Figure 11. Overall research flow diagram, including the simulation of the TRITON ground track and corresponding reflection events, the proposed GNSS-R measurements and the observables, and the ground truth data generated for the performance evaluation.

3. Results

In this section, the performance of the retrieved wind speed using three observables from the proposed GNSS-R measurement is assessed using the test data set and the GMD derived from the dual model. Figure 12 shows the density scatter plot of the true wind speed and retrieved wind speed from different observables. As shown in the figure, the difference between the retrieved wind speed from the three observables is not noticeable. Nonetheless, we could still find that the use of CDP_1 caused less outlier branch behavior than when using CDW_1 and $CDDMA_1$, particularly with low to moderate wind speeds. Table 4 summarizes the overall statistical results, including the bias error, root-mean-square error, and the correlation coefficient of the retrieved wind speed using three observables. The statistical analysis indicated that the cDP-derived wind speed yielded the best performance, with an unbiased error, an RMSE less than 1 m/s, and a correlation coefficient of 0.96. This result is consistent with the data in Figure 12. However, the difference between wind speeds retrieved from the different observables is almost ignorable. The retrieved wind speeds all had an unbiased error, an RMSE value less than 1.1 m/s, and a correlation coefficient of more than 0.96.

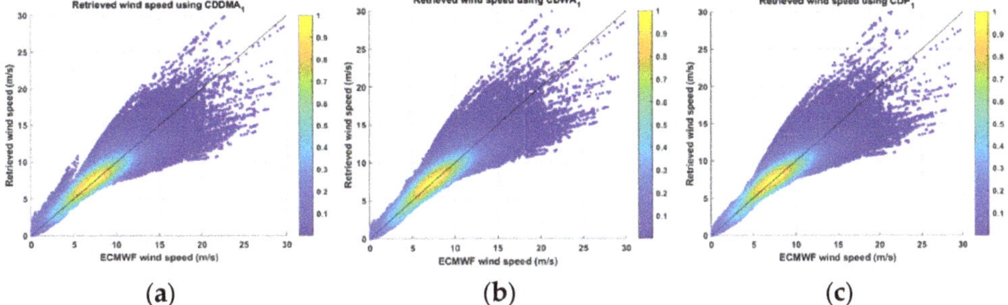

Figure 12. Scatter density plot of true wind speeds versus retrieved wind speed using (**a**) $cDDMA_1$, (**b**) $cDWA_1$, and (**c**) cDP_1, respectively, extracted from the composite delay-Doppler map test data. The solid black line represents the 1:1 agreement wind speed.

Table 4. Retrieved wind speed performance statistics for the different observables using the test data set. Bias and root-mean-square error (RMSE) are expressed in m/s. R represents the correlation coefficient between the ground truth wind speeds and the retrieved wind speeds.

Observable	Bias	RMSE	R
cDDMA	−0.0057	1.0294	0.9603
cDWA	0.0002	0.9811	0.9640
cDP	0.0001	0.9657	0.9652

The obvious benefit from using cDP to retrieve wind speed is the higher geometric resolution, which is valuable when monitoring typhoon information. In this paper, however, we cannot affirm that using cDP with the proposed GNSS-R measurement is the best choice to retrieve wind speed, because the simulation is not able to reveal real ocean surface situations. Therefore, it is necessary to collect real data to test and evaluate the performance using different observables.

Additionally, the retrieval error under different wind speeds for all observables was also analyzed, as shown in Figure 13. The error plots were calculated using wind speed intervals of width ±1 m/s. It is clear that retrieving accuracy decreases as wind speed increases when the wind speed is more than 25 m/s. The figure also shows that the RMSE increases with increases in the wind speed when the wind speed is more than 15 m/s. Overall, the proposed method can provide unbiased retrieved wind speed with an RMSE of less than 2 m/s for wind speeds lower than 23 m/s. The reason for the worse retrieval accuracy and precision at high wind speeds is that the sensitivity of the observables on wind speed decreases, as shown in Figure 12.

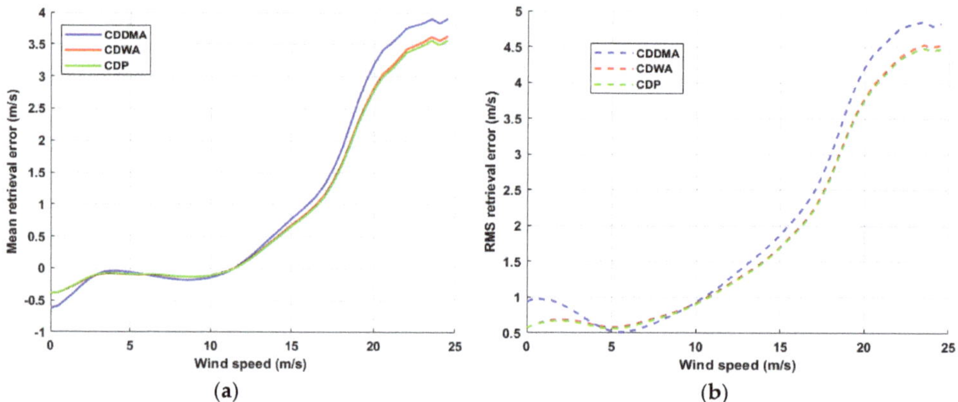

Figure 13. (a) Mean and (b) RMS retrieval error versus ground-truth wind speed for (blue) cDDMA, (red) composite delay-waveform average (cDWA), and (green) cDDM peak (cDP).

Figure 14 shows the probability density function (PDF) for the ground truth wind speeds and retrieved winds using different observables based on the proposed method. Obviously, regardless of which observables are used, the proposed method produces a PDF coincident with the reference PDF. In conclusion, the results obtained from the three observables have similar performance and yield impressive retrieval capabilities.

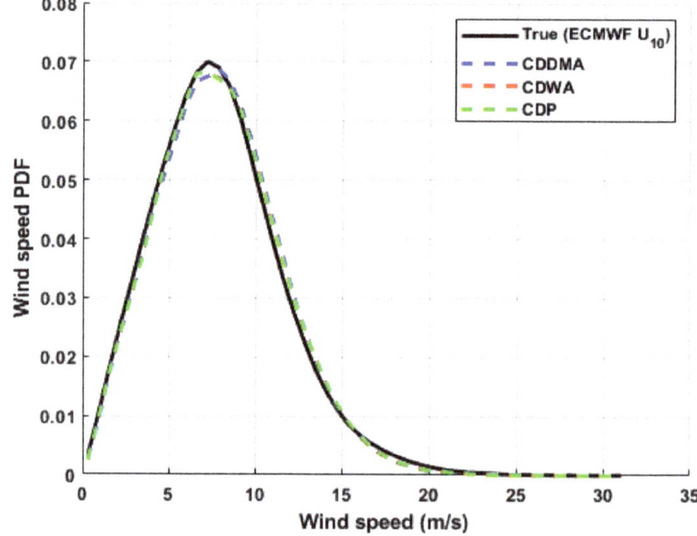

Figure 14. Probability density function (PDF) of reference winds (solid back) and retrieved winds using the proposed method with cDDMA (broken blue line), cDWA (broken red line), cDP (broken green line).

4. Discussion

This research describes a remote sensing method for deriving the ocean wind speed for the upcoming TRITON mission. On the other hand, as the United States proceeds to deploy GPS III satellites, the new signals are expected to benefit not only positioning services but also remote sensing performance. Therefore, we proposed a new GNSS-R measurement, composite delay-Doppler map (cDDM), by utilizing the properties in the new signals and used the proposed GNSS-R measurement to develop a wind speed retrieval algorithm specifically for the TRITON. In addition, we also compared the performance of wind speed retrieval between different models. The comparison results indicated that the retrieval difference between different models is less. It could be concluded that the relationship between the observable and the geophysical parameter dominates the retrieval performance. Nonetheless, the proposed dual model yields better retrieval accuracy and precision than other models at low wind speed.

In addition, the performance evaluation of wind speed retrieval by using different observable was conducted. The results show that the observables extracted from the cDDM all could provide unbiased and high precision wind speed estimations. Furthermore, the derivation of the cDDM suggests that system calibration may be disregarded. This contribution was based on the hypothesis that different signal components (i.e., L1 C/A and L1C) suffer the same effects from noise and system gain. The algorithm has been tested with simulated DDMs that represent the expected features of the TRITON in-orbit measurements. However, many effects cannot be verified through simulations, such as ocean surface conditions around a tropical cyclone. It should be noted that the allocated transmission power for L1 C/A and L1C may vary among different GPS satellite vehicles. Future studies should account for the effect of L1 C/A to L1C transmission power differences between different navigation satellites when implementing the proposed method.

Compared with previous studies, as described in Section 1, this research is aimed at investigating the potential benefit of new generation GNSS signal characteristics for GNSS-R remote sensing. A new GNSS-R measurement, composite DDM, is proposed by processing and integrating BPSK-modulated signal and BOC-modulated signal. Through the derivation and simulation analysis, it seems that the proposed method is insensitive to the variation of the transmitter power and antenna gain, the factors which are of great concern in most of the conventional methods. The present study also compared the retrieval performance by using different fitting functions to develop GMF. In comparison, most of the previous work only develop their GMF using a specific fitting function. The reason may be limited by the relationship between the derived observable based on the conventional GNSS-R measurement and the remote sensing parameter. In this regard, only a slight difference is shown by modeling the relationship between observables that are extracted from the cDDM and wind speed using different fitting functions.

5. Conclusions

In this study, we propose and evaluate the effectiveness of a new GNSS-R measurement, the composite delay-Doppler map (cDDM), from the perspective of retrieving sea wind speed, by utilizing the signal characteristics of the next-generation GPS through a simulation. The ground track for the upcoming TRITON satellite, which is designed to perform GNSS-R mission, is simulated. The ECMWF data are used to provide the ground truth 10-m reference wind speed so that the DDMs including the proposed cDDMs which are regarded as the main observables in a GNSS-R mission can be generated. The cDDM is a combination of the DDM from BPSK signals and the DDM from BOC-modulated signals. As modern GNSS satellites broadcast navigation signals under these two modulations, the cDDM can be constructed. The paper emphasizes on the benefits of using cDDM for data retrieval. To this end, it is shown that the cDDM based method is less sensitive to the variations of the transmitted power and antenna gain. Three observations, namely, cDDMA, cDWA, and cDP, are extracted from the cDDM and associated with wind speed through a dual-model geophysical model function (GMF). The statistical analysis shows that the retrieved wind speed using cDP exhibits the best performance in comparison with cDDMA and cDWA. The paper provides a new and potential GNSS-R observable and processing approach on GNSS-R remote sensing. In the future, the cDDM based retrieval methods will be further investigated and its applications to TRITON will be studied.

Author Contributions: Conceptualization, H.-Y.W.; Formal analysis, H.-Y.W.; Methodology, H.-Y.W.; Project administration, J.-C.J.; Supervision, J.-C.J.; Writing—original draft, H.-Y.W.; Writing—Creview & editing, J.-C.J. All authors have read and agreed to the published version of the manuscript.

Funding: This research was funded by the National Space Organization, Taiwan, grant number NSPO-S-107151.

Acknowledgments: The authors would like to thank ECMWF for sharing the reanalysis data for the public. The authors would like to thank the reviewers for their reviews and comments that have helped to improve this article.

Conflicts of Interest: The authors declare no conflict of interest.

Abbreviations

The following abbreviations are used in this manuscript:

BOC	Binary Offset Carrier
BPSK	Binary Phase Shift Keying
cDDM	Composite Delay Doppler Map
cDDMA	Composite Delay Doppler Map Average
cDWA	Composite Delay Waveform Average
cDP	Composite Delay Doppler Map Peak
CYGNSS	Cyclone Global Navigation Satellite System
DDM	Delay Doppler Map
DDMA	Delay Doppler Map Average
DW	Delay Waveform
ECMWF	European Centre for Medium-Range Weather Forecasts
FDS	Fully Developed Sea
GMF	Geophysical Model Function
GNSS	Global Navigation Satellite System
GPS	Global Positioning System
LES	Leading Edge Slope
LHCP	Left Hand Circular Polarized
NDW	Normalized Delay Waveform
NOAA	National Oceanic and Atmospheric Administration
NSPO	National Space Organization
PVT	Position, Velocity, Time
QZSS	Quasi-Zenith Satellite System
RO	Radio Occultation
TDS	TechDemoSate
U_{10}	Ocean Surface Wind Speed reference to a 10 m Height
YSLF	Young Sea/Limited Fetch

References

1. Rodriguez-Alvarez, N.; Bosch-Lluis, X.; Camps, A.; Vall-Llossera, M.; Valencia, E.; Marchan-Hernandez, J.F.; Ramos-Perez, I. Soil moisture retrieval using GNSS-R techniques: Experimental results over a bare soil field. *IEEE Trans. Geosci. Remote Sens.* **2009**, *47*, 3616–3624. [CrossRef]
2. Rodriguez-Alvarez, N.; Camps, A.; Vall-Llossera, M.; Bosch-Lluis, X.; Monerris, A.; Ramos-Perez, I.; Valencia, E.; Marchan-Hernandez, J.F.; Martinez-Fernandez, J.; Baroncini-Turricchia, G. Land geophysical parameters retrieval using the interference pattern GNSS-R technique. *IEEE Trans. Geosci. Remote Sens.* **2010**, *49*, 71–84. [CrossRef]
3. Chew, C.; Shah, R.; Zuffada, C.; Hajj, G.; Masters, D.; Mannucci, A.J. Demonstrating soil moisture remote sensing with observations from the UK TechDemoSat-1 satellite mission. *Geophys. Res. Lett.* **2016**, *43*, 3317–3324. [CrossRef]
4. Cardellach, E.; Fabra, F.; Nogués-Correig, O.; Oliveras, S.; Ribó, S.; Rius, A. GNSS-R ground-based and airborne campaigns for ocean, land, ice, and snow techniques: Application to the GOLD-RTR data sets. *Radio Sci.* **2011**, *46*, 1–16. [CrossRef]

5.	Yan, Q.; Huang, W. Spaceborne GNSS-R sea ice detection using delay-Doppler maps: First results from the UK TechDemoSat-1 mission. *IEEE J. Sel. Top. Appl. Earth Obs. Remote Sens.* **2016**, *9*, 4795–4801. [CrossRef]

6.	Alonso-Arroyo, A.; Zavorotny, V.U.; Camps, A. Sea ice detection using UK TDS-1 GNSS-R data. *IEEE Trans. Geosci. Remote Sens.* **2017**, *55*, 4989–5001. [CrossRef]

7.	Gleason, S.; Hodgart, S.; Sun, Y.; Gommenginger, C.; Mackin, S.; Adjrad, M.; Unwin, M. Detection and processing of bistatically reflected GPS signals from low earth orbit for the purpose of ocean remote sensing. *IEEE Trans. Geosci. Remote Sens.* **2005**, *43*, 1229–1241. [CrossRef]

8.	Rodriguez-Alvarez, N.; Akos, D.M.; Zavorotny, V.U.; Smith, J.A.; Camps, A.; Fairall, C.W. Airborne GNSS-R wind retrievals using delay–Doppler maps. *IEEE Trans. Geosci. Remote Sens.* **2012**, *51*, 626–641. [CrossRef]

9.	Foti, G.; Gommenginger, C.; Jales, P.; Unwin, M.; Shaw, A.; Robertson, C.; Rosello, J. Spaceborne GNSS reflectometry for ocean winds: First results from the UK TechDemoSat-1 mission. *Geophys. Res. Lett.* **2015**, *42*, 5435–5441. [CrossRef]

10.	Zavorotny, V.U.; Voronovich, A.G. Scattering of GPS signals from the ocean with wind remote sensing application. *IEEE Trans. Geosci. Remote Sens.* **2000**, *38*, 951–964. [CrossRef]

11.	Zavorotny, V.U.; Gleason, S.; Cardellach, E.; Camps, A. Tutorial on remote sensing using GNSS bistatic radar of opportunity. *IEEE Geosci. Remote Sens. Mag.* **2014**, *2*, 8–45. [CrossRef]

12.	Clarizia, M.P.; Ruf, C.S.; Jales, P.; Gommenginger, C. Spaceborne GNSS-R minimum variance wind speed estimator. *IEEE Trans. Geosci. Remote Sens.* **2014**, *52*, 6829–6843. [CrossRef]

13.	Clarizia, M.P.; Ruf, C.S. Wind speed retrieval algorithm for the Cyclone Global Navigation Satellite System (CYGNSS) mission. *IEEE Trans. Geosci. Remote Sens.* **2016**, *54*, 4419–4432. [CrossRef]

14.	Ruf, C.S.; Balasubramaniam, R. Development of the CYGNSS geophysical model function for wind speed. *IEEE J. Sel. Top. Appl. Earth Obs. Remote Sens.* **2018**, *12*, 66–77. [CrossRef]

15.	Ruf, C.S.; Gleason, S.; McKague, D.S. Assessment of CYGNSS wind speed retrieval uncertainty. *IEEE J. Sel. Top. Appl. Earth Obs. Remote Sens.* **2018**, *12*, 87–97. [CrossRef]

16.	Park, J.; Said, F.; Katzberg, S.J.; Soisuvarn, S.; Jelenak, Z.; Chang, P.S. Analysis of CYGNSS wind characteristics with NOAA L2 retrievals and TES method. In Proceedings of the IGARSS 2019–2019 IEEE International Geoscience and Remote Sensing Symposium, Yokohama, Japan, 28 July–2 August 2019; pp. 8043–8045.

17.	Gao, H.; Yang, D.; Wang, F.; Wang, Q.; Li, X. Retrieval of Ocean Wind Speed Using Airborne Reflected GNSS Signals. *IEEE Access* **2019**, *7*, 71986–71998. [CrossRef]

18.	Wang, Q.; Zhu, Y.; Kasantikul, K. A Novel Method for Ocean Wind Speed Detection Based on Energy Distribution of Beidou Reflections. *Sensors* **2019**, *19*, 2779. [CrossRef]

19.	Juang, J.-C.; Lin, C.-T.; Tsai, Y.-F. Comparison and Synergy of BPSK and BOC Modulations in GNSS Reflectometry. *IEEE J. Sel. Top. Appl. Earth Obs. Remote Sens.* **2020**, *13*, 1959–1971. [CrossRef]

20.	Kaplan, E.; Hegarty, C. Undestanding GPS/GNSS: Principles and applications. In *GNSS Technology and Applications Series*; Artech House Publisher: Norwood, MA, USA, 2017.

21.	Juang, J.-C.; Ma, S.-H.; Lin, C.-T. Study of GNSS-R techniques for FORMOSAT mission. *IEEE J. Sel. Top. Appl. Earth Obs. Remote Sens.* **2016**, *9*, 4582–4592. [CrossRef]

22.	Juang, J.-C.; Tsai, Y.-F.; Lin, C.-T. FORMOSAT-7R Mission for GNSS Reflectometry. In Proceedings of the IGARSS 2019–2019 IEEE International Geoscience and Remote Sensing Symposium, Yokohama, Japan, 28 July–2 August 2019; pp. 5177–5180.

23.	Gleason, S. Algorithm Theoretical Basis Document Level 1A DDM Calibration. Available online: https://clasp-research.engin.umich.edu/missions/cygnss/reference/148-0136_ATBD_L1A_DDMCalibration_Rev2_Aug2018_release.pdf (accessed on 20 August 2020).

24.	Gleason, S.; Ruf, C.S.; O'Brien, A.J.; McKague, D.S. The CYGNSS Level 1 calibration algorithm and error analysis based on on-orbit measurements. *IEEE J. Sel. Top. Appl. Earth Obs. Remote Sens.* **2018**, *12*, 37–49. [CrossRef]

25.	Gleason, S.; Ruf, C.S.; Clarizia, M.P.; O'Brien, A.J. Calibration and unwrapping of the normalized scattering cross section for the cyclone global navigation satellite system. *IEEE Trans. Geosci. Remote Sens.* **2016**, *54*, 2495–2509. [CrossRef]

26.	Fabra, F.; Cardellach, E.; Li, W.; Rius, A. WAVPY: A GNSS-R open source software library for data analysis and simulation. In Proceedings of the 2017 IEEE International Geoscience and Remote Sensing Symposium (IGARSS), Fort Worth, TX, USA, 23–28 July 2017; pp. 4125–4128.

27. Unwin, M.; Jales, P.; Tye, J.; Gommenginger, C.; Foti, G.; Rosello, J. Spaceborne GNSS-reflectometry on TechDemoSat-1: Early mission operations and exploitation. *IEEE J. Sel. Top. Appl. Earth Obs. Remote Sens.* **2016**, *9*, 4525–4539. [CrossRef]

28. Rodriguez-Alvarez, N.; Garrison, J.L. Generalized linear observables for ocean wind retrieval from calibrated GNSS-R delay–Doppler maps. *IEEE Trans. Geosci. Remote Sens.* **2015**, *54*, 1142–1155. [CrossRef]

 *remote sensing*

Article

First Evidences of Ionospheric Plasma Depletions Observations Using GNSS-R Data from CYGNSS

Carlos Molina [1,2,*] and Adriano Camps [1,2]

[1] CommSensLab-UPC, Department of Signal Theory and Communications, UPC BarcelonaTech, c/Jordi Girona 1-3, 08034 Barcelona, Spain; camps@tsc.upc.edu

[2] Institut d'Estudis Espacials de Catalunya-IEEC/CTE-UPC, Gran Capità, 2-4, Edifici Nexus, Despatx 201, 08034 Barcelona, Spain

* Correspondence: carlos.molina@upc.edu

Received: 10 October 2020; Accepted: 14 November 2020; Published: 18 November 2020

Abstract: At some frequencies, Earth's ionosphere may significantly impact satellite communications, Global Navigation Satellite Systems (GNSS) positioning, and Earth Observation measurements. Due to the temporal and spatial variations in the Total Electron Content (TEC) and the ionosphere dynamics (i.e., fluctuations in the electron content density), electromagnetic waves suffer from signal delay, polarization change (i.e., Faraday rotation), direction of arrival, and fluctuations in signal intensity and phase (i.e., scintillation). Although there are previous studies proposing GNSS Reflectometry (GNSS-R) to study the ionospheric scintillation using, for example TechDemoSat-1, the amount of data is limited. In this study, data from NASA CYGNSS constellation have been used to explore a new source of data for ionospheric activity, and in particular, for travelling equatorial plasma depletions (EPBs). Using data from GNSS ground stations, previous studies detected and characterized their presence at equatorial latitudes. This work presents, for the first time to authors' knowledge, the evidence of ionospheric bubbles detection in ocean regions using GNSS-R data, where there are no ground stations available. The results of the study show that bubbles can be detected and, in addition to measure their dimensions and duration, the increased intensity scintillation (S_4) occurring in the bubbles can be estimated. The bubbles detected here reached S_4 values of around 0.3–0.4 lasting for some seconds to few minutes. Furthermore, a comparison with data from ESA Swarm mission is presented, showing certain correlation in regions where there is S_4 peaks detected by CYGNSS and fluctuations in the plasma density as measured by Swarm.

Keywords: ionosphere; scintillation; GNSS-R; equatorial plasma depletions; bubbles; CYGNSS; swarm

1. Introduction

1.1. Physics of the Ionosphere

The ionosphere is a layer of the atmosphere that plays a very important role in satellite communications and Earth Observation. This layer, which ranges from around 60 km to more than 500 km altitude, contains free electrons and ions, making a sort of "electric conductor" that interacts with the electromagnetic waves crossing it.

The shape and density in electron density profile, are highly affected by many factors in a balance between production and destruction processes driven, mainly, by solar irradiation. Also, its dynamics is influenced by the movements of the inner atmospheric layers and the outer magnetosphere. Because of all these factors, the ionospheric fluctuations take place in a wide range of amplitudes and duration, from kilometers to several meters and from a few seconds or minutes to days.

It has been observed that there is an increase in the activity, related to solar irradiation, in the transitions (sunrise and sunset), with a rearrangement of the layers. Seasonal dependence is also noticed, with more activity during the periods around the Spring and Autumn Equinoxes. And, on top of that, there is a positive correlation with solar activity following the 11-year solar cycle, with a minimum happening this year 2020.

Electromagnetic waves crossing the ionosphere may suffer from those perturbations, making the signal to change its direction of propagation, polarization, or creating fluctuations in intensity and phase, the so-called, ionospheric scintillation. This phenomenon is known since the first satellites put in orbit and it has been a subject of study until today. Some models have been developed to try to describe and predict its behavior. During the past years, some ionospheric scintillation models have appeared, for example:

- The Global Ionospheric Scintillation propagation Model (GISM) [1] provides time series of intensity and phase scintillation models using a turbulent ionosphere, and an electromagnetic waves propagator in turbulent media, using the Multiple Phase Screen theory (MPS) [2]. GISM is the model accepted by the ITU-R (International Telecommunication Union - Radiocommunication Sector) for the ionospheric communications [3].
- The WideBand MODel (WBMOD) [4] is also based on electron-density irregularities and uses the MPS theory to compute the scintillation effects in an statistical sense. In this model it is possible to set-up a communication scenario (time, location, and other geophysical conditions).
- The Wernik-Alfonsi-Materassi Model (WAM) [5] also uses the MPS theory, but generates its statistics from in situ measurements on ionization fluctuations measured by the Dynamics Explorer 2 satellite. It can also predict the ionospheric S_4 index along a defined path given some environmental conditions.

More recently, in the context of a European Space Agency (ESA) project, the GISM functionalities were extended in the UPC/OE/RDA (Universitat Politecnica de Catalunya/Observatori de l'Ebre/Research and Development in Aerospace) SCIONAV model [6], to reproduce more realistically the behavior of ionospheric scintillation at both low and high latitudes, taking into account the different physical phenomena that create them, such as the bubbles and depletions in equatorial regions and it is based on lookup tables resulting from extensive analyses.

However, all these models require further improvements:

- In polar and auroral regions the models still need a better description in terms of velocity, distribution, duration and intensity of the ionospheric effects.
- A better modelling (3D) of the equatorial plasma bubbles (EPBs) can be incorporated in the UPC/OE/RDA SCIONAV model, relating its 3D properties to the altitude, and the amount of plasma depletion with the S_4 produced [7].
- In GNSS-R applications, some anomalous fluctuations have been reported in equatorial regions in regions over calm ocean, which are supposed to come from ionospheric scintillation [8].

1.2. Using GNSS-R to Study the Ionosphere

The goal of this study is to provide a way to improve the ionospheric models, by means of GNSS-R. This idea was firstly proposed in 1996 by S. J. Katzberg and J. L. Garrison [9], but not really explored the following years because the reflection over ocean (rough) surfaces is mostly incoherent, and sensitive to the winds, as reported in an airborne experiment of the same authors [10]. It was not until 2016 that it was shown [8] that the large fluctuations in the peak of the measured Delay-Doppler Map of GNSS-R instruments around the geomagnetic equator in calm sea conditions (wind speeds < 3 m/s), where reflections are more coherent, could be due to ionospheric scintillation.

The main purpose of GNSS-R missions is to study geophysical parameters of the Earth by studying the reflection of GNSS signals on the Earth's surface. Applications include sea altimetry, soil moisture measurement or ice detection. The signals used to apply this technique cross the ionosphere in their down-welling and up-welling paths, that is, first from the GNSS satellite to the specular reflection point on the Earth's surface, and then from this point to the GNSS-R receiver. Because of the height of the satellites involved, the down-welling path always crosses all the ionospheric layers, but the up-welling one may only cross some of them, depending on the altitude of the GNSS-R instrument.

Furthermore, GNSS signals would only allow the study of ionospheric scintillation in the GNSS bands, usually the L1/E1/B1, L2, and L5/E5. Other bands object of interest nowadays in Earth Observation, such as P-band (to be used in ESA BIOMASS Mission) are also affected by these perturbations, but they cannot be detected and/or quantified by this technique, although it could be applied using other signals of opportunity such as those from MUOS (Mobile User Objective System) satellites [11].

2. Materials and Methods

The current study uses open data from NASA CYGNSS (Cyclone Global Navigation Satellite System) mission [12,13], which continuously delivers GNSS-R data since March 2017. NASA CYGNSS is a microsatellite constellation led by the University of Michigan and the Southwest Research Institute, launched and operated by NASA in December 2016. The mission main goal is to provide a better forecast of hurricanes by studying the interaction between the sea and the atmosphere.

2.1. CYGNSS Dataset Description

The open-access CYGNSS database is provided by the Physical Oceanography Distributed Active Archive Center (PO.DAAC) [14]. The data is recorded continuously since 17 March 2017, for each of the 8 satellites in orbit. The data includes the DDM (Delay-Doppler Map), that is the cross-correlation of the reflected signal with a replica of the transmitted signal for different delay lags, and Doppler frequencies [15]. CYGNSS provides DDMs from up to the 4 channels tracking the signals received from 4 different GPS satellites, with a sample rate of 1 Hz. For each sample, auxiliary data such as the timestamp, the position of both CYGNSS satellite and the emitting GPS spacecraft are provided, together with the post-computed position of the specular point for the reflected signal on the Earth's ellipsoid.

In any case, the data from reflections is only available from around 40°N to 40°S around the equator, as the satellite orbit inclination is around 35°. The orbital period is ∼95 min but, as the constellation is formed by 8 satellites distributed along the orbit, the overall revisit time is much less. Figure 1 shows the average number of samples measured during one day by the whole constellation as a function of the latitude per different cell sizes. CYGNSS has an almost circular Low Earth Orbit (LEO) at an altitude of ∼520 km, at an average speed of 7.6 km/s, much larger than the GPS satellites linear velocity (3.89 km/s), which complete an orbit every 12 h. Table 1 summarizes the mission parameters.

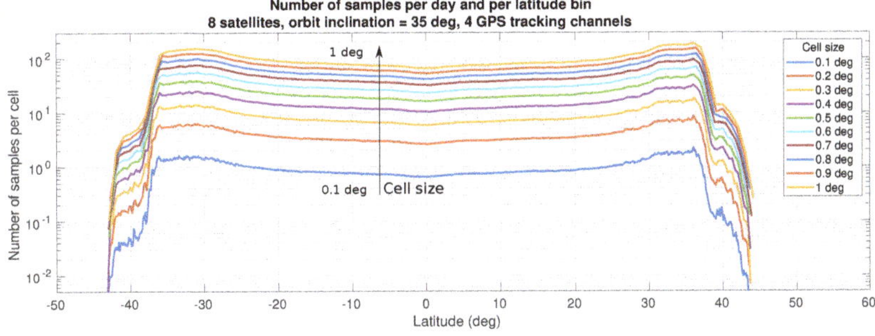

Figure 1. Average number of samples per day and per cell, for different cell sizes (from 0.1° × 0.1° to 1° × 1°) for the 8 Cyclone Global Navigation Satellite System (CYGNSS) satellites tracking up to 4 GPS satellites simultaneously from 90 min period and 35° of inclination orbit.

Table 1. Summary of CYGNSS mission.

Parameter	Value
Orbital altitude	~520 km
Inclination	~35°
Period	95 min
Linear speed	7.6 km·s^{-1}
Covered region	40°N to 40°S
Number of satellites	8
Number of GNSS satellites tracked per receiver	4
DDM sample rate	1 Hz
DDM size	11 × 17 (Doppler, delay)
DDM Doppler bin resolution	200 Hz
DDM Delay bin resolution	0.2552 chips
Total number of DDMs per day	2.7 × 10^6
Dataset volume per day	10.4 GB

2.2. Data Processing

The data processing is performed to derive the CYGNSS observables of ionospheric activity. Source data comes from the DDMs. DDMs provided by CYGNSS has a size of 11 Doppler frequency bins × 17 delay bins, with a resolution of 200 Hz per Doppler bin, and 0.2552 chips delay bin. Since 1 C/A chip is equal to 1/1,023,000 s, it is equivalent to 293.3 m. Therefore, the resolution per delay bin is 74.8 m. The images below present two examples of CYGNSS's DDM over two different surfaces.

In order to assess the conditions of sea surface during the acquisition of these measurements, sea wave height is overlaid with the specular reflection points where the DDMs were taken, and it is shown for both cases in Figure 2, at points labeled as A and B. In the case of the DDM for rough water surface (Figure 3a) the *SNR* (Signal-to-Noise Ratio) is 2.3 dB and the wave height is around 3 m, marked in the map as A. For the calm ocean in Figure 3b, the *SNR* is much higher, around 15.3 dB, and the waves are around 0.6 m, marked as B in the map.

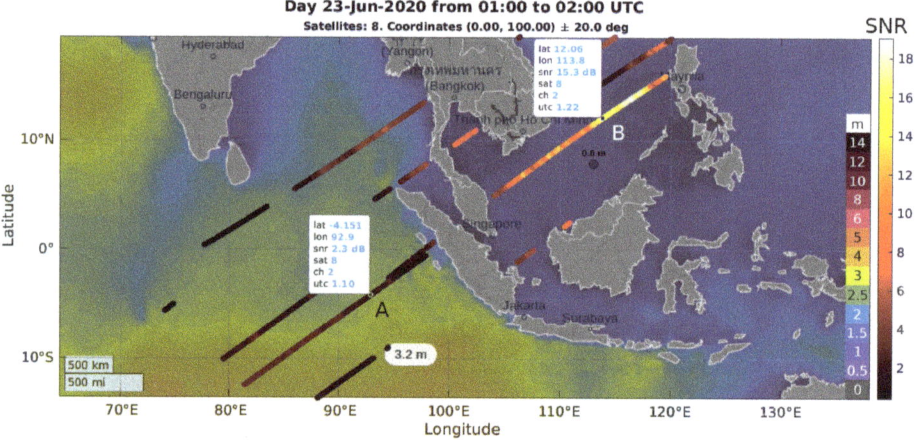

Figure 2. Sea surface roughness shown as the wave height for the selected DDMs in Figure 3 is underlaid the *SNR* (Signal-to-Noise Ratio) values (in dB) for some samples around the exact coordinates of the DDM, marked with labels A and B, respectively.

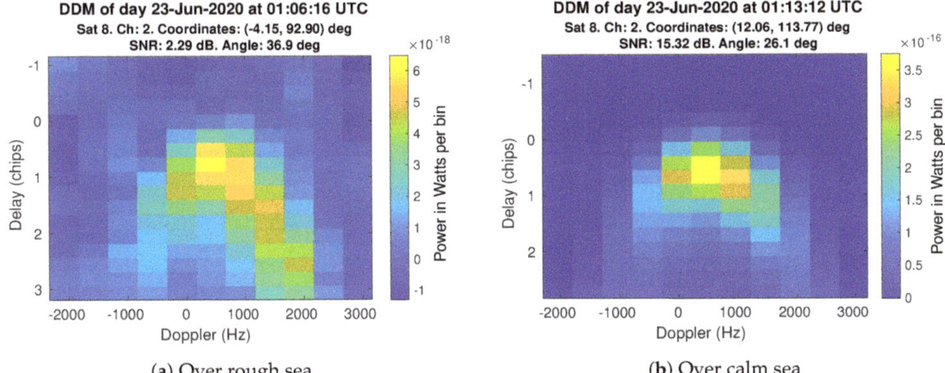

(**a**) Over rough sea   (**b**) Over calm sea

Figure 3. CYGNSS Delay-Doppler Maps (DDMs) examples of specular reflections over two different surfaces: (**a**) rough sea surface and (**b**) calm sea surface. Plots represent the equivalent analog power in Watts received per each delay/Doppler bin, distributed around the adjusted zero delay/Doppler for the particular geometry configuration. Note the difference in the color scale.

To perform this study, the main source of information is taken from the *SNR* (Signal-to-Noise Ratio) measurements available in the CYGNSS database. The *SNR* is post-computed on the ground using the DDM values as follows:

$$SNR = 10 \log_{10} \frac{DDM_{max}}{N_{avg}}, \tag{1}$$

where DDM_{max}, is the value of the *DDM* bin with maximum signal and N_{avg} is the average noise per bin, computed from the delay-Doppler bins before the DMM peak (i.e., delay < 0, as in Reference [8]. The value of the *SNR* is then given in decibels (dB). The maximum number of samples per day, considering the 8 satellites and their 4 channels, is around 2.7×10^6 measurements. An example of this data is plotted in a world map in Figure 4, where both the measurement over land and oceans are displayed. Colorbar indicates the *SNR* of reflected signal in dB.

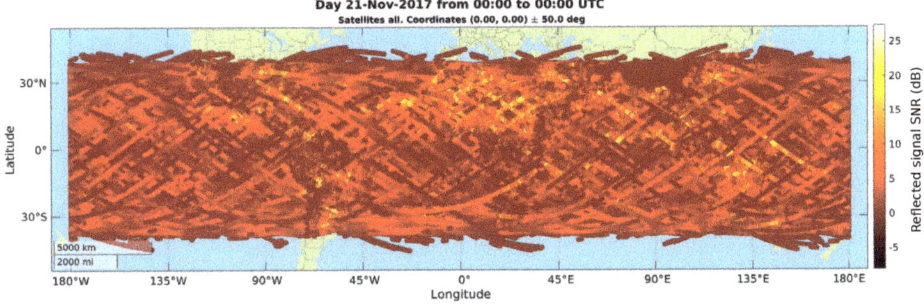

Figure 4. SNR values on specular reflection points for the full day 21 November 2017, on all regions (sea and land) from 40°S to 40°N. Values in dB.

As the aim of the study is to observe the possible fluctuations in the intensity of the signal after crossing the ionosphere, the following computation is performed for every vector of data in a day. Having the 4 channels of SNR values during a whole day, it is chopped from a user-input "start_time" to an "end_time". Then, using the *quality flags* that the CYGNSS database provides, the vector is chopped using only the points over the oceans farther away than 25 km from the coastline, and within a rectangular area that the user can define.

The output is a vector of data with internal discontinuities in time. In each per-channel vector, there are also discontinuities in the GPS satellite tracked. These discontinuities can be detected by the change of the PRN (Pseudo-Random Number) that identifies the tracked GNSS transmitter. Both types of jumps must be avoided in the computation of S_4, because they may create false S_4 peaks. So, the vector, per each of the 4 channels, is divided into those periods between jumps (PRN or time jumps), and then the computation of the S_4 value is done using:

$$S_4 = \sqrt{\frac{\langle I^2 \rangle - \langle I \rangle^2}{\langle I \rangle^2}},\tag{2}$$

where I is intensity in linear units, computed from the *SNR* value in dB, and the average is computed using a 12-samples-moving window along the vector.

In Figure 5, the S_4 values using the SNR data from Figure 4 are plotted. For the sake of clarity in the interpretation, the points where S_4 is lower than 0.1 are not shown, and for the rest, their point size and color are proportional to the S_4.

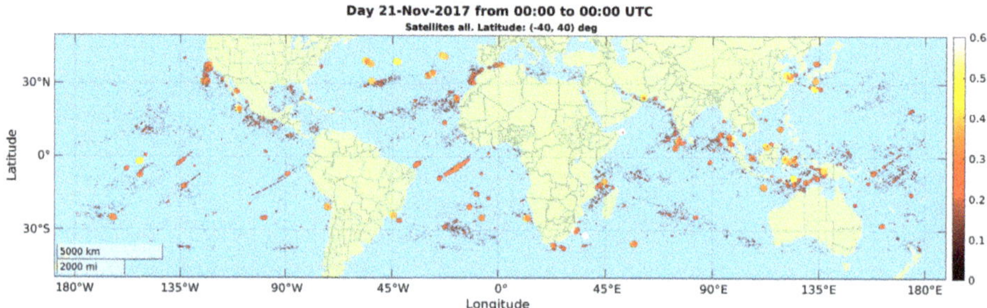

Figure 5. Computed and filtered S_4 values over all sea regions from 40°S to 40°N for 21 November 2017.

Additionally, the roughness of the water surface could make the reflection to be diffuse instead of specular. This would happen when there are high waves in the area, as opposed to having a flat ocean surface. In practice, this is making a sort of filter in which the possible scintillation happening over rough sea surfaces, is hidden, keeping only high S_4 values over regions with a calm ocean.

To check the wind and wave height for day and hour under study, ICON [16,17] model meteorological online data were used [18]. As an example, continuing with day 21 November 2017, in Figure 6, the colormap of wave-height along the Atlantic ocean is underlaid the S_4 values, only showing the ones that are larger than 0.05. It can be check that most of S_4 peaks only appear over regions with quite calm ocean.

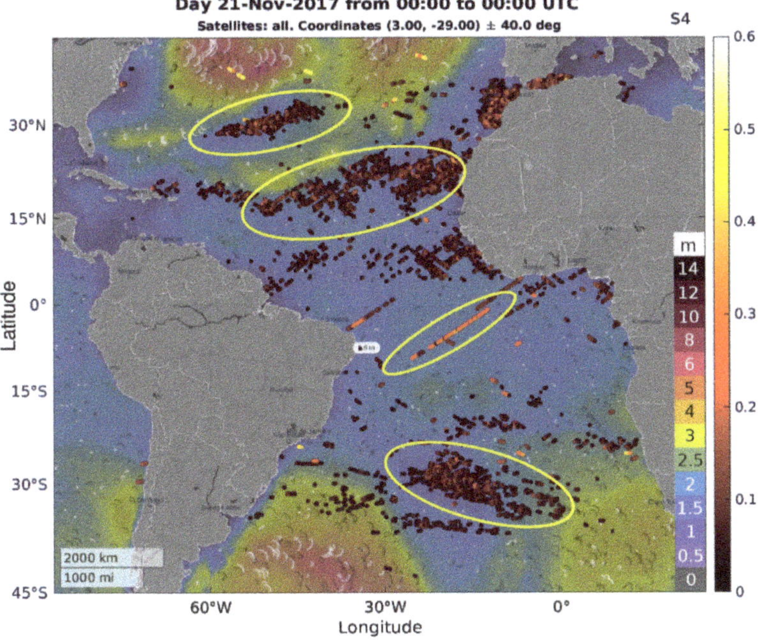

Figure 6. S_4 values above 0.05 over Atlantic ocean for 21 November 2017. It can be see that most of the S_4 points appear only when the ocean is relatively calm.

3. Results

3.1. Bubbles and Depletions

Using the data available from the year 2017 until present, after applying the processing method described above, several interesting results found are shown. The research is mainly focused on the detection of Equatorial Plasma Bubbles (EPBs) or depletions. Given the characteristics of this experiment, these bubbles are expected to take the shape of transient peaks in S_4 curves as the reflected signal path crosses the bubble, while the GNSS receiver is moving along its orbit.

A first visual inspection of some days, like the one depicted before, shows that there are many peaks no matter which day is analyzed. Some of them are isolated, but others occur along with other peaks in slightly different times and other satellites or channels but in a relatively small region. One of these peaks in S_4 is located in mid the Atlantic during 21 November 2017, shown in Figure 7, and it may serve as an example to study its origin.

This isolated event in the Atlantic is the consequence of the rapid fluctuation of the *SNR* of the signal arriving to channel 1 of CYGNSS satellite 2, at around 4:26 UTC. The approximate length of the peak is 1400 km, and it lasts for about 5 minutes. As it can be seen in the *SNR* plot (Figure 8a), channel 1 oscillates rapidly from 1 dB to 7 dB, starting at 2:13 h to almost 2:18 h Local Time (LT), according to the longitude of the region. During this event, the ocean is calm, with wave heights of at most 2 m, as seen in Figure 6.

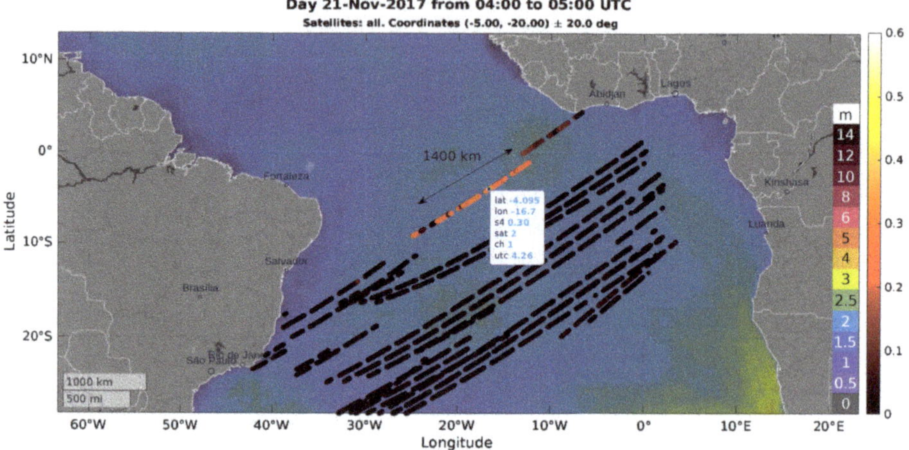

Figure 7. S_4 peak in middle Atlantic.

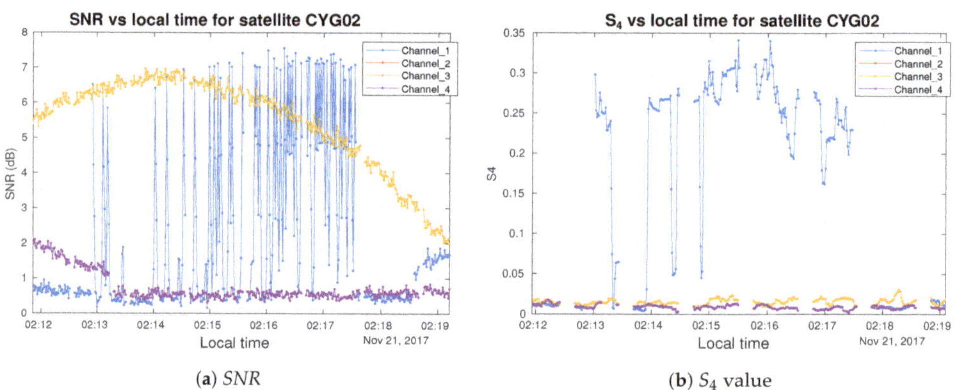

(a) *SNR*

(b) S_4 value

Figure 8. *SNR* fluctuations and S_4 value for 21 November 2017 peak vs LT (Local Time) corresponding to previous Figure 7.

Also, these plots help to explain that, given the way the S_4 is computed, only the *SNR* fluctuations occurring in a timescale of 12 s can appear as high/moderate S_4 values. In channel 3 of the same satellite, a slow drift of the *SNR* is observed at the same time as channel 1 is fluctuating quickly (Figure 8a). Despite this, the S_4 value in channel 3 keeps being very small, less than 0.05 (Figure 8b). On the other hand, the singular parabolic shape that it exhibits is due to the change in the incident angle of the reflected signal as the receiver satellite moves in the orbit capturing the signal of the much slower and higher GPS transmitter. In this case, the angle of incidence is almost vertical ($\sim$4°), when the signal is maximum ($\sim$7 dB), and increases up to around 25° at the end of the plot. In the case of channel 1, the angle of incidence varies less than 1° around 70° during all the fluctuation period.

The previous example was an isolated peak in just one of the 4 channels of one satellite, but in other days analyzed, there are concentrations of these events in a small region, occurring during a finite time period. Figure 9 shows data from 24 August 2017. The points in red mark moderate scintillation measurements around 0.2–0.3, and they appear in some concentrated regions for different channels and satellites crossing above those regions. During this day, almost all the North Atlantic was very calm, with wave heights less than 2 m, as seen in Figure 10.

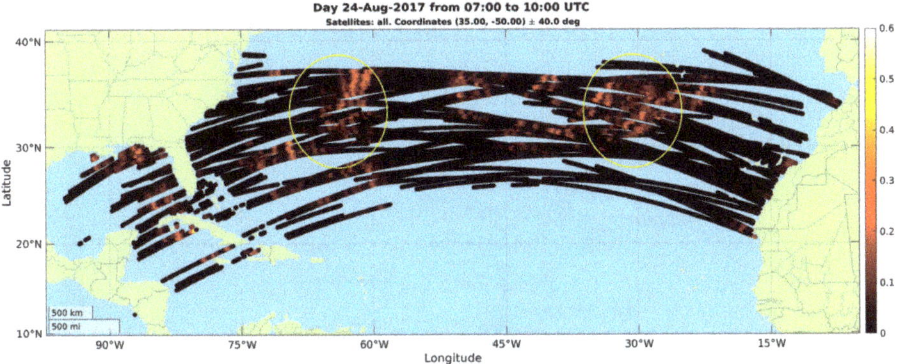

Figure 9. S_4 values over the Northern Atlantic for 24 August 2017, showing two regions of moderate scintillation.

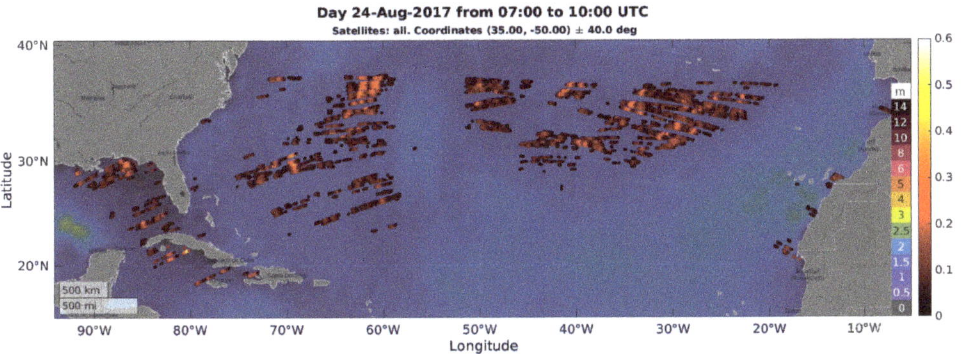

Figure 10. Filtered high S_4 values overlaid with the wave height for 24 August 2017, showing the calm sea surface.

To have a better view of these events, two detail plots are shown.

In the first one in the Western region, (Figure 11), located about 300 km North of Bermuda island, a large number of moderate S_4 points are closely spaced in a zone that extends approximately 250 km × 320 km across different CYGNSS satellites and channels. All the measurements occur during the period between 6 h and 10 h UTC, even lasting during different passes of CYGNSS satellites over this zone.

In this region, the perturbation is particularly interesting as the peaks occur roughly at the same positions for several passes or channels, with similar values of S_4 in each of them. The event starts around 7:00 h UTC and ends around 10:00 h UTC, so given that the timezone is UTC-4, the local time is from 3 h to 6 h in the morning, coinciding with the hours immediately before the sunrise, that, in this place, on 24 August 2017, took place at 5:36 h LT.

Focusing on this peak, the values of SNR and the derived S_4 are plotted in Figure 12a,b versus LT. The sudden increase of SNR values is clearly visible in two of the channels of the satellite CYG06. In other satellites (not shown in these graphs), there are also similar events, that constitute the peak in S_4 shown in Figure 11. It is important to note that the angle of the incident signal over the ocean is smoothly varying around 28° for channel 3, and from 14° to 10° in the case of channel 1, so the change in the reflection angle is not the cause of the rapid fluctuation of the signal.

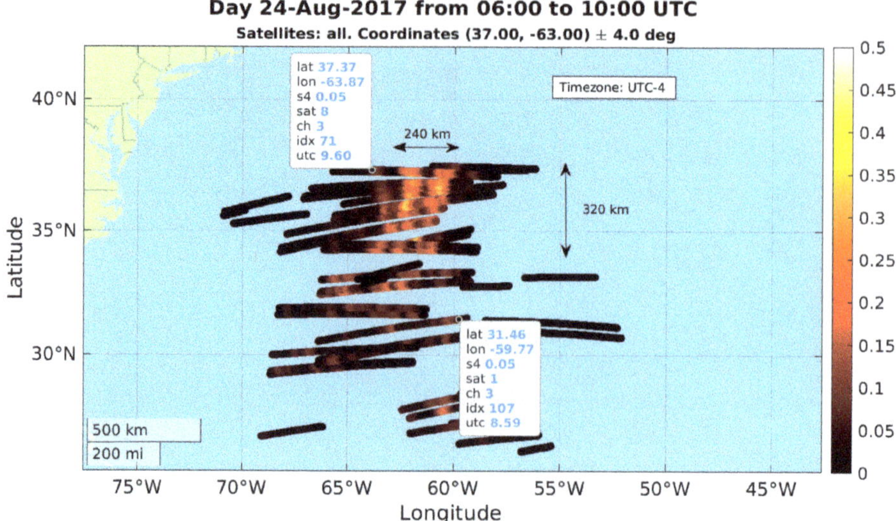

Figure 11. Peak in S_4 value in Western part of North Atlantic on 24 August 2017.

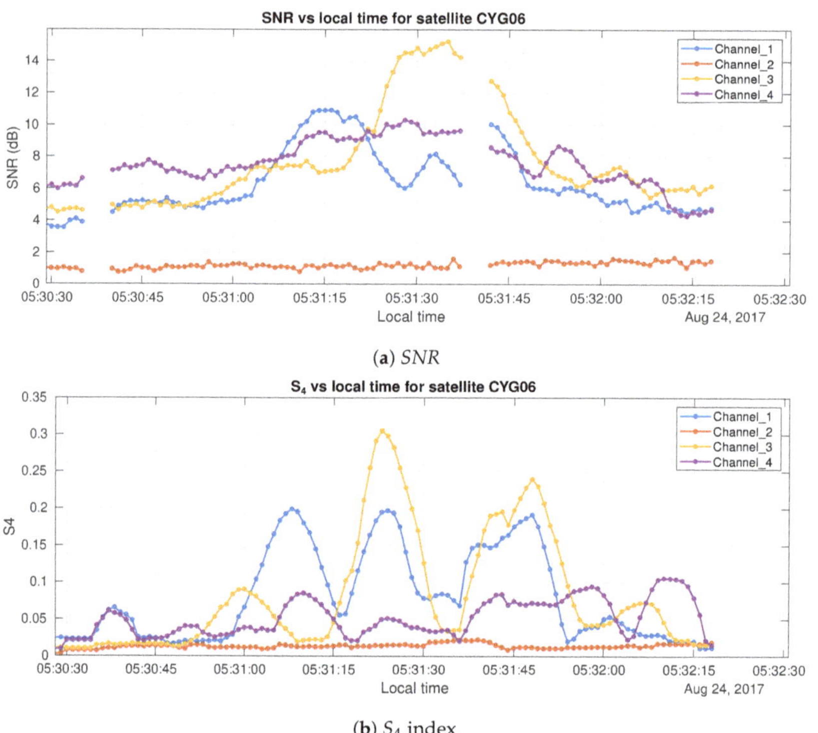

(**a**) *SNR*

(**b**) S_4 index

Figure 12. *SNR* and S_4 values plotted against LT for western peak in Figure 11 during 24 August 2017, for the four channels of one of CYGNSS satellites. Note that the S_4 rises with the transitions of SNR. For channel 3, it appears 2 peaks of S_4, one before and one after the SNR peak, and the same happens for channel 1.

In the S_4 map for the Eastern peak in North Atlantic (Figure 13), a region of about 900 km $\times$ 600 km of moderate scintillation is depicted. As in the previous case, different passes of CYGNSS satellites in several channels show values of S_4 larger than 0.2. This region is around 200 km to 600 km southwest from the Azores islands, and the water surface was also in calm, as shown in previous Figure 10 for the whole day in almost all the North Atlantic ocean.

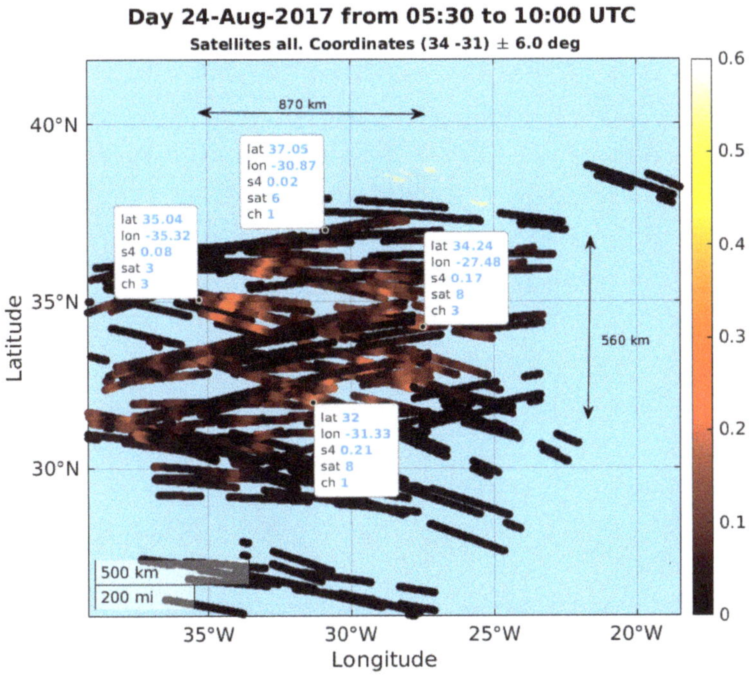

Figure 13. Peak in S_4 value in eastern part of northern Atlantic on 24 August 2017.

The information on these and other S_4 peaks is summarized in tables, recording for every event its date and time, duration, length, and magnitude of the S_4 value. In the data analysis these peaks are identified and characterized using the mentioned parameters. An event to be recorded must meet that the peak of S_4 must be larger than 0.2 during at least 5 s. Given that the integration window for computing S_4 is 12 s width, it means that a reported peak implies at least 17 seconds of fluctuations in the SNR of the received signal. Events shorter than this duration (or dimension) cannot be detected with the current coherent and incoherent integration times of the GNSS-R payload onboard. Table 2 shows the recorded peaks for 24 August 2017, from 7 h to 10 h (UTC), in all the North Atlantic region.

Table 2 shows the date and LT, the satellite and channel in which the peak was detected, the coordinates where the event started, its duration and length, the inclination of the incident signal on the sea (measured from the vertical), and the maximum value of S_4 achieved during the peak. The length of the event is computed measuring the distance between the first and the last geolocated points belonging to the peak.

Note that the highlighted row in Table 2 corresponds to the peak measured by satellite 6 in channel 3 on (37.27°N, 62.21°W) at 5:31 LT, which is the one plotted in yellow in Figure 12a,b. As it can be seen, the peak lasts for 7 seconds over the threshold of 0.2, archiving a maximum height of 0.31.

Another day of interesting results, this case in the Indian Ocean during 20 May 2019, is shown in Figure 14. As seen in the plot, the Indian Ocean was calm during that day in regions from the Equator to the North. The plot only shows the S_4 values greater than 0.05, and they are distributed in regions relatively delimited between the Equator and latitude 15°N.

Table 2. Table of peaks with S_4 larger than 0.2 during more than (or equal to) 5 s on 24 August 2017, in the North Atlantic region.

Date (dd-mm-yy)	LT	Sat	Ch	Lat (°)	Lon (°)	Duration (s)	Length (km)	Inclination (°)	Max S_4
24-08-2017	4:34	1	1	36.53	−60.96	9	67.3	29.2	0.31
24-08-2017	4:59	2	1	36.63	−62.13	7	52.8	21.1	0.25
24-08-2017	7:05	2	2	35.02	−34.24	6	44.1	26	0.28
24-08-2017	4:21	3	2	34.96	−50.14	6	43.8	23.8	0.27
24-08-2017	4:23	3	2	35.88	−41.51	7	52.1	17.3	0.24
24-08-2017	2:16	3	3	29.71	−69.37	7	48	10.3	0.36
24-08-2017	6:31	5	1	32.88	−29.54	5	35.6	18	0.22
24-08-2017	2:45	6	1	20.66	−82.03	5	32	53.2	0.25
24-08-2017	1:45	6	2	30.04	−87.14	152	1044.6	50.8	0.42
24-08-2017	5:31	6	3	37.27	−62.21	7	53.3	27.9	0.31
24-08-2017	7:14	7	2	33.22	−29.58	7	50.1	26.6	0.26
24-08-2017	2:24	7	3	30.42	−69.66	5	34.5	12.1	0.37
24-08-2017	2:22	7	4	19.67	−76.12	62	396.2	59.5	0.45
24-08-2017	3:56	8	1	34.56	−61.95	7	51.1	18.6	0.32
24-08-2017	1:50	8	2	30.05	−86.84	144	988.1	48.8	0.54
24-08-2017	6:02	8	3	34.94	−34.41	7	51.3	13.7	0.31
24-08-2017	2:50	8	4	21.28	−81.78	76	490.4	51.3	0.57

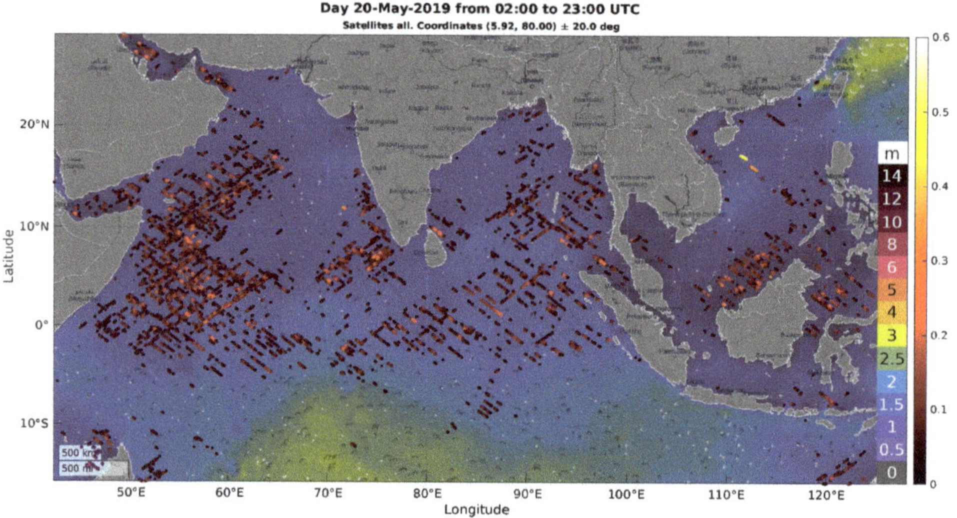

Figure 14. Filtered high S_4 values overlaid with the wave height for 20 May 2019, in the whole Indian Ocean.

Table 3 shows S_4 peaks above 0.2 lasting for at least 5 s during that day, indicating its properties.

Table 3. Table of peaks with S_4 higher than 0.2 during more than (or equal to) 5 s on 20 May 2019, at Indian Ocean.

Date (dd-mm-yy)	LT	Sat	Ch	Lat (°)	Lon (°)	Duration (s)	Length (km)	Inclination (°)	Max S_4
20-05-2019	19:05	1	3	−1.25	58.34	7	42.6	39.9	0.32
20-05-2019	06:08	1	3	−16.87	121.40	6	37.7	58.7	0.28
20-05-2019	08:59	2	1	15.00	41.64	325	2069.9	14.5	0.51
20-05-2019	16:40	2	1	17.24	111.29	11	68.7	33.3	0.47
20-05-2019	08:21	2	2	6.79	55.99	6	36.2	27.1	0.28
20-05-2019	09:00	2	3	18.41	40.30	245	1555.9	33.7	0.41
20-05-2019	08:21	2	4	2.89	58.37	6	36.0	59.3	0.27
20-05-2019	17:18	2	4	11.00	97.45	65	396.6	20.0	0.63
20-05-2019	17:30	4	1	8.86	97.49	146	881.6	20.2	0.27
20-05-2019	07:55	4	3	7.04	74.61	8	48.2	16.3	0.32
20-05-2019	08:32	4	3	9.07	56.36	7	42.4	33.1	0.33
20-05-2019	16:59	4	3	12.92	53.16	27	166.9	66.0	0.26
20-05-2019	08:05	5	2	13.69	97.45	192	1186.6	25.4	0.74
20-05-2019	08:39	5	2	9.66	75.84	106	645.0	36.3	0.52
20-05-2019	09:15	5	2	11.98	52.58	38	235.0	22.9	0.35
20-05-2019	07:35	5	3	3.05	57.99	6	36.2	34.2	0.30
20-05-2019	09:16	5	3	9.80	56.21	6	36.4	53.2	0.29
20-05-2019	17:06	5	4	12.01	71.56	20	126.0	68.3	0.34
20-05-2019	07:09	6	3	5.67	73.95	9	53.9	52.4	0.27
20-05-2019	18:48	6	3	7.98	116.57	32	197.5	37.3	0.48
20-05-2019	09:16	6	4	7.13	98.00	53	326.0	65.4	0.28
20-05-2019	07:45	6	4	0.33	57.62	6	35.6	45.6	0.30
20-05-2019	17:13	7	2	16.22	112.29	12	75.5	38.1	0.36
20-05-2019	08:17	7	3	6.87	75.01	6	36.7	49.5	0.29
20-05-2019	08:43	8	2	16.92	41.11	265	1673.2	41.9	0.23

Better statistics can be extracted by obtaining these tables for a large number of days. In the following lines, the study of 25 days equally distributed every 15 days from March 2017 to March 2018 is presented. The methodology used is to parse the whole day in all the oceanic regions from 40°N to 40°S recording the events with S_4 above 0.2 lasting more than 5 s. The number of peaks recorded per day in the year is shown in Figure 15.

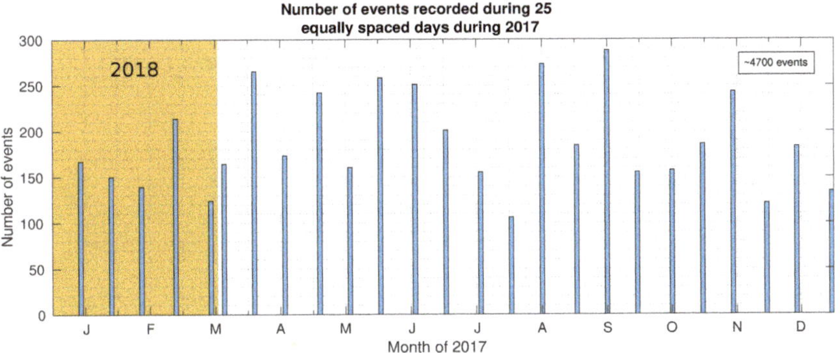

Figure 15. Histogram of events recorded during 25 equally during the first year of operation of CYGNSS, starting in March 2017, and ending in March 2018. Note that bars over the orange background belongs to 2018. The height of each peak represents the number of events recorded this day. The total number of counts is around 4700.

Figure 16 shows the histogram of those peaks in terms of the LT in which they were recorded, according to the timezone. It shows a peak around 6 h in the morning, which matches to the mean sunrise time. Figure 17 shows the same data regarding the maximum S_4 value reached during the peak.

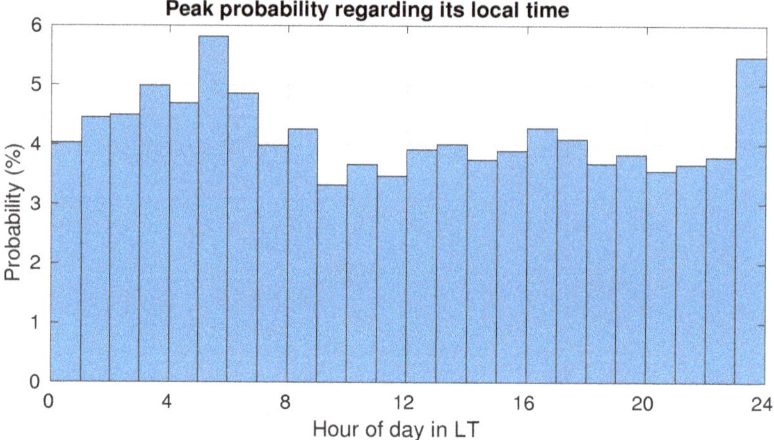

Figure 16. Histogram of number of events wrt its LT.

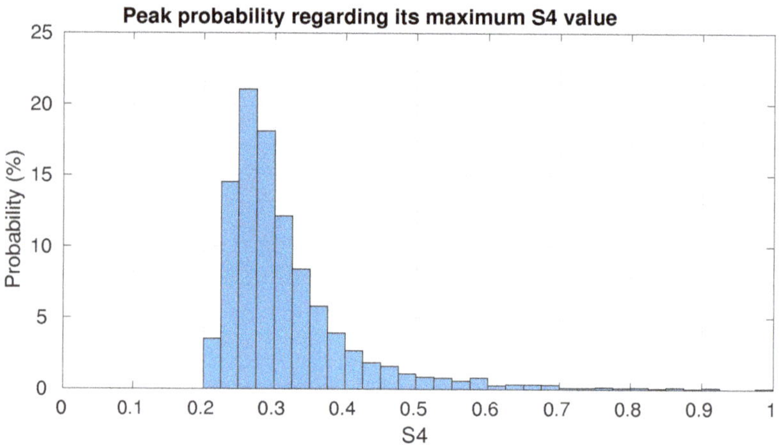

Figure 17. Histogram of number of events wrt the maximum S_4 value reached.

3.2. S_4 Statistics

Another analysis conducted, is the study of the probability to find points with certain S_4 at a particular hour in LT, averaging the data over long periods of time. This way, it can be studied if there is any correlation between the S_4 obtained with this technique and the LT, as previous models and experimental evidence shows. The set of histograms in Figure 18 shows all the data points from 0 h to 24 h (UTC) during selected days in the year 2017.

The horizontal axis of the histograms represents the local time in hourly bins, and the vertical scale is the S_4 value in bins of 0.1. The color represents the number of counts per each of the two-dimensional bins. The color scale is adjusted up to 500 counts to increase the contrast for S_4 values larger than 0.1. This means that all the bins in red on the bottom side of the histogram are saturated.

The data plotted in the histograms correspond to all the specular reflections over open oceans and seas (further away than 50 km from any coastline), within the latitudes of CYGNSS coverage

$(40°N, 40°S)$. In total, every day there are approximately 1.9×10^6 samples. The registered values include all reflections on the cited areas without any other filter (i.e., no sea surface roughness filter).

In these histograms, a small dependency with LT can be observed, that is more visible in April and August, with a slightly higher probability of higher S_4 during sunrise and sunset, around 6 h and 17 h during 21st August, and 9 h and 21 h during both days in April. Also, a similar correlation can be observed on 20th November.

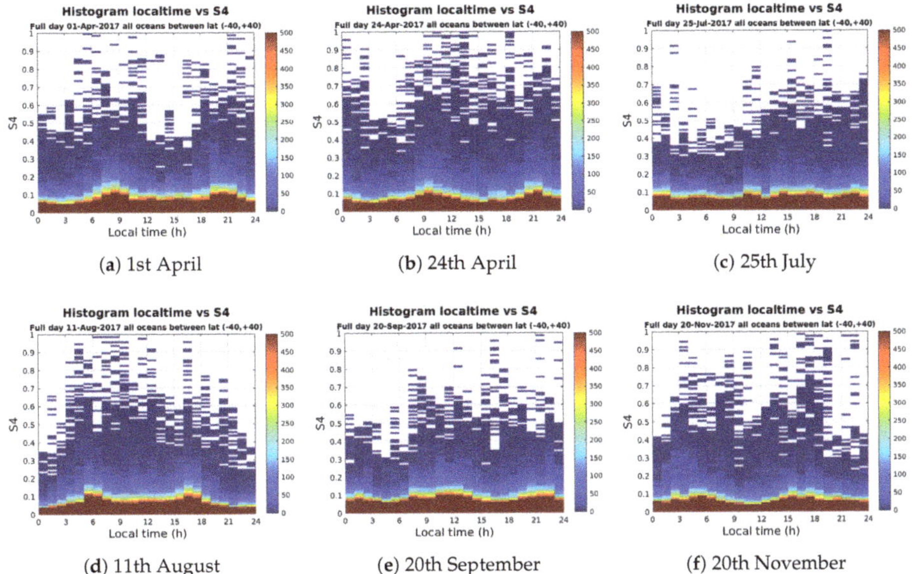

(**a**) 1st April   (**b**) 24th April   (**c**) 25th July

(**d**) 11th August   (**e**) 20th September   (**f**) 20th November

Figure 18. Histogram count of specular reflection points over all oceans between 40°N and 40°S for each S_4 value and LT hour, from 0 h to 24 h (UTC) during selected days in 2017.

4. Discussion

The results presented in this study are consistent with previous observations and studies. As a general starting point, several peaks in S_4 are found in many days during the years of operations of the CYGNSS mission.

One of the main concerns during the study was the need to filter the GNSS reflections only for coherent reflections, as ionospheric scintillation may only be "visible" when a coherent wave crosses the ionosphere. In practice, this means that the signal has to be reflected over calm water. That is the reason why in every particular case, the wave height and winds speed maps were checked before a more in-depth analysis was performed to the data. For example, two of the days shown, 24 August 2017 (Figure 10) and 21 November 2017 (Figure 6), exhibit calm sea on the regions of interest where some scintillation was found.

4.1. Bubbles Study

Obtained results match with previous studies and, in some cases, they can be clearly interpreted as the EPBs described, for example, in Reference [19]. In this work, a method for detecting and measure equatorial plasma depletions is explained. These depletions in the electron content of the ionosphere highly affect the radio wave signals crossing them, and they can rapidly appear, travel some distance, and then vanish. The way to study them in Reference [19] is by the use of data provided by ground-based GNSS receivers belonging to the International GNSS Service (IGS), available online since, at least, 2002, including different solar cycles.

GNSS satellites orbit at 20,000 km altitude (GPS), with periods of 12 h, which means that from a ground station they are in view during long times, making it possible to study the region of the ionosphere that crosses the vector from the satellite to the receiver. As this vector moves slowly compared to the estimated drift velocity of the plasma bubbles, they can cross the signal path during the tracking of a single GNSS satellite. This method is applied systematically to measure the duration and depth of these bubbles. The results show that these bubbles could have very different duration, lasting from 10 to almost 100 min during a solar maximum, or shorter periods during a solar minimum, as shown in Figure 19.

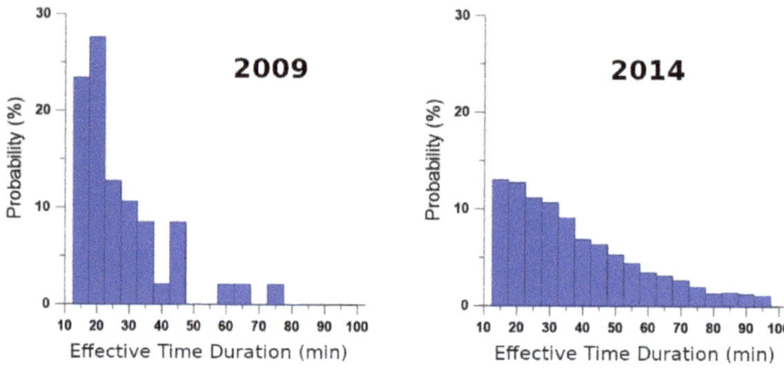

Figure 19. Probability to detect equatorial plasma bubbles (EPBs) with a particular effective time duration, for both years studied in Reference [19]: 2009 (**left**) during solar minimum and 2014 (**right**) during the last solar maximum (adapted from Reference [19]).

In order to make a comparison between both studies, the equivalent length of the bubbles has been computed. Regarding the measurement from ground stations and neglecting the movement of the GNSS satellites during the transition of the bubble, the length of the bubbles can be estimated using the mean drift velocity of the ionospheric plasma, $\sim$100 m/s. That transformation linearly translates the horizontal scale of Figure 19, from 10–100 min to 60–600 km, respectively.

From the recorded CYGNSS data, using equally spaced days every 15 days from March 2017 to March 2018 –25 days in total–, all the peaks lasting more than 5 s with an S_4 value above 0.2 have been computed to display the histogram shown in Figure 20. The extension of the bubble, in this case, is computed from the coordinates of the first and last specular reflection considered part of the peak and it is shown in Figure 21. The total number of peaks analyzed in these histograms is around 4700 and they are distributed along the 25 days in 2017 using the data shown in the previous section represented in Figure 15.

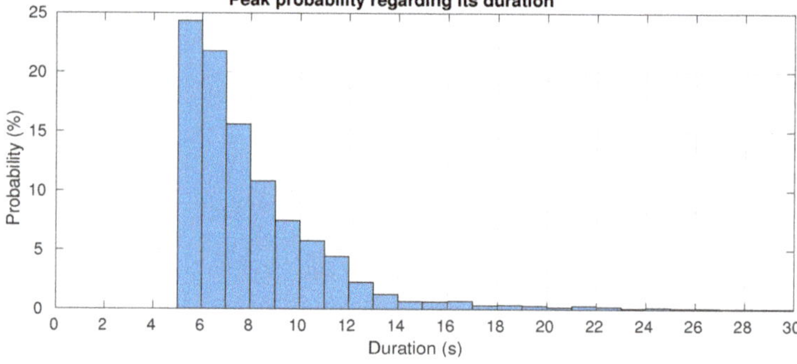

Figure 20. Histogram of peak duration during 25 equally spaced days in 2017 representing the probability in bins of 1 s with a total number of samples of around 4700.

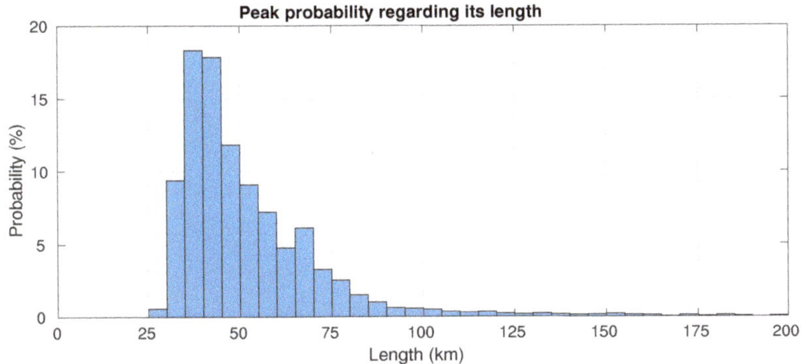

Figure 21. Histogram of bubble length during 25 equally spaced days in 2017 representing the probability in bins of 5 km with a total number of samples of around 4700.

Comparing CYGNSS data with the one studied in Reference [19], many similarities can be extracted. The shape of the distribution fits very well in terms of bubble extension, ranging from around 40 km to 200 km. In both plots, an exponential-like shape can be observed.

4.2. Correlation to Local Time

Another aspect to highlight is the correlation between local time and the occurrence of high S_4 events, as shown in the set of histograms in Figure 18. Some of them show a higher probability of finding high S_4 values during hours around the sunrise and sunset. In particular, plots on 1 and 24 April, which are also very similar to each other, reveal peaks around 9 h in the morning and 8–9 h in the evening. It is also important to remark that the sunrise and sunset do not occur at the same time in all the regions covered in this graphs, which are all the oceans and seas with latitudes from 40°N to 40°S. Considering April is close to the equinox, the span between local sunrise and sunset is not very large from 40°N to 40°S: around 30 min. It would be exactly zero on the equinoxes. This difference is much higher for 25 July, around 2 h 20 min.

In this way, being the sunrise and sunset the most active periods in terms of ionospheric activity, in the equinoxes the peak in these histograms should be even higher than in the solstices because, during the equinoxes, sunrise and sunset happen at the same time for all latitudes. This can be observed in the July histogram (Figure 18c), but it is not evident in the November's one (Figure 18f). However, it is important to consider that the measurements used for these histograms have no sea roughness filter, and they use all the data over oceans within CYGNSS available latitudes.

4.3. Comparison with Plasma Density Data from Swarm

ESA Swarm mission is a constellation of 3 satellites launched in 2013 that globally provides data on the geomagnetic field and plasma density. Satellites named as Swarm A and C fly together in a polar orbit of around 460 km altitude and 87.4° inclination, so they can sense the plasma density on the ionosphere. Swarm B is orbiting at 530 km, within another polar orbit plane, so it can sense a bit higher layer of the ionosphere. Swarm instruments can measure geomagnetic field intensity and direction, and also, local electron plasma density.

In order to compare with previous results from CYGNSS, Swarm measurements during 24 August 2017 have been analyzed. In Figure 22, the plasma density during some hours of this day is plot along with an index that the authors have defined to indicate the intensity of its fluctuations, the Normalized N_e Fluctuation Index (NNeFI). This index is obtained from the plasma density given by Swarm (N_e), once per second. Ne measures the number of electrons per cubic centimeter with values that are in the order of 10^5 e$^-$/cm^3, the typical electron density in the ionosphere at this height. Using N_e, the NNeFI is computed as:

$$NNeFI = \sqrt{\frac{\langle N_e^2 \rangle - \langle N_e \rangle^2}{\langle N_e \rangle^2}}, \tag{3}$$

where N_e is the electron density measured by Swarm and the average is computed using a 12-samples moving window.

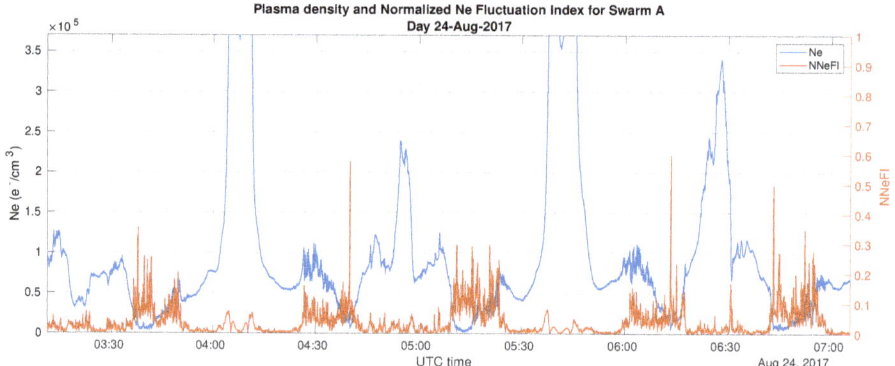

Figure 22. Plasma density (Ne) and Normalized Ne Fluctuation Index for Swarm A during some hours on day 24 August 2017. Note the values of NNeFI when there is fluctuations in the plasma density.

This index is plotted in a map in Figure 23, for the whole day and the three Swarm satellites, and it is underlaid the mean occurrence of scintillation events, using the WBMOD model predictions of the 90*th* percentile S_4 index. Note that only values of NNeFI above 0.05 are shown. It can be observed a correlation between the appearance of plasma density fluctuations and the predicted S_4 map.

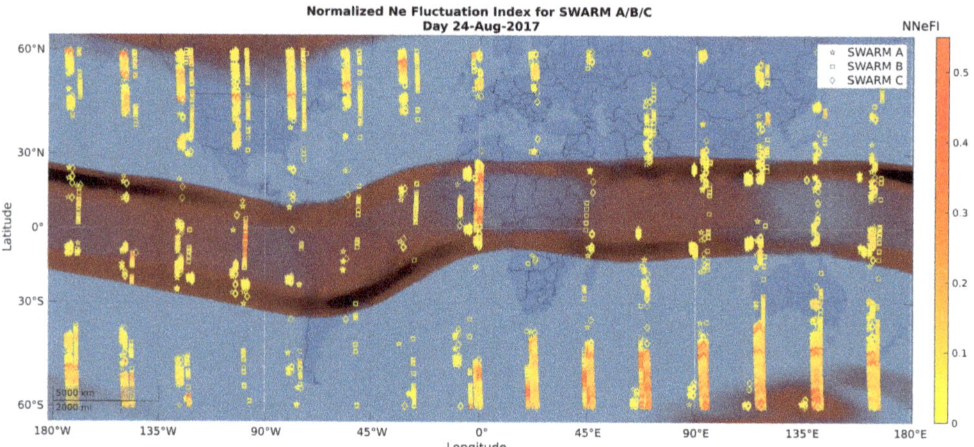

Figure 23. Normalized Ne Fluctuation Index for Swarm A, B and C during 24 August 2017 with the predicted mean occurrence of S_4 scintillation using WBMOD model. Background image adapted from Reference [20] and overlaid with Swarm data.

In Figure 24 it is plotted the same data from Swarm NNeFI along with the peaks of S_4 reported by CYGNSS during the same day that was studied in previous sections of this work. S_4 values plotted are filtered to be higher than 0.12. Swarm computed NNeFI exhibits high occurrence in high latitudes, above 60°, both Northern and Southern, but they are out of the region covered by CYGNSS and they will be not mentioned here.

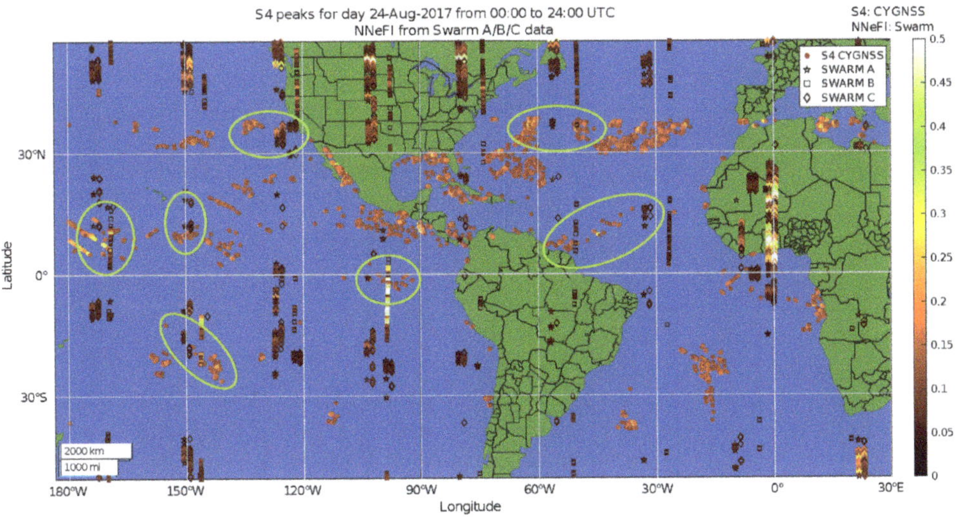

Figure 24. NNeFI for Swarm A, B and C above 0.05 during day 24*th* August, along with the S_4 peaks above 0.12 detected by CYGNSS during the same day. Note the regions marked with a yellow circle where it can be observed a spatial correlation between both sources of data. Note also, the color scale is shared for S_4 and NNeFI.

Around meridian 0° there is a large region of fluctuations observed by Swarm A and C, over Sahara and Ghana, which, in the part that enters to the Atlantic ocean, there is also some S_4 peaks detectable by CYGNSS. Another example is found in the line of S_4 peaks that goes from The Guianas to Cabo Verde that seems to be completed with some peaks in the NNeFI index from Swarm B. There is also some cases along the Pacific Ocean that are marked with circles, where there is a cluster of S_4 peaks at the same time as NNeFI peaks. There are also other regions in which there is mostly S_4 peaks but few or none Swarm indicators, for example in the South of Mexico, in the region between Bermudas and Cuba or in the Middle South Atlantic

It is important to note that the availability of data is much larger in the CYGNSS constellation than the Swarm one. For one day, CYGNSS has 8 satellites with 4 channels each in a orbit with less inclination; Swarm is only three satellites in polar orbit. This fact could difficult the correlation between both databases, but here, there are some evidences that motivates further study.

5. Conclusions

This work has presented the first experimental evidence that GNSS-R can be used as a global ionospheric scintillation monitor, and in particular, of bubble and depletions. Future work on the search of ionospheric activity at all latitudes and during different solar activity periods could help in the improvement of existing ionospheric models, providing them with more complementary data than ground stations are currently providing and with a denser spatial coverage than Swarm mission.

The study shows the result of scintillation events during the years 2017 and 2019, including tables with peaks that correspond to EPBs that are consistent with previous studies in terms of duration, LT, and also latitudes. The duration of the events shown here is in the order of seconds (from 5 s to 30 s), which means bubbles of around 25 km to 150 km length. The depth of these bubbles in terms of maximum S_4 is in the order of 0.2 to 0.5.

This way to study the ionospheric scintillation could be very beneficial because it can cover large areas all over the Earth, including remote regions in the middle of oceans, where no ground stations could be placed. However, there is also has some limitations and they should be carefully identified and refined. One of them is that only calm ocean surfaces ensure coherent specular reflections. If the water surface has ripples, the signal would reflect from a larger glistening zone, arriving at the receiver with different phases, hiding the possible fluctuations in amplitude and phase. In further studies continuing the analysis of CYGNSS and other GNSS-R missions, an automated data filtering algorithm must be implemented to eliminate data affected by sea roughness.

On the other hand, it is important to take into account that the results presented in the study come from years 2017 to 2019, and solar activity cycle should reach its minimum in 2020. That implies that less activity is expected, and consequently, fewer scintillation events. It would be very interesting to study GNSS-R data from future missions when approaching the solar maximum.

A correlation study between these results and others obtained from different sources, for example, the Swarm geomagnetic and plasma density products is being performed to find possible confirmations of this new proposed technique and also to check what new possibilities could GNSS-R technique bring to the ionospheric monitoring field.

The results presented in this study are a small sample of the total amount of information that can be obtained using this technique. CYGNSS generates around 10.4 GB per day, and further exhaustive analyses of it could bring new information or improve current models of the ionosphere, which is especially interesting regarding oceanic regions.

Author Contributions: Conceptualization, C.M. and A.C.; methodology, C.M. and A.C.; software, C.M.; validation, C.M. and A.C.; formal analysis, C.M. and A.C.; investigation, C.M.; resources, A.C.; data curation, C.M.; writing–original draft preparation, C.M.; writing–review and editing, C.M. and A.C.; visualization, C.M.; supervision, A.C.; project administration, A.C.; funding acquisition, A.C. All authors have read and agreed to the published version of the manuscript.

Funding: This work has been partially funded by the Spanish MCIU and EU ERDF project (RTI2018-099008-B-C21/AEI/10.13039/501100011033) "Pioneering Opportunistic Techniques", grant to "CommSensLab-UPC" Excellence Research Unit Maria de Maeztu (MINECO grant MDM-2016-600), and ESA project AO 1-8862/17/NL/AF/eg "RADIO CLIMATOLOGY MODELS OF THE IONOSPHERE: STATUS AND WAY FORWARD".

Acknowledgments: Thanks to the CommsSensLab administrative and research personnel and to the NanoSatLab members for all the support. Special thanks to Badr-Eddine Boudriki for his help in Swarm data retrieval and pre-processing, and to Hyuk Park for his wise advises and comments on this study.

Abbreviations

The following abbreviations are used in this manuscript:

EPB	Equatorial Plasma Bubble
ESA	European Space Agency
GISM	Globa Ionospheric Scintillation Model
GNSS	GeoNavigation Satellite System
GNSS-R	GNSS Reflectometry
GPS	Global Positioning System
ITU-R	International Telecommunications Union, Radiocommunication Sector
LEO	Low Earth Orbit
LT	Local Time
NNeFI	Normalized Ne Fluctuation Index
PO.DAAC	Physical Oceanography Distributed Active Archive Center
SNR	Signal-to-Noise Ratio
UTC	Coordinated Universal Time

References

1. Béniguel, Y. Global Ionospheric Propagation Model (GIM): A propagation model for scintillations of transmitted signals. *Radio Sci.* **2002**, *37*, 4-1–4-13, [CrossRef]
2. Knepp, D.L. Multiple phase-screen calculation of the temporal behavior of stochastic waves. *Proc. IEEE* **1983**, *71*, 722–737. [CrossRef]

3. ITU. *Ionospheric Propagation Data and Prediction Methods Required for the Design of Satellite Services and Systems. Recommendation ITU-R531-10*; International Telecommunication Union, Geneva, Switzerland, 2017. Available online: https://www.itu.int/rec/R-REC-P.531-10-200910-S/en (accessed on 18 October 2020).

4. Fremouw, E.J.; Secan, J.A. Modeling and scientific application of scintillation results. *Radio Sci.* **1984**, *19*, 687–694, [CrossRef]

5. Wernik, A.W.; Alfonsi, L.; Materassi, M. Scintillation modeling using in situ data. *Radio Sci.* **2007**, *42*, 1–21. [CrossRef]

6. Camps, A.; Barbosa, J.; Juan, M.; Blanch, E.; Altadill, D.; González, G.; Vazquez, G.; Riba, J.; Orús, R. Improved modelling of ionospheric disturbances for remote sensing and navigation. In Proceedings of the 2017 IEEE International Geoscience and Remote Sensing Symposium (IGARSS), Fort Worth, TX, USA, 23–28 July 2017; pp. 2682–2685.

7. Tulasi Ram, S.; Ajith, K.K.; Yokoyama, T.; Yamamoto, M.; Niranjan, K. Vertical rise velocity of equatorial plasma bubbles estimated from Equatorial Atmosphere Radar (EAR) observations and HIRB model simulations. *J. Geophys. Res. Space Phys.* **2017**, *122*, 6584–6594, [CrossRef]

8. Camps, A.; Park, H.; Foti, G.; Gommenginger, C. Ionospheric Effects in GNSS-Reflectometry From Space. *IEEE J. Sel. Top. Appl. Earth Obs. Remote Sens.* **2016**, *9*, 5851–5861. [CrossRef]

9. Katzberg, S.J.; Garrison, J.L. Utilizing GPS to determine ionospheric delay over the Ocean. *NASA Technical Memorandum 4750*; Hampton, Virginia, 1996.

10. Garrison, J.L.; Katzberg, S.J. Detection of ocean reflected GPS signals: Theory and experiment. In Proceedings of the IEEE SOUTHEASTCON '97, 'Engineering the New Century', Blacksburg, VA, USA, 12–14 April 1997; pp. 290–294.

11. Ye, Z.; Satorius, H. Channel modeling and simulation for mobile user objective system (MUOS)—Part 1: Flat scintillation and fading. In Proceedings of the IEEE International Conference on Communications, 2003, ICC'03, Anchorage, AK, USA, 11–15 May 2003. [CrossRef]

12. Ruf, C.; Gleason, S.; Jelenak, Z.; Katzberg, S.; Ridley, A.; Rose, R.; Scherrer, J.; Zavorotny, V. The NASA EV-2 Cyclone Global Navigation Satellite System (CYGNSS) mission. In Proceedings of the 2013 IEEE Aerospace Conference, Big Sky, MT, USA, 2–9 March 2013; pp. 1–7.

13. Ruf, C.; Gleason, S.; Ridley, A.; Rose, R.; Scherrer, J. The Nasa CYGNSS Mission: Overview and Status Update. In Proceedings of the 2017 IEEE International Geoscience and Remote Sensing Symposium (IGARSS), Fort Worth, TX, USA, 23–28 July 2017; pp. 2641–2643.

14. CYGNSS. CYGNSS Level 1 Science Data Record Version 2.1. 2018. Available online: https://podaac.jpl.nasa.gov/dataset/CYGNSS_L1_V2.1 (accessed on 8 September 2020).

15. Zavorotny, V.U.; Gleason, S.; Cardellach, E.; Camps, A. Tutorial on Remote Sensing Using GNSS Bistatic Radar of Opportunity. *IEEE Geosci. Remote Sens. Mag.* **2014**, *2*, 8–45. [CrossRef]

16. Korn, P. Formulation of an unstructured grid model for global ocean dynamics. *J. Comput. Phys.* **2017**, *339*, 525–552. [CrossRef]

17. Giorgetta, M.A.; Brokopf, R.; Crueger, T.; Esch, M.; Fiedler, S.; Helmert, J.; Hohenegger, C.; Kornblueh, L.; Köhler, M.; Manzini, E.; et al. ICON-A, the atmosphere component of the ICON Earth System Model: I. Model description. *J. Adv. Model. Earth Syst.* **2018**, *10*, 1613–1637. [CrossRef]

18. InMeteo. Ventusky—Weather Forecast Maps. Available online: www.ventusky.com (accessed on 11 June 2020).

19. Blanch, E.; Altadill, D.; Juan, J.M.; Camps, A.; Barbosa, J.; González-Casado, G.; Riba, J.; Sanz, J.; Vazquez, G.; Orús-Pérez, R. Improved characterization and modeling of equatorial plasma depletions. *J. Space Weather. Space Clim.* **2018**, *8*, A38. [CrossRef]

20. Australian Goverment Bureau of Meteorology, S.W.S. About Ionospheric Scintillation. Available online: https://www.sws.bom.gov.au/Satellite/6/3 (accessed on 4 November 2020).

Publisher's Note: MDPI stays neutral with regard to jurisdictional claims in published maps and institutional affiliations.

Letter

First Evidence of Mesoscale Ocean Eddies Signature in GNSS Reflectometry Measurements

Mostafa Hoseini [1,2,*], Milad Asgarimehr [2,3], Valery Zavorotny [4], Hossein Nahavandchi [1], Chris Ruf [5] and Jens Wickert [2,3]

[1]   Department of Civil and Environmental Engineering, Norwegian University of Science and Technology NTNU, 7491 Trondheim, Norway
[2]   German Research Centre for Geosciences GFZ, 14473 Potsdam, Germany
[3]   Technische Universität Berlin, 10623 Berlin, Germany
[4]   Cooperative Institute for Research in Environmental Sciences, University of Colorado Boulder, Boulder, CO 80309, USA
[5]   Climate and Space Department, University of Michigan, Ann Arbor, MI 48109, USA
*   Correspondence: mostafa.hoseini@ntnu.no

Received: 28 November 2019; Accepted: 3 February 2020; Published: 6 February 2020

Abstract: Feasibility of sensing mesoscale ocean eddies using spaceborne Global Navigation Satellite Systems-Reflectometry (GNSS-R) measurements is demonstrated for the first time. Measurements of Cyclone GNSS (CYGNSS) satellite missions over the eddies, documented in the Aviso eddy trajectory atlas, are studied. The investigation reports on the evidence of normalized bistatic radar cross section (σ^0) responses over the center or the edges of the eddies. A statistical analysis using profiles over eddies in 2017 is carried out. The potential contributing factors leaving the signature in the measurements are discussed. The analysis of GNSS-R observations collocated with ancillary data from the European Centre for Medium-Range Weather Forecasts (ECMWF) Reanalysis-5 (ERA-5) shows strong inverse correlations of σ^0 with the sensible heat flux and surface stress in certain conditions.

Keywords: GNSS Reflectometry; Mesoscale ocean eddies; Bistatic Radar Cross Section; CYGNSS

1. Introduction

Mesoscale ocean eddies can drive atmosphere response at mesoscales mainly through heat fluxes [1] and they have a local influence on near-surface wind, cloud properties, and rainfall [2]. Analysis of mesoscale eddy-atmosphere interactions from general circulation models suggests significant intermodel differences mainly stemming from two factors: surface wind strength and marine atmospheric boundary layer adjustments to mesoscale heat flux anomalies [3]. Several Earth-observing satellites have been aiding these models for decades with their data products.

Global Navigation Satellite System Reflectometry (GNSS-R) is a relatively new Earth observation technique for monitoring a large variety of geophysical parameters (see [4,5] for a review). This technique exploits the GNSS signals of opportunity after being reflected from the Earth's surface, both over lands and oceans. The signals are intercepted by low-cost, low-power and low-mass GNSS-R receivers and are processed to extract geophysical information. These receivers onboard small low Earth-orbiting satellites offer cost-effective Earth observations with high coverage and unprecedented sampling rate. Cyclone GNSS (CYGNSS) is the satellite constellation consisting of eight microsatellites with the main science objective of ocean wind speed monitoring especially during hurricane events, launched in December 2016 [6].

Ocean monitoring is one of the most mature spaceborne GNSS-R applications, with a proven capability of surface wind measurement [7–9]. Insignificant level of sensitivity to rain attenuation [10]

and cost-effective observation frequency are the main advantageous characteristics motivating researchers to develop new ideas for additional applications over oceans [11–13], and for the development of future novel GNSS-R missions [14,15].

Remote sensing of oceanic features, e.g., eddies, based on high precision GNSS-R altimetric measurements, are being pursued. For instance, [16] deduced sea surface topography observations from the GNSS-R phase measurements onboard the German High Altitude Long Range (HALO) research aircraft. In an air-borne GNSS-R study, the so-called "Eddy Experiment", the capabilities of the technique for ocean altimetry [17] and scatterometry [18] were additionally demonstrated. Nevertheless, the response of the measurements over mesoscale eddies is not yet characterized and documented, despite the available large datasets from recent GNSS-R satellite missions.

A high number of observations are provided by CYGNSS offering a possibility to study the feasibility of observing ocean eddies using GNSS-R measurements. This research focuses on the GNSS-R scatterometric observations (rather than in an altimetry configuration) and tries to characterize eddy signatures in those measurements for the first time. The data are empirically analyzed and the signatures and physical explanations are discussed. Following this introduction, Section 2 describes the datasets and the method. The results are reported and discussed in Section 3. Finally, concluding remarks are given in Section 4.

2. Data and Method

Four datasets are used for the analysis covering the period from March to December 2017. The main dataset consists of the CYGNSS GNSS-R measurements. The eight CYGNSS microsatellites are dispersed in $35°$ inclined orbits with an altitude of $\approx$520 km. The onboard GNSS-R receivers are equipped with distinct channels measuring up to four simultaneous GPS signals after reflection from the ocean surface [19]. The corresponding data are available at different processing levels. Level 1 (L1) provides a variety of parameters including the calibrated measurements of bistatic radar cross section (BRCS) as well as the Normalized BRCS (NBRCS) σ^0. The L1 data are further processed into the 10 m referenced wind speed above the ocean surface at Level 2 (L2). For the analysis in this study, σ^0 product is extracted from the Version 2.1 (v2.1) dataset [20,21].

CYGNSS measurements over the documented mesoscale eddies in Aviso's trajectory atlas version 2.0 are extracted. The atlas is a multi-mission altimetry-derived product with a daily temporal resolution [22]. Eddy characteristics, including the position and radius, spinning speed, and the type (cyclonic/anticyclonic) are extracted from the atlas.

Near-surface ocean current estimates from the Ocean Surface Current Analysis Real-time dataset (OSCAR) are also used in this study [23]. The ocean current data are provided with a spatial resolution of one-third degree. Nevertheless, they are spatially interpolated along the CYGNSS tracks. Due to the five-day temporal resolution of the OSCAR dataset, the tracks on those days, on which OSCAR current estimates are available, are collected for the analysis.

The analysis also uses ancillary data retrieved from the European Centre for Medium-Range Weather Forecasts (ECMWF) Reanalysis-5 (ERA-5) product. The ERA5 is a global atmospheric reanalysis based on an ECMWF model assimilating observations from various sources including satellite and ground-based measurements [24]. The retrieved parameters include surface wind-field, Sea Surface Temperature (SST), Sensible Heat Flux (SHF), and turbulent surface stress field. These data products offer a possibility to discuss potential interactions of the geophysical parameters with the GNSS-R σ^0. The reanalysis measurements are provided hourly with a spatial resolution of $0.25°$. The estimates are spatiotemporally interpolated along with the CYGNSS tacks being used in the study.

The eddy trajectory atlas detects an eddy as the outermost closed-contour of Sea Level Anomaly (SLA) encompassing a single extremum [22]. The area enclosed by the contour of maximum circum-average speed is considered as the eddy radius R. The CYGNSS tracks overpassing the eddy with a maximum distance of $2R$ from the eddy center are collected and transformed into a local coordinate system (Figure 1). The local coordinate system has the origin at the center of the moving eddy with x- and y-axes oriented toward geographical east and north, respectively. Observations marked with a poor quality flag in the CYGNSS dataset (L1, v2.1) and tracks with more than 10% data loss are excluded from the collocated dataset.

The methodology of this study is based on the following steps. First, the signatures in the CYGNSS σ^0 are visually sought. The observed behavior in several cases can be the first evidence on the possibility of an eddy-left signature in the GNSS-R measurements. This examination is followed by statistical analyses to quantitatively characterize the signatures. We investigate the collocated dataset consisting of more than 2.7×10^5 NBRCS profiles over $\approx$ 6000 mesoscale eddies. The profiles in the along-track coordinate system are normalized using the radius of each eddy and gridded between $-1.1 \times R$ to $+1.1 \times R$ (Figure 1).

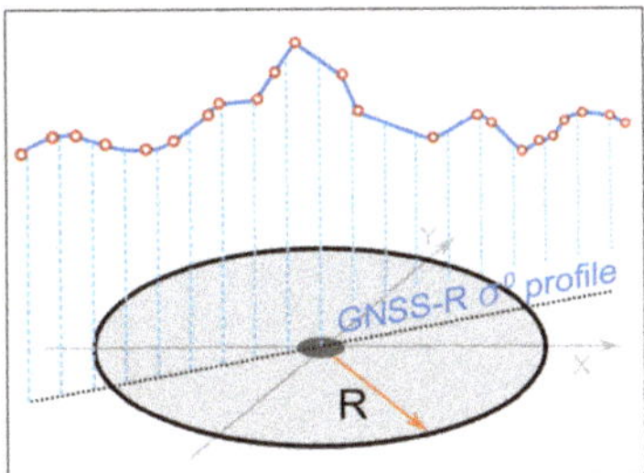

Figure 1. A sketch of the gridded GNSS-Reflectometry profile of Cyclone GNSS (CYGNSS) over an eddy and the local coordinate system with x- and y-axes oriented toward east and north, respectively.

The visually observed behaviors of the σ^0 profiles show noticeable changes over the central region or the edges of the eddies. These patterns are along with some linear and nonlinear changes in different scales. To extract the main nonlinear anomalies over the center or at the edges of eddies within the profiles, linear and small scales fluctuations of σ^0 should be filtered out. We apply Principal Component Analysis (PCA) [25] to reduce the dimensionality of the dataset while preserving most of the information within the σ^0 profiles. To this end, a data matrix $\mathbf{X}_{\mathbf{m} \times \mathbf{n}}$ is formed using n profiles, each of which with m gridded observation points. The profiles are centered by subtracting the mean values. Using Singular Value Decomposition (SVD), the data matrix X can be written as:

$$\mathbf{X} = \mathbf{U}\mathbf{L}\mathbf{V}^{\mathbf{T}} \tag{1}$$

where the columns of $\mathbf{U}$ and $\mathbf{V}$ are the left and right singular vectors, respectively. $\mathbf{L}$ is a diagonal matrix with non-negative elements, the singular values λ. A proper group of singular values and corresponding singular vectors is selected to reconstruct the data matrix. Columns of the reconstructed matrix contain the filtered σ^0 profiles. Assuming the set $I = \{i, i+1, ..., k\}$ whose elements are the indices of the selected group, the reconstructed data matrix, $\hat{\mathbf{X}}$ is:

$$\hat{\mathbf{X}} = \mathbf{X}_{\mathbf{i}} + \mathbf{X}_{\mathbf{i+1}} + ... + \mathbf{X}_{\mathbf{k}}, \quad \mathbf{X}_{\mathbf{i}} = \lambda_i \mathbf{U}_{\mathbf{i}} \mathbf{V}_{\mathbf{i}}^{\mathbf{T}} \tag{2}$$

where $\mathbf{U}_{\mathbf{i}}$ and $\mathbf{V}_{\mathbf{i}}$ are the left and right singular vectors associated with the singular value λ_i. Columns of the matrices $\mathbf{X}_{\mathbf{i}}$ represent uncorrelated features of the σ^0 profiles. The quality of each principal component (PC) can be measured by:

$$\Lambda_i = \frac{\lambda_i}{\sum_{l=1}^{d} \lambda_l} \tag{3}$$

where Λ_i represents the proportion of total variance explained by the principal component i. The parameter d ($d \leq min\{m, n\}$) is the number of non-zero singular values.

The investigation is followed seeking the conditions, in which the σ^0 response is more pronounced. To this end, the correlation coefficient between σ^0 and surface sensible heat flux is calculated at different wind speeds. Similarly, the correlation coefficient between σ^0 and the mean turbulent surface stress is obtained in a range of angular differences between the CYGNSS observational track and the turbulent surface stress. The results are presented in the following section.

3. Results and Discussion

Generally, two prominent anomalies are observed in our investigation as responses of σ^0 to the presence of the eddies: one jump at the eddy center (single-jump behavior) or two jumps at the eddy edges with a lower value at the center (double-jump behavior). Figure 2 demonstrates the double- (a–c) and single-jump (d–f) behaviors in different exemplary cases. The sudden increase in σ^0 is significant enough to be easily discerned in the measurements.

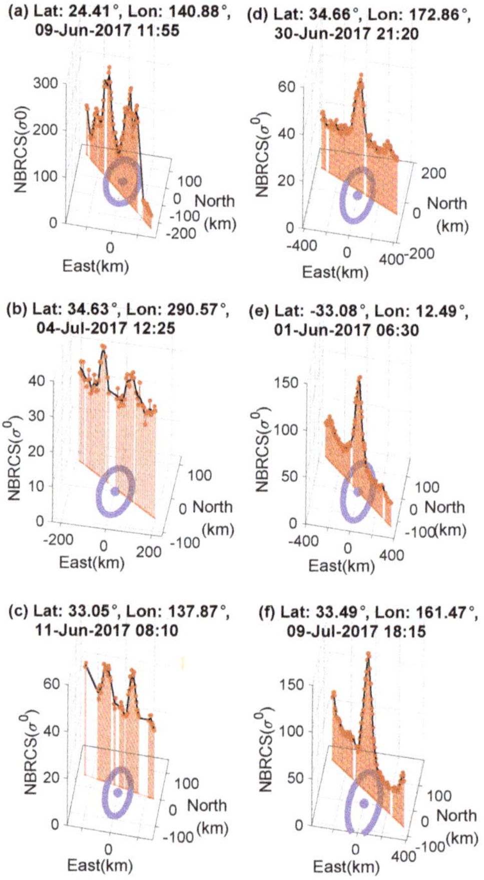

Figure 2. Exemplary cases of GNSS-Reflectometry σ^0 double-jump (**a–c**) and single-jump (**d–f**) behaviors observed in Cyclone GNSS (CYGNSS) tracks.

Additional exemplary cases are shown along with the collocated ancillary data in Figures 3–5. In Figure 3, clear fluctuations are repeatedly demonstrated over the eddy edges (similar to Figure 2a–c). Once the track enters the eddy-affected area, σ^0 increases significantly and then drops quickly at the center followed by another jump once the track leaves the eddy.

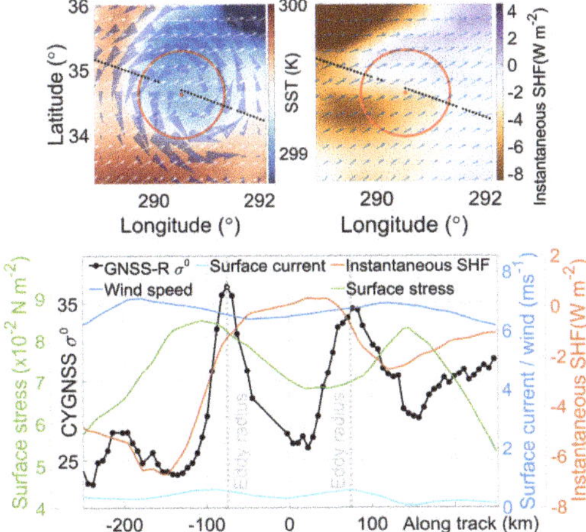

Figure 3. A track of Cyclone GNSS (CYGNSS) overpassing an eddy on 4 July 2017, 12:24. The top-left panel displays sea surface temperature, surface wind (white arrows) and current (blue cones). On the top-right, instantaneous surface sensible heat flux (SHF) as well as surface stress (blue arrows) are visualized. The bottom panel profiles CYGNSS σ^0 along with the wind and current velocity, instantaneous SHF and surface stress magnitudes.

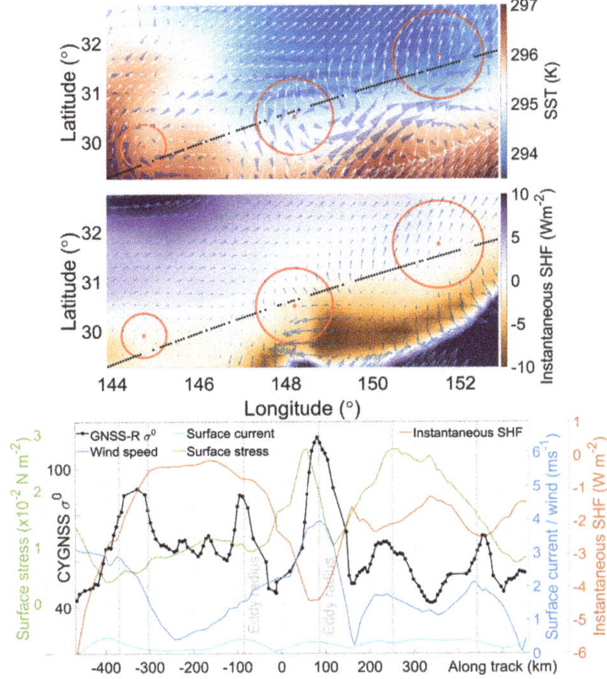

Figure 4. A track of Cyclone GNSS (CYGNSS) overpassing three eddies on 4 June 2017, 08:11. The top panel displays sea surface temperature, surface wind (white arrows) and current (blue cones). In the middle, instantaneous surface sensible heat flux (SHF) as well as surface stress (blue arrows) are visualized. The bottom panel profiles CYGNSS σ^0 along with the wind and current velocity, instantaneous SHF and surface stress magnitudes, referenced at the center of the middle eddy.

Figure 4 shows a CYGNSS track which is long enough to overpass three cyclonic eddies. The σ^0 behaves similarly to Figures 2a–c and 3. The track does not cross the first eddy center. This causes an increase in the value of σ^0 when it passes the eddy outer lying area. A remarkable fact is that σ^0 remains almost at the same level moving over the eddy edges and again drops to lower values once it leaves the affected region. Reaching the second eddy, the track sweeps also the areas close to the eddy center and σ^0 responds with a lower value at the center and two considerable increases at the edges. The behavior of σ^0 is similar over the third eddy, however, the peaks stand at lower values.

Figure 5 shows another CYGNSS track overpassing three eddies. Similar to Figure 2d–f, σ^0 shows a single peak at the center. The track enters the core region with a sudden increase in σ^0 which again drops to its initial level once the track moves off the center. Similar behavior of σ^0 is observed reaching the central region of the second and third eddies.

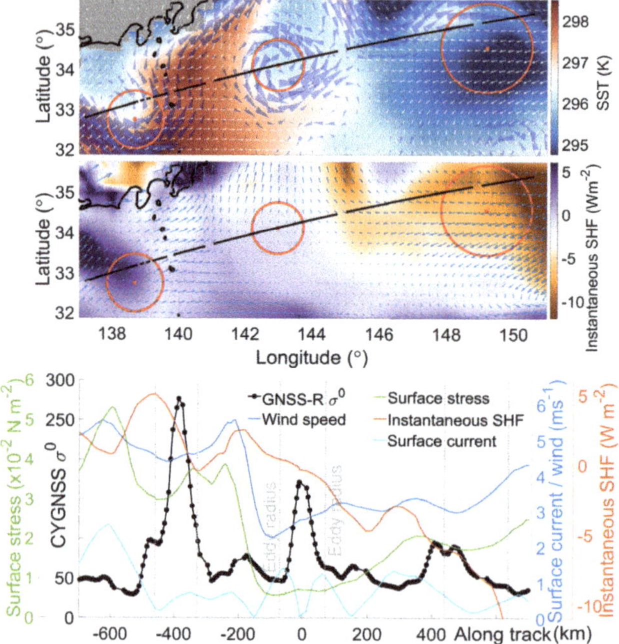

Figure 5. A track of Cyclone GNSS (CYGNSS) overpassing three eddies on 29 June 2017, 20:45. The top panel displays sea surface temperature, surface wind (white arrows) and current (blue cones). In the middle, instantaneous surface sensible heat flux (SHF) as well as surface stress (blue arrows) are visualized. The bottom panel profiles CYGNSS σ^0 along with the wind and current velocity, instantaneous SHF and surface stress magnitudes, referenced at the center of the second eddy.

Figure 6 shows the PCA results where the first nine principal components of the dataset preserve more than 95% of the statistical information in the dataset. The PCs represent low to high-fluctuating patterns within the profiles. The first PC mainly reflects the overall linear trend of the σ^0 profile. The other PCs capture the remaining non-linear variations of the profiles over the eddies. We reconstruct the profiles using the eight components PC2-PC9 and calculate the correlation coefficient of each reconstructed profile with synthetic templates of the two observed patterns. Since the peaks over the edges or at the center of the eddies could be slightly displaced from the exact expected location, we consider up to $\pm 0.1 \times R$ lag for the calculation of the correlation.

The analysis reveals that about 12.7% (15.9%) of profiles demonstrate a correlation coefficient of 0.7 or more with the single (double) peak template. We also carried out the same statistical analysis over a new set of profiles collected regardless of the presence of eddies. In a reverse approach, the profiles demonstrating a high correlation with the templates (≥ 0.7) are investigated. About 45% of these profiles are either located on the eddies (according to the Aviso's trajectory atlas) or show a high correlation (≥ 0.7) with the surface current.

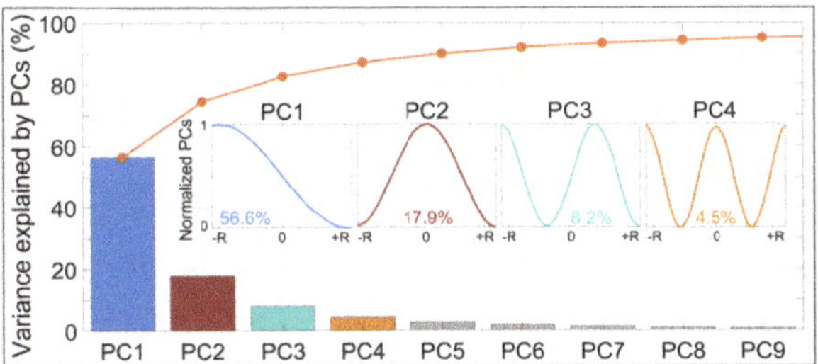

Figure 6. Principal components of the profiles and the total variance of the data explained by each principal component.

Results of the next statistical analysis over the collocated dataset reveal a strong negative correlation of CYGNSS σ^0 observations with both SHF and surface stress under certain conditions. Figure 7 provides insights into the favorable conditions, in which CYGNSS is more likely to sense surface stress and SHF over the eddies.

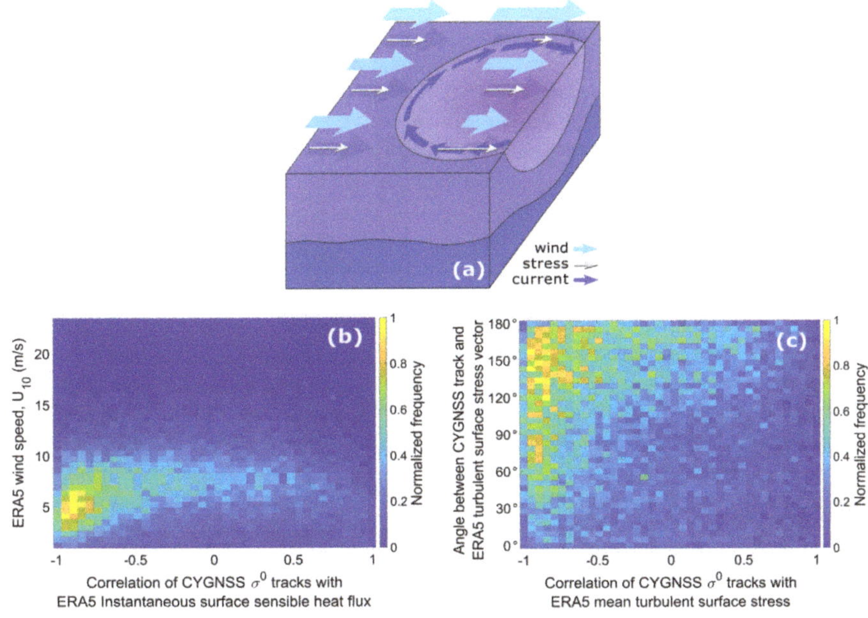

Figure 7. Schematic representation of surface stress change due to the interaction of an eastward uniform wind with the surface current associated with an anticyclonic eddy (**a**), Correlation of the σ^0 profiles of Cyclone GNSS (CYGNSS) with anomalies of instantaneous surface sensible heat flux at different wind speeds (**b**), the impact of different angular distances of the CYGNSS tracks with surface stress vector on the correlation between the σ^0 profiles and mean turbulent surface stress (**c**).

Figure 7a illustrates a simplified model of changing surface stress due to the interaction between the eddy surface current and wind speed. In Figure 7b, the behavior of σ^0 is highly correlated with SHF over the eddies at wind speeds between ≈ 3 m/s and 7 m/s, where the values of the correlation coefficients are mainly between -0.8 to -0.95. According to the theory, at high enough wind speed ($\approx > 5$ m/s), the surface parameter that controls the intensity of GNSS reflections from the ocean surface, or σ^0, is the low-pass mean square slope, MSS_{LP}, of the ocean surface [26]. It is determined by the part of the wave slope spectrum that resides at wavenumbers smaller than $k_* = k\,cos\theta_{inc}/3$ where θ_{inc} is an incidence angle and k is the wavenumber $(2\pi/\lambda)$ of the L-band GNSS signal [27]. The σ^0 is inversely proportional to MSS_{LP}. The largest contribution to the MSS_{LP} originates from the short-wave portion of the spectrum near k_*. From classic works of [28,29], it is known that there are two main mechanisms affecting that part of the wave spectrum: the varying wind surface stress and interaction of short waves with the current gradients. At low enough wind speed, the scattering of GNSS signals does not follow a pure quasi-specular scattering and there is a coherent scattering component that tends the mechanism to a higher-order Bragg scattering, driven by Rayleigh parameter [30]. Rayleigh parameter is proportional to waves at any wavenumbers. So, at this regime of wind speed, GNSS-R measurements could be more sensitive to surface state, even to small-scale roughness modifications [12]. Figure 7c shows the impact associated with the angular difference of CYGNSS tracks and surface stress field direction. The direction of the CYGNSS track with respect to the surface stress vector can increase the sensitivity of σ^0 to surface stress anomalies within the eddies. This means the GNSS-R measurements are highly likely to sense the stress field with a direction against the moving GNSS-R specular points. It can be also seen that for the absolute angular distances in the range of about 60 to 180 degrees the wind stress would be more pronounced in the CYGNSS measurements.

Atmospheric boundary layer change associated with the eddy-induced SST anomalies results in a varying wind field [31]. The modified local surface wind influenced by marine boundary layer dynamics [32,33] can partially explain the GNSS-R σ^0 patterns. The enhanced local wind over the warm core of the eddy can lead to the abrupt change in the GNSS-R σ^0 values. Since the improvement in the weather and climate projections require detailed observations and understanding of warm eddy-atmosphere interactions [34], this possible promising contribution by the GNSS-R technique should be investigated.

The first cold-core eddy shown in Figure 5 can cause a strong dampening of wind intensity due to downward transport of wind momentum, decelerating local surface wind. The sharp peak of GNSS-R σ^0 resides at the core region of the eddy where the SST has a lower value. This deceleration could also happen when a tropical cyclone reaches a strong cold-core eddy. Such eddies can broaden the eye size of the storm during its passage and reduce its intensity [35]. For instance, an unforeseen rapid weakening was demonstrated when the category 4 hurricane Kenneth passed over a cold-core eddy on 19–20 September 2005 [36].

The discussed air-sea interactions over the eddies could explain the response of GNSS-R observations to SHF at the ocean-atmosphere interface through the modified surface stress. In Figure 3, a local minimum of ERA5 surface stress values takes place almost over the core region of the eddy. The peaks of the stress values approximately reside over the rotating current of the eddy. The impact of the surface stress on the profile of CYGNSS σ^0 is evident where sudden fluctuations are seen over the edges and in the core. Larger SHF values with negative sign, i.e. upward direction of the flux, are well synchronized with two σ^0 minima at -150 and 150 km along with track coordinates.

In Figure 4, the most prominent change in the σ^0 profile can be seen over the middle eddy. The possible signature of this eddy could be explained by a high value of stress approximately at the eddy center where an increment of upward SHF is observed. The ERA5 could be subjected to deficiencies in resolving local sudden changes and It seems that it does not reveal the same level of details over the left eddy as those provided by the CYGNSS measurements. The behavior of σ^0 over the right eddy in this figure can be described by the expected behavior of σ^0 at very low wind speeds.

According to [37], at very low wind speeds (< 2.5 m/s), the bistatic radar cross section is directly proportional to the roughness (unlike the inverse correlation at higher wind speeds). Therefore, the clear correspondence between the magnitude of upward SHF and wind speed over this eddy closely matches the similar pattern in σ^0 while the wind speed values are mainly below 2 m/s.

The surface current associated with eddies is another factor that can affect surface stress. Considering surface stress as a function of wind and ignoring the surface current in the oceanic numerical modeling, can result in the overestimation of the total energy input of wind to the ocean [38]. Wind stress (τ) can be calculated as [39]:

$$\tau = \rho_a C_D \left(W - U \right) \left| W - U \right| \tag{4}$$

where ρ_a is the density of the air, C_D is the drag coefficient, and W and U are the wind and surface current, respectively.

The behavior of σ^0 in Figure 5 can be partially attributed to the modified surface stress at the eddy currents. Eddy-induced current can amplify or decrease the wind stress (Figure 7a) or alter its direction which can in turn change the level of σ^0 sensitivity to surface stress. Over the left eddy in Figure 5, the similar directional orientation of the CYGNSS track with respect to the surface stress field can lead to the weaker impact of stress on the σ^0 values (see Figure 7c). Interaction of eddy-induced current with surface stress can increase the σ^0 sensitivity over the edges resulting in lower σ^0 values. Therefore, the vanishing current at the core region would lead to the less pronounced impact of stress on σ^0. Although the stress field over the middle eddy is not as strong, the angular difference of the CYGNSS track with the stress field intensifies the impact. The strong current velocity on the edges enhances the stress on the left side and decreases the stress on the right side of the eddy (see Figure 7a), resulting slightly higher σ^0 values on the right edge compared to the left edge. The low magnitude of SHF over this cold-core eddy together with almost zero current velocity at the center cause a sudden peak in the σ^0 value. The higher SHF magnitudes and stress values between the two eddies keep the σ^0 values at a lower level.

It is worth mentioning that concentrated biogenic films from natural life in the ocean can potentially play a role in the power of reflected GNSS-R signals. The turbulence associated with the eddies brings the natural biogenic surfactants released from plankton and fishes to the surface, where the concentration of the surfactant molecules can generate a surface tension. This phenomenon could inhibit the development of Bragg waves [40]. Such areas are discerned as dark regions in the synthetic aperture radar images since the signal is mainly forward scattered rather than being backscattered. In a bistatic forward scattering configuration, the wide-enough smoothed regions can increase the power of GNSS signals after reflection from the ocean. Therefore, a dramatic increase in σ^0 over these regions can be expected. The characterization of biogenic surfactants' role in the signal forward scattering is recommended for future studies.

4. Conclusions

In this study, it is shown that spaceborne GNSS-R measurements can respond to the existence of eddies. Different characteristics of eddies can impact the local wind as well as surface stress which can, in turn, affect GNSS-R measurements. The normalized bistatic radar cross section (NBRCS) exhibits a clear inverse correlation with surface heat flux and surface stress under certain conditions. Nevertheless, characterization of the observed signatures requires further study considering other potential factors such as the effect of biogenic surfactants and the eddy-induced currents in the surface stress and ocean state. Many factors produce NBRCS changes. The complexity of oceanic and atmospheric mechanisms controlling the GNSS scattering demands further sophisticated analyses in future studies. There are still open questions such as the conditions of occurrences or the measurements specific behaviors over cyclonic or anticyclonic eddies. This study initiates the development of the novel GNSS-R technique for studying ocean mesoscale eddies, the feasibility of which has been demonstrated for the first time.

Author Contributions: Conceptualization, M.H., H.N., M.A.; Data curation, M.H.; Formal analysis, M.H., M.A., V.Z. and C.R.; Funding acquisition, H.N.; Investigation, M.H. and M.A.; Methodology, M.H., M.A.; Software, M.H.; Supervision, H.N. and J.W.; Validation, M.H. and V.Z.; Visualization, M.H.; Writing–original draft, M.H. and M.A.; Writing–review and editing, M.H., M.A., V.Z., H.N., C.R. and J.W. All authors have read and agreed to the published version of the manuscript.

Funding: This research was funded by Norwegian University of Science and Technology grant number 81771107.

Acknowledgments: Authors would like to thank the teams in charge of CYGNSS, ECMWF, Aviso and OSCAR data products which made this study possible. All the data used in this study are publicly available and free of charge at the associated repositories. The CYGNSS and OSCAR datasets can be found at the NASA Physical Oceanography Distributed Active Archive Center, PO.DAAC (https://podaac.jpl.nasa.gov). ERA5 dataset from ECMWF can be downloaded from https://cds.climate.copernicus.eu and the Aviso trajectory atlas is available on https://www.aviso.altimetry.fr.

Conflicts of Interest: The authors declare no conflict of interest.

References

1. Small, R.D.; DeSzoeke, S.; Xie, S.; O'Neill, L.; Seo, H.; Song, Q.; Cornillon, P.; Spall, M.; Minobe, S. Air–sea interaction over ocean fronts and eddies. *Dyn. Atmos. Ocean.* **2008**, *45*, 274–319. [CrossRef]
2. Frenger, I.; Gruber, N.; Knutti, R.; Münnich, M. Imprint of Southern Ocean eddies on winds, clouds and rainfall. *Nat. Geosci.* **2013**, *6*, 608. [CrossRef]
3. Yang, P.; Jing, Z.; Wu, L. An Assessment of Representation of Oceanic Mesoscale Eddy-Atmosphere Interaction in the Current Generation of General Circulation Models and Reanalyses. *Geophys. Res. Lett.* **2018**, *45*, 11–856. [CrossRef]
4. Jin, S.; Cardellach, E.; Xie, F. *GNSS Remote Sensing*; Springer: Berlin/Heidelberg, Germany, 2014.
5. Zavorotny, V.U.; Gleason, S.; Cardellach, E.; Camps, A. Tutorial on remote sensing using GNSS bistatic radar of opportunity. *IEEE Geosci. Remote. Sens. Mag.* **2014**, *2*, 8–45. [CrossRef]
6. Ruf, C.S.; Atlas, R.; Chang, P.S.; Clarizia, M.P.; Garrison, J.L.; Gleason, S.; Katzberg, S.J.; Jelenak, Z.; Johnson, J.T.; Majumdar, S.J.; et al. New ocean winds satellite mission to probe hurricanes and tropical convection. *Bull. Am. Meteorol. Soc.* **2016**, *97*, 385–395. [CrossRef]
7. Foti, G.; Gommenginger, C.; Jales, P.; Unwin, M.; Shaw, A.; Robertson, C.; Rosello, J. Spaceborne GNSS reflectometry for ocean winds: First results from the UK TechDemoSat-1 mission. *Geophys. Res. Lett.* **2015**, *42*, 5435–5441. [CrossRef]
8. Ruf, C.S.; Gleason, S.; McKague, D.S. Assessment of CYGNSS wind speed retrieval uncertainty. *IEEE J. Sel. Top. Appl. Earth Obs. Remote. Sens.* **2018**, *12*, 87–97. [CrossRef]
9. Asgarimehr, M.; Wickert, J.; Reich, S. TDS-1 GNSS Reflectometry: Development and Validation of Forward Scattering Winds. *IEEE J. Sel. Top. Appl. Earth Obs. Remote. Sens.* **2018**, *11*, 4534–4541. [CrossRef]
10. Asgarimehr, M.; Wickert, J.; Reich, S. Evaluating Impact of Rain Attenuation on Space-borne GNSS Reflectometry Wind Speeds. *Remote. Sens.* **2019**, *11*, 1048. [CrossRef]
11. Alonso-Arroyo, A.; Zavorotny, V.U.; Camps, A. Sea ice detection using UK TDS-1 GNSS-R data. *IEEE Trans. Geosci. Remote. Sens.* **2017**, *55*, 4989–5001. [CrossRef]
12. Asgarimehr, M.; Zavorotny, V.; Wickert, J.; Reich, S. Can GNSS Reflectometry Detect Precipitation Over Oceans? *Geophys. Res. Lett.* **2018**, *45*, 12–585. [CrossRef]
13. Clarizia, M.P.; Ruf, C.; Cipollini, P.; Zuffada, C. First spaceborne observation of sea surface height using GPS-Reflectometry. *Geophys. Res. Lett.* **2016**, *43*, 767–774. [CrossRef]
14. Wickert, J.; Cardellach, E.; Martín-Neira, M.; Bandeiras, J.; Bertino, L.; Andersen, O.B.; Camps, A.; Catarino, N.; Chapron, B.; Fabra, F.; et al. GEROS-ISS: GNSS reflectometry, radio occultation, and scatterometry onboard the international space station. *IEEE J. Sel. Top. Appl. Earth Obs. Remote. Sens.* **2016**, *9*, 4552–4581. [CrossRef]
15. Cardellach, E.; Wickert, J.; Baggen, R.; Benito, J.; Camps, A.; Catarino, N.; Chapron, B.; Dielacher, A.; Fabra, F.; Flato, G.; et al. GNSS Transpolar Earth Reflectometry exploriNg System (G-TERN): Mission Concept. *IEEE Access* **2018**, *6*, 13980–14018. [CrossRef]
16. Semmling, A.; Beckheinrich, J.; Wickert, J.; Beyerle, G.; Schön, S.; Fabra, F.; Pflug, H.; He, K.; Schwabe, J.; Scheinert, M. Sea surface topography retrieved from GNSS reflectometry phase data of the GEOHALO flight mission. *Geophys. Res. Lett.* **2014**, *41*, 954–960. [CrossRef]

17. Ruffini, G.; Soulat, F.; Caparrini, M.; Germain, O.; Martín-Neira, M. The Eddy Experiment: Accurate GNSS-R ocean altimetry from low altitude aircraft. *Geophys. Res. Lett.* **2004**, *31*. [CrossRef]

18. Germain, O.; Ruffini, G.; Soulat, F.; Caparrini, M.; Chapron, B.; Silvestrin, P. The Eddy Experiment: GNSS-R speculometry for directional sea-roughness retrieval from low altitude aircraft. *Geophys. Res. Lett.* **2004**, *31*. [CrossRef]

19. Ruf, C.S.; Chew, C.; Lang, T.; Morris, M.G.; Nave, K.; Ridley, A.; Balasubramaniam, R. A new paradigm in earth environmental monitoring with the CYGNSS small satellite constellation. *Sci. Rep.* **2018**, *8*, 8782. [CrossRef]

20. Gleason, S.; Ruf, C.S.; O'Brien, A.J.; McKague, D.S. The CYGNSS Level 1 calibration algorithm and error analysis based on on-orbit measurements. *IEEE J. Sel. Top. Appl. Earth Obs. Remote. Sens.* **2018**, *12*, 37–49. [CrossRef]

21. Ruf, C.; Asharaf, S.; Balasubramaniam, R.; Gleason, S.; Lang, T.; McKague, D.; Twigg, D.; Waliser, D. In-Orbit Performance of the Constellation of CYGNSS Hurricane Satellites. *Bull. Am. Meteorol. Soc.* **2019**, *100*, 2009–2023. [CrossRef]

22. Faghmous, J.H.; Frenger, I.; Yao, Y.; Warmka, R.; Lindell, A.; Kumar, V. A daily global mesoscale ocean eddy dataset from satellite altimetry. *Sci. Data* **2015**, *2*, 150028. [CrossRef] [PubMed]

23. Bonjean, F.; Lagerloef, G.S. Diagnostic model and analysis of the surface currents in the tropical Pacific Ocean. *J. Phys. Oceanogr.* **2002**, *32*, 2938–2954. [CrossRef]

24. Hersbach, H.; Dee, D. ERA5 reanalysis is in production. *ECMWF Newsl.* **2016**, *147*, 5–6.

25. Jolliffe, I.T.; Cadima, J. Principal component analysis: A review and recent developments. *Philos. Trans. R. Soc. Math. Phys. Eng. Sci.* **2016**, *374*, 20150202. [CrossRef]

26. Zavorotny, V.U.; Voronovich, A.G. Scattering of GPS signals from the ocean with wind remote sensing application. *IEEE Trans. Geosci. Remote. Sens.* **2000**, *38*, 951–964. [CrossRef]

27. Zavorotny, V.U.; Voronovich, A.G. Validity of the Kirchhoff-Geometric Optics Approach for Modeling of Ocean Bistatic Radar Scattering. In Proceedings of the IGARSS 2019–2019 IEEE International Geoscience and Remote Sensing Symposium, Yokohama, Japan, 28 July–2 August 2019; pp. 668–671. [CrossRef]

28. Plant, W.J. A relationship between wind stress and wave slope. *J. Geophys. Res. Ocean.* **1982**, *87*, 1961–1967. [CrossRef]

29. Phillips, O. On the response of short ocean wave components at a fixed wavenumber to ocean current variations. *J. Phys. Oceanogr.* **1984**, *14*, 1425–1433. [CrossRef]

30. Voronovich, A.G.; Zavorotny, V.U. Bistatic radar equation for signals of opportunity revisited. *IEEE Trans. Geosci. Remote. Sens.* **2017**, *56*, 1959–1968. [CrossRef]

31. Johannessen, J.A.; Kudryavtsev, V.; Akimov, D.; Eldevik, T.; Winther, N.; Chapron, B. On radar imaging of current features: 2. Mesoscale eddy and current front detection. *J. Geophys. Res. Ocean.* **2005**, *110*. [CrossRef]

32. Wallace, J.M.; Mitchell, T.; Deser, C. The influence of sea-surface temperature on surface wind in the eastern equatorial Pacific: Seasonal and interannual variability. *J. Clim.* **1989**, *2*, 1492–1499. [CrossRef]

33. Samelson, R.; Skyllingstad, E.; Chelton, D.; Esbensen, S.; O'Neill, L.; Thum, N. On the coupling of wind stress and sea surface temperature. *J. Clim.* **2006**, *19*, 1557–1566. [CrossRef]

34. Sugimoto, S.; Aono, K.; Fukui, S. Local atmospheric response to warm mesoscale ocean eddies in the Kuroshio-Oyashio Confluence region. *Sci. Rep.* **2017**, *7*, 11871. [CrossRef]

35. Ma, Z.; Fei, J.; Liu, L.; Huang, X.; Cheng, X. Effects of the Cold Core Eddy on Tropical Cyclone Intensity and Structure under Idealized Air–Sea Interaction Conditions. *Mon. Weather. Rev.* **2013**, *141*, 1285–1303. [CrossRef]

36. Walker, N.D.; Leben, R.R.; Pilley, C.T.; Shannon, M.; Herndon, D.C.; Pun, I.F.; Lin, I.I.; Gentemann, C.L. Slow translation speed causes rapid collapse of northeast Pacific Hurricane Kenneth over cold core eddy. *Geophys. Res. Lett.* **2014**, *41*, 7595–7601. [CrossRef]

37. Voronovich, A.G.; Zavorotny, V.U. The transition from weak to strong diffuse radar bistatic scattering from rough ocean surface. *IEEE Trans. Antennas Propag.* **2017**, *65*, 6029–6034. [CrossRef]

38. Renault, L.; Molemaker, M.J.; McWilliams, J.C.; Shchepetkin, A.F.; Lemarié, F.; Chelton, D.; Illig, S.; Hall, A. Modulation of Wind Work by Oceanic Current Interaction with the Atmosphere. *J. Phys. Oceanogr.* **2016**, *46*, 1685–1704. [CrossRef]

39. Seo, H.; Miller, A.J.; Norris, J.R. Eddy–Wind Interaction in the California Current System: Dynamics and Impacts. *J. Phys. Oceanogr.* **2016**, *46*, 439–459. [CrossRef]

40. Gagliardini, D.A., Medium Resolution Microwave, Thermal and Optical Satellite Sensors: Characterizing Coastal Environments Through the Observation of Dynamical Processes. In *Remote Sensing of the Changing Oceans*; Tang, D., Ed.; Springer: Berlin/Heidelberg, Germany, 2011; pp. 251–277. [CrossRef]

Article

Experimental Evidence of Swell Signatures in Airborne L5/E5a GNSS-Reflectometry

Joan Francesc Munoz-Martin [1,*], Raul Onrubia [1], Daniel Pascual [1], Hyuk Park [1], Adriano Camps [1], Christoph Rüdiger [2], Jeffrey Walker [2] and Alessandra Monerris [3]

[1] CommSensLab—UPC, Universitat Politècnica de Catalunya—BarcelonaTech, and IEEC/CTE-UPC, 08034 Barcelona, Spain; onrubia@tsc.upc.edu (R.O.); daniel.pascual@upc.edu (D.P.); park.hyuk@tsc.upc.edu (H.P.); camps@tsc.upc.edu (A.C.)

[2] Department of Civil Engineering, Monash University, Clayton, VIC 3800, Australia; chris.rudiger@monash.edu (C.R.); jeff.walker@monash.edu (J.W.)

[3] Department of Infrastructure Engineering, The University of Melbourne, Parkville, VIC 3010, Australia; alessandra.monerris@unimelb.edu.au

[*] Correspondence: joan.francesc@tsc.upc.edu; Tel.: +34-626-253-955

Received: 22 April 2020; Accepted: 27 May 2020; Published: 29 May 2020

Abstract: As compared to GPS L1C/A signals, L5/E5a Global Navigation Satellite System-Reflectometry (GNSS-R) improves the spatial resolution due to the narrower auto-correlation function. Furthermore, the larger transmitted power (+3 dB), and correlation gain (+10 dB) allow the reception of weaker reflected signals. If directive antennas are used, very short incoherent integration times are enough to achieve good signal-to-noise ratios, allowing the reception of multiple specular reflection points without the blurring induced by long incoherent integration times. This study presents for the first time experimental evidence of the wind and swell waves signatures in the GNSS-R waveforms, and it performs a statistical analysis, a time-domain analysis, and a frequency-domain analysis for a unique data set of waveforms collected by the UPC MIR instrument during a series of flights over the Bass Strait, Australia.

Keywords: GNSS-R, waveform, GPS, Galileo, sea, swell, waves

1. Introduction

The Microwave Interferometric Reflectometer (MIR) is an airborne multi-constellation, dual-beam, and dual-band (L1/E1 and L5/E5a), conventional and interferometric GNSS-R [1] instrument (Figure 1). It has two very directive antenna arrays to pick the direct and reflected GNSS signals [2,3]. MIR was conceived for real-time processing, but raw data is also stored at 32.768 MS/s at 1 bit I/Q for offline processing. One of its maiden flights was over the Bass Strait that separates Australia and Tasmania. The large antenna directivity (18.1 dB for the down-looking array @ L5/E5a) allows a clean detection of the reflected signal even with short incoherent integration times (40–300 ms). In these conditions, the peak of the reflected signal is not blurred, and re-tracking is not required. This has allowed that multiple peaks systematically appear in the Delay-Doppler Maps (DDM) and waveforms (WF), provided they occur within the antenna footprint. As it is presented in this study, these multiple peaks are related to sea wave information, produced by multiple reflections on consecutive waves, producing a forward Bragg scattering in consecutive wave crests.

(a)   (b)

Figure 1. (**a**) Microwave Interferometric Reflectometer (MIR) instrument and up-looking array mounted inside the airplane, and (**b**) down-looking array covered with a radome hanging from the airplane's fuselage [2].

Wind-driven and swell period or wavelengths have both been measured using other remote sensing techniques such as High Frequency (HF) radar [4], where the HF signal is back Bragg scattered over multiple swell crests, and the signal retrieved contains a modulation in frequency based on the scattered signal. Swell retrieval has been also conducted using microwaves signals, as the approach using a Doppler radar at S-Band [5], where both wind sea and swell components can be measured through spectral analysis. Moreover, Synthetic Aperture Radar (SAR) instruments have also retrieved swell parameters from space based on spectral estimation, as in Sentinel-1 [6]. As seen, most of those measurements are based on a back Bragg scattering mechanism in consecutive wave crests (Chapter 10 of [7]).

The same physical principle is presented in this study, but using GNSS-R signals in a forward scattering configuration. As stated in [8], swell components can only be seen using large bandwidth signals with narrow auto-correlation functions. Examples are given for resolutions up to 30 m. However, this is not the case for current GNSS-R instruments, which are nearly all designed to work with GPS L1 C/A signals, with 300 m spatial resolution (width of the auto-correlation function). However, the use of a higher spatial resolution signal, such as the GNSS L5 (with a spatial resolution of 30 m), opens the possibility to measure the swell spectra, as the spatial resolution provided by these signals is comparable to the swell wavelength, as will be covered in the next section.

2. Data set Description and Validation Data

On 6th June 2018 MIR flew over the Bass Strait, Australia, installed on an airplane flying at $\sim$1500 m height, and at speed of 74 m/s. During the post-processing of the waveforms generated by MIR (as shown in [9]), it has been found that most L5/E5 waveforms have second peak. In order to illustrate it, two different tracks have been selected, as shown in Figures 2 and 3. Each track has two beams directly pointing to a GNSS satellite reflection, as detailed in Table 1, which summarizes the metadata information of the reflection, including the satellite PRN, constellation, and the reflection incidence angle to each of the tracks shown in Figure 3. Incidence angle is defined as the angle between transmitter-to-specular point ray and the zenith vector perpendicular to the reflection surface, which is assumed flat for the sake of simplicity.

Table 1. Track, beam, PRN, and mean incidence angle of the measurement.

Track ID	Beam ID	Constellation and PRN	Incidence Angle (°)
1	1	GPS #1	40
1	2	GPS #32	52
2	1	Galileo #3	20
2	2	GPS #3	42

The data acquired by MIR is sampled at 32.768 MS/s, packetized, and time-tagged every 300 ms. In order to ease the data processing scheme, a single 300 ms packet is used for a single waveform, but a much shorter incoherent integration time of 40 ms is used to avoid blurring of the waveform due to the plane movement. Taking into account the plane speed (74 m/s), and the incoherent integration time (40 ms), the blurring corresponds to an integration over ∼3 m. In addition, the generation rate of the waveform (i.e., the final observable) is at 3.33 Hz, or one every 300 ms, hence each waveform is separated ∼22 m.

In order to understand the origin of all the secondary peaks observed in the waveforms (see Section 3), the sea state conditions during the flight are studied. As there is a lack of in situ buoy information in the area, the ICON model [10] wind wave period [s], swell wave period (s), and wind speed over the sea [m/s] are shown in Figure 2. The waves shown in Figure 2a have a period of $T_{wind} \approx 5$ s, and they are moving with a look angle with respect to the plane trajectory of ∼50°. The swell waves (Figure 2b) have a period of $T_{swell} \approx 9$ s, and the look angle with respect to the plane trajectory is ∼120°. Both wind-driven and swell wave period are related to the wave speed, and therefore to the waves wavelength (Λ_{waves}) [11] (i.e., distance between the crests of the wave) by means of Equation (1).

$$C_{waves} = 1.56 \cdot T_{waves},$$
$$\Lambda_{waves} = C_{waves} \cdot T_{waves}, \tag{1}$$

where C_{waves} is the sea wave celerity (or speed (m/s)), and T_{waves} is either the swell period T_{swell} or the wind-driven period T_{wind} as defined in the previous section. In deep water conditions, the relationship between them is the above closed formula.

Note that the nomenclature used in this study is X_{waves} (X stands for any parameter as C, T, or Λ) for any measurement that is not directly related to either wind-driven or swell waves, and X_{wind} or X_{swell} for any measurement directly related to the wind or the swell respectively, retrieved from the ICON model.

In the case under study, the wavelengths of the wind and swell waves are between 39 m and 126 m. Note that, the aggregation of the different waves may cause a different set of wavelengths or periods right with a minimum wavelength of 39 m and a maximum wavelength of 126 m. The following sections show a complete analysis of this second peak and its relations to the sea state conditions. As stated in [12,13], wind-driven and swell waves are combined forming different waves with different periods and heights. The wind and swell components (or even multiple swell components) can be separated by means of spectral analysis [14].

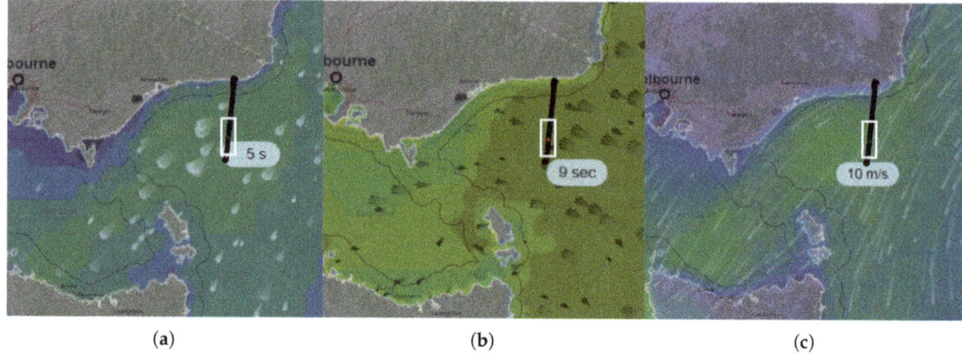

Figure 2. Position of the plane superposed to (**a**) wind-driven wave period, (**b**) swell wave period, and (**c**) wind speed over sea of the data used for the secondary specular reflection analysis. The plane trajectories framed in white box is zoomed in Figure 3. The wind-driven period, the swell period, and the wind speed are provided below the white box [15]

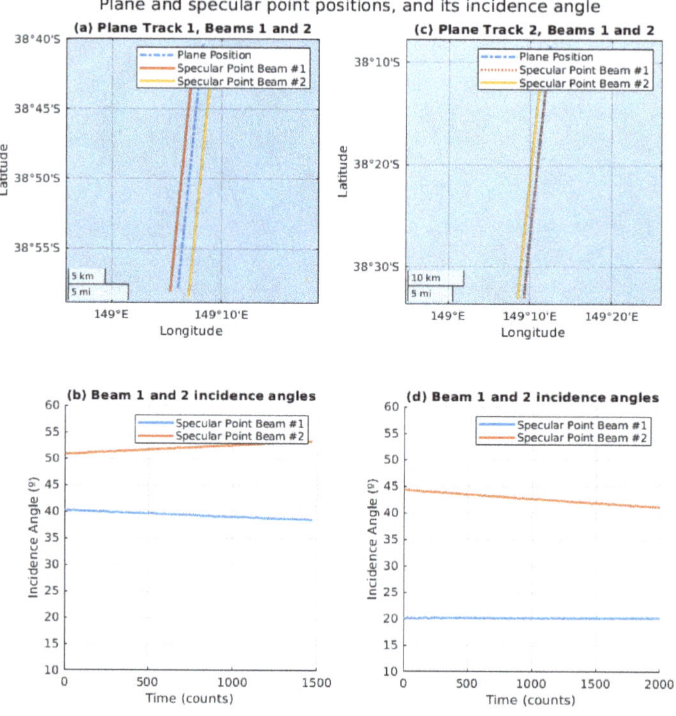

Figure 3. Top frame—plane position and specular point position of Track 1 (**a**) and Track 2 (**c**). Bottom frame—incidence angle to the specular point of Track 1 (**b**) and Track 2 (**d**). The plane positions correspond to the white frame box in Figure 2. Note that, the plane position and specular point of Beam 1 of Track 2 are superposed due to the very low incidence angle (i.e., almost nadir).

3. Sea Wavelength Retrieval from Waveforms

As in most GNSS scenarios, and due to the sea surface roughness, the combination of both swell waves and wind-driven waves cause multiple reflections, which are then captured by the down-looking antenna of the GNSS-R instrument. Despite that, and as seen in other works [9,16], the reflection over the sea surface presents a notable coherent component, which depends on the geometry of the reflection and the wind speed. Most of the waveforms retrieved in this work present a very high coherent component, the same data set used in [9] is now used to analyze the presence of secondary peaks in the retrieved waveforms, which turns to be linked to two strong reflections over nearby wave crests.

The reflection scenario is illustrated in Figure 4. Note that, the semi-radius of the ellipse formed by the first Fresnel zone [17] (i.e., the area that contains the specular reflection) in the described scenario is limited by the plane altitude as shown in Equation (2).

$$l_{Fr} = \frac{\sqrt{\lambda R_r}}{cos(\theta_{inc})}, \ where \ R_r = \frac{h}{cos(\theta_{inc})} \tag{2}$$

where $\lambda = 25$ cm for L5, $h = 1500$ m, and θ_{inc} is the wave incidence angle. Therefore, at $\theta_{inc} = 45$, $l_{Fr_{\angle 45}} = 33$ m, whereas for an almost-nadir reflection as Beam 1 of Track 2, $l_{Fr_{\angle 20}} = 21$ m.

In this case, second and third reflections occurring in consecutive wave crests in the first Fresnel zone produce different signal wave fronts that are added constructively or destructively in the receiver antenna. However, thanks to the large bandwidth, and hence the very narrow auto-correlation [18] function of the L5/E5a GNSS signal (i.e., 30 m in space), reflections from nearby crests with a distance between them larger than 30 m can produce second peaks in the retrieved waveform.

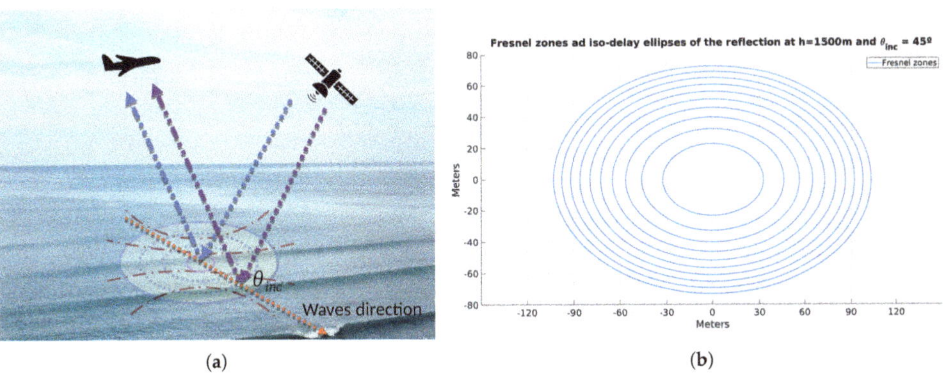

(a) (b)

Figure 4. (a) An example of the reflection scenario, where two wavefronts are reflected over two consecutive wave crests, where θ_{inc} is the incidence angle of the reflected signal. (b) The first 10 Fresnel zones associated to the reflection scenario described in (a), computed at $\theta_{inc} = 45°$. Note that each Fresnel zone corresponds to a 180° phase rotation.

As an example, two wavefronts are reflected over two successive wave crests, and hence two different specular points, resulting in the addition of the two signals in the receiver antenna, which is detected as a second peak in the processed waveform, as detailed in Figure 5. Sometimes even a third peak appears.

Due to the variability and the surface roughness of the ocean, the distance between two crests varies. In this case, where wind wave and swell wave components are both strong and also with different directions of propagation, the peak-to-peak distance is not preserved from one realization to the next one.

As explained in [19], wind-driven or swell wave period is not a fixed magnitude, but the average of the swell waves component period for the swell waves, and the average of the wind-driven period for the wind-driven waves. As an example, a swell period of 9–10 s generates a larger span of waves, which have a period between them from 8 s to 11 s. Therefore, a relative large data set analysis is required to correctly retrieve the swell waves period or wind-driven waves period.

Finally, the peak-to-peak distance does not only reproduce the crest-to-crest between two consecutive waves, but also a modulation depending on the height between the two waves where the reflection has occurred.

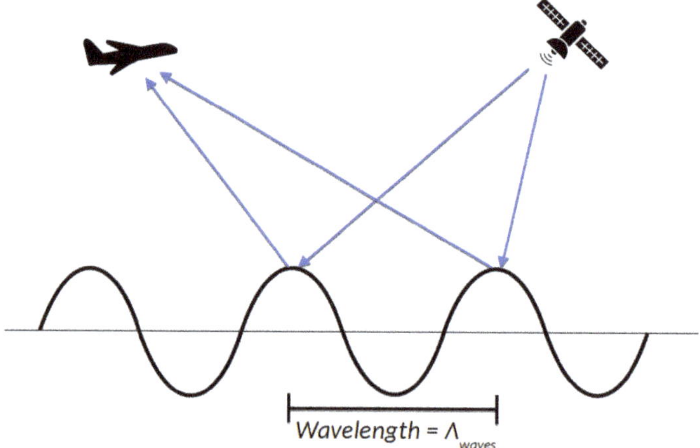

Figure 5. Forward scattering over two wave crests.

3.1. Waveform Simulation

As opposed to reflections in L1 C/A, where the spatial resolution is mostly limited by the chip length (i.e., 300 m), the new GPS L5 or Galileo E5a signals have 10 times higher spatial resolution, as the auto-correlation function is 10 times narrower (i.e., 30 m in space), allowing de reception of multiple specular points with a distance larger than 30 m. In this case, any reflection from any point separated by more than 30 m produces a secondary peak on the retrieved waveform instead of a blurring on the retrieved wavelength.

This section shows different simulation scenarios of the retrieved waveforms modulated by a forward scattering occurring at the crests of the waves, belonging to different Fresnel zones (Figure 4). To do so, a given PRN of a GPS signal at L5 ($\tau_{coh} = \tau_{inc} = 1ms$) is sampled at 32.768 MHz (to have better granularity and mimic the MIR sampling ratio) which is then delayed and added as in Equation (3). Note that, due to the sampling used, the distance from the peak to the zero of the auto-correlation (i.e., half the width of the whole auto-correlation function) is ~3.2 samples.

$$S'[k] = \sum_{i=0}^{N-1} \Phi[i] \cdot S[k-i] \tag{3}$$

where $S'[k]$ is the resulting PRN sequence at the kth sample, sampled at 32.768 MHz, Φ is the amplitude coefficient applied to each sample of the clean PRN sequence, S, and N is the length of the Φ vector, which represents the reflection scenario, where for this example is set to 10, causing a delay up to three L5 chips.

In order to illustrate this concept, three simulated waveforms are computed with different c coefficients. In the first example, using $\Phi = [1, 0.5, 0.5, 0.4, 0.4, 0.3, 0.3, 0.3, 0.2, 0.1]$ a reflection scenario is computed with lots of small contributions from all the Fresnel zones near the specular point. This produces a waveform as in Figure 6a, where the power decreases, and no secondary peaks are found. However, using $\Phi = [0.1, 1.0, 0.1, 0.1, 0.1, 0.1, 0.1, 0.1, 1.0, 0.1]$ produces two identical peaks which are easily distinguished. Note that, the separation between the peaks in samples is seven samples. In order to convert the sample distance into a tangible magnitude, the same method as in Equation (10) from [9] is used here in Equation (4).

$$\Delta_m = \Delta_{samples} \cdot \frac{c}{f_s},\tag{4}$$

where c is the speed of light, $\sim 3 \cdot 10^8$ [m/s], and f_s is the receiver sampling rate, $32.768 \cdot 10^6$ [S/s].

Applying it to Figure 6b, the distance between the two specular reflections (identified as $\Phi[1]$ and $\Phi[8]$ in previous example), a distance of 64 meters is retrieved.

However, reflections are not always as perfect as in Figure 6b. A third simulation (Figure 6c) is generated using a more realistic case, with $\Phi = [0.1, 0.1, 0.4, 1.0, 0.6, 0.1, 0.1, 0.2, 0.8, 0.4, 0.1, 0.1, 0.1]$, trying to reproduce two crests separated by five samples or 45 meters. Note that, the amplitudes of the neighbor areas have been also included, but they do not affect the high amplitude specular reflection contribution.

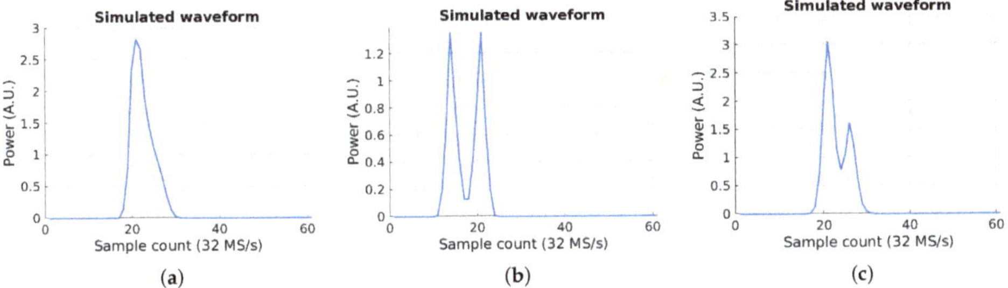

Figure 6. Three simulated waveforms different contributions from nearby Fresnel zones, producing secondary peaks driven by strong specular and coherent reflections over two hypothetical consecutive wave crest. (**a**) Simulated waveform with small contributions from all the nearby Fresnel zones, (**b**) simulated waveform with two ideal peak reflections separated by seven samples, and (**c**) simulated waveform with realistic contributions from a secondary strong reflection separated by five samples.

3.2. Evolution of Complex Waveforms in a Single Beam

The previous section has shown three different simulated waveforms synthetically generated to illustrate the concept of a second specular reflection in a consecutive wave crest around the second to sixth Fresnel zones.

In this section, three consecutive measured waveforms (see Figure 7) have been selected to illustrate these effects. The separation in time between each waveform is 600 ms, and they have been selected because of their similitude to the ones simulated in the previous section. Note that, the three selected waveforms are from Track 1 Beam 1, and both amplitude and phase evolution of the waveform are shown.

The first waveform (Figure 7a) shows an example similar to the one in Figure 6a, where very small contributions from all nearby Fresnel zones causes a blurring of the tail of the waveform.

The main peak is around sample count 17, with a phase approximately constant around +100°. Furthermore, there is a second small peak (in sample 25) around eight samples (i.e., 73 m in space) after the main one (in sample 17). The distance between the peaks indicates that small reflections are coming from the sixth to eighth Fresnel zones.

The second waveform (Figure 7b) shows a example similar to the one in Figure 6b, but with the reflection occurring in a closer Fresnel zone. In this case, the distance between these two peaks is three samples or ~ 30 m, which corresponds to a reflection in the second Fresnel zone. Compared to the first waveform, the reflections occurred within the first Fresnel zone causing a blurring and a widening of the L5 auto-correlation function are now split in two different peaks.

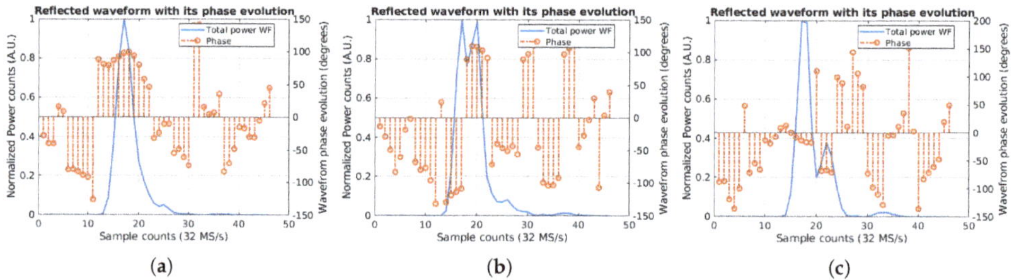

Figure 7. Three sample waveforms containing different contributions from different Fresnel zones, producing secondary peaks with different shapes. (**a**) Retrieved waveform with small contributions from all the nearby Fresnel zones, (**b**) retrieved waveform with two big peaks of the same amplitude separated three samples one from other, and (**c**) retrieved waveform with two peaks with different amplitudes and separated by five samples.

As seen, the first peak (in sample 17) has a phase of $\sim-110°$, and the second peak (in sample 20) has a phase of $\sim110°$ (i.e., close to the phase of the peak in Figure 7b), with a relative phase difference $\sim220°$, corresponding to reflection in the second Fresnel zone, as both are separated more than $180°$. Note that, there is a strong phase jump between the two peaks, indicating that both reflections are coherent (i.e., the phase information is preserved in both cases). As in the previous case, this third peak (in sample count 26) has an approximately constant phase around $-50°$. In addition and similar to the first waveform, a third small peak is also present nine samples (i.e., 82 m in space) away from the very first one. In this case, the reflection is taking place between the eighth and the tenth Fresnel zones.

Finally, the third waveform (Figure 7c) shows a similar shape as the simulation in Figure 6c, with a relative high second peak (in sample 22) in the reflected waveform which in this case is five samples from the main one (in sample 17). In this last case, the phase evolution is approximately constant in both peaks with a strong jump in the transition between them (i.e., very small incoherent reflection from other Fresnel zones). Note that in this case the relative phase difference is $\sim50°$, but as the delay in time is ~45 m, the reflection takes place further away than the second Fresnel zone, the phase with respect to the first peak has advanced at least $180°$. Therefore, the absolute phase shift between the two peaks is $\sim315°$.

3.3. First Waveform Analysis: Determination of the Crest-to-Crest Distance from Peak-to-Peak Distance

The last section showed three examples of three different waveforms retrieved from Track 1 Beam 1, where the location of the second peak varies with time. Furthermore, it has been shown that this second reflection comes from a different Fresnel zone, and in some cases present a coherent component (i.e., a single specular reflection over a wave crest). Figure 8 illustrates the example of different consecutive peaks present in the GNSS waveforms originated by the forward scattering on two consecutive wave crests, but now in the two beams for the two tracks of the flight.

Note that, the two waveforms for each track (i.e., beams 1 and 2 for Track 1 and beams 1 and 2 for Track 2) are taken at the same moment. As seen, the four waveforms present, at least, one second reflection.

Analyzing each of the tracks, Figure 8a,c present a single second reflection at six samples from the specular one. For the beam 2 case, Figure 8b presents a very large second reflection at six samples from the specular one. In addition, two small reflections are taking place 10 and 13 samples from the specular one. Finally, Figure 8d shows a very low secondary reflection point at nine samples from the main peak.

Applying Equation (4) to the waveform shown in Figure 8, the distance between the peaks in meters are: 54.9 m for both Figure 8a,c, 54.9 m for the first peak of Figure 8b, and 91.5 m for the third peak. Finally for Figure 8d, the distance between the peaks is 82.4 m.

Comparing those retrieved waveforms to the simulated waveforms from Figure 6, it is possible to identify a clear similitude between them, where the second peak (i.e., the one in Figure 8b) reminds the simulated in Figure 6c, due to a specular reflection over a wave crest.

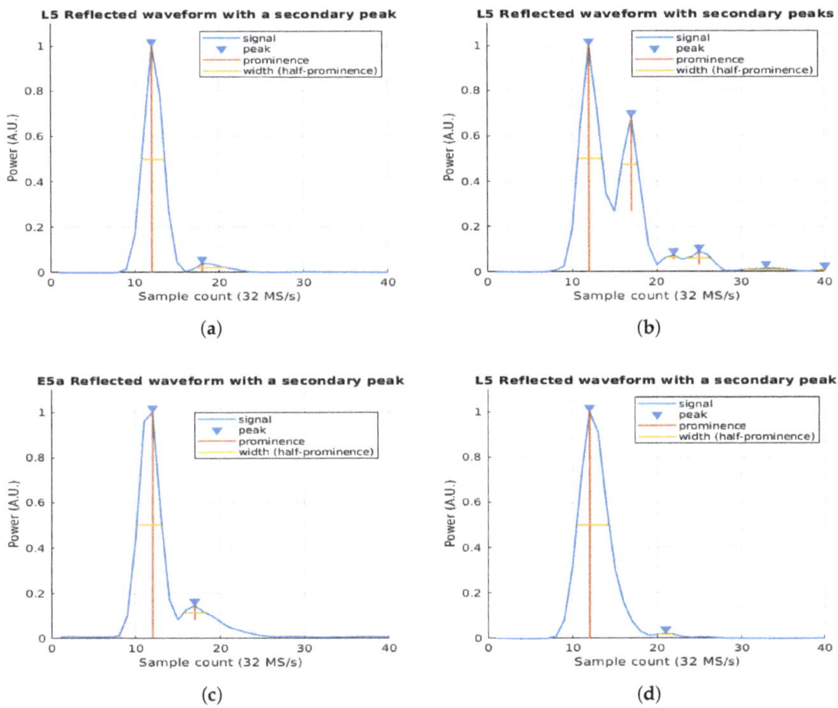

Figure 8. Reflected waveform for (**a**) Track 1, Beam 1; (**b**) Track 1, Beam 2; (**c**) Track 2, Beam 1; and (**d**) Track 2, Beam 2 of GPS L5 and Galileo E5a signals over the sea surface with an incoherent integration time of 40 ms.

In order to improve the estimations of the separation between peaks in the waveform, since raw data was acquired satisfying the Nyquist criteria at baseband for L5/E5a signals (i.e., $f_s > 10$ MHz), therefore the signal can be re-sampled without loss of information. Therefore, waveforms are re-sampled using the Fourier interpolation method (i.e., by inverse FFT of the zero-padded FFT of the waveform), and the local maxima of the interpolated waveform is located. The first two maxima of the waveform are the ones used to retrieve the distance between two consecutive wave crests. The difference in samples between the two peaks position is then converted into meters following the same approach as in previous section, but with a sampling rate f_s scaled by the interpolation rate K.

The selection of this parameter is a trade-off between computational requirements and accuracy when selecting the final position of the multiple peaks. In order to efficiently apply the Fourier interpolation using the Fast Fourier Transform, the signal shall be a multiple of 2, and hence the interpolation rate K, shall be a multiple of 2 as well. In addition, the larger the K parameter is, the lower is the error determination of the peak position of the reflected waveform. For this case an interpolation rate of $K = 2$ has a peak position uncertainty of 4.5 meters, and a $K = 8$ has a peak position uncertainty of $\sim$1 m.

In this case, an interpolation rate of $K = 8$ has been selected (i.e., now $f_s = 256$ MHz). Furthermore, Figure 9 shows the re-sampled version of the waveforms in Figure 8 following the Fourier interpolation method.

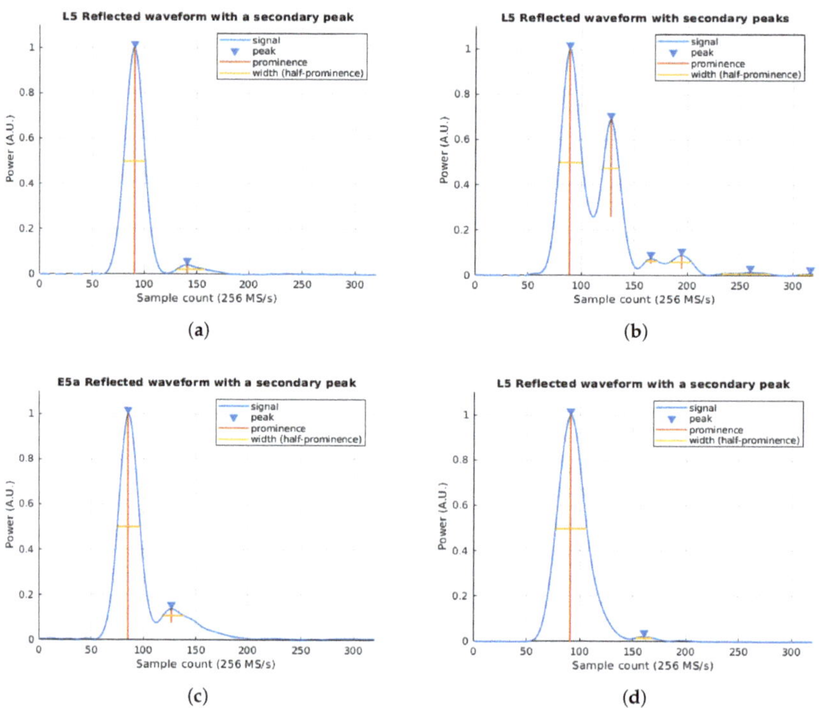

Figure 9. Reflected waveforms for (**a**) Track 1, Beam 1; (**b**) Track 1, Beam 2; (**c**) Track 2, Beam 1; and (**d**) Track 2, Beam 2 from Figure 8 now sampled at 256 MHz. Note that, thanks to the re-sampling the determination of the peak position is enhanced.

The re-sampled versions of the waveforms show a better spatial resolution when estimating the peak-to-peak distance. In this case the distance in samples is: 50, 39, 42, and 70 respectively form (a) to (d). Moreover, the third peak of Figure 9b is located at 77 samples from the specular one.

Applying Equation (4) to the above sample distances, the retrieved crest-to-crest distance (Λ_{waves}) in meters is: 57.2, 44.6, 48, and 80.1 m respectively for (a) to (d), and 88.1 m for the third peak of (b). Table 2 compares the values prior to the re-sampling with the re-sampled values. Note that, the estimation of both the first peak position is enhanced thanks to the FFT interpolation. In addition, the estimation of the position of the second peaks is also enhanced. A proper estimation of the position of both peaks is crucial to better determine the distance between the two reflections (i.e., the two wave crests). Note the difference in Beam 2 of Track 1 and Beam 1 of Track 2, where the first approach does not show the proper crest-to-crest distance, introducing an error of almost 10 m in the measurement.

This section has covered the simulation and its comparison with real data of different scenarios where different specular reflections are generating second and even third peaks on the reflected waveform. The study on the phase evolution in this secondary peaks shows that the second reflection is coherent, and hence is coming from a different specular point, as was detailed in Figure 5. Finally, the use of FFT interpolation enhances the determination of the peak-to-peak distance, showing a better granularity when determining this magnitude.

Table 2. Track, beam, crest-to-crest estimated distance using the original waveform, and the crest-to-crest distance after re-sampling the original waveform using the FFT interpolation.

Track ID	Beam ID	Original Crest-to-Crest Distance [m]	Re-Sampled Crest-to-Crest Distance [m]
1	1	54.5	57.2
1	2	54.5	44.6
2	1	54.5	48.0
2	2	82.4	80.1

4. Sea Wavelength Retrieval Using Large Data Set

The study of a larger data set is required to validate the relationship between sea waves wavelength Λ_{waves} or wave period T_{waves} to the distance between the peaks in the waveforms. Tracks identified in Table 1 of Section 2 are composed by 1470 waveforms, and 1994 waveforms respectively, generated once per 30 ms (i.e., 3.33 Hz), which corresponds to $\sim$8 min of data for each of the two beams for the first track, and 11 minutes of data for the second track. In order to analyze the data set, three different approaches are considered:

- Statistical analysis: statistics of the retrieved crest-to-crest distance and presence or not of third peaks.
- Time-series analysis: analysis in the time domain of the retrieved crest-to-crest distance.
- Spectral analysis: analysis in the frequency domain of the time-series data.

4.1. Statistical Analysis

Reviewing the histogram of all the 1470 measurements for Track 1 (see Figure 10), most of the results are condensed between 30 m (the wind-driven wavelength from the model is 39 m) and 80 m (the swell wavelength from the model is 126 m).

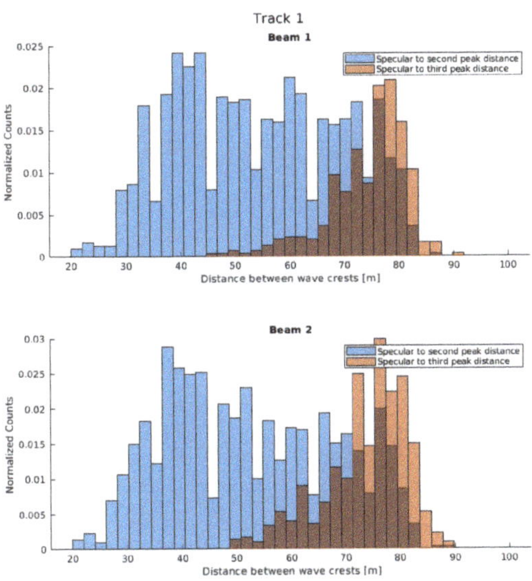

Figure 10. Histogram of the peak-to-peak distances for Track 1. Top figure corresponds to Beam 1 (GPS L5) and bottom figure to Beam 2 (GPS L5).

In addition, the distance to a third peak (if is exists) is condensed just at the end of the histogram, at the end of the histogram of the distance to the second peak. As seen, the measurements are condensed mostly between 40 and 90 m for both beams.

Both beams are following similar statistics for the second peak distance, with an average value of 54.3 m for Beam 1, and an average value of 52.6 m for Beam 2. The third peak has an average value of 74.5 m for Beam 1, and 73.5 m for Beam 2. Note that, the average of the measurements is a mixture of all possible wave crest distance contributions. Note that, the least wave crest distance is $\sim$ 30 m, which coincides with the spatial resolution due to the chip bandwidth of the L5/E5a signal, 30 m.

In addition, the same histogram of the second-track is shown in Figure 11. The first beam corresponds to a Galileo E5a signal, which has an incidence angle of $\sim$20°. The mean peak-to-peak distance in this case is 50.8 m. The second beam has an incidence angle of 42°, with an average value of 53.5 m.

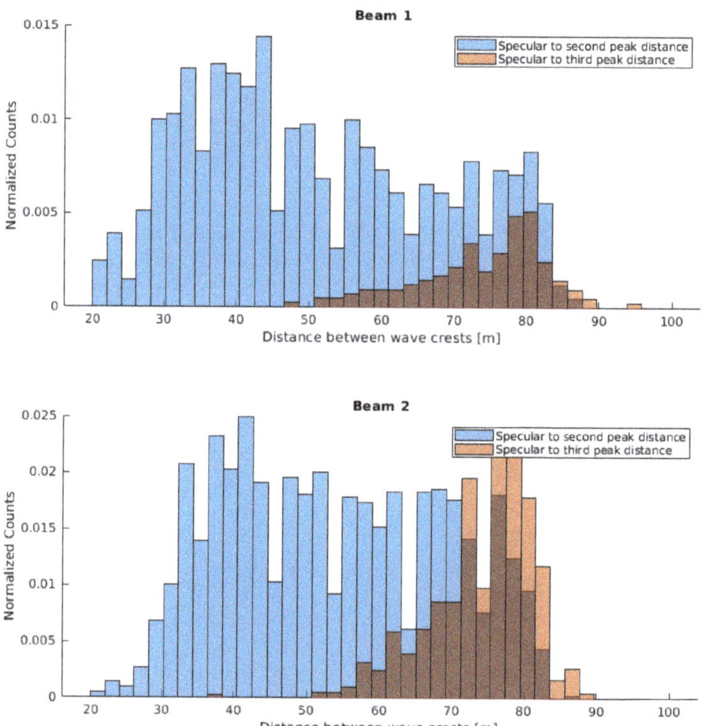

Figure 11. Histogram of the peak-to-peak distances for Track 2. Top figure corresponds to Beam 1 (Galileo E5a) and bottom figure to Beam 2 (GPS L5).

In case of this first beam, as the incidence angle is lower, the semi-radius of the first Fresnel zone is smaller (from 33 m with $\theta_{inc} = 45$ to 21 m with $\theta_{inc} = 20$). As the overall sizes of the Fresnel zones are smaller, the area where the possible strong reflection occurs over a wave crest is also smaller, hence some waves with a large distance between them can not be captured.

Note that, in this case, the third peak position for Beam 1 is poorly determined because of the small incidence angle, and the size of the Fresnel zones for that case. As seen in the histogram, the amount of third peak reflections in Beam 1 with respect to Beam 2 is in the order of 0.25.

As seen in the two tracks, the histogram analysis shows that multiple reflections are present in most of the waveforms, with an average value ~50 m for the distance between the first two peaks. From this first analysis, the wind-driven waves wavelength can be estimated (i.e., the model set this distance to 40 m), and from it the wind-driven wave period.

4.2. Time-Series Analysis

The statistical analysis through the histogram visualization shows that the crest-to-crest distances are in the same range than in the ICON model (Section 2), the analysis of the time evolution of the peak-to-peak distances shows a wavy behavior.

In order to reduce the noise of the measurement, four different averaging windows have been applied to the peak-to-peak distance. In this case, the averaging selected is 5, 10, 15, and 30 samples. As seen in Figure 12, this wavy behavior is clearly identified in the measurement, which is present even with large averaging windows. As seen with the highest averaging time, the mean of the crest-to-crest distance is ~55 m for Beam 1, and ~52.5 m for Beam 2. As seen in [14,19], the average wave period, or the average wavelength, Λ_{waves}, tends to the combination of all the different wave contributions.

A similar behavior is also found in Track 2, as shown in Figure 13. The Galileo signal (i.e., Beam 1), has a mean value ~50 m, and Beam 2 a mean value on the order of ~ 52.5 m, as in Track 1. Note that, a large portion (~ 40%) of the Beam 1 reflections do not have a secondary peak because of the reflection geometry. As the incidence angle is very close to 0° (i.e., very close to nadir), the possible reflections are coming from closer wave crests, therefore the receiver is only able to infer consecutive crests which are closer one to each other.

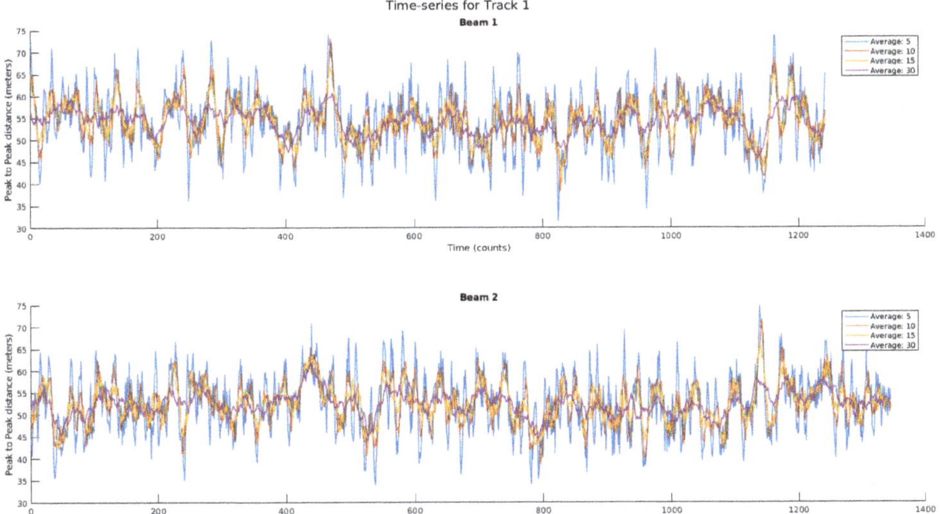

Figure 12. Time-series evolution of the Track 1 peak-to-peak calibrated distance with different moving average windows. Top figure corresponds to Beam 1, and bottom figure to Beam 2.

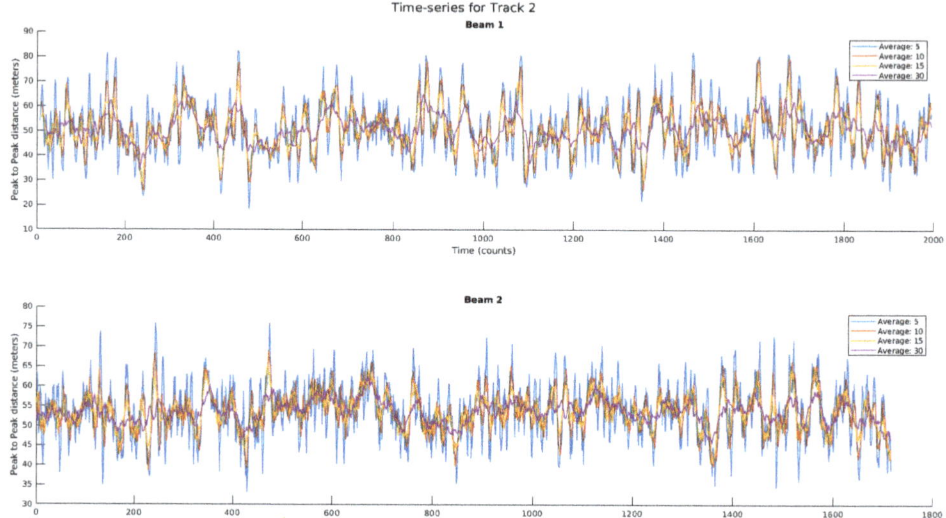

Figure 13. Time-series evolution of the Track 2 peak-to-peak calibrated distance with different moving average windows. Top figure corresponds to Beam 1, and bottom figure to Beam 2.

4.3. Spectral Analysis

The time-series analysis together with the statistical one show that there is a vast range of retrieved sea wave wavelengths Λ_{waves}, with some apparent periodicity. As detailed in [12,14], a spectral analysis over the magnitude under study helps to separate the different swell or wind-driven wave contributions.

In order to do that, the power spectral density of the time-series data from previous section is computed. Figure 14 shows the power spectral density of the distance to the second peak (or the first encountered crest in the sea wave surface). In all four figures, there are some common frequencies at $\sim$0.06 Hz, 0.1 Hz, 0.15 Hz, and also a 0.2 Hz component in Beam 1 of both Tracks 1 and 2.

Converting those frequencies to seconds, the wave periods of each of the wavelengths are $T_1 = 16.6$ s, $T_2 = 10$ s, and $T_3 = 6.66$ s. Comparing the three periods to the swell and wind-driven wave periods to the ones of the ICON model in Figure 2, $T_2 \simeq T_{swell}$ and $T_3 \simeq T_{wind}$.

In addition, the 0.2 Hz component, corresponding to a period of $T_4 = 5$ s, which is only present in Beam 1 of both Track 1 and Track 2, exactly matches the wind-driven prediction by the model, where $T_{wind} = 5$ s.

As already stated, the lower the incidence angle of the beam, the shorter waves can be retrieved (i.e., waves with shorter period), which is mostly the case for Figure 14c, where the incidence angle is closer to $0°$ (i.e., nadir), and hence wind-wave periods can be retrieved rather than swell-driven wave periods.

Furthermore, in case of analyzing the power spectral density (Figure 15) for the distance to the third peak, the fundamental frequencies are below the previous case, with a first fundamental frequency consistent in (a), (c), and (d) around $\sim$0.03 Hz, 33 s of period.

The spectrum shape of this third peak indicates a long period sea wavelength which is not contemplated in the model. As stated in [19], this high order period is present on swell waves, and also modulates the amplitude of the swell waves, it is known as second swell.

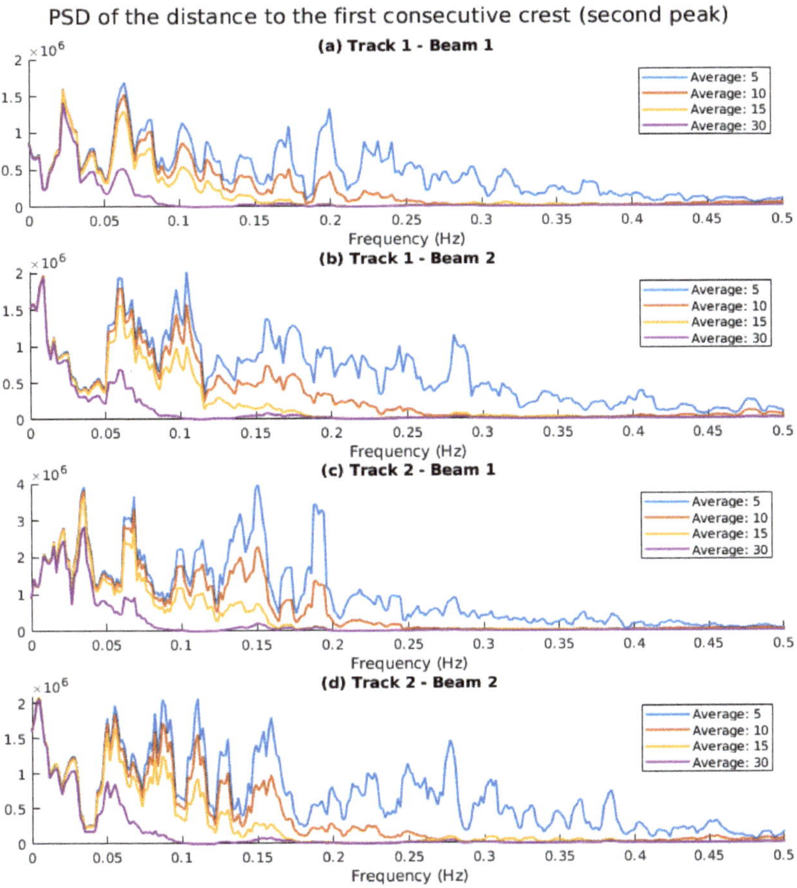

Figure 14. Fourier Transform of the time series evolution of (Figures 12 and 13), for the second peak distances.

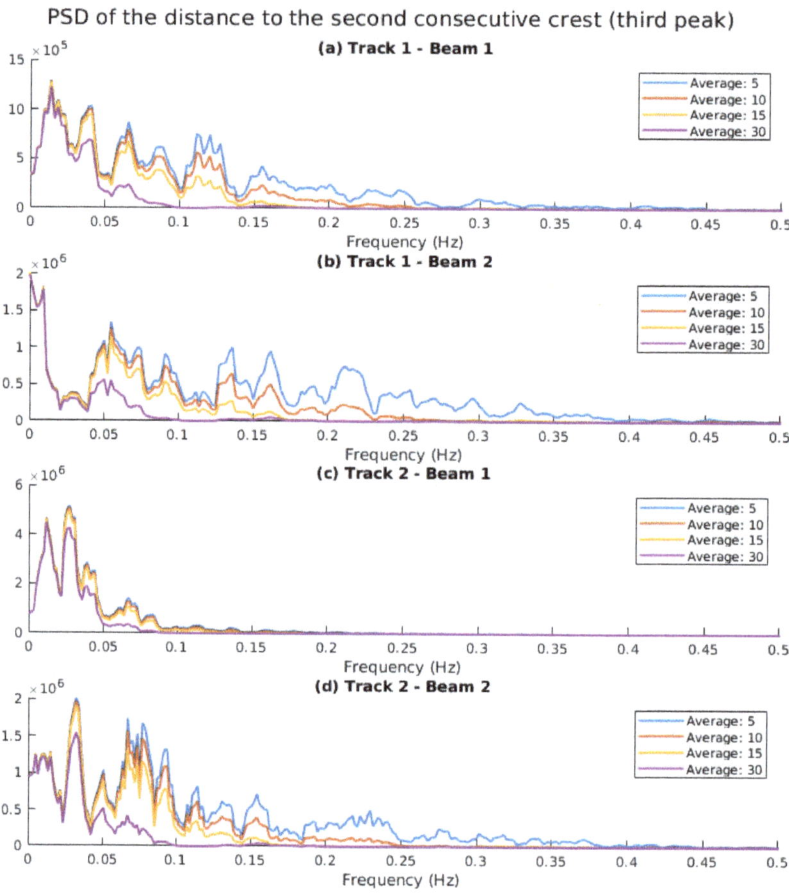

Figure 15. Fourier Transform of the time series evolution of (Figures 12 and 13), for the third peak distances.

In this case, the third peak spectrum does not give as much information as the second peak, from where it is possible to infer the wave period for different wave sources, as swell, wind-driven wind, or secondary swell waves.

5. Discussion

Each analysis provides a different perspective and hence a different parameter of the ocean can be estimated, and they are all summarized and compared to the values predicted by the ICON model in Table 3. Note that, the wave periods in the statistical and time-series approaches have been determined from the average crest-to-crest distance and assuming $\Lambda_{waves} = 1.56 \cdot T_{waves}^2$, as in (1).

Table 3. Estimated wave period from the three different analysis applied and its comparison ot the ICON model prediction.

Measurement	Est. Wave Period	ICON Model Estimation
Statistical analysis	5.8 s	Wind-driven = 5 s Swell = 9 s
Time-series analysis	5.6 s	Wind-driven = 5 s Swell = 9 s
Spectral analysis	Four components: 5 s, 6.3 s, 10 s, 16.6 s	Wind-driven = 5 s Swell = 9 s

In this case, both the statistical and the time-series analysis are providing a good estimation of the wind-driven wavelength, as the period of the wind-driven waves is smaller, the amount of wave crests per unit of area is larger, and hence the mean value of the waves is biased towards the wind-driven waves.

It is also important to remark that the time analysis serves to illustrate the wavy behavior of the crest-to-crest distance with respect to time. In this case, this analysis shows in a qualitative way how crest-to-crest distance is a mixture of different components, as explained in [12,14].

Last but not least, the spectral analysis clearly shows the different periods that can be identified in the time analysis. Multiple periods can be identified, among them τ_{swell} and τ_{wind} are identified as the main components of the spectrum of the crest-to-crest distance.

Furthermore, the results from the spectral analysis and the statistical analysis confirm that due to the geometry of the reflection, the lower the incidence angle is, the closer the reflection of the second crest occurs.

Finally, the spectral analysis also shows very low frequency components due to secondary swell or high period gravitational waves, which are interesting for other oceanographic studies, but still need further investigation.

6. Conclusions

This work has presented the first experimental evidence of wind and swell waves signatures in L5/E5a GNSS-R waveforms from second and third peaks. An algorithm has been presented to retrieve the peak-to-peak distance. Evidences for both the first and the second peaks are presented, and a detailed analysis from a large data set $N > 1000$ waveforms has been presented. The analysis in three domains (statistical, time, and spectrum) shows that additional ocean parameters can be retrieved from GNSS-R measurements, e.g., wind driven waves period and swell period, which have not been measured yet using GNSS-R technique.

This application still requires further refinements and a more exhaustive analysis with in-situ data. However, this study shows the promising results for the use of L5/E5a GNSS-R signals to infer new geophysical parameters as the wave period or wavelength of the sea thanks to the sharper shape of the auto-correlation function.

Author Contributions: Conceptualization, J.F.M.-M., H.P. and A.C.; methodology, J.F.M.-M.; software, J.F.M.-M.; validation, J.F.M-M. and A.C.; formal analysis, J.F.M.-M. and A.C.; investigation, J.F.M.-M.; resources, R.O., D.P., H.P., A.C., C.R., J.W. and A.M.; data curation J.F.M.-M., R.O., D.P.; visualization J.F.M-M.; supervision, H.P., A.C.; project administration, A.C.; funding acquisition, A.C.; writing—original draft preparation, J.F.M.-M.; writing—review and editing, J.F.M.-M., H.P., A.C. All authors have read and agreed to the published version of the manuscript.

Funding: This work was founded by the Spanish Ministry of Science, Innovation and Universities, "Sensing with Pioneering Opportunistic Techniques", grant RTI2018-099008-B-C21, and the grant for recruitment of early-stage research staff FI-DGR 2015 of the AGAUR - Generalitat de Catalunya (FEDER), Spain, and the grant for recruitment of early-stage research staff FI 2018 of the AGAUR - Generalitat de Catalunya (FEDER), Spain, and Unidad de Excelencia María de Maeztu MDM-2016-060.

Acknowledgments: Thanks to the CommsSensLab administrative and research personnel and to the NanoSatLab members for the support, specially to Adrian Pérez to help us mounting the database system that helped the authors to better analyze the retrieved data.

Conflicts of Interest: The authors declare no conflict of interest.

References

1. Zavorotny, V.U.; Gleason, S.; Cardellach, E.; Camps, A. Tutorial on Remote Sensing Using GNSS Bistatic Radar of Opportunity. *IEEE Geosci. Remote. Sens. Mag.* **2014**, *2*, 8–45, doi:10.1109/MGRS.2014.2374220.

2. Onrubia, R.; Pascual, D.; Park, H.; Camps, A.; Rüdiger, C.; Walker, J.; Monerris, A. Satellite Cross-Talk Impact Analysis in Airborne Interferometric Global Navigation Satellite System-Reflectometry with the Microwave Interferometric Reflectometer. *Remote. Sens.* **2019**, *11*, 1120, doi:10.3390/rs11091120.

3. Onrubia, R. ;Pascual, D.; Park, H.; Camps, A. *Preliminary Altimetric and Scatterometric Results with the Microwave Interferometric Reflectometer (MIR) during Its first Airborne Experiment*; ARSI-KEO 2019; ESA: Noordwijk, The Netherlands.

4. Bathgate, J.S.; Heron, M.L.; Prytz, A. A Method of Swell-Wave Parameter Extraction From HF Ocean Surface Radar Spectra. *IEEE J. Ocean. Eng.* **2006**, *31*, 812–818.

5. Chen, Z.; Zhang, L.; Zhao, C.; Chen, X. Wind sea and swell meansurements using S-band Doppler radar. In Proceedings of the OCEANS 2016, Shanghai, China, 19–23 September 2016; pp. 1–5.

6. Johnsen, H.; Husson, R.; Vincent, P.; Hajduch, G. *Sentinel-1 Ocean Swell Wave Spectra (OSW) Algorithm Definition*; Northern Research Institute (Norut): Tromso, Norway, 2020.

7. Robinson, I.S. *Measuring the Oceans from Space: The Principles and Methods of Satellite Oceanography*; Springer: Berlin, Germany, 2004.

8. Ward, K.D.; Tough, R.J.A.; Watts, S. Sea clutter: Scattering, the K distribution and radar performance. *Waves Random Complex Media* **2007**, *17*, 233–234, doi:10.1080/17455030601097927.

9. Munoz-Martin, J.F.; Onrubia, R.; Pascual, D.; Park, H.; Camps, A.; Rüdiger, C.; Walker, J.; Monerris, A. Untangling the Incoherent and Coherent Scattering Components in GNSS-R and Novel Applications. *Remote. Sens.* **2020**, *12*, 1208, doi:10.3390/rs12071208.

10. (DWD), D.W. ICON model description by DWD. Available online: https://www.dwd.de (accessed on 7 Januray 2020).

11. of Engineers, U.A.C. *Coastal Engineering Manual Part II: Coastal Hydrodynamics (EM 1110-2-1100)*; Books Express Publishing: Newbury Berkshire, UK, 2012.

12. Wang, D.W.; Hwang, P.A. An Operational Method for Separating Wind Sea and Swell from Ocean Wave Spectra. *J. Atmos. Ocean. Technol.* **2001**, *18*, 2052–2062, doi:10.1175/1520-0426(2001)018<2052:AOMFSW>2.0.CO;2.

13. Portilla, J.; Ocampo-Torres, F.J.; Monbaliu, J. Spectral Partitioning and Identification of Wind Sea and Swell. *J. Atmos. Ocean. Technol.* **2009**, *26*, 107–122, doi:10.1175/2008JTECHO609.1.

14. Hwang, P.A.; Ocampo-Torres, F.J.; García-Nava, H. Wind Sea and Swell Separation of 1D Wave Spectrum by a Spectrum Integration Method. *J. Atmos. Ocean. Technol.* **2012**, *29*, 116–128, doi:10.1175/JTECH-D-11-00075.1.

15. InMeteo. Ventusky. Available online: https://ventusky.com (accessed on 7 January 2020).

16. Voronovich, A.G.; Zavorotny, V.U. The Transition From Weak to Strong Diffuse Radar Bistatic Scattering From Rough Ocean Surface. *IEEE Trans. Antennas Propag.* **2017**, *65*, 6029–6034.

17. Giacobazzi, J. 50 - Line of sight radio systems. In *Telecommunications Engineer's Reference Book*; Mazda, F., Ed.; Butterworth-Heinemann: Oxford, UK, 1993; pp. 50-1–50-17, doi:10.1016/B978-0-7506-1162-6.50056-3.

18. Saleem, T.; Usman, M.; Elahi, A.; Gul, N. Simulation and Performance Evaluations of the New GPS L5 and L1 Signals. *Wirel. Commun. Mob. Comput.* **2017**, *2017*, 1–4, doi:10.1155/2017/7492703.

19. Government, A. Waves. Available online: http://www.bom.gov.au/marine/knowledge-centre/reference/waves.shtml (accessed on 14 March 2020).

Article

Detecting Targets above the Earth's Surface Using GNSS-R Delay Doppler Maps: Results from TDS-1

Changjiang Hu [1,*], Craig Benson [1], Hyuk Park [2], Adriano Camps [3], Li Qiao [1] and Chris Rizos [4]

[1] School of Engineering and Information Technology, University of New South Wales, Canberra 2612, Australia; c.benson@adfa.edu.au (C.B.); l.qiao@adfa.edu.au (L.Q.)

[2] Department of Physics, Univeristat Politecnica de Catalunya, 08034 Barcelona, Spain; park.hyuk@tsc.upc.edu

[3] Department of Signal Theory and Communications, Universitat Politecnica de Caltalunya and IEEC/UPC, 08034 Barcelona, Spain; camps@tsc.upc.edu

[4] School of Civil and Environmental Engineering, University of New South Wales, Sydney 2052, Australia; c.rizos@unsw.edu.au

* Correspondence: changjiang.hu@student.adfa.edu.au; Tel.: +61-0415-365-889

Received: 12 August 2019; Accepted: 4 October 2019; Published: 7 October 2019

Abstract: Global Navigation Satellite System (GNSS) reflected signals can be used to remotely sense the Earth's surface, known as GNSS reflectometry (GNSS-R). The GNSS-R technique has been applied to numerous areas, such as the retrieval of wind speed, and the detection of Earth surface objects. This work proposes a new application of GNSS-R, namely to detect objects above the Earth's surface, such as low Earth orbit (LEO) satellites. To discuss its feasibility, 14 delay Doppler maps (DDMs) are first presented which contain unusually bright reflected signals as delays shorter than the specular reflection point over the Earth's surface. Then, seven possible causes of these anomalies are analysed, reaching the conclusion that the anomalies are likely due to the signals being reflected from objects above the Earth's surface. Next, the positions of the objects are calculated using the delay and Doppler information, and an appropriate geometry assumption. After that, suspect satellite objects are searched in the satellite database from Union of Concerned Scientists (UCS). Finally, three objects have been found to match the delay and Doppler conditions. In the absence of other reasons for these anomalies, GNSS-R could potentially be used to detect some objects above the Earth's surface.

Keywords: GNSS reflectometry; delay Doppler map; target detection

---

1. Introduction

The use of Global Navigation Satellite System reflectometry (GNSS-R) to remotely sense the Earth's surface was originally proposed more than three decades ago [1]. GNSS-R works as a bistatic radar with receivers having two antennas: one up-looking antenna receiving the direct signals, and one down-looking antenna receiving the reflected signals. GNSS-R can be ground-based, airborne or spaceborne. During the first two decades of GNSS-R, a range of low-altitude (ground or airborne) campaigns were carried out to verify its feasibility [2–4]. In recent years, spaceborne GNSS-R has attracted increasing attention for its advantages of global data collection and data coverage capability. The first mission equipped with a GNSS-R payload was UK Disaster Monitoring Constellation (UK-DMC) which was in operation from 2003 to 2011 [5]. The follow-on mission of the UK-DMC was TechDemoSat-1 (TDS-1), which ceased operations in 2018 [6]. NASA's Cyclone GNSS (CYGNSS) mission is now operating [7]. Other missions being planned or in preparation can be found in [8–11].

GNSS-R applications known to date are broadly divided into two categories: retrieval of surface parameters, and object detection. The former can be further divided to ocean and land parameters retrieval. GNSS-R ocean altimetry was first proposed in 1993 [12], it could achieve good performance when the reflecting surface is smooth, such as lake surfaces [13], and ice [14,15]. It is still challenging to obtain accurate results with open ocean reflections. Using GNSS reflected signals to determine land moisture was proposed in 2000 [16], and experimental results have been reported in [17] where the moisture of bare and vegetated soil is estimated using GNSS interferometric signals. In 1998 the effect of ocean roughness on reflected GNSS signals was detected [18], and sea roughness was retrieved using a model to fit the observed delay Doppler maps (DDMs) in an aircraft experiment in 2004 [19]. Wind speed retrieval using GNSS reflected signals was theoretically studied in [20], and in recent years ocean wind speed was retrieved using space-based GNSS-R data, such as [21] using TDS-1 data. In [21] precision around 2.2 m/s was achieved, which demonstrates the potential of ocean wind speed using GNSS-R. CYGNSS is another mission dedicated to study wind speed. GNSS-R can retrieve other parameters, such as forest biomass [22], snow depth [23], and ionospheric delay and scintillation [24]. For object detection, objects on the Earth's surface that can be detected by GNSS-R include sea ice, sea surface targets, oil slicks, and tsunamis, etc. The detection of sea ice, sea surface target and oil slicks is based on the reflection coefficient differences between the targets and the reflection bodies around them. For example, the signals reflected from ice are very strong which exhibits different features from ocean reflections in DDMs. An example of using TDS-1 DDMs to detect sea ice over Arctic and Antarctic regions can be found in [25]. A technique using two separate antennas to detect oil slicks was presented in [26]. Different from the detection of the sea ice, sea surface targets and oil slicks, tsunami detection uses the high spatial resolution of GNSS-R altimetric measurements [27]. The target detection mentioned above utilises space-based GNSS-R which has the advantage of short revisit time compared to traditional synthetic aperture radar and optical systems [28]. In 2018, it was found that GNSS-R can be used to detect traffic flow [29], which is based on phase difference information of the reflected signals. In [29], a ground-based experiment is employed, and if the space-based GNSS-R is feasible to detect the traffic flow has not been reported yet.

Different from the aforementioned applications, this paper proposes a new application of GNSS-R. The previous studies usually use signals reflected from the Earth's surface. However, this study concludes that the GNSS-R technique may be able to detect objects above the Earth's surface, such as spacecraft. Traditionally such detection has been done using ground-based radar technologies [30,31], which are regional systems. The traditional techniques have limited detection areas. However, GNSS-R has several advantages over the traditional techniques, such as global coverage, large volume of data, and short revisit time. Detecting targets above the Earth's surface not only extends the applications of GNSS-R, but also provides an alternative approach to solve this issue.

This work presents experimental evidence and demonstrates the process to verify the feasibility of target detection. Firstly, Section 2 of the study presents 14 DDMs from the UK TDS-1 mission which contain "anomalous" reflections. In Section 3 five probable causes for the anomaly are analysed, reaching the finding that the anomalous reflections are likely from objects above the Earth's surface based on their path delays. Section 4 describes the determination of an object's position using delay and Doppler frequency information. In order to find the suspected objects, in Section 5 the satellite database from the Union of Concerned Scientists (UCS) [32] has been searched, and three satellites satisfying the delay and Doppler conditions have been identified. Section 6 includes conclusions and remarks.

2. TDS-1 Satellite GNSS-R Data

The TDS-1 satellite was launched in 2014, and ceased its operations at the end of 2018. Although its initial purpose was to study wind speed, its GNSS-R data has been studied for many other applications, such as altimetry [14,15,33], ice detection [25], and sea target detection [28], to name a few. TDS-1 provides three kinds of data products according to processing level: L0 raw sampling data, L1 DDMs data, and L2 wind speed product [34]. A comprehensive introduction to the TDS-1 mission and its data products can be found in [6,34].

The data used in this paper are the L1 DDMs which are generated on-board by 1 ms coherent integration, and 1000 ms non-coherent integration. Figure 1 is a typical TDS-1 DDM over the ocean. The TDS-1 DDMs have 20 pixels in the Doppler axis, and 128 pixels in the delay axis. The resolutions (one pixel in Doppler and delay) are 500 Hz in Doppler, and 244 ns in delay. The Doppler of column 11 is zero which compensates the Doppler frequency of specular reflection. The specular point through which the reflected signal has the shortest path length is around the black dot marked in Figure 1. The "horseshoe" pattern of Figure 1 indicates reflected signals away from the specular reflection point, and the blue background above the black dot are noisy pixels without a return [28].

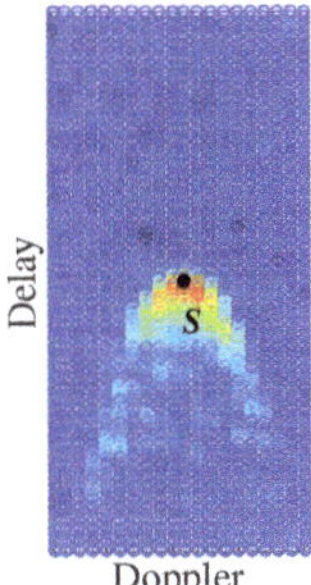

Figure 1. An example of delay Doppler map of TDS-1 satellite. *S* is specular point.

Fourteen anomalous DDMs that are used in this study are shown in Figure 2. They were all collected on 14 and 15 September 2018, and they are found from a total of 224,805 DDMs. Each DDM in Figure 2 is referred to as an "event". The location of the specular point of each event is marked by a red star in Figure 3. The reflecting bodies of the 14 events include water and land. The water reflections usually produce a clear "horseshoe" pattern in DDMs, such as Events 11 and 14 in Figure 2. Land reflections usually do not exhibit such a clear "horseshoe" pattern, such as Events 5 and 13 in Figure 2. The anomalies discussed in this paper are the bright points in the red circles of Figure 2. Table 1 gives the values of two important parameters (elevation angle and "*DirectSignalInDDM*") of the 14 DDMs, which are used in following analyses. The "*DirectSignalInDDM*" is a system parameter of the TDS-1 mission which indicates if the DDM contains direct signals. More details about this parameter can be found in [34]. In addition, the locations of the 14 DDMs in TDS-1 dataset are listed in Table 1 by track ID and observation time. For example, the DDM of Event 1 is in Group 05, Folder H18 of the metadata collected on 14 September 2018, and the observation time is 15:19:26.999 of the corresponding date. The third column of Table 1 gives the column number of the bright point in the DDM.

Table 1. Information of the 14 DDMs.

	Elevation Angle	Column No. of Bright Point	"*DirectSignal InDDM*"	Track ID in Dataset	Observation Time
Event 1	71°	11	False	20180914H18Group05	15-19-26.999
Event 2	72°	11	False	20180914H18Group05	15-20-07.999
Event 3	73°	11	False	20180914H18Group05	15-20-58.999
Event 4	64°	17	False	20180914H18Group08	15-33-15.999
Event 5	85°	12	False	20180914H18Group51	18-51-51.999
Event 6	59°	8	False	20180914H18Group82	20-21-29.999
Event 7	67°	8	False	20180915H00Group27	23-09-16.999
Event 8	67°	11	False	20180915H00Group52	01-05-40.999
Event 9	77°	11	False	20180915H00Group32	23-30-54.999
Event 10	76°	11	False	20180915H00Group49	00-51-30.999
Event 11	61°	17	False	20180915H06Group53	06-22-39.999
Event 12	66°	20	False	20180915H06Group78	07-46-53.999
Event 13	73°	9	False	20180915H12Group25	10-51-09.999
Event 14	59°	8	False	20180915H18Group30	16-39-04.999

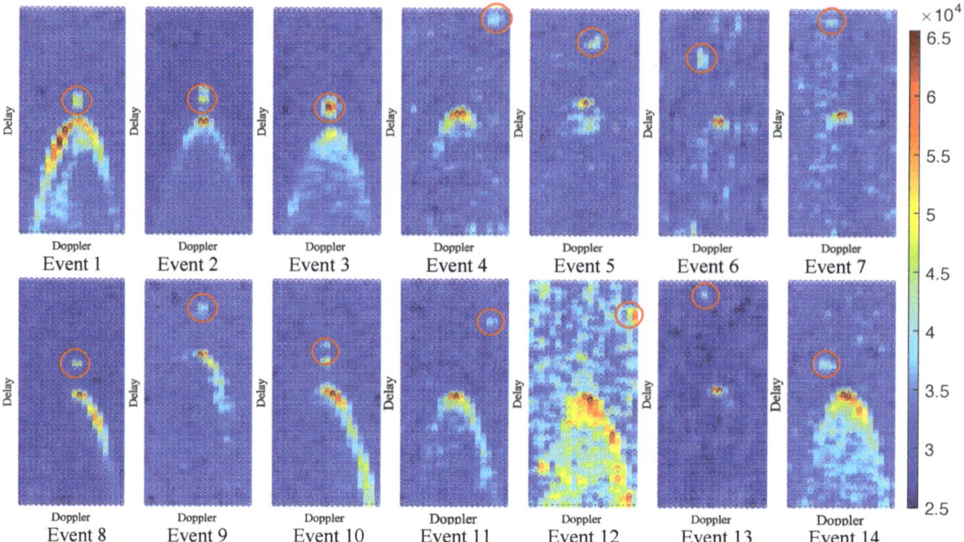

Figure 2. Fourteen delay Doppler maps (DDMs) of TDS-1 containing anomalous features that are marked by red circles.

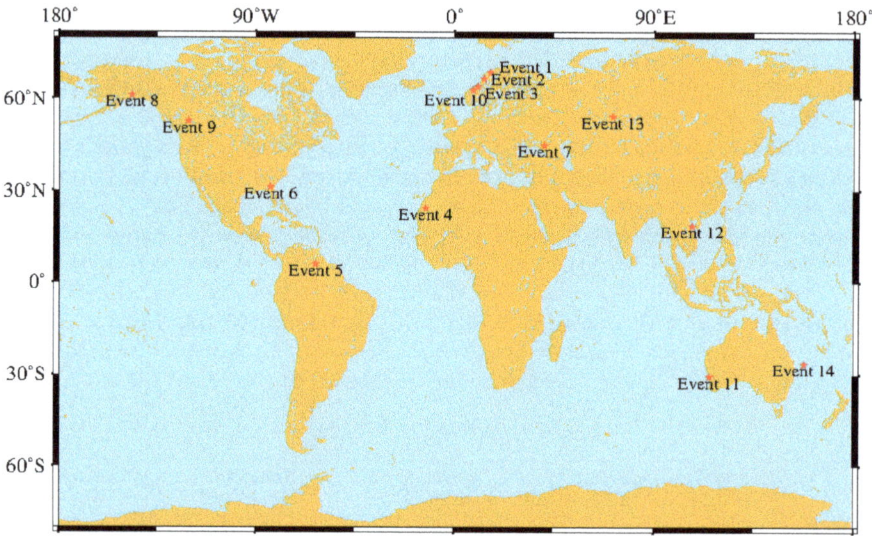

Figure 3. The distribution of specular points of the 14 events. The specular points are marked by red stars.

3. Analyses of Possible Causes

The generation of DDMs involves three parts from the receiver to transmitter, such as receiver hardware, signal processing, and signal propagation. For typical DDMs the bright points circled in Figure 2 are unexpected. The problems in each part of DDMs generation may cause the bright points. Seven possible causes of the bright points are considered in this work. Three of them relate to receiver hardware, such as (1) random noise, (2) DDM generator losing tracking, and (3) leakage of direct signals. The possible cause relating to the signal processing is aliasing. The remaining three causes are associated with the signal propagation, such as (1) reflections from an Earth's surface target, (2) GNSS radio-occultation, and (3) reflection from targets above the Earth's surface and under the nadir-looking antenna of the receiver. In the following sections, the first six possible causes are investigated, and evidence found for these not being the most likely causes.

3.1. Random Noise

The blue background above the "horseshoe" pattern is utilised to investigate the noise properties for each event. Then a 99% confidence interval is given based on the property information of the noises, which is further used to decide if the bright points are random noise. The bright points of Figure 2 are excluded in noise samples, and there are about 1000 noise samples for each event. This investigation is carried out as follows. First the probability distribution function, mean and standard deviation of the noise samples are obtained. Then the noise samples are assumed to obey normal distribution considering the probability distribution function of the noise samples, and the mean and standard deviation of the noise samples are used to generate a normal distribution function. The next step computes the 99% confidence interval (a, b) using the normal distribution function obtained in the last step. The final step is to determine if the correlation power value of the bright point ($Bval$) is located in the 99% confidence interval (a, b).

Figure 4 gives the probability distribution function for each event, as shown by black dots. The 99% confidence interval (a, b) is indicated by red lines. Since the volume of the noise samples is small and the noise samples may not exactly obey normal distribution, the normal distribution function determined by the mean and standard deviation of the noise samples can not fit the black dots well. Hence the normal distribution functions are normalised to fit the black dots, which are represented by the blue lines of Figure 4. The values for a, b and $Bval$ are given in the top of each sub-plot. It can be seen that the values for $Bval$ are far outside the 99% confidence interval. Therefore, the bright points are not random noises.

3.2. DDM Generator Losing Tracking

The investigation of DDM generator is to check if the 14 events happened just after the DDM generator recovers from loss of track. Differential operation is applied to DDMs observation time for the tracks involved in the 14 events to find if and when DDM generator lost tracking. Then the observation time of each event is compared with the DDM generation recovery time. Figure 5 is the differential results of the DDMs observation time of Track ID "20180914H18Group05". This track involves Events 1, 2, and 3, which can be seen from Table 1. The horizontal axis is DDM index in the track, which is of convenience to indicate the observation time. TDS-1 outputs DDMs every second, so the differential results of Figure 5 are mostly 1. However, there are seven outliers which means that the DDM generator loses tracking for seven times during this track. The DDM indexes corresponding to the outliers are shown in the figure. The DDM indexes for Event 1, 2 and 3 are 407, 447, and 498 respectively, which are not around the time that the DDM generator loses tracking. The same results are obtained for the other 11 events, and details are not shown here.

Figure 6 gives the two DDMs that are just before and after the DDM generator loses tracking at DDM index 213 of Figure 5. The DDM indexes are marked by blue rectangles, and observation time by red rectangles. It can be seen from Figure 5 that the DDM generator loses tracking for seven seconds at DDM index 213, which is confirmed by the observation time difference between 15:16:5.999 and 15:16:12.999 of Figure 6. It can be seen from Figure 6 that the two DDMs show a very clear "horseshoe" pattern without any strange patterns. This is understandable because TDS-1 receiver uses on-board-calculated specular point to estimate the delay and Doppler of the reflected signals, to keep tracking the reflected signals. All the tracks relating to the 14 events are checked and it is found that there are no outliers in the positions of the on-board-calculated specular points, even if the DDM generator loses tracking. Therefore, the bright points are not caused by the DDM generator losing tracking.

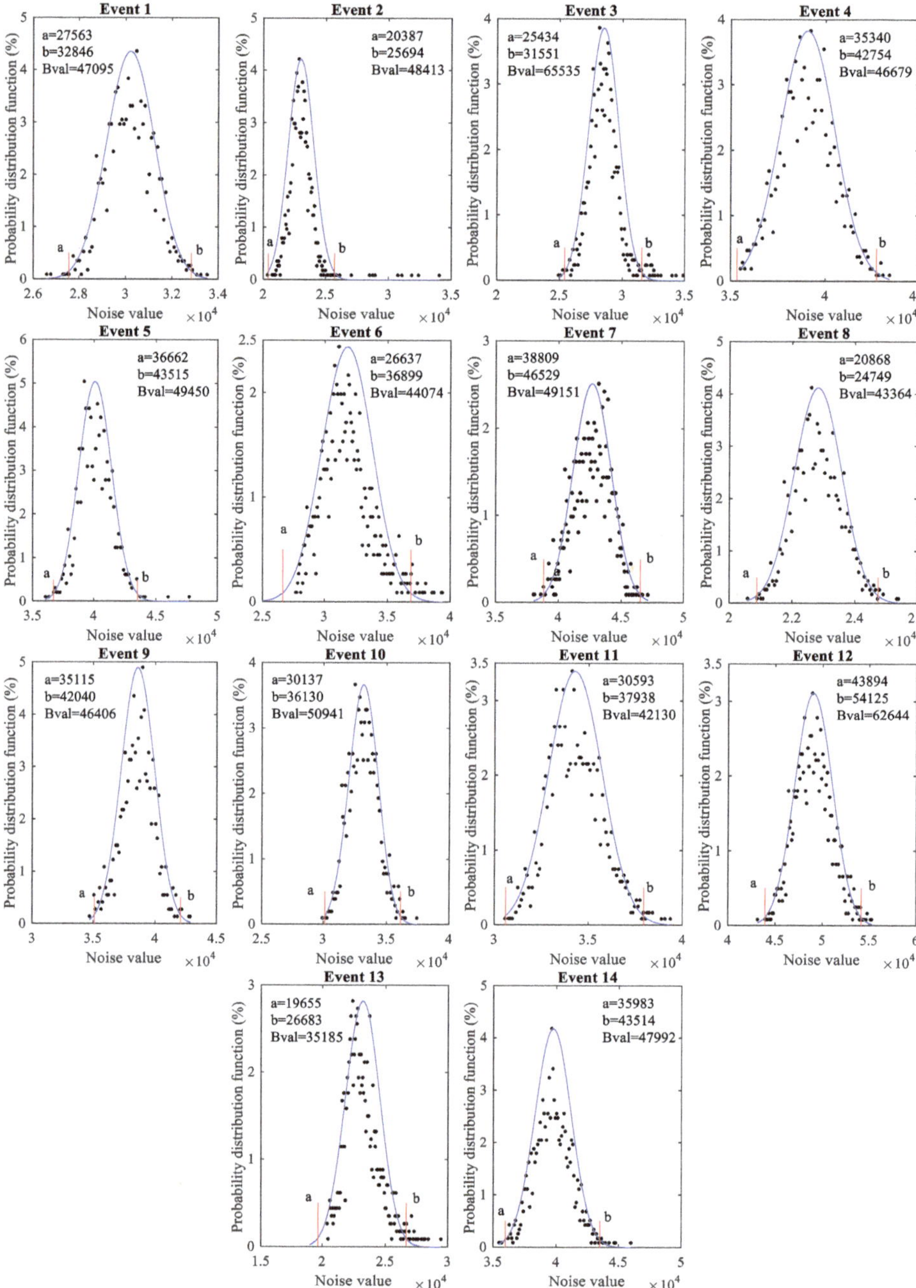

Figure 4. Probability distribution functions of the background noises for the 14 events, which are presented by black dots. The noises are assumed to obey normal distributions, and the normal distribution functions with the means and standard deviations corresponding to these of the noises are obtained. (a, b) is 99% confidence interval of the normal distribution functions. The blue curves are the results of normalising the normal distribution functions to fit the black dots. *Bval* is the correlation power value for the bright point.

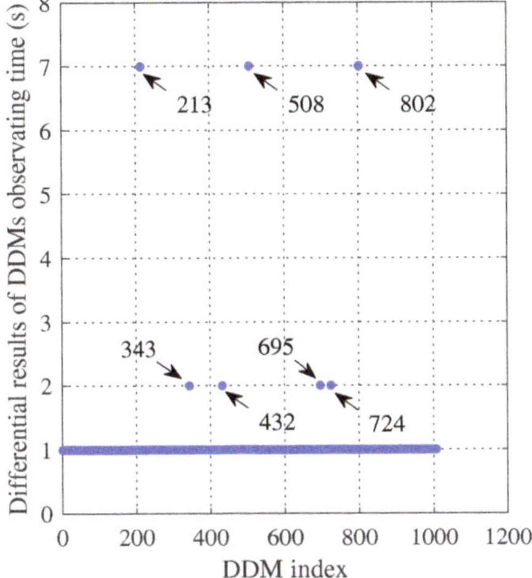

Figure 5. Differential results of DDMs observation time of Track ID "20180914H18Group05". The horizontal axis is DDM index, and the DDM indexes corresponding to the seven outliers are indicated by the arrows around them.

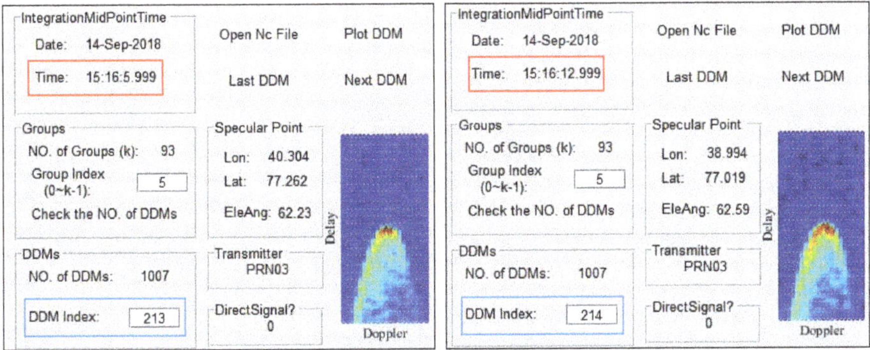

Figure 6. Two DDMs that are just before and after DDM generator loses tracking at DDM index 213 of Figure 5.

3.3. Leakage of Direct Signal

Two aspects are examined to decide if the bright points correspond to a leakage of the direct signal. One is to analyse the path length and Doppler of the direct signal, and another is to check the system parameter "*DirectSignalInDDM*" as listed in the fourth column of Table 1.

The path length and Doppler frequency difference between the bright points and reflected signals can be extracted from the corresponding DDMs, while the path length and Doppler frequency differences between direct and reflected signals are not directly available from DDMs and other data products. Simulations have been conducted to generate the path length and Doppler frequency of the direct and reflected signals. The simulations use TDS-1 orbit, precise orbit of transmitter, WGS-84 ellipsoid model, corrections for the ionospheric and tropospheric delays, and correction for the mean sea surface height using the DTU13 model [35]. Two parameters, ρ_{rem} for the path length difference between direct and reflected signals and D_d^S for the Doppler frequency difference between direct and reflected signals, are obtained from the simulations:

$$\rho_{\mathrm{rem}} = L_{sp} - L_d - \mu \cdot \tau \cdot c \tag{1}$$

$$D_d^S = D_{dir}^S - D_{ref}^S \tag{2}$$

where L_{sp} is the reflected path length through the specular point, L_d is the path length of the direct signal, τ is the time period of the C/A code (1 ms), μ is an integer ambiguity to ensure that ρ_{rem} has a value between 0 and $\tau \cdot c \approx 300$ km, D_{dir}^S and D_{ref}^S are the simulated Doppler frequencies of the direct and reflected signals, respectively. The superscript S of (2) indicates simulated variables, which is used to separate them from observation variables which are marked by the superscript O, such as the D_d^O of (4).

The upper delay window of the DDM is defined as:

$$L_{sp} - L_x < Dp \cdot Dr \cdot c \cdot \eta \tag{3}$$

where L_x is the reflected path length from any point, Dp is the number of delay pixels (128 for TDS-1 DDMs), Dr is the delay resolution of DDM (244 ns for TDS-1 DDMs), c is the speed of light, and η is the ratio of the number of pixels above the specular point to the number of delay pixels. The upper delay window is the necessary condition that a reflected signal is above the "horseshoe". Given that the specular point is basically located in the mid delay axis of the 14 DDMs, η is set to 0.5. Therefore the upper delay window is less than 4682 m, e.g., $0 < L_{sp} - L_x < 4682$ m, and the path length difference between the bright points and reflected signals follows this relation. The Doppler frequency difference between the bright points and reflected signals D_d^O can be obtained from the DDMs [34]:

$$D_d^O = (N - 11) \cdot Dre \tag{4}$$

where N is the column number of the bright point, and Dre is the Doppler resolution (500 Hz in TDS-1). The column numbers of the bright points are listed in Table 1. The specular reflection (DDM peak) is set to be in the 11th column in the TDS-1 DDM.

Table 2 lists the values of ρ_{rem}, D_{ref}^S, D_d^S and D_d^O for the 14 DDMs. It can be seen that the ρ_{rem} of the DDMs ranges from 15.1 km to 292.6 km, which are considerably beyond the upper delay window. Therefore, the direct signals cannot be above the "horseshoe" of the DDMs, despite the fact that some DDMs have small difference between D_d^S and D_d^O (Events 3, 7 and 13). Furthermore, the path difference between the direct and reflected signals seems to be random. Therefore, the bright points in Figure 2 do not appear to be the leakages of direct signals.

Table 2. The values of ρ_{rem}, D_{ref}^S, D_d^S and D_d^O for the 14 DDMs.

	ρ_{rem} (km)	D_{ref}^S (Hz)	D_d^S (Hz)	D_d^O (Hz)
Event 1	273.1	6703	1701	0
Event 2	284.4	4726	1177	0
Event 3	292.6	2242	518	0
Event 4	186.7	2081	453	3000
Event 5	54.7	−3114	−745	500
Event 6	116.3	15812	3826	−1500
Event 7	221.1	−6864	−1595	−1500
Event 8	224.1	5715	1418	0
Event 9	22.4	−5668	−1622	0
Event 10	15.1	−8527	−2079	0
Event 11	156.7	2087	730	3000
Event 12	211.8	−14858	−3780	4500
Event 13	290.8	−3627	−1052	−1000
Event 14	119.4	−80	−173	−1500

The system parameter *"DirectSignalInDDM"* indicates if the reflected signals are separated from direct signals. It did not catch all cases in early version data set, but the problem was solved in data products after February 2018 [34]. It can be seen from Table 1 that the *"DirectSignalInDDM"* of the 14

DDMs are "False", which means that the DDMs do not contain any direct signal, which confirms the previous analysis.

3.4. Aliasing

Aliasing is the signal ambiguity in the frequency domain [36]. An appropriate sampling frequency could avoid this issue according to the Nyquist-Shannon sampling theorem. In TDS-1 signal processing, two processes involve sampling [34]. One is the sampling of the centre frequency signal (4.188 MHz) using a sampling frequency of 16.367 MHz. The other one is the decimation which makes the effective sampling frequency drop to 32 kHz. More details about the two processes can be found in [34]. It is known that DDMs are the correlation results in two domains (time and frequency), which means that the bright points or the "horseshoe" of Figure 2 are produced with appropriate delay and Doppler frequency. Given that aliasing is just the ambiguity in the frequency domain, it cannot produce the bright points without appropriate delays. Therefore, the bright points of Figure 2 do not appear to be due to aliasing.

3.5. Reflections from an Earth's Surface Target

It can be seen from Figure 2 that the bright points are located in delay bins before the "horseshoe" pattern, i.e., before the specular reflection from the Earth's surface arrives. These regions are often called the "forbidden zone" of the DDM because they correspond to delays shorter than the delay associated with the points over the Earth's surface. Therefore, the bright points of Figure 2 cannot be caused by reflections from an Earth's surface target. Moreover, the height difference corresponding to the path length difference between the bright point and specular point is obtained to investigate if the bright point is caused by properties on the Earth's surface. The height difference can be obtained by [12]

$$\Delta h = -\frac{\Delta \rho}{2 \sin \Theta} \tag{5}$$

where $\Delta \rho$ is the path length difference between the bright point and specular point, and Θ the elevation angle which is given in Table 1. The negative sign means that height measurements decrease with the path lengths of reflected signals. The C/A code integer ambiguity μ in (1) is set to zero in this investigation to obtain the minimum value of the height difference for each event. It is obtained that the height differences for the 14 events are greater than 445 m (Event 1), and can reach to thousands of metres such as Event 7. Skyscrapers are the properties of which the heights are hundreds of metres, but it is found that the reflection areas for the 14 events have no skyscrapers that match the corresponding height differences. Hence the bright points of Figure 2 are not caused by the properties on the Earth's surface.

3.6. GNSS Radio-Occultation

GNSS radio occultation (GNSS-RO) event happens when GNSS satellites are setting or rising behind the Earth's limb as viewed by a receiver fixed on LEO satellite [37], which is shown by the red line of Figure 7. In addition, Figure 7 gives an illustration of GNSS-R which is presented by black lines. The GNSS-RO uses refracted signals, while the GNSS-R uses reflected signals. It can imagine that the GNSS-R may be close to the GNSS-RO when the elevation angle is very low, such as about $0°$. However, in the case of the 14 events, their elevation angles which are listed in Table 1 are considerably high, ranging from $59°$ to $85°$ which precludes the occurrence of GNSS-RO events.

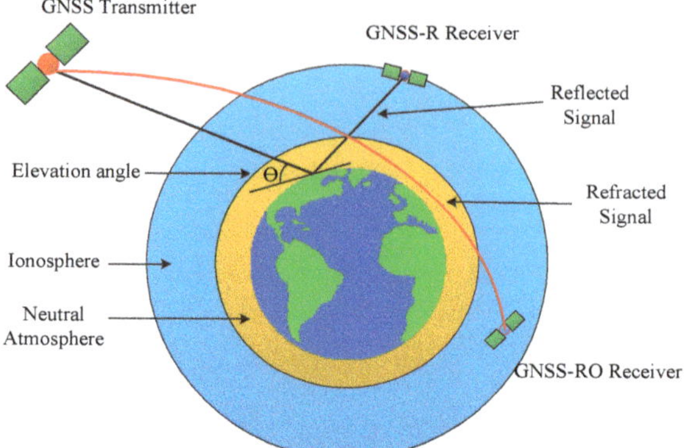

Figure 7. Illustration GNSS-R and GNSS-RO, which are shown by black and red lines respectively. Θ is elevation angle.

4. Object Positioning

The bright points of Figure 2 appear to be the reflection from objects above the Earth's surface because their path lengths are shorter than those of the specular points. This section describes the method used to calculate their position using the delay, Doppler frequency and appropriate geometry, and gives the conditions that the suspected objects have to fulfil.

The retrieval of target position using delay and Doppler frequency usually has no unique solution even if the target is on the Earth's surface. This issue is more complicated for space target reflections. Figure 8 gives an illustration of the geometry for the retrieval of object position which are shown by black lines. The solid lines represent direct signals, and the dashed lines the reflected signals. This geometry has two main assumptions, (1) the altitude of the target is lower than that of the receiver, and (2) the normal vector of the reflecting surface is perpendicular to the Earth's surface. The Earth's surface is assumed to be the WGS-84 ellipsoid model. It can be seen from Figure 8 that the target could be at different altitudes due to the integer ambiguity μ of (1). It is possible that the reflection could happen under other conditions, with a typical example shown in Figure 8 by blue lines. Since other reflection conditions are unpredictable, they are not considered in the retrieval of target position.

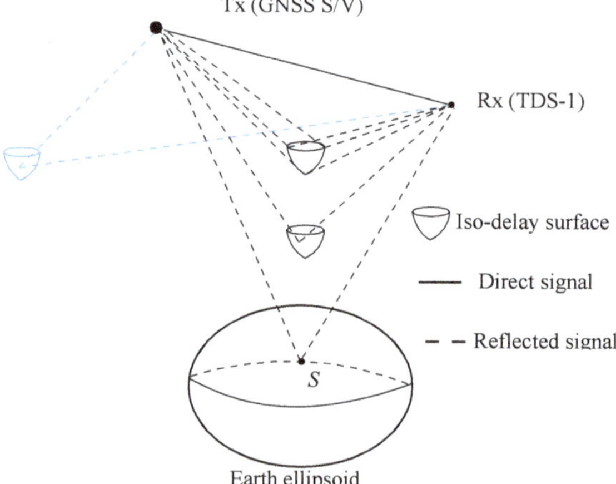

Figure 8. Illustration of geometry model used in the retrieval of the target position, as shown by black lines. The blue lines give a typical example of unpredictable reflection conditions. *S* is specular point on the Earth's surface.

Three steps are used to retrieve the target positions. The first step obtains the path lengths and Doppler frequency of the bright points. The delay and Doppler frequency of the TDS-1 DDMs are relative to those of the on-board-calculated specular point to keep the specular reflection in the middle of the DDM. The Doppler frequency pixel corresponding to the on-board specular point is the 11th column, which is set to zero, and the delay pixel corresponding to the on-board specular point can be obtained using the system parameters *"TrackingOffsetDelayNs"* and *"SpecularPathRangeOffset"* [34]. The delay and Doppler frequency of the bright point relative to those of the on-board specular point can be estimated from the number of pixels between them. The TDS-1 dataset does not provide the path length and Doppler frequency of the on-board specular point directly. The path length of the on-board specular point is obtained using the on-board receiver position, on-board transmitter position, and on-board specular point position. Similarly the Doppler frequency of the on-board specular point is simulated using on-board trajectory information of the receiver, transmitter, and the position of on-board specular point.

The second step is positioning the target using path length information. It is known that a given path delay (except for specular delay) corresponds to a branch of reflection points, which form an iso-delay curve in the two-dimensional domain and an iso-delay surface in the three-dimensional domain. The iso-delay surface is shaped like a "top", as shown in Figure 9. In this investigation the "top" is sliced into 11 iso-delay curves, and the specular point of each iso-delay curve is marked by blue dots in Figure 9. The path length difference of the specular points of two adjacent iso-delay curves is 30 m. Considering the C/A code integer ambiguity μ, the possible positions of the target will have several iso-delay surfaces in one event.

The last step is based on the results of the second one, and uses Doppler frequency information to further constrain the target position. Two assumptions are made on the motion of the target: (1) the velocity of the target is perpendicular to the radial vector of the target, and (2) the target describes a nearly circular orbit when $\mu > 0$, and the speed v is determined from Kepler's equation. The target is assumed to be static when $\mu = 0$. These assumptions are reasonable for satellite orbits with very small eccentricities, but have limitations when the eccentricities are too large to produce nearly circular orbits. In addition, the assumptions do not consider the situations that the targets are moving when $\mu = 0$ because in this case the moving direction is arbitrary. Hence the assumptions work in the normal reflection as shown by black lines of Figure 8. For a given target position such as the red dot of Figure 9, the Doppler frequency corresponding to each moving direction is calculated, and compared with the Doppler frequency from the first step. If the Doppler frequencies of all directions are considerably different from the Doppler frequency of the first step, the given position will be eliminated.

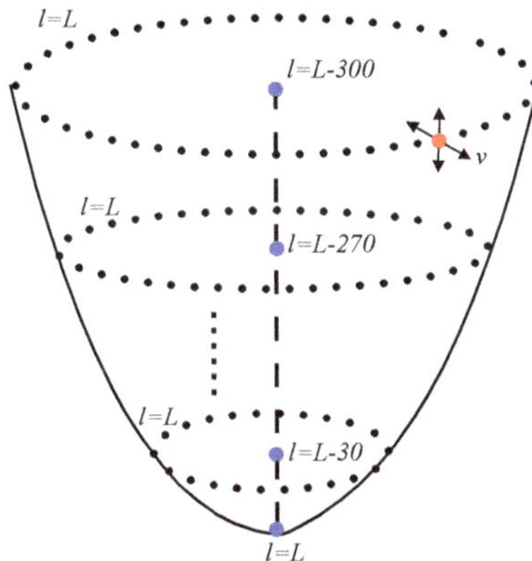

Figure 9. Illustration of iso-delay surface in three-dimensional domain. *l* indicates path length, and *L* is a given path length. The black dots are the possible positions of the target, and the blue dots are the specular points of the corresponding iso-delay curve.

These three conditions (delay, Doppler and geometry) are considered in the verification of a target candidate. The path delay d^T, and Doppler frequency D^T of a given target candidate can be calculated using the position and velocity information of the transmitter, receiver, and the target candidate. In addition, the path delay d^O, and Doppler frequency D^O of the bright points of Figure 2 were already obtained in the previous part. Then, the path delay difference d^{OT} between a particular target candidate and the bright point can be obtained, and similarly, the Doppler frequency difference D^{OT} can also be obtained. It should be noted that d^{OT} is between 0 and $\tau \cdot c \approx 300$ km, which eliminates the impact of C/A code ambiguity. The delay and Doppler frequency conditions are that d^{OT} and D^{OT} should be smaller than particular thresholds, which are discussed in the next section. The geometry condition is that the target candidate can be seen by the transmitter, and that its altitude is less than that of the TDS-1.

5. Results Analyses

5.1. Target Position Retrieval

Figure 10 gives the results of target positions of Event 11, with four sub-plots for C/A code ambiguity $\mu =$ 0, 1, 2 and 3. The blue dots are the results considering only the delay, and the red dots considering both the delay and Doppler frequency. The dots show a "top" pattern in all sub-plots, which confirms the analyses of Figure 9. It can be seen that in sub-plot (a) where the "target" is near the Earth's surface ($\mu = 0$), there are no red dots, which means there are no solutions under the geometry assumption of Figure 9 and the assumption that the "target" is static. This is understandable because the Doppler frequency difference between the bright point and specular point of Event 11 is 3000 Hz according to Table 2, which is large because the Doppler frequency of a scattering point of which path delay is less than 300 m is usually less than 1500 Hz as compared to that of a specular point on the Earth's surface. When $\mu > 0$, there are an increasing number of red dots as shown in sub-plots (b) to (d). The results for target positions of other events are similar to Figure 10, which are not presented here.

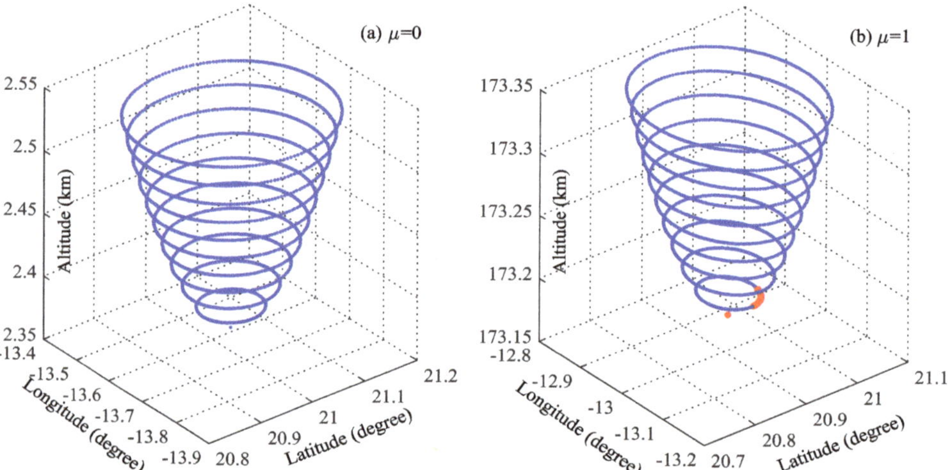

Figure 10. *Results of target positions of Event 11. Sub-plots (a) and (b) correspond to $\mu = 0$ and $\mu = 1$, respectively. The blue dots are the results only considering the delay condition, and the red dots are the results further considering the Doppler frequency condition.*

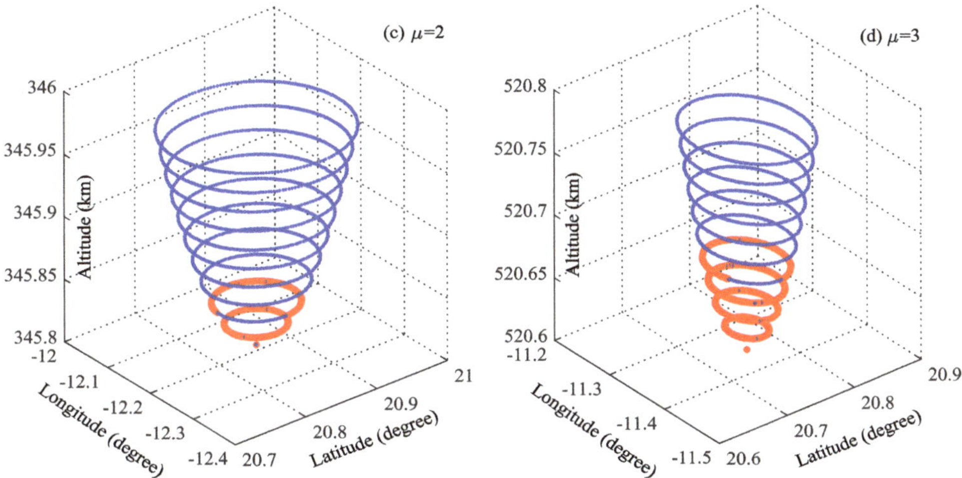

Figure 10. Results of target positions of Event 11. Sub-plots (c) and (d) correspond to $\mu = 2$ and $\mu = 3$, respectively. The blue dots are the results only considering the delay condition, and the red dots are the results further considering the Doppler frequency condition.

5.2. Verification of Suspected Objects

The UCS provides the list of operational satellites orbiting the Earth, and updates the database regularly. The database used in this investigation was issued on 1 December 2018. There are 1957 satellites in the database, of which 834 remain after applying such filter conditions as: (1) the class of orbit is LEO, (2) perigee is less than 650 km considering that TDS-1 altitude is around 640 km, and (3) launch date is before 15 September 2018. Other objects such as debris and aircraft are not considered in this investigation, which is explained Section 6.

The position and velocity of the 834 satellites are obtained by simplified general perturbations 4 (SGP4) method [38] using two line elements (TLE) [39]. The TLE information can be accessed from Space-Track [39]. The TLE of 746 satellites at the time around the observing period of the 14 DDMs were obtained from the Space-Track, while the other 85 satellites were not available. Therefore, 746 satellites are the target candidates, and the d^{OT} and D^{OT} were obtained by calculation. The position precision of SGP4 is around 2.5 km for the TDS-1, by comparison with the TDS-1 on-board position, which could be a reference for the target candidates.

The threshold value of d^{OT} is set to 10 km considering the SGP4 orbit error, TLE age, and other error sources including path delay error caused by the DDMs, and orbit errors of the transmitter and receiver. The threshold value of D^{OT} is set to around 1000 Hz considering the DDM Doppler resolution is 500 Hz. Three events have valid candidates after applying the criteria that $d^{OT} < 10$ km, $D^{OT} < or \approx 1000$ Hz, and the geometry condition. The relevant information of the three events is listed in Table 3, including d^{OT}, D^{OT}, and the altitude of the target candidate. It can be seen from Table 3 that the valid candidates for Events 2, 4 and 5 are Kondor, Step Cube Lab and Deimos 2, respectively. The values of d^{OT} for Kondor, Step Cube Lab and Deimos 2 are around 500 m which is small. This is understandable because (1) the reflected path length is usually insensitive to the position of scattering point because the semi-major and semi-minor of an iso-delay curve is from several kilometres to tens of kilometres in space-based scenario [40], (2) the path delay extracted from the DDMs is of the order of tens of metres, which is related to the delay resolution (244 ns), and (3) the errors induced by other factors such as the orbit of the transmitter and receiver are much smaller, and considered to be negligible. In addition, the D^{OT} of the three candidates are 607 Hz, 568 Hz and 1046 Hz respectively, which are also reasonable considering the Doppler resolution of 500 Hz for the TDS-1 DDMs. Therefore, Kondor, Step Cube Lab and Deimos 2 can be said to meet the delay and Doppler frequency conditions, so they are possibly targets of the bright points for Events 2, 4 and 5, respectively.

Table 3. Three events with valid target candidates.

	Name and NORAD ID	d^{OT} (km)	D^{OT} (Hz)	Altitude (km)
Event 2	Kondor (39194)	0.495	607	459.5
Event 4	Step Cube Lab (43138)	0.458	568	505.9
Event 5	Deimos 2 (40013)	0.653	1046	616.4

In addition, the sub-satellite positions of the transmitter, receiver and valid target candidates of the three events are shown in Figure 11. The upper, middle and lower sub-figures are for the Event 2, 4 and 5 respectively. In normal reflection of GNSS-R, such as the dashed line through S in Figure 8, the sub-satellite positions of transmitter, receiver and reflection point (S) are on a line. It can be seen from Figure 11 that the sub-satellite point of the target candidate is far away from the line formed by those of the transmitter and receiver for the three events. This means the reflection conditions of the three events are like the one shown by the blue lines of Figure 8.

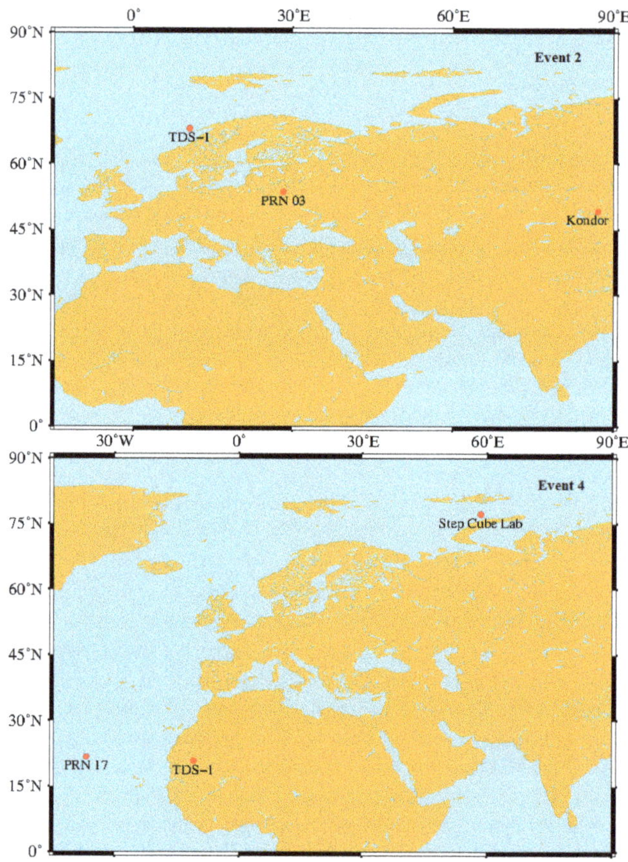

Figure 11. *The sub-satellite locations of transmitter, receiver and valid target candidates for Event 2 (upper figure) and 4 (lower figure), which are shown by red dots.*

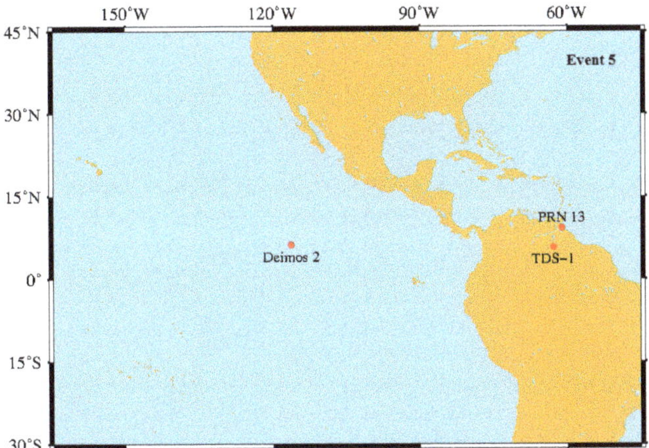

Figure 11. The sub-satellite locations of transmitter, receiver and valid target candidates for Event 5, which are shown by red dots.

6. Conclusions and Remarks

This work proposed a new application of GNSS-R to detect objects above the Earth's surface such as satellites. This finding was derived from the analysis of 14 DDMs of the TDS-1 mission, which contain anomalous reflections, as the bright points (Figure 2) occurring in the "forbidden" zone of the DDMs. Seven possible causes of the unusually bright points were analysed, including random noise, DDM generator losing tracking, leakage of direct signals, aliasing, reflection from an Earth's surface target, GNSS radio-occultation, and reflections from objects above the Earth's surface. The conclusion was that the bright points are likely coming from the reflection of targets above the Earth's surface according to their delays. Two parts of work were done based on this conclusion, target positioning by theoretical analyses and target search in satellite database. It is challenging to theoretically retrieve the target positions because there are unpredictable reflections, such as the reflection shown by the blues lines in Figure 8. Hence this work investigated the target positions under normal reflection conditions (black lines of Figure 8). Moreover, the UCS satellite database has been searched, and it was found that Kondor, Step Cube Lab, and Deimos 2 may be responsible for Events 2, 4 and 5, as they meet the delay and Doppler frequency conditions, with delay error d^{OT} around 500 m and Doppler error D^{OT} less than 1050 Hz.

These 14 DDMs containing anomalies were found in a time period of less than two days, which indicates that these anomalies are not rare. If there were no other reasons for the anomalies, GNSS-R will be possible to detect targets above the Earth's surface. This work is a preliminary study of this new application, and there are two limitations which leave space for future work. First, the location distributions of the 14 DDMs are in land or coastal areas, and none of them is in open ocean. This is due to the limited number of events, if more events could be found, a random distribution pattern would be seen. The future work will use both TDS-1 and CYGNSS dataset to find more events. Second, three events have valid target candidates, while no candidates are found for the other 11 events. In the search for target candidates, one satellite database was used, and other databases such as aircraft and debris are not included because it is challenging to gather the database which is effective in the observing period, and that provides accurate position information. The future work will gather a more effective database to search the target candidates.

Author Contributions: Conceptualization, C.H. and C.B.; methodology, C.H.; formal analysis, C.H., C.B., H.P. and A.C.; resources, C.H.; writing–original draft preparation, C.H.; writing–review, C.B., A.C., L.Q. and C.R.; supervision, C.B., L.Q. and C.R.

Funding: This research received no external funding.

Conflicts of Interest: The authors declare no conflict of interest.

References

1. Hall, C.; Cordey, R. Multistatic scatterometry. In Proceedings of the IGARSS, Edinburgh, UK, 12–16 September 1988; pp. 561–562.
2. Martin-Neira, M.; Caparrini, M.; Font-Rossello, J.; Lannelongue, S.; Vallmitjana, C.S. The PARIS concept: An experimental demonstration of sea surface altimetry using GPS reflected signals. *IEEE Trans. Geosci. Remote Sens.* **2001**, *39*, 142–150.
3. Garrison, J.L.; Komjathy, A.; Zavorotny, V.U.; Katzberg S.J. Wind speed measurement using forward scattered GPS signals. *IEEE Trans. Geosci. Remote Sens.* **2002**, *40*, 50–65.
4. Rius, A.; Aparicio, J.M.; Cardellach, E.; Martín-Neira, M.; Chapron, B. Sea surface state measured using GPS reflected signals. *Geophys. Res. Lett.* **2002**, *29*, 37-1–37-4.
5. Gleason, S. Remote Sensing of Ocean, ICE and land Surfaces Using Bistatically Scattered GNSS Signals from Low Earth Orbit. Ph.D Thesis, University of Surrey, London, UK, 2006.
6. Unwin, M.; Jales, P.; Tye, J.; Gommenginger, C.; Foti, G.; Rosello, J. Spaceborne GNSS-reflectometry on TechDemoSat-1: Early mission operations and exploitation. *IEEE J. Sel. Top. Appl. Earth Observ. Remote Sens.* **2016**, *9*, 4525–4539.
7. Ruf, C.S.; Gleason, S.; Jelenak, Z.; Katzberg, S.; Ridley, A.; Rose, R.; Scherrer, J.; Zavorotny, V. The CYGNSS nanosatellite constellation hurricane mission. In Proceedings of the IGARSS, Munich, Germany, 22–27 July 2012; pp. 214–216.
8. Wickert, J.; Cardellach, E.; Martín-Neira, M.; Bandeiras, J.; Bertino, L.; Andersen, O.B.; Camps, A.; Catarino, N.; Chapron, B.; Fabra, F. GEROS-ISS: GNSS reflectometry, radio occultation, and scatterometry onboard the international space station. *IEEE J. Sel. Top. Appl. Earth Observ. Remote Sens.* **2016**, *9*, 4552–4581.
9. Martín-Neira, M.; Li, W.; Andrés-Beivide, A.; Ballesteros-Sels, X. "Cookie": A satellite concept for GNSS remote sensing constellations. *IEEE J. Sel. Top. Appl. Earth Observ. Remote Sens.* **2016**, *9*, 4593–4610.
10. Høeg, P.; Fragner, H.; Dielacher, A.; Zangerl, F.; Koudelka, O.; Beck, P.; Wickert, J. PRETTY: Grazing altimetry measurements based on the interferometric method. In Proceedings of the 5th Workshop on ARSI, Noordwijk, The Netherlands, 12–14 September 2017.
11. Cardellach, E.; Wickert, J.; Baggen, R.; Benito, J.; Camps, A.; Catarino, N.; Chapron, B.; Dielacher, A.; Fabra, F.; Flato, G. GNSS transpolar earth reflectometry exploring system (G-TERN): Mission concept. *IEEE Access* **2018**, *6*, 13980–14018.
12. Martin-Neira, M. A passive reflectometry and interferometry system (PARIS): Application to ocean altimetry. *ESA J.* **1993**, *17*, 331–355.
13. Treuhaft, R.N.; Lowe, S.T.; Zuffada, C.; Chao, Y. 2-cm GPS altimetry over Crater Lake. *Geophys. Res. Lett.* **2001**, *28*, 4343–4346.
14. Hu, C.; Benson, C.; Rizos, C.; Qiao, L. Single-pass sub-meter space-based GNSS-R ice altimetry: Results from TDS-1. *IEEE J. Sel. Top. Appl. Earth Observ. Remote Sens.* **2017**, *10*, 3782–3788.
15. Li, W.; Cardellach, E.; Fabra, F.; Rius, A.; Ribó, S.; Martín-Neira, M. First spaceborne phase altimetry over sea ice using TechDemoSat-1 GNSS-R signals. *Geophys. Res. Lett.* **2017**, *44*, 8369–8376.
16. Masters, D.; Zavorotny, V.; Katzberg, S.; Emery, W. GPS signal scattering from land for moisture content determination. In Proceedings of the IGARSS, Honolulu, HI, USA, 24–28 July 2000; pp. 3090–3092.
17. Chew, C.; Small, E.E.; Larson, K.M. An algorithm for soil moisture estimation using GPS-interferometric reflectometry for bare and vegetated soil. *GPS Solut.* **2015**, *20*, 525–537.
18. Garrison, J.L.; Katzberg, S.J.; Hill, M.I. Effect of sea roughness on bistatically scattered range coded signals from the Global Positioning System. *Geophys. Res. Lett.* **1998**, *25*, 2257–2260.
19. Germain, O.; Ruffini, G.; Soulat, F.; Caparrini, M.; Chapron, B.; Silvestrin, P. The Eddy Experiment: GNSS-R speculometry for directional sea-roughness retrieval from low altitude aircraft. *Geophys. Res. Lett.* **2004**, *31*, L21307.
20. Zavorotny, V.U.; Voronovich, A.G. Scattering of GPS signals from the ocean with wind remote sensing application. *IEEE Trans. Geosci. Remote Sens.* **2000**, *38*, 951–964.

21. Foti, G.; Gommenginger, C.; Jales, P.; Unwin, M.; Shaw, A.; Robertson, C.; Roselló J. Spaceborne GNSS reflectometry for ocean winds: First results from the UK TechDemoSat-1 mission. *Geophys. Res. Lett.*, **2015**, *42*, 5435–5441.

22. Carreno-Luengo, H.; Camps, A.; Querol, J.; Forte, G. First results of a GNSS-R experiment from a stratospheric balloon over boreal forests. *IEEE Trans. Geosci. Remote Sens.* **2016**,

23. Yu, K.; Li, Y.; Chang, X. Snow depth estimation bsed on combination of pseudorange and carrier phase of GNSS dual-frequency signals. *IEEE Trans. Geosci. Remote Sens.* **2019**, *57*, 1817–1828.

24. Camps, A.; Park, H.; Foti, G.; Gommenginger, C. Ionospheric effects in GNSS-reflectometry from space. *IEEE J. Sel. Top. Appl. Earth Observ. Remote Sens.* **2016**, *9*, 5851–5861.

25. Alonso-Arroyo, A.; Zavorotny, V.U.; Camps, A. Sea ice detection using UK TDS-1 GNSS-R data. *IEEE Trans. Geosci. Remote Sens.* **2017**, *55*, 4989–5001.

26. Li, C.; Huang, W.; Gleason, S. Dual antenna space-based GNSS-R ocean surface mapping: Oil slick and tropical cyclone sensing. *IEEE J. Sel. Top. Appl. Earth Observ. Remote Sens.* **2015**, *8*, 425–435.

27. Yu, K. Tsunami-wave parameter estimation using GNSS-based sea surface height measurement. *IEEE Trans. Geosci. Remote Sens.* **2015**, *53*, 2603–2611.

28. Di Simone, A.; Park, H.; Riccio, D.; Camps, A. Sea target detection using spaceborne GNSS-R delay-Doppler maps: Theory and experimental proof of concept using TDS-1 data. *IEEE J. Sel. Top. Appl. Earth Observ. Remote Sens.* **2017**, *10*, 4237–4255.

29. Gao, C.; Yang, D.; Hong, X.; Xu, Y.; Wang, B.; Zhu, Y. Experimental results about traffic flow detection by using GPS reflected signals. *IEEE J. Sel. Top. Appl. Earth Observ. Remote Sens.* **2018**, *11*, 5076–5087.

30. Howard, D.; Roberts, S.; Brankin, R. Target detection in SAR imagery by genetic programming. *Adv. Eng. Softw.* **1999**, *30*, 303–311.

31. Bekkerman, I.; Tabrikian, J. Target detection and localization using MIMO radars and sonars. *IEEE Trans. Signal Process.* **2006**, *54*, 3873–3883.

32. UCS Satellite Database. 2019. Available online: https://www.ucsusa.org/nuclear-weapons/space-weapons/satellite-database (accessed on 1 April 2019).

33. Mashburn, J.; Axelrad, P.; Lowe, S.T.; Larson, K.M. Global ocean altimetry with GNSS reflections from TechDemoSat-1. *IEEE Trans. Geosci. Remote Sens.* **2018**, *56*, 4088–4097.

34. MERRByS Product Manual-GNSS Reflectometry on TDS-1 with the SGR-ReSI. 2019. Available online: http://merrbys.co.uk/wp-content/uploads/2018/02/MERRByS-Product-Manual-V4.pdf (accessed on 3 December 2018).

35. Andersen, O.; Knudsen, P.; Kenyon, S.; Factor, J.; Holmes, S. The DTU13 global marine gravity field—First evaluation. In Proceedings of the OSTST, Boulder, CO, USA, 8–11 October 2013; pp. 1–14.

36. Lyons, R.G. Periodic sampling. In *Understanding Digital Signal Processing*, 3rd ed.; Pearson Education, Inc.: Boston, MA, USA, 2011; pp. 48–110.

37. Yu, K.; Rizos, C.; Burrage, D.; Dempster, A. G.; Zhang, K.; Markgraf, M. An overview of GNSS remote sensing. *EURASIP Journal on Advances in Signal Processing*, **2014**, *1*, 134.

38. Vallado, D.; Crawford, P. SGP4 orbit determination. In Proceedings of the AIAA/AAS Astrodynamics Specialist Conference and Exhibit, Honolulu, HI, USA, 18–21 August 2008; doi:10.2514/6.2008-6770.

39. Two Line Element Data. 2019. Available online: www.space-track.org/ (accessed on 30 April 2019).

40. Hajj, G.A.; Zuffada, C. Theoretical description of a bistatic system for ocean altimetry using the GPS signal. *Radio Sci.* **2003**, *38*, 1089.

 *remote sensing*

Letter

Can GNSS-R Detect Abrupt Water Level Changes?

Sajad Tabibi * and Olivier Francis

Faculty of Science, Technology, and Medicine, Belval Campus, University of Luxembourg, 4365 Esch-sur-Alzett, Luxembourg; olivier.francis@uni.lu
* Correspondence: sajad.tabibi@uni.lu

Received: 21 September 2020; Accepted: 31 October 2020; Published: 3 November 2020

Abstract: Global navigation satellite system reflectometry (GNSS-R) uses signals of opportunity in a bi-static configuration of L-band microwave radar to retrieve environmental variables such as water level. The line-of-sight signal and its coherent surface reflection signal are not separate observables in geodetic GNSS-R. The temporally constructive and destructive oscillations in the recorded signal-to-noise ratio (SNR) observations can be used to retrieve water-surface levels at intermediate spatial scales that are proportional to the height of the GNSS antenna above the water surface. In this contribution, SNR observations are used to retrieve water levels at the Vianden Pumped Storage Plant (VPSP) in Luxembourg, where the water-surface level abruptly changes up to 17 m every 4-8 h to generate a peak current when the energy demand increases. The GNSS-R water level retrievals are corrected for the vertical velocity and acceleration of the water surface. The vertical velocity and acceleration corrections are important corrections that mitigate systematic errors in the estimated water level, especially for VPSP with such large water-surface changes. The root mean square error (RMSE) between the 10-min multi-GNSS water level time series and water level gauge records is 7.0 cm for a one-year period, with a 0.999 correlation coefficient. Our results demonstrate that GNSS-R can be used as a new complementary approach to study hurricanes or storm surges that cause abnormal rises of water levels.

Keywords: GPS; GNSS; reflectometry; GNSS-R; SNR; water level

---

1. Introduction

Global navigation satellite systems (GNSS) signal have been used for unanticipated purposes, such as remote sensing of the environment [1]. In the last two decades, GNSS reflectometry (GNSS-R) has emerged as a reliable method to estimate water levels, with many studies highlighting the technique's potential [2–7]. Geodetic GNSS-R uses near-surface or ground-based signals of opportunity at the L-band frequency to retrieve environmental variables such as snow depth, soil moisture, and water level at intermediate spatial scales that are proportional to the height of the antenna above the surface being observed (i.e., 1000 m^2 for a GNSS antenna installed on 2-m-tall monument) [4,8–12]. This technique is based on the simultaneous reception of the direct or line-of-sight transmissions, along with their coherently reflected signals.

GNSS-R can retrieve absolute water levels by taking the vertical land motion into account. It should be noted that the vertical land motion of the tide gauge benchmarks ought to be corrected to isolate tide gauges from the land motion [13]. Furthermore, the international workshop on sea-level measurements in hostile conditions that was held from 13 to 15 March 2018 and was cosponsored by the Intergovernmental Oceanographic Commission (IOC) of the United Nations Educational, Scientific and Cultural Organization (UNESCO) has recommended the investigation of GNSS reflectometry (GNSS-R) "if conditions are too difficult for conventional observing methods"; this was in addition to the general recommendation that "tide gauges should be connected to benchmarks and co-located with GNSS as required by the Global Sea Level Observing System (GLOSS) standards". Therefore, geodetic GNSS-R would be especially valuable in the harsh environments where conventional tide gauges are very hard to maintain, and therefore, sea level observations are very sparse (i.e., sea-ice-covered regions).

The potential of GPS L1 reflections from a ground-based GNSS-R experiment first was demonstrated for sea-state monitoring in [14]. They used an upright right-handed circularly polarized (RHCP) antenna and an upside-down left-handed circularly polarized (LHCP) antenna to generate a times series of the interferometric complex field. The authors of [15] further developed the concept of the GNSS-R using GPS L1 phase delays from two geodetic-quality GNSS antennas.

The excess delay of water-surface reflections with respect to the line-of-sight transmission that is recorded with a commercial off-the-shelf GNSS receiver can be used to estimate the water level. The signal-to-noise ratio (SNR) observations from a geodetic-quality receiver/upright antenna at Kachemak Bay, Alaska were first demonstrated by [16] to retrieve the water levels. GNSS-R water levels using carrier-phase and SNR observations of Global Positioning System (GPS) and Russian global navigation satellite system (GLObal'naya NAvigatsionnaya Sputnikovaya Sistema; GLONASS) from a GNSS station in Sweden were compared with those of a co-located tide gauge sensor; the results showed correlation coefficients of 0.86 to 0.97 [17]. Using the GNSS station on top of the Cordouan Lighthouse for a three-month period, [18] reported a root means square error (RMSE) of 0.87 m between GNSS-R water levels and tide gauge (TG) records that were located ~9 km from the lighthouse, taking the dynamic of the reflection surface into account.

In this contribution, we used SNR observations for about a one-year period (from 27 June 2017 to 28 July 2018) at the Vianden Pumped Storage Plant (VPSP) in Luxembourg (Figure 1a) to report the water-surface levels to investigate if geodetic GNSS-R can detect sudden and large water-surface changes caused by nonastronomical forces. We corrected the estimated GNSS-R heights for tropospheric propagation [19]. Then, we corrected the reflector heights for vertical velocity and acceleration of the water surface [6]. The water level of the upper reservoir can abruptly change up to about 17 m every 4–8 h in either generating or pumping modes. This process repeats daily but at different times of the day when the energy demand is high. Therefore, the dynamic satellite and surface corrections are very important, as the water surface at VPSP has a unique water-surface regime. In the last decade, the velocity and vertical velocity corrections were estimated by fitting a daily sinusoidal fit based on the mean frequencies of the dominant tides [16]. However, the traditional method cannot be applied, as the water-surface changes are not caused by diurnal and semidiurnal tides at VPSP. Here, the estimated GNSS-R water levels are corrected using centered finite differences. Finally, the 10-min GNSS-R water levels with a two-hour window width are compared with those of a co-located water level gauge to assess the performance of the technique.

(a)

Figure 1. *Cont.*

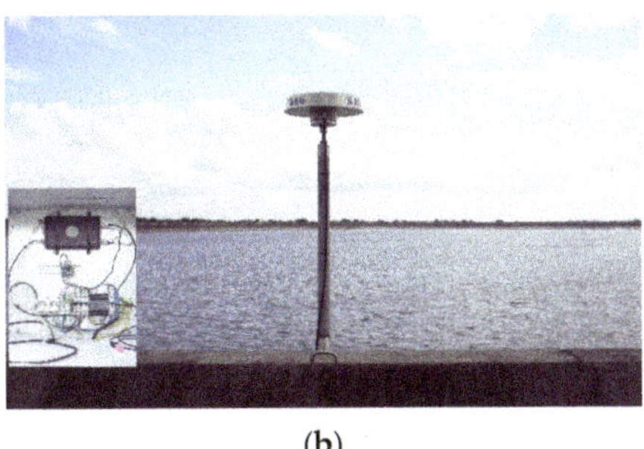

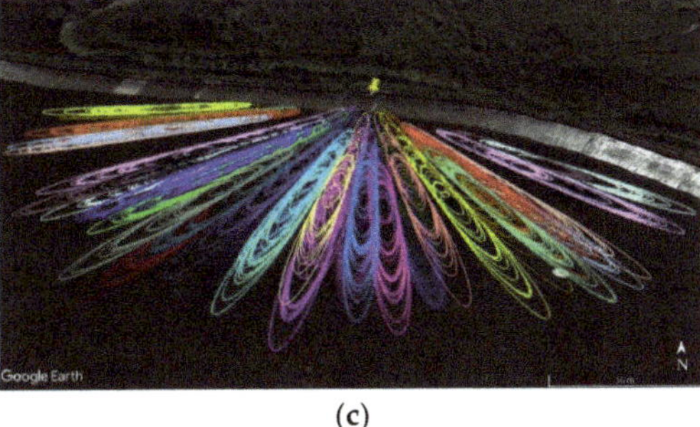

(b)   **(c)**

Figure 1. Vianden-pumped storage plant (**a**) (courtesy of the SEO, [20]), global navigation satellite system (GNSS) station VPSP at the Vianden pumped storage plant (**b**), and (**c**) occurrences of GS1C first Fresnel zones (FFZs) of July 2018 at the VPSP. Each color depicts each GPS satellite, and the location of the GNSS station is shown as a yellow pin.

2. Experiment Setup

The Vianden power station that is located north of Vianden in Diekirch District, Luxembourg is a pumped storage power station that is used to store excess energy and generate peak current. The first four sets of pump generators in the Vianden pumped storage plant were commissioned in the winter of 1962/1963 [20]. The second phase of work was completed in 1964 by installing another five reversible turbine-pump generators. At times when the power consumption is low, water is pumped into the upper reservoir from the lower reservoir dam at VPSP. When the energy demand increases, the stored water is used to generate a peak current. The elevation difference between the upper and lower reservoirs is 282.8 m [20]. The VPSP hydroelectric plant generates an average of 1650 gigawatt-hours electricity every year. The upper reservoir with the total capacity of 7,340,000 m^3 is equipped with ten pump-turbine generators. The 1st to 9th reversible turbine-pump generators with the turbine-runner diameter of 2.45 m can discharge 39.5 m^3/s of water when electricity is generated. The 10th machine with a diameter of 4.40 m has a water flow of 77 m^3/s and 74 m^3/s in the turbine operation and pumping mode, respectively [20]. The 11th pump generator with the turbine capacity of 200 megawatts was commissioned in December 2014.

The GNSS station VPSP (latitude: 49.94°, longitude: 6.18°, altitude: 561.96 m) was installed in late-June 2017 at VPSP. It was equipped with a Septentrio PolaRx5 receiver and a Trimble TRM59800.00 antenna with no external radome (Figure 1b). The antenna is installed ~11.0 m above the mean water level, and the receiver was set at a 1-Hz sampling rate. As the station VPSP was installed specifically for GNSS-R, it is positioned towards the south direction in order to maximize the number of satellite tracks (Figure 1c).

VPSP is also equipped with Rittmeyer high-precision pressure gauge MPW2Q sensors. MPW2Q provides hydrostatic or pneumatic pressure measurements every 1 min. MPW2Q reports high-precision water levels (WL) with a resolution better than 1 mm [21].

An example of SNR observations of a rising arc at VPSP station is shown in Figure 2. The protected/encrypted signals (P(Y)) on both the L1 and L2 bands are ignored, as they produce nosier SNR fringes [12]. It can be seen that SNR fringes are very clear for all the seven signals, especially for low-elevation angle observations, i.e., observations below a 20° elevation angle. Second, GS5Q have the highest direct power, as no military signal is planned for L5 [11]. Third, there are up to 10-dB different direct power levels between the strongest (GS5Q) and the weakest (RS2P) signals.

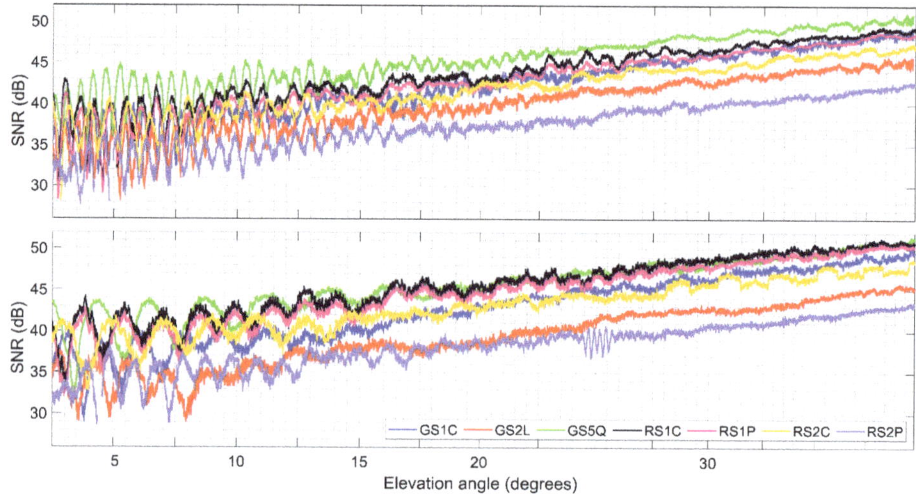

Figure 2. Signal-to-noise ratio (SNR) observations for a rising arc at VPSP for two different epochs when the reservoir had different water levels. Top panel is for decimal year 2018.0024 when the energy demand increased, and bottom panel shows SNR observations for decimal year 2018.0034 when the power consumption was low.

3. GNSS-R Data Processing

SNR observations can be decomposed into a trend ($tSNR$) that is a monotonically increasing function of the elevation angle (e) and detrended interference fringes ($dSNR$). The detrended interference fringes that are superimposed on the trend can be expressed in terms of the detrended interference fringes' amplitude ($aSNR$) and reflection excess phase (ϕ_i) with respect to the direct phase [22]:

$$dSNR = aSNR \cos \phi_i \tag{1}$$

The interferometric phase is a function of the coupled surface-antenna properties (ϕ_X), geometric interference component (ϕ_I), and right-handed circularly polarized (RHCP) antenna phase pattern evaluated in the direct path (Φ_d^R) [23]:

$$\phi_i = \phi_X + \phi_I - \Phi_d^R \tag{2}$$

The compositional component of the interferometric phase can be expressed as a function of the complex-valued Fresnel reflection coefficients for same- $\left(R^S\right)$ and cross-sense circular polarization $\left(R^X\right)$ and antenna complex vector effective length ($\overline{L}$) [11,22]:

$$\phi_X = \arg(R^S L_r^R + R^X L_r^L) \tag{3}$$

The geometric component of the interferometric phase can be modeled as a function of the wavenumber (k):

$$\phi_I = k\tau_i \tag{4}$$

Assuming a flat and smooth surface, the interferometric propagation delay (τ_i) can be modeled as a function of the satellite elevation angle and an a priori reflector height (H)—the vertical distance between the GNSS antenna phase center and the reflecting surface [11]:

$$\tau_i = 2H \sin e \tag{5}$$

As the GNSS antenna surroundings (i.e., water level) are an unknown condition, the GNSS-R inverse model is parameterized by a few biases assuming a priori values for the physical parameters. Therefore, the $tSNR$ and $dSNR$ in the electromagnetic forward model are defined as $tSNR\left(K^{-2} + K^{-2}B^2\right)$ and $dSNR\left(K^{-2}|B|^{-1}\right)$, respectively. The real-valued noise power K and complex-valued interferometric bias B with magnitude $|B|^2 = 10^{B_{dB}/10}$ and phase ϕ_B are introduced for the interferometric model as [12]:

$$K_{dB} = \sum_{j=0}^{n} K_{dB}^{(j)} \sin^j e \tag{6}$$

$$B_{dB} = \sum_{j=0}^{n'} B_{dB}^{(j)} \sin^j e \tag{7}$$

$$\phi_B = \sum_{j=0}^{n''} \phi^{(j)} \sin^j e \tag{8}$$

where the superscript in parentheses is an index; otherwise, it is a power exponent. While the direct and interferometric power biases are introduced only to improve the sinusoid fit to the SNR observations, the interferometric phase bias is used to retrieve the reflector height.

The inverse model can use the forward model internally to simulate SNR observations to estimate parameter corrections by fixing them to their a priori values and corresponding covariance matrices. For more details, the reader is referred to [12] and [23]. Therefore, the interferometric phase can be updated in the forward model as:

$$\phi_i = \phi_i - \phi_B^{(0)} - \phi_B^{(1)} \sin e \ldots \tag{9}$$

Finally, the linear phase bias can be used to estimate the total reflector height:

$$H_{total} = H - \phi_B^{(1)} \lambda / 4\pi. \tag{10}$$

where λ is the carrier wavelength.

After inversion, a number of post-processing measures are applied to detect and reject outliers resulting from unexpected situations (i.e., secondary reflections, diminished interferometric powers) of each individual GNSS signal. However, SNR observations should be first clustered so retrieved biases will be compared under comparable environmental conditions. To do that, SNR observations are partitioned into satellite tracks, between rising/setting near the horizon, and culmination near the zenith. The satellite tracks are then clustered by taking advantage of GNSS ground tracks repeatability. Therefore, satellite tracks are discriminated by whether the satellite is ascending or descending near the same azimuth. Intra-signal and intra-cluster quality control (QC) compare the estimated linear phase biases within each individual cluster using robust estimators and statistical intervals and reject outliers resulting from anomalous conditions [9]. In the next step, as phase biases are highly correlated to one another, the constrained least-squares (LS) adjustment is implemented to retrieve constrained constant phase biases to improve the precision of the water level estimates [6].

The water level at VPSP abruptly changes with time. Therefore, the estimated water levels must be corrected for the dynamic water surface level [16]. The interferometric phase rate can be defined as a function of the stationary surface height (H) and height rate correction ($\partial H / \partial k_z$):

$$\partial \phi_i / \partial k_z = H + k_z \partial H / \partial k_z \tag{11}$$

where $k_z = 2k \sin e$ is the vertical wavenumber [6]. The vertical velocity of the water surface level is an important correction, as the water level at VPSP varies up to 17 m every 4–8 h. Therefore, the height-rate correction can be modeled as:

$$\partial H / \partial k_z = \dot{H} \tan e / \dot{e} + \ddot{H} \Delta t \tan e / \dot{e} \tag{12}$$

The first term on the right-hand side represents the angular delay rate, and the second is the integrated delay rate. The height rate and velocity rate can be solved iteratively using the centered finite differences [6]. The elevation angel rate ($\partial e / \partial t$) is derived from the GNSS precise ephemerides. The authors of [6] reported a height rate and velocity rate of about 40 cm/h and 20 cm/h^2, respectively, for a GNSS station with about 3-m daily tidal envelopes. Therefore, the height-rate correction cannot be neglected especially at VPSP; otherwise, GNSS-R water levels would be biased. It should be noted that vertical acceleration would be also beneficial, especially when sub-hourly water levels (10 min) with such a window width (2 h) are reported [6].

Tropospheric delays as a function of the satellite elevation angle and the reflector height were found as an important source of error in GNSS-R studies [19,24]. Therefore, here, the estimated water levels are corrected for atmospheric refraction after intra-signal and intra-cluster QC. The Vienna Mapping Function 3 (VMF3) that is based on the ray-traced delays of the ray-tracer RADIATE, together with the Global Pressure and Temperature 3 model (GPT3) on a $1° \times 1°$ global grid that represents a comprehensive troposphere model [25], is used to estimate the tropospheric delay corrections. Using the minimum and maximum elevation angle observations of each individual satellite track, the along-path tropospheric delays can be estimated as:

$$H_t = 2H(\bar{n} - 1) / \sin e \tag{13}$$

where $\bar{n}$ is the index of refraction along the instantaneous excess delay path [19,24].

All cluster-by-cluster tracks are combined to produce a regularly spaced signal-specific epoch-by-epoch water level time series together with their respective uncertainties based on a weighted moving average. The moving average returns water levels every 10 min using a 2-h window width [6]. It should be noted that the moving regression is not a simple moving average, as it also estimates the height and velocity rates. Finally, sub-hourly water level time series are reported using the signal-specific variance factors. The statistical inter-signal combination that offers a single multi-GNSS time series was found to improve the precision of geodetic GNSS-R retrievals [12]. The post-inversion algorithm of the water level retrieval at VPSP is summarized in Figure 3.

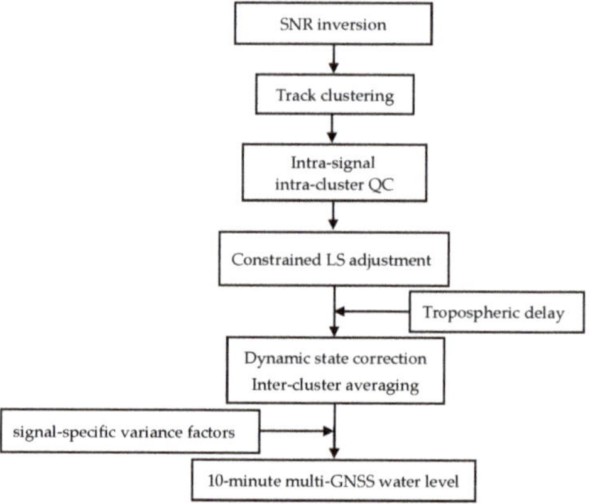

Figure 3. Flowchart of the water-surface level retrieval algorithm. QC: quality control.

Before we combined all the independent GNSS-R water levels to provide a single multi-GNSS water level time series, first we compared all the signal-dependent water levels together. Here, the GS1C water level time series is used to evaluate the inter-signal consistency among all the GNSS-R retrievals. A systematic error at a decimeter level between the GPS-L1 and GPS-L2 signal water-surface level retrievals was reported in [26]. In this study, we found linear relationships between different GNSS-R water level estimates and GS1C, with correlations that exceeded 0.98 (Figure 4).

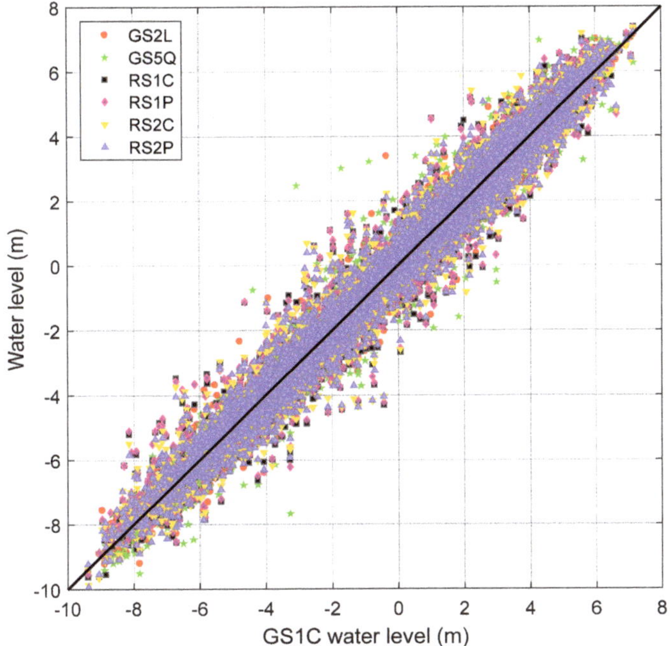

Figure 4. Scatterplots of the GNSS-derived water level show an excellent agreement between GS1C retrieval and other GNSS-derived water level retrievals; the 1:1 diagonal is shown as a black line.

4. Results

Figure 5 shows the tropospheric delays at station VPSP. Due to the variations of the water-surface level, the atmospheric delay changes with respect to the satellite elevation angle. The tropospheric delay difference between the two modes of VPSP when the water-surface level is at its minimum and its maximum is 29.40 cm. The regression slope 0.984 m/m ± 0.001 m/m between the GNSS-R water level time series and WL records are statistically significant vis-à-vis its 95% confidence interval if the tropospheric delay corrections are neglected.

SNR observations from a GNSS station in Sweden, [17] demonstrated that GS2W and RS2P should be avoided for GNSS-R water level studies. The authors of [12] found that legacy GPS signals are worse in terms of accuracy, while modernized GPS and GLONASS signals are comparable. Therefore, signal-specific variance factors are used to report multi-GNSS GNSS-R time series, together with their formal errors (Figure 6).

The GNSS-R water level time series are compared with those of the WL records. First, the WL records are interpolated at the GNSS sampling epochs and vice versa by disabling the interpolations across the data outage periods. The RMSE error between the sub-hourly GNSS-R time series and WL records is 7.0 cm. The agreement between the WL records and GNSS-R water level retrievals is excellent, with a correlation coefficient that exceeds 0.999 (Figure 7).

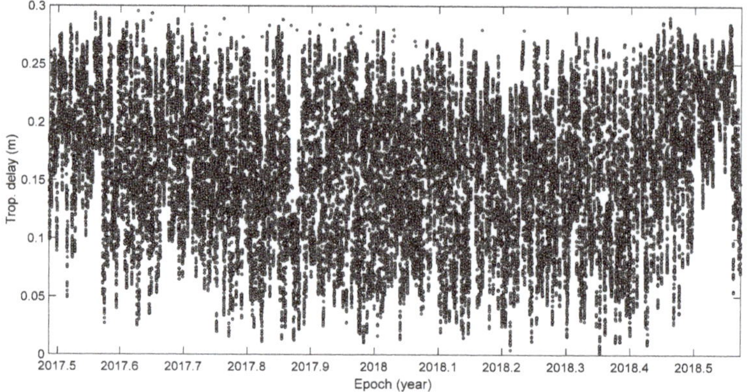

Figure 5. Tropospheric delays at station VPSP.

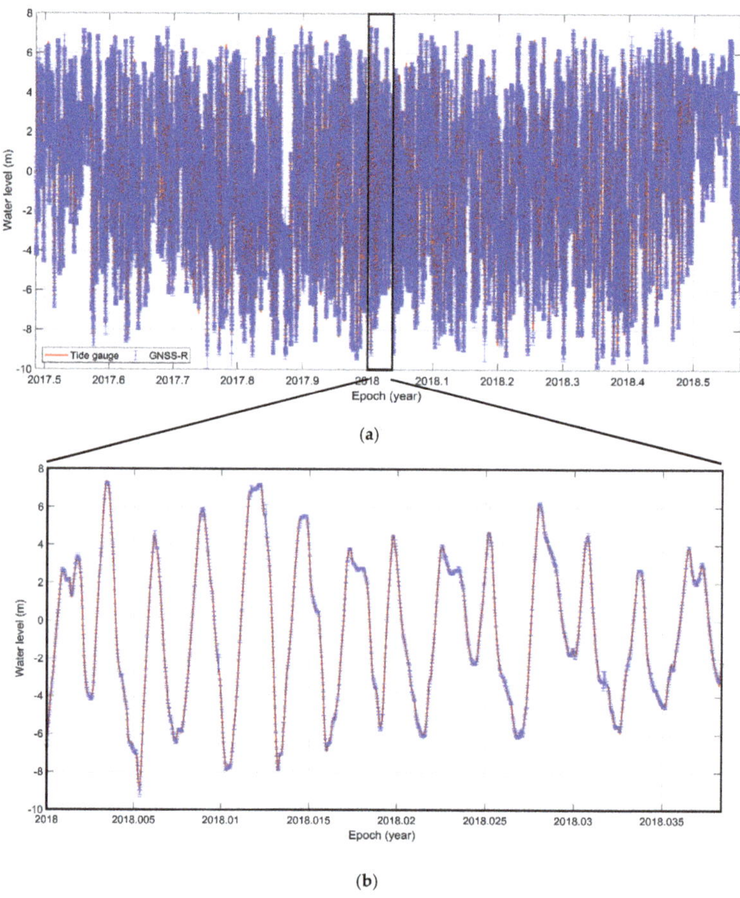

Figure 6. Water level time series from Rittmeyer high-precision pressure gauge sensor (red) and geodetic GNSS-R (blue) at the Vianden pumped storage plant (**a**) and zoom in of the first 15 days of 2018 (**b**).

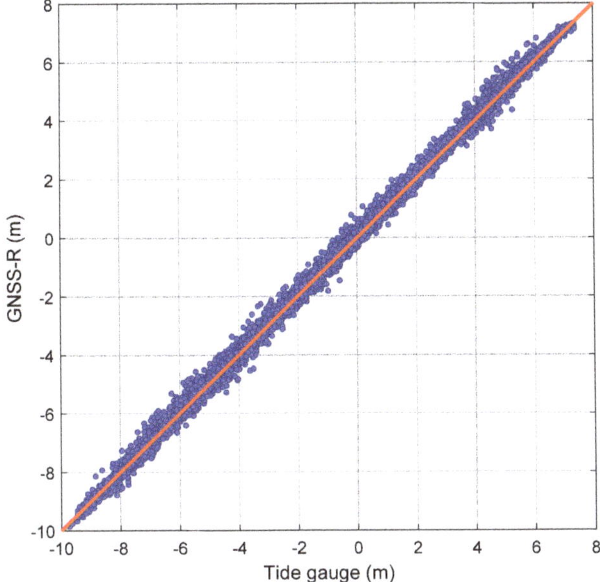

Figure 7. Scatterplot of the water levels show an excellent agreement between GNSS-R retrievals and water level (WL) records.

There is no scale error between the GNSS-R water level time series and WL records that can be seen in the van de Casteele diagram in Figure 8. The *x*-axis presents the difference between geodetic GNSS-R water levels as the test sensor, and the y-axis represents the WL records as the reference sensor. The two water levels deviate by −0.27% from a one-to one relationship. The authors of [4] reported a scale error of 0.84% for a GNSS site with ~2.5-m tidal amplitudes. We found that the RMS reduction at VPSP is 49% when the vertical acceleration is applied. In addition, the GNSS-R water level time series would be deviated by 1.97% from the WL records if they were only corrected for the height rate.

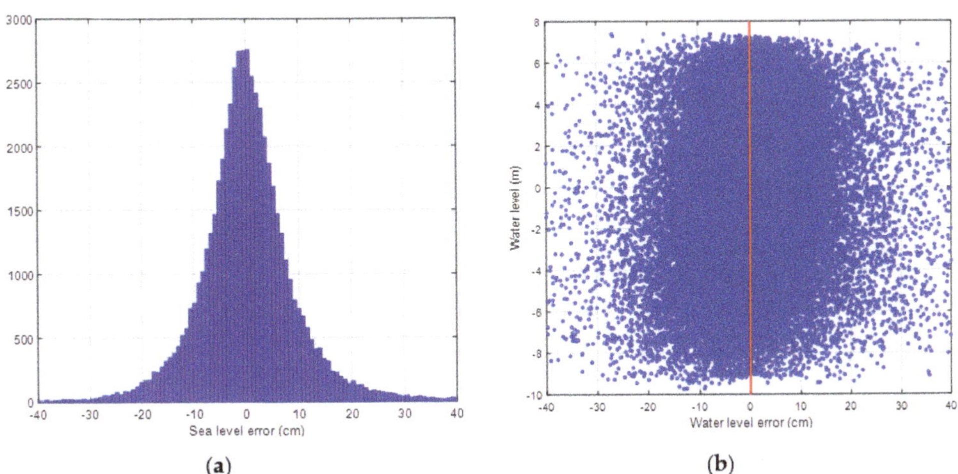

(a)   (b)

Figure 8. Histogram of the water level error between the GNSS-R and WL sensor at the VPSP (**a**), and van de Casteele diagrams for the GNSS-R water levels over 1 year at the VPSP (**b**).

We also compared the sub-hourly maximum and minimum GNSS-R water level retrievals and water level gauge records (Figure 9). The RMSE between the maximum GNSS-R retrievals and water level records is 6.5 cm, but it is 11.1 cm between the minimum GNSS-R estimates and water level measurements.

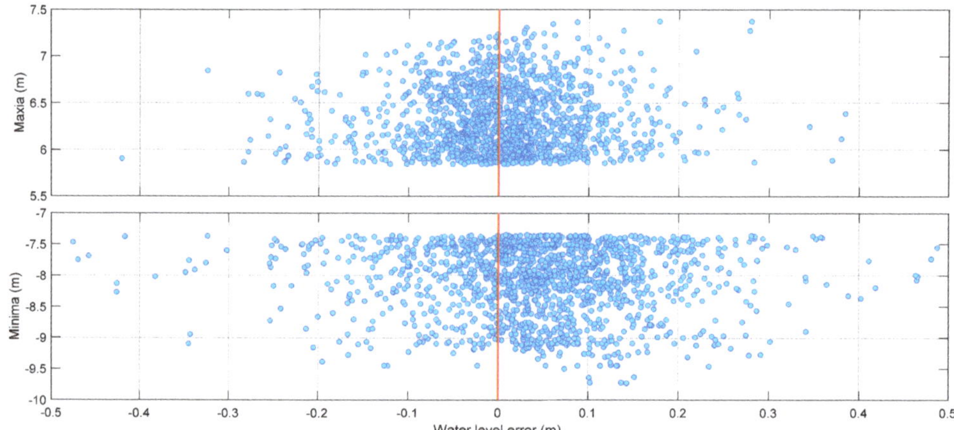

Figure 9. Van de Casteele diagrams for the sub-hourly water level extremes measured by the GNSS-R and water level gauge.

Carrying out a regression between maxima measurements from the two sensors, the intercept −0.03 m ± 0.16 m and regression slope 0.99 m/m ± 0.03 m/m are not statistically significant vis-à-vis their 95% confidence intervals. The regression intercept of the minima GNSS-R retrievals and water level records is −0.02 m ± 0.19 m, and the regression slope is 1.01 m/m ± 0.02 m/m. While the median offset between the maximum measurements is 0.5 cm, it is 4.7 cm for the minimum retrievals. On one hand, GNSS-R retrievals are less precise than those of the water level. On the other hand, the resolution of the Rittmeyer high-precision pressure gauge sensor is 10 times higher than the GNSS-R, and GNSS-R retrievals are averaged over the FFZs over 10 min. In addition, the offset error between minimum retrievals can be the result of the ramp slope of the dam. When electricity is generated, the water level goes down. Therefore, the vertical and horizonal distances between the GNSS antenna and water surface increase. There would be reflections from the ramp slope for higher elevation angle observations where FFZs are close to the antenna.

5. Conclusions

The geodetic GNSS-R is used to estimate water levels at the Vianden pumped storage plant (VPSP), where the water-surface level changes vertically ~17 m several times a day to generate a peak current. The retrieved heights are corrected for tropospheric delay using VMF3/GPT3 models before signal-specific water levels are estimated based on a weighted moving average using a 10-min post-spacing and a 2-h averaging window width. The inter-cluster combination is implemented in a way to take the dynamic state of the sea surface into account by correcting the retrieved heights for the velocity and vertical velocity corrections. Finally, GNSS-R sub-hourly time series are reported using the signal-specific variance factors. The standard deviation between the sub-hourly GNSS-R water levels and co-located tide gauge records for about one year at the VPSP is 7.0 cm, with a correlation coefficient exceeding 0.999.

The results confirm that GNSS-R can detect sudden and large water surface changes. Therefore, geodetic GNSS-R can be used to study water-surface changes caused by nonastronomical forces such as hurricanes or storm surges. The performance of GNSS-R water levels can be improved in terms of temporal sampling by tracking all GNSS signals from other global navigation systems (i.e., Russian GLONASS, European Galileo, and the Chinese BeiDou global). In the future, the atmospheric bias in he -R water levels will be solved using an atmospheric ray-tracing procedure [24].

Author Contributions: Conceptualization, S.T. and O.F.; methodology, S.T.; software, S.T.; validation, S.T. and O.F.; formal analysis, S.T.; investigation, S.T.; resources, S.T. and O.F.; data curation, S.T and O.F.; writing—original draft preparation, S.T.; writing—review and editing, S.T. and O.F.; visualization, S.T.; project administration, O.F. All authors have read and agreed to the published version of the manuscript.

Funding: This research was carried out at the Geophysics Laboratory of the University of Luxembourg, and received no external funding.

Acknowledgments: The authors would like to thank Société Electrique de l'Our (SEO) for their support in the execution of the VPSP field experiment. We thank F.G. Nievinski for the valuable discussions. We would like to thank Gilbert Klein and Marc Seil for the technical support of the project.

Conflicts of Interest: The authors declare no conflict of interest.

References

1. Jin, S.; Cardellach, E.; Xie, F. *GNSS Remote Sensing: Theory, Methods and Applications*; Remote Sensing and Digital Image Processing; Springer: Dordrecht, The Netherlands, 2014; ISBN 978-94-007-7481-0.
2. Anderson, K.D. Determination of Water Level and Tides Using Interferometric Observations of GPS Signals. *J. Atmos. Ocean. Technol.* **2000**, *17*, 1118–1127. [CrossRef]
3. Geremia-Nievinski, F.; Makrakis, M.; Tabibi, S. Inventory of published GNSS-R stations, with focus on ocean as target and SNR as observable. *Zenodo* **2020**. [CrossRef]
4. Larson, K.M.; Ray, R.D.; Williams, S.D.P. A 10-Year Comparison of Water Levels Measured with a Geodetic GPS Receiver versus a Conventional Tide Gauge. *J. Atmos. Ocean. Technol.* **2016**, *34*, 295–307. [CrossRef]
5. Strandberg, J.; Hobiger, T.; Haas, R. Improving GNSS-R sea level determination through inverse modeling of SNR data. *Radio Sci.* **2016**, *51*, 1286–1296. [CrossRef]
6. Tabibi, S.; Geremia-Nievinski, F.; Francis, O.; van Dam, T. Tidal analysis of GNSS reflectometry applied for coastal sea level sensing in Antarctica and Greenland. *Remote Sens. Environ.* **2020**, *248*, 111959. [CrossRef]
7. Geremia-Nievinski, F.; Hobiger, T.; Haas, R.; Liu, W.; Strandberg, J.; Tabibi, S.; Vey, S.; Wickert, J.; Williams, S. SNR-based GNSS reflectometry for coastal sea-level altimetry: Results from the first IAG inter-comparison campaign. *J. Geod.* **2020**, *94*, 70. [CrossRef]
8. Larson, K.M. GPS interferometric reflectometry: Applications to surface soil moisture, snow depth, and vegetation water content in the western United States. *Wires Water* **2016**, *3*, 775–787. [CrossRef]
9. Nievinski, F.G.; Larson, K.M. Inverse Modeling of GPS Multipath for Snow Depth Estimation—Part II: Application and Validation. *IEEE Trans. Geosci. Remote Sens.* **2014**, *52*, 6564–6573. [CrossRef]
10. Small, E.E.; Larson, K.M.; Chew, C.C.; Dong, J.; Ochsner, T.E. Validation of GPS-IR Soil Moisture Retrievals: Comparison of Different Algorithms to Remove Vegetation Effects. *IEEE J. Sel. Top. Appl. Earth Obs. Remote Sens.* **2016**, *9*, 4759–4770. [CrossRef]
11. Tabibi, S.; Nievinski, F.G.; van Dam, T.; Monico, J.F.G. Assessment of modernized GPS L5 SNR for ground-based multipath reflectometry applications. *Adv. Space Res.* **2015**, *55*, 1104–1116. [CrossRef]
12. Tabibi, S.; Nievinski, F.G.; van Dam, T. Statistical Comparison and Combination of GPS, GLONASS, and Multi-GNSS Multipath Reflectometry Applied to Snow Depth Retrieval. *IEEE Trans. Geosci. Remote Sens.* **2017**, *55*, 3773–3785. [CrossRef]
13. Wöppelmann, G.; Marcos, M. Vertical land motion as a key to understanding sea level change and variability. *Rev. Geophys.* **2016**, *54*, 64–92. [CrossRef]
14. Soulat, F.; Caparrini, M.; Germain, O.; Lopez-Dekker, P.; Taani, M.; Ruffini, G. Sea state monitoring using coastal GNSS-R. *Geophys. Res. Lett.* **2004**, *31*. [CrossRef]
15. Löfgren, J.S.; Haas, R.; Scherneck, H.-G.; Bos, M.S. Three months of local sea level derived from reflected GNSS signals. *Radio Sci.* **2011**, *46*. [CrossRef]
16. Larson, K.M.; Ray, R.D.; Nievinski, F.G.; Freymueller, J.T. The Accidental Tide Gauge: A GPS Reflection Case Study From Kachemak Bay, Alaska. *IEEE Geosci. Remote Sens. Lett.* **2013**, *10*, 1200–1204. [CrossRef]
17. Löfgren, J.S.; Haas, R. Sea level measurements using multi-frequency GPS and GLONASS observations. *EURASIP J. Adv. Signal Process.* **2014**, *2014*, 50. [CrossRef]

18. Roussel, N.; Ramillien, G.; Frappart, F.; Darrozes, J.; Gay, A.; Biancale, R.; Striebig, N.; Hanquiez, V.; Bertin, X.; Allain, D. Sea level monitoring and sea state estimate using a single geodetic receiver. *Remote Sens. Environ.* **2015**, *171*, 261–277. [CrossRef]

19. Williams, S.D.P.; Nievinski, F.G. Tropospheric delays in ground-based GNSS multipath reflectometry—Experimental evidence from coastal sites. *J. Geophys. Res. Solid Earth* **2017**, *122*, 2310–2327. [CrossRef]

20. Société Electrique de l'Our. Available online: http://www.seo.lu/ (accessed on 25 May 2020).

21. Rittmeyer, A.G. MPW2Q High-Precision Pressure Gauge. Available online: https://rittmeyer.com/en/instrumentation/applications/overview/ (accessed on 24 June 2020).

22. Nievinski, F.G.; Larson, K.M. Forward Modeling of GPS Multipath for Near-surface Reflectometry and Positioning Applications. *GPS Solut.* **2014**, *18*, 309–322. [CrossRef]

23. Nievinski, F.G.; Larson, K.M. Inverse Modeling of GPS Multipath for Snow Depth Estimation—Part I: Formulation and Simulations. *IEEE Trans. Geosci. Remote Sens.* **2014**, *52*, 6555–6563. [CrossRef]

24. Nikolaidou, T.; Santos, M.C.; Williams, S.D.P.; Geremia-Nievinski, F. Raytracing atmospheric delays in ground-based GNSS reflectometry. *J. Geod.* **2020**, *94*, 68. [CrossRef]

25. Landskron, D.; Böhm, J. VMF3/GPT3: Refined discrete and empirical troposphere mapping functions. *J Geod* **2018**, *92*, 349–360. [CrossRef]

26. Santamaría-Gómez, A.; Watson, C.; Gravelle, M.; King, M.; Wöppelmann, G. Levelling co-located GNSS and tide gauge stations using GNSS reflectometry. *J. Geod.* **2015**, *89*, 241–258. [CrossRef]

Publisher's Note: MDPI stays neutral with regard to jurisdictional claims in published maps and institutional affiliations.

 *remote sensing*

Technical Note

Flood Inundation Mapping by Combining GNSS-R Signals with Topographical Information

S L Kesav Unnithan [1,2,3,*], Basudev Biswal [2,4] and Christoph Rüdiger [3]

[1] IITB-Monash Research Academy, Mumbai 400076, India
[2] Department of Civil Engineering, Indian Institute of Technology Bombay, Mumbai 400076, India; basudev@civil.iitb.ac.in
[3] Department of Civil Engineering, Monash University, Clayton, VIC 3168, Australia; chris.rudiger@monash.edu
[4] Interdisciplinary Program in Climate Studies, Indian Institute of Technology Bombay, Mumbai 400076, India
* Correspondence: kesav.sreekuttanlakshmidevi@monash.edu; Tel.: +91-8281343572

Received: 27 June 2020; Accepted: 19 August 2020; Published: 17 September 2020

Abstract: The Cyclone Global Navigation Satellite System (CYGNSS) mission collects near-global hourly, pseudo-randomly distributed Global Navigation Satellite System - Reflectometry (GNSS-R) signals in the form of signal-to-noise ratio (SNR) point data, which is sensitive to the presence of surface water, due to their operating frequency at L-band. However, because of the pseudo-random nature of these points, it is not possible to obtain continuous flood inundation maps at adequately high resolution. By considering topological indicators, such as height above nearest drainage (HAND) and slope of nearest drainage (SND), which indicate the probability of a certain area being prone to flooding, we hypothesize that combining static topographic information with the dynamic GNSS-R signals can result in large-scale, high-resolution flood inundation maps. Flood mapping was performed and validated with flood extent derived using available Sentinel-1A synthetic aperture radar (SAR) data for flooding in Kerala during August 2018, and North India during August 2017. The results obtained after thresholding indicate that the model exhibits a flooding accuracy ranging from 60% to 80% for lower threshold values. We observed significant overestimation error in mapping inundation across the flooding period, resulting in an optimal critical success index of 0.22 for threshold values between 17–19.

Keywords: CYGNSS; HAND; flood inundation mapping; Sentinel-1A SAR

1. Introduction

Floods are among the worst natural disasters, affecting millions of people across different parts of the world. The magnitude of these flood events is projected to increase [1] because of exacerbated climate change and rapid urbanization, thus making flood control and management an important agenda on the Sendai Framework for Disaster Risk Reduction [2]. To alleviate the impact of floods on the surrounding environment, and to enhance decision making for emergency response and recovery, a comprehensive global flood risk assessment must be carried out, quantifying flood hazard, exposure, and vulnerability. Flood hazard [3] is defined as the occurrence probability of a flood event within a region. Hence, in order to quantify the probability of flooding of a region, we need accurate flood area maps with high temporal and spatial resolutions that are associated with a given rainfall and discharge volume.

Earlier methods of producing spatial flood maps were based on manual observation of flood extents, which are not only laborious but also affect the safety of people undertaking observations in flooding conditions. Satellite-based remote sensing observation is currently used for flood inundation mapping, as it provides a synoptic view of a large area [4–8]. However, the presence of clouds acts as a hindrance to the use of optical remote sensing images. Since most flood events are characterized by heavy rainfall, this limits the detailed mapping of the inundation below clouds. Synthetic aperture radar (SAR)-based flood mapping techniques have the advantage of an all-weather, all-time capability that has found a lot of applications during floods [9–11]. However, most of the open data source SAR satellites have poor revisit times (in the order of ~10–14 days), which are larger than the typical flood event durations that last around 3–7 days. Hence, unless in a constellation, space-borne SAR missions generally fail to capture the spatial evolution of the flooding event, even though they may capture the flooding for a given day more accurately when compared to optical remote sensing methods.

In the case of rare flood events, where the magnitude of the flood peak is such that the swath width of the satellite in one day cannot fully cover the flooded region, SAR-derived flood maps may not be able to capture the entire extent of flooding, in particular for extensive flood events such as in Chennai during December 2015 [12], and the nearby state of Kerala during August 2018 [13], both located in western India. In both instances, the catchments received their highest rainfall in a century [14,15]. Both flood events were characterized by heavy rainfall in the catchment area, followed by dams in the basin area being filled to their maximum capacity, and dam authorities eventually having to release enormous quantities of water to prevent dam failure [16]. Subsequently, large areas were inundated across the floodplain region, resulting in the deaths of over 400 people and 100,000 heads of livestock, while displacing millions from their homes and damaging crops spread across an area of 4000 km^2 [15,17]. The extent was so large that the available SAR images during the flooding period did not cover the flooded region entirely. It must be noted that, in the case of the Chennai floods, much of the dam operations were performed during the night, catching the people unaware of the rising flood levels, thereby aggravating the problem of rescue and recovery [18]. Also, there exist significant challenges in identifying flooding under vegetation and for urban areas from SAR imagery, which are currently overcome by considering other ancillary information, including land cover maps and optical imagery [10]. However, current and future SAR missions, including NISAR, TerraSAR/TanDEM-L, CSK-2, the Radarsat constellation, and open data access policies between organizations in charge of their data cataloguing, can lead to sufficiently extensive temporal SAR inventory of the Earth's surface that can, in turn, result in comprehensive remote observations of flood extents across the world. However, until those missions are fully operational, there still exist significant challenges in reliably estimating the spatial extent of flooding.

Global Navigation Satellite System Reflectometry (GNSS-R) is the technique of measuring the reflected signals from navigation satellites, the most popular being the Global Positioning System (GPS), which provides daily, near-global point information over land, ocean, and atmosphere [19]. The Cyclone Global Navigation Satellite System (CYGNSS) mission was originally planned to improve the understanding of cyclonic behavior with sub-daily wind speed measurements. Recently, this data has found land surface applications in the form of large-scale flood inundation mapping [20,21] through change detection approaches. The signals operating in the L-band penetrate clouds, and are less affected by signal attenuation through heavy rainfall and vegetation than observations in other frequencies, and thus provide an all-time and -weather observational capability. The inland areas, when submerged in water, have a distinctive increase in signal strength when compared to non-water surfaces. Water, being a smooth scatterer, reflects most of the incoming GPS signals, and hence the signals received by the 8-satellite CYGNSS constellation from a water-covered surface have enhanced signal strength. Thus, GNSS-R acts, in this case, as a dynamic indicator of the presence of water on a near-global daily scale. However, the reflected GPS data points are pseudo-randomly distributed as a function of the satellite positions, and hence the same location may not be mapped periodically, effectively resulting in a coarse daily spatial footprint for CYGNSS over land surfaces. Hence, it is hypothesized that upon integrating fine resolution static topographic information with highly frequent signal-to-noise ratio (SNR) data, it will be feasible to obtain actionable flood inundation maps that provide useful information regarding how susceptible to inundation a given region will be for a given flood event.

Height above nearest drainage (HAND) is a topography-based conceptual flood terrain index that quantifies the floodplains along the river reach into pixels, which are identified by their relative height to the nearest drainage point along the flow direction from the corresponding pixels [22–24]. However, it is a static flood terrain index that does not provide any information about the propagation speed of the flood wave, or if an area will be inundated or not. The index also requires a flood height cut-off, which acts as a threshold for the floodplain area, which in turn may then be used to derive the flood inundation extent. Studies have shown that there exists a large amount of overestimation of the floodplain area, depending on the threshold value chosen and the quality of the digital elevation model (DEM) used for estimating HAND values [25–28]. In addition to the HAND index, which provides information on the floodplain, further information is therefore needed to characterize the flooding potential of a channel pixel, which can be obtained by the slope of nearest drainage (SND), which is a function of the height of water in it. Thus, HAND and SND together act as static indicators for deriving flood extent, while CYGNSS provides dynamic information.

In this work, complimentary dynamic CYGNSS and static HAND-SND datasets are combined to map the extent of floodplain areas arising in the case of riverine floods. The proposed model was analyzed and compared against Sentinel-1A SAR-derived flood maps [7]. The error associated with the corresponding threshold values of the proposed method was determined. The potential of future GNSS-R-enabled satellite system missions in flood inundation mapping is also explored. The present paper is organized as follows: Section 2 describes the dataset used for model development and comparison, as well as the study area; Section 3 explains the methodology undertaken; Section 4 illustrates the results obtained and discussion therein; and finally, Section 5 provides the conclusion.

2. Data Used and Study Regions

The CYGNSS satellite constellation is an eight-satellite bi-static radar system launched by NASA in December 2016, collecting near-global (between 38°N and 38°S latitudes) daily reflected L-1 coarse acquisition (C/A) GPS signals in the form of delay-Doppler maps (DDMs), with four simultaneous point observations measured by a single satellite. The DDMs capture the behavior of the reflecting surface, characterized by the specular point of reflection and reflectivity of the surface itself, and strength of the reflected signal as indicated by the signal-to-noise ratio (SNR). For the model development, SNR data was obtained from the Level-1 bistatic radar cross-section (BRCS) of Earth surface data collection, hosted at the University of Michigan [29] (Supplementary Table S1). The globally available open-access Advanced Spaceborne Thermal Emission and Reflection Radiometer (ASTER) digital elevation model (DEM) Version 3 (GDEM 003) dataset was also used in this study, which is provided as $1° \times 1°$ tile at 30 m spatial resolution, and referenced to the WGS84/EGM96 geoid [30]. The proposed model is applied to two regions suffering from flood events: (a) the southern Indian state of Kerala with an area of 38,000 km^2 (August 2018); and (b) Bangladesh and parts of North and Northeast India, including the states of Uttar Pradesh, Bihar, West Bengal, Sikkim, Assam, and Arunachal Pradesh, totaling an area of 531,000 km^2, which witnessed widespread and heavy flooding during the monsoon season of 2017 [31]. Both study regions are having contrasting land use/land cover, with the state of Kerala characterized by uneven, hilly terrain, while parts of North India are among the flattest delta regions in the world. This makes it prone to regular flooding, while Kerala has witnessed sporadic periods of intensive flooding. North India is also home to the most fertile basins of Ganga and Brahmaputra, because of the sedimentation following its floods. The proposed model was evaluated for its robustness over those two regions featuring contrasting topographic and flooding behaviors. Sentinel-1 synthetic aperture radar (SAR) ground range detected (GRD) imagery [32] acquired in vertical (VV) polarization during the flood events of Kerala in August 2018 and over parts of North India across August 2017 (Supplementary Table S2) was used to derive the flood inundation extent that was used for validating against the inundation extent obtained from the proposed model.

3. Methodology

The proposed model workflow for the conceptual near-global flood inundation mapping technique is shown in Figure 1. The multi-temporal SNR data available from CYGNSS as point information, although useful in providing information regarding the underlying surface, has considerable spatial gaps in their daily ground footprint, which prevents mapping floods consistently across inland areas. Therefore, in order to obtain spatially contiguous flood inundation maps, the daily SNR point data was accumulated across a temporally moving window of 30 days length, traversing a two-month period before and during the flood event for the given region; i.e., SNR values from the (1 July to 1 August) window are used cumulatively to obtain an SNR spatial map by following the inverse distance weighted (IDW) interpolation technique, before proceeding to the next window iteration (2 July to 2 August), until the (30 July to 30 August) iteration. Extreme rainfall was observed from 1 August in both the case studies; hence the study period was considered for a total of two months, i.e., one month prior to and one month after the start of the rainfall event. The temporally moving window of a duration of roughly 30 days was taken as a baseline for mapping observed flood events of about 2-week duration. Since there existed no meaningful relationship between the semi-variance of SNR data as a function of the search radius, the value at a particular pixel was interpolated from the values of the three nearest neighboring SNR data points using IDW at 30 m spatial resolution to produce continuous surface values. The maximum value among the SNR spatial maps generated for each pixel was recorded and combined with the topographical information in the form of the HAND-SND terrain index to enhance the spatial extent of maximum inundation that can be mapped.

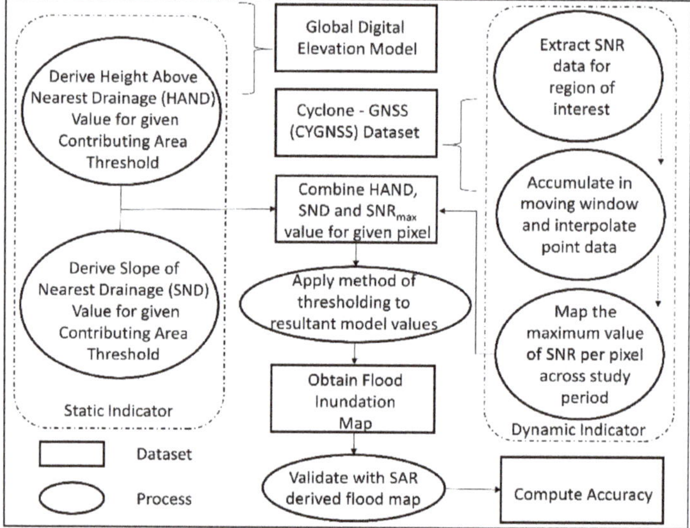

Figure 1. Workflow of the conceptual flood inundation mapping technique undertaken in this study.

Calculation of the HAND value for a given pixel was performed as follows (Figure 2):

1. ASTER DEM was filled, to avoid sinks and to make it hydrologically consistent.
2. The D8 flow direction algorithm (Figure 2c) was applied to the filled DEM.
3. Flow accumulation raster was obtained from flow direction, to find how many pixels drain to neighboring downstream pixels.
4. Stream raster was delineated by choosing the appropriate cumulative area (CA) threshold. In this case, the CA value of 1.000 was chosen.

5. Based on the flow direction raster and the stream raster delineated in the earlier step, the nearest drainage pixel for every pixel in the study area was found.

6. Finally, the HAND value of a pixel at 30 m spatial resolution was calculated as the height of pixel relative to its nearest drainage pixel.

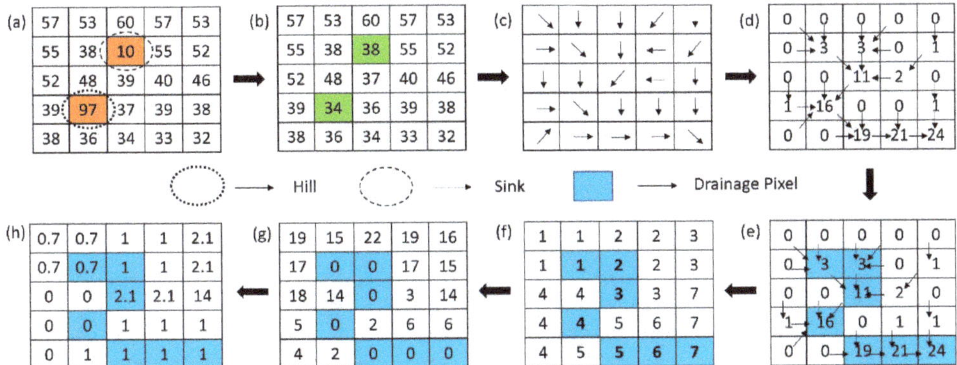

Figure 2. (**a**) Hydrologically inconsistent sample digital elevation model (DEM) with hill and sink indicated; (**b**) with DEM corrections undertaken (green); (**c**) D8 flow direction applied on corrected DEM; (**d**) flow accumulation calculated; (**e**) sample contributing area threshold of 3 applied to obtain drainage pixels; (**f**) nearest drainage pixel mapped for every other pixel in study region; (**g**) height above nearest drainage (HAND) value calculated with respect to nearest drainage pixel; (**h**) slope of nearest drainage (SND) assigned to corresponding floodplain pixels.

The effect of critical parameters that affect the flood extent is determined for the integration of CYGNSS and HAND-SND datasets. A flooded pixel is associated with higher SNR value, and hence the maximum SNR value mapped in the window period, w, can be assumed to have a directly proportional relationship with the likelihood of flooding P(F), where F represents the flood variable, i.e.,

$$P(F) \propto \text{Max (SNR)}|_w \tag{1}$$

Furthermore, based on the HAND flood terrain model, it can be concluded that the higher the HAND value of a pixel, the less likely it is to be flooded than a pixel with a relatively lower HAND value, and hence,

$$P(F) \propto \frac{1}{\text{HAND}} \tag{2}$$

Similarly, once we have the nearest drainage map derived from step 5 mentioned above, the slope of the nearest drainage (SND) is then assigned to its corresponding floodplain pixel (Figure 2h). This step was performed because drainage pixels with the same HAND value and nearest channel pixels with higher slope values are less likely to inundate than those drainage pixels with lower slope values. Also, Manning's Equation shows that a given discharge amount (Q) in the channel is proportional to hydraulic radius (R) and channel slope (S), where $Q \propto R^{5/3}S^{1/2}$. Furthermore, for wide rectangular channels, ($B \gg H$) R simplifies to $BH^{5/3}$, where B is the channel width and H is the discharge height. If the value of B/η is assumed constant (where η denotes the roughness coefficient of the riverbed), it can be concluded that for a given discharge, the water level in the stream is inversely related to the slope of channel pixel given as $H \propto S^{-3/10}$, and therefore,

$$P(F) \propto \frac{1}{\text{SND}^{0.3}} \tag{3}$$

In order to characterize the static potential of flooding, the following points are required: (a) the height of a given pixel relative to the nearest drainage pixel height; and (b) the height of the water level in the channel, which is inversely related to the slope of the channel pixel, as shown above. This characterization is in contrast to [24], which considers the slope of a floodplain pixel relative to the nearest drainage pixel as an extension of HAND. Thus, combining (1), (2), and (3), the dynamic (CYGNSS) and static (HAND and SND) dataset available for the study region under consideration was formalized as follows:

$$P(F) \propto \frac{MAX(SNR)|_w}{\left(HAND \times SND^{0.3}\right)} \tag{4}$$

The product of HAND and SND can result in null values in the case of pixels located within channels, leading to undefined model output. We consider that $1 + (HAND \times SND^{0.3}) \cong (HAND \times SND^{0.3})$, and thus the model variable F from (4) was further simplified without considering any additional parameters in the following form:

$$F = \frac{MAX(SNR)|_w}{1 + \left(HAND \times SND^{0.3}\right)} \tag{5}$$

In addition, the proposed model output raster was subjected to different model output threshold values to obtain the respective maximum extent of inundation across the flood duration at 30 m resolution. The model was compared against co-located Sentinel-1 GRD VV-polarized images. Those images were pre-processed by applying standard steps, including (a) applying the orbit file; (b) sub-setting of the images; (c) thermal noise removal; (d) border noise removal; (e) radiometric calibration; (f) speckle filtering using a Lee filter of 7×7 window size; and (g) range-Doppler terrain correction, followed by image binarization using Otsu's method [33]. The model sensitivity analysis was performed by introducing parameter β to (5), yielding

$$F = \frac{MAX(SNR)|_w}{1 + \left(HAND \times SND^{0.3}\right)^{\beta}} \tag{6}$$

and analyzing the flood accuracy (FA) and critical success index (CSI) parameters for different threshold values. These parameters are given by

$$FA = \frac{h}{h + m} \tag{7}$$

$$CSI = \frac{h}{h + m + z} \tag{8}$$

where h represents the number of pixels which have an agreement (hit) between model and reference pixels, m represents the number of pixels that the model has failed (miss) to identify as inundated when compared with the reference inundation map, and z represents the number of pixels that the model has overestimated as inundated when compared to the reference.

4. Results and Discussion

After processing the ASTER DEM, the HAND flood terrain model for Kerala was obtained, as shown in Figure 3a. The figure shows that Kerala is enveloped by the mountainous Western Ghats to the East, which is associated with high HAND and SND values (Figure 3b), and the relatively flat flood plains along the east coast on the left. The CYGNSS Level-1 BRCS SNR daily data was extracted over the southern Indian state of Kerala for the August 2018 flood event, accumulated across the 30-day window period before and after the flood event and spatially interpolated, yielding the maximum SNR values, as shown in Figure 3c. The coastal areas are clearly associated with higher SNR values than the mountainous Western Ghats. These values are indicative of surface water present in the pixel and coincide with ground observations of flood points reported at the time [34]. However, the SNR data alone cannot accurately map the inundation at the river channel scale, because of the mixed pixel response from the water surface along the coastlines. It can only provide overall information on which places could have been inundated across a large spatial extent of the study region. Hence, the proposed merged model output from CYGNSS, HAND, and SND data yields possible inundation, as shown in Figure 3d. The model output indicates areas that have higher values in the low-lying parts of the central and southern coastal areas, which witnessed extensive flooding [34]. In the case of the August 2017 flood event across North India and its neighboring regions, most of the terrain, except the mountainous Arunachal Pradesh, is flat, and associated with low HAND (Figure 4a) and SND (Figure 4b) values across the floodplains of the Ganga and Brahmaputra rivers, making it prone to extensive flooding. The processed CYGNSS data indicates large areas with significant SNR values (<15) across the study region, as shown in Figure 4c. The combined output also shows a significant increase in model value across large portions of the floodplain areas, given in Figure 4d. It must be noted that HAND, and SND as its extension, were chosen based on a better conceptual representation of hydrologic similarity of floodplain regions than other topographic indices [35].

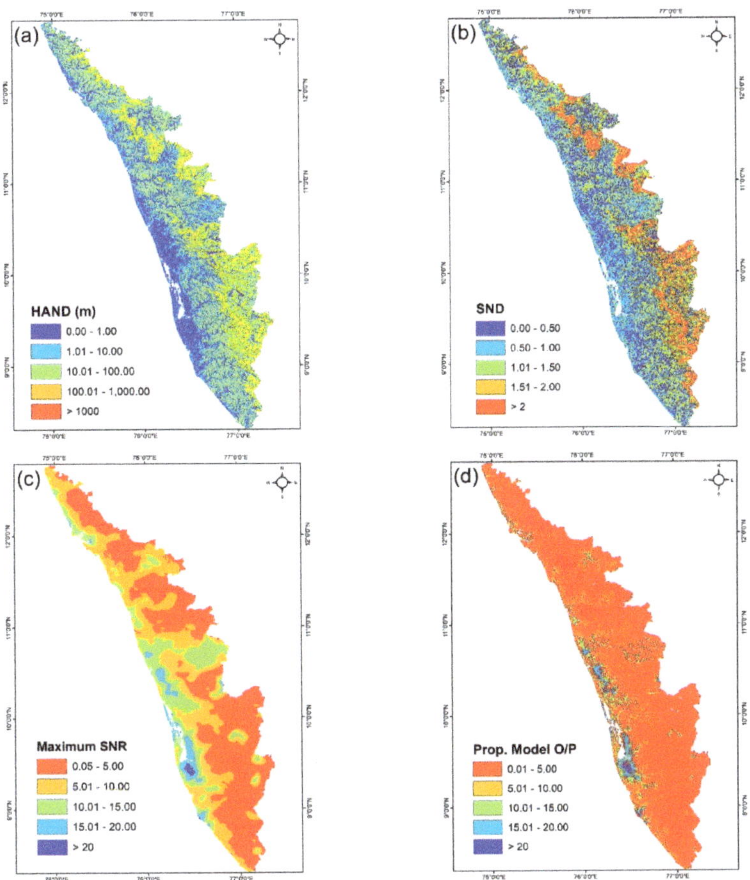

Figure 3. Case I (**a**) HAND flood terrain model for Kerala; (**b**) SND obtained for Kerala; (**c**) maximum SNR obtained from the Cyclone Global Navigation Satellite System (CYGNSS) for August 2018; (**d**) proposed model output for Kerala during floods of August 2018.

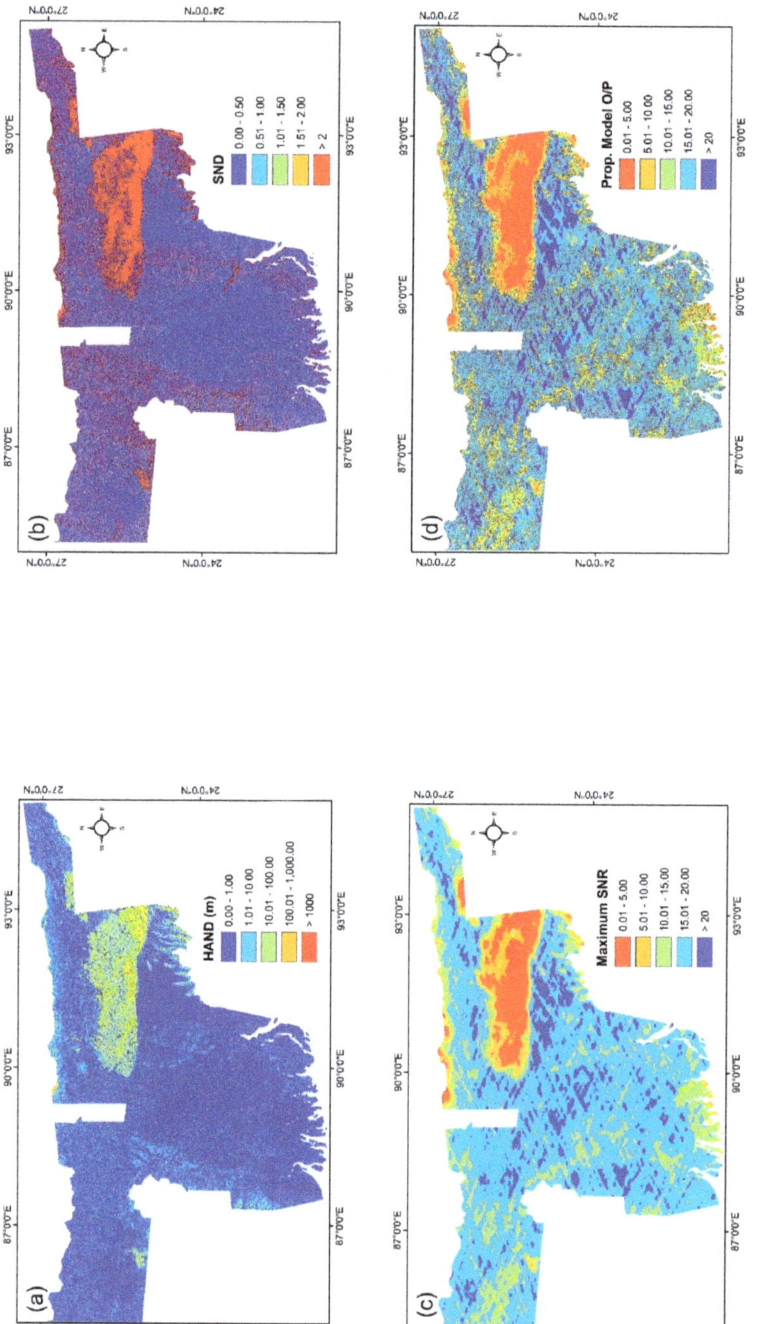

Figure 4. Case II (**a**) HAND flood terrain model for North India; (**b**) SND obtained for North India; (**c**) maximum SNR obtained from CYGNSS for August 2017; (**d**) model output for North India during floods of August 2017.

In the case of the Kerala flood event, Sentinel-1A could capture the flood inundation only on August 21, 2018, and hence imagery available (three scenes) over parts of Kerala state was used for deriving inundation extent by thresholding the pre-processed SAR image using Otsu's method of binarization. The inundation extent was derived at 30 m spatial resolution (shown in Figure 5a) to make it comparable with the proposed model result. The reference inundation map exhibits a similar inundation pattern, as shown by the model output inundation extent.

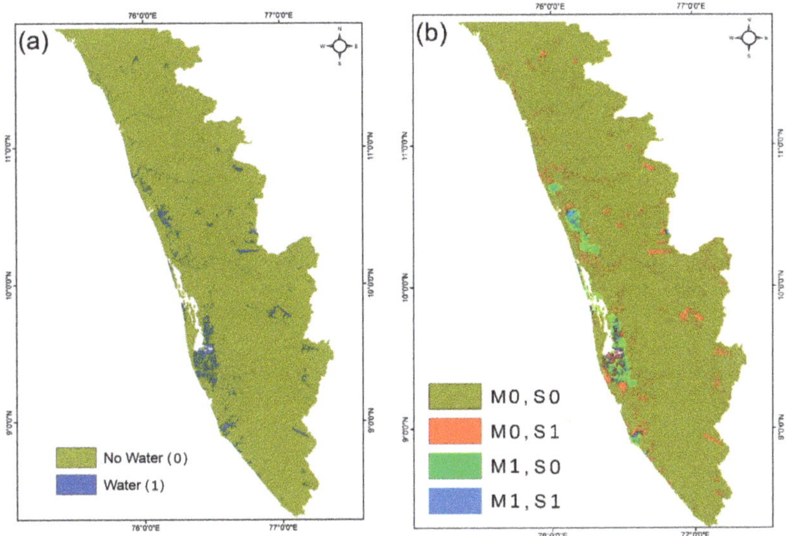

Figure 5. (**a**) Flood inundation derived from Sentinel-1 synthetic aperture radar (SAR) imagery; (**b**) model (M) output for threshold 15 and Sentinel (S) flood inundation maps (water—1, and no water—0) overlaid on each other over Kerala on 23 August 2018.

The model output was then subjected to different threshold values, to compare how the extent of inundation matched with the validation inundation map obtained from SAR imagery. Figure 5b shows the maximum extent of inundation obtained after considering a model output threshold value of 15 overlaid with the validation inundation map with the percentage overlap of flooded and non-flooded areas, as well as the over- and underestimation by the proposed model indicated. Figure 6 shows the variation of the CSI calculated (from (7)) between both the inundation maps with the threshold value.

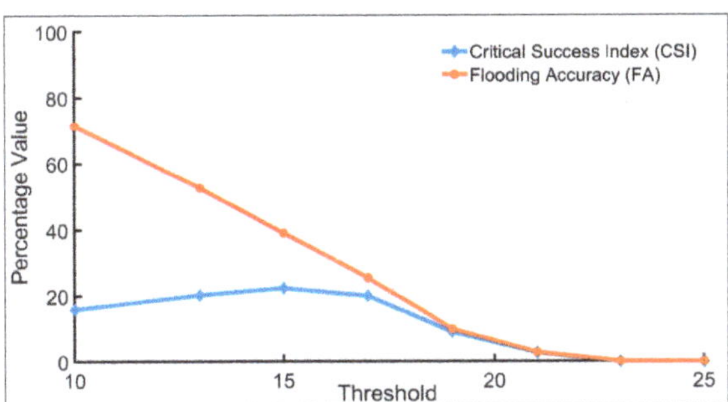

Figure 6. Mapping critical success index (CSI) and flood accuracy (FA) of proposed model output, compared with the Sentinel-derived flood inundation map for the Kerala flood event of August 2018.

   The model results were found to be most sensitive around values of 15–19, after which FA and CSI decreased considerably. For lower values of model threshold, high values of FA were observed, but the corresponding value of CSI indicates the presence of model overestimation. In the case of the North India flood event of August 2017, several Sentinel-1 SAR scenes were collected across different days (24–29 August) of the flood period, subjected to thresholding, and mosaicked to obtain a rough extent of inundation across such a large study region (Figure 7a).

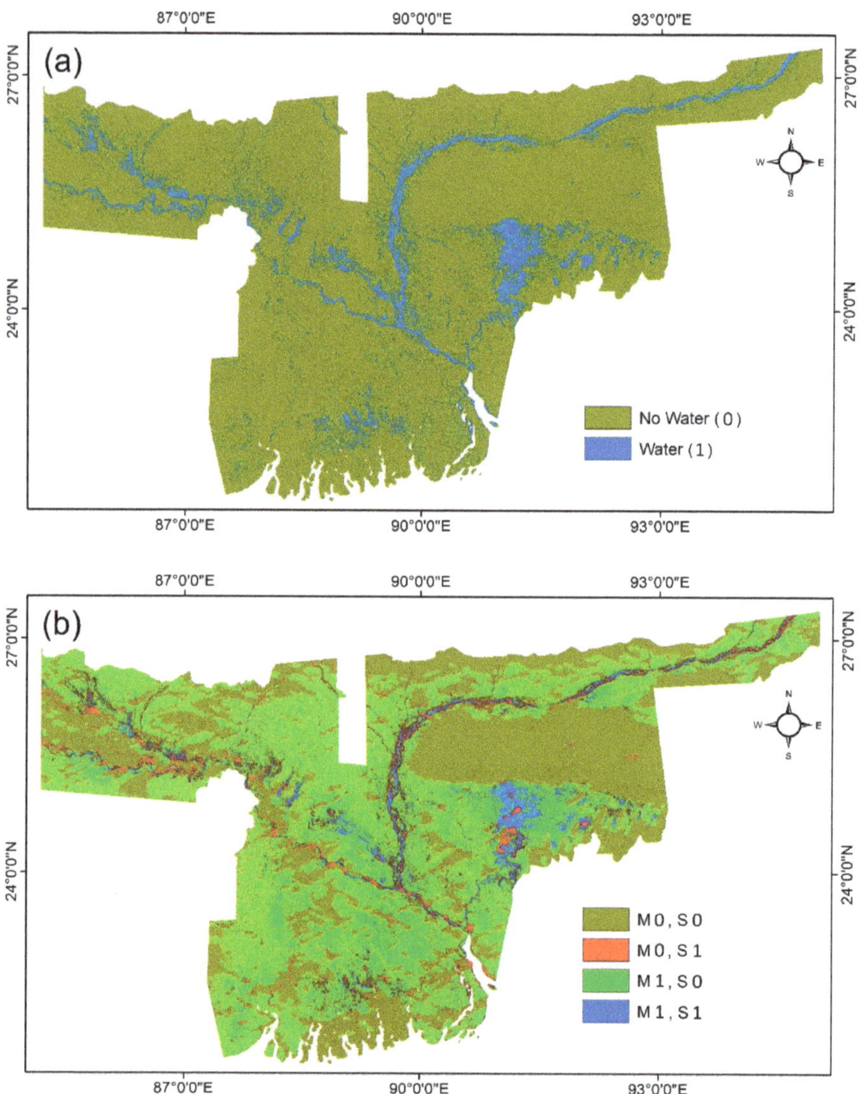

Figure 7. (**a**) Flood inundation derived from Sentinel-1 SAR imagery; and (**b**) model (M) output for threshold 15 and Sentinel (S) flood inundation maps (water—1, and no water—0) overlaid on each other over North India for August 2017.

Similar to the case of the flood event in Kerala, the model output was subjected to thresholding for different values. The overlapped inundation map obtained for a sample threshold value of 15 with the reference inundation map from Sentinel-1 SAR shows that there exists a large amount of overestimation error by the model (Figure 7b). This overestimation by the model was found to be mainly caused by:

1. Coarse spatial resolution of the CYGNSS SNR points, which are spaced widely apart, whose signal strength also depends spatially on the first Fresnel zone of the coherently reflecting surface.
2. Poor vertical accuracy and coarse spatial resolution of the ASTER DEM in resolving the floodplain areas, which caused the HAND flood terrain model being unable to distinguish between the river channel and the delta regions.
3. Sentinel-1 SAR imagery suffering from larger temporal latency and therefore failing to capture the maximum inundation, which will be complemented by the daily available CYGNSS dataset.
4. Presence of many irrigated paddy fields in the fertile area of the Ganga and Brahmaputra delta region, as well as the presence of dense vegetation in the case of Kerala, can result in the proposed model identifying those high-moisture regions as being inundated due to flood event.

Also, Figure 8 underlines that the model result was sensitive around 17–21, much as the earlier CSI plot obtained for the Kerala flood event. At higher threshold model values (>20), substantial underestimation errors in inundation mapping exist, which shows that the proposed model was highly sensitive around model values of 17–19 when considering both cases.

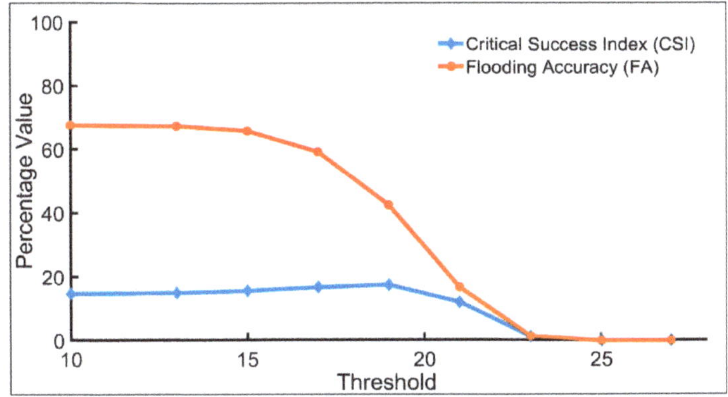

Figure 8. Mapping CSI and FA of proposed model output compared with Sentinel-derived flood inundation map for Kerala flood event of August 2018.

However, the model threshold value cannot definitively be fixed for a given flood event to obtain the flooding extent, because of the varied land-use/land-cover usage (including interference due to non-flooding attributes in the case of inundated paddy fields dominating the North India study region) and flooding behaviors associated with the study regions. Furthermore, the model structure was modified by including an exponential parameter, β, to the static (HAND and SND) part of the term which was varied from 0.1 to 10, resulting in similar CSI values for a given model threshold value for both the cases. This insensitivity to model structure variation is due to the over-/underestimation of the inundation extent as obtained by the HAND/SND threshold, as this is more sensitive to the channel initiation value (contributing area) [27] than assigning more/less weight to the HAND/SND values in itself.

This model was applied to flood events arising due to extreme rainfall events, and specifically, the SND component is applicable only in case of floods arising with respect to channel pixel slope values. The model cannot be applied in its current form to other forms of flooding, including pluvial or coastal flooding, which can be possible only by considering other relevant factors.

As such, there are several limitations to the validation of the proposed model. These include the choice of Sentinel-1A SAR data as the only other reliable source that can provide the spatial extent of a flood inundation. However, the Sentinel satellite overpass over the study regions was mostly during the receding phase of both the flood events; hence the flood peak extent could have probably been missed. Furthermore, the FA and CSI values are also affected by the amount of observational overlap between CYGNSS and Sentinel-1 datasets, as they do not have the same coverage extent and daily repeat rates. The presence of vegetation in both study regions greatly affects the accuracy in deriving the flood extent from SAR images, while rain and wind also affect the omission of inundation retrievals from single SAR imagery. In addition, the presence of smooth surfaces in the reference scene, including roads and rooftops, cause commission errors [36].

In the future, by adopting a probabilistic framework [37], through which it will be possible to generate a probabilistic flood inundation map by exploiting the histogram of the combined model output and separating the non-water pixels, which will result in better mapping capabilities of the inundation uncertainty for a region. With the availability of additional improved GNSS-R SNR datasets, from other navigation satellite constellations, there are active navigation signals that are already available, which are sensitive to the presence of inland surface water that can have downstream applications [38], including inundation mapping in the case of flood events. Those include satellites operated by different space agencies around the world, such as the Global Navigation Satellite System (GLONASS) of Russia), European Space Agency's GALILEO, and the Navigation with Indian Constellation (NavIC), launched by the Indian Space Research Organisation (ISRO).

5. Conclusions

GNSS-R Signals of Opportunity (SOOP) are sensitive to the presence of surface water. The study proceeded, considering these signals to be highly temporal (sub-daily); however, because of the poor spatial resolution (order of kilometers) available for a given region from CYGNSS [39], SNR data points were accumulated and interpolated spatially, employing a moving window approach before and during a flood event. The maximum observed SNR value during the flooding period provides an opportunity to estimate the maximum flood inundation extent while integrating those with topographical information. Thus, this paper explored the potential of combining the coarse resolution, but temporally dynamic, GNSS–R signals with the static, but spatially continuous, HAND-SND-based flood terrain index for high-resolution global flood inundation mapping. CYGNSS SNR daily data was extracted for two flood events, one over the southern Indian state of Kerala during August 2018, and for an extensive event during August 2017 across Bangladesh and parts of North and Northeast India. HAND and SND values were calculated for the respective study regions and were integrated with CYGNSS SNR data using the proposed approach. The model was evaluated with flood inundation maps derived from Sentinel-1 SAR imagery available during the flood duration. It was found that the model-derived inundation extent was sensitive to values around 17–19 on thresholding with a maximum CSI of 0.22, although there was high flooding accuracy (FA) in mapping 60–80% of flood extent for lower model threshold values. The significant over-/underestimation is mainly attributed to the spatially-coarse pseudo-randomly distributed SNR data, and to the limited availability of the Sentinel-1 SAR imagery. This method thus gives an advantage over existing SAR satellite-derived flood inundation methods, because (a) SAR-derived flood maps provide only an incomplete snapshot, not covering the entire extent of flooding, whereas in the case of CYGNSS, a near-global daily coverage is available; and (b) because of low revisit frequency in the case of SAR satellites, they are not able to provide information about the spatial dynamics of the flood event. Thus, there exists no single standard dataset which can satisfy all the conditions that are necessary for the temporal evolution of flood inundation extent to be captured with reasonable accuracy. With a larger amount of GNSS–R data available in the future, this approach provides an opportunity that would greatly benefit peak flood extent estimation, and also help in mapping floods at the sub-daily temporal resolution, which can provide useful information when a flood intensifies and recedes.

Supplementary Materials: The following are available online at http://www.mdpi.com/2072-4292/12/18/3026/s1, Table S1: Details of the CYGNSS SNR dataset used along with observed flood extent considered in this study; Table S2: Details of the Sentinel—1 SAR imagery used in this study.

Author Contributions: Conceptualization: S.L.K.U. and B.B.; methodology: S.L.K.U., B.B., and C.R.; validation: S.L.K.U. and C.R.; formal analysis: S.L.K.U., B.B., and C.R.; resources: S.L.K.U.; data curation: S.L.K.U.; writing—original draft preparation: S.L.K.U.; writing—review and editing: B.B. and C.R.; visualization: S.L.K.U., B.B., and C.R.; supervision: B.B. and C.R. All authors have read and agreed to the published version of the manuscript.

Funding: This research received funding from CSIRO Data61, Australia.

Acknowledgments: The first author acknowledges the support from IITB-Monash Research Academy through the joint Ph.D. program between the Indian Institute of Technology Bombay, India, and Monash University, Australia.

Conflicts of Interest: The authors declare no conflict of interest.

References

1. UNISDR Global Assessment Report on Disaster Risk Reduction 2015 Making Development Sustainable: The Future of Disaster Risk Management. Available online: https://www.preventionweb.net/english/hyogo/gar/2015/en/gar-pdf/GAR2015_EN.pdf (accessed on 1 December 2019).
2. UNISDR Sendai Framework for Disaster Risk Reduction 2015–2030. Available online: https://www.preventionweb.net/files/43291_sendaiframeworkfordrren.pdf (accessed on 1 December 2019).
3. UNISDR Global Assessment Report on Disaster Risk Reduction 2011 Revealing Risk, Redefining Development. Available online: https://www.preventionweb.net/english/hyogo/gar/2011/en/bgdocs/GAR-2011/GAR2011_Report_Chapter1.pdf (accessed on 1 December 2019).
4. Yan, K.; Di Baldassarre, G.; Solomatine, D.P.; Schumann, G.J.P. A review of low-cost space-borne data for flood modelling: Topography, flood extent and water level. *Hydrol. Process.* **2015**, *29*, 3368–3387. [CrossRef]
5. Fayne, J.; Bolten, J.; Lakshmi, V.; Ahamed, A. Optical and Physical Methods for Mapping Flooding with Satellite Imagery. In *Remote Sensing of Hydrological Extremes*; Lakshmi, V., Ed.; Springer: Berlin/Heidelberg, Germany, 2017; pp. 83–103. ISBN 978-3-319-43743-9.
6. Schumann, G.; Bates, P.D.; Apel, H.; Aronica, G.T. Global Flood Hazard Mapping, Modeling, and Forecasting: Challenges and Perspectives. In *Global Flood Hazard: Applications in Modeling, Mapping, and Forecasting*; Schumann, G.J.-P., Bates, P.D., Apel, H., Aronica, G.T., Eds.; John Wiley & Sons, Inc.: Hoboken, NJ, USA, 2018; pp. 239–244.
7. Schumann, G.J.P.; Brakenridge, G.R.; Kettner, A.J.; Kashif, R.; Niebuhr, E. Assisting flood disaster response with earth observation data and products: A critical assessment. *Remote Sens.* **2018**, *10*, 1230. [CrossRef]
8. Notti, D.; Giordan, D.; Caló, F.; Pepe, A.; Zucca, F.; Galve, J.P. Potential and limitations of open satellite data for flood mapping. *Remote Sens.* **2018**, *10*, 1673. [CrossRef]
9. Schumann, G.J.P.; Moller, D.K. Microwave remote sensing of flood inundation. *Phys. Chem. Earth* **2015**, *83–84*, 84–95. [CrossRef]
10. Dasgupta, A.; Grimaldi, S.; Ramsankaran, R.A.A.J.; Pauwels, V.R.N.; Walker, J.P. Towards operational SAR-based flood mapping using neuro-fuzzy texture-based approaches. *Remote Sens. Environ.* **2018**, *215*, 313–329. [CrossRef]
11. Shen, X.; Wang, D.; Mao, K.; Anagnostou, E.; Hong, Y. Inundation extent mapping by synthetic aperture radar: A review. *Remote Sens.* **2019**, *11*, 879. [CrossRef]
12. Bremner, L. Planning the 2015 Chennai floods. *Environ. Plan. E Nat. Sp.* **2019**. [CrossRef]
13. Hunt, K.M.R.; Menon, A. The 2018 Kerala floods: A climate change perspective. *Clim. Dyn.* **2020**, *54*, 2433–2446. [CrossRef]
14. World Weather Attribution Chennai Floods. 2015. Available online: https://www.worldweatherattribution.org/chennai-floods-december-2015/ (accessed on 1 January 2019).
15. News Minute The Scale of Kerala's Largest Flood in a Century. Available online: https://www.thenewsminute.com/article/8-charts-scale-keralas-largest-flood-century-86948 (accessed on 1 January 2019).

16. Mishra, V.; Shah, H.L. Hydroclimatological perspective of the Kerala flood of 2018. *J. Geol. Soc. India* **2018**, *92*, 645–650. [CrossRef]

17. The Hindu Business Line Northeast Monsoon Claimed 470 Lives in Tamil Nadu: Jayalalithaa. Available online: https://www.thehindubusinessline.com/news/national/northeast-monsoon-claimed-470-lives-in-tamil-nadu-jayalalithaa/article8064661.ece (accessed on 14 July 2020).

18. Arabindoo, P. Unprecedented natures?: An anatomy of the Chennai floods. *City* **2016**, *20*, 800–821. [CrossRef]

19. Ruf, C.S.; Chew, C.; Lang, T.; Morris, M.G.; Nave, K.; Ridley, A.; Balasubramaniam, R. A new paradigm in earth environmental monitoring with the CYGNSS small satellite constellation. *Sci. Rep.* **2018**, *8*, 1–13. [CrossRef] [PubMed]

20. Chew, C.; Reager, J.T.; Small, E. CYGNSS data map flood inundation during the 2017 Atlantic hurricane season. *Sci. Rep.* **2018**, *8*, 1–8. [CrossRef] [PubMed]

21. Chew, C.; Small, E. Estimating inundation extent using CYGNSS data: A conceptual modeling study. *Remote Sens. Environ.* **2020**, *246*, 111869. [CrossRef]

22. Nobre, A.D.; Cuartas, L.A.; Hodnett, M.; Rennó, C.D.; Rodrigues, G.; Silveira, A.; Waterloo, M.; Saleska, S. Height above the nearest drainage—A hydrologically relevant new terrain model. *J. Hydrol.* **2011**, *404*, 13–29. [CrossRef]

23. Nobre, A.D.; Cuartas, L.A.; Momo, M.R.; Severo, D.L.; Pinheiro, A.; Nobre, C.A. HAND contour: A new proxy predictor of inundation extent. *Hydrol. Process.* **2016**, *30*, 320–333. [CrossRef]

24. Vasconcelos, V.V.; De Engenharia, C.; Follador, M.; Amoni, M.; Alves, S.; Horizonte, B. Floodable cross-sectional area and slope to the nearest drainage as extensions of the hand model: Mapping flood susceptibility in the region of Lucas Do Rio Verde, Mato Grosso State, Brazil. *Rev. GeoAmazonia* **2017**, *5*, 3–25.

25. Rennó, C.D.; Nobre, A.D.; Cuartas, L.A.; Soares, J.V.; Hodnett, M.G.; Tomasella, J.; Waterloo, M.J. HAND, a new terrain descriptor using SRTM-DEM: Mapping terra-firme rainforest environments in Amazonia. *Remote Sens. Environ.* **2008**, *112*, 3469–3481. [CrossRef]

26. Jafarzadegan, K.; Merwade, V. A DEM-based approach for large-scale floodplain mapping in ungauged watersheds. *J. Hydrol.* **2017**, *550*, 650–662. [CrossRef]

27. Speckhann, G.A.; Borges Chaffe, P.L.; Fabris Goerl, R.; de Abreu, J.J.; Altamirano Flores, J.A. Flood hazard mapping in Southern Brazil: A combination of flow frequency analysis and the HAND model. *Hydrol. Sci. J.* **2018**, *63*, 87–100. [CrossRef]

28. Jafarzadegan, K.; Merwade, V. Probabilistic floodplain mapping using HAND-based statistical approach. *Geomorphology* **2019**, *324*, 48–61. [CrossRef]

29. CYGNSS Level 1 Science Data Record Version 2.1. 2017. Available online: https://podaac.jpl.nasa.gov/dataset/CYGNSS_L1_V2.1 (accessed on 7 December 2019).

30. ASTER Digital Elevation Model V003. Available online: https://dx.doi.org/10.5067/ASTER/AST14DEM.003 (accessed on 30 January 2019).

31. Siddique, H. South Asia Floods Kill 1200 and Shut 1.8 Million Children Out of School. Available online: https://www.theguardian.com/world/2017/aug/30/mumbai-paralysed-by-floods-as-india-and-region-hit-by-worst-monsoon-rains-in-years (accessed on 30 September 2018).

32. ESA Sentinel-1 User Handbook. Available online: https://sentinel.esa.int/documents/247904/349449/S1_SP-1322_1.pdf (accessed on 1 January 2019).

33. Otsu, N. A threshold selection method from gray-level histograms. *IEEE Trans. Syst. Man. Cybern.* **1979**, *9*, 62–66. [CrossRef]

34. Dynamical Downscalingof Regional Climate: Simulation of Extreme RainfallEvents and Their Impacts over the State of Kerala in the Near-Future. Available online: https://sdma.kerala.gov.in/wp-content/uploads/2019/08/KSDMA_Project_report_2017-2018.pdf (accessed on 14 December 2019).

35. Gao, H.; Birkel, C.; Hrachowitz, M.; Tetzlaff, D.; Soulsby, C.; Savenije, H.H.G. A simple topography-driven and calibration-free runoff generation module. *Hydrol. Earth Syst. Sci. Discuss.* **2018**, 1–42. [CrossRef]

36. Grimaldi, S.; Xu, J.; Li, Y.; Pauwels, V.R.N.; Walker, J.P. Flood mapping under vegetation using single SAR acquisitions. *Remote Sens. Environ.* **2020**, 111582. [CrossRef]

37. Giustarini, L.; Hostache, R.; Kavetski, D.; Chini, M.; Corato, G.; Schlaffer, S.; Matgen, P. Probabilistic flood mapping using synthetic aperture radar data. *IEEE Trans. Geosci. Remote Sens.* **2016**, *54*, 6958–6969. [CrossRef]

38. Gerlein-Safdi, C.; Ruf, C.S. A CYGNSS-based algorithm for the detection of inland waterbodies. *Geophys. Res. Lett.* **2019**, *46*, 12065–12072. [CrossRef]
39. Bussy-Virat, C.D.; Ruf, C.S.; Ridley, A.J. Relationship between temporal and spatial resolution for a constellation of GNSS-R satellites. *IEEE J. Sel. Top. Appl. Earth Obs. Remote Sens.* **2019**, *12*, 16–25. [CrossRef]

Article

Geolocation, Calibration and Surface Resolution of CYGNSS GNSS-R Land Observations

Scott Gleason [1,*], Andrew O'Brien [2], Anthony Russel [3], Mohammad M. Al-Khaldi [2] and Joel T. Johnson [2]

1. Constellation Observing System for Meteorology, Ionosphere and Climate (COSMIC), University Corporation for Atmospheric Research, Boulder, CO 80301, USA
2. Department of Electrical and Computer Engineering, The Ohio State University, Columbus, OH 43210, USA; obrien.200@osu.edu (A.O.); al-khaldi.2@buckeyemail.osu.edu (M.M.A.-K.); johnson.1374@osu.edu (J.T.J.)
3. Department of Climate and Space Sciences and Engineering, The University of Michigan, Ann Arbor, MI 48109, USA; russelan@umich.edu
*   Correspondence: gleason@ucar.edu

Received: 26 March 2020; Accepted: 18 April 2020; Published: 22 April 2020

Abstract: This paper presents the processing algorithms for geolocating and calibration of the Cyclone Global Navigation Satellite System (CYGNSS) level 1 land data products, as well as analysis of the spatial resolution of Global Navigation Satellite System Reflectometry (GNSS-R) coherent reflections. Accurate and robust geolocation and calibration of GNSS-R land observations are necessary first steps that enable subsequent geophysical parameter retrievals. The geolocation algorithm starts with an initial specular point location on the Earth's surface, predicted by modeling the Earth as a smooth ellipsoid (the WGS84 representation) and using the known transmitting and receiving satellite locations. Information on terrain topography is then compiled from the Shuttle Radar Topography Mission (SRTM) generated Digital Elevation Map (DEM) to generate a grid of local surface points surrounding the initial specular point location. The delay and Doppler values for each point in the local grid are computed with respect to the empirically observed location of the Delay Doppler Map (DDM) signal peak. This is combined with local incident and reflection angles across the surface using SRTM estimated terrain heights. The final geolocation confidence is estimated by assessing the agreement of the three geolocation criteria at the estimated surface specular point on the local grid, including: the delay and Doppler values are in agreement with the CYGNSS observed signal peak and the incident and reflection angles are suitable for specular reflection. The resulting geolocation algorithm is first demonstrated using an example GNSS-R reflection track that passes over a variety of terrain conditions. It is then analyzed using a larger set of CYGNSS data to obtain an assessment of geolocation confidence over a wide range of land surface conditions. Following, an algorithm for calibrating land reflected signals is presented that considers the possibility of both coherent and incoherent scattering from land surfaces. Methods for computing both the bistatic radar cross section (BRCS, for incoherent returns) and the surface reflectivity (for coherent returns) are presented. a flag for classifying returns as coherent or incoherent developed in a related paper is recommended for use in selecting whether the BRCS or reflectivity should be used in further analyses for a specific DDM. Finally, a study of the achievable surface feature detection resolution when coherent reflections occur is performed by examining a series of CYGNSS coherent reflections across an example river. Ancillary information on river widths is compared to the observed CYGNSS coherent observations to evaluate the achievable surface feature detection resolution as a function of the DDM non-coherent integration interval.

Keywords: land processes; calibration; GNSS; GPS; reflectometry; bistatic radar; CYGNSS

---

1. Introduction

Global Navigation Satellite System (GNSS) Reflectometry (GNSS-R) performs Earth remote sensing by measuring reflections off the Earth's surface by signals transmitted from various GNSS constellations, including the U.S. Global Positioning System (GPS), the European Galileo constellation, and others. All of the satellites within a GNSS constellation typically transmit within the same frequency bands, but use spread spectrum techniques to distinguish the transmissions of different space vehicles [1]. This allows an appropriately designed GNSS receiver to track multiple GNSS signals simultaneously, all of which can potentially be used for surface remote sensing. a history and overview of GNSS-R and its applications can be found in [2].

The potential of GNSS-R to perform land Earth observations was first demonstrated from a space platform in 2007 using a small amount of raw sampled data from the UK-DMC satellite [3]. Previously, the sensitivity of GPS reflections to surface water and soil moisture was demonstrated using aircraft experiments [4]. The launch of the NASA Cyclone GNSS (CYGNSS) mission in Dec 2016 marked a significant opportunity to further validate and develop GNSS-R land applications. The primary mission of the eight satellite CYGNSS constellation is to measure ocean near surface winds for hurricane research [5]. However, multiple studies have recently demonstrated the use of CYGNSS land observations in surface water monitoring applications, including flood inundation [6] and near surface soil moisture [7,8].

The CYGNSS Level 1 ocean calibration is based on an inversion of the bistatic radar equation, and includes corrections for all instrument related and non-geophysical parameters affecting the received power levels [9,10]. However, GNSS-R signals received from land surfaces require several non-trivial modifications to the existing ocean calibration algorithms to enable land parameter retrievals. These modifications are a result of the fundamental differences in the reflecting surface and the associated surface scattering processes between ocean and land observations.

The ocean surface (to first order) can be approximated by a Gaussian distribution of directional surface wave slopes, which are indirectly related to the near surface wind speed [11]. By contrast, land surfaces exhibit more diverse and irregular distributions of reflecting surfaces, as land surfaces often do not lend themselves to simple surface distribution approximations (although exponential and other land slope approximations have shown some utility [12]). In addition, the typically large RMS heights of the ocean surface relative to the 19 cm electromagnetic wavelength result in an incoherent scattering process (with occasional exceptions in low wind/wave conditions). This allows the CYGNSS Level 1 ocean calibration to focus on incoherent scattering alone. In constrast, CYGNSS land observations exhibit both coherent reflection and incoherent scattering, due to the underlying variability in both topography and surface cover [6]. The presence of inland water bodies typically having very smooth surfaces also increases the frequency of very strong coherent or mixed reflection returns [13]. Previous studies on GNSS-R land retrievals, coherent and incoherent scattering and the impact of topography on GNSS-R land observations can be found in [14–16].

In addition to the fundamental differences in surface scattering mechanisms, the geolocation of the dominant scattering region on the Earth's surface is complicated by large scale variations in land surface topography. The existing CYGNSS ocean geolocation calculation [17] uses the WGS84 ellipsoid to define the Earth's surface with a small correction for the mean sea surface height. The geolocation of land reflections requires additional consideration of local terrain variations, which can significantly shift the dominant scattering region. In some instances (e.g., extreme mountain environments) a land geolocation can become infeasible or ambiguous.

Additionally, of great importance to land remote sensing applications is the achievable surface resolution under different observation conditions. When the land surface is relatively smooth, with respect to the L-band incident wavelength, the received signal at the spacecraft will be coherent. In theory, this will result in a surface reflection footprint of approximately the first Fresnel zone, integrated over the motion of the surface reflection point during instrument processing. In this paper, the actual minimum feature detection resolution for coherent reflections will be investigated using an ancillary data set of river widths.

These fundamental differences between ocean and land GNSS-R observations necessitate more detailed consideration of the geolocation, calibration, and spatial resolution of land observations. Following this introduction, Section 2 presents a land geolocation algorithm that uses three criteria for identifying the dominant scattering location on Earth's surface. Section 3 then describes the calibration of GNSS-R land observations, including the calculation of both diffuse and coherent observation data products, including land σ_0, reflectivity, and coherence estimate. Section 4 presents the results of analysis to estimate the spatial resolution achievable by CYGNSS 1 Hz and 2 Hz coherent land observations, while Section 5 provides general discussion on the algorithms and on-going work and Section 6 provides a summary of the results of the paper.

2. Geolocating GNSS Reflections from Land

2.1. Surface Criteria for Forward Reflection

The overall smoothness and uniformity of the ocean allows for the identification of a single specular reflection point at the minimum path difference between the transmitter, surface and receiver using a WGS84 ellipsoidal representation of the Earth, corrected for the mean sea surface height [17]. This technique often works well for flat land surfaces whose heights are not far from the WGS84 approximation [18].

In these land cases, using a local surface topography model [19] to adjust the estimated WGS84 specular point to the land surface can be adequate for most applications. Generally, for reflections from land with topographic variations, there is often no singular unique specular point as specular reflections may occur from multiple places separated in delay Doppler space. In such conditions, additional criteria are often required to assure that the geolocation estimate is sufficiently accurate for science applications. Empirical criteria are based on matching the CYGNSS-observed delay and Doppler values to values predicted using knowledge of local terrain topography. Additional geometric criteria include the degree to which local incidence and scattering angles are suitable to specular reflection.

For a given CYGNSS land Delay Doppler Map (DDM), these three conditions (matching of Delay and Doppler values to CYGNSS-observed values, and "nearness" to a specular geometry) are evaluated over a local digital elevation map grid centered at the specular reflection point initially estimated using the WGS84 surface model. Points within the grid that satisfy specified delay, Doppler and geometric conditions are then selected as representing the highest probability surface scattering points. For complicated terrain, multiple (even discontinuous) regions may exist for which all the three conditions are met. Additional information on the presence of water bodies within the local surface grid is also relevant to assist in interpreting the likelihood of coherent reflection.

2.2. Local Surface Delay Calculation

CYGNSS-observed delay (τ_{max}) and Doppler (D_{max}) values are selected as those corresponding to the peak power value in the DDM in what follows. These values are then compared to predicted values mapped across a local surface. The expected code phase time delay at any given surface point can be expressed as

$$\tau_s = \tau_{dir} - \delta P \tag{1}$$

where τ_s is the code phase delay at surface point s, τ_{dir} is the code phase delay of the tracked direct signal at the same time epoch and δP is the additional code phase delay between the direct and surface reflected code phase delays. The additional code phase delay can be calculated as

$$\delta P = P_{T2S2R} - P_{T2R} = (P_{S2R} + P_{T2S}) - P_{T2R} \tag{2}$$

where P_{T2R} is the length of the path from the transmitter to the receiver, P_{T2S2R} is the sum of the path from the transmitter to the surface point (P_{T2S}) plus the path from the surface point to the receiver (P_{S2R}). The additional code phase delay can also be expressed as a function of the transmitter, receiver, and surface point locations,

$$\delta P = (|\vec{T} - \vec{S}| + |\vec{S} - \vec{R}|) - |\vec{T} - \vec{R}| \tag{3}$$

where $\vec{R}$ is the receiver position, $\vec{T}$ is the transmitter position and $\vec{S}$ is the surface position in a specified reference frame. The difference between the CYGNSS-observed code phase delay (τ_{max}) and the code phase delay at a given surface location (τ_s) can be estimated over the three dimensional local land surface grid,

$$\delta\tau = \tau_{max} - \tau_s \tag{4}$$

Note that the GPS signal code phase is a continually repeating length of psuedo-random sequence of a finite length [1]. Therefore, τ_{max} and τ_s need to be corrected based on the repeating length of the specific GNSS code sequence. For the GPS L1 C/A code, the code phase repeats every 1024 chips. The approximate length of one GPS L1 C/A code chip is 293.0522561 m.

The value of $\delta\tau$ is calculated across the local three dimensional surface area centered on the initial WGS84 specular point location. In this process, the grid point location is determined using

$$\vec{S} = \vec{S}^{W84}_{lat',lon'} + \frac{\vec{S}^{W84}_{lat',lon'}}{\left| \vec{S}^{W84}_{lat',lon'} \right|} \Delta H \tag{5}$$

where $\vec{S}^{W84}_{lat',lon'}$ is a vector from the Earth center to the WGS84 ellipsoid position at an offset latitude (lat') and longitude (lon') [17], and ΔH is the DEM height above the WGS84 ellipsoid at the offset latitude and longitude. The local surface delay difference map ($\delta\tau$) is produced using Equation (4) for a grid extending 100 km in all compass directions from the center reference in steps of 1 km. Points on the local surface contour map which satisfy the condition $\delta\tau = 0$ represent the iso-range contour (or point) where the CYGNSS-observed code phase delay is in agreement with the reflecting surface code phase delay.

Local Surface Doppler Estimate

A similar surface contour map can be generated using the frequency dimension of GNSS-R land observations. The expected Doppler frequency at any given local surface point ($\vec{S}$) can be calculated as

$$D_s = D_R + D_T + D_{clk} \tag{6}$$

where D_R is the Doppler component due to the receiver velocity, D_T is the Doppler due to the transmitter velocity and D_{clk} is the Doppler bias from the drift of the satellite instrument clock. The individual Doppler terms induced by the receiver and transmitter are calculated as

$$D_R = -1(\vec{R}_v \cdot \hat{D}_{RS})(\frac{f}{c}) \tag{7}$$

$$D_T = -1(\vec{T}_v \cdot \hat{D}_{TS})(\frac{f}{c}) \tag{8}$$

where $\vec{R}_v$ is the receiver velocity, $\vec{T}_v$ is the transmitter velocity, f is the GNSS signal transmit frequency (for GPS L1 = 1.57542 GHz), and c is the speed of light (299,792,458 m/s). Also $\hat{D}_{RS}$ and $\hat{D}_{TS}$ are unit vectors directed between the receiver and surface point and transmitter and surface point, respectively:

$$\hat{D}_{RS} = \frac{\vec{R} - \vec{S}}{\left|\vec{R} - \vec{S}\right|} \tag{9}$$

$$\hat{D}_{TS} = \frac{\vec{T} - \vec{S}}{\left|\vec{T} - \vec{S}\right|} \tag{10}$$

where $\vec{R}$ is the receiver position, $\vec{T}$ is the transmitter position, and $\vec{S}$ is a point on the local three dimensional surface grid. The predicted Doppler frequencies at every point on the local surface grid are then differenced with the Doppler frequency of the received GNSS-R observations (D_{max}) to obtain

$$\delta D = D_{max} - D_s \tag{11}$$

The Doppler difference contour map is then calculated over the local three dimensional surface, and the iso-Doppler contour which satisfies the condition $\delta D = 0$ represents the physical surface region in agreement with the CYGNSS-observed Doppler.

2.3. Local Surface Snell Reflection Criteria

The final check on the land surface geo-location is purely geometrical and is not linked to the signal peak in the received GNSS-R DDM. This check is often the most robust as it is sometimes the case that the CYGNSS-observed delay and Doppler values can be difficult to estimate with certainty due to low receiver antenna gain or adverse Earth surface properties. a reference frame to facilitate the calculation of the incidence and scattering angles from a local surface was defined and is illustrated in Figure 1. With respect to this figure, the following points are defined: S is the center pixel in a 3 by 3 group on the local surface grid, and S1, S2, S3 and S4 are points at pixels directly to the north, south, east and west, respectively, spaced typically at 1 km. The points T and R are the GNSS transmitter and instrument receiver satellite, respectively, while P_1 and P_2 are the line-of-sight paths from the transmitter to a given surface point and from the surface point to the receiver respectively. The θ_i, ϕ_i, θ_r, and ϕ_r angles represent the incidence and scattering polar and azimuthal angles, respectively, and are the variables to be calculated at every local surface point ($\vec{S}$).

In the example in Figure 1, all 9 points are shown above the WGS84 reference ellipsoid (shown in green). From the 5 local reference points, incidence and scattering angles can be calculated based on the incoming line-of-sight rays to the center pixel from the transmitter and receiver. We define local east, north, up (ENU) unit vectors from the center reference pixel, S, as follows:

$$\hat{E}_r = \frac{(\vec{S3} - \vec{S4})}{\left|\vec{S3} - \vec{S4}\right|}, \hat{N}_r = \frac{(\vec{S1} - \vec{S2})}{\left|\vec{S1} - \vec{S2}\right|}, \hat{U}_r = \hat{E} \times \hat{N} \tag{12}$$

Subsequently, the projected line-of-sight ray paths from the transmitter to the center surface pixel from the center pixel to the receiver (P_1 and P_2, respectively, in Figure 1) are projected to the local surface ENU surface reference frame directions:

$$\begin{aligned}
P'_{1,E} &= (\vec{S} - \vec{T}) \cdot \hat{E}_r \\
P'_{1,N} &= (\vec{S} - \vec{T}) \cdot \hat{N}_r \\
P'_{1,U} &= (\vec{S} - \vec{T}) \cdot \hat{U}_r
\end{aligned} \tag{13}$$

$$\begin{aligned}
P'_{2,E} &= (\vec{R} - \vec{S}) \cdot \hat{E}_r \\
P'_{2,N} &= (\vec{R} - \vec{S}) \cdot \hat{N}_r \\
P'_{2,U} &= (\vec{R} - \vec{S}) \cdot \hat{U}_r
\end{aligned} \tag{14}$$

From these unit vectors it is possible to calculate the polar and azimuth angles shown in Figure 1 as

$$\theta_i = atan(P'_{1,U}, \sqrt{(P'_{1,E})^2 + (P'_{1,N})^2}) \tag{15}$$

$$\theta_r = atan(P'_{2,U}, \sqrt{(P'_{2,E})^2 + (P'_{2,N})^2}) \tag{16}$$

$$\phi_i = atan(P'_{1,N}, P'_{1,E}) \tag{17}$$

$$\phi_r = atan(P'_{2,N}, P'_{2,E}) \tag{18}$$

We are interested in the difference between the calculated angles and an ideal forward specular reflection from the given surface point. The resulting angle differences from a specular reflecting geometry are then

$$\delta\theta = \theta_i - \theta_r \tag{19}$$

$$\delta\phi = atan(\sin(\phi_r - (\phi_i + \pi)), \cos(\phi_r - (\phi_i + \pi))) \tag{20}$$

We define the total angular error metric as the sum of the absolute errors in the azimuth and elevation angles relative to that of an ideal specular reflection,

$$\delta\Psi = |\delta\theta| + |\delta\phi| \tag{21}$$

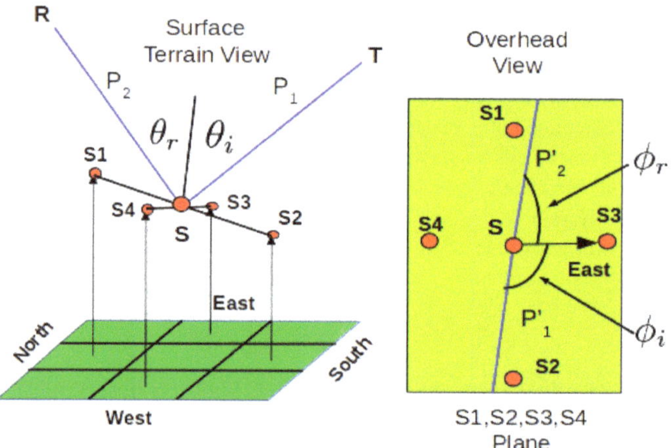

Figure 1. (**left**) Surface terrain view of point on local surface grid showing incident and reflection angles with respect to the surface normal. (**right**) Overhead view with incident and reflected vectors projected to a local two dimensional surface with forward scattering azimuth angles shown.

2.4. Summary of Geolocation Estimation Criteria

The three geolocation parameter checks described above are computed over the local surface grid around the initial estimate of the specular reflection point. If each criteria check is within limits for a given grid point, it is likely that this location is contributing to the received GNSS-R observation. The limits are determined by the geolocation requirements for a given application. Table 1 summarizes the three criteria and the limits used during the analysis that follows.

The limits in Table 1 were selected to achieve an approximate geolocation accuracy of 1 km or less (although results will vary depending on the amplitude of the terrain variations). The algorithm also assigns a confidence level to each geolocation, as specified in Table 2 based on whether any points can be identified within the local surface grid that satisfy the three conditions and based on the DDM SNR.

Table 1. Summary of land Geo-location criteria parameters and thresholds used during performance analysis.

Criteria	Threshold	Verification
$|\delta\tau|$	2.5 C/A Code Chips	Maximum delay
$|\delta D|$	200 Hz	Maximum Doppler
$\delta\Psi$	2 Degrees	Reflection geometry

Table 2. Summary of land geolocation confidence flag logic.

SNR Thold (2 dB)	Delay, Doppler, Snell	Confidence	Land Percentage	Comment
Above	Invalid	0	1.3 (0 to 1)	Most likely incorrect
Below	Invalid	1	21.5 (1 to 2)	Probably incorrect
Below	Valid	2	28.4 (2 to 2.5)	Likely correct
Above	Valid	3	48.8 (2.5 to 3)	High probability correct

2.5. Land Geolocation Example

An example land surface reflection track is shown in Figure 2. The track starts in South Texas (the western most point of the track) and extends toward the east to approximately Lake Pontchartrain in Louisiana. Five example observation locations are shown along the track from west to east at relative seconds; 11, 33, 50, 60 and 120, respectively, indicated as consecutive green markers. The corresponding 5 DDMs at these seconds are shown below the track map. These DDMs start as predominantly non-coherent reflections to the west at higher altitude and more varied topography and proceed down in altitude toward sea level and a region with a significant surface water and wetlands coverage. The geolocation results of each of the 5 example DDMs can be briefly summarized as follows:

1. Second 11: Relatively high terrain with a varied surface topography. The reflected power appears incoherent, with high confidence in the geolocation.
2. Second 33: a "double peaked" DDM, which complicates geolocation determination. Delay and Doppler location checks are based on the highest of the DDM power peaks, which in this case is ambiguous. This "double peak" DDM (with one peak at delay bin 6, and another at delay bin 10) is due to the irregular topography of the surface reflection at second 33, where two distinct and spatially separated surface locations are meeting the geometric conditions suitable for directing power toward the receiver. Attempts at surface retrievals using this DDM would be problematic.
3. Seconds 50 and 60: Surface reflections transition from rougher more varied topography to the flatter terrain of western Louisiana. The corresponding DDMs transition from noncoherent at second 50 to largely coherent at second 60.
4. Second 120: An area containing multiple surface water bodies within and near the estimated geolocation region. The DDM itself shows a very strong coherent reflection within a limited region around the estimated specular point.

The local surface contour maps used in the geolocation at Second 11 are shown in Figure 3. The top-left plot shows how the delay difference of the Second 11 DDM (peak roughly in bin 8 in Figure 2 (bottom-left)) maps across the 205 km × 205 km local surface grid. The dark area around the initial prediction center clearly indicates that the surface region that corresponds to the peak delay is near the estimated center. The Doppler difference map shown on the top-right further narrows the surface region of interest. Finally, the reflection geometry difference angle (bottom-left) isolates the likely scattering region. These three criteria combined result in the estimated surface geolocation region shown in the bottom-right, where the estimated geolocation for this DDM is predicted to extend over several 1 km surface pixels at and around the center of the local area grid. This geolocation estimate provides the center location of a larger diffuse scattering region (10 km or more).

These local surface maps are calculated for each estimated land reflection point using a global (1 km resolution) SRTM surface height map, cropped around the initial geolocation estimate. This adds considerably to the processing overhead for geolocating each measurement, but is necessary to provide confidence in the predicted surface location of the received signal power, especially as the terrain becomes more varied. In some extreme cases (i.e., mountain regions), the local area maps often reveal highly irregular and discontinuous surface regions which are not always in agreement between each of the three confidence checks. In these cases the geolocation is classified as low confidence and a detailed individual DDM analysis would be required to recover the observations at these locations.

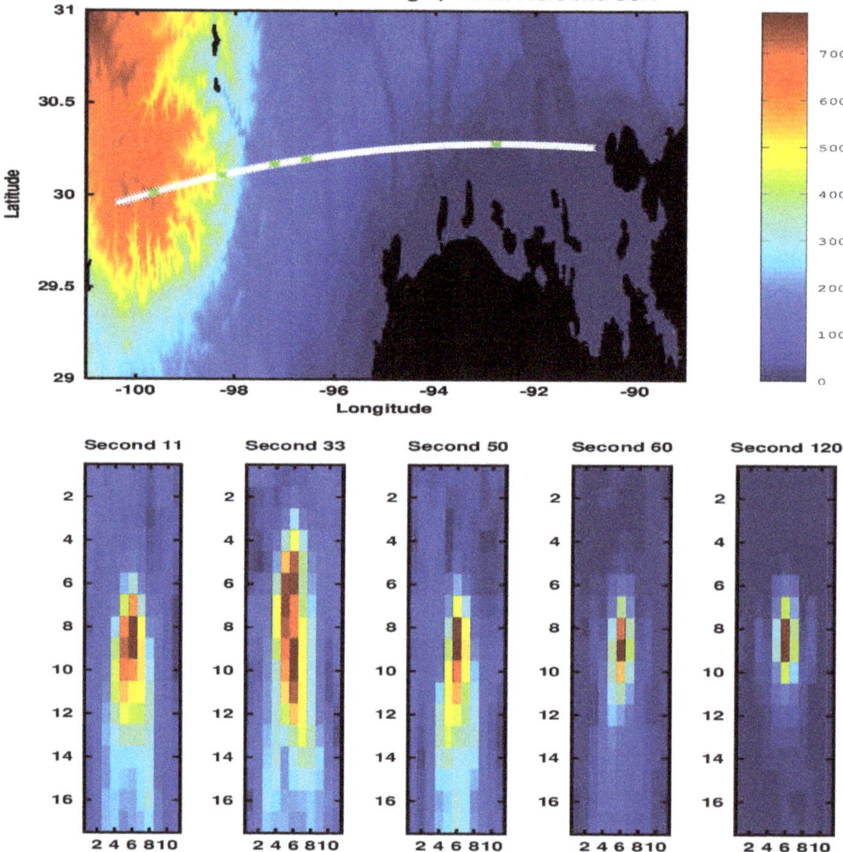

Figure 2. (**top**) Example Cyclone Global Navigation Satellite System (CYGNSS) reflection track over changing terrain. Image background is Shuttle Radar Topography Mission (SRTM) height map in meters. CYGNSS track shown as white arch traveling from west to east, green stars along track indicate Delay Doppler Map (DDM) locations. Second 11 is the western most green star, Second 120 is the eastern most. Water regions shown in black. (**bottom**) Example DDMs at points along the track from various terrain conditions.

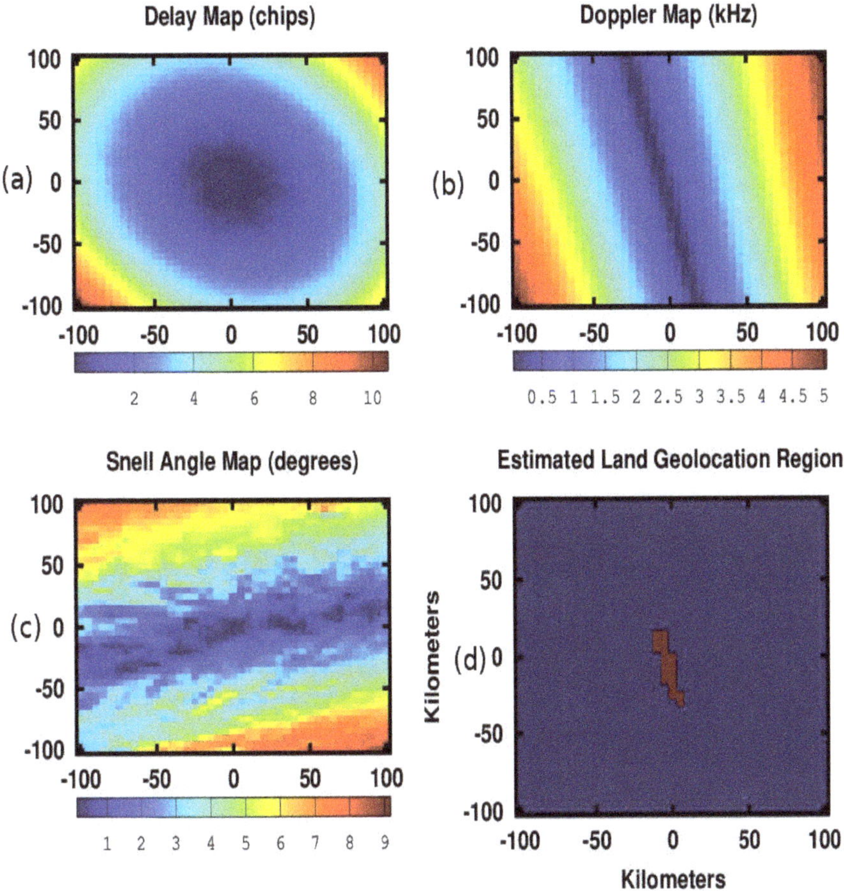

Figure 3. Contour maps of the three geolocation checks across local surface grids at second 11 of Figure 2. (**a**) Empirical delay difference ($\delta\tau$) contour map, (**b**) Empirical Doppler difference (δD) contour map, (**c**) Geometric Snell reflection angle difference ($\delta\Psi$), (**d**) Logical "AND" of the delay, Doppler and Snell area checks, indicating region which satisfies the tolerance limits of all three forward reflection geolocation criteria.

For the DDMs showing coherent reflection (such as seconds 60 and 120), the received power is received from a small area (approx. 1 km) near the predicted geolocation center pixel. The actual surface resolution in the case of coherent reflections is investigated further in Section 4.

2.6. Global Assessment of Land Geolocation Confidence

GNSS-R observation geometries for which at least one local grid point satisfying the delay, Doppler and Snell criteria is located are defined as valid geolocation while other cases are classified as invalid geolocation. Figure 4 is a map resolved into 0.1 by 0.1 degree pixels of the fraction of valid geolocations using 12 days of data from all 8 CYGNSS observatories. Note that mean validity values between zero and one are possible due to the multiple looks within a map pixel from varying CYGNSS look angles.

The geolocation confidence flag (which includes the effects of DDM SNR as defined in Table 2) is shown in Figure 5. The results show the frequent high confidence values in most regions excepting those areas having high topographic variations. The percentage of the Earth's land surface in each confidence category are also included in Table 2). The global Earth land coverage percentage of observations estimated to be valid (confidence level greater than 2) was found to be 77.2%.

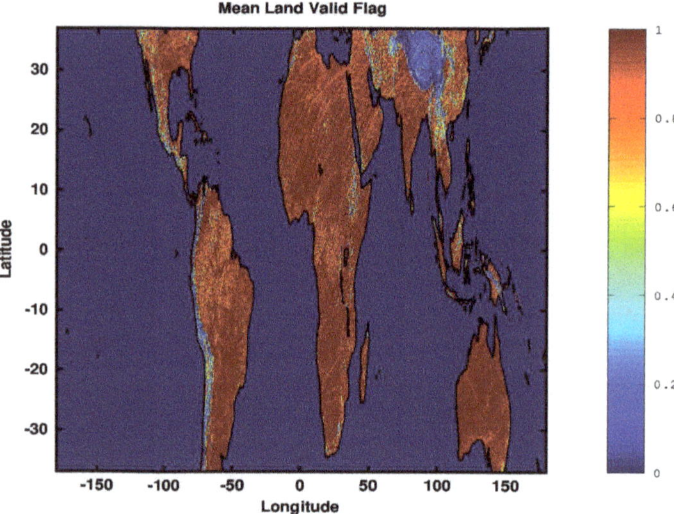

Figure 4. Map of mean land geolocation validity flag (0 or 1). Areas of lower validity correlate strongly with regions of high terrain variability. CYGNSS is unable to capture DDMS from the Tibetan Plateau and Andes Mountain regions.

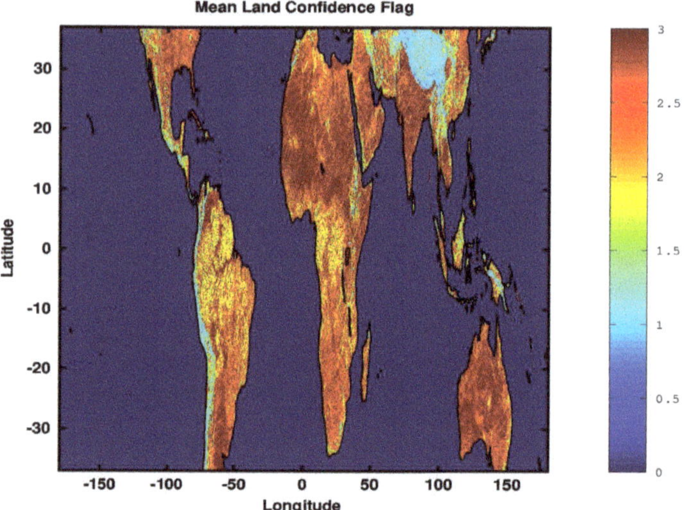

Figure 5. Map of land geolocation confidence flag. Confidence values range between 0 (lowest) and 3 (highest). As expected the general trend is that the estimated geolocation is of higher confidence over lower lying and more moderate terrain. It can also be observed that the geolocation often more difficult in dense forest regions (South America and Central Africa) due to the relatively lower SNR from these surfaces, which make the empirical signal peak checks more difficult.

3. Calibration of GNSS-R Land Reflections

The conversion of CYGNSS-observed count values into calibrated bistatic radar cross section (BRCS) or surface reflectivity information also requires reconsideration for land surfaces.

3.1. Level 1 Normalized Bistatic Radar Cross Section (NBRCS) Land Calibration

The Level 1a conversion from instrument counts to received power (P in Watts) is performed using the same formulation as for the CYGNSS ocean L1a calibration [9]:

$$P = \frac{(C - C_N)(P_B + P_r)}{C_B} \tag{22}$$

where C and C_N are the total Level 0 counts and noise counts, C_B and P_B are the black body calibration load counts and external noise power, respectively, and P_r is the instrument noise power. All of these parameters are calculated exactly as in the ocean L1a calibration with the exception of the DDM noise counts C_N. The value of C_N is obtained from portions of the DDM that occur at times prior to the specular reflection point, so that only thermal noise is present. Because the delay and Doppler location of ocean DDM returns can be reliably predicted using the WGS84 ellipsoid, a fixed set of time offsets from the WGS84 delay is used for computing C_N for ocean observations. Over land, high surface elevations above the WGS84 ellipsoid can cause surface scattering to impinge upon the standard noise time offsets. Therefore, the delays used for computing C_N over land surfaces exclude any points that approach the land surface delay, calculated as

$$\tau_L = \tau_O - 2\cos(\theta)\Delta H \tag{23}$$

where ΔH is the DEM height above the WGS84 ellipsoid specular point, θ is the WGS84 specular point incidence angle, and τ_O is the delay pixel at the ocean surface, or roughly the WGS84 ellipsoid. Whereas C_N for ocean observations is calculated using noise pixels at delays prior to τ_O, C_N for land observations is computed using only delays less than τ_L. Due to the repeating nature of the GNSS code delay sequences, code phase roll-overs should be corrected accordingly.

The Level 1b conversion from power in Watts (P) to an estimate of the BRCS can be calculated using the same algorithm as for an ocean reflected signal, using the estimated transmitter power and antenna gain, receiver antenna gain and propagation path losses associated with the reflection geometry determined by the land geolocation point [9]:

$$\sigma = \frac{P(4\pi)^3 R_R^2 R_T^2}{Y \lambda^2 G^R} \tag{24}$$

where Y is the effective isotropic radiated power (EIRP), including the transmitter power and antenna gain combined, G^R is the receive antenna gain, R_R and R_T are the ranges from the surface to the receiver and transmitter, respectively. The constant $\frac{4\pi}{\lambda^2}$ accounts for the diffuse scattering spreading terms. The Level 1 BRCS is calculated over a range of delay (τ) and Doppler frequency (f) bins. The delay and frequency bins containing the maximum received BRCS are used to evaluate the estimated surface geolocation point.

3.2. Level 1 Reflectivity Calculation

The surface footprint returning power to the receiver is significantly altered in the case of coherent reflection, as explored in Section 4. In these conditions the surface is most appropriately described using the surface reflectivity rather than BRCS, where reflectivity (Γ) is the standard definition used in remote sensing [20]. Under these conditions, the reflection process is modeled using the Friis transmission equation:

$$P^{coh} = \frac{Y G^R \lambda^2 \Gamma}{(4\pi)^2 (R_R + R_T)^2} \tag{25}$$

The key difference with the diffuse bistatic equation [11]) is the scaling by the total path length squared as opposed to the squares of the separate transmit and receive path lengths, as the signal reflects and does not diffusely scatter at the surface. The above equation can be rearranged into an expression of surface reflectivity as

$$\Gamma(\tau, f) = \frac{P^{coh}(4\pi)^2(R_R + R_T)^2}{Y G^R \lambda^2} \tag{26}$$

where $P^{coh}(\tau, f)$ is the level 1a DDM of received power in Watts. Considering the smaller reflection area, it is often useful to consider only the maximum power pixel in the DDM, representing the power level received within a small area near the estimated geolocation point, such that,

$$\Gamma_{max} = max[\Gamma_{\tau, f}] \tag{27}$$

3.3. Distinguishing Coherent and Incoherent Returns

It is widely known from surface reflection and scattering theory that the surface properties determine the physical mechanism of electromagnetic radiance scattered from the surface toward the instrument [11,20]. Generally, if the surface is smooth with respect to the incident radiation (in this case the GPS L1 C/A code signal, at approximately 19 cm wavelength) the result will be a coherent reflection. Conversely, relatively rough surfaces with respect to the EM wavelength will result in diffuse scattering toward the instrument. Each of these cases result in distinctly different equations to estimate the surface parameter of interest: for coherent reflection the surface reflectivity, for diffuse scattering the NBRCS.

Considering this duality of possible physical reflection/scattering mechanisms, for land GNSS-R observations it is advisable to calculate both the NBRCS and the surface reflectivity for every CYGNSS DDM. To assist the user in determining which is most appropriate for a given DDM, a flag for identifying the presence of coherence was developed. This flag is based on assessment of the degree of "power spreading" in a DDM, and is described in detail in [13]. It is acknowledged that this flag retains some uncertainty, motivating the reporting of both the NBRCS and reflectivity for all land observations.

3.4. Level 1 Land Calibration Results

The specified land calibration algorithms were tested using 12 days of CYGNSS data. Two regions were analyzed at finer resolution to better illustrate the differences in NBRCS and reflectivity over known wetlands and regions of higher likelihood of coherent reflection: the southern U.S. and Gulf of Mexico region and the greater Indian Peninsula. Maps of the mean surface reflectivity, mean NBRCS, and coherence flag are shown in Figure 6. Regions of high coherent reflection correlate well with the regions of high reflectivity and NBRCS and with known regions having a high prevalence of inland water bodies.

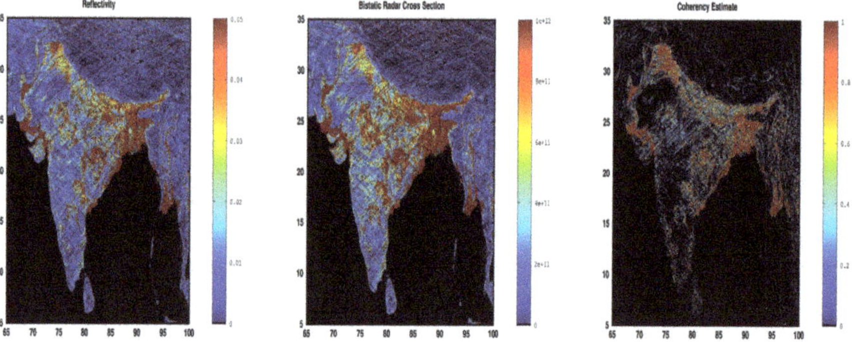

Figure 6. *Cont.*

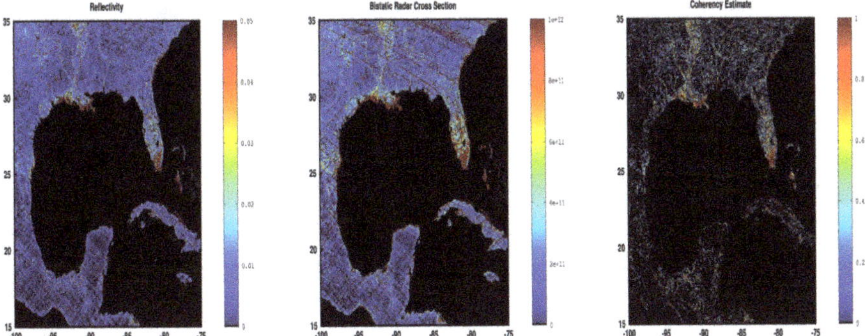

Figure 6. (**left**) Mean surface reflectivity from 12-days of CYGNSS observations. (**middle**) Mean surface peak bistatic radar cross section (BRCS) over the same 12-days of CYGNSS observations. (**right**) Estimated mean reflected signal coherency estimate over same 12-day interval (1 = coherent, 0 = non-coherent) for the Indian Peninsula and the Gulf of Mexico regions.

4. Analysis of Coherent Surface Resolution Using Ancillary River Width Data

To assess CYGNSS's spatial resolution under the conditions of coherent reflection quantitatively, we have performed a statistical study using the North American River Width Data Set (NARWidth) [21]. This dataset provides a reference for locations with greater expectation of generating a coherent reflection. The RMS error of the NARWidth river data set, compared to in-situ measured river widths, is estimated to be approximately 38 m [21]. a detailed coherency analysis using raw sampled data from CYGNSS (from which coherent phase information can be estimated) was previously presented in [22]. However, this work presents CYGNSS results in its normal processing mode (including the non-coherent instrument integration which eliminates all phase information in the observable).

Using the available CYGNSS measurements together with reported river widths, the mean Level-1 power ratio PR coherence metric with respect to varying river widths is examined in Figure 7. The PR is an estimate of the ratio of the amount of received power near the peak to the integrated power away from the peak, derived in [13]. Increasing river widths were found to be associated with a steady increase in the Level-1 coherence metric and the probability of detecting the river, with the detection probability approaching 100% for rivers with widths exceeding ≈ 1.2 km. This is to be expected as rivers with larger widths present a target more conducive to result in identifiable coherence characterized by a significant concentration of power about the DDM peak. They are also less susceptible to ambiguities introduced by the L1 integration time(s) as well as perturbations associated with more heterogeneous scenes. It is also noted that operationally, a DDM measurement is declared as being the result of a dominantly coherent reflection process if its power ratio exceeds the detection threshold $\rho_0 = 2.0$.

From the analysis presented in Figure 7 it is observed that this occurs, on average, for rivers with a minimum width exceeding 250 m and 200 m for 1 Hz and 2 Hz CYGNSS data, respectively. This places a minimum bound on the finest spatial features CYGNSS measurements are able to resolve, on average, suggesting that this is on the order of a first Fresnel zone and provides empirical support for previous theoretical treatments of this question [23]. However, there are several caveats to this analysis which assumes that rivers will generally satisfy the conditions to produce coherent CYGNSS forward reflections:

1. River widths are known to change dynamically, seasonally and during flooding events. The NARWidth data set provides an estimate of the river width at mean discharge, thus it does not account for natural seasonal variations. The analysis undertaken here focuses on using CYGNSS data over two eight month periods, the first starting with July 2018 and ending with February

2019 and the second starting with July 2019 and ending with February 2020. This included a total of 50 million measured CYGNSS specular points, within the NARWidth dataset's coverage, providing ample data to generate statistically significant results. While variations in the size of the coherent reflection surface due to natural fluctuations in river widths will contribute errors into the analysis, the mean correspondence of measures of interest to varying river widths is expected to remain indicative of CYGNSS's spatial resolution under conditions of coherent reflection.

2. Previous analysis of CYGNSS data has shown that river surfaces are generally smooth enough to result in coherent forward reflections [6]. However, there will be cases when a river surface is rough relative to the reflecting L-band wavelength (19 cm) and will not generate coherent forward reflections. In these circumstances, the area of surface scattering will be significantly larger than an integrated Fresnel zone and not result in coherent detection of the river crossing.

3. After (approximately) July 2019, the CYGNSS output data rate was increased from 1 Hz to 2 Hz Level 0 observations. The 2 Hz data output rate corresponds to a 0.5 s instrument integration interval, which reduces the length that the single look surface footprint is integrated across the surface at the specular point velocity. Generally, CYGNSS 1 Hz observations result in a surface integration on the order of 6 km in the along track direction due to satellite motion, while 2 Hz observations will span roughly half of that distance. Note that short intervals of coherent reflection (such as from a small water bodies) within the larger integration footprint often dominate the received power, resulting in detections of small water bodies within the longer integrated surface footprint.

4. Horizontal errors in the surface geolocation of reflections near the Courantyne river will potentially introduce small errors (less than ≈ 1 km) in the river width estimates.

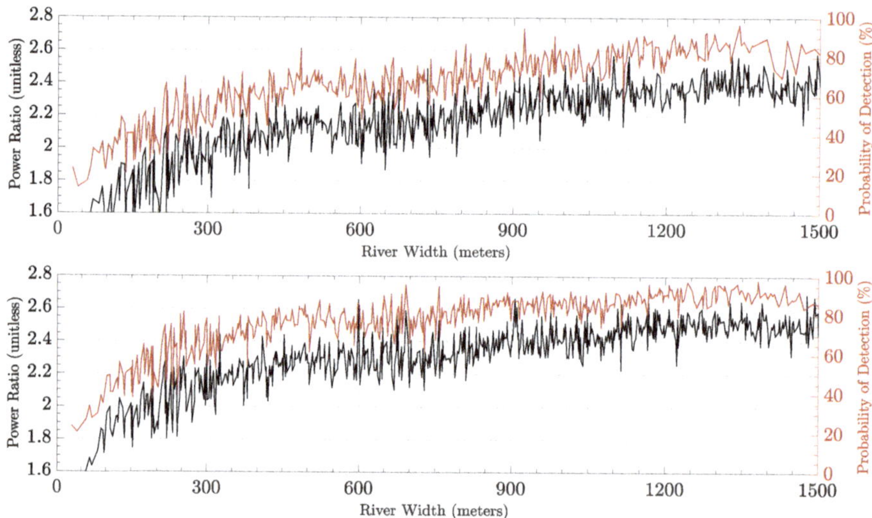

Figure 7. Analysis deriving from the NARWidth data set illustrating effects of varied river widths on the CYGNSS Level-1 coherence metric and probability of detection. (**Top**) Across a sixth month period from 1 September 2018 to 28 February 2019, exclusively comprising 1 Hz data. (**Bottom**) Across a sixth month period from 1 September 2019 to 2 February 2020, exclusively comprising 2 Hz data.

The upper bound of the constellation's ability to resolve features on the surface, under a coherent reflection regime, is expected to be governed by the incoherent integration time. The reduction of the integration interval reduces the along track smearing of L1 DDMs, thereby improving the overall spatial resolution. To illustrate this, CYGNSS measurements from over a (mean) 500 m wide section of the Courantyne River were analyzed, illustrated in Figure 8, over 8 month periods when the constellation operated exclusively in 1 Hz and 2 Hz modes. Details of the selected test region and CYGNSS track crossings are described in Table 3.

As illustrated in Figure 9, it can be observed that the 1 Hz CYGNSS data mean power ratios remain above the detection threshold for distances up to 5 km, slightly less than the ≈6 km distances traveled during the 1 s integration time. Within this 5 km radius, the *PR* is maintained above the detection threshold as the result of dominant coherent reflections arising from the (mean) 500 m wide river. The contributions of the coherent reflector (the river) get 'smeared' along track thereby degrading the surface resolution of the individual observation. This is contrasted with the behavior of measurements made after the switch to 2 Hz sampling over the same region where it is observed that due to the halving of the incoherent integration time, the along track smearing is reduced to approximately the distance traveled along track between two consecutive specular points of ≈3 km. As a result, the power ratios are estimated as 'dominantly coherent' up to 2.4 km away from the river center resulting in an average improvement in spatial resolution of ≈48% brought about by the reduction in integration time. a summary of the observation statistics based on the data shown in Figure 9, is included in Table 4.

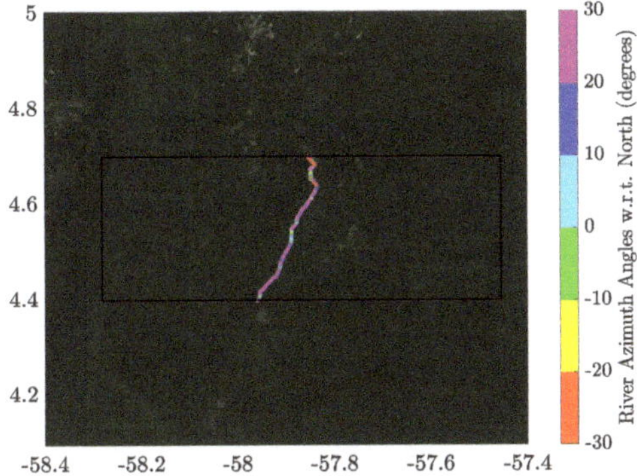

Figure 8. Test region latitude/longitude used in CYGNSS integration time analysis. Test region was centered around a 35 km long section of the Courantyne River between Guyana and Suriname.

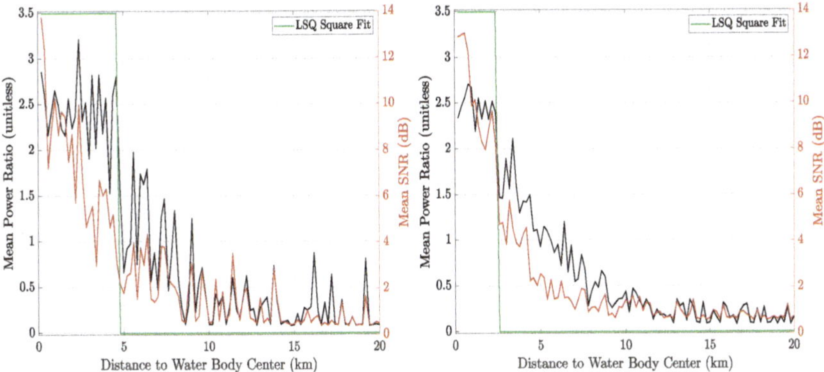

Figure 9. Effects of varying the CYGNSS mission's Level-1 product integration time for observations in the vicinity of the Courantyne River on the CYGNSS coherence metric and Signal-to-Noise Ratio (SNR) at varying distances from the water body. (**Left**) From 3 July 2018 to 28 February 2019, exclusively comprising 1 Hz data. (**Right**) From 3 July 2019 to 28 February 2020, exclusively comprising 2 Hz data. Green traces indicates Least-Squares estimated transition distance between coherence and non-coherence *PR* threshold.

Table 3. Description of selected test region for river crossing coherency analysis.

Courantyne River Region (Guyana/Suriname)		
Parameter	**Mean Value**	**Comment**
River Width	500 m	NARWidth [21]
North Angle	29 deg	Figure 8
Test Region	**Min**	**Max**
Latitude	4.4	4.7
Longitude	−58.3	−57.45
Track Direction	**North Angle**	**River Angle**
Ascending	53 deg	24 deg
Descending	127 deg	97 deg

Table 4. Observation statistics and estimated LSQ fit of coherence detection threshold distances (4600 m at 1 Hz, and 2400 m at 2 Hz) for 1 Hz and 2 Hz DDM frequencies.

Parameter	Distance	Mean	STD
1 Hz DDM Frequency (Obs. Spacing: 5990m)			
Power Ratio	500 m	2.38	0.77
Power Ratio	4600 m	2.41	1.54
SNR	500 m	9.75 dB	5.35 dB
SNR	4600 m	6.63 dB	5.23 dB
Detection Prob.	500 m	91.67 %	N/A
Detection Prob.	4600 m	77.67 %	N/A
2 Hz DDM Frequency (Obs. Spacing: 3010m)			
Power Ratio	500 m	2.43	0.87
Power Ratio	2400 m	2.46	1.13
SNR	500 m	12.91 dB	6.48 dB
SNR	2400 m	10.05 dB	5.97 dB
Detection Prob.	500 m	92.16 %	N/A
Detection Prob.	2400 m	87.12 %	N/A

5. Discussion

This manuscript addresses several of the underlying requirements unique in the processing of GNSS-R land observations: geolocating the observations in diverse topography, calibration the observations into both reflectivity and NBRCS estimates, and assessing the surface resolution for the coherent reflection case.

We have shown that in a majority of CYGNSS land observations we were able to predict the surface observation point with high confidence. This can generally be attributed to many regions of the Earth having relatively low and moderate surface topographic variations, and those variations often being more or less consistent over kilometer scales. However, the exceptional cases in extreme environments will often be incompatible with a generalized algorithm and will require detailed per-DDM analysis across larger regions to identify the locations generating the received power at the instrument.

As discussed above, the increased likelihood of complexities associated with the presence of coherent (and mixed) observations as well as diffuse scattering from land surfaces, can be mitigated with a dual approach for land observations that estimates both surface reflectivity and NBRCS, accompanied by an additional estimate as to the coherence in the level 0 observation. This approach assumes that individual users working on higher level geophysical parameter estimation will need to consider carefully how they apply the different surface observations in conjunction with the provided (or independently estimated) likelihood of signal coherency.

We feel this initial study on the detectable surface resolution of small coherent surface features provides a useful first quantitative analysis of the impact of the non-coherent instrument integration and the resulting blurring of surface footprint and how this impacts the detection of small features within a larger integrated surface area. The surface resolution in the case of diffusely scattered land observations is a subject of further study and expected to be more on the order of the surface resolution of a typical ocean observation (on the order of 15 km or greater).

6. Results

This paper presents an algorithm for the geolocation of CYGNSS observations over diverse terrain conditions. Three criteria were used in the assessment of surface regions for the likelihood of directing power toward the GNSS-R receiver: delay iso-surface range agreement, Doppler iso-surface frequency agreement, and forward reflection Snell angle geometry. It was found that it was possible to estimate the geo-location point of the land reflected signal accurately (mean confidence level of greater than 2) over more than 77.2% of the Earth's land surface. It was also observed that in more complicated terrain environments it is possible for multiple surface areas to contribute to a single GNSS-R observation, thus making observation geolocation inherently problematic.

Additionally, a dual approach level 1 land calibration algorithm was presented which calculates both the BRCS and the surface reflectivity for every land DDM. It was proposed that the CYGNSS level 1 DDM coherency estimator could be used to assist in the determination of the physical scattering or reflection conditions and guide which surface parameter to apply (reflectivity or NBRCS) during land geophysical parameter retrievals. Reflectivity and BRCS maps over regions of varied coherent reflection were presented using 12 days of data from the NASA CYGNSS mission.

Finally, an investigation was performed to quantify the the CYGNSS surface resolution for coherently reflected land signals, and to assess the probability of small water body detection, using the NARWidth Data Set of mean river widths. The resulting analysis showed that (for a 500 m mean width river), the CYGNSS level-1 coherence detector could be used to estimate the bounds of the surface detection resolution with respect to the incoherent integration interval. The resulting detection resolution was determined to be 4600 m for 1 Hz observations (1 s intergation) and 2400 m for 2 Hz observations (0.5 s integration). This clearly demonstrates the advantage (48% improvement) with the shorter integration intervals in the detection of land coherent water bodies.

Author Contributions: S.G.: methodology, validation, writing, review and editing, all Sections. A.O.: methodology, Section 2. A.R.: software, Sections 2 and 3. M.M.A.-K.: methodology and formal analysis, Section 4. J.T.J.: methodology and supervision, all Sections. All authors have read and agreed to the published version of the manuscript.

Funding: This research was funded by the NASA Earth Ventures Cyclone Global Navigation Satellite System (CYGNSS), Extended Mission.

Conflicts of Interest: The authors declare no conflict of interest.

Abbreviations

The following abbreviations are used in this manuscript:

CYGNSS	Cyclone Global Navigation Satellite System
GNSS	Global Navigation Satellite System
GPS	Global Positioning System
GNSS-R	Global Navigation Satellite System Reflectometry
DDM	Delay Doppler Map

PR	Power Ratio
BRCS	Bistatic Radar Cross Section
NBRCS	Normalized Bistatic Radar Cross Section
NARWidth	North American River Width Data Set
DEM	Digital Elevation Model
ENU	East, North, Up
SNR	Signal to Noise Ratio
SRTM	Shuttle Radar Topography Mission
WGS84	World Geodetic System 1984

References

1. Misra, P.; Enge, P. *Global Positioning System: Signals, Measurements, and Performance*; Ganga Jamuna Press: Lincoln, MA, USA, 2001; ISBN 0-9709544-0-9.
2. Gebre-Egziabher, D.; Gleason, S. *GNSS Applications and Methods*; Artech House: Norwood, MA, USA, 2009.
3. Gleason, S. Remote Sensing of Ocean, Ice and Land Surfaces Using Bistatically Scattered GNSS Signals From Low Earth Orbit. Ph.D. Thesis, University of Surrey, Guildford, UK, 2006.
4. Masters, D.; Katzberg, S.; Axelrad, P. Airborne GPS bistatic radar soil moisture measurements during SMEX02. In Proceedings of the 2003 IEEE International Geoscience and Remote Sensing Symposium (IGARSS 2003), Toulouse, France, 21–25 July 2003; Volume 2, pp. 896–898.
5. Ruf, C.S.; Asharaf, S.; Balasubramaniam, R.; Gleason, S.; Lang, T.; McKague, D.; Twigg, D.; Waliser, D. In-Orbit Performance of the Constellation of CYGNSS Hurricane Satellites. *Bull. Am. Meteorol. Soc.* **2019**, *100*, 2009–2023. [CrossRef]
6. Chew, C.; Reager, J.T.; Small, E. CYGNSS data map food inundation during the 2017 Atlantic hurricane season. *Sci. Rep.* **2018**, *8*, 9336. [CrossRef] [PubMed]
7. Chew, C.C.; Small, E.E. Soil Moisture Sensing Using Spaceborne GNSS Reflections: Comparison of CYGNSS Reflectivity to SMAP Soil Moisture. *Geophys. Res. Lett.* **2018**, *45*, 4049–4057. [CrossRef]
8. Al-Khaldi, M.M.; Johnson, J.T.; O'Brien, A.J.; Balenzano, A.; Mattia, F. Time-Series Retrieval of Soil Moisture Using CYGNSS. *IEEE Trans. Geosci. Remote. Sens.* **2019**, *57*, 4322–4331. [CrossRef]
9. Gleason, S.; Ruf, C.S.; Clarizia, M.P.; O'Brien, A.J. Calibration and Unwrapping of the Normalized Scattering Cross Section for the Cyclone Global Navigation Satellite System. *IEEE Trans. Geosci. Remote. Sens.* **2016**, *54*, 2495–2509. [CrossRef]
10. Gleason, S. *CYGNSS Algorithm Theoretical Basis Documents, Level 1A and 1B*; University of Michigan: Ann Arbor, MI, USA, 2018.
11. Zavorotny, V.; Voronovich, A. Scattering of GPS signals from the ocean with wind remote sensing application. *IEEE Trans. Geosci. Remote. Sens.* **2000**, *38*, 951–964. [CrossRef]
12. Yardim, C.; Johnson, J.T.; Burkholder, R.; Teixeira, F.L.; Ouellette, J.D.; Chen, K.-S.; Brogioni, M.; Pierdicca, N. An intercomparison of models for predicting bistatic scattering from rough surfaces. In Proceedings of the 2015 IEEE International Geoscience and Remote Sensing Symposium (IGARSS), Milan, Italy, 26–31 July 2015; pp. 2759–2762. [CrossRef]
13. Al-Khaldi, M.M.; Johnson, J.T.; Gleason, S.; Loria, E.; Yi, Y. An Algorithm for Detecting Coherence in Cyclone Global Navigation Satellite System Mission Level 1 Delay Doppler Maps. *IEEE Trans. Geosci. Remote Sens.* **2019**, under review.
14. Carreno-Luengo, H.; Luzi, G.; Crosetto, M. First Evaluation of Topography on GNSS-R: An Empirical Study Based on a Digital Elevation Model. *Remote. Sens.* **2019**, *11*, 2556. [CrossRef]
15. Camps, A.; Park, H.; Pablos, M.; Foti, G.; Gommenginger, C.; Liu, P.-W.; Judge, J. Sensitivity of GNSS-R Spaceborne Observations to Soil Moisture and Vegetation. *IEEE J. Sel. Top. Appl. Earth Obs. Remote. Sens.* **2016**, *9*, 4730–4742. [CrossRef]
16. Comite, D.; Ticconi, F.; Dente, L.; Guerriero, L.; Pierdicca, N. Bistatic Coherent Scattering From Rough Soils with Application to GNSS Reflectometry. *IEEE Trans. Geosci. Remote. Sens.* **2020**, *58*, 612–625. [CrossRef]
17. Gleason, S. A Real-Time On-Orbit Signal Tracking Algorithm for GNSS Surface Observations. *Remote. Sens.* **2019**, *11*, 1858. [CrossRef]

18. Gleason, S.; Ruf, C.S.; O'Brien, A.J.; McKague, D.S.; OrBrien, A.J. The CYGNSS Level 1 Calibration Algorithm and Error Analysis Based on On-Orbit Measurements. *IEEE J. Sel. Top. Appl. Earth Obs. Remote. Sens.* **2019**, *12*, 37–49. [CrossRef]

19. Shuttle Radar Topography Mission (SRTM). Available online: https://www.usgs.gov/centers/eros/ (accessed on 16 November 2018).

20. Ulaby, F.T.; Long, D.G. *Microwave Radar and Radiometric Remote Sensing*; University Michigan Press: Ann Arbor, MI, USA, 2014.

21. Allen, G.H.; Pavelsky, T.M. Patterns of river width and surface area revealed by the satellite-derived North American River Width data set. *Geophys. Res. Lett.* **2015**, *42*, 395–402. [CrossRef]

22. Loria, E.; O'Brien, A.; Gupta, I.J. Detection and Separation of Coherent Reflections in GNSS-R Measurements Using CYGNSS Data. In Proceedings of the 2018 IEEE International Geoscience and Remote Sensing Symposium 2018 (IGARSS 2018), Valencia, Spain, 22–27 July 2018; pp. 3995–3998. [CrossRef]

23. Camps, A. Spatial Resolution in GNSS-R Under Coherent Scattering. *IEEE Geosci. Remote. Sens. Lett.* **2020**, *17*, 32–36. [CrossRef]

Article

Untangling the Incoherent and Coherent Scattering Components in GNSS-R and Novel Applications

Joan Francesc Munoz-Martin [1,*], Raul Onrubia [1], Daniel Pascual [1], Hyuk Park [1], Adriano Camps [1,*], Christoph Rüdiger [2], Jeffrey Walker [2] and Alessandra Monerris [3]

[1] CommSensLab—UPC, Universitat Politècnica de Catalunya—BarcelonaTech, and IEEC/CTE-UPC, 08034 Barcelona, Spain; onrubia@tsc.upc.edu (R.O.); daniel.pascual@upc.edu (D.P.); park.hyuk@tsc.upc.edu (H.P.)

[2] Department of Civil Engineering, Monash University, Clayton, VIC 3800, Australia; chris.rudiger@monash.edu (C.R.); jeff.walker@monash.edu (J.W.)

[3] Department of Infrastructure Engineering, The University of Melbourne, Parkville, VIC 3010, Australia; alessandra.monerris@unimelb.edu.au

* Correspondence: joan.francesc@tsc.upc.edu (J.F.M.-M.); camps@tsc.upc.edu (A.C.); Tel.: +34-626-253-955 (J.F.M.-M.)

Received: 26 February 2020; Accepted: 4 April 2020; Published: 9 April 2020

Abstract: As opposed to monostatic radars where incoherent backscattering dominates, in bistatic radars, such as Global Navigation Satellite Systems Reflectometry (GNSS-R), the forward scattered signals exhibit both an incoherent and a coherent component. Current models assume that either one or the other are dominant, and the calibration and geophysical parameter retrieval (e.g., wind speed, soil moisture, etc.) are developed accordingly. Even the presence of the coherent component of a GNSS reflected signal itself has been a matter of discussion in the last years. In this work, a method developed to separate the leakage of the direct signal in the reflected one is applied to a data set of GNSS-R signals collected over the ocean by the Microwave Interferometer Reflectometer (MIR) instrument, an airborne dual-band (L1/E1 and L5/E5a), multi-constellation (GPS and Galileo) GNSS-R instrument with two 19-elements antenna arrays with 4 beam-steered each. The presented results demonstrate the feasibility of the proposed technique to untangle the coherent and incoherent components from the total power waveform in GNSS reflected signals. This technique allows the processing of these components separately, which increases the calibration accuracy (as today both are mixed and processed together), allowing higher resolution applications since the spatial resolution of the coherent component is determined by the size of the first Fresnel zone (300–500 meters from a LEO satellite), and not by the size of the glistening zone (25 km from a LEO satellite). The identification of the coherent component enhances also the location of the specular reflection point by determining the peak maximum from this coherent component rather than the point of maximum derivative of the incoherent one, which is normally noisy and it is blurred by all the glistening zone contributions.

Keywords: GNSS-R; sea; coherent scattering; incoherent scattering

1. Introduction

During the last years, Global Navigation Satellite System-Reflectometry (GNSS-R) has been implemented mainly by performing the incoherent integration (the sum of the modulus square) of a set of coherently integrated GNSS codes over short integration times (1–4 ms at most from space). This integration basically removes any coherency present in the reflected signal. The presence of a coherent component in a GNSS reflection can actually be translated to a surface property or a geophysical parameter. Coherent reflections containing a ratio of a potential coherent component has been both studied [1], and they have been captured in different surfaces [2,3]. As an example, sea-ice reflections contain a strong coherent component, where the famous K-shape of a Delay-Doppler Map (DDM) is almost negligible [4], thus the reflection occurs mostly in the first Fresnel zone.

Coherent reflections have been found from low height ground-based instruments based on the coherent interference between the direct and the reflected signals [5]. However, as shown in [6], the maximum interferometric delay is limited to about half the chip length, which puts a trade-off limit on the receiver height and on the satellite elevation angle ($\sim$150 m for L1CA; $\sim$15 m for L5/E5a). This coherent component is almost negligible (but still present) in many reflections for airborne, high stratospheric balloons [7], and spaceborne instruments [8]. The use of a larger antenna (i.e., dish antenna) or an array of antennas (i.e., multiple microstrip antennas) allows a higher gain on the receiver side, and therefore a higher signal-to-noise ratio. Therefore, using shorter integration times increases even further the signal-to-noise ratio of the coherent component of the reflected wave. This work analyzes in more depth the presence of a coherent component with new data acquired by the Microwave Interferometer Reflectometer (MIR) instrument.

The Microwave Interferometer Reflectometer (MIR) [9,10] is an airborne GNSS-R instrument conceived to perform cGNSS-R and iGNSS-R using dual-band (L1/E1 and L5/E5a) high directive up-looking and down-looking antenna arrays ($\sim$21 dB at L1; $\sim$18 dB at L5). Despite the instrument being conceived for real-time processing, the 1-bit raw data sampled at 32 MS/s is also stored as part of the observables, to test other processing techniques offline. MIR maiden flights were conducted in Victoria, Australia in 2018. One of the flights was conducted over the Bass Strait, the area that separates Australia and Tasmania. The large directivity of the MIR antennas allows a very clear detection of the GNSS reflected signal over the ocean with short incoherent integration times (40–300 ms). Despite the evidences shown in the phase of the Delay-Doppler Map (DDM) over the ocean [11] and over land [7], the presence or not of a coherent component in the GNSS reflected signal and its magnitude has been the object of discussion during the last years.

2. MIR Data Description

The data under analysis corresponds to a flight over the Bass Strait on 6 June 2018. The plane followed three passes over a straight line going from 37.9°S, 149.23°E to 38.9°S, 149.1°E, as shown in Figure 1. The GNSS-R data used include both L1/E1 and L5/E5a bands and also contain data from both GPS and Galileo constellations at different incidence angles and coming from different azimuths.

The plane flew at a height of $h \sim 1500$ meters at an average speed of 74 m/s. In that case, the specular reflection occurs in the first Fresnel zone, which is limited by the plane altitude by Equation (1),

$$l_{Fr} = \frac{\sqrt{\lambda R_r}}{cos(\theta_{inc})}, where R_r = \frac{h}{cos(\theta_{inc})} \tag{1}$$

where $\lambda = 19$ cm for L1, $\lambda = 25$ cm for L5, $h = 1500$ m, and θ_{inc} the wave incidence angle. Thus, at nadir $\theta_{inc} = 0°$ and $R_r = h$, which lead to a semi-major axis of the first Fresnel zone for L1 $l_{Fr_{L1}} = 17$ m, and for L5 $l_{Fr_{L5}} = 19$ m, and a Fresnel zone of $l_{Fr_{L1}} = 28$ m, and for L5 $l_{Fr_{L5}} = 33$ m for an incidence angle of $\theta_{inc} = 45°$.

Considering the plane height and both L1 and L5 antenna 3 dB beam-width, $\theta_{L1} = 18°$ and $\theta_{L5} = 25.5°$, the footprint projection (in one direction) over the Earth surface is given by Equation (2),

$$L = R_r \cdot \left(cotg\left(\theta_{inc} - \frac{\theta_{3dB}}{2} \right) - cotg\left(\theta_{inc} + \frac{\theta_{3dB}}{2} \right) \right) \tag{2}$$

where for an incidence angle $\theta_{inc} = 0°$ and $\theta_{3dB} = \theta_{L1}$, $L = 475$ m, for a $\theta_{inc} = 0°$ and $\theta_{3dB} = \theta_{L5}$ the coverage length is $L = 678$ m, for $\theta_{inc} = 45°$ and $\theta_{3dB} = \theta_{L1}$ the coverage length is $L = 975$ m, and for $\theta_{inc} = 45°$ and $\theta_{3dB} = \theta_{L5}$ the coverage length is $L = 1430$ m.

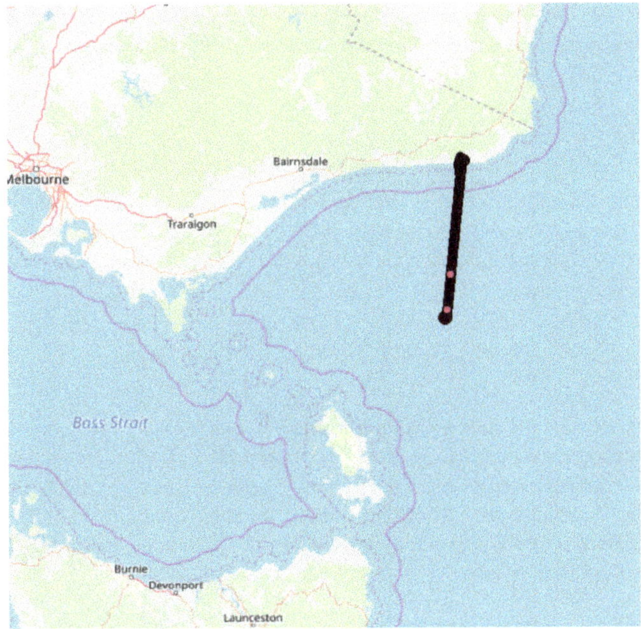

Figure 1. Flight path on 6 June 2018 in which the data used for the coherency analysis were acquired.

3. Coherency of the cGNSS-R Signal

Conventional GNSS-R (cGNSS-R) is based on the correlation of the reflected signal $x(t)$ with a Doppler-shifted (v) clean replica of the GNSS code $y(t)$ [12]. In most GNSS-R instruments [13–15], the correlation time is limited by the code length used by the GNSS signal (i.e., $\tau_c = 1$ ms for GPS L1 C/A, $\tau_c = 1$ ms for GPS L5 without secondary codes, and $\tau_c = 20$ ms for GPS L5 including secondary codes),

$$Y_i(\tau, v) = \frac{1}{T_c} \int_{iT_c}^{(i+1)T_c} x(t)y^*(t-\tau)e^{-j2\pi vt}dt \tag{3}$$

There are several techniques used to perform this cross-correlation [16]: serial search, parallel phase search, or the most used in modern GNSS receivers, the parallel code-phase search (PCPS) algorithm. PCPS is based on the use of the Fast Fourier Transform (FFT) as large as 1 period of the code length (i.e., $\tau_c = 1$ ms for GPS L1 C/A). Therefore, Y_i is obtained as the Doppler cut containing the maximum absolute value of the signal.

In order to increase the signal-to-noise ratio (SNR), the resulting cross-correlation is incoherently averaged with subsequent correlations. The incoherent integration has been used for GNSS signal acquisition and also to perform DDMs, as it increases the observable SNR, but destroys the coherent part of the signal, which is the one coming from the specular reflection point, around the first Fresnel zone.

In [17], a method was proposed to detect and eliminate the direct signal. In this work, this technique is applied to detect the coherent component present in a GNSS-R signal. The technique explained in [17] consists of the computation of the variance of the coherently integrated DDM (Y in Equation (3)), i.e., prior to the incoherent averaging, as in Equation (4).

$$Var\,(Y) = E[|Y|^2] - |E\,[Y]|^2, \tag{4}$$

where $E[|Y|^2]$ is the incoherently averaged DDM. In practice, the variance term, $Var(Y)$ is computed as the mean square of the N_{inc} samples (amount of samples incoherently averaged) minus the arithmetic mean of the samples (μ), as in Equation (5). Note that Y is a complex value vectors, therefore any sum, multiplication, or mean calculus has to follow the complex arithmetic (i.e., mean of a complex vector is $mean(I) + j \cdot mean(Q)$)

$$Var(Y) = \frac{1}{N_{inc}} \sum_{i=1}^{N_{inc}} |Y_i - \mu|^2 \tag{5}$$

$$\mu = \frac{1}{N_{inc}} \sum_{i=1}^{N_{inc}} Y_i \tag{6}$$

The difference of Equations (3) and (5) leads to the coherent component averaged over N_{inc} samples, as shown in Equation (7).

$$|E[Y]|^2 = \frac{1}{N_{inc}} \sum_{i=1}^{N_{inc}} |Y_i|^2 - \frac{1}{N_{inc}} \sum_{i=1}^{N_{inc}} |Y_i - \mu|^2. \tag{7}$$

The implementation of the coherent integration as in Equation (7) opens many possibilities and analysis methods for signal processing. As an example, the ratio of both coherent component and total power waveform is defined as proposed in [18] as the degree of coherency (DOC), as in Equation (8). This ratio represents how coherent is the GNSS signal, for instance a direct GNSS signal has a DOC very close to 1. A reflected GNSS signal may have a large DOC in case of a quasi-specular reflection, but in general over land, it does not.

$$DOC = \frac{|E[Y]|^2}{E[|Y|^2]} \tag{8}$$

4. Data Processing

The raw data is processed in different steps. First of all, each waveform Y_i is used to perform both the coherent (7) and the variance calculations following the PCPS algorithm, as detailed in Figure 2. Note that, the PRN clean replica length, and therefore the final waveform length, is set by the PRN code length, which is 1 ms for GPS L1 C/A, 1 ms for the GPS L5 without secondary codes, and 20 ms for GPS L5 including the secondary codes.

The cross-correlation process is repeated N_{inc} times for the amount of total integration time (both total power waveform and coherent component through the variance calculus), from now on, each of the Y_i realizations will be identified as an integration period.

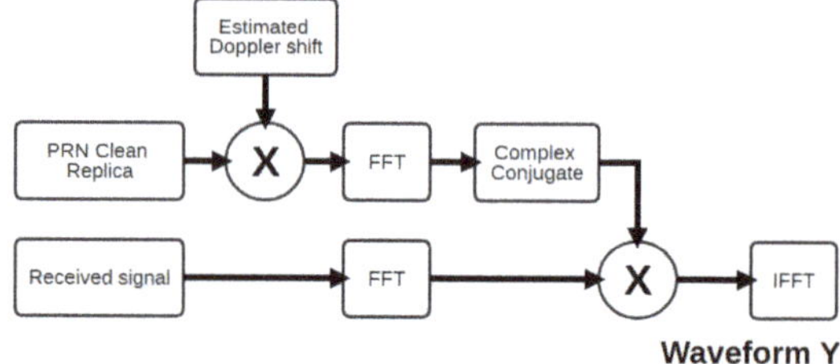

Figure 2. Cross-correlation process of the Microwave Interferometer Reflectometer (MIR) data based on the parallel code-phase search (PCPS) algorithm.

Once each Y_i waveform for each of the N_{inc} integration periods has been retrieved, the algorithm in Figure 3 is followed in order to provide the three products: total power waveform; variance part (which is the incoherent part); its difference, which corresponds to the coherent component of the signal; and finally the phase evolution for each integration period.

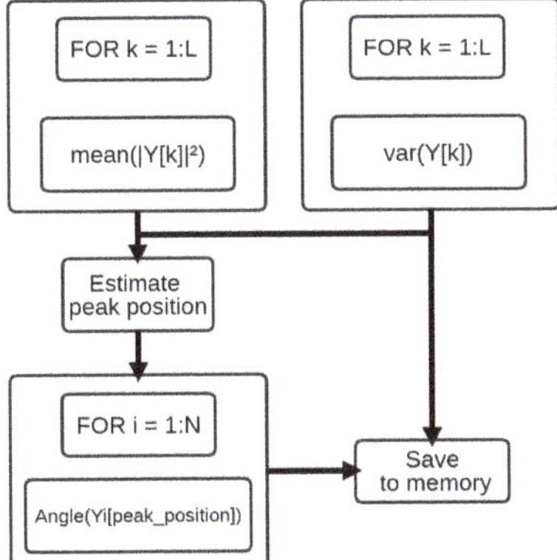

Figure 3. Total power waveform and coherent component processing algorithm of MIR data, including the phase retrieval of the peak.

Note that, mean() and var() functions are calculated over Yi, where $i = 1$ to N, and N is the integration time.

4.1. Navigation Bit Transitions during the Coherent Integration

As described in [19], the bit transition in GNSS needs to be handled in case of large coherent integration periods. The variance method is also sensitive to the bit change, therefore a bit change in the middle of the integration process causes that the coherent term in Equation (5) drops to zero. The bit transition effect can be compensated by means of retrieving the navigation bit sign and multiplying each of the resulting waveforms (Y_i), by the corresponding sign. The navigation bit sign is retrieved by looking at the phase evolution during the integration period.

The Navigation bit transition can also affect the reflected GNSS signal in case the coherency is preserved in the reflection. However, this is not actually the case, and thus the navigation bit cannot be retrieved and compensated as easy as in the direct signal case. In such cases, the navigation bit can be compensated using the direct signal information. The algorithm to retrieve the coherent part for both direct and reflected signals, compensating the navigation bit transition, is detailed in Figure 4.

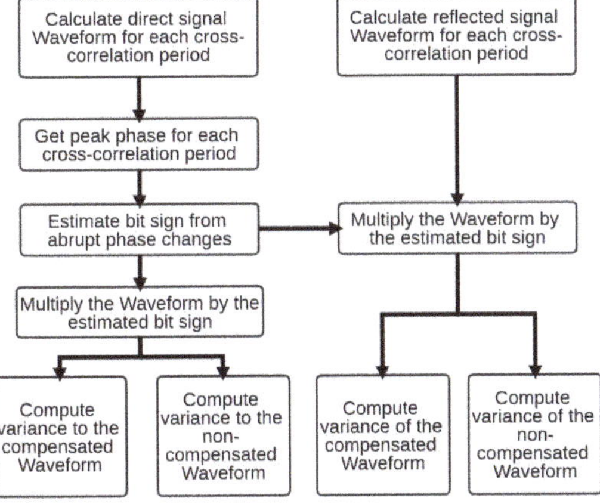

Figure 4. Variance calculus algorithm in the presence of bit transitions for both direct and reflected waveforms.

In order to estimate the bit sign ($\hat{x}$) in the direct signal, the discriminator in Equation (9) is used.

$$\hat{x} = atan2\,(Q, I)\,, \tag{9}$$

where Q is the imaginary component of the WAF (Y_i) at its peak (for both code and Doppler), I the real component, and *atan2* is the four quadrant arctangent function.

4.2. Open-Loop Tracking of the Coherent Part of the Reflected Signal

The described algorithm allows for an open-loop tracking of the reflected signal with a variable coherent integration time compensating the bit transitions because of the information of the direct signal. In addition, this algorithm stores both the bit-compensated and the non-compensated coherent integrated waveforms, which are useful to evaluate the coherency characteristics of the signal (i.e., the reflected signal can contain a coherent part, but no bit information can be retrieved).

Figure 5 illustrates the bit change effect on the impact on the coherency part in a GPS L1 C/A GNSS signal captured by the MIR instrument and integrated during 40 ms. A data set containing two bit transitions (periods 12 and 32 as seen in the phase plot) in the middle of the integration has been selected. As seen, the coherent part in case the non-compensated case (dash-dot black line) for the direct signal goes down to zero, as the coherency is lost due to the navigation bit change. However, as the bit transition is detected and compensated, the coherent (dashed blue line) part goes almost as high as the total power waveform; therefore, the direct signal DOC for this example is $\sim$0.9.

The reflected case is quite different. First of all, as expected, the coherency of the signal is much lower than the direct one, but also the navigation bit compensation does not make any difference in terms of the coherency. Note that the phase evolution in the reflected peak presents the 180° jump at the 12th integration period, but after the 22nd period, it completely losses the phase and hence the coherency of the signal. Despite that, the signal presents a coherent component with a DOC $\sim$0.2.

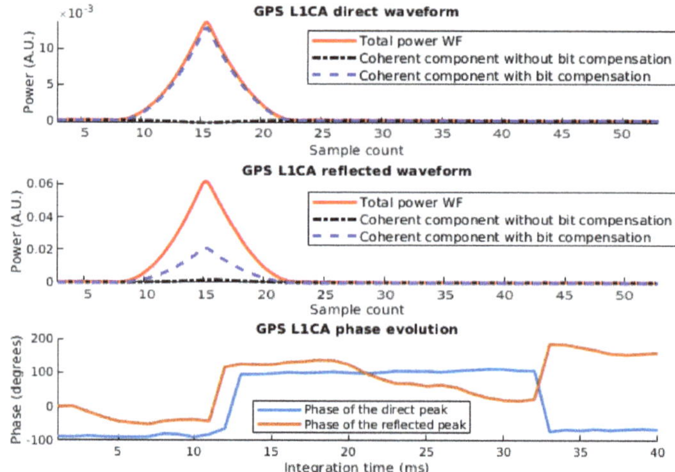

Figure 5. Coherency of a GPS L1CA direct signal (top figure) within 40 ms of integration with a two bit transitions, coherency of the same GPS signal once reflected over the sea (middle figure), and phase evolution of the peak for each integration sample (bottom figure). All figures with $f_s = 32.768$ MHz, 1 sample = 30 nanoseconds.

Note that in the case of reducing the integration time to 20 ms, to avoid the signal coherency degradation present at the 22nd period, the DOC of the reflected signal increases (as shown in Figure 6) up to ∼0.8, as expected. Therefore, this increase on the DOC for 20 ms of integration that the reflection is almost coherent in this period.

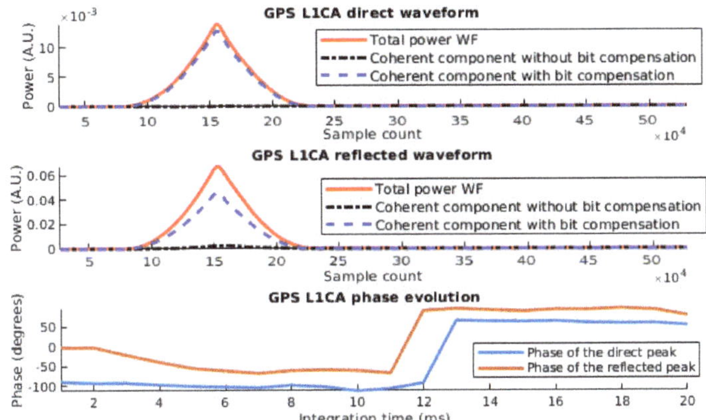

Figure 6. Coherency of a GPS L1CA direct signal (top figure) within 20 ms of integration with a bit transition in the middle, coherency of the same GPS signal once reflected over the sea (middle figure), and phase evolution of the peak for each integration sample (bottom figure). All figures with $f_s = 32.768$ MHz, 1 sample = 30 nanoseconds.

4.3. Coherency in the Presence of Secondary Codes

The secondary codes present in GPS L5 signal produce a similar behavior on the coherent part than the navigation bit. As the navigation information, the secondary code is a pseudorandom sequence of $+/-1$, thus multiple sign changes occur in a 1 ms integration period. As the repetition period of the secondary code for GPS L5 is 20 ms, performing the coherent integration as in GPS L1 C/A without the secondary code produces a degradation on both direct and reflected signals, as shown in Figure 7, with a DOC for both direct and reflected <0.05. Note that the integration has been performed over the pilot component, as in that way we are able to remove the entropy of the navigation bit sign change.

However, performing the cross-correlation, but now including the secondary code, and preserving the integration time (40 ms) has a direct impact on the DOC. As seen in Figure 8, the DOC of both direct and reflected signal goes up to ∼1, which means that, for the selected waveform, the coherency of the sea spectrum at L5 is preserved within 40 ms. Note as well the typical elongation of the trailing edge of the waveform (Figure 7, central panel), and that, in the reflected signal case, two coherent peaks can be identified. In addition, the phase evolution of the L5Q signal with secondary codes included can only be represented once every 20 ms, as the cross-correlation process and waveform retrieval is taken in multiples of the code length.

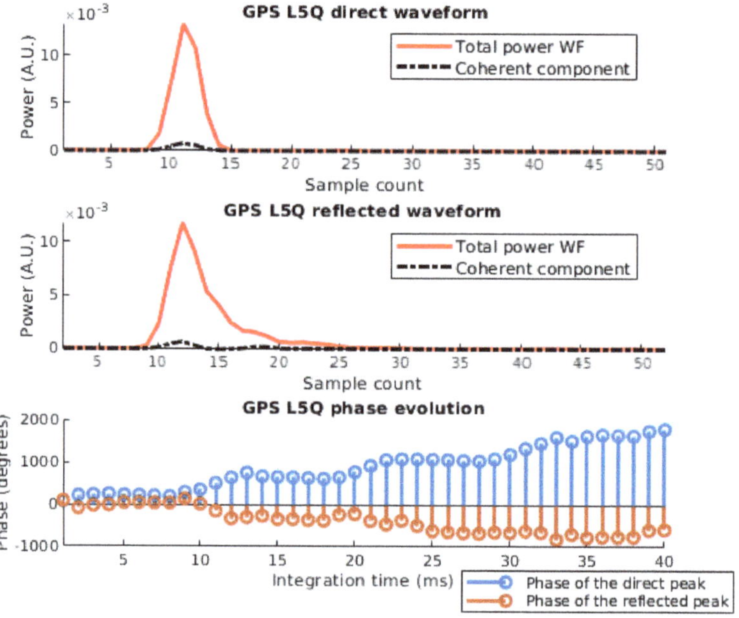

Figure 7. Coherency of a GPS L5Q without secondary codes direct signal (top figure) within 40 ms of integration with a two bit transitions, coherency of the same GPS signal once reflected over the sea (middle figure), and phase evolution of the peak for each integration sample (bottom figure). All figures with $f_s = 32.768$ MHz, 1 sample = 30 nanoseconds

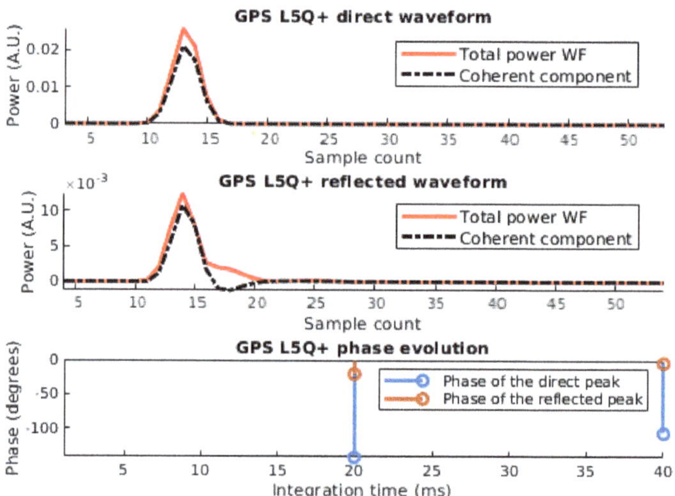

Figure 8. L5QCoherency of a GPS L5Q with secondary codes direct signal (top figure) within 40 ms of integration with a two bit transitions, coherency of the same GNSS signal once reflected over the sea (middle figure), and phase evolution of the peak for each integration sample (bottom figure). All figures with $f_s = 32.768$ MHz, 1 sample = 30 nanoseconds

The second reflection peak seen in this figure is placed seven samples away from the first peak. Converting the sample distance to meters (as in Equation (10)), we clearly see that this second peak is out of the first Fresnel zone (27 m for L5); therefore, it came from the reflection glistening zone, but presenting a coherent component.

$$\Delta_m = c \cdot \frac{\Delta_{samples}}{f_s}$$

$$\Delta_m = 64m$$

(10)

where c is the light speed, f_s = 32.768 MHz is the sampling rate, and $\Delta_{samples}$ corresponds to the peak-to-peak distance in samples.

5. Reflected Signal Coherency Analysis

This section covers an analysis of different GPS L1 and L5 signals for different integration times.

As seen in Equation (10) in [20] the coherency time of a given surface for this flight is for both L1 and L5, and assuming an incidence angle of 0° and 45°.

$$\tau_s = \frac{\lambda}{2 \cdot v} \sqrt{\frac{h}{2 \cdot c \cdot \tau_c \cdot \cos(\theta_i)}},$$

$$\tau_{s_{L1}}(\theta_{inc} = 0°) = 2\,ms,\ \tau_{s_{L1}}(\theta_{inc} = 45°) = 2.8\,ms,$$

$$\tau_{s_{L5}}(\theta_{inc} = 0°) = 7.7\,ms,\ \tau_{s_{L5}}(\theta_{inc} = 45°) = 10\,ms$$

(11)

This section analyzes a set of different waveforms for GPS L1 C/A and GPS L5Q with secondary codes. Those waveforms have been selected and reproduced for a set of integration times, which are multiples (up to 10 times) of the sea coherence time. In addition, the analysis does not only cover the waveform shape, but the phase of the signal at all the integration steps, which helps to understand why the signal has a given coherency or not.

5.1. GPS L1 C/A Reflected Signal Signatures

The GNSS reflected signal contains a portion of coherency, which is easily untangled from the incoherent one thanks to the proposed algorithm. A first example for GPS L1 C/A is provided in Figure 9 showing four different integration times: 5 ms, 10 ms, 20 ms, and 40 ms. As it can be seen, the DOC of the reflected waveform decreases as the coherent integration time increases. In addition, the phase of the peak is detailed for each 1 ms cross-correlation interval. Note that, the scale of the phase evolution has been set the same for the four measurements to ease its visualization.

Note that, the phase for the direct signal is flat and does not present 180° jumps, as the navigation bit has the same sign for the 40 ms. Note also that the phase unwrapping function has been used to ease the visualization of the phase evolution.

Analyzing the DOC for the four cases, the DOC decreases as the integration time increases, which is logical due to the coherence time of the sea as shown in Equation (11).

In addition, analyzing the phase variation for the four cases (and in particular the 20 and 40 ms case), the coherency loss is coming from a change of the reflected surface, as, for instance, the reflection has gone from the wave crest to the wave valley. As seen, the peak phase is very smooth and follows a shape which can be identified as the wave evolution with time. This was first observed in a controlled experiment in a water tank (Figures 7 and 10 of [21]). This phase precision can be actually used to perform phase altimetry measurements as in [22,23].

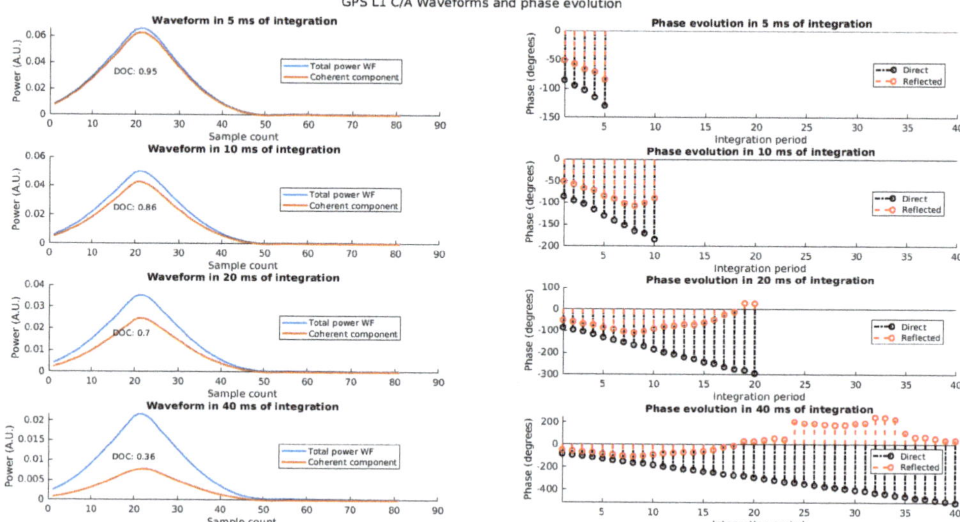

Figure 9. Left column: Total power waveform and coherent component for different integration times (from 5 to 40 ms). Right column: phase evolution of GPS L1 direct (black) and reflected (red) signals for the 1 ms coherently integrated waveforms used for different integration times (from 5 to 40 ms).

The second example (Figure 10) shows a *noisier* environment, where the coherency of the signal is almost lost after 10 ms of integration. As seen in the 10 ms and 20 ms examples, the phase evolution of the reflected signal is very noisy, which causes a coherence loss. However, despite the phase is not constant in this case, the coherent component can be still untangled from the incoherent one, allowing for that a better estimation of the first Fresnel zone characteristics and the surface roughness associated to the illuminated area.

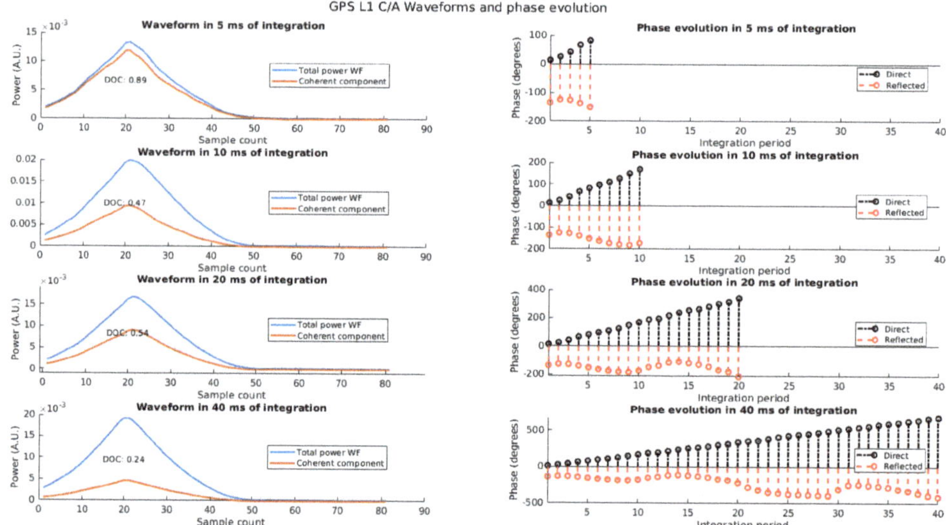

Figure 10. Left column: Total power waveform and coherent component for different integration times (from 5 to 40 ms). Right column: phase evolution of GPS L1 direct (black) and reflected (red) signals for the 1-ms coherently integrated waveforms used for different integration times (from 5 to 40 ms).

5.2. GPS L5Q Reflected Signal Signatures

As presented in Section 4.3, the coherency of the L5 signal (either I or Q components) is lost if the secondary codes are not taken into account. Two examples for GPS L5Q with secondary codes are provided for five different integration time: 40 ms, 80 ms, 120 ms, 160 ms, and 200 ms.

The first example is shown in Figure 11. In the 40 ms case of this figure, it is seen that the coherent component presents a positive and a negative part, as it is a real correlation and the coherent integration is not based on the absolute value operation. This negative part can be identified as an out-of-phase addition of the reflected signal paths. Despite that, the signal presents a high coherency, with a DOC ~0.22 for 40 ms of integration. In addition, the coherent peak due to the incoherent integration is not located in exactly the same chip delay than the incoherent one. The position of this coherent peak is also very useful for altimetry applications, as, from one side, the coherent reflection is linked only to the first Fresnel zone (i.e., better spatial resolution), and from the other side, the peak position estimation is better than in the incoherent case.

As seen in the 80, 120, 160, and 200 ms cases, increasing the integration time also increases the blurring of the total power waveform (no retracking applied [24]), also providing evidence that the peak position for the total power case is changing. Despite that, the coherent peak is not moving from its original position, as it is only identifying the coherent reflection, and not all the contributions from the glistening zone. Even though the coherent component is still, its amplitude decreases as the integration time increases due to the surface changes among time. Take into account that, after 200 ms, the plane has been moved ~15 m, which is half of the the first Fresnel zone size at L5.

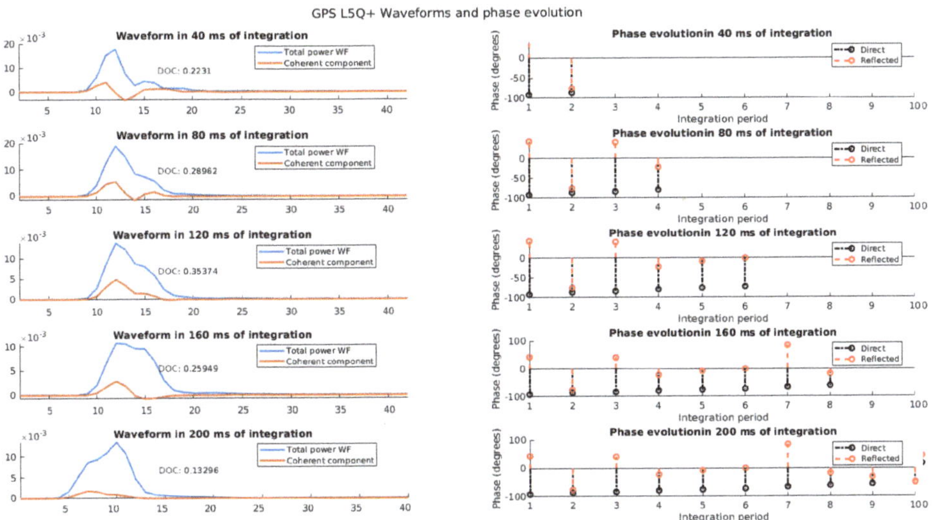

Figure 11. Left column: Total power waveform and coherent component for different integration times (from 5 to 40 ms). Right column: phase evolution of GPS L5Q direct (black) and reflected (red) signals for the 20 ms coherently integrated waveforms used for different integration times (from 40 to 200 ms).

In addition to the waveform analysis, the phase evolution and its difference with respect to the direct signal one gives very useful information. The difference between both is changing as the integration time increases. In this case, the direct phase is still, and the reflected phase changes within ±100°. The abrupt changes in phase is indicating two different phenomena: on one hand, the signal coherency is not 100% preserved between one integration step and the next one, which is also reflected in the DOC parameter. On the other hand, the phase evolution follows a *half-sine* slope, which may be linked with the sea surface shape at the specular point. Therefore, further processing this phase evolution, taking into account the peak position enhancement thanks to the coherent integration will help to enhance phase altimetry precision [23].

Finally, the second example for L5Q shows a very different shape as the previous example. As seen in Figure 12, the larger the integration time, the larger the change of the shape of the coherent component. First of all, the DOC for the 5 cases is very similar, on the order of ~ 0.1, which means that even a small coherency is preserved for even large integration times.

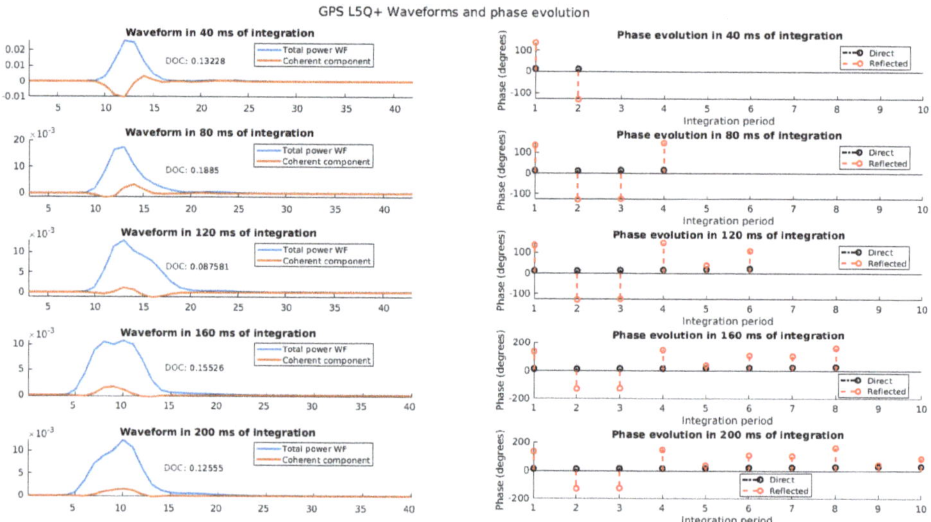

Figure 12. Left column: Total power waveform and coherent component for different integration times (from 5 to 40 ms). Right column: phase evolution of GPS L5Q direct (black) and reflected (red) signals for the 20 ms coherently integrated waveforms used for different integration times (from 40 to 200 ms).

The first case (40 ms of integration) shows a first negative peak on the coherent part, followed by a positive peak. Comparing it to the total power waveform, which uses the modulus operator, the presence of this negative peak tells us that the specular reflection (and hence the coherent one) is coming from the positive coherent component, which is two lags away from the peak estimation of the total one.

Looking at the evolution of both the coherent and total power peak positions with respect to the integration time, it is clear that as the total one gets blurred and its maximum moves within 20 samples, the coherent one is almost frozen at the same sample, and starts blurring when the integration time is too large. Looking at the phase evolution for this example, it is clear that the coherency of the signal is not high, as the phase performs $\sim 180°$ jumps from one integration period to the next one, while the direct phase is the same. In this case, a *half-sine* shape can be identified for large integration times, which can be linked to the sea slope, and hence indicating that phase altimetry may be feasible and will be enhanced thanks to the coherent component peak position determination.

6. Potential Applications from the Coherent Component Analysis of the GNSS-R Waveform

The previous sections have shown that the coherent component of the GPS L1 C/A and GPS L5, including secondary codes, is not negligible. This component decreases as the integration time increases, and its decreasing ratio depends on the roughness of the reflected surface, as it is linked to the surface coherence time. The application of the coherent component untangling technique opens a number of potential applications from its use, which are highlighted below.

6.1. Scatterometry Using the Coherent Component

As seen from the GPS L1 C/A case, the coherent component is very strong for short integration times. In addition, knowing that the coherent component reflection comes from the specular point, scatterometry measurements are much easier to compute and also with a better spatial resolution.

From now, the scattering model used to compute scatterometry measurements assumes a normalized bistatic scattering cross section (σ_0) over the full glistening zone of the reflected surface [25]. In this case, the reflected area is very large, therefore having a poor spatial resolution. However, as the coherent component contains the reflection from the specular point, and hence the first Fresnel zone, the spatial resolution of the reflected signal is improved. At this point, scatterometry measurements are as easy as computing the power ratio between the direct and the reflected signal coherent components, as in Equation (12).

$$\Gamma = \frac{P_{ref}}{P_{dir}} = \left(\frac{R_{T-SP} + R_{SP-R}}{R_{T-R}}\right)^2 \cdot \frac{G_{zenith}(\theta_{dir}, \phi_{dir})}{G_{nadir}(\theta_{ref}, \phi_{ref})} \cdot \frac{G_T(\theta_1, \phi_1)}{G_T(\theta_2, \phi_2)} \tag{12}$$

where R_{T-SP}, R_{SP-R}, and R_{T-R} are the distances from the transmitter to the specular reflection point, from the specular reflection point to the receiver, and from the transmitter to the receiver, respectively; G_{zenith} and G_{nadir} are the gain of the receiver antennas; and G_T is the gain of the transmitter antenna in the direction of the specular point and the receiver position. Note that the G_T term can be neglected for receivers and specular points that are very close (i.e., the specular point is almost at nadir), or for low altitude receivers (i.e., a plane flying at 1500 meters height). In addition, the range correction term is $\simeq 1$ and can be neglected for low altitude platforms, but it can be a fraction of a dB for LEO satellites.

6.2. Precise Altimetry from Precise Peak Position Estimation in L5Q+ Waveforms

From the examples provided in Section 5.2, it is clear that the coherent part of the GNSS L5Q signal with secondary codes provides additional information of the peak position, which can be useful to estimate the lag corresponding to the maximum position of the waveform. The better the estimation of this peak, the better the altimetry resolution. The example shown in Figure 13 is a zoom of a L5Q+ reflected waveform. Both total power and coherent parts of the reflected signal have been interpolated by 8 (i.e., $f_s = 262.144$ MHz) using the FFT interpolation method. In addition, the derivative of the total power part is calculated. Note that all three signals have been normalized with respect to its maximum value in order to ease its visualization.

For the four cases, 40, 60, 80, and 100 ms of integration time, the coherent component peak position is found in five samples for the 40 ms case, and seven samples for the next three cases before the peak of the total power averaged.

In addition, the peak of the coherent component is placed between 9 and 13 samples after the maximum of the derivative of the total power waveform, which means that the actual specular point (in terms of lag delay) is placed between the point of maximum slope of the total power waveform and the actual maximum of the waveform. This behavior was studied in [26], which for an ideal conditions of a rough surface, and a theoretically infinite incoherent integration time, the delay corresponding to the specular point was placed right on the maximum of the derivative of the total power waveform. In the example shown in this study, the delay corresponding to the specular point is identified as the coherent component peak.

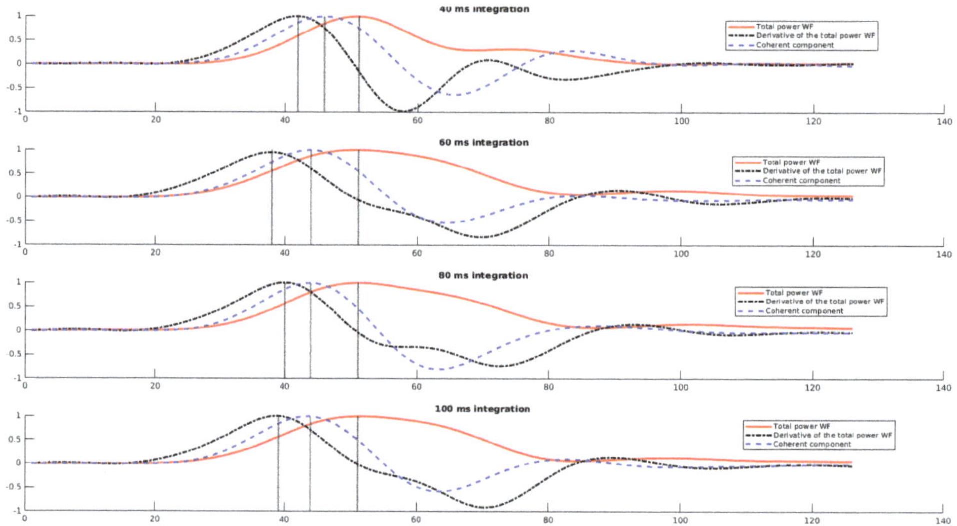

Figure 13. Peak position estimation depending on the integration time (40 to 100 ms) from the coherent component at L5Q with secondary codes, $f_s = 262.144$ MHz, 1 sample = 3.81 nanoseconds. Note all functions are normalized for the sake of clarity.

A proper estimation of the specular reflection point gives an improved accuracy of the altimetry measurement, as seven samples of difference in the estimation of this peak leads to an error up to 3 m (i.e., for an incidence angle of 45°, from Equation (34) in [12]).

Swell Wave Period from Secondary Peaks in L5Q+ Waveforms

As seen in Figure 13, apart from the waveform peak, there are several secondary peaks, mainly in the 40 ms of integration case, whose second reflection is identified in both coherent component and total power processed reflected waveforms. The 60, 80, and 100 ms cases present a secondary peak in the coherent component, but not in the total one. Despite the incoherent integration has blurred up the waveform for relative large integration times, the coherent component presents the same *secondary peak*, indicating that the incoherent integration is preventing other applications of the GNSS-R. Note that the secondary peaks that are always positive in the total power waveform, are negative in the coherent one, but still it represents a coherent reflection with negative sign (i.e., 180° rotation with respect to the specular peak).

Taking as a reference the coherent component maximum, the first secondary peaks (i.e., the negative ones) are placed ∼20 samples away from the maximum one. Transforming the sample distance to meters following Equation 10 but with $f_s = 262.144$ MHz, the Δ_m is ∼23 m. Moving to the third, which is almost negligible in the figure, it is placed ∼37 samples from the maximum, therefore giving a Δ_m ∼42 m.

Looking to the sea state conditions during the flight in [27] (see Figure 14), it is found that swell waves had a period of T_{waves} ∼9 s, which corresponds to a sea speed (C_{waves}) and a wavelength (i.e., distance between crests) as in Equation (13) of [28].

$$C_{waves} = 1.56 \cdot T_{waves} \simeq 14\,m/s,$$
$$\Lambda_{waves} = C_{waves} \cdot T_{waves} \simeq 126\,m. \tag{13}$$

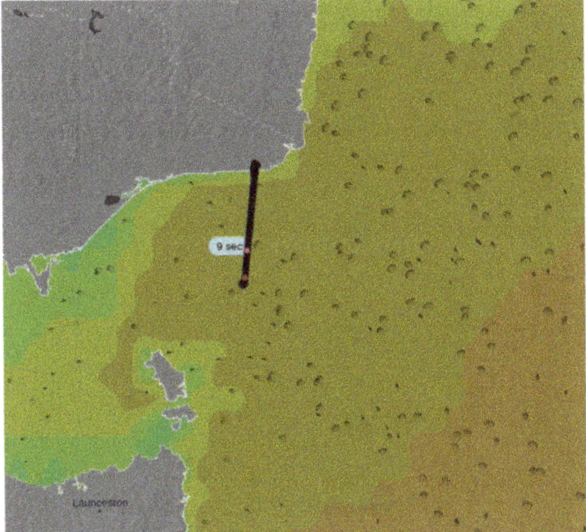

Figure 14. Swell period in the flight path the 6th of June, 2018. Downloaded from Ventusky [27].

In the exposed case, the maximum distance found is $\sim$57 m, which is approximately $\frac{\lambda_{waves}}{2}$. This distance corresponds to a two reflections on consecutive wave crests. Therefore, the *negative* coherent peak corresponds to the sum of the components reflected between the wave crest and the wave trough.

7. Conclusions

This study has presented a technique to untangle the coherent signal from the total power waveform on reflected GNSS signals from the sea surface, for both L1 and L5 bands from GPS. The results presented confirm the presence of a non-negligible coherent component in a reflection over the sea surface from an aircraft at 1500 m height.

A waveform and phase evolution analysis has been presented, allowing the processing of the coherent and total power components separately. The application of this technique to any GNSS-R signal opens new possible applications. Including a better resolution for altimetry products thanks to a better estimation of the peak position, the estimation of the wave period thanks to the identification of the secondary peaks present on the coherent component, and other derived products such as the sea state (U_{10}), and the sea surface height with improved resolution.

Finally, the combination of both precise altimetry measurements and secondary peak identification can provide a more complete analysis of the sea state, not only the swell period, but also the swell amplitude.

Author Contributions: Conceptualization, J.F.M.-M., H.P. and A.C.; methodology, J.F.M.-M.; software, J.F.M.-M.; validation, J.F.M.-M. and A.C.; formal analysis, J.F.M.-M.; investigation, J.F.M.-M.; resources, R.O., D.P., H.P., A.C., C.R., J.W. and A.M.; data curation J.F.M.-M., R.O., D.P.; visualization J.F.M.-M.; supervision, H.P., A.C.; project administration, A.C.; funding acquisition, A.C.; writing—original draft preparation, J.F.M-M.; writing—review and editing, J.F.M-M., H.P., A.C. All authors have read and agreed to the published version of the manuscript.

Funding: This work was founded by the Spanish Ministry of Science, Innovation and Universities, "Sensing with Pioneering Opportunistic Techniques", grant RTI2018-099008-B-C21, and the grant for recruitment of early-stage research staff FI-DGR 2015 of the AGAUR—Generalitat de Catalunya (FEDER), Spain, and Unidad de Excelencia María de Maeztu MDM-2016-060.

Acknowledgments: Thanks to the CommsSensLab administrative and research personnel and to the NanoSatLab members for the support.

Conflicts of Interest: The authors declare no conflict of interest.

References

1. Voronovich, A.G.; Zavorotny, V.U. The Transition From Weak to Strong Diffuse Radar Bistatic Scattering From Rough Ocean Surface. *IEEE Trans. Antennas Propag.* **2017**, *65*, 6029–6034. [CrossRef]
2. Carreno-Luengo, H.; Camps, A. Unified GNSS-R formulation including coherent and incoherent scattering components. In Proceedings of the 2016 IEEE International Geoscience and Remote Sensing Symposium (IGARSS), Beijing, China, 10–15 July 2016. [CrossRef]
3. Camps, A. Spatial Resolution in GNSS-R Under Coherent Scattering. *IEEE Geosci. Remote Sens. Lett.* **2020**, *17*, 32–36. [CrossRef]
4. Alonso-Arroyo, A.; Zavorotny, V.; Camps, A. Sea Ice Detection Using U.K. TDS-1 GNSS-R Data. *IEEE Trans. Geosci. Remote Sens.* **2017**, *55*, 4989–5001. [CrossRef]
5. Alonso-Arroyo, A.; Camps, A.; Park, H.; Pascual, D.; Onrubia, R.; Martin, F. Retrieval of Significant Wave Height and Mean Sea Surface Level Using the GNSS-R Interference Pattern Technique: Results From a Three-Month Field Campaign. *IEEE Trans. Geosci. Remote Sens.* **2015**, *53*, 3198–3209. [CrossRef]
6. Rodriguez Alvarez, N.; Camps, A.; Vall-llossera, M.; Bosch, X.; Monerris, A.; Ramos-Perez, I.; Valencia, E.; Marchan, J.; Martínez-Fernández, J.; Baroncini Turricchia, G.; et al. Land Geophysical Parameters Retrieval Using the Interference Pattern GNSS-R Technique. *IEEE Trans. Geosci. Remote Sens.* **2011**, *49*, 71–84. [CrossRef]
7. Carreno-Luengo, H.; Amézaga, A.; Vidal, D.; Olivé, R.; Martin, J.M.; Camps, A. First Polarimetric GNSS-R Measurements from a Stratospheric Flight over Boreal Forests. *Remote Sens.* **2015**, *7*, 13120–13138. [CrossRef]
8. Gerlein-Safdi, C.; Ruf, C.S. A CYGNSS-Based Algorithm for the Detection of Inland Waterbodies. *Geophys. Res. Lett.* **2019**, *46*, 12065–12072. [CrossRef]
9. Onrubia, R.; Pascual, D.; Park, H.; Camps, A.; Rüdiger, C.; Walker, J.; Monerris, A. Satellite Cross-Talk Impact Analysis in Airborne Interferometric Global Navigation Satellite System-Reflectometry with the Microwave Interferometric Reflectometer. *Remote Sens.* **2019**, *11*, 1120. [CrossRef]
10. Pascual, D.; Onrubia, R.; Alonso-Arroyo, A.; Park, H.; Camps, A. The microwave interferometric reflectometer. Part II: Back-end and processor descriptions. In Proceedings of the International Geoscience and Remote Sensing Symposium (IGARSS), Quebec City, QC, Canada, 13–18 July 2014; pp. 3782–3785. [CrossRef]
11. Valencia, E.; Camps, A.; Marchan-Hernandez, J.F.; Bosch-Lluis, X.; Rodriguez-Alvarez, N.; Ramos-Perez, I. Advanced architectures for real-time Delay-Doppler Map GNSS-reflectometers: The GPS reflectometer instrument for PAU (griPAU). *Adv. Space Res.* **2010**, *46*, 196–207. [CrossRef]
12. Zavorotny, V.U.; Gleason, S.; Cardellach, E.; Camps, A. Tutorial on Remote Sensing Using GNSS Bistatic Radar of Opportunity. *IEEE Geosci. Remote Sens. Mag.* **2014**, *2*, 8–45. [CrossRef]
13. Jales, P. Spaceborne Receiver Design for Scatterometric GNSS Reflectometry. Ph.D. Thesis, University of Surrey, Guildford, UK, 2012.
14. eoPortal Directory. Cyclone GNSS Mission Description Website. Available online: https://directory.eoportal.org/web/eoportal/satellite-missions/c-missions/cygnss (accessed on 2 January 2019).
15. Munoz-Martin, J.F.; Fernandez, L.; Ruiz-de-Azua, J.; Camps, A. The Flexible Microwave Payload -2: A combined GNSS-R and L-band radiometer with RFI mitigation payload for CubeSat-based Earth Observation Missions. In Proceedings of the 2019 IEEE International Geoscience and Remote Sensing Symposium (IGARSS), Yokohama, Japan, 28 July–2 August 2019.
16. Khan, R.; Khan, S.U.; Zaheer, R.; Khan, S. Acquisition strategies of GNSS receiver. In Proceedings of the International Conference on Computer Networks and Information Technology, Abbottabad, Pakistan, 11–13 July 2011; pp. 119–124. [CrossRef]
17. Martin, F.; Camps, A.; Fabra, F.; Rius, A.; Martin-Neira, M.; D'Addio, S.; Alonso, A. Mitigation of direct signal cross-talk and study of the coherent component in GNSS-R. *IEEE Geosci. Remote Sens. Lett.* **2015**, *12*, 279–283. [CrossRef]
18. Holzer, J.A.; Sung, C.C. Scattering of electromagnetic waves from a rough surface. II. *J. Appl. Phys.* **1978**, *49*, 1002–1011. [CrossRef]
19. Foucras, M.; Ekambi, B.; Bacard, F.; Julien, O.; Macabiau, C. Optimal GNSS acquisition parameters when considering bit transitions. In Proceedings of the 2014 IEEE/ION Position, Location and Navigation Symposium—PLANS 2014, Monterey, CA, USA, 5–8 May 2014; pp. 804–817. [CrossRef]

20. Camps, A.; Park, H.; Valencia I Domenech, E.; Pascual, D.; Martin, F.; Rius, A.; Ribo, S.; Benito, J.; Andres-Beivide, A.; Saameno, P.; et al. Optimization and performance analysis of interferometric GNSS-R altimeters: Application to the PARIS IoD mission. *IEEE J. Sel. Top. Appl. Earth Obs. Remote Sens.* **2014**, *7*, 1436–1451. [CrossRef]

21. Carreno-Luengo, H.; Camps, A. Empirical Results of a Surface-Level GNSS-R Experiment in a Wave Channel. *Remote Sens.* **2015**, *7*, 7471–7493. [CrossRef]

22. Fabra, F.; Cardellach, E.; Rius, A.; Ribo, S.; Oliveras, S.; Nogues-Correig, O.; Belmonte Rivas, M.; Semmling, M.; D'Addio, S. Phase Altimetry With Dual Polarization GNSS-R Over Sea Ice. *IEEE Trans. Geosci. Remote Sens.* **2012**, *50*, 2112–2121. [CrossRef]

23. Fabra, F.; Cardellach, E.; Ribo, S.; Li, W.; Rius, A.; Arco-Fernández, J.; Nogués-Correig, O.; Praks, J.; Rouhe, E.; Seppänen, J.; et al. Is Accurate Synoptic Altimetry Achievable by Means of Interferometric GNSS-R? *Remote Sens.* **2019**, *11*, 505. [CrossRef]

24. Park, H.; Camps, A.; Valencia, E.; Rodriguez Alvarez, N.; Bosch, X.; Ramos-Perez, I.; Carreno-Luengo, H. Retracking Considerations in Spaceborne GNSS-R Altimetry. *GPS Solut.* **2012**, *16*, 507–518. [CrossRef]

25. Zavorotny, V.U.; Voronovich, A.G. Scattering of GPS signals from the ocean with wind remote sensing application. *IEEE Trans. Geosci. Remote Sens.* **2000**, *38*, 951–964. [CrossRef]

26. Rius, A.; Cardellach, E.; Martin-Neira, M. Altimetric analysis of the sea-surface GPS-reflected signals. *IEEE Trans. Geosci. Remote Sens.* **2010**, *48*, 2119–2127. [CrossRef]

27. InMeteo. Ventusky. Available online: https://ventusky.com (accessed on 7 January 2020).

28. U.S. Army Corps of Engineers. *Coastal Engineering Manual Part II: Coastal Hydrodynamics (EM 1110-2-1100)*; U.S. Army Corps of Engineers: Washington, DC, USA, 2012.

remote sensing

MDPI

Article

Towards a Topographically-Accurate Reflection Point Prediction Algorithm for Operational Spaceborne GNSS Reflectometry—Development and Verification [†]

Lucinda King [1,*], Martin Unwin [2], Jonathan Rawlinson [2], Raffaella Guida [1] and Craig Underwood [1]

[1] Surrey Space Centre, University of Surrey, Guildford GU2 7HX, UK; r.guida@surrey.ac.uk (R.G.); c.underwood@surrey.ac.uk (C.U.)
[2] Surrey Satellite Technology Ltd., Guildford GU2 7YE, UK; m.unwin@sstl.co.uk (M.U.); j.rawlinson@sstl.co.uk (J.R.)
*   Correspondence: l.s.king@surrey.ac.uk
†   This paper is an extended version of our paper published in IGARSS 2020, 26 September–2 October 2020.

Abstract: GNSS Reflectometry (GNSS-R), a method of remote sensing using the reflections from satellite navigation systems, was initially envisaged for ocean wind speed sensing. In recent times there has been significant interest in the use of GNSS-R for sensing land parameters such as soil moisture, which has been identified as an Essential Climate Variable (ECV). Monitoring objectives for ECVs set by the Global Climate Observing System (GCOS) organisation include a reduction in data gaps from spaceborne sources. GNSS-R can be implemented on small, relatively cheap platforms and can enable the launch of constellations, thus reducing such data gaps in these important datasets. However in order to realise operational land sensing with GNSS-R, adaptations are required to existing instrumentation. Spaceborne GNSS-R requires the reflection points to be predicted in advance, and for land sensing this means the effect of topography must be considered. This paper presents an algorithm for on-board prediction of reflection points over the land, allowing generation of DDMs on-board as well as compression and calibration. The algorithm is tested using real satellite data from TechDemoSat-1 in a software receiver with on-board constraints being considered. Three different resolutions of Digital Elevation Model are compared. The algorithm is shown to perform better against the operational requirements of sensing land parameters than existing methods and is ready to proceed to flight testing.

Keywords: GNSS-R; topography; data compression; on-board data processing

Citation: King, L.; Unwin, M.; Rawlinson, J.; Guida, R.; Underwood, C. Towards a Topographically-Accurate Reflection Point Prediction Algorithm for Operational Spaceborne GNSS Reflectometry—Development and Verification. *Remote Sens.* **2021**, *13*, 1031. https://doi.org/10.3390/rs13051031

Academic Editors: Nereida Rodriguez-Alvarez and Mary Morris

Received: 20 January 2021
Accepted: 3 March 2021
Published: 9 March 2021

1. Introduction

GNSS reflectometry (GNSS-R) is a method of remote sensing which uses navigation signals such as those from GPS as "Signals of Opportunity" in a system similar to bistatic radar. The navigation satellites transmit L-band signals which are incident all over the Earth, that interact with and reflect from the surface (thereby acquiring information about the geometric, textural and dielectric properties of the surface) and are collected by a receiver, in this context a low Earth orbit spacecraft. The method has to-date most typically been used for sensing ocean wind-speed [1], but recent work has focused on the possibility of using the method to target land parameters such as soil moisture (e.g., [2,3]).

The primary observable of GNSS-R is the delay-Doppler map (DDM) which is built up by correlating the reflected signal with another copy (either the direct signal or a local replica—in this paper the latter is considered) shifted to have a range of different delays and Doppler values, thereby building up a map of reflected power centred around the delay and Doppler of the reflection point. This point is the specular point (SP) for forward-scattering GNSS-R, which is the most common form currently, although the issues addressed herein also apply to backscattering mode. As the reflected signals are very weak, the reflection points must be predicted in advance and tracked using an open loop method to direct the correlators to look for the reflected signal in the appropriate delay and Doppler search space [4].

The prediction of the reflection points requires the positions of the transmitter (the GNSS satellite), the receiver (the spacecraft in LEO receiving reflections—in this paper data from TechDemoSat-1 (TDS-1) is used [5])—and a model of the Earth's surface. Current GNSS-R spaceborne instrumentation uses a quasi-spherical Earth approximation, in which the Earth model used is the WGS-84 ellipsoid [6]. The quasi-spherical algorithm works by first scaling the Earth ellipsoid to a unit sphere, and applying the same transformation to the receiver and transmitter positions. The specular point is then determined in the transformed co-ordinates, using a method described in [7], before being scaled back to the ellipsoid. A full description of the method, including comparison with other methods and justification for its use on current instrumentation, can be found here [4].

The quasi-spherical method does not account for topography, but it has been sufficient for ocean scatterometry applications because the ocean surface approximately follows the geoid, with deviations less than ~100 m [8], and is unaffected by topography (see Figure 1a). For land sensing such an approximation will not always be sufficient due to the impact of terrestrial topography, which could result in inaccurate prediction of the reflection point, as shown in Figure 2. The centre of the DDM, or "cross-hairs", of both DDMs in Figure 1 corresponds to the predicted specular reflection point (using the quasi-spherical method). It should be noted that the quasi-spherical method does have its own errors in prediction of the specular point, even over the ocean, which have been shown in some cases to impact the inversion of the DDMs to ocean wind-speed values due to an asymmetry in power distribution in the bins around the SP [9]. As inversion mechanisms for land parameters typically use the peak reflectivity (derived from the peak power bin only) the impact of these errors is lessened but they should still be considered in future analysis, as any errors from the initial smooth-Earth estimate will be incorporated into the topographically-corrected result.

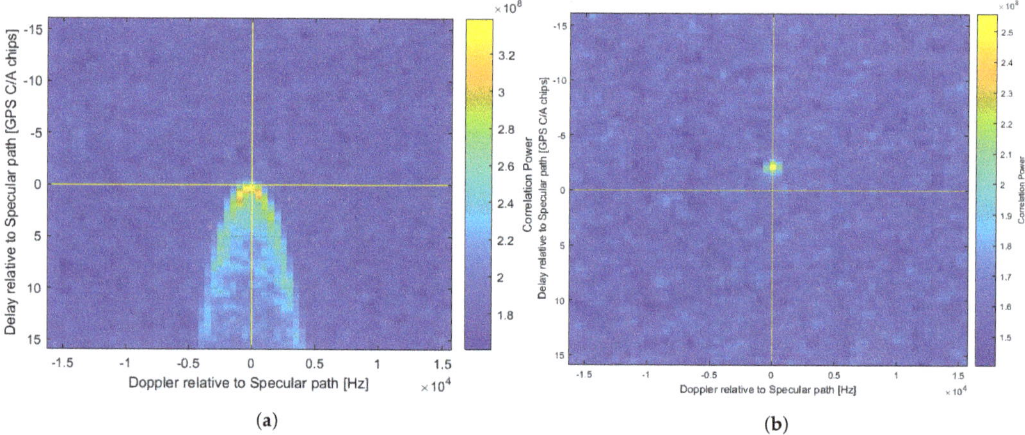

Figure 1. Example DDMs from different Earth surfaces. (**a**) A typical DDM collected over the ocean, with the characteristic horseshoe shape showing the spreading of power over the surface. The peak power is located on the DDM cross-hairs as the specular point has been predicted with sufficient accuracy by the quasi-spherical model. (**b**) A DDM from a dataset over low-elevation mountains in South Sudan (approximate elevation 500 m). The cross-hairs represent the predicted reflection point, showing the effect of topography on the location of the peak reflected power in the DDM.

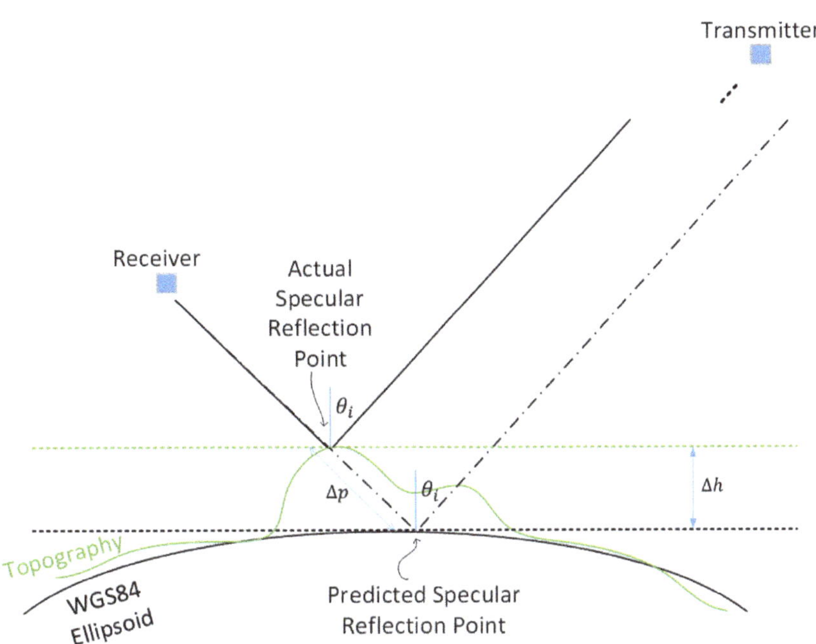

Figure 2. Diagram with simplified geometry demonstrating the influence of topography on specular reflection point prediction. The transmitter is considered to be far enough removed (20,200 km altitude for GPS) that the incoming direction of the incident signals at two nearby points can be assumed parallel. This diagram assumes that the elevated surface is flat, which is not often the case, but is a simplification to allow the most basic form of the algorithm to be implemented (see Section 2). Reproduced from [10] with permission.

This paper presents an algorithm for predicting GNSS-R reflection points in the presence of topography, which has been specifically designed for operational use on-board a small satellite. A recent paper by Gleason [11] has introduced a similar specular point prediction algorithm, developed independently, which includes the addition of a surface height term; however this is stated to only account for "low-lying land" and an operational limit is not given. In addition the algorithm in question was developed for a software receiver and although the paper mentions the possibility of its use on-board a reflectometry instrument, the actual implementation is not discussed. Other recent works include a method for modelling the effect of topography on DDMs [12], and a post-processing algorithm for accurately geolocating the reflection points in the presence of topography [13]. Whilst both these methods are valuable, they are designed for use on the ground. The objective of this paper is to introduce an algorithm that can work on-board, enabling reflected power to be captured by a GNSS-R instrument no matter the elevation of the reflecting surface—which is not currently possible. Should more accurate geolocation be required, this can be achieved on the ground using algorithms such as those referenced previously. In this way the algorithm presented here and the previously proposed post-processing algorithms are complementary.

The structure of this paper is as follows: in Section 2 there is a discussion of the motivation for using GNSS-R for sensing land parameters and the associated need to correct for topography when predicting reflection points. Section 3 contains a description of the proposed topographically-accurate reflection point prediction algorithm (TARPP) and the Digital Elevation Model chosen. The test procedures used to analyse the algorithm performance are given in Section 4 and results of these tests, with discussion, are presented in Section 5. The paper finishes with a conclusion in Section 7.

The study presented in this paper is a continuation of work introduced at the IGARSS conference in 2020 [10].

2. Motivation and Requirements

Land parameters such as soil moisture and Above Ground Biomass (AGB) have been designated as Essential Climate Variables (ECVs) by an international organisation, the Global Climate Observing System (GCOS) [14,15]. This confirms that measurement of these variables is critical for monitoring and modelling climate change. In addition, soil moisture, for example, is important for forecasting local weather and flood risks [16].

GCOS has identified monitoring objectives for the ECVs which include a call to reduce the spatial and temporal gaps in datasets collected from space [17]. GNSS-R is ideally placed to meet this challenge. The L-band signals are sensitive to parameters such as soil-moisture, freeze-thaw status and AGB [3,18–20]; and the low size, weight and power (SWaP) platforms used for the method enable—in a complementary way to other technologies— single satellites or even constellations to be built and launched more quickly, cheaply and flexibly than large traditional radar missions. GNSS-R technology must be adapted to optimally target land parameters and this includes developing means of accurately predicting the reflection points over the land.

An associated objective, which also requires accurate knowledge of the reflection points over the Earth's surface is that of data compression via windowing, which will also help improve instrumentation and support GNSS-R land sensing missions.

For both the objectives—land sensing over all elevations, and data compression— accurate prediction of the reflection points is required, for the following reasons. These are particularly relevant for downlink of on-board processed DDMs—rather than "raw" sampled IF (intermediate-frequency) data—as would be the case for an operational mission. These reasons can be translated into objectives for the topography algorithm, as shown in Table 1.

- The DDM is built up through correlation of the received reflected signals with a clean replica that is systematically shifted through the various combinations of delay and Doppler (d, D) to calculate the received power at every pixel. If the prediction of (d, D) of the specular reflection is correct, this reflected power will be present at the centre of the DDM. However, if the prediction is wrong, this power can appear at other positions in the DDM or sometimes may not appear at all (see Equation (3)). An example of the latter can be seen in Figure 1b, where the peak reflected power is offset from the centre due to topography. If the reflected power is not captured within the DDM at all then that measurement is lost.

- For forward-scattered DDMs, the centring of the reflected power allows the definition of a "Noise Box" in the null space at the top of the DDM, which corresponds to delays shorter than the predicted specular point (SP) delay—over the ocean such delays are typically not physically possible and so any power in this region can be assumed to be noise. This Noise Box is used to help measure noise power and thus calculate absolute received signal power from measured Signal-to-Noise Ratio (SNR). Reflections from unaccounted-for topography can contaminate the area resulting in inaccurate measurements or an inability to use the method at all [21].

- The minimum requirement for successful recording of reflected power on the space-craft is that the peak is located somewhere within the DDM (the effect on the Noise Box and thus calibration notwithstanding). If necessary, users can perform more accurate geo-location on the ground through re-processing. However, a stronger requirement is for the peak reflected power to be recorded as close to the centre of the DDM (delay offset = Doppler offset = 0) as possible at the on-board processing stage. This allows more efficient methods of data compression (i.e., windowing the DDM) to be used, which can be an enabler for usable data to be disseminated to users on the ground faster. This is key for soil moisture in particular. For certain missions e.g., those using very low power platforms, significant compression may be essential to close the data downlink budget. In general, regardless of the application or size of the platform, enabling better data compression is desirable for more efficient operation. For example, to enable 24/7 Level 1b DDM downlink operations on DoT-1 (an SSTL spacecraft on which this algorithm will be tested) it has been calculated that DDMs should be reduced to a window with 64 pixels in area. An 8×8 box has been chosen for initial testing (over e.g., 16×4) as this places a stronger test on the path length prediction requirement, and will also capture more of the DDM width, which may help to record multiple peaks if they should occur in e.g., mountainous terrain. Figure 3 demonstrates the impact of this desired windowing on DDMs generated with and without a topographically accurate algorithm.

- Some mission concepts that have recently been developed have proposed a so-called "coherent channel" to track the peak reflected power with longer coherent integration times, which allows carrier phase information to be preserved. As this results in twice as much data (both I and Q components are stored), the channel should ideally monitor one pixel of the DDM only, and therefore the peak reflected power must be located accurately in this coherent pixel. An accurate topography algorithm is part of the process for achieving this.

Table 1. Spectrum of Objectives for Topography Algorithm. The stretch objective for coherent pixel tracking is not targeted by the TARPP v1 algorithm tested as part of this study, but is included to give the full context of the current work and future directions.

Algorithm Outcome	Land Sensing Enabled Over Whole Globe	Noise Box Preserved	Data Compression Enabled (Windowing)	Coherent Pixel Tracking
Baseline Outcome Requirement: SP power captured within 128×20 pixel window	✓	Not guaranteed	×	×
Good Outcome Requirement: SP power captured within 8×8 pixel window	✓	✓	✓	×
Best Outcome—"Stretch Objective" Requirement: Location of SP power in DDM window known to 1 pixel.	✓	✓	✓	✓

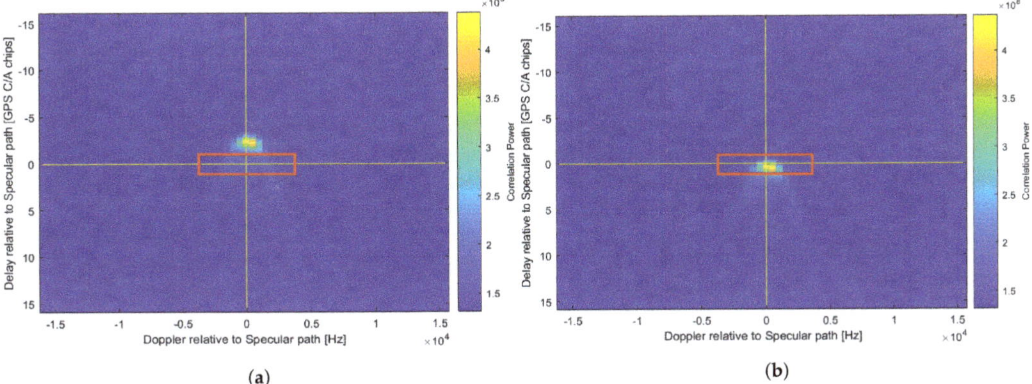

Figure 3. These DDMs demonstrate the impact of topography on the DDM and the impact of compressing the DDM using windowing. The red box represents an 8×8 pixel window. In this case, without the topography algorithm the peak power data would be lost as it falls outside the window. The surface elevation where this DDM was collected was approximately 400 m. (**a**) Land DDM generated with quasi-spherical algorithm. (**b**) The same DDM generated with the topography algorithm.

In a TechDemoSat-1 (TDS-1) DDM the centre, or cross-hairs, corresponds to the delay-Doppler (d, D) value of the specular reflection point. The DDM is normalised so that this point has $d = D = 0$, and other delays and Dopplers are measured relative to this. In a TDS-1 Level 1b DDM the delay dimension extends over 128 pixels, equivalent to 16 chips (each chip is equivalent to 293 m, or 977 ns, for the GPS L1 C/A code) on either side of the zero-delay point, and 10 Doppler pixels each way. The raw TDS-1 data used in this study (discussed in Section 4) has been processed by a software receiver and has 64 Doppler pixels in total (slightly more than the L1b DDMs so more of the search space is visible for analysis- it does not effect the processing of the DDMs otherwise). These extend to approximately ±15 kHz relative to the specular point (see Figure 1b). Considering the delay dimension, if the path length of the reflection is predicted with an inaccuracy greater than 16 chips due to omitted topography, the reflected power will not appear in the DDM. Using a simplified view of the Earth's surface (not accounting for surface slopes), as shown in Figure 2, an elevation above the ellipsoid greater than ~2300 m is sufficient to produce these path length differences or greater. This is calculated using the following:

$$\frac{2\Delta h}{cos(\theta_i)} = \Delta p \tag{1}$$

$$16 \text{ chips} = 16 \times 293 \text{ m}$$
$$= 4688 \text{ m} \tag{2}$$

$$\Delta h > \frac{4688}{2} cos(\theta_i) \text{ m} \tag{3}$$

where Δh is the elevation above the ellipsoid, θ_i is the incidence angle between reflection point and receiver and Δp is the resulting path length change.
For the limiting case of $\theta_i = 0$ (nadir),

$$\Delta h > 2344 \text{ m} \tag{4}$$

For high incidence angles, smaller height deltas above the ellipsoid can result in the same path length change, which will increase the required prediction accuracy. For example, assuming 70° as the incidence angle limit means that elevations over 800 m will result in unacceptable path length changes, unless the elevation above the ellipsoid is accounted for.

According to hypsometric curves of the Earth's surface ([22]) ~10% of the terrestrial surface is over 2300 m elevation above sea level. This includes regions such as the Tibetan Plateau (average elevation over 4000 m), an area which has a well-documented impact on the Asian climate and weather and for which monitoring of soil moisture is therefore very important [23]. This is just one example of an area requiring denser spatio-temporal sampling from satellite measurements, which could be addressed by GNSS-R technology— but only if the topography requirements are addressed.

The percentage of surface potentially creating reflections outside of the DDM is substantially greater than 10% once data compression is considered. For example, in standard mode CYGNSS compresses the downlinked Level 1 DDMs produced on-board to 1 delay chip "above" and 3 "below" the nominal specular point (17 delay × 11 Doppler pixels) [24]. If 70° is again assumed as the incidence angle limit, then heights above 50 m will cause the reflections to fall outside the downlinked DDM. This corresponds to ~90% of the terrestrial surface.

It is acknowledged that the assumption that the elevated surface is flat, and that the effect of topography does not shift the reflection point laterally, is a simplification which could lead to inaccuracies in the specular point estimation. However, this assumption has been made for this algorithm as the intention is to operate on-board small satellites where non-iterative algorithms are preferred. The intention is to explore what can be achieved with a very simple algorithm, which will open up the possibilities for the implementation of the technology. Future work (discussed in Section 6 will include development of more accurate algorithms including e.g., surface slopes, and the possibilities of multiple reflection points.

3. Algorithm

3.1. Baseline Algorithm (Elevation Only)—TARPP v1

The baseline algorithm that has been developed is described in Figure 4 and is referred to as TARPP v1—a Topographically Accurate Reflection Point Predictor (version 1). This algorithm is non-iterative, which is preferable for inclusion in flight software, and applies a height correction extracted from an on-board Digital Elevation Model (DEM) to an initial "smooth-Earth" estimate—"smooth" here meaning that the estimate uses a purely mathematical model of the Earth which does not include topography. In this case the initial estimate is generated using the existing quasi-spherical method, which is described elsewhere [4,21]. The algorithm refers to a "Reflection Point" rather than a "Specular Point" because although this study concerns forward-scattering GNSS-R, in general the algorithm could be applied to any predicted reflection point, including for example backscattered reflection points, which will be addressed in future work.

Highly accurate geolocation in both horizontal and vertical dimensions, such as for altimetry, is not the primary focus of the algorithm, however it will be ensured that future operational implementation of the algorithm will be transparent so any effects can be undone by users, should they wish to recalculate a more accurate reflection point on the ground.

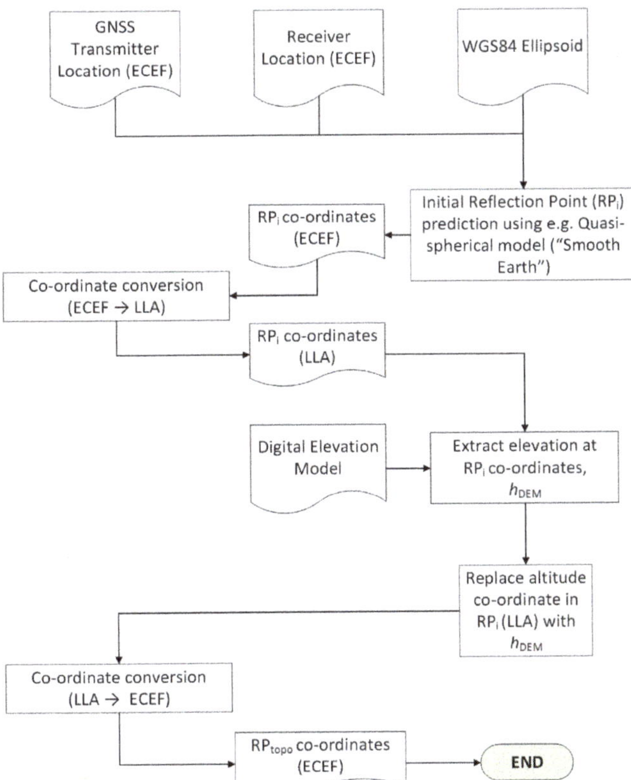

Figure 4. Diagram of the baseline version of the Topographically Accurate Reflection Point Prediction algorithm (TARPP v1).

The initial estimate is calculated in Earth-Centred-Earth-Fixed (ECEF) co-ordinates using the receiver and transmitter positions and the WGS84 ellipsoid. The point co-ordinates are then transformed into Latitude-Longitude-Altitude (LLA) format and used to index into a DEM and extract the elevation at that point. The altitude co-ordinate, originally zero, is then set to this elevation value and the point transformed back to ECEF.

The TARPP v1 algorithm has been developed in MATLAB as part of a software-defined GNSS-R receiver that processes raw TDS-1 data. It has undergone testing in this environment, the method and results of which are discussed in Sections 4 and 5 respectively.

3.2. Digital Elevation Models and On-Board Constraints

A key component of the algorithm is the Digital Elevation Model (DEM) from which the elevation is extracted, and therefore the choice of DEM should be evaluated carefully. Criteria to be considered include the resolution, extent over the Earth and data sources. The EarthEnv suite of topographic variables [25] provides several DEMs of various horizontal resolutions and aggregation methods. DEMs from this suite have been chosen for use as they are near-global in scope (84N - 56S), and the availability of other resolution DEMs and slope data within the suite which have been generated using the same source data will make comparison of different resolutions, and the development of the next stage of the algorithm, simple and consistent. The EarthEnv topographic variables, including the DEMs, use the 250 m GMTED2010 [26] as the primary source, with the 90 m SRTM4.1dev [27] used for validation. The accuracy of GMTED2010 is affected by the presence of vegetation, however this is non-trivial to resolve and beyond the scope of this study. Approximately 10.9 m of the global mean error of SRTM, on which GMTED is based, is due to vegetation [28], however in the worst case this would result in a few pixels of error in the reflection point location in the DDM. This means it would need to be accounted for to achieve the "stretch objective" of coherent pixel tracking shown in Table 1, but this is not a goal of the TARPP v1 algorithm and so is not considered further here.

There are different options available for the method used to aggregate the source data to the different spatial grains. The mean was chosen, rather than the minimum or maximum, to reduce the possibility that outliers would result in a significant over or under compensation of the elevation at a certain point. It should be noted that the EarthEnv DEM is referenced to the EGM96 geoid surface but the positions of the transmitter and receiver, which are calculated using GNSS solutions, are referenced to the WGS84 ellipsoid. As discussed in Section 1, the geoid and the ellipsoid do not deviate by more than ±100 m. Initial testing on a selected dataset in the Taklamakan desert region which showed a bias demonstrated that including a geoid correction in the DEM did not show an improvement to the bias in this case. Therefore due to this, and to expedite the analysis, the geoid undulations have not been included in the DEMs for the purposes of this study.

The operational constraints affecting the choice of DEM for a spaceborne platform are:

- Uplink rate, affecting file size.
- On-board storage, also affecting file size.
- Processing requirements for unpacking and using on-board.

The larger the file to be uploaded the more passes required, which can stretch over several days. For future missions using on-board topography algorithms, there will be the option to upload the DEM on the ground before launch, which mitigates this constraint. However there are several GNSS-R capable spacecraft currently in orbit that could be adapted to address the land sensing monitoring objectives and so consideration of uplink constraints is important. The file size limit will constrain the horizontal and vertical resolutions of the DEM chosen for use.

The other aspect of the resolution trade-off is the comparison of the horizontal resolution with the spatial resolution of GNSS-R scatterometry over the land. As discussed in recent literature [29,30], scattering over the land surface is a complex mix of coherent and incoherent scattering, affected by many variables such as surface roughness, vegetation cover and large-scale topography. Purely coherent scattering, with spatial resolution comparable to the first Fresnel Zone ($\sim$1 km) is possible, but is mostly limited to inland water bodies, a very small fraction of the land surface. Primarily, the scattering over the land surface has a large diffuse component and so has a lower spatial resolution. However, the horizontal resolution of the DEM should be chosen to correspond with the best possible case of spatial resolution, that of coherent scattering, so that in these cases the DEM resolution is not broader than the scattering resolution. This instantaneous spatial resolution, of the order of 1 km, is combined with the motion of the reflection point over the surface during the period of incoherent integration (in this case 1 s), which is dominated by the motion of the LEO receiver. Therefore the resolution is of the order of 7 km in the along-track dimension, and the chosen horizontal DEM resolution should be comparable with this.

Regarding the vertical resolution, if a 1 m level of precision is assumed (storing the values as 16 bit signed integers) this gives a maximum error in a stored value of 0.5 m. Using Equation (1) this gives a worst case error in the path length of 57 m. This is $\sim$1/5 of a chip which is acceptable with respect to the requirement to centre the reflected power in the DDM window. Storing the DEM values as integers significantly reduces the size of the DEM. The EarthEnv DEM values range from -412 m to 7919 m which can be adequately represented using 16 bit signed integers. There is no benefit to switching to unsigned integers as the range is more than large enough, and it would require adding a constant factor to every DEM value. Although this is trivial to implement in software, it is an unnecessary complication. On other platforms, if the available memory on-board is severely limited the DEM values could be represented by unsigned 8 bit integers and similarly converted using additive and multiplicative factors, however this would have an impact on the vertical accuracy and result in an increased path-length error.

This algorithm has testing planned on-board the SSTL DoT-1 technology demonstration satellite and so this spacecraft has been used as an example for the development and testing of the algorithm. DoT-1 is a 17.5 kg satellite in a 530 km polar orbit, with an S-band up and downlink. The uplink rate is the dominant operational constraint in this case, as the satellite is already in orbit. Flight software on DoT-1 is stored in non-volatile memory, with 128 MB available, and the DEM must fit within these bounds. The EarthEnv 5 km and 10 km resolution DEMs meet the on-board memory constraints, and so these will be chosen for test. The 1 km DEM will not feasibly fit on-board but has been selected for comparison. Therefore the DEM provisionally selected for use on-board is the 5 km resolution, and will be compared with the higher and lower resolution DEMs, 1 km and 10 km.

4. Test Procedure

4.1. Datasets

This study has made use of data from the TDS-1 satellite, which are generated in two forms on-board. Firstly, Level 1a DDMs, which are combined with metadata to form Level 1b DDMs which are then disseminated via the MERRByS online portal [31]. These are 128 by 20 pixels in delay and Doppler respectively—details of how the on-board instrument generated these can be found in [32]. Secondly, Level 0 "raw" data—sampled IF data which were not processed into DDMs on-board. A small number of sample Level 0 datasets are available from MERRByS, however this study had access to a larger catalogue of raw data.

The advantage of using raw data is that it allows simulation of on-board processing using real satellite data, with the ability to apply different processing methods to those used on-board currently. A disadvantage is that the results are dependent on the quality and characteristics of the real data.

There were approximately 60 datasets containing land or partial land reflections in the TDS-1 raw data catalogue, six of which were selected for intensive testing of the algorithm, as described in Table 2. Each dataset contained reflection tracks from several GNSS satellites (only GPS reflections have been considered in this study) and these are distinguished by their PRN number, which refers to the unique Pseudo-Random Noise code which modulates the carrier wave for that satellite [33].

Table 2. Raw TDS-1 datasets used in this study.

Dataset Test ID	Collection Date and Timeslot	No. of DDMs	Approx. Location	Land Cover	PRN	Track Start			Track End		
						Co-Ordinate	Elevation (m)	El. angle at SP (°)	Co-Ordinate	Elevation (m)	El. angle at SP (°)
TT1	2015-01-11, H18	126	Texas, USA	Open shrubland and grassland. Some mountainous regions.	15	35.585, −97.421	316	65.3	30.627, −98.897	385	61.1
					18	35.650, −103.721	1259	59.7	30.596, −104.891	994	59.1
					21	36.702, −101.193	861	69.3	31.637, −102.525	849	64.3
TT2	2015-06-05, H12	129	Sahara Desert (Algeria to Mali)	Desert	12	25.665, −2.526	298	66.1	18.709, −4.045	284	66.7
					15	24.408, 0.665	332	74.0	17.309, −1.029	288	79.0
					24	29.132, −0.741	342	56.0	22.345, −2.368	352	48.3
TT3	2015-11-04, H06	130	Far Eastern China (Taklamakan Desert) to N. India.	Desert and mountains	5	41.255, 81.677	985	61.2	34.417, 79.510	5504	52.9
					20	37.967, 76.976	1522	67.3	30.982, 75.161	216	75.9
					29	40.964, 76.881	3910	69.0	33.818, 74.953	1631	64.1
TT4	2015-11-19, H18	130	Arizona to Californian coast, USA	Desert and shrubland, mountains (Sierra Nevada). Some tracks cross the ocean.	15	36.069, −110.878	1713	53.1	29.163, −113.060	0	49.9
					18	35.530, −117.960	948	67.9	28.472, −119.648	0	70.9
					20	38.840, −111.737	2050	50.9	32.107, −113.652	388	42.9
					21	38.055, −116.500	2060	67.5	31.018, −118.364	0	60.0
TT5	2018-10-06, H12	128	South Sudan to DR Congo	Savanna and forests. Some rivers.	8	1.589, 27.740	682	46.6	−5.211, 26.360	584	54.2
					11	6.872, 27.991	466	75.9	−0.195, 26.493	585	71.4
					18	7.140, 30.227	414	73.4	0.097, 28.772	1166	66.2
					23	6.338, 25.489	649	54.9	−0.560, 24.021	458	55.0
TT6	2018-10-11, H12	129	DR Congo	Forest, some mountainous regions, savanna. Some lakes.	8	−7.152, 27.347	639	49.3	−14.058, 25.878	1115	57.2
					11	−2.067, 27.266	939	77.4	−9.205, 25.726	795	71.7
					18	−1.813, 29.517	2158	70.9	−8.937, 28.073	1452	64.0

4.2. Method—MATLAB Testing

The aim of the algorithm is to centre the peak reflected power within the DDM window to ensure that the data is captured despite high elevation reflection surfaces and compression constraints. Therefore analysing the position of the peak power pixel relative to the centre of the DDM is a good test for the performance of the algorithm. Whilst it is accepted that in the case of diffuse scattering the peak of the DDM can lag in the delay axis behind the true position of the specular point [34], the testing will proceed with the peak power being used as a proxy for the specular point, as this is currently accurate enough for our purposes.

For testing purposes the TARPP v1 algorithm was implemented in MATLAB as a module of an existing GNSS-R software receiver. This software receiver is capable of processing raw TDS-1 data into DDMs using a variety of methods, including the quasi-spherical method currently used on-board (described in Section 1). The six test datasets were processed first using this method and then using the TARPP algorithm, with three different DEM horizontal resolutions −1, 5 and 10 km. A 5 km resolution was the provisional value chosen for on-board use as discussed in Section 3.2, and a resolution either side was chosen for comparison.

4.3. Analysis 1a—Peak Power Offset Graphs

The first analysis assessed the offset of the peak power from the centre of the DDM, using all pixels within ±10 Doppler pixels of the delay axis to simulate the current on-board DDM window (128 × 20 pixels). Within the test datasets there were several DDMs of reflections which had very weak coherent components and in these cases the DDMs had a noise-like quality and the highest power pixel was randomly distributed, as can be seen in the Results section. Preliminary work using a Gaussian filter to smooth the DDMs before running the analysis did not show an improvement in the results. At this stage a consistent method for automatically removing the noise-like DDMs has not been resolved and is still under development. Not all of the large delay offsets seen in the results are due to noise, as some coherent reflections can be seen at offsets away from the centre, even after the topography algorithm has been applied. This will be addressed further in the Results and Discussion section. This analysis was conducted on each test dataset separately.

4.4. Analysis 1b—Peak Power Offset Histograms

This analysis used histograms to display the distribution of the peak power offset from the centre. It was conducted on both the test datasets indivdually and a consolidated set of data from all PRNs from all datasets, to give an overall assessment of the algorithm performance.

4.5. Analysis 2—Compression Boxes

This analysis tested the ability of the algorithm to place the peak power pixel within various sized boxes around the centre of the DDM, simulating different window sizes that could be used for data compression. The box sizes tested were 4 × 4 pixels, 8 × 8 pixels and 20 × 20 pixels. The 8 × 8 pixel box could theoretically enable 24/7 operations on DoT-1, as discussed in Section 2. This analysis was conducted on each test dataset separately and the results were also combined to form an overall score for each DEM resolution for the consolidated dataset.

5. Results and Discussion

This section contains the test results of the TARPP v1 algorithm, using the analyses described in Section 4 on the datasets given in Table 2. Each section includes a comparison of the different DEM resolutions.

5.1. Analysis 1a

Analysis 1a presents the offset of the peak power from the DDM centre throughout the full DDM track. For conciseness, a graph for one PRN from each dataset has been selected for presentation, which is representative of the results from that test dataset. The results, as shown in Figure 5, show that the TARPP algorithm for all DEM resolutions is successful at placing the peak power pixel close to the centre of the DDM window, significantly better than the quasi-spherical model. There is some noise in the results due to some DDMs presenting incoherent scattering as discussed in Section 4.3, however the overall impression is that the algorithm improves the placement of the peak.

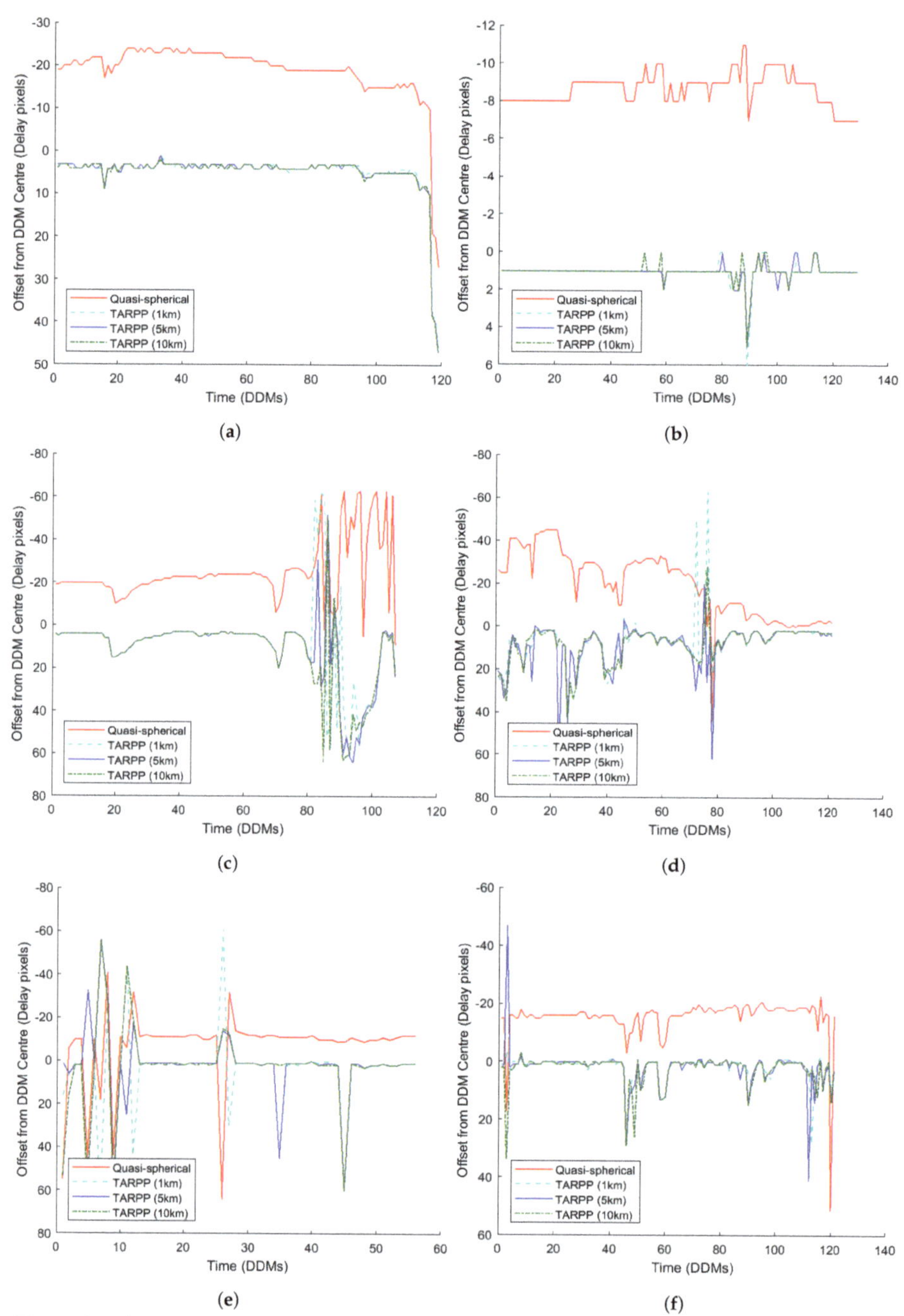

Figure 5. The results of Analysis 1a for an example PRN from each test dataset—plots of peak power offset from the centre of the DDM. (**a**) TT1, PRN 21; (**b**) TT2, PRN 15; (**c**) TT3, PRN 05; (**d**) TT4, PRN 20; (**e**) TT5, PRN 08; (**f**) TT6, PRN 11.

There are several points of interest: firstly, in many of the datasets there appears to be a constant bias in the location of the peak pixel below the centre of the DDM. Whilst this is preferable to a bias above the DDM centre, which could lead to contamination of the noise box, the cause should still be discovered and mitigated. The exact value of the bias is not common to all datasets and PRNs and so is unlikely to be a systematic, instrument-based bias however this cannot be said for sure without further tests. As discussed in Section 3.2, a similar bias found in another dataset was predicted to be caused by geoid undulations. However, analysis found that this was not the case, and so it is not predicted to be the reason here. Future analysis will investigate this possibility more thoroughly.

Secondly, taking TT1 as an example, there is a feature present in the data from all three resolutions towards the end of the DDM track which shows the peak power pixel shifting smoothly to greater delay values. This behaviour could be explained by a gradual decrease in elevation along the track of the specular point which is not accounted for by the DEM, however this is not corroborated by elevation maps of the area. Both these phenomena will be subject to further investigation.

Thirdly, in some instances, such as in the "Time (DDMs) > 80" portion of Figure 5c, the errors of the quasi-spherical and TARPP series appear to trend in opposite senses. In this case, this is believed to be due to very high elevations towards the end of the track (it ends at ∼5500 m). This would cause the peak reflected power to fall a large number of delay pixels outside the DDM—corresponding visually to the space "above" the DDM. The extended "arms" of the DDM horseshoe (see Figure 1a), which although lower in power than the peak are still higher than the noise background, could then contaminate the top portion of the DDM. This could cause the peak value to be recorded at low delay pixel values more often—corresponding to the top half of the graphs in Figure 5. Conversely, the TARPP method corrects the peak pixel on or close to the centre of the DDM, leaving the top part of the DDMs uncontaminated, other than the effects of incoherent scattering and noise. The simplifying assumptions of the TARPP v1 algorithm, which will be addressed in further developments, do not account for the possibilities of multiple reflection points, which may be contaminating the lower portion of the DDMs following application of the TARPP algorithm.

5.2. Analysis 1b

Analysis 1b was conducted on the six test datasets both individually and as a consolidated set, however for conciseness here only the consolidated results are presented. Figure 6 contains histograms describing the position of the peak power pixel relative to the centre of the DDM for a combination of the data from all six datasets. Figure 6a–c show that the TARPP algorithm, for all resolutions of DEM, is much better at placing the peak power pixel at, or close to, the centre of the DDM window than the quasi-spherical model. Figure 6d shows that the performance of the three DEM resolutions is very similar, with the 5 km resolution showing a very slight advantage over the others. The distribution is skewed to the right meaning it is less likely that there will be contamination of the noise box even if the peak pixel is not placed exactly at the centre.

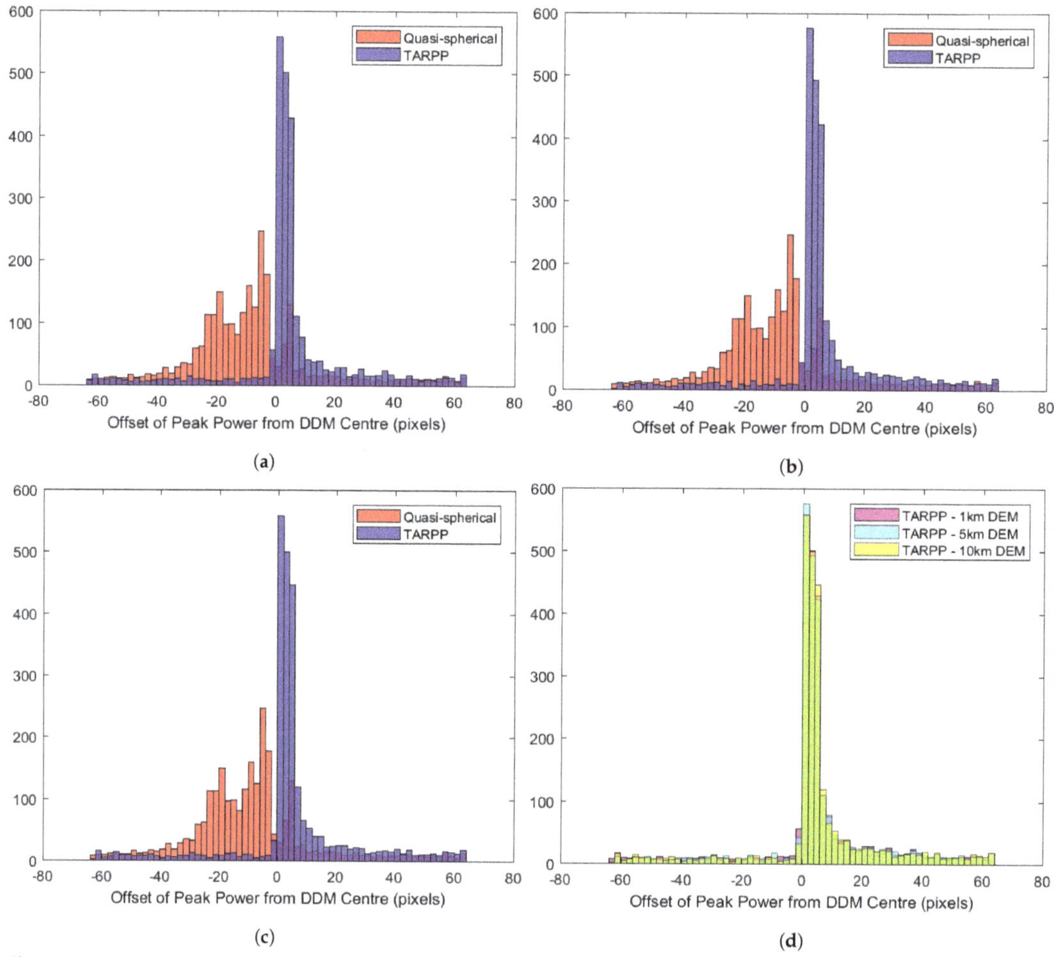

Figure 6. Histograms, generated using the combined data of all DDMs from all six datasets, showing the offset of the peak power pixel from the centre of the DDM. The three DEM resolutions are shown and compared with the quasi-spherical model in (**a**–**d**) shows the three DEM resolutions directly compared to each other. All three resolutions have a similar performance and thus the histograms nearly overlay each other, and so appear green. (**a**) 1 km DEM; (**b**) 5 km DEM; (**c**) 10 km DEM; (**d**) DEM Comparison.

5.3. Analysis 2

Analysis 2 tested how often the peak pixel was located within various sized boxes around the centre of the DDM, and was conducted on each test dataset individually and a consolidated set. Table 3 contains the results of Analysis 2 on the TT1 dataset, which is provided as an example to show the significant variation between the PRNs (which have different tracks on the Earth's surface). Reasons for the differences may include different surface roughness and different elevation angles to each of the transmitting satellites. In addition, as discussed in Section 4.3, many of the DDMs in all of the test datasets had a noise-like quality. This is due to a weak coherent scattering component, which can be caused by surface roughness or attenuation by vegetation. This could account for some of the differences between the PRNs, and is also a contributing reason to why the scores are lower than might be expected when considered in absolute terms, particularly when compared with the histograms in Figure 6 for example. However, for the purposes of this study, a comparison is sufficient to demonstrate the improvement offered by the TARPP algorithm. In future work, a robust method for cleaning the datasets to leave coherent DDMs, which will lend themselves better to testing using these methods, will be developed.

Table 3. Results of Analysis 2 on the TT1 dataset.

Dataset and PRN	No. of DDMs	Box Size (Pixels)	Score (% Success Rate of Placing Pixel in Box)			
			Quasi-Spherical	TARRP—1 km	TARPP—5 km	TARPP—10 km
TT1, 15	126	4 × 4	1.6	4.8	4.8	4.8
		8 × 8	23.8	68.3	72.2	74.6
		20 × 20	88.9	85.7	85.7	85.7
TT1, 18	126	4 × 4	0	5.6	11.9	8.7
		8 × 8	1.6	55.6	54.0	51.6
		20 × 20	4.8	81.7	81.0	79.4
TT1, 21	126	4 × 4	0	1.6	0.8	0.8
		8 × 8	2.4	77.0	72.2	72.2
		20 × 20	4.8	92.9	92.9	92.9

Table 4 shows the results of Analysis 2 with the data from all the datasets and PRNs combined using a weighted average, taking into account that there were slightly different numbers of DDMs for each dataset and PRN (due to different track lengths). This corroborates the results shown in Figure 6—that the three DEM resolutions each show very similar performances, with the 5 km resolution showing a very slight advantage (though this could be negligible when noise is considered). The 8 × 8 box size, which could theoretically allow 24/7 operations on DoT-1 through DDM windowing and thus data compression, has been tested. The results show that with the current test method there is a five-fold improvement over the quasi-spherical method. A 55% success rate of capturing the peak reflected power with this window size is not adequate for a fully operational service however it is a significant step closer than using the current methods. It should be noted that the lack of a 100% score, for example, does not necessarily mean that the algorithm stops working at certain points in the track, but rather that the signal may have been attenuated as described above and thus the maximum power no longer comes from the specular point pixel. More advanced test methods to be undertaken in future work should give a better indication of success rate.

Table 4. Results of Analysis 2 on all datasets, combined as a weighted average using number of DDMs for each dataset and PRN.

Box Size (Pixels)	Weighted Average Score (% Success Rate of Placing Pixel in Box)			
	Quasi-Spherical	TARRP—1 km	TARPP—5 km	TARPP—10 km
4 × 4	1.05	30.18	30.77	30.36
8 × 8	10.35	55.52	55.09	54.46
20 × 20	35.20	67.72	67.76	67.19

None of the three DEM resolutions shows a significant improvement over the others. It was expected that the higher the resolution, the better the performance, however this has not been demonstrated in the results. This may be explained by incoherent scattering dominating in the test datasets, causing the spatial resolution of the scattering to be larger than all three of the DEM resolutions, so that there is not significant distinction between them.

6. Future Work

This study has shown that although the baseline algorithm TARPP v1 is simple, it is able to improve the placement of the peak power pixel in the DDM window over existing methods. However, it makes the assumption that the land surface is flat (as well as elevated), when this is not very often the case. Therefore development has now begun on the next stage of the algorithm, TARPP v2.

The TARPP v2 algorithm will include surface slopes to more accurately assess the equality of incidence and reflection angles, a criteria for determining the specular reflection point. Two components of slope from DEM data will be used—slope data products are typically provided along N-S and E-W lines, however the slope that is important to the reflection process is the slope projected in the incidence plane, defined for bistatic radar as the plane containing the transmitter, receiver and specular point positions (providing a smooth-Earth approximation is used). Complications arise when it is considered that the effect of topography may now mean that the shortest path to the ground is not the one that lays in the nominal incidence plane, and this will be considered during the development of the algorithm.

A trade-off will be conducted as to whether the slope components will be stored on-board, or calculated in real-time from the existing elevation values. Two components of slope plus elevation at every point would result in a trebling of the required data, affecting uplink times and storage, if the storage option is chosen. There will also be consideration of multiple reflection points, which can occur in regions of mountainous terrain or near cliffs.

It is expected that the TARPP v2 algorithm will show improved results compared with TARPP v1, including demonstrating the predicted improvement in performance with resolution (to a point—once the resolution is comparable to the best-possible spatial resolution of reflectometry, i.e., 1 km, it is unlikely that further improvements will be gained by increasing the resolution). As the envelope of GNSS reflectometry technology continues to be pushed and the use cases expanded, there will be a need for even more accurate specular point prediction. Therefore incorporated into the future work on the TARPP algorithm will be a consideration of the error terms introduced by the initial smooth-Earth estimate, including a comparison with the ellipsoidal Earth model.

7. Conclusions

This paper has presented the motivations for sensing land parameters with GNSS-R and demonstrated that a reflection point prediction method which takes topography into account is required to enable this. The TARPP v1 algorithm, which has been developed to meet this need on-board operational GNSS-R satellites, has been described along with planned future developments. Testing of this algorithm has been conducted using raw data from TDS-1 in a software receiver, and three horizontal resolutions of DEM have been tested with the algorithm -1 km, 5 km and 10 km.

The results show that the TARPP v1 algorithm performs better than the quasi-spherical model at centring the peak reflected power in a DDM, meeting the operational requirements significantly more often. It has also been shown to be suitable for operational use on-board a small reflectometry satellite. An interesting result is that no one of the DEM resolutions demonstrated significantly better results than the others, which is possibly explained by the largely incoherent nature of land surface scattering. The flight stage of testing will proceed with the 5 km resolution DEM, which has been shown to fit on-board the DoT-1 satellite. TARPP v1 has been written into C code ready for deployment to the satellite at a suitable time and testing will be undertaken in the near future.

Demonstrating that a topography algorithm can be used successfully on-board an operational spacecraft is a significant step towards 24/7 land sensing operations and the deployment of GNSS-R missions targeting Essential Climate Variables. This will help to achieve the GCOS goals of reducing satellite-sourced ECV data gaps.

Author Contributions: Conceptualization, L.K., M.U. and J.R.; methodology, L.K.; software, L.K., J.R., and M.U.; validation, L.K.; formal analysis, L.K.; investigation, L.K.; resources, M.U., J.R. and L.K.; data curation, M.U., J.R. and L.K.; writing—original draft preparation, L.K.; writing—review and editing, R.G., M.U., C.U. and L.K.; visualization, L.K.; supervision, C.U., M.U. and R.G.; project administration, L.K.; funding acquisition, R.G., C.U., M.U. and L.K. All authors have read and agreed to the published version of the manuscript.

Funding: This research was funded by the Natural Environment Research Council via Lucinda King's SCENARIO PhD studentship, grant number NE/L002566/1. The PhD stipend was topped-up by Surrey Satellite Technology Ltd. The APC was funded by the University of Surrey.

Data Availability Statement: The data presented in this study are available in Figshare at https://doi.org/10.6084/m9.figshare.14171153.v1. They are available under a CC BY 4.0 license, and if used should be attributed to SSTL and Lucinda King.

Acknowledgments: Lucinda King would like to thank Surrey Satellite Technology Ltd. for providing the raw TDS-1 data used in this research.

Conflicts of Interest: The authors state that funding comes in part from Surrey Satellite Technology Ltd. (SSTL), owner of the TDS-1 and DoT-1 satellites and co-sponsor of the PhD study. SSTL offered support in design, collection, analysis and interpretation of data. The majority of PhD funding is from the Natural Environment Research Council who played no role in the design of the study; in the collection, analyses, or interpretation of data; in the writing of the manuscript, or in the decision to publish the results.

References

1. Foti, G.; Gommenginger, C.; Jales, P.; Unwin, M.; Shaw, A.; Robertson, C.; Rosellõ, J. Spaceborne GNSS reflectometry for ocean winds: First results from the UK TechDemoSat-1 mission. *Geophys. Res. Lett.* **2015**, *42*, 5435–5441, doi:10.1002/2015GL064204.
2. Camps, A.; Park, H.; Pablos, M.; Foti, G.; Gommenginger, C.P.; Liu, P.W.; Judge, J. Sensitivity of GNSS-R Spaceborne Observations to Soil Moisture and Vegetation. *IEEE J. Sel. Top. Appl. Earth Obs. Remote. Sens.* **2016**, *9*, 4730–4742, doi:10.1109/JSTARS.2016.2588467.
3. Chew, C.C.; Small, E.E. Soil Moisture Sensing Using Spaceborne GNSS Reflections: Comparison of CYGNSS Reflectivity to SMAP Soil Moisture. *Geophys. Res. Lett.* **2018**, *45*, 4049–4057, doi:10.1029/2018GL077905.
4. Jales, P. Spaceborne Receiver Design for Scatterometric GNSS Reflectometry. Ph.D. Thesis, University of Surrey, Guildford, UK, 2012.
5. Unwin, M.; Jales, P.; Tye, J.; Gommenginger, C.; Foti, G.; Rosello, J. Spaceborne GNSS-Reflectometry on TechDemoSat-1: Early Mission Operations and Exploitation. *IEEE J. Sel. Top. Appl. Earth Obs. Remote. Sens.* **2016**, *9*, 4525–4539, doi:10.1109/JSTARS.2016.2603846.
6. DMA WGS84 Development Committee. *Department of Defense World Geodetic System 1984, Its Definition and Relationships with Local Geodetic Systems*, 2nd ed.; Technical Report 8350.2; Department of Defense: Fort Lee, VA, USA, 1991.
7. Martin-Neira, M. PARIS: Application to Ocean Altimetry. *ESA J.* **1993**, *17*, 331–355.
8. Fraczek, W. *Mean Sea Level, GPS, and the Geoid*; ArcUser: Redlands, CA, USA 2003.
9. Grieco, G.; Stoffelen, A.; Portabella, M.; Belmonte Rivas, M.; Lin, W.; Fabra, F. Quality Control of Delay-Doppler Maps for Stare Processing. *IEEE Trans. Geosci. Remote. Sens.* **2019**, *57*, 2990–3000, doi:10.1109/TGRS.2018.2879059.
10. King, L.; Unwin, M.; Rawlinson, J.; Guida, R.; Underwood, C. A Topographically-Accurate GNSS-R Reflection Point Predictor For On-Board Operational Processing. In Proceedings of the IGARSS 2020—2020 IEEE International Geoscience and Remote Sensing Symposium, Waikoloa, HI, USA, 26 Septmber–2 October 2020; pp. 6194–6197, doi:10.1109/IGARSS39084.2020.9324677.
11. Gleason, S. A Real-Time On-Orbit Signal Tracking Algorithm for GNSS Surface Observations. *Remote. Sens.* **2019**, *11*, 1858, doi:10.3390/rs11161858.
12. Campbell, J.D.; Melebari, A.; Moghaddam, M. Modeling the Effects of Topography on Delay-Doppler Maps. *IEEE J. Sel. Top. Appl. Earth Obs. Remote. Sens.* **2020**, *13*, 1740–1751, doi:10.1109/JSTARS.2020.2981570.
13. Gleason, S.; O'Brien, A.; Russel, A.; Al-Khaldi, M.M.; Johnson, J.T. Geolocation, Calibration and Surface Resolution of CYGNSS GNSS-R Land Observations. *Remote. Sens.* **2020**, *12*, 1317, doi:10.3390/rs12081317.
14. Global Climate Observing System (GCOS). *Essential Climate Variables*; GCOS: Geneva, Switzerland, 2010.
15. Global Climate Observing System (GCOS). *The GCOS Story*; GCOS: Geneva, Switzerland, 2010.
16. Brocca, L.; Ciabatta, L.; Massari, C.; Camici, S.; Tarpanelli, A. Soil moisture for hydrological applications: Open questions and new opportunities. *Water* **2017**, *9*, 140, doi:10.3390/w9020140.
17. Global Climate Observing System (GCOS). *GCOS Monitoring Principles*; GCOS: Geneva, Switzerland, 2003.
18. Camps, A.; Vall·Llossera, M.; Park, H.; Portal, G.; Rossato, L. Sensitivity of TDS-1 GNSS-R Reflectivity to Soil Moisture: Global and Regional Differences and Impact of Different Spatial Scales. *Remote. Sens.* **2018**, *11*, 1856, doi:10.3390/rs10111856.
19. Pierdicca, N.; Mollfulleda, A.; Costantini, F.; Guerriero, L.; Dente, L.; Paloscia, S.; Santi, E.; Zribi, M. Spaceborne GNSS Reflectometry Data For Land Applications: An Analysis Of TechDemoSat Data. In Proceedings of the IGARSS 2018—2018 IEEE International Geoscience and Remote Sensing Symposium, Valencia, Spain, 22–27 July 2018; pp. 3343–3346.
20. Comite, D.; Cenci, L.; Colliander, A.; Pierdicca, N. Monitoring Freeze-Thaw State by Means of GNSS Reflectometry: An Analysis of TechDemoSat-1 Data. *IEEE J. Sel. Top. Appl. Earth Obs. Remote. Sens.* **2020**, *13*, 2996–3005, doi:10.1109/JSTARS.2020.2986859.
21. Surrey Satellite Technology Ltd. *MERRByS Product Manual—GNSS Reflectometry on TDS-1 with the SGR-ReSI*, 7 ed.; Surrey Satellite Technology Ltd.: Guildford, UK, 2019.

22. Eakins, B.W.; Sharman, G.F. *Hypsographic Curve of Earth's Surface from ETOPO1*; NOAA National Geophysical Data Center: Boulder, CO, USA, 2012.

23. Wu, G.; Liu, Y. Impacts of the Tibetan Plateau on Asian Climate. *Meteorol. Monogr.* **2016**, *56*, 7.1–7.29, doi:10.1175/AMSMONOGRAPHS-D-15-0018.1.

24. Gleason, S. *Algorithm Theoretical Basis Document—Level 1B DDM Calibration*; CYGNSS, rev 2 ed.; NASA: Washington, DC, USA, 2018.

25. Amatulli, G.; Domisch, S.; Tuanmu, M.N.; Parmentier, B.; Ranipeta, A.; Malczyk, J.; Jetz, W. A suite of global, cross-scale topographic variables for environmental and biodiversity modeling. *Sci. Data* **2018**, *5*, 1–15. doi:10.1038/sdata.2018.40.

26. Danielson, J.; Gesch, D. *Global Multi-Resolution Terrain Elevation Data 2010 (GMTED2010)*; U.S. Geological Survey Open-File Report 2011-1073; US Geological Survey: Sunrise Valley, VA, USA, 2011.

27. Jarvis, A.; Reuter, H.; Nelson, A. *Hole-Filled Seamless SRTM Data V4*; International Centre for Tropical Agriculture (CIAT): Cali, Colombia, 2008.

28. O'Loughlin, F.; Paiva, R.; Durand, M.; Alsdorf, D.; Bates, P. A multi-sensor approach towards a global vegetation corrected SRTM DEM product. *Remote. Sens. Environ.* **2016**, *182*, 49–59, doi:10.1016/j.rse.2016.04.018.

29. Loria, E.; O'Brien, A.; Zavorotny, V.; Downs, B.; Zuffada, C. Analysis of scattering characteristics from inland bodies of water observed by CYGNSS. *Remote. Sens. Environ.* **2020**, *245*, 111825, doi:10.1016/j.rse.2020.111825.

30. Al-Khaldi, M.M.; Johnson, J.T.; Gleason, S.; Loria, E.; O'Brien, A.J.; Yi, Y. An Algorithm for Detecting Coherence in Cyclone Global Navigation Satellite System Mission Level-1 Delay-Doppler Maps. *IEEE Trans. Geosci. Remote. Sens.* **2020**, 1–10., doi:10.1109/TGRS.2020.3009784.

31. SSTL; NOC. *Measurement of Earth Reflected Radio-Navigation Signals By Satellite*; SSTL: Guildford, UK, 2015.

32. De Vos Van Steenwijk, R.; Unwin, M.; Jales, P. Introducing the SGR-ReSI: A next generation spaceborne GNSS receiver for navigation and remote-sensing. In Proceedings of the Programme and Abstract Book—5th ESA Workshop on Satellite Navigation Technologies and European Workshop on GNSS Signals and Signal Processing, NAVITEC 2010, Noordwijk, The Netherlands, 8–10 December 2010; doi:10.1109/NAVITEC.2010.5708063.

33. Kwan, P. *NAVSTAR GPS Space Segment/User Segment Interfaces*; IS-GPS-200; GPS Directorate, Space & Missile Systems Center (SMC): Los Angeles, CA, USA, 2019.

34. Rius, A.; Cardellach, E.; Martin-Neira, M. Altimetric Analysis of the Sea-Surface GPS-Reflected Signals. *IEEE Trans. Geosci. Remote. Sens.* **2010**, *48*, 2119–2127, doi:10.1109/TGRS.2009.2036721.

Article

First Evaluation of Topography on GNSS-R: An Empirical Study Based on a Digital Elevation Model

Hugo Carreno-Luengo *, Guido Luzi and Michele Crosetto

Centre Tecnòlogic de Telecomunicacions de Catalunya (CTTC/CERCA), 08860 Castelldefels (Barcelona), Spain
* Correspondence: hugo.carreno@cttc.cat; Tel.: +34-93-645-29-00

Received: 16 September 2019; Accepted: 29 October 2019; Published: 31 October 2019

Abstract: Understanding the effects of Earth's surface topography on Global Navigation Satellite Systems Reflectometry (GNSS-R) space-borne data is important to calibrate experimental measurements, so as to provide accurate soil moisture content (SMC) retrievals. In this study, several scientific observables obtained from delay-Doppler maps (DDMs) $\left\langle \left| Y_{r,topo}(\tau, f) \right|^2 \right\rangle$ generated on board the Cyclone Global Navigation Satellite System (CyGNSS) mission were evaluated as a function of several topographic parameters derived from a digital elevation model (DEM). This assessment was performed as a function of Soil Moisture Active Passive (SMAP)-derived SMC at grazing angles $\theta_e \sim$ [20,30] ° and in a nadir-looking configuration $\theta_e \sim$ [80,90] °. Global scale results showed that the width of the trailing edge (TE) was small TE~ [100, 250] m and the reflectivity was high $\Gamma \sim$ [−10, −3] dB over flat areas with low topographic heterogeneity, because of an increasing coherence of Earth-reflected Global Positioning System (GPS) signals. However, the strong impact of several topographic features over areas with rough topography provided motivation to perform a parametric analysis. A specific target area with little vegetation, low small-scale surface roughness, and a wide variety of terrains in South Asia was selected. A significant influence of several topographic parameters i.e., surface slopes and curvatures was observed. This triggered our study of the sensitivity of TE and Γ to SMC and topographic wetness index (TWI). Regional scale results showed that TE and Γ are strongly correlated with the TWI, while the sensitivity to SMC were almost negligible. The Pearson correlation coefficients of TE and Γ with TWI are ~ 0.59 and $r_{TE} \sim -0.63$ at $\theta_e \sim$ [20, 30] ° and $r_\Gamma \sim$ 0.48 and $r_{TE} \sim$ −0.50 at $\theta_e \sim$ [80, 90] °, respectively.

Keywords: GNSS-R; CyGNSS; SMAP; topography; digital elevation model; elevation angle; soil moisture content; topographic wetness index; bistatic scattering

---

1. Introduction

Soil moisture content (SMC) is an important component in the global water and energy balance because it determines the flux of water that infiltrates to groundwater or drains via surface water [1]. SMC determines the water and energy available for evapotranspiration. The dependence of SMC on topographic convergence suggests that topography is an important element in the spatial distribution of SMC [2]. The state and pattern of SMC affect the hydrological response of a catchment area [1]. Temporal dynamics in the state of SMC depend on rates of outgoing horizontal and vertical fluxes, e.g., drainage and evaporation. On the other hand, the dependence of SMC on topographic convergence [2] suggests that topography is an important factor that partially determines the spatial distribution of SMC. As such, processes that lead to spatial organization in SMC could be represented by topographic parameters derived from a digital elevation model (DEM). Indeed, topography-based wetness indices can be used to predict the spatial distribution of wetlands because topography is one of the most important elements that influences the spatial patterns of saturated areas. Alternatively, wetlands mapping and spatial patterns of wetness can be generated in catchments under the assumption that groundwater tables follow topography holds.

The future of Earth observation in SMC determination is strongly based on the ability of the next generation of space-borne sensors to improve the Earth's surface spatio-temporal sampling properties. Global Navigation Satellite Systems Reflectometry (GNSS-R) appears in this framework, with an intrinsic capability to advance over more classical remote sensing approaches. GNSS-R is a sort of L-band passive bistatic radar (as many transmitters as navigation satellites are in view) that provides a wide swath up to ~ 1500 km. It was first conceived as an interferometric technique or iGNSS-R to provide higher precision in ocean altimetry measurements [3,4]. However, it can also be used for land surfaces applications. In this scenario, the conventional GNSS-R technique or cGNSS-R is more appropriate because lower coherent and incoherent integration times are required, so that the associated spatial resolution is improved [5]. At present, several works have demonstrated the feasibility of SMC estimation using space-borne cGNSS-R data from the United Kingdom (U.K.) TechDemoSat-1 [6–8], Soil Moisture Active Passive (SMAP) [9], and Cyclone Global Navigation Satellite System (CyGNSS) [10,11] missions.

CyGNSS was selected by the National Aeronautics and Space Administration (NASA) in 2012 as a highly innovative Earth Venture space system, and it was launched into space on 15 December 2016 [12]. CyGNSS is an 8-microsatellite constellation. Each single CyGNSS satellite carries on board two down-looking antennas with left hand circular polarization (LHCP). Both antennas point to the Earth's surface with an elevation angle θ_e~ 62 ° on each side of the satellite ground track (θ_e ~ 90 °at the normal direction to the Earth's surface). The gain of the antennas is ~ 14.7 dB (antenna boresight).

Topographic inhomogeneities can distort GNSS-R measurements [13,14]. They have to be compensated to improve the accuracy in the retrieval of SMC. In a bistatic radar geometry, GNSS scattered signals can be collected along the specular direction (elevation angle of the incident wave equals elevation angle of the scattered wave i.e., $\theta_{e,i}=\theta_{e,s}=\theta_e$). The scattering is only strong over an area around the nominal specular point. On the other hand, the scattering area depends on the local topography as well as on the GNSS satellites' elevation angles. Local slopes modify the scattering area as compared to that corresponding to a flat Earth assumption. The scattering area may be a complex and sometimes non-contiguous area on ground. However, it remains uncertain to what extent topography influences GNSS-R data and how these effects can be corrected. The main purpose of the present investigation was to further understand the effects of rough topography using data collected by CyGNSS. The counterpart of the radar pulse width in CyGNSS is ~ 300 m (1 C/A code chip). As such, reflected delay Doppler maps (DDMs) are potentially sensitive to the relative orientation of the scattering surfaces, but also to the vertical relief distribution. In the first part of this study, the influence of six different topographic parameters derived from the Global Multi-resolution Terrain Elevation Data 2010 (GMTED2010) model on DDMs derived from CyGNSS Level 1 Science Data Record Version 2.1 [12,15,16] was evaluated as a function of SMC derived from SMAP-enhanced L3 Radiometer Global Daily 9 km Level L3 SPL3SMP_E Version 1.0 product. In the last part, the following question was addressed: Can topographic features determine the main spatial pattern of Earth's surface reflectivity as it is measured by CyGNSS?

The remainder of this paper is organized as follows. Section 2 provides a theoretical background on the effects of topography on GNSS-R. Section 3 describes the methodology. Section 4 describes the effects of topography on CyGNSS-derived DDMs based on several parameters derived form a DEM. Additionally, it investigates the relationship between DDMs and a topography-based surface wetness index. Section 5 includes final discussions, and the main conclusions are d in Section 6.

2. Theoretical Background

CyGNSS allows sampling of the Earth's surface along 32 tracks simultaneously, within a wide range of satellites' elevation angles θ_e~ [20, 90] ° over the tropical latitudinal band Latitude ~ [−40, 40]°. Geophysical parameters retrieval from CyGNSS requires that the GNSS-R payload cross-correlates directly the collected signals against a known replica of the GPS L1 C/A code, after compensating for the Doppler shift. Consequently, 32 complex reflected DDMs, Y_r, are generated on board at a time, using a short coherent integration time, $T_c \sim 1$ ms. Complex DDMs, Y_r, are incoherently averaged along ~ 1 s to finally generate the so-called power DDMs.

Power DDMs, $\left\langle |Y_r(\tau, f)|^2 \right\rangle$, can be modelled using geometrical and scattering related parameters as follows [17]:

$$\left\langle |Y_r(\tau, f)|^2 \right\rangle = \frac{T_c^2 P_T \lambda^2}{(4\pi)^3} \iint \frac{G_T G_R |\chi(\tau - (R_T + R_R)/c, f - f_c)|^2 \sigma^0}{R_T^2 R_R^2} d^2\bar{\rho}, \tag{1}$$

where P_T is the power of the right HCP(RHCP) transmitted signals, $\lambda \sim 19$ cm is the wavelength of the signals at L1, G_T and G_R are the transmitting and receiving antenna gains, R_T and R_R are the ranges from the transmitter and the receiver to the specular point, respectively, χ is the Woodward ambiguity function (WAF), τ is the delay of the signal from the transmitter to the receiver, f is the Doppler shift of the reflected signal, f_c is aimed to compensate the Doppler shift of the signal, σ^0 is the bistatic scattering coefficient at LHCP, and $\bar{\rho}$ is the positioning vector of the scattering point.

The bistatic scattering coefficient can be defined as follows [18]:

$$\sigma^0 = \sigma^{coh,0} + \sigma^{incoh,0}, \tag{2}$$

where $\sigma^{coh,0}$ and $\sigma^{incoh,0}$ are the coherent and the incoherent scattering terms respectively. Consequently, DDMs consist on a sum of two terms as follows [19–21]:

$$\left\langle |Y_r(\tau, f)|^2 \right\rangle = \left\langle |Y_{r,coh}(\tau, f)|^2 \right\rangle + \left\langle |Y_{r,incoh}(\tau, f)|^2 \right\rangle. \tag{3}$$

$\left\langle |Y_{r,coh}(\tau, f)|^2 \right\rangle$ accounts for coherent reflections from the surface, while $\left\langle |Y_{r,incoh}(\tau, f)|^2 \right\rangle$ is responsible for the diffuse scattering.

$\left\langle |Y_{r,coh}(\tau, f)|^2 \right\rangle$ can be expressed analytically as follows [10]:

$$\left\langle |Y_{r,coh}(\tau, f)|^2 \right\rangle = \frac{T_c^2 P_T \lambda^2 G_T G_R |\chi(\tau, f)|^2}{(4\pi)^2 (R_T + R_R)^2} |R(\theta_e)|^2 \gamma \exp(-(2k\sigma \sin \theta_e)^2), \tag{4}$$

where R is the LHCP Fresnel reflection coefficient, k is the signal angular wavenumber, σ is the surface height standard deviation (related to small-scale surface roughness), and γ is the transmissivity of the vegetation.

$\left\langle |Y_{r,incoh}(\tau, f)|^2 \right\rangle$ includes contributions from vegetation, small-scale surface roughness, and topography. Further details on the incoherent scattering over rough surfaces can be found in Reference [22]. The impact of vegetation should be also considered to establish a model of the incoherent scattered field. A comprehensive method can be found in References [23,24].

Over regions without rough topography, DDMs can be expressed as follows [25]:

$$\left\langle |Y_r(\tau, f)|^2 \right\rangle = |\chi(\tau, f)|^2 ** |\sigma^0(\tau, f)|^2, \tag{5}$$

where $**$ is the 2-D convolution in both domains, delay and Doppler. The WAF can be defined as follows [17]:

$$|\chi(\tau, f)| = \Lambda(\tau)S(f), \tag{6}$$

where Λ is the autocorrelation function of the pseudo-random noise (PRN) codes and S is the system impulse response in the frequency domain.

Regions with rough topography are characterized with a root mean square error (RMSE) of the surface height variation higher than the length of the GNSS code under study. The effects of rough topography in the DDMs can be understood as follows:

$$\left\langle \left| Y_{r,topo}(\tau, f) \right|^2 \right\rangle = \left\langle \left| Y_r(\tau, f) \right|^2 \right\rangle * * \xi(\tau, f), \tag{7}$$

where the function ξ represents the mean density of scattering points as a function of the delay, and the Doppler shift.

3. Data and Methodology

3.1. Topographic Parameters

Several topographic parameters were derived from the ~250 m GMTED2010 model based on a 3×3 moving window (Figure 1). This window was composed of the focal cell and its eight surrounding cells. The DEM surface was approximated by a bivariate quadratic equation [26,27]:

$$Z = aX^2 + bY^2 + cXY + dX + eY + f, \tag{8}$$

where Z represents the height of the DEM surface and X and Y are the horizontal coordinates. The coefficients in Equation (8) were solved within the 3×3 window using simple combinations of neighboring cells, based on the so-called Wood's approach [28].

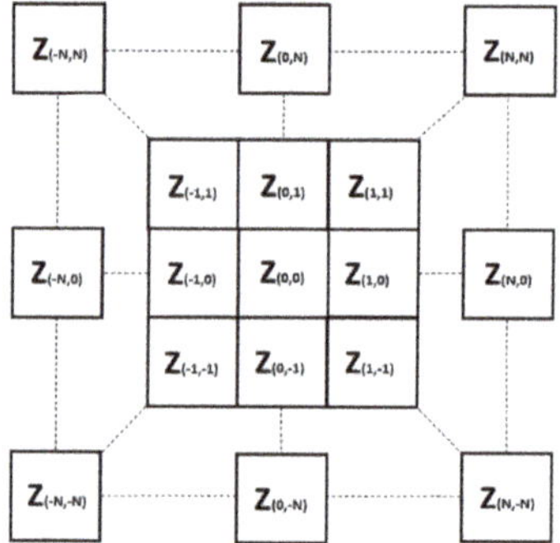

Figure 1. Image of the raster grid. Numbering system is shown. Z is the value of the raster. Notation is simplified. $N = (n - 1)/2$ is used for any $n \times n$ analysis window, where n may be any odd integer smaller than the number of cells in the shortest side of the raster.

The topographic parameters were used to investigate the impact of different topographic features on $\left\langle \left| Y_{r,topo}(\tau, f) \right|^2 \right\rangle$. In particular, three heterogeneity variables (topographic position index or TPI, terrain ruggedness index or TRI, vector ruggedness measure or VRM), slope (β), and two curvature variables (tangential (TC) and profile curvature (PC) were used in the parametric study. They have the potential to enable a better understanding on the properties of the Earth's surface than raw DEM data. Topographic heterogeneity is described as the variability of surface elevations within an area. On the other hand, curvature attributes are based on the change of slope in a particular direction. More specifically, the selected topographic parameters (Figure 2) were defined as follows [26–28]:

- TPI is the difference between the elevation of a focal cell and the mean of its eight surrounding cells. Positive and negative values correspond to ridges and valleys, respectively. Zero values correspond to flat areas. It was computed as follows:

$$
\text{TPI} = Z_{(0,0)} - \sum_{i=1}^{m} Z_i / m,
\tag{9}
$$

where m is total number of surrounding points employed in the evaluation.

- TRI is the mean of the absolute differences in elevation between a focal cell and its eight surrounding cells. It quantifies the total elevation change across the 3×3 moving window. Flat areas have a value of zero and mountain areas with steep ridges have positive values. It was computed as follows:

$$
\text{TRI} = \frac{\sum_{i=-N}^{N} \sum_{j=-N}^{N} \left| Z_{(i,j)} - Z_{(0,0)} \right|}{(n^2 - 1)},
\tag{10}
$$

where $N = (n-1)/2$ and n may be any odd integer smaller than the number of cells in the shortest side of the raster.

- VRM quantifies the terrain ruggedness based on the measurements of the dispersion of vectors orthogonal to the surface. VRM quantifies the local variation of slope in the terrain more independently than TPI and TRI. VRM values range from 0 over flat areas to 1 over rough areas. It was computed using vector analysis [26].

- β is defined as the rate of change of elevation in magnitude for the steepest descent vector. It influences hydraulic gradients driving any surface flows and also sub-surface flows when the water table has a similar slope to the ground surface. Slope and curvature serve as useful input variables in erosion and hydrological models. It was computed as follows:

$$
\beta = \arctan\left(\sqrt{d^2 + e^2} \right).
\tag{11}
$$

- TC measures the rate of change perpendicular to the slope gradient. It is related to the convergence and divergence of lateral flows across a surface. It was computed as follows:

$$
\text{TC} = \frac{200 \left(bd^2 + ae^2 - cde \right)}{\left(e^2 + d^2 \right)^{1.5}}.
\tag{12}
$$

- PC measures the rate of change of slope along a flow line. It affects the acceleration or deceleration of surface flows along the surface and thus it influences erosion and deposition of soils. Positive and negative values indicate convex and concave surfaces, respectively. It was computed as follows:

$$
\text{PC} = \frac{-200 \left(ad^2 + be^2 + cde \right)}{\left(e^2 + d^2 \right)\left(1 + e^2 + d^2 \right)^{1.5}}.
\tag{13}
$$

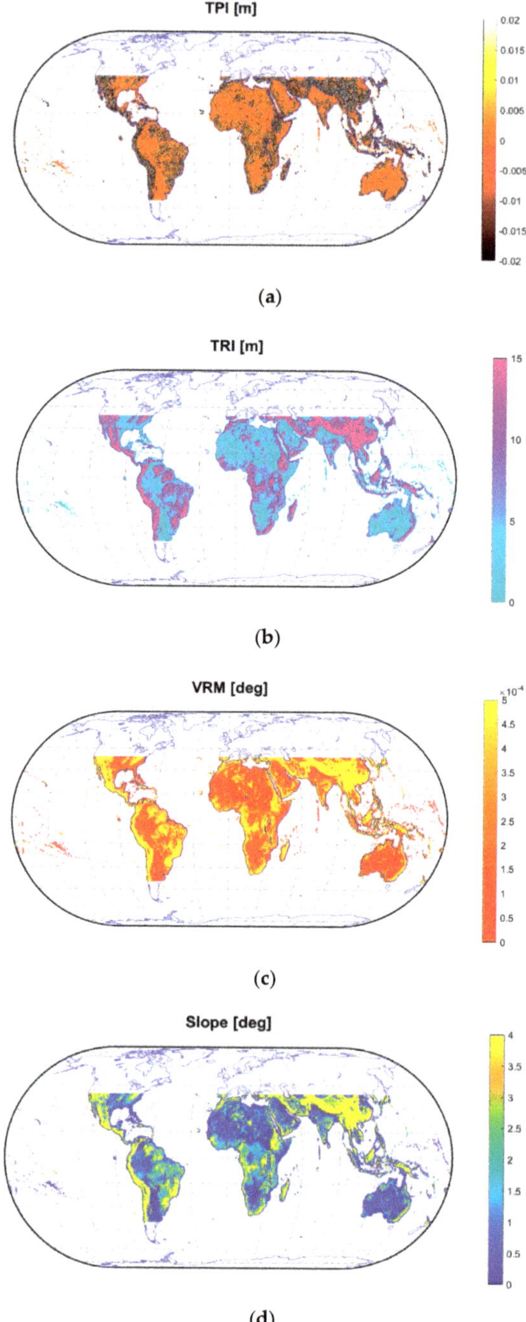

Figure 2. *Cont.*

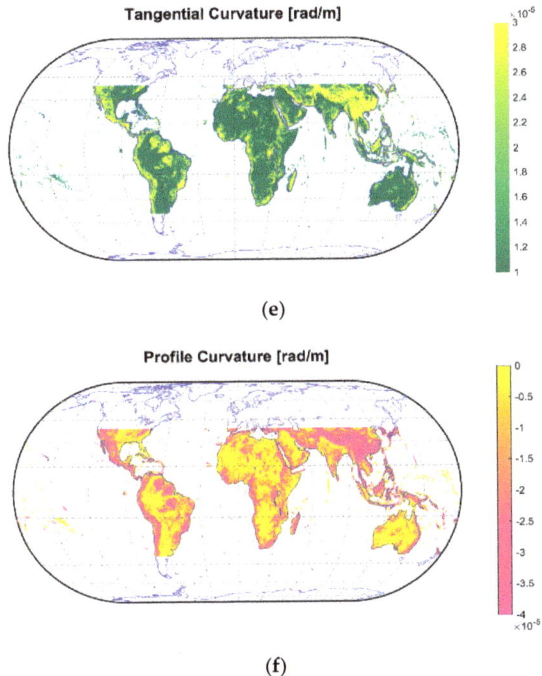

(e)

(f)

Figure 2. Global Multi-resolution Terrain Elevation Data 2010 (GMTED2010) model-based topographic parameters. (**a**) Topographic position index (TPI) (m), (**b**) terrain ruggedness index (TRI) (m), (**c**) vector ruggedness measure (VRM) (deg), (**d**) Slope β (deg), (**e**) tangential curvature (TC) (rad/m), and (**f**) profile curvature (PC) (rad/m). Parameters were displayed over the Cyclone Global Navigation Satellite System (CyGNSS) coverage over land surfaces.

3.2. Topographic Wetness Index

The topographic wetness index (TWI) is a parameter used to match runoff- producing elements in the landscape [29]. Different target areas with similar TWI values should have similar hydrological dynamics [30]. As such, TWI is an indicator of the relative propensity of soils to become saturated to the surface because of the local topography. In this work, TWI values were calculated based on the GA2 algorithm [31]. GA2 calculates the outflow gradient of each target area and uses precalculated up-land values from HydroSHEDS for the catchment area [32]. It can be expressed as follows [31]:

$$\text{TWI} = \ln\!\left(\frac{A}{\tan \beta}\right)\!, \tag{14}$$

where A is the specific catchment area [33,34]. TWI values are low at ridges and high in valleys. Generally, humid areas generate higher TWI values, although there are some exceptions such as desert areas where high TWI values do not correlate with high flow accumulation.

3.3. Delay Doppler Maps Parameters

CyGNSS Level 1 Science Data Record Version 2.1 [12,15,16] was selected for this study. Original data were first filtered out using an equivalent "CyGNSS overall quality flag" over land surfaces [35]. Reflected delay waveforms $\text{WF}_{\text{r,raw}}$ were obtained from the original DDMs $\left\langle \left|Y_{\text{r,topo}}(\tau, f)\right|^2 \right\rangle$ at zero Doppler frequency:

$$\text{WF}_{\text{r,raw}} = \left\langle \left|Y_{\text{r,topo}}(\tau, f = 0)\right|^2 \right\rangle. \tag{15}$$

The delay bin resolution of the original 17 lag waveforms $WF_{r,raw}$ was ~ 0.2552 GPS C/A chips. A re-sampling and interpolation of the resulting 1700 lag waveforms was performed using a spline method to increase the accuracy of the waveforms, before applying the algorithms to extract the observables.

The trailing edge (TE) width was defined as the lag difference between the 70% power threshold of the high resolution waveforms $WF_{r,threshold}$, and that corresponding to the maximum power of the waveforms $WF_{r,peak}$ [9]:

$$TE = \tau_{WF_{r,threshold}} - \tau_{WF_{r,peak}}. \tag{16}$$

Different power thresholds were tested. The 50% power threshold was found to cut off some lags, when this threshold was out of the original 17 lag waveforms. On the other hand, the 90% power threshold provided a lower dynamic range as compared to the 70%. The incoherent scattering term $\sigma^{incoh,0}$ was the main contribution to $WF_{r,threshold}$, while both the coherent $\sigma^{coh,0}$ and the incoherent $\sigma^{incoh,0}$ scattering terms significantly contributed to the peak power of the waveforms $WF_{r,peak}$.

The reflectivity Γ was estimated as the ratio of the reflected $WF_{r,peak}$ and the direct $WF_{d,peak}$ power waveforms' peaks, after compensating for the noise power floor and the antennas' gain patterns as a function of the elevation angle θ_e as in [10]:

$$\Gamma = WF_{r,peak}/WF_{d,peak}. \tag{17}$$

The antennas' gain patterns were compensated versus the gain at the corresponding boresight direction, and the difference of both antennas at boresight [10]. The nominal mission lifetime was 2 years. This milestone was achieved in March 2019. At present, the mission is operating nominally, and the automatic gain control (AGC) was disabled from August 2018 [12]. This fact improves the accuracy in the estimation of Γ using Equation (17). On the other hand, Γ could also be inverted from scattering models. This approach relies on the assumption that the scattering is totally coherent [35] or totally incoherent [36]. The assumption in [35] is valid over smooth surface areas with little vegetation. The assumption in [36] is valid if inland water bodies are removed. However, in a more general scenario, the total scattered electromagnetic field is composed of both a coherent and an incoherent contribution [18] in different proportions depending on the dielectric and geometrical properties of the scattering medium, and the directions of incoming and outgoing electromagnetic waves. Both, the coherent $\sigma^{coh,0}$ and the incoherent $\sigma^{incoh,0}$ scattering terms contribute to the peak power of the waveforms. This was the main motivation to use Equation (17) in this work.

4. Understanding the Effects of Rough Topography on GNSS-R Space-Borne Data

4.1. Global Scale

The main goal of this work was to study the impact of rough topography on CyGNSS-derived TE & Γ as a function of SMAP-derived SMC. As a first step, the selected topographic parameters (Figure 2), TWI (Figure 3a), CyGNSS-derived TE and Γ (Figure 3b,c), and SMAP-derived SMC (Figure 3d) were displayed over the complete coverage of the Earth's surface enabled by CyGNSS. SMC data were derived based on the V-pol single channel algorithm [37,38]. Corrections for roughness, effective surface temperature, and vegetation water content (VWC) were also applied. Six months of CyGNSS and SMAP data were selected from 1 August 2018 to 31 January 2019. This specific time period was selected because the AGC was disabled in July 2018, so as to improve the performance in the estimation of Γ (Equation (17)). The selected temporal length was as large as possible at the time of starting this work. All the parameters were averaged using a $0.1° \times 0.1°$ latitude/longitude grid with a moving window of $0.2°$ in steps of $0.1°$. This gridding strategy enabled the analysis using auxiliary data and different sensors. The resulting spatial resolution was ~ 20 km $\times$ ~ 20 km at equatorial latitudes, which was wide enough to account for the across-track spreading ~ 20 km of the DDMs due to $\left\langle \left| Y_{r,incoh}(\tau, f) \right|^2 \right\rangle$.

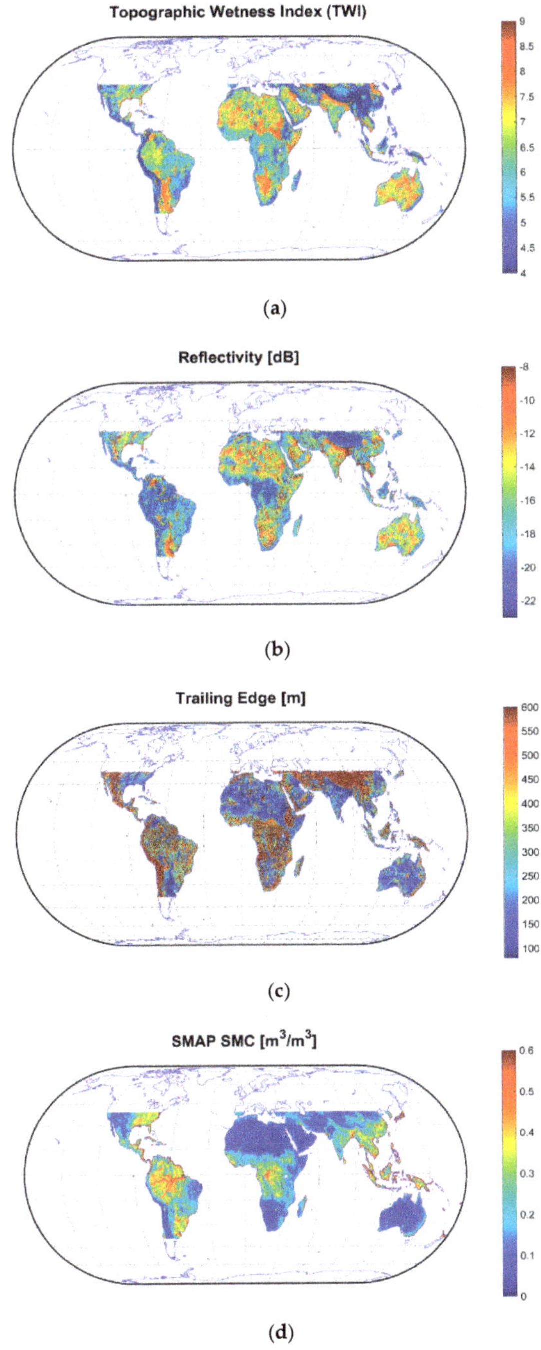

Figure 3. (**a**) TWI, (**b**) CyGNSS-derived Γ (dB), (**c**) CyGNSS-derived TE (m), and (**d**) Soil Moisture Active Passive (SMAP)-derived SMC (m³/m³). CyGNSS and SMAP data correspond to a temporal window of six months (1 August 2018 to 31 January 2019). The parameters were displayed over the CyGNSS coverage over land surfaces.

Overall, it was found that the spreading of TE was low ~ [100, 250] m and Γ is high ~ [−10, −3] dB over flat areas with low topographic heterogeneity and small surface elevation gradients (Figures 2 and 3). Over these areas, the scattering was quite specular and the coherence of the Earth-reflected GPS signals could be assumed to be high [35]. Under these circumstances, $\left\langle \left| Y_{r,coh}(\tau, f) \right|^2 \right\rangle \gg \left\langle \left| Y_{r,incoh}(\tau, f) \right|^2 \right\rangle$. Thus, the power of the reflected signals could be assumed to be roughly independent of the platform's height (Equation.4). This point justifies the strong values of Γ and the reduced spreading of the waveforms. SMC was high over regions that do not always correspond to flat areas (Figure 3d). On the other hand, the spatial patterns of TE & Γ showed a certain agreement with TWI. An exception is found over e.g., tropical rainforests. Here, the spreading of TE was high and Γ was low because specular reflection is weakened by vegetation attenuation, γ, and vegetation structure contributes to diffuse scattering, both of which significantly modify the shape of $\left\langle \left| Y_{r,topo}(\tau, f) \right|^2 \right\rangle$ [39]. A more quantitative analysis is performed at a regional scale over an area with little vegetation and low small-scale surface roughness. Over desert areas e.g., Sahara, Kalahari, and Australia it was found that Γ was also quite high, despite SMC being almost negligible. This aspect was previously attributed to the effect of sub-surface scattering over rich sand content areas [6,9]. In very dry conditions, GNSS-R observables have shown a certain sensitivity to the soil moisture at deeper levels [8].

In addition, specular scattering over areas with high TWI could significantly increase Γ, despite SMC being low (Figure 3). Indeed, TE was rather low over areas with high TWI, which is an additional indication of the specular nature of the scattering. TWI accounted for saturation of SMC in local topographic converging areas and was used to predict zones of surface saturation and therefore organized spatial fields of SMC. It could be expected that inland water bodies, i.e., rivers, are predominant in regions with high TWI, which could justify the specular nature of the scattering [29]. On the other hand, topographic heterogeneity (TPI, TRI, and VRM), β, and curvature parameters (TPI, TRI) are high over the Andes and Himalayan mountains, where TE is high because of the effect of diffuse scattering $\left\langle \left| Y_{r,incoh}(\tau, f) \right|^2 \right\rangle$.

4.2. Regional Scale

4.2.1. Introduction

A specific target area over South Asia was selected to further investigate topographic effects in $\left\langle \left| Y_{r,topo}(\tau, f) \right|^2 \right\rangle$, without the influence of high levels of σ, above ground biomass (AGB), and canopy height (CH). The coordinates of the selected target area were as follows: latitude = [5, 36] ° and longitude = [64, 94] °. The characteristics of this region enabled this study to be performed over a rich variety of terrains, including the Tibetan Plateau, Great Himalaya, Indo-Gangetic Plains, and Hindustan. As such, the impact of different topographic features in TE and Γ can be properly evaluated (Figures 4–6). In so doing, $\left\langle \left| Y_{r,topo}(\tau, f) \right|^2 \right\rangle$ were classified in two groups to evaluate the effects of topography over two different scenarios. At θ_e ~ [20, 30] °, the main contribution to the generation of the reflected signal in a specular manner (elevation angle of the incident wave equals elevation angle of the scattered wave i.e., $\theta_{e,i} = \theta_{e,s} = \theta_e$) came from surface facets over high elevation mountains. Some low elevation areas could be occulted by the top of the mountains. On the other hand, at θ_e ~ [80, 90] °, both valleys and mountain peaks could reflect signals in a specular manner towards the CyGNSS. This study was additionally complemented with SMAP-derived SMC and its coefficient of variation (CV). A wide range of SMC levels could be observed. The highest SMC values were associated with both the Indo-Gangetic Plains and mountains regions in the West of Hindustan, while the lowest SMC levels were found over the Tibetan Plateau and the Thar Desert. The spatial patterns of SMC did not appear correlated with the topographic parameters. Significant SMC levels could be found over mountains but also over rivers basins.

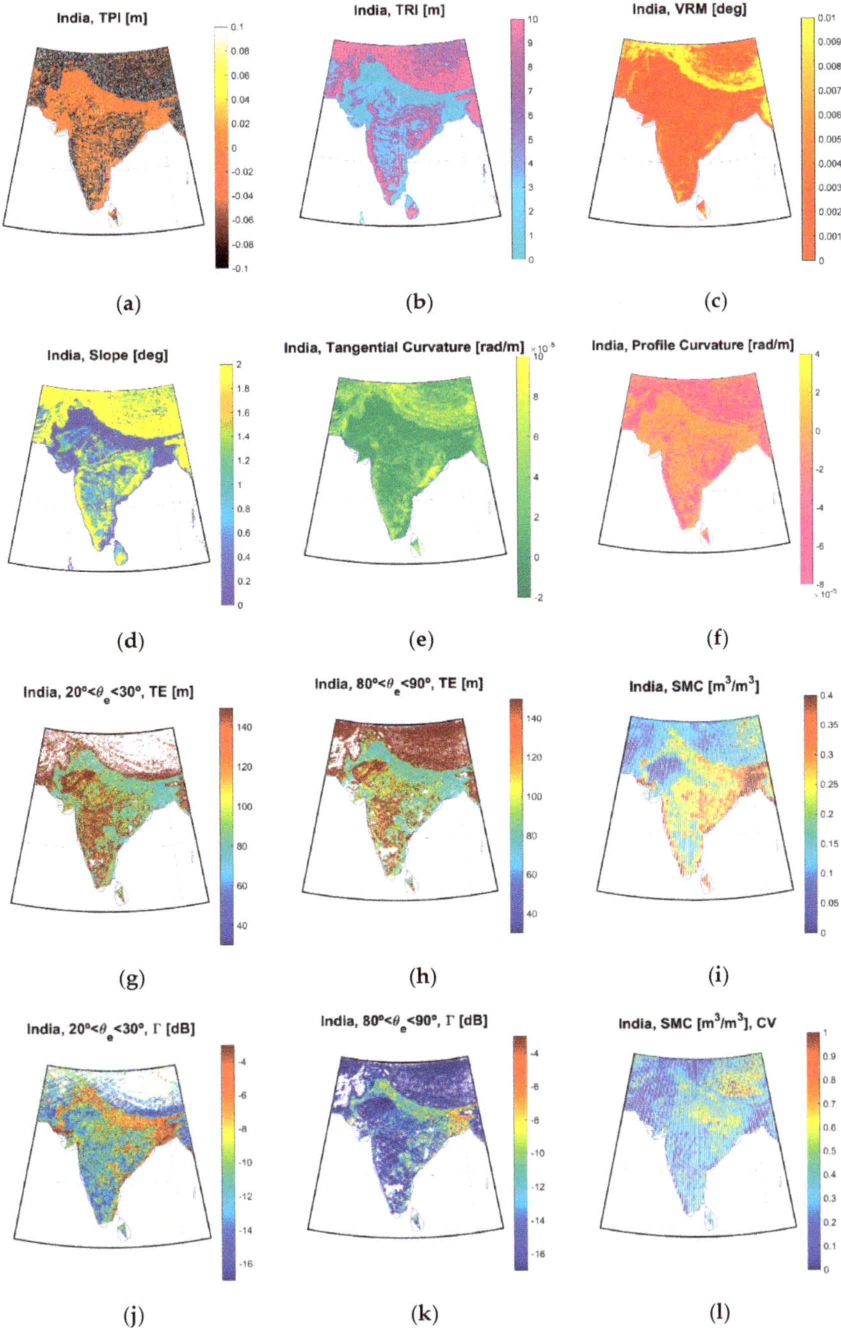

Figure 4. Regional scale study over a target area in South Asia. Different scientific observables and auxiliary data were depicted: (**a**) TPI, (**b**) TRI, (**c**) VRM, (**d**) β, (**e**) TC, (**f**) PC, (**g**) CyGNSS-derived TE at $\theta_e \sim [20, 30]\,°$, (**h**) TE at $\theta_e \sim [80, 90]\,°$, (**i**) SMAP-derived SMC, (**j**) CyGNSS-derived Γ at $\theta_e \sim [20, 30]\,°$, (**k**) Γ at $\theta_e \sim [80, 90]\,°$, and (**l**) SMC CV.

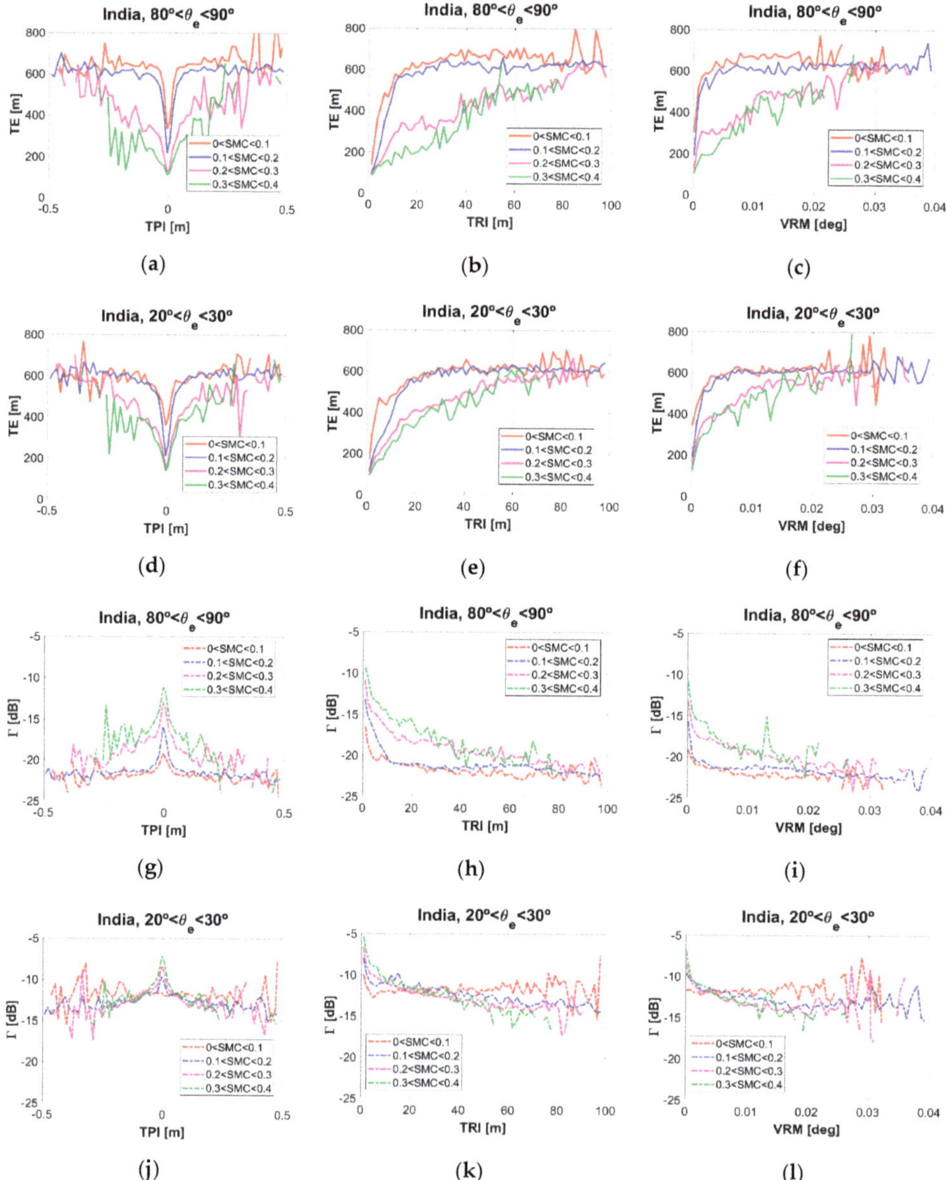

Figure 5. Parametric evaluation as a function of SMAP-derived SMC: (**a**,**d**,**g**,**j**) TPI, (**b**,**e**,**h**,**k**) TRI, and (**c**,**f**,**i**,**l**) VRM. Analysis of CyGNSS-derived TE at $\theta_e \sim [80, 90]°$ (**a**–**c**) and $\theta_e \sim [20, 30]°$ (**d**–**f**). Analysis of CyGNSS-derived Γ at $\theta_e \sim [80, 90]°$ (**g**–**i**) and $\theta_e \sim [20, 30]°$ (**j**–**l**).

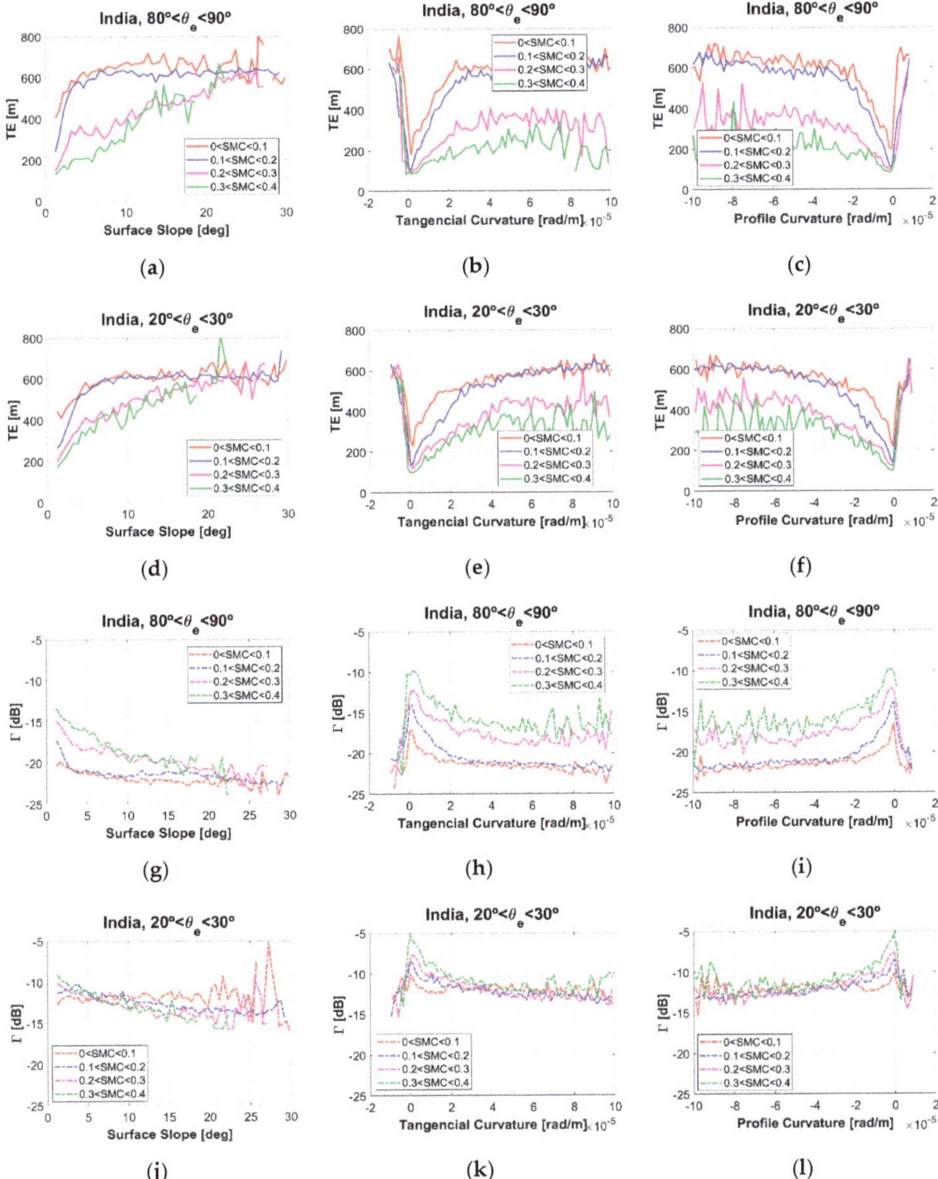

Figure 6. Parametric evaluation as a function of SMAP-derived SMC: (**a,d,g,j**) β, (**b,e,h,k**) TC, and (**c,f,i,l**) PC. Analysis of CyGNSS-derived TE at θ_e ~ [80, 90] ° (**a–c**) and θ_e ~ [20, 30] ° (**d–f**). Analysis of CyGNSS-derived Γ at θ_e ~ [80, 90] ° (**g–i**) and θ_e ~ [20, 30] ° (**j–l**).

SMAP SMC data were used as an auxiliary product to help in the interpretation of the experimental results. The SMAP baseline accuracy requirement of 4% is not guaranteed over areas with very rough topography. SMC data are here provided with a lower accuracy. Nonetheless, 6 months of SMC data were averaged using a 0.1 ° × 0.1 ° latitude/longitude grid with a moving window of 0.2 ° in steps of 0.1 °. This strategy mitigated the effects of potential artefacts in SMAP-derived SMC. Additionally, this moving filter smoothed potential residual errors in the GMTED2010. Some errors have been reported over areas above Latitude ~ 60 ° [26]. However, this study was focused on areas within CyGNSS coverage Latitude ~ [−40, 40] °.

4.2.2. Interpretation of the Empirical Results

Figure 4 shows that TE increases and Γ decreases with increasing values of the heterogeneity variables i.e.,TPI (Figure 4a), TRI (Figure 4b), and VRM (Figure 4c). Additionally, the measurements of β (Figure 4d), TC (Figure 4e), and PC (Figure 4f) have a clear impact on both observables. The ranges of TE and Γ are the same in all the plots. This strategy was assumed to provide intercomparable plots, and at the same time show full variability. Over the Himalayan mountains, the spreading of TE was very high (up to ~ 600 m) and Γ was very low (down to ~ −25 dB). On the other hand, an inverse behavior appeared over the Indo-Gangetic Plains. Here, the scattered electromagnetic field became mostly coherent at grazing angles because the exponential factor (Equation (4)) increased significantly with decreasing angles [40]. As a consequence, there was an increment of Γ, which was distributed in a rather homogeneous manner (Figure 4j), despite the strong SMC gradient from east to west (Figure 4i). In Reference [40] it was previously found that the Γ dynamic range decreased at grazing angles. In other words, the sensitivity to SMC reduced at this geometry, which helps to justify the homogeneous distribution of Γ. On the other hand, Γ appeared more associated with the variability of SMC in a nadir-looking configuration, despite being lower than at grazing angles [8,40].

Figures 5 and 6 show the relationship of TE and Γ with the selected topographic parameters as a function of SMAP-derived SMC, over the two angular ranges used in this study $\theta_e \sim [20, 30]$ ° and $\theta_e \sim [80, 90]$ °: TPI (Figure 5a,d,g,j), TRI (Figure 5b,e,h,k), VRM (Figure 5c,f,i,l), β (Figure 6a,d,g,j), TC (Figure 6b,e,h,k), and PC (Figure 6c,f,i,l). The influence of SMC was evaluated at four differentiated ranges from ~0 m^3/m^3 to ~0.4 m^3/m^3 in steps of ~0.1 m^3/m^3.

For low SMC ~ [0, 0.2] m^3/m^3, the impact of the heterogeneity variables on both TE and Γ was quite strong in the ranges TPI~ [−0.01, 0.01] m, TRI~ [0, 20] m, and VRM~ [0, 0.005] °. This point could be just justified because for low SMC levels: (a) the geometrical properties of the scattering medium play a stronger role than the dielectric properties and (b) the GNSS-R observables become saturated for a smaller topographic roughness.

Increasing SMC levels reduced the gradients because of the waveforms' peak power increment. Consequently, the 70% power threshold appeared at smaller TE values as compared to dry soils. The variability of TE with SMC was higher in a nadir-looking configuration than at grazing angles because of the lower topographic influence. On the other hand, Γ increased with increasing SMC in a nadir-looking configuration. However, it was roughly independent of SMC at grazing angles [8]. As such, the spreading of TE and the decrease of Γ showed a much more gradual evolution for increasing heterogeneity levels up to TE ~ 600 m and $\Gamma \sim -22$ dB at $\theta_e \sim [80, 90]$ ° and TE ~ 600 m and $\Gamma \sim -15$ dB at $\theta_e \sim [20, 30]$ °.

Finally, it is pointed out that the topographic dynamic range of both GNSS-R parameters TE and Γ was higher for SMC ~ [0, 0.2] m^3/m^3 than for SMC ~ [0.2, 0.4] m^3/m^3. This observation is also attributed to the stronger influence of the geometrical properties for decreasing SMC levels. A similar behavior was also pointed out in the case of the small-scale surface roughness [40].

The impact of surface slopes and curvature variables on both observables, TE and Γ, was quite strong at low SMC ~ [0, 0.2] m^3/m^3, over the following ranges: β~ [0, 5] °, TC~ [-1, 2] rad/m, and PC~ [−2, 1] rad/m. Increasing SMC levels reduced the gradients in agreement with the findings using the heterogeneity variables (Figure 5). The spreading of TE and the decrease of Γ showed a much more gradual evolution for increasing curvature levels up to (a) $\theta_e \sim [20, 30]$ °: TE~ 600 m and $\Gamma \sim -13$ dB for SMC ~ [0.3, 0.4] m^3/m^3, TE~ 600 m and $\Gamma \sim -13$ dB for SMC ~ [0.2, 0.3] m^3/m^3, TE ~ 400 m and $\Gamma \sim -13$ dB for SMC ~ [0.1, 0.2] m^3/m^3, and TE~ 300 m and $\Gamma \sim -13$ for SMC ~ [0, 0.1] m^3/m^3 and (b) $\theta_e \sim [80, 90]$ °: TE~ 600 m and $\Gamma \sim -16$ dB for SMC ~ [0.3, 0.4] m^3/m^3, TE~ 600 m and $\Gamma \sim -18$ dB for SMC ~ [0.2, 0.3] m^3/m^3, TE~ 375 m and $\Gamma \sim -22$ dB for SMC ~ [0.1, 0.2] m^3/m^3, and TE~ 200 m and $\Gamma \sim -23$ dB for SMC ~ [0, 0.1] m^3/m^3.

Surface slopes modify the scattering area with respect to that corresponding to a flat Earth assumption. They are a key parameter in GNSS-R studies over land surfaces, which have a certain degree of correlation with TRI and VRM because increasing topographic heterogeneity belongs to increasing surface slopes. On the other hand, the PC and TC have a much clearer differentiated impact on TE and Γ. These topographic variables are directly related to water accumulation over the surface. Over flat surfaces, Γ is high and TE is low because the curvature is negligible, and the scattering is quite specular. However, the impact of SMC is lower compared to areas with higher curvatures. In these areas, TE and Γ have a clearer differentiated response to different SMC levels. Finally, it was found that the variability with SMC is higher in a nadir-looking configuration than at grazing angles. It increases for negative PC levels (concave), while for positive curvature levels (convex) there is a negligible of influence of SMC.

4.2.3. Sensitivity of CyGNSS to TWI vs. SMC

The strong influence of primary terrain attributes on GNSS-R observables (Figures 5 and 6) triggered the sensitivity study of TE and Γ with TWI vs. SMC. To do so, a specific target area (Latitude ~ [20, 27] ° and Longitude ~ [72, 84] °) was selected with a strong gradient of SMC, and a wide topographic heterogeneity (Figure 7). Figure 8 shows the density scatter plots of Γ vs. TWI at θ_e~ [20, 30] ° (Figure 8a) and θ_e~ [80, 90] ° (Figure 8b), Γ vs. SMC at θ_e~ [20, 30] ° (Figure 8c) and θ_e~ [80, 90] ° (Figure 8d), TE vs. TWI at θ_e~ [20, 30] ° (Figure 8e) and θ_e~ [80, 90] ° (Figure 8f), and TE vs. SMC at θ_e~ [20, 30] ° (Figure 8g) and θ_e~ [80, 90] ° (Figure 8h).

Both observables TE and Γ are more correlated with TWI than with SMC (Table 1). This point could be justified because the potential presence of inland water bodies over areas with high TWI have a stronger impact on GNSS-R observables than SMC. In this analysis, areas with high elevation and very rough topography (i.e., Himalayan mountains) were excluded (Figure 7) because SMAP-derived SMC could have some uncertainties over areas with very rough topography. As such, SMC is delivered within the baseline accuracy in this target area.

The Pearson correlation coefficients increased at grazing angles because of the stronger impact of topography at this geometry as compared to a nadir-looking configuration. On the other hand, the correlation of TE and Γ with SMC is almost negligible despite the moderate-to-high SMC dynamic range ~ [0, 0.4] m^3/m^3. The slopes of the linear regressions were also included in Table 1, for both angular ranges. It is found that the sensitivity of TE to TWI and SMC increased at grazing angles, while that of Γ was stronger in a nadir-looking configuration in agreement with the results in Reference [8]. This empirical evidence suggests that topographic features can determine the main spatial pattern of Earth's surface reflectivity, at least over some specific target areas.

Figure 7. Close up of the area selected for the sensitivity analysis of CyGNSS observables to TWI and SMAP-derived SMC.

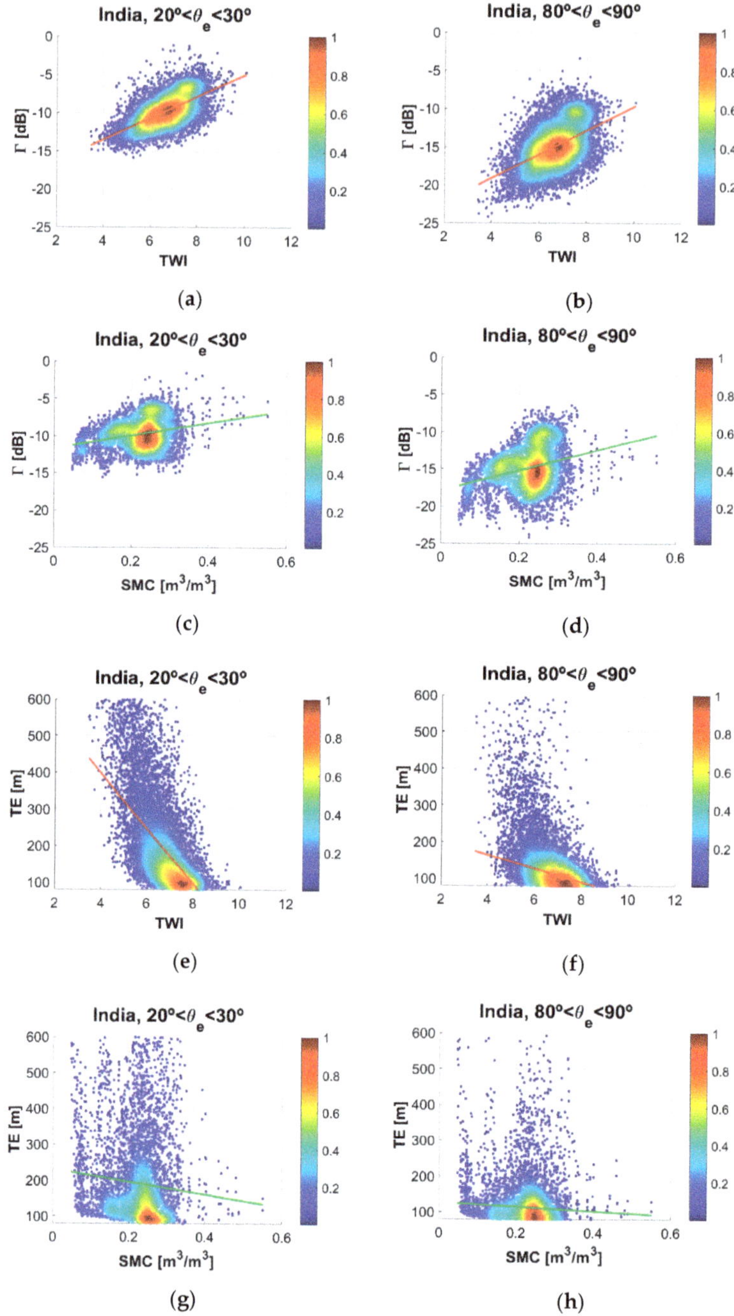

Figure 8. Density scatter plots over North India target area: CyGNSS-derived Γ vs. TWI at (**a**) $\theta_e \sim [20, 30]\,°$ and (**b**) $\theta_e \sim [80, 90]\,°$. Γ vs. SMAP-derived SMC at (**c**) $\theta_e \sim [20, 30]\,°$ and (**d**) $\theta_e \sim [80, 90]\,°$. TE vs. TWI at (**e**) $\theta_e \sim [20, 30]\,°$ and (**f**) $\theta_e \sim [80, 90]\,°$. TE vs. SMC at (**g**) $\theta_e \sim [20, 30]\,°$ and (**h**) $\theta_e \sim [80, 90]\,°$.

Table 1. Statistics (Pearson correlation coefficient r, RMSE, and slope of the linear fits) of the scatter plots of CyGNSS-derived TE and Γ with TWI and SMAP-derived SMC over the selected target area over North India.

	Topographic Wetness Index (TWI)				SMC [m^3/m^3]			
	TE [m]	Γ [dB]	TE [m]	Γ [dB]	TE [m]	Γ [dB]	TE [m]	Γ [dB]
	[20, 30]°	[20, 30]°	[80, 90]°	[80, 90]°	[20, 30]°	[20, 30]°	[80, 90]°	[80, 90]°
r	−0.63	0.59	−0.50	0.48	−0.08	0.27	−0.06	0.30
RMSE	95.3	1.7	68.5	2.7	121.9	2.1	80.7	2.8
Slope	−75.13	1.40	−18.65	1.57	−175.21	8.4078	−63.47	13.55

Finally, it is worth pointing out that these results at a regional scale support qualitative results obtained at a global scale (Figure 1). As such, attention should be paid when analyzing GNSS-R data because soils saturated by the effect of topography could introduce confounding effects with respect to the impact of SMC.

5. Final Discussion

CyGNSS space-borne GNSS-R data appear to have a strong dependence on topography. This dependence was found to be different from that corresponding to small-scale surface roughness. The impact of topographic roughness increased with decreasing angles θ_e, while the impact of surface roughness increased with increasing θ_e. This latter point can be justified based on the Rayleigh criterion that establishes that a surface (without vegetation and rough topography) could be considered smooth if the phase difference between two reflected electromagnetic waves is lower than $\pi/2$ rad. On the other hand, it was found that topographic roughness was a much more complex phenomenon, which depended on several topographic features in a differentiated manner. Some Earth observation missions such as the Soil Moisture Ocean Salinity (SMOS) mission used a method to flag the pixels according to the relative impact of topography on the brightness temperature [41]. The GNSS-R scenario appears more complex because the bistatic scattering includes both a coherent $\sigma^{coh,0}$ and an incoherent $\sigma^{incoh,0}$ component, in different proportions depending on the dielectric and geometrical properties of the scattering medium, and the directions of incoming and outgoing electromagnetic waves. On the other hand, the spatial resolution depends on the coherent-to-incoherent scattering ratio. This point complicates the topographic masking because of the impact of θ_e.

The scattering over surfaces with rough topography could be assumed to be totally incoherent. However, experimental results showed that Γ increased with decreasing θ_e even over regions with very rough topography. This angular behavior is classically associated with the coherent scattering term $\sigma^{coh,0}$ because the effective surface roughness is much lower at grazing angles [40]. At this point, it could be hypothesized that the scattering could be locally coherent over specific mountains facets. The total scattered electromagnetic field could be coherent if only a single mountain facet contributes in a specular manner to the reflected signal as collected by CyGNSS.

6. Conclusions

Several topographic parameters, TPI, TRI, VRM, β, TC, and PC were derived from the ~250 m GMTED2010 DEM. The impact of these parameters on different GNSS-R observables TE and Γ was assessed over land surfaces as a function of SMAP-derived SMC and θ_e. A parametric study was then performed over a target area in South Asia. For low SMC levels, TE and Γ presented a strong fluctuation from flat surfaces to areas with moderate topographic roughness. On the other hand, increasing SMC levels reduced the gradients. Additionally, it was found that the sensitivity of both observables to SMC increased in a nadir-looking configuration as compared to grazing angles. As a final remark, TE and Γ converged with increasing levels of TPI, TRI, VRM, and β; while they diverge with increasing levels of TC and PC for different SMC values.

In the last part of this study, the sensitivity of TE and Γ to TWI and SMC was evaluated over a specific target region in North India, with a strong SMC gradient and a wide topography heterogeneity. Areas with very rough topography were excluded. The Pearson correlation coefficients of TE and Γ with TWI were rather high. On the other hand, the correlation with SMC was almost negligible. This empirical evidence suggests that topographic features could determine the main spatial patterns of Γ, at least over some areas. TWI provided an indication of the relative propensity of soils to become saturated to the surface due to the effect of the local topography.

Present and future applications of GNSS-R for Earth Observation over land surfaces could benefit from the results of this work. Accuracy on geophysical parameters retrieval e.g., SMC, could be improved after properly calibrated reflected DDMs. The estimation of SMC based on GNSS-R observables relies on their ability to represent the dielectric constant of the soil. However, topography significantly distorts these observables. The empirical relationships that were found between the selected topographic parameters and the observables pave the way to compensate underestimated SMC levels due to topographic roughness. Future work should include ground truth data to further validate these results.

Author Contributions: Conceptualization, methodology, software, investigation, and writing-original draft preparation by H.C.-L.; supervision, by G.L. and M.C.

Funding: This research received no external funding.

Acknowledgments: This research was carried out with the support grant of a Juan de la Cierva postdoctoral research fellowship from the Spanish Ministerio de Ciencia, Innovación y Universidades, reference FJCI-2016-29356.

Conflicts of Interest: The authors declare no conflict of interest.

References

1. Dunne, T.; Black, R.D. Partial area contributions to storm runoff in a small New England watershed. *Water Resour. Res.* **1970**, *6*, 1296–1311. [CrossRef]

2. Western, A.W.; Grayson, R.B.; Bloschl, G.G.; Willgoose, R.; McMahon, T.A. Observed spatial organization of soil moisture and its relation to terrain indices. *Water Resour. Res.* **1999**, *35*, 797–810. [CrossRef]

3. Martín-Neira, M. A passive reflectometry and interferometry system (PARIS): Application to ocean altimetry. *ESA J.* **1993**, *17*, 331–355.

4. Martín-Neira, M.; Li, W.; Andrés-Beivide, A.; Ballesteros-Sels, X. Cookie: A satellite concept for GNSS remote sensing constellations. *IEEE J. Sel. Top. Appl. Earth Obs. Remote Sens.* **2016**, *9*, 4593–4610. [CrossRef]

5. Cardellach, E.; Rius, A.; Martín-Neira, M.; Fabra, F.; Nogués-Correig, O.; Ribó, S.; Kainulainen, J.; Camps, A.; D'Addio, S. Consolidating the precision of interferometric GNSS-R ocean altimetry using airborne experimental data. *IEEE Trans. Geosci. Remote Sens.* **2014**, *52*, 4992–5004. [CrossRef]

6. Camps, A.; Hyuk, H.; Pablos, M.; Foti, G.; Gommerginger, C.P.; Liu, P.W.; Judge, J. Sensitivity of GNSS-R spaceborne observations to soil moisture and vegetation. *IEEE J. Sel. Top. Appl. Earth Obs. Remote Sens.* **2016**, *9*, 4730–4732. [CrossRef]

7. Chew, C.; Shah, R.; Zuffada, C.; Hajj, G.; Masters, D.; Mannucci, A.J. Demonstrating soil moisture remote sensing with observations from the UK TechDemoSat-1 satellite mission. *Geophys. Res. Lett.* **2016**, *43*, 3317–3324. [CrossRef]

8. Camps, A.; Vall·llossera, M.; Park, H.; Portal, G.; Rossato, L. Sensitivity of TDS-1 GNSS-R reflectivity to soil moisture: Global and regional differences and impact of different spatial scales. *Remote Sens.* **2018**, *10*, 1856. [CrossRef]

9. Carreno-Luengo, H.; Lowe, S.T.; Zuffada, C.; Esterhuizen, S.; Oveisgharan, S. Spaceborne GNSS-R from the SMAP mission: First assessment of polarimetric scatterometry over land and cryosphere. *Remote Sens.* **2017**, *9*, 362. [CrossRef]

10. Carreno-Luengo, H.; Luzi, G.; Crosetto, M. Sensitivity of CyGNSS bistatic reflectivity and SMAP microwave brightness temperature to geophysical parameters over land surfaces. *IEEE J. Sel. Top. Appl. Earth Obs. Remote Sens.* **2018**, *12*, 107–122. [CrossRef]

11. Chew, C.; Small, E.E. Soil moisture sensing using spaceborne GNSS reflections: Comparison of CYGNSS reflectivity to SMAP soil moisture. *Geophys. Res. Lett.* **2018**, *45*, 4049–4057. [CrossRef]

12. Ruf, C.; Cardellach, E.; Clarizia, M.P.; Galdi, C.; Gleason, S.T.; Paloscia, S. Foreword to the special issue on Cyclone Global Navigation Satellite System (CYGNSS) early on orbit performance. *IEEE J. Sel. Top. Appl. Earth Obs. Remote Sens.* **2018**, *12*, 1–4. [CrossRef]

13. Park, H.; Camps, A.; Pascual, D.; Alonso-Arroyo, A.; Querol, J.; Onrubia, R. Improvement of PAU/PARIS end-to-end performance simulator (P2EPS): Land scattering including topography. In Proceedings of the IEEE International Geoscience and Remote Sensing Symposium, Beijing, China, 10–15 July 2016; pp. 5607–5610.

14. Carreno-Luengo, H.; Luzi, G.; Crosetto, M. An experimental assessment of rough topography on spaceborne delay Doppler maps. In Proceedings of the IEEE GNSS+R Specialist Meeting on Reflectometry using GNSS and other Signals of Opportunity, Benevento, Italy, 20–22 May 2019; pp. 1–4.

15. Ruf, C.; Atlas, R.; Chang, P.S.; Clarizia, M.P.; Garrison, J.L.; Gleason, S.; Katzberg, S.J.; Jelenak, Z.; Johnson, J.T.; Majumdar, S.J.; et al. New ocean winds satellite mission to probe hurricanes and tropical convection. *Bull. Amer. Meteorol. Soc.* **2015**, *97*, 385–395. [CrossRef]

16. CYGNSS. CYGNSS Level 1 Science Data Record. Ver. 2.1. PO.DAAC, CA, USA. 2017. Available online: http://dx.doi.org/10.5067/CYGNSL1X20 (accessed on 11 May 2019).

17. Zavorotny, V.U.; Voronovich, A.G. Scattering of GPS signals from the ocean with wind remote sensing applications. *IEEE Trans. Geosci. Remote Sens.* **2000**, *38*, 951–964. [CrossRef]

18. Ulaby, F.T.; Long, D.G. *Microwave Radar and Radiometric Remote Sensing*; Univ. Michigan Press: Ann Arbor, MI, USA, 2014; p. 252.

19. Pierdicca, N.; Guerriero, L.; Brogioni, M.; Egido, A. On the coherent and non-coherent components of bare soil and vegetated terrain bistatic scattering: Modelling the GNSS-R signal over land. In Proceedings of the IEEE International Geoscience and Remote Sensing Symposium, Munich, Germany, 22–27 July 2012; pp. 3407–3410.

20. Carreno-Luengo, H.; Camps, A.; Querol, J.; Forte, G. First results of a GNSS-R experiment from a stratospheric balloon over boreal forests. *IEEE Trans. Geosci. Remote Sens.* **2015**, *54*, 2652–2663. [CrossRef]

21. Voronovich, A.G.; Zavorotny, V.U. Bistatic radar equation for signals of opportunity revisited. *IEEE Trans. Geosci. Remote Sens.* **2017**, *56*, 1959–1968. [CrossRef]

22. Tsang, L.; Newton, R.W. Microwave emissions from soils with rough surfaces. *J. Geophys. Res.* **1982**, *87*, 9017–9024. [CrossRef]

23. Pierdicca, N.; Guerriero, L.; Giusto, R.; Brogioni, M.; Egido, A. SAVERS: A Simulator of GNSS Reflections from Bare and Vegetated Soils. *IEEE Trans. Geosci. Remote Sens.* **2014**, *52*, 6542–6554. [CrossRef]

24. Zribi, M.; Guyon, D.; Motte, E.; Dayau, S.; Wigneron, J.-P.; Baghdadi, N.; Pierdicca, N. Performance of GNSS-R GLORI data for biomass estimation over the Landes forests. *Int. J. Earth Obs. Geoinf.* **2019**, *74*, 150–158. [CrossRef]

25. Gleason, S. Reflecting on GPS sensing land and ice from low earth orbit. *GPS World Innov. Column.* **2007**, *18*, 44–49.

26. Amatulli, G.; Domisch, S.; Tuanmu, M.-N.; Parmentire, B.; Ranipeta, A.; Malczyk, J.; Jetz, W. A suite of global cross-scale topographic variables for environmental and biodiversity modelling. *Nat. Sci. Data* **2018**, 180040. [CrossRef] [PubMed]

27. Wilson, M.F.J.; O'Connell, B.; Brown, C.; Guinan, J.C.; Grehan, A.J. Multiscale terrain analysis of multibeam bathymetry data for habitat mapping on the continental slope. *Taylor Fr. Mar. Geod.* **2007**, *40*, 3–35. [CrossRef]

28. Wood, J. The Geomorphological Characterization of Digital Elevation Models. Ph.D. Thesis, University of Leicester, Leicester, UK, 1996.

29. Beven, K.J.; Kirkby, M.J. A physically based, variable contributing area model of basin hydrology. *Hydrol. Sci. Bull.* **1979**, *24*, 43–69. [CrossRef]

30. Wolock, D.M. *Simulating the Variable-Source-Area Concept of Streamflow Generation with the Watershed Model TOPMODEL*; United States Geological Survey Water-Resources Investigations Report 93-4124; US Geological Survey: Lawerence, KS, USA, 1993.

31. Marthews, T.R.; Dadson, S.J.; Lehner, B.; Abele, S.; Gedney, N. High-resolution global topographic index values for use in large-scale hydrological modelling. *EGU Hydrol. Earth Syst. Sci.* **2015**, *19*, 91–104. [CrossRef]

32. Lehner, B.; Grill, G. Global river hydrography and network routing: Baseline data and new approaches to study the world's large river systems. *Hydrol. Process.* **2013**, *27*, 2171–2186. [CrossRef]

33. Quinn, P.; Beven, K.; Chevalier, P.; Planchon, O. The prediction of hillslope flow paths for distributed hydrological modelling using digital terrain models. *Hydrol. Process.* **1991**, *5*, 59–79. [CrossRef]
34. Quinn, P.; Beven, K.; Lamb, R. The ln (a/tanB) index: How to calculate it and how to use it within the TOPMODEL framework. *Hydrol. Process.* **1995**, *9*, 161–182. [CrossRef]
35. Clarizia, M.-P.; Pierdicca, N.; Constantini, F.; Floury, N. Analysis of CyGNSS data for soil moisture retrieval. *IEEE J. Sel. Top. Appl. Earth Obs. Remote Sens.* **2019**, *12*, 2227–2235. [CrossRef]
36. Al-Khaldi, M.; Johnson, J.T.; O'Brien, A.J.; Balenzano, A.; Mattia, F. Time-series retrieval of soil moisture using CYGNSS. *IEEE Trans. Geosci. Remote Sens.* **2019**, *57*, 4322–4331. [CrossRef]
37. Entekhabi, D.; Yueh, S.; O'Neill, P.E.; Kellogg, K.H.; Allen, A.; Bindlish, R.; Brown, M.; Chan, S.; Colliander, A.; Crow, W.; et al. SMAP Handbook. Soil Moisture Active Passive. Available online: https://nsidc.org/data/SPL3SMP_E/versions/1 (accessed on 6 April 2019).
38. Chan, S.K.; Bindlish, R.; O'Neill, P.; Jackson, T.; Njoku, E.; Dunbar, S.; Chaubell, J.; Piepmeier, J.; Yueh, S.; Entekhab, D.; et al. Development and assessment of the SMAP enhanced passive soil moisture product. *Remote Sens. Environ.* **2018**, *204*, 931–941. [CrossRef]
39. Carreno-Luengo, H.; Luzi, G.; Crosetto, M. Sensitivity of CyGNSS to above ground biomass and canopy height over tropical forests. In Proceedings of the IEEE GNSS+R Specialist Meeting on Reflectometry using GNSS and other Signals of Opportunity, Benevento, Italy, 20–22 May 2019; pp. 1–4.
40. Carreno-Luengo, H.; Luzi, G.; Crosetto, M. Impact of the elevation angle on CyGNSS GNSS-R bistatic reflectivity as a function of effective surface roughness over land surfaces. *Remote Sens.* **2018**, *10*, 1749. [CrossRef]
41. Mialon, A.; Coret, L.; Kerr, Y.H.; Secherre, F.; Wigneron, J.-P. Flagging the topographic impact on the SMOS signal. *IEEE Trans. Geosci. Remote Sens.* **2008**, *46*, 689–694. [CrossRef]

 *remote sensing*

Article

Analytical Computation of the Spatial Resolution in GNSS-R and Experimental Validation at L1 and L5

Adriano Camps [1,2,*] and Joan Francesc Munoz-Martin [1,2]

[1] CommSensLab-UPC, Department of Signal Theory and Communications, UPC BarcelonaTech, c/Jordi Girona 1-3, 08034 Barcelona, Spain; joan.francesc@tsc.upc.edu

[2] Institut d'Estudis Espacials de Catalunya-IEEC/CTE-UPC, Gran Capità, 2-4, Edifici Nexus, despatx 201, 08034 Barcelona, Spain

* Correspondence: camps@tsc.upc.edu

Received: 3 November 2020; Accepted: 24 November 2020; Published: 28 November 2020

Abstract: Global navigation satellite systems reflectometry (GNSS-R) is a relatively novel remote sensing technique, but it can be understood as a multi-static radar using satellite navigation signals as signals of opportunity. The scattered signals over sea ice, flooded areas, and even under dense vegetation show a detectable coherent component that can be separated from the incoherent component and processed accordingly. This work derives an analytical formulation of the response of a GNSS-R instrument to a step function in the reflectivity using well-known principles of electromagnetic theory. The evaluation of the spatial resolution then requires a numerical evaluation of the proposed equations, as the width of the transition depends on the reflectivity values of two regions. However, it is found that results are fairly constant over a wide range of reflectivities, and they only vary faster for very high or very low reflectivity gradients. The predicted step response is then satisfactorily compared to airborne experimental results at L1 (1575.42 MHz) and L5 (1176.45 MHz) bands, acquired over a water reservoir south of Melbourne, in terms of width and ringing, and several examples are provided when the transition occurs from land to a rough ocean surface, where the coherent scattering component is no longer dominant.

Keywords: GNSS-R; spatial resolution; diffraction; experiment; airborne; L1; L5

---

1. Introduction

Global navigation satellite systems reflectometry (GNSS-R) is a relatively new remote sensing technique that uses navigation signals as signals of opportunity in a multi-static radar configuration [1].

The spatial resolution of GNSS-R instruments was first analyzed in [2], and it was estimated from the dimensions of the area determined by the intersection of the first iso-delay and iso-Doppler lines. In this case, for a Low Earth Orbit (LEO) satellite at 700 km, the predicted spatial resolution for an instrument using the GPS L1 C/A (Coarse/Acquisition) code would be in the range of 36 to 54 km, for elevation angles from 90° to 40°.

Later on, the definition of spatial resolution was revised and "an effective spatial resolution derived as a function of measurement geometry and delay–Doppler (DD) interval, and as a more appropriate representation of resolution than the geometric resolution previously used in the literature" was proposed in [3]. This effective spatial resolution was defined following an approach similar to the definition of effective bandwidth used in filter theory. An increase in the effective spatial resolution is observed for increasing incidence angle, receiver altitude, and delay–Doppler window. With this definition, the effective spatial resolution is estimated to range from 25 to 37 km for a 700 km height receiver (Figure 4, [3]).

The above definitions implicitly assume that the scattering is incoherent, as the radar equation used is the one that corresponds to surface scattering. However, the development of advanced GNSS-R receivers, capable of processing the coherent component and separating it from the incoherently scattered component, showed that much finer features can be distinguished [4,5]. In fact, when coherent scattering occurs, the GNSS reflectivity values have a very high spatial resolution, as most of the power is mainly coming from the first Fresnel zone [6], an ellipse of semi-major axis a (Equation (1a)) and semi-minor axis b (Equation (1b)):

$$a = \sqrt{\lambda \frac{R_T \cdot R_R}{R_T + R_R}} \tag{1a}$$

$$b = \frac{a}{cos(\theta_i)}, \tag{1b}$$

where λ is the electromagnetic wavelength, R_T and R_R are the distances from the transmitter and receiver to the specular reflection point, and θ_i is the incidence angle. This means that even from an LEO satellite, the spatial resolution can be as small as ~400–500 m.

In [7] the spatial resolution under coherent scattering was theoretically analyzed in detail, including the effect of the different Fresnel zones. It was shown that up to ~45° most of the power is coming from an area determined by the first Fresnel zone, although a power equivalent to the one received in free-space conditions is actually coming from a region ~0.58–0.62 times the first Fresnel zone. However, significant contributions from regions farther away than the first Fresnel zone also contribute to the received power, making its interpretation more difficult. This region was ~800 m away at a 60° elevation angle for an LEO satellite at 500 km height.

In this work, we have gone one step further and analyzed the spatial resolution from a theoretical point of view, using known principles of electromagnetism. The use of these results makes it possible to compute in a very compact form the expression of the ringings and oscillations observed when the specular reflection point transitions from the land into the ocean or vice versa. Analytical results are validated with airborne experimental data gathered at L1 and L5 with the Microwave Interferometric Reflectometer (MIR) instrument when crossing the coastline [8].

2. Methodology

The scattering geometry is shown in Figure 1, where the receiver (R) and transmitter (T) lie in the YZ plane, and are at heights h_R and h_T over a flat surface. R_T and R_R are the distances to the origin of coordinates where the specular reflection point is, and θ_i is the incidence angle, which for the specular reflection point (origin of coordinates) is equal to the reflected angle θ_r. The different disks represent the Fresnel zones, with the odd ones (1, 3, 5, etc.) colored dark gray, and the even ones (2, 4, 6, etc.) colored white.

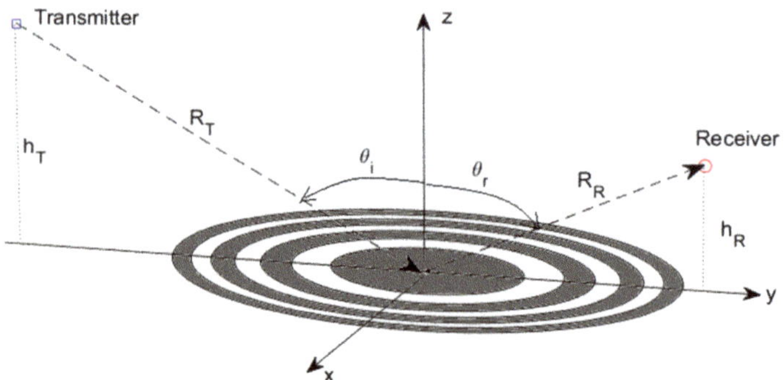

Figure 1. Global navigation satellite systems reflectometry (GNSS-R) scattering geometry showing the transmitter, receiver, specular reflection point (origin of coordinates), and the different Fresnel zones.

In what follows, a few principles of electromagnetism are reviewed. They will be used to compute a closed-form solution of the GNSS-R response to a step function, such as in the transition between water and the ocean.

2.1. Review of Some Basic Principles of Electromagnetism

2.1.1. Principle of Equivalence and Image Theory

The principle of equivalence [9] allows us to compute the solutions in the region of interest (half-space in front of the conducting plane) by replacing the plane conductor with the images of the dipoles. The image sources must satisfy the boundary condition that the tangential electric field at the conducting surface should be zero. The total electric field above the conducting plane (Figure 2) is then computed as the sum of the one created by the original currents and the image ones, while the total electric field below the conducting plane is zero. Applying this principle to our problem, the transmitter can be replaced by its "image" underneath the plane, as illustrated in Figure 3.

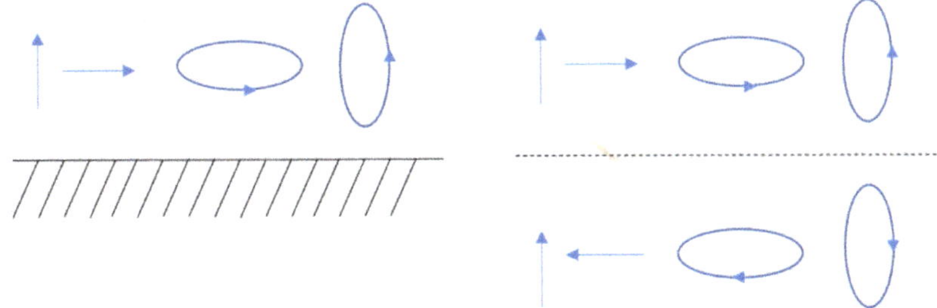

Figure 2. Left: linear and loop currents over a perfect conductor. Right: the conductor has been replaced by the image currents. Note that the sign reversal of the currents parallel to the conducting plane are responsible for the polarization change of the "reflected" wave.

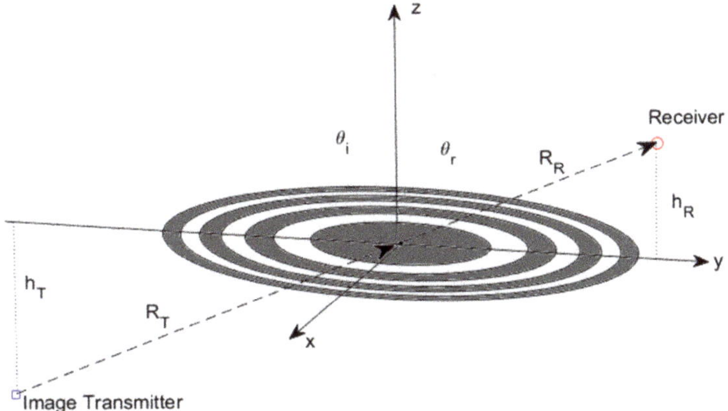

Figure 3. GNSS-R scattering geometry in Figure 1 modified according to the principle of equivalence: the transmitter is replaced by its image. Computed electric fields are only valid in the region $z > 0$.

2.1.2. The Huygens Principle and Knife-Edge Diffraction

The Huygens principle states that every point (red dots in Figure 4) on a primary wave front (blue curve in Figure 4) may be regarded as a new source of spherical secondary waves, and that at each point in space these waves propagate in every direction at a speed and frequency equal to that of the primary wave, and in such a way that the primary wave front at some later time is the envelope of these secondary waves (red curve in Figure 4).

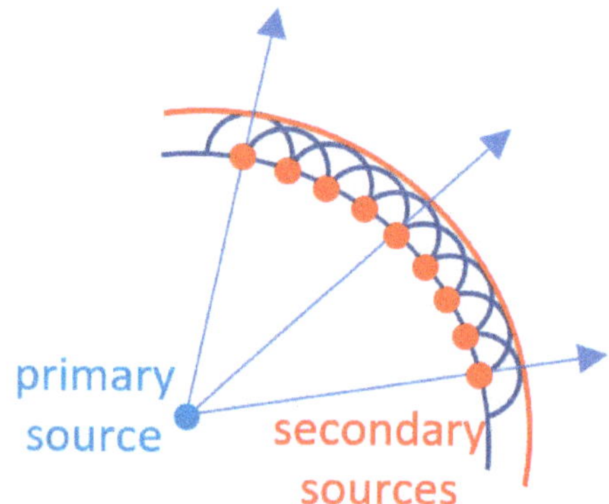

Figure 4. Graphical representation of the Huygens diffraction principle.

The Huygens principle can be applied to compute the electric fields when part of the trajectory is partially obstructed by a conducting plane as illustrated in Figure 5, leaving a clearance h with respect to the line of sight. This academic case is known as the "knife-edge diffraction", and it is used in radio communications to assess the propagation losses with respect to the free space caused by the obstruction by a mountain, building, etc. In the latter this will represent the transition from one medium to another [6].

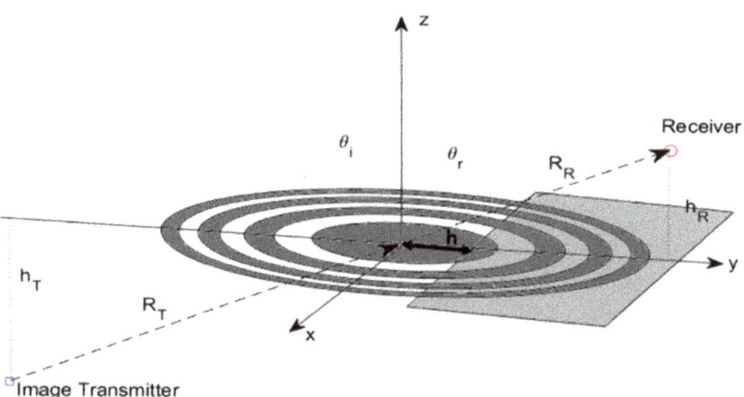

Figure 5. Huygens principle applied to the "knife-edge diffraction" case. The light-gray area partially blocks the propagation of the signal, hiding parts of some Fresnel zones. h is the clearance of the obstruction.

The development of an analytical solution can be found in many text books, and only the final result is provided below:

$$E_{diff} = E_{free\ space} \cdot F(v) \tag{2}$$

where:

$$F(v) \triangleq \frac{1+j}{2} \left[\int_v^\infty \cos\left(\frac{\pi}{2}t^2\right)dt - j \int_v^\infty \sin\left(\frac{\pi}{2}t^2\right)dt \right] = \frac{1+j}{2}\left\{\left[\frac{1}{2} - C(v)\right] - j\left[\frac{1}{2} - S(v)\right]\right\}, \tag{3}$$

$C(v)$ and $S(v)$ are the so-called Fresnel integrals, which are given be:

$$C(v) = \int_0^v \cos\left(\frac{\pi}{2}t^2\right)dt, \tag{4a}$$

$$S(v) = \int_0^v \sin\left(\frac{\pi}{2}t^2\right)dt, \tag{4b}$$

and v is the Fresnel–Kirchhoff parameter:

$$v = h \cdot \sqrt{\frac{2 \cdot (R_T + R_R)}{\lambda \cdot R_T \cdot R_R}} \tag{5}$$

Figure 6 shows the absolute value of $F(v)$ in Equation (3) in linear units. When $h = 0$, half of the electric field is blocked, which translates into a power loss of 6 dB with respect to the free-space conditions. Note the ringing for $h < 0$ due to the different Fresnel zones being blocked or not and contributing to the total electric field with different phases.

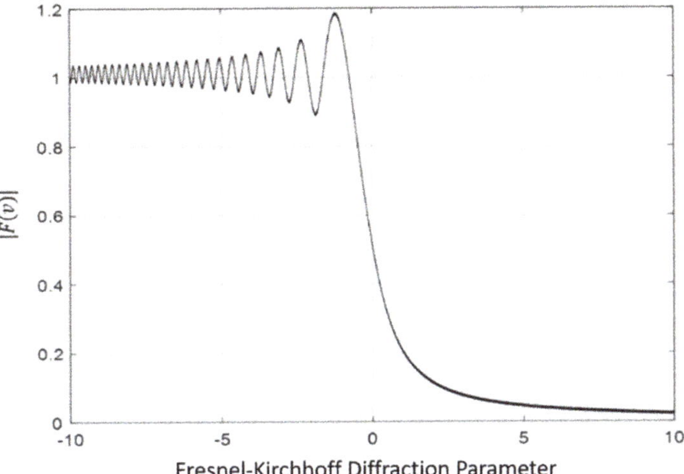

Figure 6. Absolute value of $F(v)$. Negative v indicates that the line-of-sight (LoS) trajectory is unobstructed, while positive v indicates an LoS obstruction (see Figure 5).

2.1.3. Babinet's Principle

Babinet's principle [10] states that the diffraction pattern from an opaque body is identical to that from a hole of the same size and shape, except for the overall forward beam intensity. The concept is expressed graphically in Figure 7 [11].

Babinet's principle can be used to compute the electric field diffracted when the signal emitted by the transmitter's image is diffracted by the light-gray opaque area in the $z = 0$ plane (the white area being "transparent"), and when it is diffracted by the white opaque area in the $z = 0$ plane (the light-gray area now being "transparent").

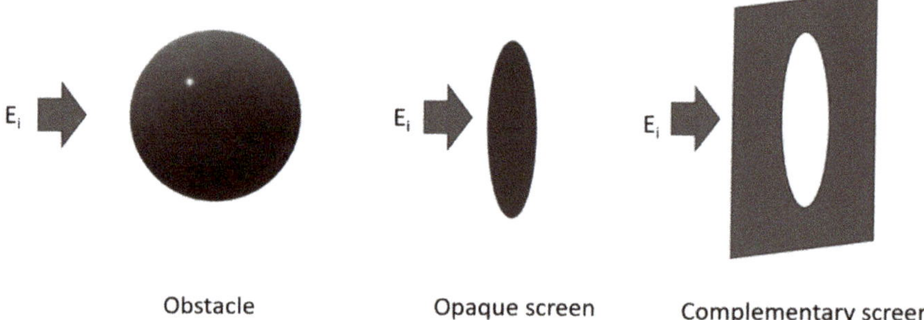

Obstacle   Opaque screen   Complementary screen

Figure 7. Graphical representation of the concept of Babinet's principle and the three equivalent obstacles: electric field incident on an arbitrary obstacle (**left**), associated opaque screen (**center**), and complementary screen (i.e., hole in the infinite plane; **right**).

2.1.4. Principle of Superposition

Once the electric fields diffracted by the semi-plane (the white area in the plane z = 0 in Figure 5) are computed, the electric fields diffracted by the complementary semi-plane (the light-gray area in the plane z = 0 in Figure 5), when it is not opaque, can be computed using Babinet's principle, and then the total electric field arriving at the receiver can be computed as the sum of both.

2.2. Computation of the Electric Field Scattered by the Discontinuity between Two Media

Assuming that an electromagnetic wave impinges over the z = 0 plane around (x,y,z) = (0,0,0), where the specular reflection point is, and:

- The extension of the Fresnel reflection zones is small enough (a few hundred meters from a low Earth orbiter) so that the variations of the local incidence angle can be neglected over a few Fresnel zones;
- The plane z = 0 is composed of two flat half-planes of different materials, and thus exhibiting different reflection coefficients; and
- The transmission coefficient for the wave emitted by the image transmitter (Figure 5) is equal to the reflection coefficient evaluated at the local incidence angle at the origin.

Then, as illustrated in Figure 8, the total electric field can be immediately computed as the sum of the following two terms:

$$E_{diff} = E_{free\ space}\cdot[F(v)\cdot\rho(\varepsilon_{r,1},\ \theta_i) + F(-v)\cdot\rho(\varepsilon_{r,2},\ \theta_i)], \tag{6}$$

where $\rho(\varepsilon_{r,n},\ \theta_i)$ is the Fresnel reflection coefficient, and $\varepsilon_{r,n}$ is the dielectric constant of the half-plane n. The received power is proportional to the square of the absolute value of Equation (6).

$$P_{diff} = P_{free\ space}\cdot\left|F(v)\cdot\rho(\varepsilon_{r,1},\ \theta_i) + F(-v)\cdot\rho(\varepsilon_{r,2},\ \theta_i)\right|^2, \tag{7}$$

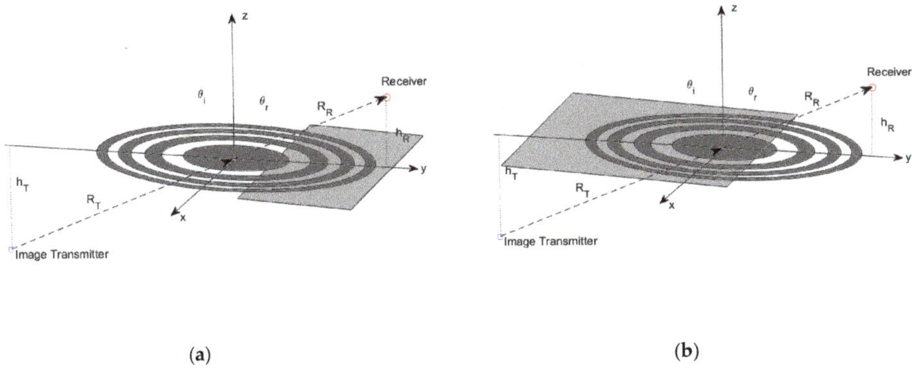

(a)   (b)

Figure 8. Graphical representation of the principle of superposition and its application to the computation of the total electric field incident. Opaque screen on the (**a**) right and (**b**) left sides block radiation.

Following Equation (7), an example is shown in Figure 9. The red and blue lines correspond to $|F(v)|$ and $|F(-v)|$ in decibels. The black line corresponds to the square of the absolute value of the term within brackets in Equation (7), computed for $\rho(\varepsilon_{r,1},\ \theta_i) = \frac{2}{3}$, and $\rho(\varepsilon_{r,2},\ \theta_i) = \sqrt{0.1}$, which corresponds to the normalized peak power that would be collected by a GNSS-R instrument.

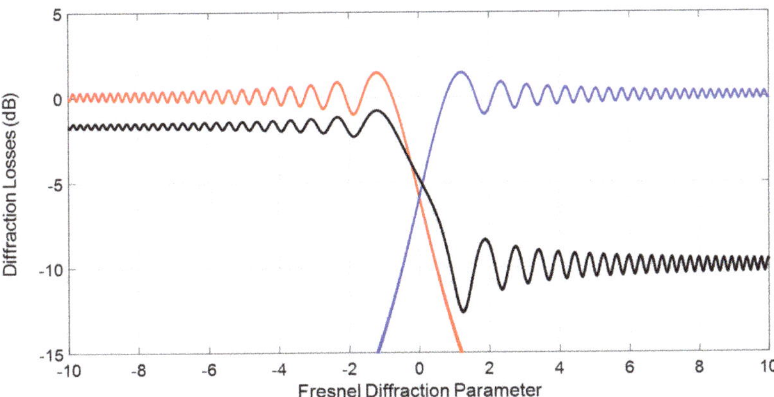

Figure 9. Graphical representation of the $|F(v)|$ (red) and $|F(-v)|$ (blue) functions, and total power loss computed as $\left| F(v) \cdot \rho\left(\varepsilon_{r,1},\ \theta_i\right) + F(-v) \cdot \rho\left(\varepsilon_{r,2},\ \theta_i\right) \right|^2$ (black) for $\rho\left(\varepsilon_{r,1},\ \theta_i\right) = \frac{2}{3}$ and $\rho(\varepsilon_{r,2},\ \theta_i) = \sqrt{0.1}$.

3. Results

3.1. Spatial Resolution Computation

Based on the above results the spatial resolution of a GNSS-R system can now be defined as the width of the transition from the 10% above $\left|\rho(\varepsilon_{r,2},\ \theta_i)\right|^2 \triangleq \left|\rho_{max}\right|^2$, and 90% of $\left|\rho(\varepsilon_{r,1},\ \theta_i)\right|^2 \triangleq \left|\rho_{min}\right|^2$

$$\left| F(v_{90\%}) \cdot \rho_{max} + F(-v_{90\%}) \cdot \rho_{min} \right|^2 = 0.9 \cdot \left|\rho_{max}\right|^2, \tag{8a}$$

$$\left| F(v_{10\%}) \cdot \rho_{max} + F(-v_{10\%}) \cdot \rho_{min} \right|^2 = 1.1 \cdot \left|\rho_{min}\right|^2 \tag{8b}$$

The solution of the above equations has to be performed numerically, and it actually depends on the values of ρ_{min} and ρ_{max}. The difference between $v_{10\%}$ and $v_{90\%}$ leads to the width of v, $\Delta v = v_{10\%} - v_{90\%}$, which is linked to the spatial resolution by means of Equation (9). For the sake of simplicity, assuming an airborne GNSS-R instrument, so that $R_T \gg R_R$, and approximating $R_R \approx h_R / cos(\theta_i)$, the following is an approximated formula for the spatial resolution:

$$\Delta v = \Delta x \sqrt{\frac{2 \cdot cos(\theta_i)}{\lambda \cdot h_R}} \Rightarrow \Delta x = \Delta v \sqrt{\frac{\lambda \cdot h_R}{2 \cdot cos(\theta_i)}}. \tag{9}$$

Figure 10 shows the evolution of $\Delta v = v_{10\%} - v_{90\%}$ as a function of $\Delta\rho = \rho_{min} / \rho_{max}$ in decibels. It varies from 0.74 for $\rho_{min} / \rho_{max} = -3$ dB to 1.7 for $\rho_{min} / \rho_{max} = -20$ dB. This means that for a plane flying at $h_R = 1000$ m height, at GPS L1/Galileo E1 ($\lambda = 19$ cm), the spatial resolution ranges from 7.2 m/cos(θ_i) to 16.6 m/cos(θ_i), depending on the contrast of the reflection coefficients in the transition.

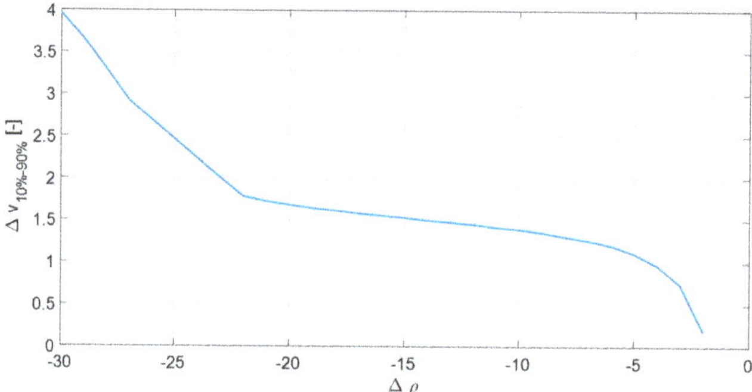

Figure 10. Width of the Δv parameter for a transition from 10% above ρ_{min} to 90% of ρ_{max}.

3.2. Experimental Validation

Figure 11 shows the measured reflectivity (Γ) in a transition between land and calm water in the Devilbend Reservoir, south of Melbourne, Victoria, Australia. The data was acquired during one of the test flights of the MIR [8] instrument on 30 April 2018. The top plot corresponds to one of the four beams at GPS L1, and the bottom plot at L5. Flight speed was 75 m/s, flight height was $h_R = 1000$ m, and GNSS-R data were coherently integrated for 1 ms (GPS L1 C/A code), and incoherently for $T_{inc} = 20$, 40, 100, and 200 ms. At 20 ms, the blurring produced by the aircraft movement is 1.5 m, which is negligible compared to the best spatial resolution computed in Section 3.1. Inspecting Figure 11, at 20 and 40 ms incoherent integration time, one can clearly observe some reflectivity ripples that are reminiscent of those seen in Figure 9. At 100 ms, the blurring is 7.5 m, which is now comparable to the best spatial resolution, and the ripples can no longer be appreciated. At 200 ms the reflectivity plot is even smoother. In the land part (left part of the plots), note that the L5 reflections exhibit a more marked fading pattern than the L1 ones. This may be because L5 penetrates a bit more in the land (and vegetation), and it is moister soil underneath the surface. Although it is not very clear, the width of the reflectivity step is approximately 0.3 s, which corresponds roughly to 22.5 m, as compared to the 20.7 m predicted value using Equation (9) and Figure 10 for a ~-15 dB reflectivity.

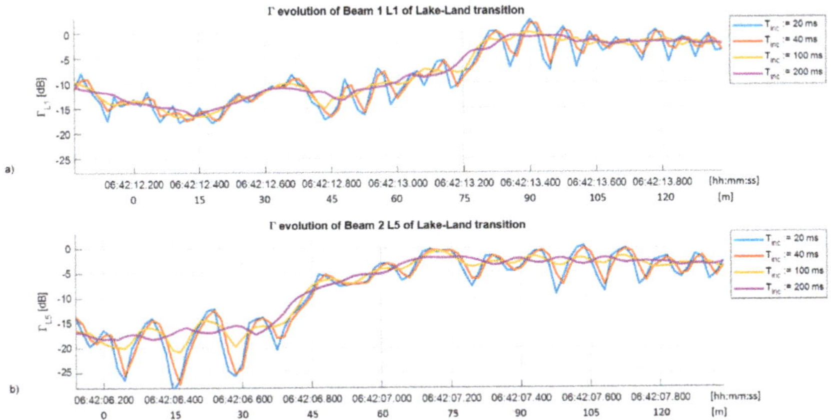

Figure 11. Land-to-water transition during a Microwave Interferometric Reflectometer (MIR) test flight over Devilbend Reservoir (38.29°S, 145.1°E): (**a**) GPS L1, (**b**) GPS L5. Horizontal axis units: time and distance. Vertical dashed line: land–water transition.

Taking a closer look to the periodicity of the ripples, it can be observed that the period in Figure 11b is larger than that in Figure 11a, as it can be inferred from Equation (5), and Figures 6 and 9, parameterized in terms of v. In Figure 11a, the separation between consecutive peaks in time is ~100 ms, while in Figure 11b it is ~130 ms, which, for an aircraft speed of $v = 75$ m/s corresponds to 7.50 and 9.75 m, respectively. More detailed results are shown in Table 1 for several peaks. Replacing $\Delta x(t)$ by $v_\perp \cdot \Delta t$ in Equation (10), where $v_\perp = v \cdot \cos(\varphi)$, φ being the angle formed between the aircraft ground-track and the coastline, one obtains:

$$\Delta v = v_\perp \cdot \Delta t \sqrt{\frac{2 \cdot \cos(\theta_i)}{\lambda \cdot h_R}}. \tag{10}$$

Table 1. Peak positions in Figure 6 and peak spacing (Δv), peak positions in time for L1 and L5 beams, time spacing and estimated peak spacing in v.

v_{peaks}	$|\Delta v_{peaks}|$	$t_{peaks,\,L1}$ [s]	$\Delta t_{peaks,\,L1}$ [ms]	$\hat{\Delta v}_{peaks,L1}$	$t_{peaks,\,L5}$ [s]	$\Delta t_{peaks,\,L5}$ [ms]	$\hat{\Delta v}_{peaks,L5}$
−1.22	-	Not visible	-	-	7.140	-	-
−2.34	1.12	13.290	-	-	7.310	170	0.95
−3.08	0.74	13.400	110	0.71	7.450	140	0.67
−3.68	0.60	13.500	100	0.65	7.570	120	0.78
−4.18	0.50	13.580	80	0.52	7.700	130	0.67

Assuming $h_R = 1000$ m, $\theta_i = 45°$, and substituting Δt for the time difference between consecutive peaks in Figure 11 (except for roundoff errors due to the discrete sampling), this ratio closely matches the squared root of the ratio of the wavelengths at L1 and L5: $\sqrt{1575.42 \text{ MHz}/1176.45 \text{ MHz}} = 1.16$. Other sources of error are variations of the flight height (approximately $h_R = 1000$ m), variations of the incidence angle, and most likely the fact that the velocity vector is not perpendicular to the coastline (as implicitly assumed in Figure 5).

These results are corroborated by many other transitions. In what follows a few more examples are presented to illustrate this effect. A minimum incoherent integration time of 40 ms is used, since blurring is still negligible, and these results are almost identical to when 20 ms is used.

Figure 12 shows a geo-located double transition from land–calm water–land, the temporal evolution of the reflection coefficient at L1, and two zooms around the transitions for different incoherent integration times. Figure 13 shows another example at L5. Note that for large incoherent integration times (i.e., 200 ms) the ringings cannot be identified for both the L1 and L5 examples.

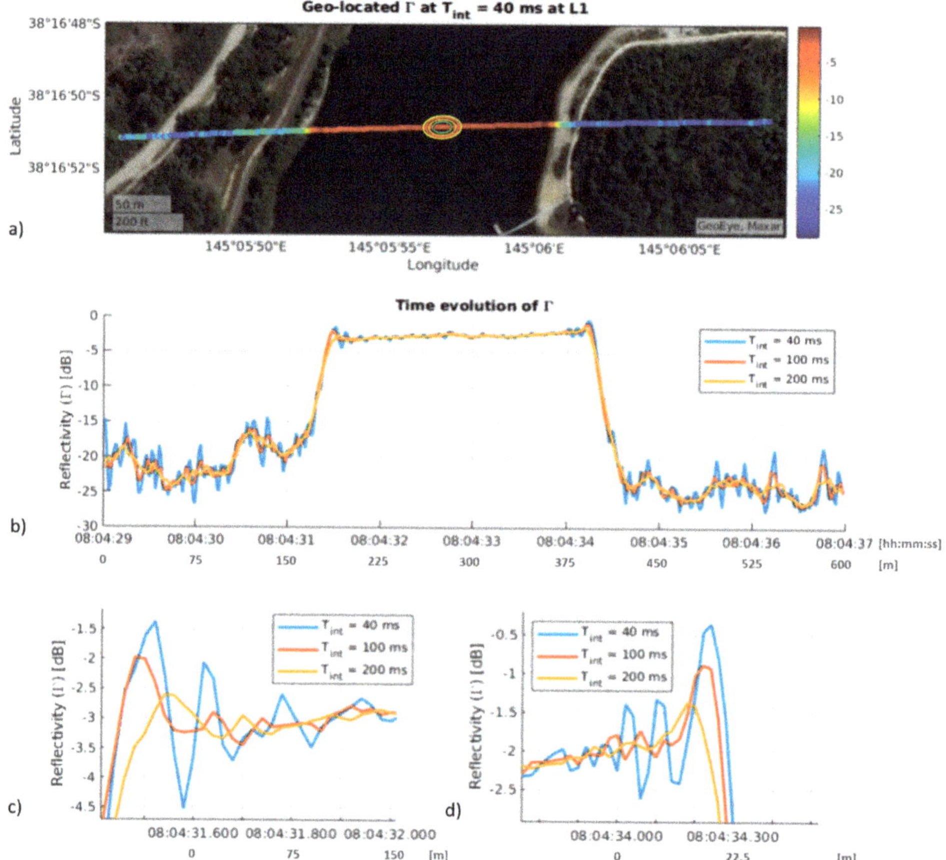

Figure 12. Land-to-water transition during an MIR at L1 test flight over Devilbend Reservoir (38.29°S, 145.1°E). (**a**) Geo-located reflectivity, and approximated representation of the first four Fresnel zones; (**b**) time evolution of the reflectivity, and a zoom-in of the step transition on the (**c**) left and (**d**) right.

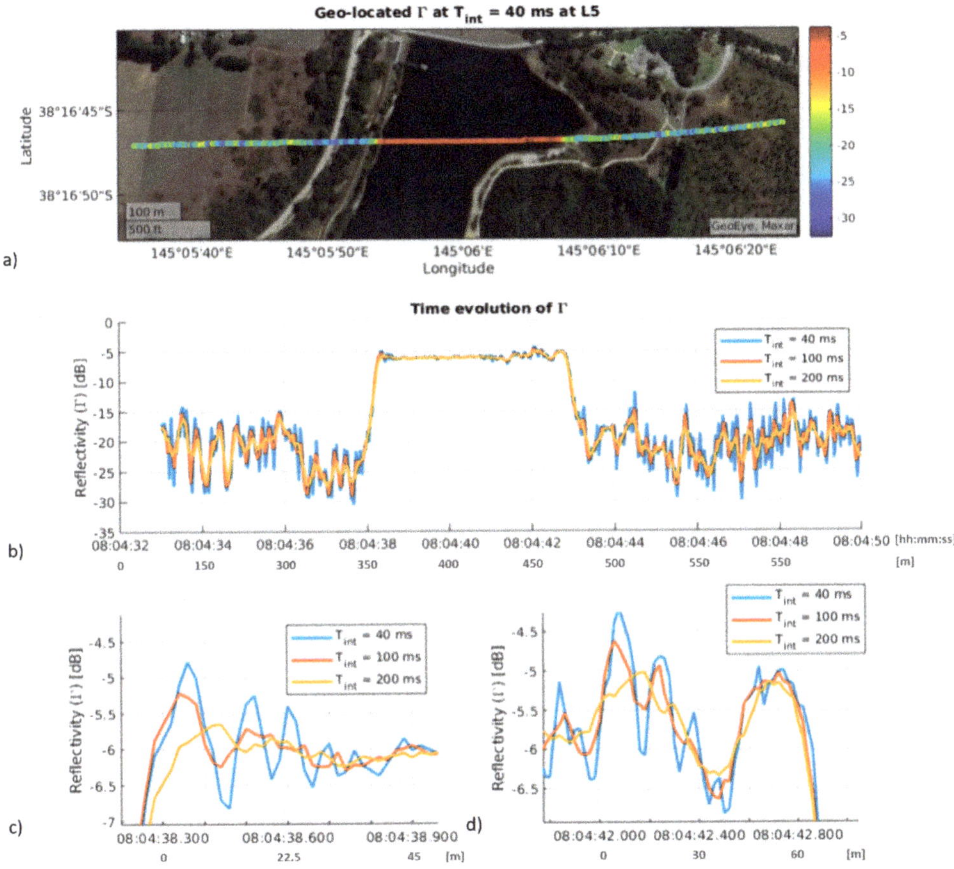

Figure 13. Land-to-water transition during a MIR at L5 test flight over Devilbend Reservoir (38.29°S, 145.1°E). (**a**) Geo-located reflectivity, (**b**) time evolution of the reflectivity, and a zoom-in of the step transition on the (**c**) left and (**d**) right.

Figure 14 presents another example on the same flight. In this case, there is an additional transition to a small portion of a peninsula being crossed by the GNSS-R track. It can be noticed that at short integration times, some ripples are produced when changing from land to water, at the beginning and the end of the track. In addition, in the central part, when passing over the peninsula the combination of the different responses produces very large amplitude ripples. In this case, large incoherent integration times (i.e., 200 ms) are not blurring the ripples in the center of the image. However, if even larger integration times were used (i.e., 1 s) no ringings could be identified.

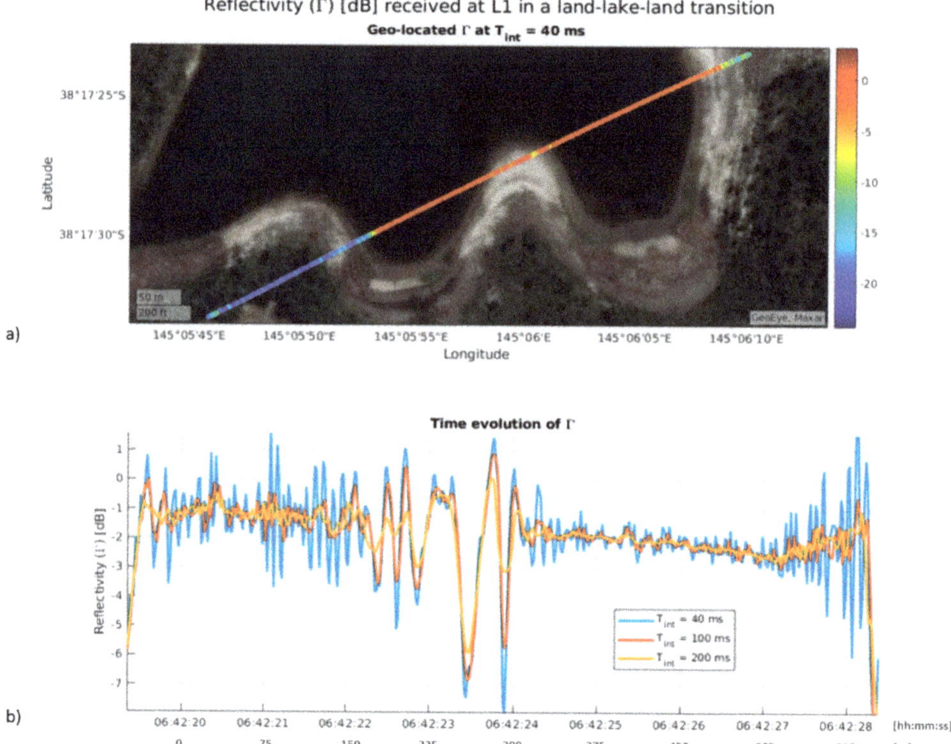

Figure 14. Land-to-water transition during an MIR at L1 test flight over Devilbend Reservoir (38.29°S, 145.1°E). (**a**) Geo-located reflectivity, and (**b**) time evolution of the reflectivity.

Finally, Figure 15 shows another double transition when passing over Sorrento, south of Melbourne. The left part corresponds to an open ocean area (Sorrento ocean beach), while the right part of the image is inside Port Phillip Bay (Sorrento front beach). Looking at the ocean–land transition in Figure 15c on the left-hand side, the reflectivity step is larger, but longer over time, and reflectivity values are noisy because of the increased surface roughness and speckle noise associated with a mostly incoherent scattering. The ripples described in the previous sections and illustrated in the previous examples can no longer be identified. However, as compared to the ocean side in Figure 15c, in Figure 15d the reflectivity step is smaller (calm water), but shorter in time (i.e., sharper), and similar ripples to the lake case can now be identified.

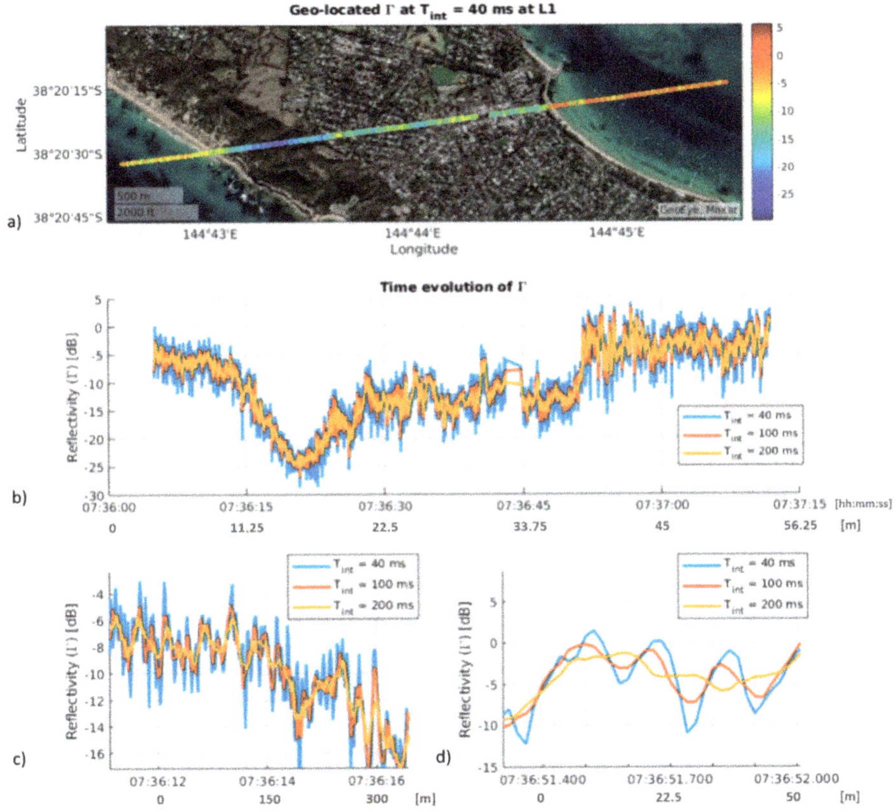

Figure 15. Land-to-water transition during an MIR at L1 test flight over Point Nepan, Melbourne (38.35°S, 144.74°E). (**a**) Geo-located reflectivity, (**b**) time evolution of the reflectivity, and a zoom-in of the step transition on the (**c**) left and (**d**) right.

4. Discussion

To the authors' best knowledge, this is the first analytical study of the response of a GNSS-R system to a reflectivity step. The analytical expression of the measured reflectivity $\hat{\Gamma}$ can be derived from Equation (7):

$$\hat{\Gamma} = \left| F(v) \cdot \rho(\varepsilon_{r,1},\ \theta_i) + F(-v) \cdot \rho(\varepsilon_{r,2},\ \theta_i) \right|^2, \tag{11}$$

with v given by Equation (5). It shows a smooth transition between $\left| \rho(\varepsilon_{r,1},\ \theta_i) \right|^2$ and $\left| \rho(\varepsilon_{r,2},\ \theta_i) \right|^2$, and ripples associated to different Fresnel zones passing from one area to the other one. The width of the transition, as measured from the 10% above the lowest reflectivity values to 90% of the highest reflectivity value, has been found to be dependent on the amplitude of the reflectivity step itself, although it is quite flat for a wide range of reflectivity steps (Figure 10): $\Delta v \approx 1.5$

$$\Delta x = \Delta v \sqrt{\frac{\lambda \cdot h_R}{2 \cdot cos(\theta_i)}} \approx 1.5 \sqrt{\frac{\lambda \cdot h_R}{2 \cdot cos(\theta_i)}}. \tag{12}$$

The spatial resolution and the position of the peaks of the ripples was validated using airborne data at the L1 and L5 bands.

It is worth mentioning that the theoretical analysis assumes that coherent scattering is the dominant mechanism, which is the case when transitioning from land to/from calm water. However, when transitioning into open ocean, where incoherent scattering is more important, the presence of the ripples is not that evident, and the reflectivity exhibits random fluctuations associated mostly to the speckle noise.

These results should allow us to improve the determination of the extent of the flooded areas using GNSS-R observations as in [12–15], and they can also be used to dynamically optimize the coherent and incoherent integration times in future GNSS-R instruments (airborne or spaceborne) depending on the application or target area, as opposed to the approach implemented in past and current missions (1 ms coherent integration time, and 1 s incoherent integration time) such as UK TDS-1 [16] and NASA CyGNSS [17] missions.

5. Conclusions

In this study, the GNSS-R response to a step function in the reflectivity was derived analytically using known principles of electromagnetism. However, the evaluation of the Fresnel integrals has to be performed numerically. Analytical/numerical predictions match well with the experimental results at L1 and L5, acquired during an airborne experiment using the MIR (Microwave Interferometric Reflectometer) instrument over a water reservoir south of Melbourne and in a transition from the ocean into Port Phillip Bay, in terms of width and ringing, if incoherent integration times multiplied by the speed of the aircraft are shorter than the width of the Fresnel zone. Larger incoherent integration times blur the step response and degrade the achievable spatial resolution. Considering the size of the first Fresnel zone (tens of meters from a plane, to a few hundred meters from a spacecraft) the analysis presented is applicable to most coastlines.

Author Contributions: Conceptualization, A.C.; methodology, A.C.; software, J.F.M.-M. and A.C.; validation, A.C.; formal analysis, A.C.; investigation, A.C.; resources, A.C. and J.F.M.-M.; data curation, J.F.M.-M.; writing—original draft preparation, A.C.; writing—review and editing, A.C. and J.F.M.-M.; visualization, A.C. and J.F.M.-M.; supervision, A.C.; project administration, A.C.; funding acquisition, A.C. All authors have read and agreed to the published version of the manuscript.

Funding: This work was funded by the Spanish MCIU and EU ERDF project (RTI2018-099008-B-C21/AEI/10.13039/501100011033) "Sensing with pioneering opportunistic techniques" and grant to "CommSensLab-UPC" Excellence Research Unit Maria de Maeztu (MINECO grant MDM-2016-600).

Acknowledgments: The authors would like to express their gratitude to R. Onrubia and D. Pascual who flew and operated the MIR instrument during the field experiment, and to C. Rüdiger, J.P. Walker, and A. Monerris for their support during the execution of the campaign.

Conflicts of Interest: The authors declare no conflict of interest.

References

1. Zavorotny, V.U.; Gleason, S.; Cardellach, E.; Camps, A. Tutorial on Remote Sensing Using GNSS Bistatic Radar of Opportunity. *IEEE Geosci. Remote Sens. Mag.* **2014**, *2*, 8–45. [CrossRef]
2. Martín-Neira, M. A passive reflectometry and interferometry system (PARIS): Application to ocean altimetry. *ESA J.* **1993**, *17*, 331–355.
3. Clarizia, M.P.; Ruf, C.S. On the Spatial Resolution of GNSS Reflectometry. *IEEE Geosci. Remote Sens. Lett.* **2016**, *13*, 1064–1068. [CrossRef]
4. Martin, F.; Camps, A.; Fabra, F.; Rius, A.; Martin-Neira, M.; D'Addio, S.; Alonso, A. Mitigation of Direct Signal Cross-Talk and Study of the Coherent Component in GNSS-R. *IEEE Geosci. Remote Sens. Lett.* **2015**, *12*, 279–283. [CrossRef]
5. Munoz-Martin, J.F.; Onrubia, R.; Pascual, D.; Park, H.; Camps, A.; Rüdiger, C.; Walker, J.; Monerris, A. Untangling the Incoherent and Coherent Scattering Components in GNSS-R and Novel Applications. *Remote Sens.* **2020**, *12*, 1208. [CrossRef]
6. Seybold, J.S. *Introduction to RF Propagation*, 8th ed.; John Wiley & Sons, Inc: Hoboken, NJ, USA, 2005. [CrossRef]

7. Camps, A. Spatial Resolution in GNSS-R Under Coherent Scattering. *IEEE Geosci. Remote Sens. Lett.* **2020**, *17*, 32–36. [CrossRef]
8. Onrubia, R.; Pascual, D.; Querol, J.; Park, H.; Camps, A. The Global Navigation Satellite Systems Reflectometry (GNSS-R) Microwave Interferometric Reflectometer: Hardware, Calibration, and Validation Experiments. *Sensors* **2019**, *19*, 1019. [CrossRef] [PubMed]
9. Kong, J.A. *Electromagnetic Wave Theory*; EMW Publishing: Cambridge, MA, USA, 2008.
10. Born, M.; Wolf, E. *Principles of Optics*; Cambridge University Press: Cambridge, UK, 1999.
11. Kubicke, G.; Yahia, Y.A.; Bourlier, C.; Pinel, N.; Pouliguen, P. Bridging the Gap Between the Babinet Principle and the Physical Optics Approximation: Scalar Problem. *IEEE Trans. Antennas Propag.* **2011**, *59*, 4725–4732. [CrossRef]
12. Nghiem, S.V.; Zuffada, C.; Shah, R.; Chew, C.; Lowe, S.T.; Mannucci, A.J.; Cardellach, E.; Brakenridge, G.R.; Geller, G.; Rosenqvist, A. Wetland monitoring with Global Navigation Satellite System reflectometry. *Earth Space Sci.* **2017**, *4*, 16–39. [CrossRef] [PubMed]
13. Chew, C.; Reager, J.T.; Small, E. CYGNSS data map flood inundation during the 2017 Atlantic hurricane season. *Sci. Rep.* **2018**, *8*, 9336. [CrossRef] [PubMed]
14. Loria, E.; O'Brien, A.; Zavorotny, V.; Downs, B.; Zuffada, C. Analysis of scattering characteristics from inland bodies of water observed by CYGNSS. *Remote Sens. Environ.* **2020**, *245*, 111825. [CrossRef]
15. Unnithan, S.L.K.; Biswal, B.; Rüdiger, C. Flood Inundation Mapping by Combining GNSS-R Signals with Topographical Information. *Remote Sens.* **2020**, *12*, 3026. [CrossRef]
16. Unwin, M.; Jales, P.; Tye, J.; Gommenginger, C.; Foti, G.; Rosello, J. Spaceborne GNSS-Reflectometry on TechDemoSat-1: Early Mission Operations and Exploitation. *IEEE J. Sel. Top. Appl. Earth Obs. Remote Sens.* **2016**, *9*, 4525–4539. [CrossRef]
17. Ruf, C.S.; Chew, C.; Lang, T.; Morris, M.G.; Nave, K.; Ridley, A.; Balasubramaniam, R. A New Paradigm in Earth Environmental Monitoring with the CYGNSS Small Satellite Constellation. *Sci. Rep.* **2018**, *8*, 8782. [CrossRef] [PubMed]

Publisher's Note: MDPI stays neutral with regard to jurisdictional claims in published maps and institutional affiliations.

Technical Note

From GPS Receiver to GNSS Reflectometry Payload Development for the Triton Satellite Mission

Yung-Fu Tsai [1],*, Wen-Hao Yeh [1], Jyh-Ching Juang [2], Dian-Syuan Yang [2] and Chen-Tsung Lin [1]

[1] National Space Organization, Hsinchu City 30078, Taiwan; whyeh@narlabs.org.tw (W.-H.Y.); tomlin@narlabs.org.tw (C.-T.L.)
[2] Department of Electrical Engineering, National Cheng Kung University, Tainan City 70101, Taiwan; juang@mail.ncku.edu.tw (J.-C.J.); n26084286@gs.ncku.edu.tw (D.-S.Y.)
* Correspondence: raymond@narlabs.org.tw

Abstract: The global positioning system (GPS) receiver has been one of the most important navigation systems for more than two decades. Although the GPS system was originally designed for near-Earth navigation, currently it is widely used in highly dynamic environments (such as low Earth orbit (LEO)). A space-capable GPS receiver (GPSR) is capable of providing timing and navigation information for spacecraft to determine the orbit and synchronize the onboard timing; therefore, it is one of the essential components of modern spacecraft. However, a space-grade GPSR is technology-sensitive and under export control. In order to overcome export control, the National Space Organization (NSPO) in Taiwan completed the development of a self-reliant space-grade GPSR in 2014. The NSPO GPSR, built in-house, has passed its qualification tests and is ready to fly onboard the Triton satellite. In addition to providing navigation, the GPS/global navigation satellite system (GNSS) is facilitated to many remote sensing missions, such as GNSS radio occultation (GNSS-RO) and GNSS reflectometry (GNSS-R). Based on the design of the NSPO GPSR, the NSPO is actively engaged in the development of the Triton program (a GNSS reflectometry mission). In a GNSS-R mission, the reflected signals are processed to form delay Doppler maps (DDMs) so that various properties (including ocean surface roughness, vegetation, soil moisture, and so on) can be retrieved. This paper describes not only the development of the NSPO GPSR but also the design, development, and special features of the Triton's GNSS-R mission. Moreover, in order to verify the NSPO GNSS-R receiver, ground/flight tests are deemed essential. Then, data analyses of the airborne GNSS-R tests are presented in this paper.

Keywords: remote sensing; GPS receiver; GNSS reflectometry; DDM; Triton

Citation: Tsai, Y.-F.; Yeh, W.-H.; Juang, J.-C.; Yang, D.-S.; Lin, C.-T. From GPS Receiver to GNSS Reflectometry Payload Development for the Triton Satellite Mission. *Remote Sens.* **2021**, *13*, 999. https://doi.org/10.3390/rs13050999

Academic Editors: Nereida Rodriguez-Alvarez and Mary Morris

Received: 28 January 2021
Accepted: 27 February 2021
Published: 5 March 2021

Publisher's Note: MDPI stays neutral with regard to jurisdictional claims in published maps and institutional affiliations.

1. Introduction

Due to the omnipresence of global navigation satellite systems (GNSS) signals and their well-maintained infrastructure, many GNSS applications are widely explored. One of most common GNSS space applications is using the direct GNSS signal to provide the position, velocity, and time (PVT) information for orbit determination (Figure 1). The other two space applications are probing the atmospheric and ionospheric profiles with the bending angle and total electron content of the GNSS signal while passing through the atmosphere and ionosphere (GNSS radio occultation, GNSS-RO) and receiving the reflected GNSS signals to sense their surface properties (GNSS reflectometry, GNSS-R). At present, equipping a space-grade global positioning system (GPS) receiver is seen as indispensable for low Earth orbit (LEO) missions because of the convenience of autonomous onboard positioning and timing. The domestic industry in Taiwan (such as MediaTek Inc. and SkyTraq Technology, Inc.) has established a strong capability for designing and manufacturing civil GPS receiver (GPSR) modules/chipsets. However, such a civil GPSR module/chipset is not able to be applied to satellite missions due to the International Traffic in Arms Regulations (ITAR) [1]. In order to overcome export control, the National Space

Remote Sens. **2021**, *13*, 999. https://doi.org/10.3390/rs13050999

Organization (NSPO) has the responsibility to play an important role in promoting the self-reliant technologies of the space industry so that the NSPO has developed the first space-grade GPSR in Taiwan (Figure 2) [2].

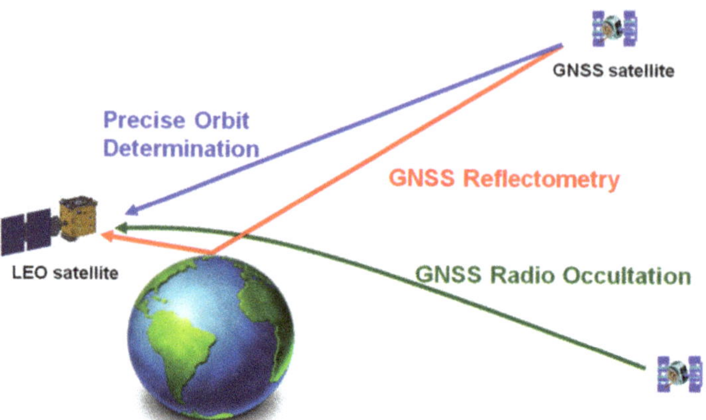

Figure 1. The space applications of global navigation satellite systems (GNSS).

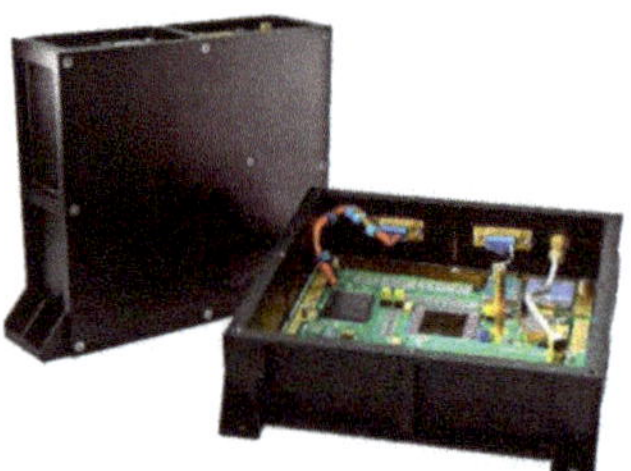

Figure 2. The first space-grade GPS receiver in Taiwan.

The first GPS RO constellation mission is the FORMOSAT-3/COSMIC (simplified as FS-3 in the following descriptions) program, which is an international collaboration program between the NSPO and the University Corporation for Atmospheric Research (UCAR) of the U.S. The FORMOSAT-7/COSMIC-2 (simplified as FS-7) is the follow-up program (launched in 2019) to the successful FS-3 program (which was decommissioned in 2020) and is equipped with the upgraded RO mission payload, which itself is a collaboration between the NSPO and the National Oceanic and Atmospheric Administration (NOAA) of the U.S. Each satellite of the FS-7 constellation is equipped with a GNSS-RO payload to receive the GNSS signal from GPS, GLONASS, and Galileo satellites. The observed RO data is then downlinked to the ground station to retrieve and process into useful atmospheric and space weather data (such as temperature, pressure, water vapor content, electron density, etc.). The RO data distribution for 24 h coverage via FS-7 constellation is shown in Figure 3, which includes about 4500 soundings from GPS and GLONASS signals. Currently, the receiving ability of the Galileo signal is under development. Therefore, it greatly increases the amount of atmospheric and ionospheric observation data available at low latitudes, including in Taiwan [3]. In addition to the RO mission, the NSPO is aggressive in the development of new GNSS remote sensing technology so that the GNSS-R mission is selected for the Triton satellite. The main objective of the Triton program is to collect GNSS signals reflected off the Earth's surface from the LEO satellite [4]. The data could be used for research in several remote sensing domains, such as ocean surface wind [5], sea surface height [6], sea ice [7], vegetation [8], and soil moisture [9]. This could be especially useful in retrieving wind speed over a sea surface for weather forecasting and typhoon path prediction, thereby benefitting people's livelihoods in Taiwan.

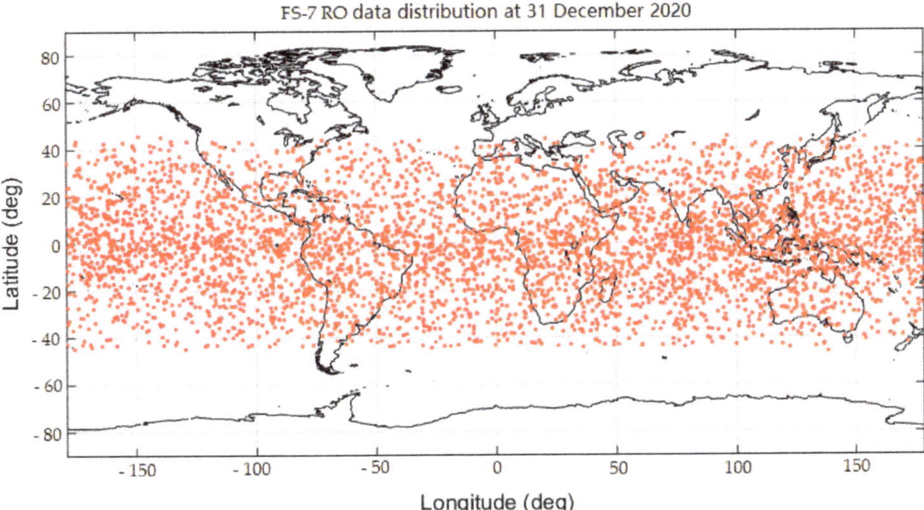

Figure 3. Data distribution for 24 h coverage by FORMOSAT-7/COSMIC-2 (FS-7) constellation.

Triton Satellite Mission

The Triton satellite is an experimental micro-satellite designed and manufactured by NSPO and was planned along with the FS-7 program. Hence, in the beginning, it continued to use the FORMOSAT serial number and subjoined a letter "R" for easier identification (FS-7R). In late 2018, the name "Triton" was officially given due to the GNSS-R mission. In addition to the GNSS-R mission, the Triton satellite mission is conceived as a technology demonstration mission, with the goal of demonstrating some key satellite bus technologies, including the onboard computer (OBC), the power control unit (PCU), H_2O_2 propulsion, a fiber optical gyro, a space-grade GPSR, a micro-stepping solar array driving module, and some associated components. These are designed and supplied by domestic industries in order to increase the technology readiness level (TRL) through flight heritage so that a baseline design of a satellite platform with self-reliant components and technologies could be provided for future NSPO missions [10]. However, due to the reduction in the FS-7 program's scope (down from 12 + 1 satellites to 6 satellites), the Triton program became less dependent on the FS-7 program, even though the Triton satellite shares some common components with FS-7 satellites, such as a solar array, an S-band transceiver, a reaction wheel, a magnetometer, a coarse sun sensor, magnetic torquers, and a battery. The mass of the Triton satellite is less than 285 kg, and its deployed and stowed configurations are depicted in Figure 4. The dimension of the stowed configuration is approximately $100 \times 120 \times 125$ cm. After separation from the launch vehicle, the solar panel will be deployed in space to generate an average power of 162 W.

In this paper, the design and development of the in-house built space-grade NSPO GPSR will be described in the following section, which also includes the performance and environmental tests. This is followed by a presentation of the design, development, and features of the GNSS reflectometry payload. Some unique features of the NSPO self-developed GNSS-R payload will also be highlighted. In order to verify the design of the NSPO's self-developed GNSS-R receiver, several airborne flight experiments have been performed. The data analysis including link budget and data comparison would be discussed as well. The conclusion and the updated status of the Triton mission would be presented in the end.

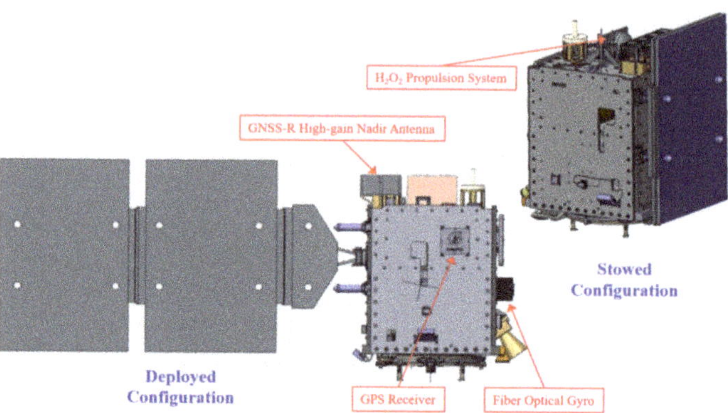

Figure 4. Triton satellite configuration.

2. GPS Receiver (GPSR) Design/Development

Figure 5 shows the block diagram of the NSPO GPSR. The MAX2769 IC [11] is the main chip of the radio frequency front-end (RFFE), which is designed for GPS L1 signal conversion and digitalization. The unpacking and filtering of the digitalized RF signal, as well as the interface between the digital signal processor (DSP) and other onboard devices, are implemented into the field programmable gate array (FPGA) unit to reduce the computational load of the DSP. The main processor (the DSP) is in charge of the fast search algorithm engine, signal tracking, almanac/ephemeris decoding, raw measurement forming, navigation solution, orbit propagation, host command and message decoding, and user data exchange.

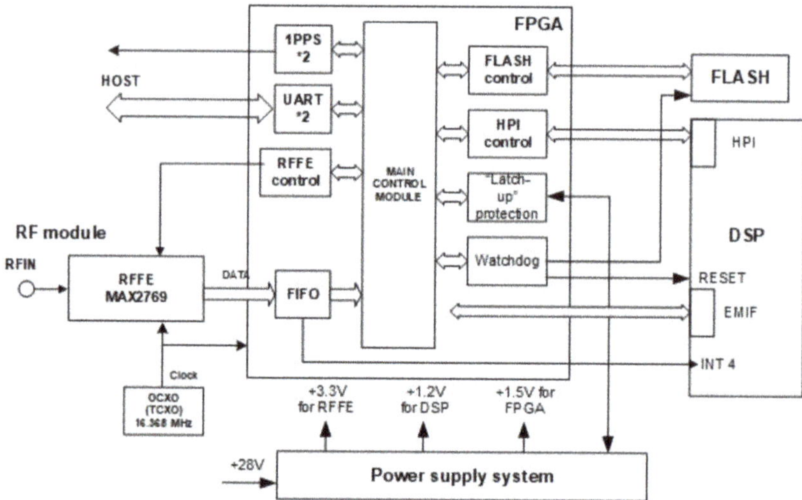

Figure 5. The National Space Organization (NSPO) GPS receiver (GPSR) block diagram.

It is noteworthy that one key feature of the common space-grade GPSRs is its capability to overcome the high Doppler shifts and Doppler shift rates of the high dynamic environment of LEO missions. Figure 6 depicts that the Doppler shift caused by a LEO satellite at an altitude of 550 km is about ± 40 kHz and the contact period is only around 40 min. Due to the implementation of the fast search algorithm, the cold start of the NSPO GPSR is within 90 s. It is faster than some of the space-grade GPS receivers, such as the SSTL SGR-07 [12], the New Space NGPS-01-422, and the Astrium MosaicGNSS single-frequency receiver.

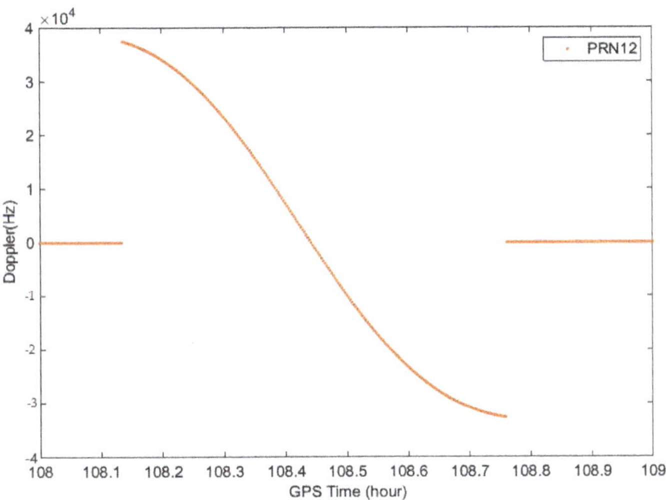

Figure 6. Doppler shift simulation of a low Earth orbit (LEO) mission.

The key specifications of the NSPO GPSR are listed in Table 1. Two types of performance tests have been performed for the NSPO GPSR, with one being a terrestrial test and the other being a high dynamic environment test. In the 4-day terrestrial test, the 3D position and velocity errors of the static GPSR performance are (4.9576, 7.9232, and 5.3995) (m) and (0.0466, 0.0602, and 0.0396) (m/s), respectively. In order to verify the capability of the NSPO GPSR, the end-to-end test platform is required to reproduce the RF signal that the receiver will experience in the high dynamic environment. There are two approaches to reproduce the designated GPS signal via Spirent GPS simulator for the high dynamic tests. One is by using pre-loaded motion data files or pre-determined scenarios in the Spirent GPS simulator. The other approach is to construct an attitude control and LEO mission dynamic environment software simulator platform, which is implemented on the NI PXI real time system to generate a corresponding satellite attitude, which is fed into the Spirent GPS simulator in a real time manner [13]. The first high dynamic test was conducted with a predetermined scenario of the orbit of altitude set at 780 km and the inclination set at 45 deg. The position and velocity errors of the nine orbits navigation performance test were (1.7637, 1.6696, and 1.30) (m) and (0.0283, 0.0277, and 0.0239) (m/s), respectively. The average time to first fix (TTFF) of the NSPO GPSR cold start for 30 times with the same high dynamic scenario was about 66 s. The other high dynamic test was conducted with the satellite attitude at a rate of 1 deg/sec to validate the NSPO GPSR's tracking stability during satellite tumbling. The upper two sub-figures of Figure 7 show the position and velocity error, respectively. The third sub-figure depicts the number of tracked GPS signals (blue line) and the number of tracked GPS signals used for navigation (red line). The lowest sub-figure indicates the information concerning the attitude of the LEO satellite, which means that the satellite is tumbling. Hence, the NSPO GPSR maintains the navigation solution (after 150 s), even if the satellite is in a tumbling situation [14]. In addition to the performance test, the environmental tests are essential to verify that the NSPO GPSR is compliant with its specification and is ready to fly with LEO missions. The NSPO GPSR has passed the entire qualification test campaign, undergoing such tests as a thermal cycle test, a vibration test, a thermal vacuum test, an electromagnetic compatibility (EMC) test, and a radiation test [15].

Table 1. The specifications of the NSPO GPSR.

Input power	28 V/3.5 W
Electrical interface	RS422/UART@115200 bps & 1 PPS output
Tracking channels	GPS L1 C/A×12
Tracking threshold	33 dB-Hz
Max Doppler shift	±65 KHz.
Position accuracy	Better than 8 m (1 sigma)
Velocity accuracy	Better than 0.1 m/s (1 sigma)
Time accuracy	Better than 1 micro-second
Cold start time	Within 90 s
Trajectory dynamic	Up to 12 g
Mass	1.8 kg
Dimension	160 × 120 × 40 mm
Mission life	5 years
Total doze	35 Krad

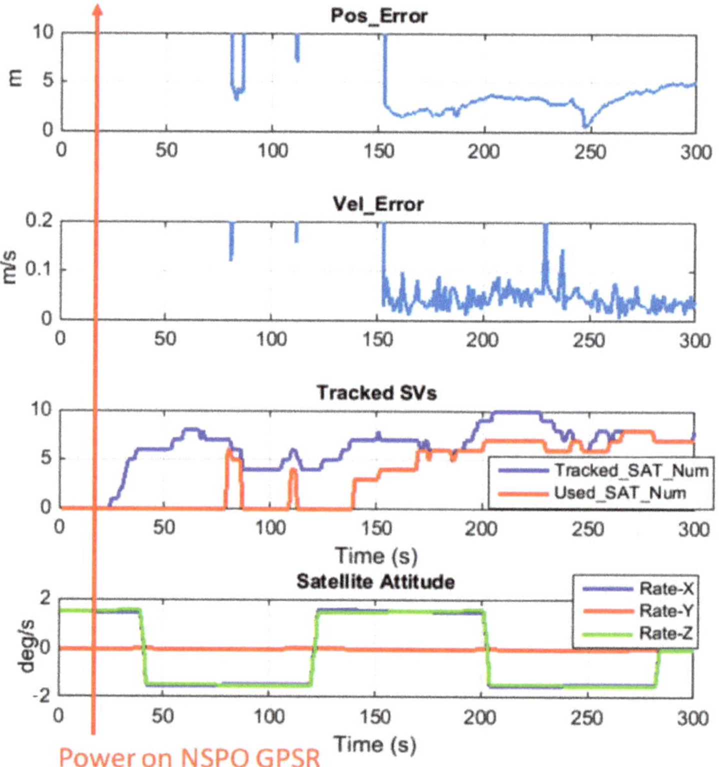

Figure 7. The tracking stability of the NSPO GPSR.

3. GNSS Reflectometry (GNSS-R) Mission Payload

3.1. GNSS-R Payload Design

The major task of the GNSS-R receiver is to receive and process reflected GNSS signals, after which the features on the surface of the earth can be retrieved. Therefore, the retrieval of the data concerning wind speed over the ocean is the main objective of the Triton mission. The design of the NSPO GNSS-R payload is based on several applicable GNSS-R processing options, which are discussed in [16]. By examining the needs and referring to the design of the SGR-ReSI (an existing GNSS-R receiver) [17], the GNSS-R payload is designed as a scatterometer, which includes the following requirements [18].

- The GNSS-R payload shall be able to process GPS L1 reflection signals and then generate the associated delay–Doppler map (DDM), either autonomously or by schedule.
- The generated DDM resolution shall be at least 128 (in code phase) * 64 (in frequency bin) and each entry of the DDM shall be better than 16 bits.
- The GNSS-R payload shall be capable of processing at least four DDMs simultaneously.
- The DDM update rate shall be at least 1 s.
- The GNSS-R payload should potentially be extended to process reflected Quasi-Zenith Satellite System (QZSS) [19] and Galileo L1 signals and generate DDMs of the aforementioned space, magnitude, and time resolution in real time.
- The GNSS-R payload shall be able to record direct line-of-sight and reflected GPS L1/L2 band signals in raw data format at the intermediate frequency (IF) for ground post-processing purposes.
- The GNSS-R payload shall include at least 512 M bytes of RAM to store the raw data.
- The GNSS-R payload shall be designed with a process-and-record mode for ground debugging purpose.
- The GNSS-R payload shall facilitate meta-data to support calibration and retrieval.

The photo of the GNSS-R payload system is depicted in Figure 8. The zenith antenna is a commercial dual-frequency (L1 and L2) right-hand circularly polarized (RHCP) antenna for the reception of the direct line-of-sight signals from GNSS satellites. The nadir antenna is a high-gain dual-frequency left-hand circularly polarized (LHCP) antenna that is used to receive the reflected/scattered signals from GNSS satellites. The low noise amplifier (LNA), installed close to the antenna, is used to provide the appropriate amplification to enhance the signal-to-noise ratio (SNR). Both the nadir antenna and the LNA are designed and supplied by the cooperation of the NSPO and domestic industries. There are a RFFE with two RF inputs, a navigation unit, a science unit, and a power unit inside the GNSS-R receiver. The RFFE is in charge of signal conditioning, down-conversion, and analog-to-digital conversion. The incoming signals are sampled at 16.368 M samples per second, and each sample is represented as 4-bit data (2-bit in-phase/ 2-bit quadrature). The core technology of the navigation unit is based on the NSPO GPSR, and it can process the zenith antenna signal to provide PVT information and GNSS satellite position/velocity information. The timing information is used to synchronize the generation of DDM. The PV information and the GNSS satellite PV information are used to compute the specular reflection point. Moreover, the tracking information provided by the navigation unit (such as code phase, code phase rate, and Doppler frequency) is used to determine the relative code shift and Doppler shift for the DDM processing. The telemetry/telecommand (TM/TC) interface of the GNSS-R payload is RS422 in order to operate and monitor the GNSS-R payload. The output data of the science unit (which includes generated DDM and/or stored raw data) are transmitted to the Triton satellite bus through the SpaceWire interface. The transmitted pulse per second (PPS) signal is used for the purpose of synchronization. The power unit regulates the input power from the Triton satellite bus to distribute regulated voltage to the other units. The kernel of the GNSS-R receiver, its science unit, is responsible for the reception and processing of reflected GNSS signals to generate DDMs in real time. The science unit of the GNSS-R receiver is implemented by using a FPGA co-processor (Zynq-7045) to meet the demands of real-time processing.

Compared with the SGR-ReSI used in the TechDemoSat-1 and CYGNSS missions [20], both reflectometry receiver designs are similar in terms of major functionality. However, there are still some upgraded/distinguishable features of the NSPO GNSS-R payload, which are listed as follows:

- A higher DDM resolution (CYGNSS: one-fourth chip/500 Hz, Triton: one-sixteenth chip/125 Hz)
- A real-time QZSS reflected signal processing capability
- The capability to generate eight DDM generations per second
- On-orbit software/firmware modification capability

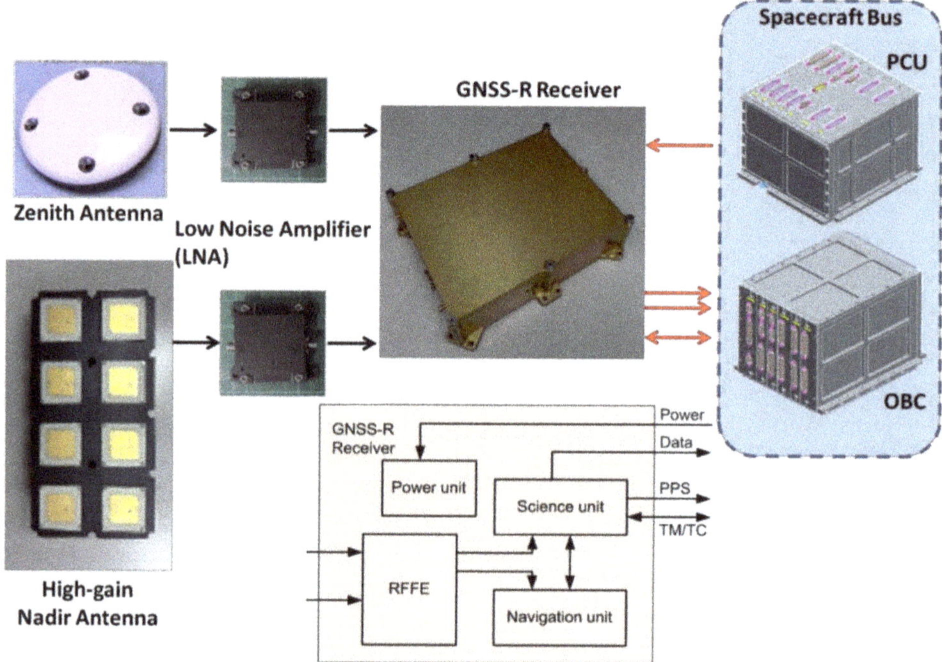

Figure 8. The GNSS-R payload system.

3.2. GNSS-R Payload Validation

The prototype of the NSPO self-developed GNSS-R receiver was designed and implemented in 2016. Figure 9 shows the ground test configuration of the GNSS-R prototype, and the test location is on the bridge over the river mouth. There is a dual-frequency zenith antenna to receive the direct line-of-sight GNSS signals, and the nadir antenna (an LHCP 2 × 2 pitch antenna) is set up to point to the surface of the river mouth to collect the reflected GNSS signals and generate the corresponding DDMs. Although several ground tests have been conducted to validate the prototype of the GNSS-R payload in the early development phase, two airborne tests have been performed with an Aerospace Industrial Development Corporation (AIDC) airplane at the end of 2016 to verify the functionality of the GNSS-R receiver [21–25]. Figure 10 depicts one snapshot of the aircraft position and four specular point positions (red points), as well as four corresponding DDMs, PRN6, PRN19, PRN17 and PRN 9, respectively. Each entry of the DDM (the correlation value) is indicated by the color as well as the x-axis and y-axis of the DDM are the Doppler frequency bin and the code delay, respectively. After that, the development processes of the engineering model (EM), the engineering qualification model (EQM), and the flight model (FM) were completed in 2017, 2018, and 2019, respectively. The entire qualification/acceptance level environmental tests, including the thermal cycle test, the vibration test, the thermal vacuum test, and the electromagnetic compatibility (EMC) test, have been performed and passed as well.

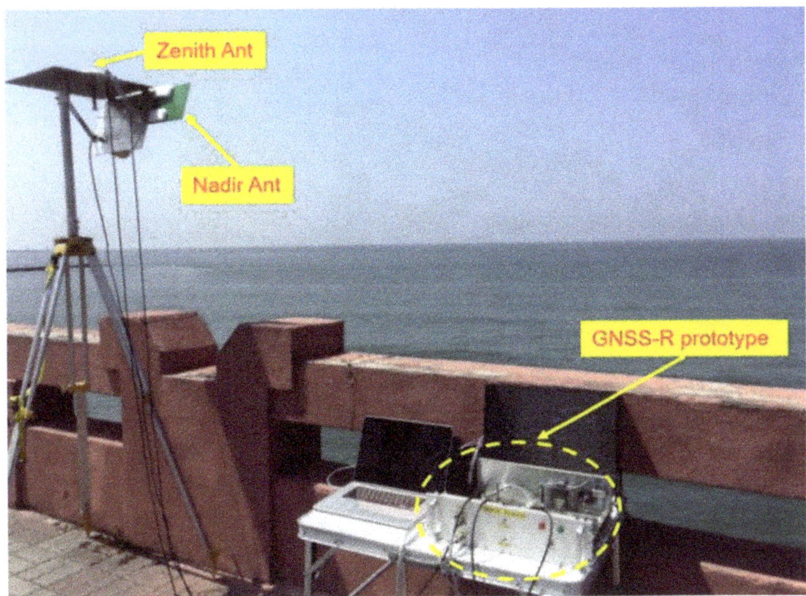

Figure 9. Ground test with the GNSS-R prototype.

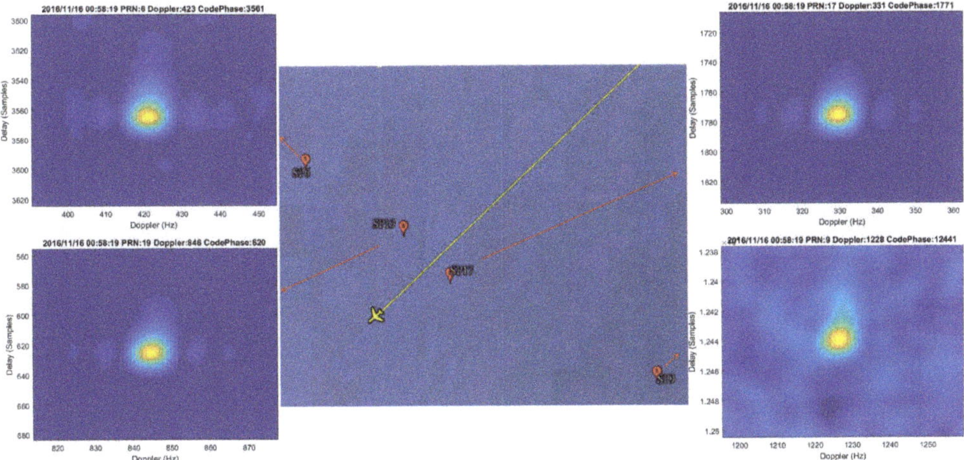

Figure 10. (a) One snapshot of the aircraft position and four specular point positions (red points) (b) four DDMs of PRN17, PRN19, PRN6, and PRN 9 (from left to right).

Owing to the flexibility of unmanned aerial vehicle (UAV) flight tests, several UAV flight tests with GNSS-R EM were performed during the EQM/FM development phase. The configuration of the UAV flight tests is depicted in Figure 11. Figure 12 shows one snapshot of the four DDMs, the corresponding specular point positions (red circle points), and the UAV flight path. Moreover, it is shown that the interference of the UAV engine is introduced into the DDMs. Hence, the objectives of the UAV flight tests were to verify the functionality, operation scheduling, status monitoring, and data science data volume validation of the NSPO GNSS-R receiver. Because there was no interference issue in the 2016 AIDC flight tests, the AIDC airplane was used to carry the GNSS-R EM to conduct the reflectometry experiments (Figure 13) in order to assess the data quality generated from the NSPO GNSS-R receiver. Three airborne reflectometry experiments were conducted on 2020/07/27, 2020/09/03, and 2020/11/03, respectively. In addition to the DDMs, the data collection of each reflectometry experiment includes the raw data for further post-processing.

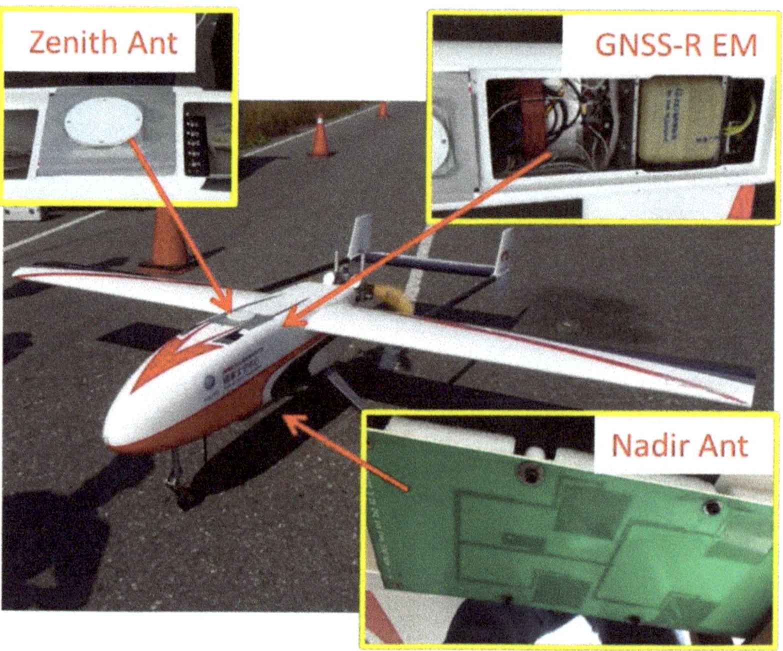

Figure 11. A unmanned aerial vehicle (UAV) flight test with GNSS-R EM.

Figure 14 shows the actual flight path of the AIDC flight test on 27 July 2020. In this figure, the black line is the flight path of the AIDC airplane. In order to avoid the influence of the signal incidence angle and the different transmission power of different GPS satellites, the reflected signal of the QZSS satellite (PRN195) in the flight test is used for analysis. The colored line in Figure 14 is the specular point distribution of PRN195, and the color bar indicates the received power of the reflected signal from the area near the specular point, which is the maximum value of the DDMs (maxDDM). The significant wave height (Hs) observation from the buoy is used for comparison with the received power of the reflected signal. In this experiment, the Hs observations of two buoys (Chimei (22°57′09″ N 119°31′06″ E) and Pratas (21°04′39″ N 118°50′50″ E)) were used. These buoys belong to the Center Weather Bureau (CWB), Taiwan. Moreover, a Drifter Technology Co., Ltd. buoy [26] was anchored at (22°55′29″ N 119°52′10″ E) to obtain more Hs observations. The diamond, triangle, and square in Figure 14 indicate the buoy positions of the Chimei, Pratas, and Drifter Technology's buoy, respectively. The comparison between Hs and maxDDM is shown in Figure 15 and the symbol indicators of the three buoys are the same as those in Figure 14. The maxDDM information comes from the DDM where the specular point (indicated with the asterisk symbol in Figure 14) is closest to the buoy position, and the red line is the linear regression result. The correlation coefficient between Hs and maxDDM is close to -0.99. According to the result, it is a well negative relation between Hs and maxDDM. This means that the maxDDM decreases as Hs increases, and a higher Hs indicates higher ocean surface roughness. When a signal is reflected by a higher roughness ocean surface, more signal power is scattered and the signal power received is lowered. Hence, it is shown that the quality of the data of the GNSS-R EM is satisfactory.

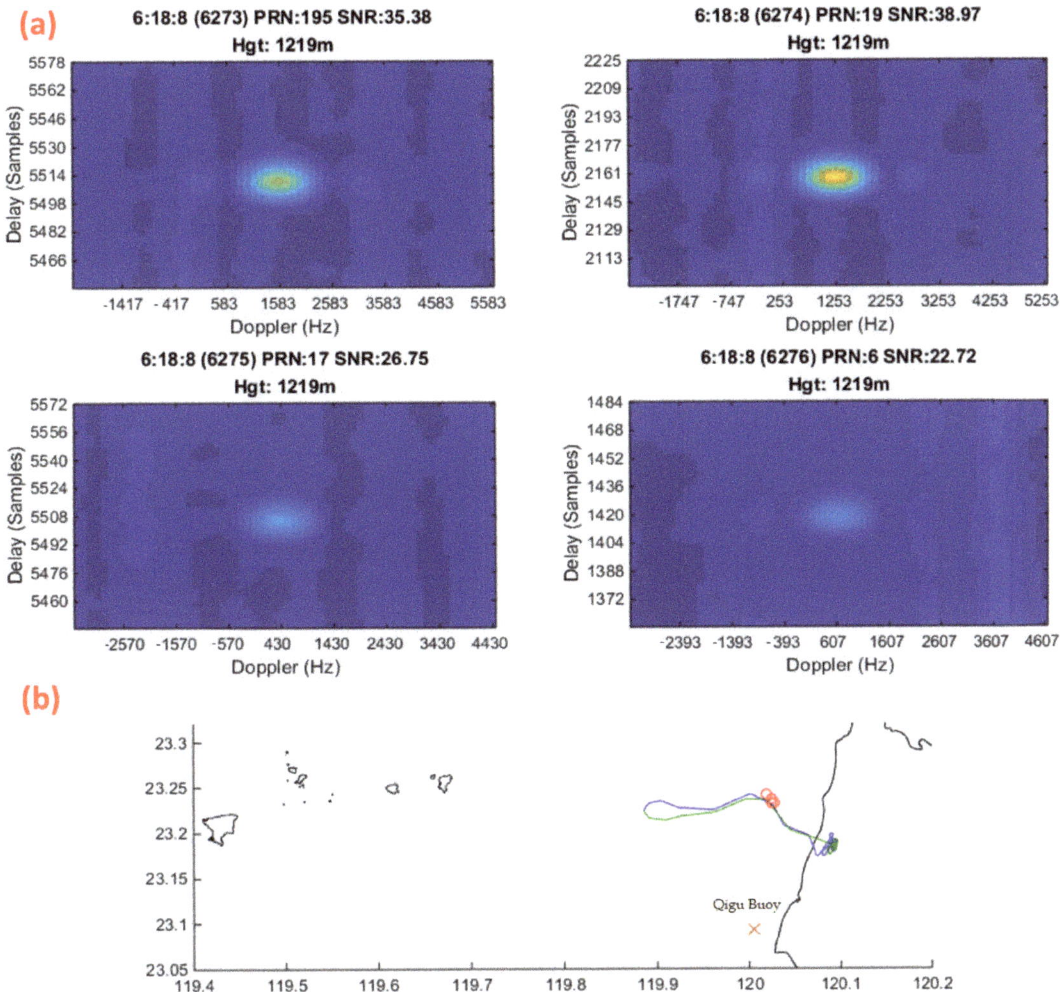

Figure 12. (**a**) One snapshot of the four DDMs (**b**) Four corresponding specular point positions (red circle points) at the UAV flight path.

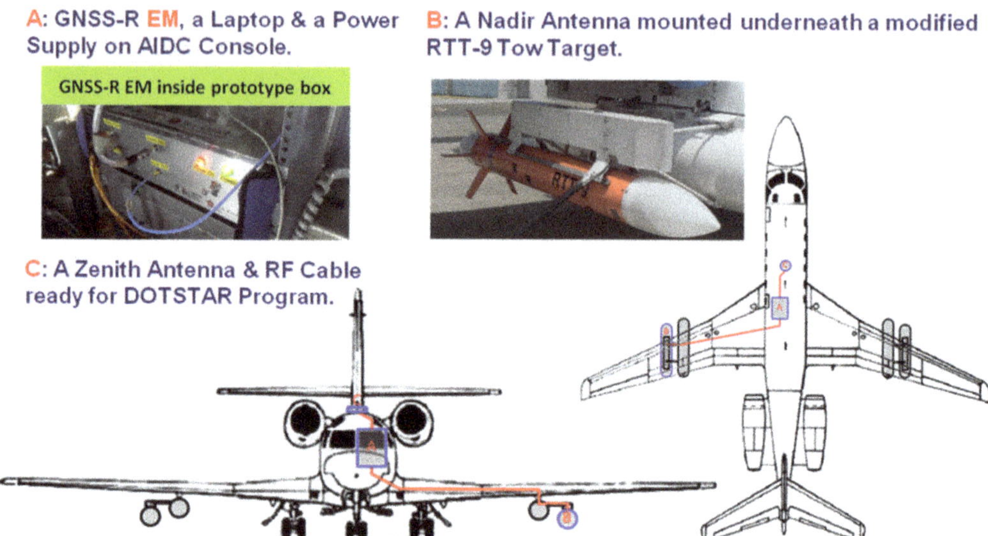

Figure 13. An Aerospace Industrial Development Corporation (AIDC) flight test with GNSS-R EM.

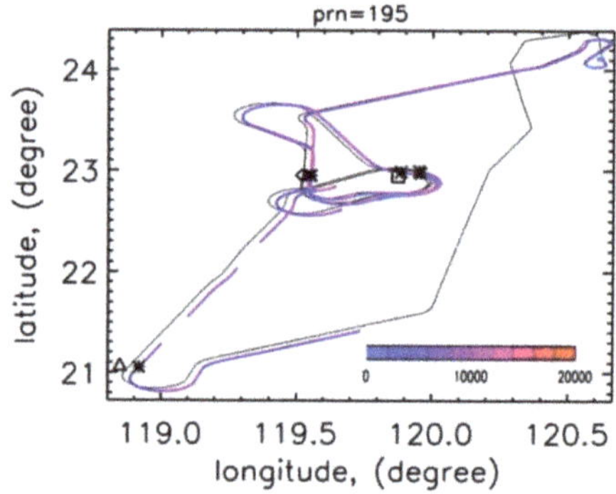

Figure 14. The actual flight path of AIDC reflectometry experiment conducted on 27 July 2020.

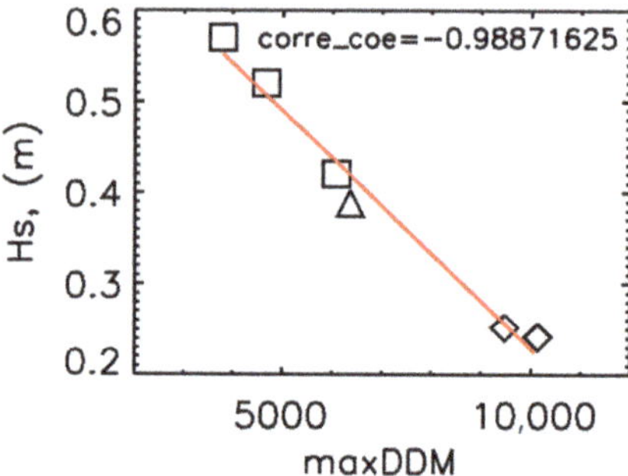

Figure 15. Comparison between the Hs of a buoy and maxDDM (Chimei, Pratas, and Drifter Technology's buoy are indicated with the diamond, triangle, and square).

Moreover, since the scattered/reflected GNSS signals for space application are extremely weak, a link budget analysis is performed and compared with real collected DDM

SNR (defined in Figure 16) to validate the design parameters, including antenna gain and coherent/non-coherent processing parameters. The theoretical model DDM could be represented by the following GPS bi-static radar equation:

$$\left\langle |Y_S(\tau,f)|^2 \right\rangle = \frac{P_T \lambda_{L1}^2}{(4\pi)^3} T i^2 \iint \frac{G_T(\theta,\phi) G_R(\theta,\phi) \sigma^0(\theta,\phi)}{R_{SR}^2(\theta,\phi) R_{ST}^2(\theta,\phi)} \Omega(\tau,f,\theta,\phi) dA(\theta,\phi) \qquad (1)$$

where the detailed parameters in the equation are defined in [27]. Therefore, Figure 17 shows the flow chart of the theoretical model SNR calculation for link budget analysis [28]. In the upper part of Figure 17, the input parameters, the GNSS satellite position, and the GNSS-R receiver position are provided by the GNSS-R EM in order to calculate the reflection point. Then, some geo-physical parameters, such as wind speed and GPS equivalent isotropically radiated power (EIRP), are assumed to be at a reasonable value according to CWB broadcast wind speed and the GNSS interface control document (ICD). Moreover, some supplemental information (such as AIDC airplane/UAV attitude, antenna gain pattern, and LAN gain/noise figure look-up table) is provided to calculate the received power and thermal noise over each grid. Finally, each grid is accumulated and processed in a non-coherent integration to obtain the theoretical model DDM SNR. The comparison of 2020/09/03 airborne reflectometry experiment is illustrated in Figure 18. The trending and value are similar so that good confidence is provided for the signal reception of the Triton GNSS-R mission based on the link budget analysis method. However, the average error of the comparison is about 0.5 dB. The reason for this is that some uncertain parameters are difficult to obtain for link budget analysis (for example, real wind speed, GNSS-R receiver/LNA temperatures, etc.).

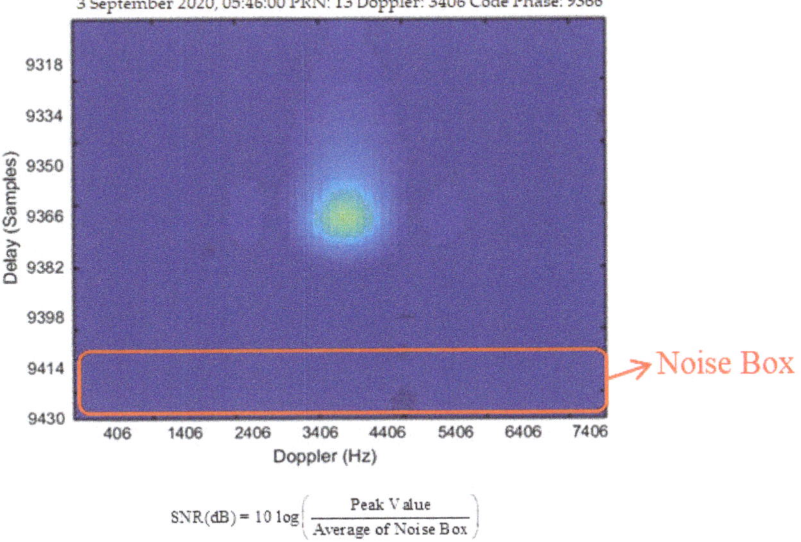

$$SNR(dB) = 10 \log\left(\frac{\text{Peak Value}}{\text{Average of Noise Box}}\right)$$

Figure 16. A signal-to-noise ratio (SNR) calculation of DDM.

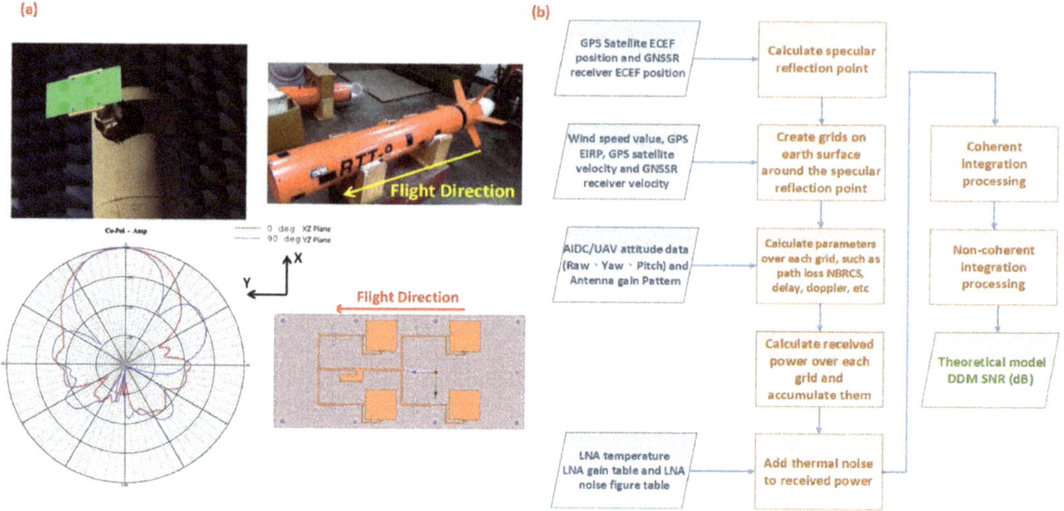

Figure 17. Link budget analysis flowchart.

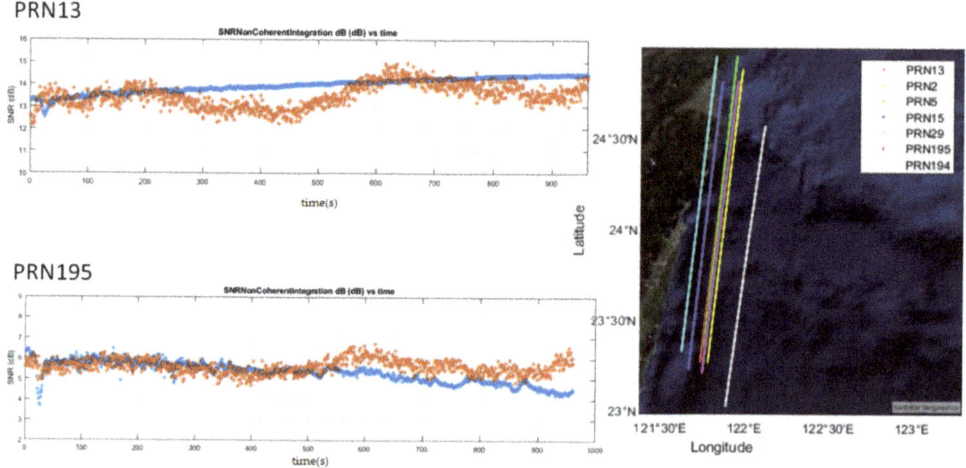

Figure 18. Comparison of the theoretical model (blue) and real collected (red) DDM SNR.

4. Conclusions

The introduction of the Triton mission is described at the beginning of the paper. The development of the NSPO GPSR, which is ready to fly onboard the Triton satellite, is then illustrated. Moreover, in order to satisfy the requirements of future NSPO missions and university CubeSats, several miniaturized versions of the NSPO in-house built GPSRs were developed without performance degradation [29]. Afterward, the paper presented not only the design and development of the Triton's GNSS-R mission payload but also the flight tests for the validation of the GNSS-R payload. Moreover, a science team has been formed to perform the tasks of data calibration, data archiving, data retrieval, and data utilization to support the Triton mission. The Triton program started in 2012 and passed several milestones, such as a system design review (SDR), a preliminary design review (PDR), a critical design review (CDR), and an integration & test review (ITR), before 2018. The NSPO team also has completed the installation of most components, harness routing, and integration. The entire environmental tests are planned to be conducted in 2021 so that the Triton satellite is ready to launch in 2022.

Author Contributions: Conceptualization, Y.-F.T. and W.-H.Y.; data curation, D.-S.Y.; writing—original draft preparation, Y.-F.T.; writing—review and editing, Y.-F.T. and W.-H.Y.; supervision, J.-C.J.; project administration, C.-T.L. All authors have read and agreed to the published version of the manuscript.

Funding: The APC was funded by the National Space Organization, Taiwan.

Acknowledgments: The authors would like to thank Center Weather Bureau (CWB), Taiwan and Drifter Technology Co., Ltd., Taiwan for providing the buoy data used in this paper.

Conflicts of Interest: The authors declare no conflict of interest.

References

1. Juang, J.C.; Lin, C.F.; Hu, C.M.; Chang, C.C.; Tsai, Y.F.; Lin, C.T. Implementation and Test of a Space-borne GPS Receiver Payload of University Microsatellite. *J. Aeronaut. Astronaut. Aviat.* **2012**, *44*, 141–148.
2. News of NSPO, Taiwan. Available online: http://www.nspo.narl.org.tw/en2016/info/news.shtml?id=000527&hid=ruEdk2nA2B (accessed on 1 March 2021).
3. FORMOSAT-7 Program, NSPO, Taiwan. Available online: https://www.nspo.narl.org.tw/inprogress.php?c=20022301&ln=en (accessed on 1 March 2021).
4. TRITON Program, NSPO, Taiwan. Available online: https://www.nspo.narl.org.tw/inprogress.php?c=20030305&ln=en (accessed on 1 March 2021).
5. Gleason, S.; Hodgart, S.; Sun, Y.; Gommenginger, C.; Mackin, S.; Adjrad, M.; Unwin, M. Detection and processing of bistatically reflected GPS signals from low earth orbit for the purpose of ocean remote sensing. *IEEE Trans. Geosci. Remote Sens.* **2005**, *43*, 1229–1241. [CrossRef]
6. Martin-Neira, M.; Caparrini, M.; Font-Rossello, J.; Lannelongue, S.; Vallmitjana, C.S. The PARIS concept: An experimental demonstration of sea surface altimetry using GPS reflected signals. *IEEE Trans. Geosci. Remote Sens.* **2001**, *39*, 142–150. [CrossRef]
7. Yan, Q.; Huang, W. Spaceborne GNSS-R sea ice detection using delay-Doppler maps: First results from the UK TechDemoSat-1 mission. *IEEE J. Sel. Top. Appl. Earth Obs. Remote Sens.* **2016**, *9*, 4795–4801. [CrossRef]
8. Small, E.E.; Larson, K.M.; Braun, J.J. Sensing vegetation growth with reflected GPS signals. *Geophys. Res. Lett.* **2010**, *37*. [CrossRef]
9. Rodriguez-Alvarez, N.; Bosch-Lluis, X.; Camps, A.; Vall-Llossera, M.; Valencia, E.; Marchan-Hernandez, J.F.; Ramos-Perez, I. Soil moisture retrieval using GNSS-R techniques: Experimental results over a bare soil field. *IEEE Trans. Geosci. Remote Sens.* **2009**, *47*, 3616–3624. [CrossRef]
10. Tsai, Y.F.; Lin, C.T.; Juang, J.C. Taiwan's GNSS Reflectometry Mission—The FORMOSAT-7 Reflectometry (FS-7R) Mission. *J. Aeronaut. Astronaut. Aviat.* **2018**, *50*, 391–404.
11. Datasheet of MAX2769 Universal GPS Receiver. Available online: https://www.maximintegrated.com/en/products/comms/wireless-rf/MAX2769.html (accessed on 1 March 2021).
12. Datasheet of Space GPS Receiver SGR-07. Available online: https://sstl-qa.azurewebsites.net/getattachment/Media-Hub/Featured/Navigation/SGR-07-Datasheet-2018-V2-Read-Only.pdf?lang=en-GB (accessed on 1 March 2021).
13. Chang, H.Y.; Chang, H.C.; Lin, C.T. Performance Demonstration of NSPO Space-borne GPS Receiver. In Proceedings of the 26th International Technical Meeting of the Satellite Division of The Institute of Navigation (ION GNSS+ 2013), Manassas, VA, USA, 21–25 September 2013.
14. Chang, H.Y.; Chiang, W.L.; Wu, K.L. A Space-Borne GNSS Receiver for Evaluation of the LEO Navigation Based on Real-Time Platform. In Proceedings of the CEAS Euro GNC Conference, Warsaw, Poland, 25–27 April 2017.
15. Cheng, C.C.; Lin, C.T.; Chen, T.L. TID Testing and SEE Mitigation Approach of a Long Mission Life GPS Receiver Using COTS Parts. In Proceedings of the 6th ESA Workshop on Satellite Navigation Technologies, Noordwijk, The Netherlands, 5–7 December 2012.
16. Zavorotny, V.U.; Gleason, S.; Cardellach, E.; Camps, A. Tutorial on remote sensing using GNSS bistatic radar of opportunity. *IEEE Geosci. Remote Sens. Mag.* **2014**, *2*, 8–45. [CrossRef]
17. Unwin, M.; de Vos Van Steenwijk, R.; Da Silva Curiel, A.; Cutter, M.; Abbott, B.; Gommenginger, C.; Mitchell, C.; Gao, S. Remote sensing using GPS signals—The SGR-ReSI instrument. In Proceedings of the 25th Annu. AIAA/USU Conf. Small Satellites, Logan, UT, USA, 8–11 August 2011.
18. Juang, J.C.; Ma, S.H.; Lin, C.T. Study of GNSS-R Techniques for FORMOSAT Mission. *IEEE J. Sel. Top. Appl. Earth Obs. Remote Sens.* **2016**, *9*, 4582–4592. [CrossRef]
19. Quasi-Zenith Satellite System (QZSS). Available online: https://qzss.go.jp/en/overview/services/index.html (accessed on 1 March 2021).
20. CYGNSS Mission. Available online: https://clasp-research.engin.umich.edu/missions/cygnss/ (accessed on 1 March 2021).

21. Juang, J.C.; Lin, C.T. Recent Development in GNSS-Reflected Signal Processing and Receiver Research in Taiwan. In Proceedings of the COSMIC/IROWG, Estes Park, CO, USA, 21–27 September 2017.
22. Juang, J.C.; Ma, S.H.; Lin, C.T. Data Analysis of Galileo Reflected Signals in an Airborne Experiment. In Proceedings of the GNSS+R, 2017 Workshop, University of Michigan Union, Ann Arbor, MI, USA, 23–25 May 2017.
23. DOTSTAR Project. Available online: http://typhoon.as.ntu.edu.tw/DOTSTAR/en/ (accessed on 1 March 2021).
24. Juang, J.C.; Tsai, Y.F.; Lin, C.T. FORMOSAT-7R Mission for GNSS Reflectometry. In Proceedings of the 2019 IEEE International Geoscience and Remote Sensing Symposium, IGARSS 2019, Yokohama, Japan, 28 July–2 August 2019.
25. Juang, J.C.; Lin, C.T.; Tsai, Y.F. Comparison and Synergy of BPSK and BOC Modulations in GNSS Reflectometry. *IEEE J. Sel. Top. Appl. Earth Obs. Remote Sens.* **2020**. [CrossRef]
26. Drifter Technology Co. Ltd. Available online: http://driftertek.com/ (accessed on 1 March 2021).
27. O'Brien, A. End-to-end Simulator Technical Memo. CYGNSS Project Doc 148-0123. Available online: https://clasp-research.engin.umich.edu/missions/cygnss/reference/148-0123_CYGNSS_E2ES_EM.pdf (accessed on 1 March 2021).
28. Yang, D.S.; Juang, J.C.; Lin, C.T.; Tsai, Y.F. Verification of GNSS-R Signals for the Triton Mission Payload. In Proceedings of the ICGPSRO 2020, Hsinchu, Taiwan, 21–23 October 2020.
29. Tsai, Y.F.; Hsieh, M.Y.; Chang, H.Y.; Lin, C.T. Development and Test of a Space Capable Miniaturized GPS/GNSS Receiver for Space Applications. In Proceedings of the ION ITM/PTTI 2018, Reston, Virginia, 29 January–1 February 2018.

MDPI

St. Alban-Anlage 66

4052 Basel

Switzerland

Tel. +41 61 683 77 34

Fax +41 61 302 89 18

www.mdpi.com

Remote Sensing Editorial Office

E-mail: remotesensing@mdpi.com

www.mdpi.com/journal/remotesensing